ISBN 978-1-5277-9795-6
PIBN 10900084

1 MONTH OF
FREE
READING

at
www.ForgottenBooks.com

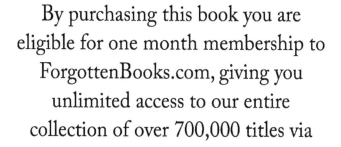

By purchasing this book you are eligible for one month membership to ForgottenBooks.com, giving you unlimited access to our entire collection of over 700,000 titles via our web site and mobile apps.

To claim your free month visit:
www.forgottenbooks.com/free900084

English
Français
Deutsche
Italiano
Español
Português

www.forgottenbooks.com

Mythology Photography **Fiction**
Fishing Christianity **Art** Cooking
Essays Buddhism Freemasonry
Medicine **Biology** Music **Ancient
Egypt** Evolution Carpentry Physics
Dance Geology **Mathematics** Fitness
Shakespeare **Folklore** Yoga Marketing
Confidence Immortality Biographies
Poetry **Psychology** Witchcraft
Electronics Chemistry History **Law**
Accounting **Philosophy** Anthropology
Alchemy Drama Quantum Mechanics
Atheism Sexual Health **Ancient History**
Entrepreneurship Languages Sport
Paleontology Needlework Islam
Metaphysics Investment Archaeology
Parenting Statistics Criminology
Motivational

THE INSTITUTION

OF

MECHANICAL ENGINEERS.

ESTABLISHED 1847.

PROCEEDINGS.

1903.

PARTS 1-2.

PUBLISHED BY THE INSTITUTION,

STOREY'S GATE, ST. JAMES'S PARK, WESTMINSTER, S.W.

CONTENTS.

1903.

PARTS 1-2.

PAST-PRESIDENTS.

GEORGE STEPHENSON, 1847–48. (*Deceased* 1848.)

ROBERT STEPHENSON, F.R S., 1849–53. (*Deceased* 1859.)

SIR WILLIAM FAIRBAIRN, BART., LL.D., F.R.S., 1854–55. (*Deceased* 1874.)

SIR JOSEPH WHITWORTH, BART., D.C.L., LL.D., F.R.S., 1856–57, 1866. (*Deceased* 1887.)

JOHN PENN, F.R.S., 1858–59, 1867–68. (*Deceased* 1878.)

JAMES KENNEDY, 1860. (*Deceased* 1886.)

THE RIGHT HON. LORD ARMSTRONG, C B., D.C.L., LL D., F.R.S., 1861–62, 1869. (*Deceased* 1900.)

ROBERT NAPIER, 1863–65. (*Deceased* 1876.)

JOHN RAMSBOTTOM, 1870–71. (*Deceased* 1897.)

SIR WILLIAM SIEMENS, D.C.L., LL.D., F.R.S., 1872–73. (*Deceased* 1883)

SIR FREDERICK J. BRAMWELL, BART., D.C.L., LL.D., F.R.S., 1874–75.

THOMAS HAWKSLEY, F.R.S., 1876–77. (*Deceased* 1893.)

JOHN ROBINSON, 1878–79. (*Deceased* 1902.)

EDWARD A. COWPER, 1880–81. (*Deceased* 1893.)

PERCY G. B. WESTMACOTT, 1882–83.

SIR LOWTHIAN BELL, BART., LL.D., F.R.S., 1884.

JEREMIAH HEAD, 1885–86. (*Deceased* 1899.)

SIR EDWARD H. CARBUTT, BART., 1887–88.

CHARLES COCHRANE, 1889. (*Deceased* 1898.)

JOSEPH TOMLINSON, 1890–91. (*Deceased* 1894.)

SIR WILLIAM ANDERSON, K.C.B , D.C.L., F.R.S., 1892–93. (*Deceased* 1898.)

PROFESSOR ALEXANDER B. W. KENNEDY, LL.D., F.R.S., 1894–95.

E. WINDSOR RICHARDS, 1896–97.

SAMUEL WAITE JOHNSON, 1898.

SIR WILLIAM H. WHITE, K.C.B., LL D., D.Sc., F.R S., 1899–1900.

WILLIAM H MAW, 1901–02.

The Institution of Mechanical Engineers.

OFFICERS.

1903.

PRESIDENT.

J. HARTLEY WICKSTEED, ... Leeds.

PAST-PRESIDENTS.

SIR LOWTHIAN BELL, BART., LL.D., F.R.S., ...	Northallerton.
SIR FREDERICK J. BRAMWELL, BART., D.C.L., LL.D., F.R.S.,	London.
SIR EDWARD H. CARBUTT, BART., ...	London.
SAMUEL WAITE JOHNSON, ...	Derby.
PROFESSOR ALEXANDER B. W. KENNEDY, LL.D., F.R S.,	London.
WILLIAM H. MAW, ...	London.
E. WINDSOR RICHARDS, ...	Caerleon.
PERCY G. B. WESTMACOTT, ...	Ascot.
SIR WILLIAM H. WHITE, K.C.B., LL.D., D.Sc., F.R.S.,	London.

VICE-PRESIDENTS.

JOHN A. F. ASPINALL, ...	Manchester.
EDWARD B. ELLINGTON, ...	London.
ARTHUR KEEN, ...	Birmingham.
EDWARD P. MARTIN, ...	Dowlais.
T. HURRY RICHES, ...	Cardiff.
A. TANNETT-WALKER, ...	Leeds.

MEMBERS OF COUNCIL.

SIR BENJAMIN BAKER, K.C.B , K.C.M.G., LL.D., D.Sc., F.R.S.,	London.
SIR J. WOLFE BARRY, K.C.B., LL.D., F.R.S., ...	London.
HENRY CHAPMAN, ...	London.
HENRY DAVEY, ...	London.
WILLIAM DEAN, ...	Folkestone.
H. GRAHAM HARRIS, ...	London.
EDWARD HOPKINSON, D.Sc., ...	Manchester.
HENRY A. IVATT, ...	Doncaster.
HENRY LEA, ...	Birmingham.
SIR WILLIAM T. LEWIS, BART., ...	Aberdare.
MICHAEL LONGRIDGE, ...	Manchester.
JAMES MANSERGH, F.R.S., ...	London.
HENRY D. MARSHALL, ...	Gainsborough.
THE RIGHT HON. WILLIAM J. PIRRIE, LL.D., ...	Belfast.
SIR THOMAS RICHARDSON, ...	Hartlepool.
JOHN F. ROBINSON, ...	London.
MARK H. ROBINSON, ...	Rugby.
JOHN W. SPENCER, ...	Newcastle-on-Tyne.
SIR JOHN I. THORNYCROFT, LL.D., F.R.S., ...	London.
JOHN TWEEDY, ...	Newcastle-on-Tyne.
HENRY H. WEST, ...	Liverpool.

HON. TREASURER.	AUDITOR.
HARRY LEE MILLAR.	ROBERT A. McLEAN, F.C.A.

SECRETARY.

EDGAR WORTHINGTON,

The Institution of Mechanical Engineers,

Storey's Gate, St. James's Park, Westminster, S.W.

Telegraphic address:—*Mech, London.*   Telephone:—*Westminster,* 264.

THE INSTITUTION OF MECHANICAL ENGINEERS.

Memorandum of Association.

AUGUST 1878.

1st. The name of the Association is "THE INSTITUTION OF MECHANICAL ENGINEERS."

2nd. The Registered Office of the Association will be situate in England.

3rd. The objects for which the Association is established are:—

(A.) To promote the science and practice of Mechanical Engineering and all branches of mechanical construction, and to give an impulse to inventions likely to be useful to the Members of the Institution and to the community at large.

(B.) To enable Mechanical Engineers to meet and to correspond, and to facilitate the interchange of ideas respecting improvements in the various branches of mechanical science, and the publication and communication of information on such subjects.

(C.) To acquire and dispose of property for the purposes aforesaid.

(D.) To do all other things incidental or conducive to the attainment of the above objects or any of them.

4th. The income and property of the Association, from whatever source derived, shall be applied solely towards the promotion of the objects of the Association as set forth in this Memorandum of Association, and no portion thereof shall be paid or transferred directly or indirectly, by way of dividend, bonus, or otherwise howsoever, by way of profit to the persons who at any time are or have been Members of the Association, or to any of them, or to any person claiming through any of them : Provided that nothing herein contained shall prevent the payment in good faith of remuneration to any officers or servants of the Association, or to any Member of the Association, or other person, in return for any services rendered to the Association, or prevent the giving of privileges to the Members of the Association in attending the meetings of the Association, or prevent the borrowing of money (under such powers as the Association and the Council thereof may possess) from any Member of the Association, at a rate of interest not greater than five per cent. per annum.

5th. The fourth paragraph of this Memorandum is a condition on which a licence is granted by the Board of Trade to the Association in pursuance of Section 23 of the Companies Act 1867. For the purpose of preventing any evasion of the terms of the said fourth paragraph, the Board of Trade may from time to time, on the application of any Member of the Association, impose further conditions, which shall be duly observed by the Association.

6th. If the Association act in contravention of the fourth paragraph of this Memorandum, or of any such further conditions, the liability of every Member of the Council shall be unlimited ; and the liability of every Member of the Association who has received any such dividend, bonus, or other profit as aforesaid, shall likewise be unlimited.

7th. Every Member of the Association undertakes to contribute to the Assets of the Association in the event of the same being wound up during the time that he is a Member, or within one year afterwards, for payment of the debts and liabilities of the Association contracted before the time at which he ceases to be a Member, and of the costs, charges, and expenses for winding up the same, and for the adjustment of the rights of the contributories amongst themselves, such amount as may be required not exceeding Five Shillings, or in case of his liability becoming unlimited such other amount as may be required in pursuance of the last preceding paragraph of this Memorandum.

8th. If upon the winding up or dissolution of the Association there remains, after the satisfaction of all its debts and liabilities, any property whatsoever, the same shall not be paid to or distributed among the Members of the Association, but shall be given or transferred to some other Institution or Institutions having objects similar to the objects of the Association, to be determined by the Members of the Association at or before the time of dissolution ; or in default thereof, by such Judge of the High Court of Justice as may have or acquire jurisdiction in the matter.

Articles of Association.

FEBRUARY 1893.

(*Article* 23 *revised March* 1902.)

INTRODUCTION.

Whereas an Association called "The Institution of Mechanical Engineers" existed from 1847 to 1878 for objects similar to the objects expressed in the Memorandum of Association of the Association (hereinafter called "the Institution") to which these Articles apply;

And whereas the Institution was formed in 1878 for furthering and extending the objects of the former Institution, by a registered Association, under the Companies Acts 1862 and 1867;

And whereas terms used in these Articles are intended to have the same respective meanings as they have when used in those Acts, and words implying the singular number are intended to include the plural number, and *vice versâ*;

Now THEREFORE IT IS HEREBY AGREED as follows :—

CONSTITUTION.

1. For the purpose of registration the number of members of the Institution is unlimited.

MEMBERS, ASSOCIATE MEMBERS, GRADUATES, ASSOCIATES, AND HONORARY LIFE MEMBERS.

2. The present Members of the Institution, and such other persons as shall be admitted in accordance with these Articles, and none others, shall be Members of the Institution, and be entered on the register as such.

3. Any person may become a Member of the Institution who shall be qualified and elected as hereinafter mentioned, and shall agree to become such Member, and shall pay the entrance fee and first subscription accordingly.

4. The qualification of Members shall be prescribed by the By-laws from time to time in force, as provided by the Articles.

5. The election of Members shall be conducted as prescribed by the By-laws from time to time in force, as provided by the Articles.

6. In addition to the persons already admitted as Graduates, Associates, and Honorary Life Members respectively, the Institution may admit such persons as may be qualified and elected in that behalf as Associate Members, Graduates, Associates, and Honorary Life Members respectively of the Institution, and may confer upon them such privileges as shall be prescribed by the By-laws from time to time in force, as provided by the Articles: provided that no Associate Member, Graduate, Associate, or Honorary Life Member shall be deemed to be a Member within the meaning of the Articles.

7. The qualification and mode of election of Associate Members, Graduates, Associates, and Honorary Life Members shall be prescribed by the By-laws from time to time in force, as provided by the Articles.

8. The rights and privileges of every Member, Associate Member, Graduate, Associate, or Honorary Life Member shall be personal to himself, and shall not be transferable or transmissible by his own act or by operation of law.

ENTRANCE FEES AND SUBSCRIPTIONS.

9. The Entrance Fees and Subscriptions of Members, Associate Members, Graduates, and Associates shall be prescribed by the By-laws from time to time in force, as provided by the Articles.

EXPULSION.

10. If any Member, Associate Member, Graduate, or Associate shall leave his subscription in arrear for two years, and shall fail to pay such arrears within three months after a written application has been sent to him by the Secretary, his name may be struck off the register by the Council at any time afterwards, and he shall thereupon cease to have any rights as a Member, Associate Member, Graduate, or Associate, but he shall nevertheless continue liable to pay the arrears of subscription due at the time of his name being so struck off: provided always that this regulation shall not be construed to compel the Council to remove any name, if they shall be satisfied the same ought to be retained.

11. The Council may refuse to continue to receive the subscriptions of any person who shall have wilfully acted in contravention of the regulations of the Institution, or who shall in the opinion of the Council have been guilty of such conduct as shall have rendered him unfit to continue to belong to the Institution; and may remove his name from the register, and he shall thereupon cease to be a Member, Associate Member, Graduate, or Associate (as the case may be) of the Institution.

GENERAL MEETINGS.

12. The General Meetings shall consist of the Ordinary Meetings, the Annual General Meeting, and of Special Meetings as hereinafter defined.

13. The Annual General Meeting shall take place in London in one of the first four months of every year. The Ordinary Meetings shall take place at such times and places as the Council shall determine.

14. A Special Meeting may be convened at any time by the Council, and shall be convened by them whenever a requisition signed by twenty Members or Associate Members of the Institution,

specifying the object of the Meeting, is left with the Secretary. If for fourteen days after the delivery of such requisition a Meeting be not convened in accordance therewith, the Requisitionists or any twenty Members or Associate Members of the Institution may convene a Special Meeting in accordance with the requisition. All Special Meetings shall be held in London.

15. Seven clear days' notice of every Meeting, specifying generally the nature of any special business to be transacted at any Meeting, shall be given to every person on the register of the Institution, except as provided by Article 35, and no other special business shall be transacted at such Meeting; but the non-receipt of such notice shall not invalidate the proceedings of such Meeting. No notice of the business to be transacted (other than such ballot lists as may be requisite in case of elections) shall be required in the absence of special business.

16. Special business shall include all business for transaction at a Special Meeting, and all business for transaction at every other Meeting, with the exception of the reading and confirmation of the Minutes of the previous Meeting, the election of Members, Associate Members, Graduates, and Associates, and the reading and discussion of communications as prescribed by the By-laws, or by any regulations of the Council made in accordance with the By-laws.

PROCEEDINGS AT GENERAL MEETINGS.

17. Twenty Members or Associate Members shall constitute a quorum for the purpose of a Meeting other than a Special Meeting. Thirty Members or Associate Members shall constitute a quorum for the purpose of a Special Meeting.

18. If within thirty minutes after the time fixed for holding the Meeting a quorum is not present, the Meeting shall be dissolved, and all matters which might, if a quorum had been present, have been done at a Meeting (other than a Special Meeting) so dissolved, may forthwith be done on behalf of the Meeting by the Council.

19. The President shall be Chairman at every Meeting, and in his absence one of the Vice-Presidents; and in the absence of all Vice-Presidents a Member of Council shall take the chair; and if no Member of Council be present and willing to take the chair, the Meeting shall elect a Chairman.

20. The decision of a General Meeting shall be ascertained by show of hands, unless, after the show of hands, a poll is forthwith demanded; and by a poll, when a poll is thus demanded. The manner of taking a show of hands or a poll shall be in the discretion of the Chairman; and an entry in the Minutes, signed by the Chairman, shall be sufficient evidence of the decision of the General Meeting. Each Member and Associate Member shall have one vote and no more. In case of equality of votes the Chairman shall have a second or casting vote: provided that this Article shall not interfere with the provisions of the By-laws as to election by ballot.

21. The acceptance or rejection of votes by the Chairman shall be conclusive for the purpose of the decision of the matter in respect of which the votes are tendered: provided that the Chairman may review his decision at the same Meeting, if any error be then pointed out to him.

BY-LAWS.

22. The By-laws set forth in the schedule to these Articles, and such altered and additional By-laws as shall be substituted or added as hereinafter mentioned, shall regulate all matters by the Articles left to be prescribed by the By-laws, and all matters which consistently with the Articles shall be made the subject of By-laws. Alterations in, and additions to, the By-laws, may be made only by resolution of the Members and Associate Members at an Annual General Meeting, after notice of the proposed alteration or addition has been announced at the previous Ordinary Meeting, and not otherwise.

COUNCIL.

23. The Council of the Institution shall be chosen from the Members only, and shall consist of one President, six Vice-Presidents, twenty-one ordinary Members of Council, and of the Past-Presidents. The President, two Vice-Presidents, and seven Members of Council (other than Past-Presidents), shall retire at each Annual General Meeting, but shall be eligible for re-election. The Vice-Presidents and Members of Council to retire each year shall, unless the Council agree otherwise among themselves, be chosen from those who have been longest in office, and in cases of equal seniority shall be determined by ballot.

24. The election of a President, Vice-Presidents, and Members of Council, to supply the place of those retiring at the Annual General Meeting, shall be conducted in such manner as shall be prescribed by the By-laws from time to time in force, as provided by the Articles.

25. The Council may supply any casual vacancy in the Council (including any casual vacancy in the office of President) which shall occur between one Annual General Meeting and another; and the President, Vice-Presidents, or Members of Council so appointed by the Council shall retire at the succeeding Annual General Meeting. Vacancies not filled up at any such Meeting shall be deemed to be casual vacancies within the meaning of this Article.

OFFICERS.

26. The Treasurer, Secretary, and other employés of the Institution shall be appointed and removed in the manner prescribed by the By-laws from time to time in force, as provided by the Articles. Subject to the express provisions of the By-laws, the officers and servants of the Institution shall be appointed and removed by the Council.

27. The powers and duties of the officers of the Institution shall, subject to any express provision in the By-laws, be determined by the Council.

POWERS AND PROCEDURE OF COUNCIL.

28. The Council may regulate their own procedure, and delegate any of their powers and discretions to any one or more of their body, and may determine their own quorum : if no other number is prescribed, three members of Council shall form a quorum.

29. The Council shall manage the property, proceedings, and affairs of the Institution, in accordance with the By-laws from time to time in force.

30. The Treasurer may, with the consent of the Council, invest in the name of the Institution any moneys not immediately required for the purposes of the Institution in or upon any of the following investments (that is to say) :—

(A) The Public Funds, or Government Stocks of the United Kingdom, or of any Foreign or Colonial Government guaranteed by the Government of the United Kingdom.

(B) Real or Leasehold Securities, or in the purchase of real or leasehold properties in Great Britain or Ireland.

(c) Debentures, Debenture Stock, or Guaranteed or Preference Stock, of any Company incorporated by special Act of Parliament, the ordinary Shareholders whereof shall at the time of such investment be in actual receipt of half-yearly or yearly dividends.

(D) Stocks, Shares, Debentures, or Debenture Stock of any Railway, Canal, or other Company, the undertaking whereof is leased to any Railway Company at a fixed or fixed minimum rent.

(E) Stocks, Shares, or Debentures of any East Indian Railway or other Company, which shall receive a contribution from His Majesty's East Indian Government of a fixed annual percentage on their capital, or be guaranteed a fixed annual dividend by the same Government.

(F) The security of rates levied by any corporate body empowered to borrow money on the security of rates, where such borrowing has been duly authorised by Act of Parliament.

31. The Council may, with the authority of a resolution of the Members and Associate Members in General Meeting, borrow moneys for the purposes of the Institution on the security of the property of the Institution, or otherwise at their discretion.

32. No act done by the Council, whether *ultra vires* or not, which shall receive the express or implied sanction of the Members and Associate Members in General Meeting, shall be afterwards impeached by any member of the Institution on any ground whatsoever, but shall be deemed to be an act of the Institution.

NOTICES.

33. A notice may be served by the Council upon any Member, Associate Member, Graduate, Associate, or Honorary Life Member, either personally or by sending it through the post in a prepaid letter addressed to him at his registered place of abode.

34. Any notice, if served by post, shall be deemed to have been served at the time when the letter containing the same would be delivered in the ordinary course of the post; and in proving such service it shall be sufficient to prove that the letter containing the notice was properly addressed and put into the post office.

35. No Member, Associate Member, Graduate, Associate, or Honorary Life Member, not having a registered address within the United Kingdom, shall be entitled to any notice; and all proceedings may be had and taken without notice to such member, in the same manner as if he had had due notice.

By-laws.

(*Last Revision, February* 1894.)

MEMBERSHIP.

1. Candidates for admission as Members must be persons not under twenty-five years of age, who, having occupied during a sufficient period a responsible position in connection with the practice or science of Engineering, may be considered by the Council to be qualified for election.

2. Candidates for admission as Associate Members must be persons not under twenty-five years of age, who, being engaged in such work as is connected with the practice or science of Engineering, may be considered by the Council to be qualified for election, though not yet to occupy positions of sufficient responsibility, or otherwise not yet to be eligible, for admission as Members. They may afterwards be transferred at the discretion of the Council to the class of Members.

3. Candidates for admission as Graduates must be persons holding subordinate situations, and not under eighteen years of age. They must furnish evidence of training in the principles as well as in the practice of Engineering. Before attaining the age of twenty-six years, those elected after 1892 must apply for election as Members, Associate Members, or Associates, if they desire to remain connected with the Institution; they may not continue Graduates after attaining the age of twenty-six.

4. Candidates for admission as Associates must be persons not under twenty-five years of age, who from their scientific attainments or position in society may be considered eligible by the Council. They may afterwards be transferred at the discretion of the Council to the class of Associate Members or of Members.

5. The Council shall have the power to nominate as Honorary Life Members persons of eminent scientific acquirements, who in their opinion are eligible for that position.

6. The Members, Associate Members, Graduates, Associates, and Honorary Life Members shall have notice of and the privilege to attend all Meetings; but Members and Associate Members only shall be entitled to vote thereat.

7. The abbreviated distinctive Titles for indicating the connection with the Institution of Members, Associate Members, Graduates, Associates, or Honorary Life Members thereof, shall be the following:—for Members, M. I. Mech. E.; for Associate Members, A. M. I. Mech. E.; for Graduates, G. I. Mech. E.; for Associates, A. I. Mech. E.; for Honorary Life Members, Hon. M. I. Mech. E.

8. Subject to such regulations as the Council may from time to time prescribe, any Member, Associate Member, or Associate may upon application to the Secretary obtain a Certificate of his membership or other connection with the Institution. Every such certificate shall remain the property of, and shall on demand be returned to, the Institution.

ENTRANCE FEES AND SUBSCRIPTIONS.

9. Each Member shall pay an Annual Subscription of £3, and on election an Entrance Fee of £2.

10. Each Associate Member shall pay an Annual Subscription of £2 10s., and on election an Entrance Fee of £1. If afterwards transferred by the Council to the class of Members, he shall pay on transference 10s. additional subscription for the current year, and £1 additional entrance fee.

11. Each Graduate shall pay an Annual Subscription of £1 10s., but no Entrance Fee. Any Graduate elected prior to 1893, if transferred by the Council to the class of Associate Members, shall pay on transference £1 additional subscription for the current year, but no additional entrance fee; if transferred direct to the class of Members, he shall pay on transference £1 10s. additional subscription for the current year, and £1 additional entrance fee.

12. Each Associate shall pay an Annual Subscription of £2 10s., and on election an Entrance Fee of £1. If afterwards transferred by the Council to the class of Associate Members, he shall pay on transference no additional subscription or entrance fee. If transferred direct to the class of Members, he shall pay on transference 10s. additional subscription for the current year, and £1 additional entrance fee; except Associates elected prior to 1893, who shall pay no additional entrance fee on transference.

13. All subscriptions shall be payable in advance, and shall become due on the 1st day of January in each year; and the first subscription of Members, Associate Members, Graduates, and Associates, shall date from the 1st day of January in the year of their election.

14. In the case of Members, Associate Members, Graduates, or Associates, elected in the last three months of any year, the first subscription shall cover both the year of election and the succeeding year.

15. Any Member, Associate Member, or Associate, whose subscription is not in arrear, may at any time compound for his subscription for the current and all future years by the payment of Fifty Pounds, if paid in any one of the first five years of his membership. If paid subsequently, the sum of Fifty Pounds shall be reduced by One Pound per annum for every year of membership after five years. All compositions shall be deemed to be capital moneys of the Institution.

16. The Council may at their discretion reduce or remit the annual subscription, or the arrears of annual subscription, of any Member or Associate Member who shall have been a subscribing member of the Institution for twenty years, and shall have become unable to continue the annual subscription provided by these By-laws.

17. No Proceedings or Ballot Lists or Certificates shall be sent to Members, Associate Members, Graduates, or Associates, who are in

arrear with their subscriptions more than twelve months, and whose subscriptions have not been remitted by the Council as hereinbefore provided.

ELECTION OF MEMBERS, ASSOCIATE MEMBERS, GRADUATES, AND ASSOCIATES.

18. A recommendation for admission according to Form A or B in the Appendix shall be forwarded to the Secretary, and by him be laid before the next Meeting of the Council. The recommendation must be signed by not less than five Members or Associate Members if the application be for admission as a Member or Associate Member or Associate, and by three Members or Associate Members if it be for a Graduate.

19. All elections shall take place by ballot, four-fifths of the votes given being necessary for election.

20. All applications for admission shall be communicated by the Secretary to the Council for their approval previous to being inserted in the ballot list for election, and the approved ballot list shall be signed by the President and forwarded to the Members and Associate Members. The name of any Candidate approved by the Council for admission as an Associate Member or an Associate shall not be inserted in the ballot list until he has signed the Form C in the Appendix. The ballot list shall specify the name, occupation, and address of the Candidates, and also by whom proposed and seconded. The lists shall be opened only in the presence of the Council on the day of election, by a Committee to be appointed for that purpose.

21. The Elections shall take place at the General Meetings only.

22. When the proposed Candidate is elected, the Secretary shall give him notice thereof according to Form D; but his name shall not be added to the register of the Institution until he shall have paid his Entrance Fee and first Annual Subscription, and signed the Form E in the Appendix.

23. In case of non-election, no mention thereof shall be made in the Minutes, nor any notice given to the unsuccessful Candidate.

24. An Associate Member desirous of being transferred to the class of Members, or an Associate to the class of Associate Members or of Members, shall forward to the Secretary a recommendation according to Form F in the Appendix, signed by not less than five Members or Associate Members, which shall be laid before the next meeting of Council for their approval. On their approval being given, the Secretary shall notify the same to the Candidate according to Form G; but his name shall not be added to the list of Members or Associate Members until he shall have signed the Form H, and shall have paid the additional entrance fee (if any), and the additional subscription (if any) for the current year.

ELECTION OF PRESIDENT, VICE-PRESIDENTS, AND MEMBERS OF COUNCIL.

25. Candidates shall be put in nomination at the General Meeting preceding the Annual General Meeting, when the Council are to present a list of their retiring Members who offer themselves for re-election; any Member or Associate Member shall then be entitled to add to the list of Candidates. The ballot list of the proposed names shall be forwarded to the Members and Associate Members. The ballot lists shall be opened only in the presence of the Council on the day of election, by a Committee to be appointed for that purpose.

APPOINTMENT AND DUTIES OF OFFICERS.

26. The Treasurer shall be a Banker, and shall hold the uninvested funds of the Institution, except the moneys in the hands of the Secretary for current expenses. He shall be appointed by the Members and Associate Members at a General or Special Meeting, and shall hold office at the pleasure of the Council.

27. The Secretary of the Institution shall be appointed, as and when a vacancy occurs, by the Members and Associate Members at a General or Special Meeting, and shall be removable by the Council upon six months' notice from any day. The Secretary shall give the same notice. The Secretary shall devote the whole of his time to the work of the Institution, and shall not engage in any other business or profession.

28. It shall be the duty of the Secretary, under the direction of the Council, to conduct the correspondence of the Institution; to attend all meetings of the Institution, and of the Council, and of Committees; to take minutes of the proceedings of such meetings; to read the minutes of the preceding meetings, and all communications that he may be ordered to read; to superintend the publication of such papers as the Council may direct; to have the charge of the library; to direct the collection of the subscriptions, and the preparation of the account of expenditure of the funds; and to present all accounts to the Council for inspection and approval. He shall also engage (subject to the approval of the Council) and be responsible for all persons employed under him, and set them their portions of work and duties. He shall conduct the ordinary business of the Institution, in accordance with the Articles and By-laws and the directions of the President and Council; and shall refer to the President in any matters of difficulty or importance, requiring immediate decision.

MISCELLANEOUS.

29. All Papers shall be submitted to the Council for approval, and after their approval shall be read by the Secretary at the General Meetings, or by the Author with the consent of the Council; or, if so directed by the Council, shall be printed in the Proceedings without having been read at a General Meeting.

30. All books, drawings, communications, &c., shall be accessible to the members of the Institution at all reasonable times.

31. All communications to the Meetings shall be the property of the Institution, and be published only by the authority of the Council.

32. None of the property of the Institution—books, drawings, &c.—shall be taken out of the premises of the Institution without the consent of the Council.

33. All donations to the Institution shall be enumerated in the Annual Report of the Council presented to the Annual General Meeting.

34. The General Meetings shall be conducted as far as practicable in the following order:—

> 1st. The Chair to be taken at such hour as the Council may direct from time to time.

> 2nd. The Minutes of the previous Meeting to be read by the Secretary, and, after being approved as correct, to be signed by the Chairman.

> 3rd. The Ballot Lists, previously opened by the Council, to be presented to the Meeting, and the new Members, Associate Members, Graduates, and Associates elected to be announced.

> 4th. Papers approved by the Council to be read by the Secretary, or by the Author with the consent of the Council.

35. Each Member or Associate Member shall have the privilege of introducing one friend to any of the Meetings; but, during such portion of any meeting as may be devoted to any business connected with the management of the Institution, visitors shall be requested by the Chairman to withdraw, if any Member or Associate Member asks that this shall be done.

36. Every Member, Associate Member, Graduate, Associate, or Visitor, shall write his name and residence in a book to be kept for the purpose, on entering each Meeting.

37. The President shall ex officio be member of all Committees of Council

38. Seven clear days' notice at least shall be given of every meeting of the Council. Such notice shall specify generally the business to be transacted by the meeting. No business involving the expenditure of the funds of the Institution (except by way of payment of current salaries and accounts) shall be transacted at any Council meeting unless specified in the notice convening the meeting.

39. The Council shall present the yearly accounts to the Annual General Meeting, after being audited by a professional accountant, who shall be appointed annually by the Members and Associate Members at a General or a Special Meeting, at a remuneration to be then fixed by the Members and Associate Members.

40. Any member wishing to have a copy of the Papers sent to him for consideration beforehand can do so by sending in his name once in each year to the Secretary; and a copy of all Papers shall then be forwarded to him as early as possible prior to the date of the Meeting at which they are intended to be read.

41. At any Meeting of the Institution any member shall be at liberty to re-open the discussion upon any Paper which has been read or discussed at the preceding Meeting; provided that he signifies his intention to the Secretary at least one month previously to the Meeting, and that the Council decide to include it in the notice of the Meeting as part of the business to be transacted.

APPENDIX.

FORM A.

Mr.   being   years of age, and desirous of admission into The Institution of Mechanical Engineers, we, the undersigned proposer and seconder from our personal knowledge, and the three other signers from trustworthy information, propose and recommend him as a proper person to belong to the Institution.

Witness our hands, this   day of

Members or Associate Members.

FORM B.

Mr.   born on   being desirous of admission into The Institution of Mechanical Engineers, we, the undersigned proposer and seconder from our personal knowledge, and the other signer or signers from trustworthy information, propose and recommend him as a proper person to become a Graduate thereof.

Witness our hands, this   day of

Members or Associate Members.

FORM C.

If elected an   of The Institution of Mechanical Engineers, I, the undersigned, do hereby engage to ratify my election by signing the form of agreement (E) and paying the Entrance Fee and Annual Subscription in conformity with the By-laws.

Witness my hand, this   day of

FORM D.

Sir,—I have to inform you that on the   you were elected a   of The Institution of Mechanical Engineers. For the ratification of your election in conformity with the rules, it is requisite that the enclosed form be returned to me with your signature, and that your Entrance Fee and first Annual Subscription be paid, the amounts of which are   and   respectively. If these be not received within two months from the present date, the election will become void.

I am, Sir, Your obedient servant,

Secretary.

FORM E.

I, the undersigned, being elected a   of The Institution of Mechanical Engineers, do hereby agree that I will be governed by the regulations of the said Institution, as they are now formed or as they may hereafter be altered; that I will advance the objects of the Institution as far as shall be in my power, and will attend the Meetings thereof as often as I conveniently can: provided that, whenever I shall signify in writing to the Secretary that I am desirous of withdrawing from the Institution, I shall (after the payment of any arrears which may be due by me at that period) be free from this obligation.

Witness my hand, this   day of

FORM F.

Mr.   being   years of age, and desirous of being transferred into the class of   of The Institution of Mechanical Engineers, we, the undersigned, from our personal knowledge recommend him as a proper person to be so transferred by the Council.

Witness our hands, this   day of

                                      Members or Associate Members.

FORM G.

Sir,—I have to inform you that the Council have approved of your being transferred to the class of   of The Institution of Mechanical Engineers. For the ratification of your transference in conformity with the rules, it is requisite that the enclosed form be returned to me with your signature, and that your additional Entrance Fee and additional Annual Subscription for the current year be paid, the amounts of which are and   respectively. If these be not received within two months from the present date, the transference will become void.

                   I am, Sir, Your obedient servant,

                                    Secretary.

FORM H.

I, the undersigned, having been transferred to the class of   of The Institution of Mechanical Engineers, do hereby agree that I will be governed by the regulations of the said Institution, as they now exist, or as they may hereafter be altered; that I will advance the objects of the Institution as far as shall be in my power, and will attend the Meetings thereof as often as I conveniently can: provided that, whenever I shall signify in writing to the Secretary that I am desirous of withdrawing from the Institution, I shall (after the payment of any arrears which may be due by me at that period) be free from this obligation.

Witness my hand, this   day of

The Institution of Mechanical Engineers.

PROCEEDINGS.

January 1903.

An ORDINARY GENERAL MEETING was held at the Institution on Friday, 16th January 1903, at Eight o'clock p.m.; WILLIAM H. MAW, Esq., President, in the chair.

The Minutes of the previous Meeting were read and confirmed.

The PRESIDENT announced that, in accordance with the Rules of the Institution, the President, two Vice-Presidents, and seven Members of Council, would retire at the ensuing Annual General Meeting; and the list of those retiring was as follows :—

PRESIDENT.

WILLIAM H. MAW, ...	London.

VICE-PRESIDENTS.

JOHN A. F. ASPINALL, ...	Manchester.
A. TANNETT-WALKER, ...	Leeds.

MEMBERS OF COUNCIL.

EDWARD B. ELLINGTON, ...	London.
HENRY LEA, ...	Birmingham.
JAMES MANSERGH, F.R.S., . . .	London.
JOHN F. ROBINSON, ...	Glasgow.
JOHN W. SPENCER, ...	Newcastle-on-Tyne.
JOHN TWEEDY, ...	Newcastle-on-Tyne.
HENRY H. WEST, ...	Liverpool.

All of the foregoing, with the exception of the President, had offered themselves for re-election, and had been nominated by the Council.

The following Nominations had also been made by the Council for the election at the Annual General Meeting :—

PRESIDENT.

J. HARTLEY WICKSTEED, ... Leeds.

MEMBERS OF COUNCIL.

Election as
Member.

1880. MICHAEL LONGRIDGE, . . . Manchester.
1887. J. ROSSITER HOYLE, . . . Sheffield.
1894. EDWARD HOPKINSON, D.Sc., . . Manchester.
1897. ALFRED MORCOM, ... Birmingham.

All of the above had consented to the Nomination.

The PRESIDENT reminded the Meeting that, according to the Rules of the Institution, any Member or Associate Member was now entitled to add to the list of candidates.

No other names being added, the President announced that the foregoing names would accordingly constitute the nomination list for the Election of Officers at the Annual General Meeting.

The PRESIDENT announced that the Ballot Lists for the election of New Members had been opened by a Committee of the Council, and that the following forty-nine candidates were found to be duly elected :—

MEMBERS.

ANGAS, ALFRED SPELMAN, . . . Woolwich.
ATTACK, EDWARD ARTHUR, . . . Romford.
BEGBIE, THOMAS, Sen., . . . Johannesburg.
CUSACK, HENRY EDWARD, . . . Dublin.

Fletcher, James Ashton, . . .	Ashton-under-Lyne.
Gelling, John Welsby, . . .	Mexico.
Grunwell, Hartley, ...	London.
May, Charles Ramsden, . . .	Assouan.
McRitchie, George Scott, . . .	Melbourne.
Nesbitt, Alexander Walter, . .	London.
Rake, George Arthur, . . .	London.
Smith, Stanley George Drew, Major	
R.A., ...	London.
Stewart, George, ...	Bangkok.
Ullmann, William Charles, . .	London.

ASSOCIATE MEMBERS.

Alexander, William, . . .	Glasgow.
Armytage, John Hawksworth, . .	London.
Bagg, Henry Arthur, . . .	Dartford.
Bennett, Henry Arthur Dillon, .	London.
Boyd, Hugh, ...	Fleetwood.
Cave, William Henry, . . .	Bolton.
Cooke, Frederick William, . .	Rugby.
Dunkerton, Ernest Charles, . .	Glasgow.
Fryer, John Edwin George, . .	London.
Goolding, Edmund, ...	Sutton, Co. Dublin.
Hawkins, Elyot Sydney, . . .	Oswestry.
Hodson, Frank Herbert, . . .	London.
Hurst, Bertram Lawrance, . .	London.
Jewson, Herbert, ...	Dereham.
Josselyn, Edward, ...	Newark.
Knowles, George Stanley, . .	Tientsin.
Koetter, Arnold, ...	Manchester.
Lancaster, John Hayes, . . .	Rangoon.
Mackay, Harry John Sutherland, .	Manchester.
Moore, James Foster, . . .	Sunderland.
Prosser, Robert Walter Ostell, .	Dublin.
Prüsmann, Carl Adolf Louis, . .	Swansea.
Shaw, William Wilson, . . .	Manchester.

THOMAS, EVAN GIBSON, ... London.
TREGONING, WYNN HAROLD, . . . London.
YOUNG, GEORGE ALBERT, ... London.

GRADUATES.

CAMPBELL, CHARLES GRAHAM, . . . London.
CHOWDRY, KRISHNA CHANDRAH ROY, . . London.
DAVIES, CYRIL ERNEST, ... Bedford.
KNOWLES, THORNTON, ... London.
PIPER, GEORGE EDMUND, ... Dublin.
ROBINSON, ISAAC VINCENT, . . . Hartlepool.
SMITH, ULRIC VIVIAN, ... Dartford.
TAYLOR, HARRY LONGFELLOW, . . . London.
YARROW, HAROLD EDGAR, . . . London.

The following two Transferences had been made by the Council since the last Meeting :—

Associate Member to Member.

KEATINGE, SHERBROOKE AUGUSTUS JOHN, . Lucknow.

Graduate to Associate Member.

MARKS, ALFRED PALLY, ... Birmingham.

The following Paper was read and discussed :—

" Cutting Angles of Tools for Metal Work, as affecting Speed and Feed "; by Mr. H. F. DONALDSON, *Member*, of Woolwich.

The Meeting terminated at Ten minutes past Ten o'clock. The attendance was 121 Members and 58 Visitors.

CUTTING ANGLES OF TOOLS FOR METAL-WORK,
AS AFFECTING SPEED AND FEED.

By Mr. H. F. DONALDSON, *Member*, of Woolwich.

The author has approached the preparation of this Paper with considerable hesitation and reluctance, because he feels that he is unable to present the subject for the consideration of the Institution with any real finality of determination by resulting rule, which would be so desirable if it could be evolved. Seeing, however, that he was invited to communicate the results of some experiments made under his direction, he did not feel that he could lightly disregard such a call; and desires to submit the following short Paper, not so much with the idea that, taken alone it will materially add to the interest of the Proceedings, but with the hope that it will fit in with other contributions on kindred subjects, and lead up to discussion and bring out some communication of experiments on similar lines, which may have been undertaken and carried out by other members. The author craves the indulgence of members for the apparent want of finality above referred to, on the ground that the experiments hereafter dealt with were undertaken to obtain sufficient data for use in an inquiry having a still wider scope, and they consequently were only carried just far enough to fulfil the requirements of that investigation, and as a result leave something to be desired when called upon for communication to the Institution. This indulgence he is counting upon.

The importance of the study of the elements which rule speed and feed cannot fail to be recognised by any one interested in matters relating to engineering manufactures, and the author believes that one of the first of these elements, and one which has by no means received the attention which it deserves, is that relating to the individual and combined angles of the tools themselves. Without a fairly clear knowledge of the effect of such combinations, it is a practical impossibility to speak with certainty about absolute speeds which might and ought to be attained under given conditions. What a tool will do, and the angles it requires, will be found to depend upon: (1) the material of which it is made; (2) upon the nature of the material to be cut; (3) upon the speed it is desired to work at; (4) the depth of cut; and (5) the feed.

In the experiments hereafter quoted the tools were all either of ordinary self-hardening steel or ordinary tool-steel, and the materials cut were such as were in daily use in the shops in which the experiments were carried out. These materials of course have varying degrees of hardness, and for purposes of comparison Tables are given (pages 12, 14–15) in which the chemical analysis, or the leading features thereof, and the mechanical tests for each specimen are set out.

It is perhaps well to mention here that, up to the present, no special trial to investigate the best cutting angles for the recently introduced rapid-cutting tool-steels have been undertaken on similar lines. A large number of tools of that nature have been tried, but, as the requirements and the conditions demand as much similarity as possible, it naturally followed that the shapes of tools were kept as uniform as possible. Incidentally, however, these experiments showed that these new steels require more obtuse angles, as well as stouter scantling. This is what would be expected, if data obtained with ordinary tool steel was used as the basis for arriving at the correct angles. Further reference will be made to this point later.

In shops where workmen grind and always have-ground their tools for themselves, it will generally be found that from long experience and by trial and error, the best workmen have arrived at a fairly close approximation to the combination of tool angles to suit the work they have to do, but it is by no means common to find one who can name the angles at which he aims, when grinding, and still more rare is it to find one who is provided with any means of checking the correctness of his grinding, if he knows exactly what he wants. This is one of the great objections to hand-grinding, but with machine-grinding accuracy in this respect can be fully assured ; and for this reason, if for no other, machine-grinding should, in the author's opinion, be adopted to the entire exclusion of all hand-grinding in the future. There are, of course, many other reasons which could be advanced in support of the plea for reform in this direction, but the present Paper does not seem the proper place to produce them ; it will suffice to point out that, in order to secure efficient results from the use of such investigations as these, machine-grinding must be used.

The variation in the tool angles is always dependent on the nature and hardness of the material to be cut ; this is specially the case in the cutting angles, but the influence and effect of the depth of cut and the fineness or coarseness of the feed must not be overlooked. There are indications in the experiments that a light cut and a fine feed should be attacked with a sharper angle than would be permissible with a heavy cut and a coarse feed. Experience, both during the investigation and subsequently, goes to show that if the work will stand it, a heavy cut and slow speed will be more efficient in production than a light cut and quick speed, even though figures would seem to show that the result should be the same.

It will be seen that the experiments covered fourteen different qualities of material, of which seven were steel, one wrought-iron, one cast-iron, and five yellow metal alloys of different degrees of hardness, and chemical constitution. It will also be seen (page 14) that the angles of the cutting edges of tools range from 55° to 75°, while the clearance angles, though showing some variation, do not appear to have the same importance. Subsequent experience indicates that these should under no circumstances be less than 3°, but should be greater as the diameter of the work in the lathe increases. The experiments were all made with pointed roughing tools, though in some cases (notably in specimens VI and VII) the point was somewhat rounded off. The angles mentioned are as follows, Fig. 1 (page 8) :—

FIG. 1.

Cutting and Clearance Angles of Tools.

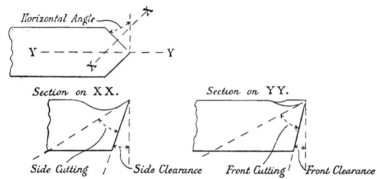

During the course of the experiments, it was thought desirable to secure some data with regard to the actual weight brought to bear upon the point of the tool when cutting, and a device was worked out by one of those to whose care the experiments hereafter detailed were entrusted, of which some short description may be of interest to Members. It is not suggested that it is in all respects perfect, or that there are not several improvements easily open to suggestion. For instance, there are parts where frictional effects are rather large, but for the purposes for which the device was produced, it was sufficiently accurate, and fulfilled its object. The apparatus may be shortly described, with the aid of the drawing, Plate 1.

It consists of an attachment, fixed by the ordinary tool studs to the slide-rest of an ordinary lathe, forming a holder for the body of the tool to be tried. Along its front edge a knife-edge A is provided, which forms the fulcrum on which the forward portion of the tool rests; at the hinder end of the tool a stop B is arranged, also beneath the tool; the tool is primarily rested upon these two points; a set-screw C is devised to be brought to bear on the top hinder end of the tool, and is adjusted when the tool is first inserted, to allow a certain amount of play, while preventing the tool from tilting an undue amount, when the cut is first put on. At a distance from the fulcrum A, about equal to that from the point of the tool to the fulcrum, is a small ram D, having a sliding block E attached to its lower end, and capable of being moved forward by the rotation of the capstan-headed screw, housed in the ram. The cylinder into which the ram fits is part of the main attachment; and the hydraulic joint is made with a small hat-leather ; into the top of the cylinder a spring gauge F is screwed. At the back of the tool a similar arrangement is provided with gauge G, the only material difference being that there is no sliding block, but the pressure is taken up on a point. The omission of a sliding block at this place, as well as at the side gauge H, is responsible for the greater portion of the frictional errors which had to be allowed for. The side gauge is arranged for on identical lines, but it may be well to eliminate this from further remark, because, owing to the friction at the end of the tool, no readings of any value were obtained. The only other moving parts are the set-screws I J disposed at each side of the tool, and in the same plane as the knife-edge fulcrum.

The process was as follows : the tool was laid upon the fulcrum A and the stop B; the set-screws I J were given a light bearing, and set-screw C was also adjusted so that the tool could not tilt much above the horizontal ; the end and side ram points were also given a firm bearing each, and the cut was put on. When the desired depth was reached, the sliding block was screwed hard down, until the axis of the tool was horizontal, and thus its contact ceased with stop B and set-screw C. The vehicle for transmitting the pressure to the gauges was oil, and the readings were direct, except that to obtain actual pressures the reading required to be multiplied by the ratio of the distance from the centre of pressure to the fulcrum, and the point of the tool to the fulcrum, seeing that it is practically impossible to fix these with exact equality.

The material operated upon was medium hard steel, the tool was an ordinary pointed one, and its angles were :—

Front cutting	60°
Front clearance	5°
Side cutting	60°
Side clearance	5°
Horizontal	9°

The following Table 1 gives results of some tests:—

TABLE 1.

Material.	Revs.	Feed.	Depth of cut.	Pressure on top in lbs. per sq. inch.	Actual pressure in lbs.	Travel.	Pressure at end in lbs.	Remarks.
		in.	in.			in.		
Medium	A 14	$\frac{1}{32}$	0·1	75·0	81·25	—	25	
hard	B 14	$\frac{1}{32}$	0·125	80·0	86·66	1	25	
steel.	C 14	$\frac{1}{20}$	0·125	150·0	162·5	1	25	At starting of cut the pressure was 250 lbs., and dropped to 150 lbs.

Speed of Material.

A = 14·65 ft per min.
B = 14·65 ft. per min.
C = 14·65 ft. per min.

It will be seen from the above that a comparatively small increase in the feed is responsible for a much more considerable increase in the pressure at the point of the tool, and that, in one experiment at any rate, there was a decided rise in the pressure at the beginning of the cut. The deductions possible from investigations of this nature will be found of great utility, in dealing not only with the steels of usual quality, but still more so if a close examination is undertaken in regard to the requisite strength of machines capable of standing up to the extra stresses likely to be brought upon them when using the newer qualities of rapid cutting tool steels. They are, moreover, valuable in considering the shape and points of tools. As stated, pointed tools were used in these tests, and it is therefore calculable, with measurements, to compare the power required if a round tool is used. With the latter more power is of course required, but for ordinary work, which will stand the strain, the efficiency, both in length of life of the tool, and in the greater amount of metal removed in a given time, seems to give a balance of advantage in its favour. When cutting the harder natures of steel this is particularly the case, because this shape of tool has necessarily more metal in it, and consequently takes up heat less quickly, and also when hot has a greater area from which to get rid of it; moreover, being stronger it is less liable to chip. Allied to this matter of strength is the subject of the method of fixing the tool in its rest; this is of much importance, for the more rigidly the tool can be fixed, the better will it cut, and the longer will it last. For this reason the nearer the point of the tool can be fixed to its rest the better; to attain this to the greatest extent possible, it will be necessary to use straight-bodied tools, and to avoid adopting tools forged with a bend or spring towards the point. "Spring" in a tool, with its attendant evils, should be avoided wherever possible. The pointed tool, however, has its distinct advantages under certain conditions, as for example, the experiments, hereafter quoted, seem clearly to indicate that it is better for light cuts and fine feeds. When using tools of rapid cutting steel, there is no doubt that the straight, heavy round-nosed tool is the best.

Turning now to Table 3 (pages 14–15), showing results of experiments with various combinations of tool angles, speeds and feeds, the author has thought it undesirable to burden this Paper with a complete list of all experiments, and therefore only enumerates the "Best Results." He gives two Tables, Table 2 (page 12) containing the chemical analysis and mechanical tests, and Table 3 (pages 14–15) the depths of cut, the speeds of material and tool, the angles of tools, the state of tools after work, the qualities of shavings, and the appearance of the work. Both Tables indicate the assessed degree of hardness.

It would be impossible, within the limits of such a Paper as this, to embark upon anything like an exhaustive analysis of the data these Tables afford, and the author therefore proposes, while remarking upon a few points, to confine himself to inviting attention to the conclusions he has drawn for himself. He has already referred to the more important of these. For instance, the tests indicate that when working with a high speed of material, a light cut, and a fine feed, different angles are required to those found most suitable for a heavier cut, a slower speed of material, and a coarser feed. Having special reference to this point, specimens II and VI are worth noticing. Both of these are similarly classed, as regards hardness, and there is only a difference in the respective cutting angles of 4°.

TABLE 2.—*Chemical Analysis and Mechanical Tests of Material Operated upon.*

Material Index No. and Degree of Hardness. (See Table 3, page 14.)	Chemical Compositions.								Mechanical Tests.	
	Carbon.	Silicon.	Manganese.	Nickel.	Phosphorus.	Sulphur.	Iron.	Manganese.	Ultimate Stress in tons per sq. inch.	Extension per cent. in a 2-inch length.
	per cent.	per cent.	per cent.	per cent.	per cent.	per cent.				per cent.
I. Steel HHH.	1·1 to 1·35	—	—	—	—	—			55 to 60	10 to 15
II. " H.	0·6 to 0·75	—	—	0·5	—	—			45 to 52	12 to 15
III. " HHN.	0·6	0·2	0·5	—	—	—			50 to 60	10 to 15
IV. " cast HH.	0·35 to 0·45	0·2 to 0·3	0·7 to 0·9	—	—	—			28 to 38	4 to 9
V. " HH.	0·3 to 0·35	0·1 to 0·2	0·5 to 0·7	—	0·04	0·02			35	30
VI. " H.	0·33	0·06	0·54	—	0·04	0·02			31	23
VII. " H.O.			0·54	—	0·1	—			33	20·5
VIII. Wrought Iron M.	0·08	trace	trace	—	—	—			23 to 25	40 to 60
IX. Cast Iron.	0·36	0·25	0·7	—		—			10 to 12	
	Copper.	Tin.	Zinc.	Lead.	Phos: Tin.	Aluminium.	Iron.	Manganese.		
	per cent.	per cent.	per cent.	per cent.	per cent.	per cent.	per cent.	per cent.		
X. G.M. hard alloy HH	90	8	2	—	2	—	—	—	19·5	30
XI. " medium " M	87 ± 2	3·5 ± 1	6 ± 2	3·5 ± 1	—	—	—	—	14·5	28
*XII. " soft " S	87 ± 2	3·5 ± 1	6 ± 2	3·5 ± 1	—	—	—	—	14·5	28
XIII. " forged " HH	60	—	40	—	—	0·25	0·25	2	27	22
XIV. " " " H.	60	—	40	2	—	0·25	—	—	32	35

The ruling factor, in fixing cutting angles, must in all cases depend to a very large extent upon the hardness of the metal to be cut ; the harder the metal, the more obtuse will be the angles of the tool. Thus in specimen II, classed as H, specimen V, classed as HH, and specimen I, classed as HHH, with a tolerably even gradation in speed of material and lightness of feed, the best cutting angles appear to be 59°, 62° and 75° respectively. In each of these examples the depth of cuts and the nature and size of tools are approximately the same. Similar evidence is forthcoming on examination of specimens VI and VII ; the material in both these cases is the same, but the latter is oil-hardened. The softer material clearly requires the sharper angle of tool, but for purposes of finish the feed is rather finer, and this supports the suggestion that, other things permitting, the variations in the angles are closely allied to speed and feed.

It will be noted, especially in connection with operations on yellow metal alloys, that the possibilities of speed of working on hard, medium, and soft material rise in accordance with the nature of the material, but that with medium and soft material, specimens XI and XII, no change of angles was necessary ; from this it may be deduced that the tendency of the soft metal to hang to the point of the tool is responsible for the obtuseness in the required angle.

The information gathered from these investigations has not been without practical results in several subsequent steps. It has tended to assist in the consideration of more than one subject connected with shop administration, with, as the author believes, very beneficial results. Among other good objects, it affords a reasonably sound basis for determining the cutting speeds with which those in authority may be content; it therefore will naturally be available and useful in the formation of estimates, whether for rate fixing, promises for delivery, or quotations to customers. It can also be used for the assistance of the workmen in the shop. Rule of thumb, and practical experience, of course have their value in dealing with such subjects, but, in the author's opinion, work-shop administration can be, and ought, in these days, to be based upon a more defined foundation. This can be much assisted by a study of anything which

ABLE 3 (*continued on next page*).

Table showing Best Results

Material, Index No., and Degree of Hardness.	Diameter.		Speed of Material.		Speed of Tool.			Tools and Angles.					
	Before.	After.	Revolutions per minute.	Feet per minute.	Feed.	Travel.	Time.	Nature of Tool.	Front Cutting.	Side Cutting.	Front Clearance.	Side Clearance.	Horizontal
	ins.	ins.			in.	ins.	mins.	inch.	deg.	deg.	deg.	deg.	de
. Steel HHH.	3·4	3·2	18	15	1/40	3	6·75	1⅛ Mu.	75	69	2	2	3
I. Steel H.	1·85	1·7	70	32	1/70	6	6	1 C.S.	59	62½	3	3	3
II. Steel HHN.	5·75	4·95	18	27	1/60	3	9	do.	62½	62½	1	·1	3
V. C. Steel HH.	6	5·15	18	28	1/40	1	2·25	do.	65	63	2½	1	4
. Steel HH.	1·5	1·25	70	27½	1/61	½	0·44	do.	62½	62½	3	3	4
I. Steel H.	8·35	7·5	5½	12	⅛	10	15	1¼ Mu.	55	52	4	5	3
II. Steel H O.	6·75	6·25	5⅓	10	1/14	18	44	do.	57½	55	8	8	4
III W. Iron M.	3	2·75	70	55	1/70	2	2	1 C.S.	57½	55	3	3	4
X. C. Iron	4·95	4·2	18	21	1/27	2	3	do.	60	64	1¼	1	3
. G.M.F.	3·37	3	140	122	0·02	4	1·25	do.	61	64	3	0	3
HH.	3	2·7	320	251	0·03	6	0·6	do.	61	64	3	0	3
I. G. Metal M.	2·6	2·37	320	217	0·03	6	0·6	do.	73	74	2	½	3
II. G.Metal S.	3·5	3·25	320	293	0·03	5	0·5	do.	73	74	2	½	3
III. G.Metal F. HH.	2·65	2·45	224	160	0·02	4	0·75	do.	78	78	1	0	3
IV. G. Metal F. H.	2·4	2·125	224	140	0·02	3	0·55	do.	62½	64	3	3	3

Very hard material indicated HHII.   Soft material indicated S.
Medium hard „ „ HH.   Forged „ „ F.
Hard „ H.   Nickel „ „ N.
Oil hardened „ O.   Mushet „ „ Mu.
Medium „ „ M.   Cast Steel „ „ C.S.

(*concluded from opposite page*) TABLE 3.

Operating Angles of Tools.

Lubricant used.	State of Tool-edge after work.			Shavings.		Surface of Specimen after Work.			Material, Index No. and Degree of Hardness.
	Not Damaged.	Slightly Damaged.	Completely Damaged.	Long.	Short.	Smooth.	Rough.	Very Rough.	
Soda Hyd.	—	Yes	—	—	Yes	Good	—	—	I. Steel HHH.
Do.	—	Yes	—	Yes	—	do.	—	—	II. Steel H.
Do. and Soap	—	{Rough on Edge}	—	—	Yes	do.	—	—	III. Steel HHN.
Do. do.	—	do.	—	Long	—	do.	—	—	IV. C. Steel HH.
Do. do.	Not	—	—	Yes	—	do.	—	—	V. Steel HH.
Do. do.	Not	—	—	Yes	—	do.	—	—	VI. Steel H..
Do.	Not	—	—	Yes	—	do.	—	—	VII. Steel H.O.
Do.	Not	—	—	Yes	—	do.	—	—	VIII. W. Iron M.
Nil.	—	Yes	—	—	Yes	do.	—	—	IX. C. Iron
{ Soda Hyd.	Not	—	—	Yes	—	do.	—	—	X. G.M.F. HH.
{ Hyd.	Not	—	—	Yes	—	do.	—	—	
Do.	Not	—	—	—	Yes	do.	—	—	XI. G. Metal M.
Nil.	Not	—	—	—	Yes	do.	—	—	XII. G. Metal S.
Soda Hyd.	Not	—	—	—	Yes	do.	—	—	XIII. G. Metal F. HH.
Do.	Not	—	—	Yes	—	do.	—	—	XIV. G. Metal F. H.

Very hard material indicated HHH.   Soft   material indicated S.
Medium hard „   „   HH.   Forged   „   „   F.
Hard   „   „   H.   ₅ Nickel   „   „   N.
Oil hardened „   „   O.   Mushet   „   „   Mu.
Medium   „   „   M.   Cast Steel „   „   C.S.

appears to contain an approach to definite laws. The author believes that a careful study of matters relating to cutting angles of tools will fully repay the trouble.

A practical illustration of how beneficial results have occurred, in one place at any rate, may perhaps be of interest, and useful. In a shop which was occupied on repair work, and equipped with a considerable number of lathes, the cutting speeds in general use were very variable, more so than the changes of the work required seemed to warrant. The workmen ground their own tools, and so long as that remained the practice it did not seem feasible to formulate definite instructions as to the cutting speeds at which the men would be expected to work. Machine grinding was introduced, and tools at approved angles were prepared for issue to the men as required. From the data of the experiments, determinations were made for the best speeds for cutting all classes of material in general use, using 1-inch diameter as the constant, and the following letters were adopted to indicate the different classes of material :—

S. V. H. means Steel very hard.	I. C. means Iron cast.
S. H. „ Steel hard.	B. H. „ Brass hard.
S. M. H. „ Steel medium hard.	B. S. „ Brass soft.
S. T. „ Steel tough.	B. F. H. „ Brass forged hard.
S. S. „ Steel soft.	B. F. S. „ Brass forged soft.
I. W. „ Iron wrought.	

A small brass plate was prepared and attached to each machine, bearing figures denoting the determinations referred to of the following form :—

Material.	Revolutions for 1 inch diameter.
S. V. H.	58
S. H.	84
S. M. H.	102
S. T.	107
S. S.	105
I. W.	210
I. C.	80
B. H.	468
B. S.	860
B. F. H.	615
B. F. S.	530

The face of each step on the cone-pulleys was stamped with the number of revolutions the spindle would make when running with single, back, and triple gear. Armed with the information which these permanent markings afford, the adjustments to be made become a very simple operation for the workman ; for all he has to do, to at once start his machine at its correct speed, is to divide the number of revolutions for 1-inch diameter by the diameter in inches of the work to be operated upon, which will give him the number of revolutions required, and he can at once arrange his belt on the necessary step of his cone pulley. Besides the advantages accruing from such a method, other benefits may be mentioned ; for instance, the foreman in charge can at a glance tell if work entrusted to him for execution is being arranged by the workman at speeds, which will ensure his estimates both for time and cost resulting. Such a display of proof of sound shop administration will be reflected on the workman, and cannot fail to act beneficially on the productions of the shop. The author has referred to the possible charge, that the results of the experiments show a want of finality, which may mar their efficiency ; he is fully alive to the possible soundness of such a criticism, but at the same time, hopes that the data offered may yet afford a basis from which others may start in the selection of tool angles most suited to their own special requirements. He believes that variation in practice in individual shops will be found to affect materially the determinations as to the best tool angles. With this in view, he has less hesitation in offering the following Table of margins based upon the experiments, even though their wideness may be considered to almost defeat the attainment of the object in view.

The criticism, that these margins are too wide for practical use, will best be met by the acknowledgment that the author has perhaps endeavoured to compress the variable requirements within too narrow limits, but he would point out that it would unduly prolong the Paper to attempt to embark upon all the possible contingencies, which he believes would arise, when considering the requirements of a particular shop. The large allowance in the margins is also made in order to cover variations in degree of toughness (as distinct from

TABLE 4.

Limits of Cutting Angles.

	Front cutting.	Side cutting.	Front clearance.	Side clearance.	Horizontal.
	deg.	deg.	deg.	deg.	deg.
For cutting mi'd steel . .	52 to 60	50 to 60	3 to 8	3 to 8	33 to 43
For cutting medium steel .	54 to 63	60 to 65	3 to 8	3 to 8	33 to 43
For cutting hard steel . .	65 to 78	60 to 70	3 to 8	3 to 8	33 to 43
For cutting soft yellow metal	62 to 74	62 to 74	3 to 8	3 to 8	33 to 33
For cutting medium yellow metal	62 to 74	70 to 75	3 to 8	3 to 8	33 to 38
For cutting hard yellow metal	60 to 80	60 to 80	3 to 8	3 to 8	33 to 38

hardness), variations in depth of cut, speed of material, speed of feed, and form of tool. The author is well aware that the subject could readily be expanded by closer examination of details, but he hopes that the remarks he has made may be the means of raising such discussion as will tend to a closer examination of the principles, which he believes underlie, and matters which depend upon the subject under review, than is, he thinks, generally given to them.

In conclusion, the author desires to express his thanks to all those, especially those in administrative control of departments and shops where the experiments were carried out, and particularly to those who brought special devices and appliances to bear, to assist in the attainment of the desired information.

The Paper is illustrated by Plate 1 and one Fig. in the letterpress, and is accompanied by an Appendix.

APPENDIX.

Some pressure has been brought to bear upon the author to afford some information regarding tests made under his direction with the newer classes of rapid-cutting tool-steels as an addendum to the foregoing Paper; he has much hesitation in doing this, because the experiments were made with a view to the selection of steels most likely to meet the requirements of the work which has to be done on the material generally used, and consequently, it is difficult to avoid reference to the firms from which the steels were procured. He does, however, feel that he cannot entirely refuse the request, as he is fully alive to the great desirability for the widest possible dissemination of information on this subject. He will therefore endeavour to formulate some remarks, which will assist in this direction, to the greatest extent possible without disclosing the names of the makers of the steels with which he has experimented up to the present.

The trials show that a rapid cutting steel tool, which gives excellent results in cutting mild material, is by no means certain to give correspondingly good results if required to cut very hard material. The converse is also true. Consequently, in examining the question, or in selecting a rapid-cutting steel, it is essential to take into consideration the nature of the material to be cut.

Twenty-seven different brands of steel have been tried and in many of these, under several sets of conditions. Table 5 (pages 20–21) shows twenty-four of these, and from them six have been selected as most satisfactorily meeting requirements. Table 6 (page 22) sets out some of the leading features; the brands are indicated alphabetically.

It will be noticed that only two brands are selected for working on hard steel, while four are found suitable for medium or mild steel. This is no doubt due to the fact that, generally speaking, medium or mild steel is dealt with in most shops, and consequently, steel-makers have aimed at supplying general requirements, and it is

D 2

TABLE 5 (*continued on opposite page*).

Results *of* Trials of High Speed Tool Steels.

Covering a period from September 1901 *to* November 1902.

Name of Steel.	Cutting Speed.	Depth of Cut.	Feed of Tool per Revolution.	Weight of Material Removed per Hour.	Average Time of Cutting.	Travel of Tool.	Condition of Tool after Test.	Order of Merit of Tool.
	feet per min.	in.	in.	lbs.	hours.	ins.		
W	17	0·25	0·046	20·48	0·116	1·43	{Required regrinding}	—
I	17	0·25	0·046	21·175	2·405	15	Do.	—
	38·6	0·35	0·1	281	7·6	—	Do.	—
	13·5	0·5	0·05	—	—	—	—	—
E	17	0·25	0·046	18·45	1·315	7·3	{Required regrinding}	—
	33	0·6	0·076	229	7·1	—	Do.	{3rd Soft Steel}
	13·5	0·93	0·05	80	9	—	Good	—
B	17	0·25	0·046	20·48	5·8	72	{Required regrinding}	{2nd Hard Steel}
D	50	0·26	0·1	269·7	8·5	—	Good	{2nd Soft Steel}
J	13·5	0·93	0·05	148	4·66	—	Good	—
K	13·5	0·93	0·05	143·5	4·25	—	Good	—
L	13·5	0·93	0·05	128·5	4·8	—	Fair	—
N	13·5	0·93	0·05	80	9	—	Bad	—
	39·5	0·228	0·098	176	2·85	—	{Required regrinding}	—
M	13·5	0·93	0·05	80	9	—	Good	—
	35	0·2	0·1	143	0·106	—	{Required regrinding}	—
O	37·3	0·215	0·1	179·2	3·95	—	Do.	—
R	28	0·15	0·08	78·3	1·62	—	Do.	—

(concluded from opposite page) T

Results of Trials of High Speed Tool Steels.
Covering a period from September 1901 to November 1902.

Nature of Material	Name of Steel	Cutting Speed	Depth of Cut	Feed of Tool per Revolution.	Weight of Material Removed per Hour.	Average Time of Cutting.	Travel of Tool.	Condition of Tool after Test.	Order of Merit of Tool.
		feet per min.	in.	in.	lbs.	hours	ins.		
Only tried on Cast Steel Soft Steel	A	44·5	0·33	0·1	313	4·12	—	Fair	
		17	0·25	0·063	25·6	12·5	214	Point broke due to accident	1st Hard Steel
Do.	S	17	0·25	0·046	19·98	3·71	41	Required regrinding	
Do.	T	17	0·25	0·046	19·75	1·37	15·5	Do.	
Do.	Y	Would not cut at all at these Feeds.							
	Q	17	0·25	0·062	25	0·35	6	Required regrinding	
		39·25	0·244	0·1	196·1	3·04	—	Edge worn away	
Suitable for Soft and Hard Steel	F	38·5	0·27	0·1	217·3	4·78	—	Cutting edge slightly worn	4th Soft Steel
		13·5	0·94	0·05	186·6	2·42	—	Good	
	V	13·5	0·94	0·05	166·8	0·83	—	Bad	
		26	0·1	0·008	4·25	0·25	—	Fair	
	X	28	0·2	0·02	24	0·116	—	Edge worn off Tool	
	C	42	0·41	0·10	350	13·5	—	Required regrinding	1st Soft Steel
	H	43·75	0·25	0·1	227·3	5·5	—	Good	
Medium Steels		21	0·2	0·047	40	1·34	26·75	Required regrinding	
Medium Steel	G	42	0·367	0·1	321·7	2·75	—	Do.	
		21	0·2	0·047	39·7	3·64	75·5	Do.	
	P	35·5	0 15	0·1	114	3·5	—	Do.	
		21	0·2	0·047	39·7	2·42	51	Do.	

TABLE 6.—*Rapid Cutting Tool Steel Trials.*

Index Mark of Brand.	Nature of Material Cut.	Cutting Speed.	Depth of Cut.	Feed of Tool per revolution.	Weight of Material Removed.	Average Time of Cutting.	Condition of Tool.
		feet per minute.	inch.	inch.	lbs. per hour.	hours.	
A*	Very hard.	17·0	0·25	0·063	25·6	12·5	Point broke due to accident.
B	Do.	17·0	0·25	0·046	20·48	5·8	{ Required re-grinding.
C	⎫	42·0	0·41	0·10	350·0	13·5	Do.
D	⎬ Hard, oil hardened.	50·0	0·26	0·1	269·7	8·5	Good.
E	⎭	33·0	0·6	0·076	229·0	7·1	{ Required re-grinding.
F		38·5	0·27	0·1	217·3	4·78	Slightly worn.

* Lathe broke down, owing to insufficient power.

therefore easier to get a greater area of supply in grades of this quality. The author does not propose to comment at any length on the foregoing Table, but thinks that he may usefully draw some comparisons, with regard to the weight of metal removed, between these tests and some of the experiments dealt with in the body of the Paper, where the relative degrees of hardness coincide.

The material in specimen I in Table 2 (page 12) was of about the same degree of hardness as was operated upon by tools A and B in Table 6 (page 22). The material removed from specimen I was equal to about $7\frac{1}{2}$ lbs. per hour, while tools A and B removed about $25\frac{1}{2}$ lbs. and $20\frac{1}{2}$ lbs. per hour respectively. Specimen VII would represent equivalent hardness of material for comparison with that operated upon by tools C, D, E, F. The metal removed from specimen VII was about $35\frac{1}{2}$ lbs. per hour, while tools C, D, E, F removed 350 lbs., 269 lbs., 229 lbs., 217 lbs. per hour respectively. This comparison brings out the interesting point that while the new rapid-cutting tool-steels show speeds of work from six to nearly ten times faster than when working with ordinary tool-steels, only about three times faster can be attained when working on the harder qualities of material. It is difficult to account for this, except by the theory that steel-makers have directed their attention more exclusively to meeting the demands for quick removal of large weights of mild or medium hard metal ; at the same time, there is reason to think that many steel-makers have not in all cases realised that tools for operating upon very hard material must possess special qualities.

Remarks in the body of the Paper (page 7) have already been made indicating that a round-nosed tool will, other things being equal, give better results in the amount of metal removed than a pointed tool. This is very marked in the case of rapid-cutting tools, for not only are the latter stiffer and stronger but they stand up to their work better and are more durable. No special experiments have been made as to cutting angles for these tools, but experience seems to show that the cutting angles should be a few degrees more obtuse than is required for tools of ordinary steel.

In conclusion, it seems desirable to draw attention to the fact that, in considering the quantities of metal removed in the two sets of experiments, true comparisons are not possible, because in the tool angle experiments no effort was made to test the tool to destruction, while in the rapid-cutting trials such an end was actually aimed at.

Discussion.

The PRESIDENT said the Paper dealt with a subject which was not only of very great importance at the present day, but was of growing importance. If the point raised by the author in the early part of his Paper was considered, that nowadays the whole tendency was towards having tools ground to certain definite angles and supplied to the workmen for their use without giving the workmen the option of changing those angles, it would at once be seen that it was particularly essential that the angles so chosen should be correct, and it was only by careful experiments, such as those carried out by the author, that the necessary angles could be determined and results obtained which would, as it were, compensate for dispensing with the individual skill of the better class of workmen which had hitherto been relied upon. The Institution was indebted to the author for the great trouble he had taken in preparing the Paper. He (the President) felt personally indebted to Mr. Donaldson to a very large extent, because it was chiefly at his request that the author was kind enough to gather together the data which had been embodied in the communication. He was sure the members would feel that the author had done excellent work for them, and he proposed a hearty vote of thanks to him for having written the Paper.

Mr. DONALDSON said he was glad to have the opportunity of saying a few words before the discussion commenced, because the issue of the Paper took place somewhat hurriedly before he was able to correct some details, and therefore there were one or two points which seemed to him to require explanation. For instance, with regard to the Tables in the Appendix (pages 20–22), originally the Appendix contained only one list, Table 6, but during the course of printing he was able to supply Table 5, of which Table 6 was a part. It was a little troublesome perhaps that two Tables should be published containing the same information. Table 6 gave the Table of best results, which were also included in Table 5. There were also some slight errors in the print of Table 1 (page 10). Instead of "Pressure on top in square inches per lb.," he thought it was tolerably evident it should be the reverse, "Pressure on top in lbs. per square inch," and in the next column the actual pressure should be only in lbs. and not per square inch. [These corrections have now been made.]

One of the main difficulties connected with the subject of the Paper was that nearly everybody did the same as was done in his case. The examination of the angles was, as he had said, a part only of another subject, and as soon as sufficient data had been collected the whole of the experimental work, so to speak, was dropped, and the main subject was proceeded with. Consequently the subject was never comprehensively examined. It was one which he thought might, could, and should be extended on its own merits, because there was a great deal of information to be gathered from a closer examination than had, so far as he knew, ever been accorded consistently throughout the whole enquiry to the subject of tool angles.

With reference to the apparatus for measuring the pressure on the point of the tool, Plate 1, so far as he knew nothing of that sort had been tried before. The measurement had never, he believed, been taken directly from the tool, as in this case. If taken in any other way, the possibility of errors of reading was much increased. Experiments to measure a certain portion of the power absorbed by the machine on the tool had been taken upon entirely different lines, and with

(Mr. Donaldson.)

entirely different results. He had hoped that Colonel Holden would
have been present that evening, but he had sent a message saying
that at the last moment he was prevented from coming. He
(Mr. Donaldson) had hoped that Colonel Holden would have said
something about a measurement which was taken out for an entirely
different purpose, with the idea of testing what would be the effect of
taking off cuts with a square-nosed tool in a machine which was now
being worked out by a committee appointed for the purpose of
preparing a standard leading screw, and a machine which would
reproduce those standard leading screws. In parenthesis, he might
say that that committee was nearly at the end of its labours, and he
hoped at no distant date the machine would be capable of reproducing
correctly standard leading screws under the direction of the National
Physical Laboratory. That, of course, dealt with another subject,
but it was the fact that an experiment had been made on different
lines which took in the frictional error to a certain extent of the
moving part being cut, and in the examination certain differences
had been brought out which at the present time he confessed he was
unable to account for; the differences existed, and he hoped they
might eventually be able to discover how those differences were
caused.

On the question of high-speed steels, Colonel Holden had told
him, in answer to a question during a conversation with him that
day, that the greatest amount of metal removed by one tool in an
hour from oil-hardened steel was 600 lbs., and that he hoped to
improve on this performance and get a ton off with two tools. Such
a result was by no means impossible.

With reference to the measurement of the weight on the point of
the tool, in order to get a common comparison he had taken the
cubic inches of the material removed per minute at a given speed, a
given feed, and a given depth, and had applied that to the high-speed
trials in order to show by calculation, based upon those premises,
what sort of pressure was taken by the point of a tool of one of the
high-speed cutting steels. Taking the examples given, he found
that in the second example of M it was 1,050 lbs. on the point of
the tool; in the first of the M's it was 955, while the highest figure

was at F, where 1,580 lbs. on the end of the tool was obtained.   He
ventured to think that an extension of experiments of that kind
might be of great use to machine builders, who were setting out,
as he was glad to see some were at present, to evolve machines
which would be capable of standing up to the work which the tools,
or some of them at any rate, were capable of doing.   At present he
had broken down some of the machines without breaking the tools.

Mr. H. F. L. ORCUTT thought the Institution was indebted to the
author for bringing out the facts contained in the Paper.   As
Mr. Donaldson said, it was to be hoped it would lead to a
development in other shops of further experiments to be carried out
on similar lines.   He wished, in the first place, to refer to the
general importance of investigating the subject, and the necessity for
manufacturers to determine upon specific angles, speeds, and feeds by
which the greatest product could be obtained from their machines
and tools.   The author very modestly refrained from bringing out
some of those points, and he hoped he would be pardoned if he
mentioned one or two very briefly.   In the first place, standardizing
the tools and definitely fixing the angles would certainly save
steel.   He thought he had mentioned to the Institution before, that
one company had gone into the question of standardizing their
angles, and found two tons of useless tool-steel in their works.
Every workman had his own kit and his own stock of tools, and
a great amount of that steel was done away with ; probably it would
average 1s. a pound.

Another important point, referring particularly to shop
administration, which the author had mentioned, was that
standardizing angles and systematizing the grinding of tools
assisted materially to maintain order in the shop ; he thought order
was of great importance in shop administration.   Another point
was that it made it possible to obtain the maximum product from
all the machines by fixing upon the angles, and not allowing the
workmen to change them.   It saved cost in the actual grinding
of the tools, a point to which the President had alluded.   A workman
who was turning, and who was supposed to be a skilful machine

(Mr. H. F. L. Orcutt.)

operator, ought not to waste his time in sharpening tools; that work should be delegated to a man who did nothing else, because with the aid of the machines now on the market it could be done by comparatively unskilled labour.

In reference to roughing cuts he took occasion to look up the results of some experiments which were made originally by Messrs. William Sellers and Co. who, he believed, were the first to study the matter systematically and thoroughly. They brought out some years ago, as the result of their experiments, a machine especially designed for grinding such tools. He noticed that they mentioned that for roughing cuts a curved edge on the tool was more efficient than a straight edge. He would like to ask the author if he had made any experiments in that direction, because the curve came in as well as the angle. Messrs. Sellers also mentioned that the tool should be arranged so as not to gouge into the work, but rather so that it would have a dragging cut; that also gave greater efficiency. He was sorry the author had not spoken of roughing tools a little more; he seemed to have confined himself principally to finishing work. He thought the roughing tool question was a most important one, for two reasons. The first was that the majority of the hard work of turning tools was the roughing; the finishing was comparatively light work. Another reason was that for a very great deal of the finishing of cylindrical surfaces the lathe disappeared altogether, on account of the increased use of the grinding machine which was now coming in, and the parts prepared for the grinding machine needed mainly to be considered with reference to the roughing tool. Some of the points brought out by the author did not quite agree with Messrs. Sellers' experiments, which were very complete, especially the statement (page 11) that when tools of rapid cutting steel were used there was no doubt that the straight, heavy, round-nosed tool was the best.

Mr. DANIEL ADAMSON said he was very much interested in the subject, on account of some experiments which were being carried out in Manchester at the new Municipal School of Technology under the auspices of their local Association of Engineers; therefore, if he appeared to criticise the Paper or to ask a number of questions, he hoped the author would understand it was with the idea of eliciting from him further information than was given in the Paper, for which his fellow-members on the committee, and also the tool-steel makers who were helping them in carrying out the experiments by sending tools, would be very grateful.

He would like to ask the author first whether the experiments were confined to lathes. The results given in the Paper appeared to refer entirely to lathes, but possibly in arriving at the deductions given the author might have had other machines in view. One very important omission he noticed, assuming all the experiments referred to work on turning lathes, was that nothing was said of the position of the tool with regard to the centres of the lathe. If that was omitted, or if no notice was taken of it, the information given about angles was, as most of the members would understand, of considerably less value, because the height of the tool in connection with the centres affected the relation of the angles of the tool to the work being cut. The author said (page 7), "even though figures would seem to show that the result should be the same." He could not understand what that meant, unless it was that, although the product of the speed and the area of the cut might by calculation show the same result in the case of a light cut and a heavy cut, the author intended them to understand that a better result was obtained with a heavy cut. There he thought the author was not very clear, because it was certain that if the work or the machine would stand it, more material could be removed by working with a heavy cut at a slow speed.

On the same page (7) the author said that the clearance should be greater as the diameter of the work in the lathe increased. That, he believed, was an error. If the tool was placed in the rest, so that the cutting edge was level with the centres of the lathe, then it was immaterial to the tool what diameter the work was. The work was passing the front of the tool in the vertical direction at right angles to the tool; therefore the clearance was just the same as if the same tool was in use in a planing machine working over a plane surface. If the tool was raised above the centre, as was usual, then the clearance ought to be increased as the work was reduced,

(Mr. Daniel Adamson.)

just the opposite to what was mentioned by the author. Perhaps the author would explain how he arrived at this finding.

The members were asked to compare specimens VI and VII (page 7); in one case the point being rounded off; but later on (page 11) the author asked them to compare No. VI with No. II, saying that the chief difference between the two experiments was the cutting angle of the tools, a difference of 4°. There were, however, other differences between specimens II and VI than the cutting angle. He would also like the author to explain more fully than he had done the meaning in Table 1 (page 10) of "the pressure on top in square inches per lb.," and also "the actual pressure in lbs." On the same page the author mentioned that the deductions possible from investigations in connection with the pressure on the tool under various cuts would be found of great utility ; but he (Mr. Adamson) suggested that, for reliable information on the point, heavier cuts ought to have been taken than a $\frac{1}{32}$-inch feed by $\frac{1}{10}$-inch cut. Those were particularly small cuts on which to base any calculation as to the amount of pressure that would come on the tool's end, with such cuts as would make the pressure an important amount.

The author also said (page 10) that it was "calculable to compare the power required if a round tool is used." He would like to ask the author to explain how he would advise them to compare the data given for the pointed tool with the round tool, which was in normal use. The author spoke favourably of the round tool ; it was therefore unfortunate that he had not given some figures of experiments with round tools. In order to ascertain for himself some idea of the comparison between the tool specified by the author and the round tools in normal use, he (Mr. Adamson) had a tool prepared of the old-fashioned cast-steel—the ordinary water-hardened steel that the author's experiments referred to—in accordance with the particulars (page 9) ; and the other end of the same tool he had forged into the normal round-nosed shape which was found suitable in their shops. He had the tool run on two forgings, one after the other, with the result that he found the round tool would run twice the length on the same bar, before destroying the cutting edge, than the tool ground to the angles given (page 9) would ; from which he concluded that under those conditions a round tool might be taken to be twice as effective as the pointed tool given in the Tables. He hoped the author would be able to inform them that he had conducted further experiments of the same character which would make that point clearer.

The author mentioned (page 11) that the pointed tool was better for light cuts and fine feeds. That, he believed, was because the pointed tool, however pointed, had a small radius on the point; therefore that small radius when cutting with a light cut was in proportion to the larger radius of the round tool on the heavier cut. The conditions geometrically were similar, supposing one had the small radius on the fine cut and the larger radius on the heavier cut. Further, the pointed tool would not, he suggested, stand with the heavier cuts, while the rounded tool was not suitable for the fine cuts; the ordinary normal rounded tool would not take fine cuts without chattering, the large radius and small depth of cut made the tool act more like a scraper than a tool.

The author said (page 11) that he enumerated only the "best results." In giving the best results he might have given some hint as to the penalty for departing from the angles which gave those "best results," in order that the angles might be appreciated at their true value. Would a slight variation from one angle to another make a serious difference in the "best results," and if so, how much? The author said, "The tests indicate that when working with a high speed of material, a light cut, and a fine feed, different angles are required to those found most suitable for a heavier cut, a slower speed of material, and a coarser feed." The author had only told them that for the harder material more obtuse angles were advisable, but had not given any recommendation as to the amount of variation under these different conditions. In his Table of hardness of different materials (page 12), No. II steel was marked H, and No. V steel was marked HH, which he took to mean that No. V was harder than No. II, but No. II contained twice the amount of carbon on analysis that No. V did, and therefore should be the harder of the two. The manganese in the higher carbon steel was not mentioned; only the carbon was given, therefore the hardness could only be

(Mr. Daniel Adamson.)

judged on the carbon, and according to the carbon the No. II should be harder than the other; but that might be a clerical error.

It was stated (page 13) " among other good objects, it affords a reasonably sound basis for determining the cutting speeds with which those in authority may be content." Again, he would ask the author, for their guidance in Manchester, how to compare the results named in this Paper with those to be expected from the modern fast cutting steels which were now in use in Lancashire shops. How could they make use of the figures given in the Paper to enable them, as the author suggested, to " determine cutting speeds with which they may be content?" Perhaps he would state his methods of deduction in arriving at that result. A Table showing the " best results " was given (pages 14 and 15). He would like to ask the author how many experiments those " best results " were based upon, in order that some basis might be obtained for judging the value of the results as " best results." Again, were the same tests repeated sufficiently often to ensure that they were the best results, and yet results which could be obtained again under the conditions specified in the Paper? One thing that made him doubtful on that point was the column marked " Lubricant used." If a lubricant was used, it was very important that it should always be applied under the same conditions, and he therefore asked the author to state how the lubricant was applied under the conditions mentioned. Was it pumped on lavishly, or was it dripped on from a can, as was done in the old days? He would also like to ask in regard to the best results whether the cuts were first cuts cr finishing cuts. Mr. Orcutt (page 28) had inferred that they were finishing cuts, but it would be advisable to have the point definitely stated. Again, how was it decided that these were " best results ; " according to the statement (page 24), the tools were not run up to the limits of their endurance.

The author referred (page 16) to five varieties of steel, but he did not give much data by means of which it was possible to compare or connect those five varieties with the seven analyses on page 12. He also spoke of four varieties of brass (page 16), and recommended certain speeds (at the foot of the page) for that brass. The speeds suggested for brass could be increased 50 per cent. with the ordinary tool steels. It seemed to him that Table 4 (page 18) was the most valuable part of the Paper; in fact, the centre round which the whole revolved. The cutting angles for steel agreed with Lancashire shop practice, but with regard to those for cutting yellow metal, he would ask the author to give some analyses of that yellow metal, or something to enable one to connect the *three* varieties of yellow metal (page 18), with the *four* varieties of brass mentioned (page 16), and the *five* varieties of gun-metal (page 12), with their analyses. At present there was nothing in the Paper to connect those various qualities with one another, therefore it was impossible to refer from one part of the Paper to the other with satisfaction. In each case where brass or yellow metal was mentioned, it should be stated whether cast or rolled was meant.

The author mentioned that in cutting hard yellow metal the front clearance should be from 3° to 8°. That, he thought, was an error. A tool for cutting brass must be more approaching a tool for cutting wood, that is, the front clearance must be more like 30° than 3°. The front *cutting* angle appeared to be about correct, so that, if the front *clearance* angle was modified, the cutting tool for brass would be obtained which was used in Lancashire practice, the only practice of which he had knowledge. In Table 5 (pages 20–21) there was nothing to indicate clearly what the material operated upon was. A Table was given showing what certain steels were recommended for, but as neither the steels nor the materials on which the tests were carried out were known, it was not much of a guide. The size of section of tool steel used was not recorded, this was most important. Nor was the subject of the Paper referred to at all, as far as he could see, in that Table; nothing was said about the angles of the tools with which the experiments were carried out; these could easily have been recorded. The comparison between new tool-steels and old tool-steels was considerably over-stated (page 23), the author stating that speeds of work from six to nearly ten times faster could be obtained. Supposing that the old tool-steel cut at 20 feet per minute, which was not a high speed with the old tool-steels, ten times that amount would be 200 feet, an amount which had never

E

(Mr. Daniel Adamson.)
been obtained with any reasonable cut. That speed was run with a finishing cut of about $\frac{1}{100}$-inch by $\frac{1}{100}$-inch ; but this was not a cut to be spoken of in testing tool-steels, and again it could be taken at higher speeds than the 20 feet with the old tool-steels. His own opinion, which might be taken for what it was worth, was that with the ordinary mild steels which the Lancashire makers were in the habit of using—they had no experience of the oil-hardened steels used in Woolwich—about four times the weight could be removed under favourable conditions than was done with the old steels. In actual workshop experience that might be modified ; for example, machines had been increased in speed two and a quarter times, and then a curious thing happened. Although the machine was running two and a quarter times faster, the man only did about $12\frac{1}{2}$ per cent. more work. The man was told that he was not taking full advantage of the tools and the increased speed offered to him. More expensive tool-steel had been bought for the man, and one would have thought it would have been to his own advantage to increase his output, being on piece-work before and after the change and at same rates ; but it was found that if the man was left alone he would do only $12\frac{1}{2}$ per cent. more work, although he was running at two and a quarter times the cutting speed. That was a problem on which another Paper could be written.

In the comparison of cast-iron, about three times the output could be obtained from the new steels.

The tool he exhibited, referred to earlier in these remarks when comparing a pointed tool with a round tool, was cutting forged steel bars containing from 0·25 per cent. to 0·3 per cent. of carbon (which was comparable with the reference No. V, page 12), $\frac{3}{16}$-inch cut and $\frac{1}{16}$-inch feed. (He chose this cut and feed because those who were at the Paris Exhibition would remember that the Taylor White tools there shown were being run at the same, and he kept to that figure as a sort of standard with which to compare other tools.) The tool was running on the first cut, that is, removing the skin, at $22\frac{1}{2}$ feet per minute ; the diameter of the work was about 9 inches, the pointed tool in one experiment ran $3\frac{1}{2}$ inches along the forging, and on a second similar forging it ran $7\frac{1}{2}$ inches. The round tool forged on the other end of the tool was hardened and tempered under similar conditions as nearly as possible, and it ran 6 inches against $3\frac{1}{2}$ inches on the first forging, and on the second forging 14 inches against $7\frac{1}{2}$, practically twice as much in each case, and left a better finish. He thought one explanation of having done twice as much on the one forging as on the other, with both of the tool shapes, was that there might be a little difference in the contents or hardness of the two forgings experimented upon.

Mr. J. HARTLEY WICKSTEED, Vice-President, said the Paper, to which he had listened with the very greatest interest, had opened to his imagination long vistas of further enquiry. It seemed to him that the subject was brought before the Institution at a most opportune moment, because at the present time so many manufacturers were removing the tools from under the planing machines and lathe benches and taking them into tool stores. No doubt the man who had been brought up all his life, as an apprentice and as a workman, to the business, and who had just as good brains as the engineer, knew perfectly well how to grind the tool by eye and without any rules to suit the work that he had to do; but if the tool was taken away from him and given to another man to grind, one must of necessity give that man precise directions as to the angle at which the tool must be ground, otherwise it would come back into the user's hands worse than if he had ground it for himself. The apparatus conceived by the author for weighing the pressure on the tool seemed to him to be most admirable; it was direct, easily read, and very fairly sensitive, so far as the author had applied it, but that was only to light cuts. He wished he could be grateful for information to come, as well as for the information already received; he also wished that the author would carry it a great deal beyond the pressure of 160 lbs. or even 200 lbs. He hoped Mr. Donaldson would make an apparatus which would carry the cut up to about 10 tons pressure; because pressure to that amount was very usually provided in large machine tools. Power was given to the belting and gearing which was equal to a pressure upon the tool of 10 tons upon a single cut. He thought that if the author made an

E 2

(Mr. J. Hartley Wicksteed.)

apparatus on a large scale for heavy cuts he would benefit by putting the tool into a carrier, and letting the carrier vibrate upon the knife-edge, because then the tool could be securely held in the carrier, and the carrier could not only see-saw upon the knife-edge, but it could also be supported sideways in a line with the knife-edge; whereas if the tool was supported otherwise, one was obliged to go higher than the part which rested upon the knife-edge, and as soon as one went higher than the axis on which the tool vibrated the moment of friction was increased between the side screws, if there was any pressure upon them.

It was stated in the Paper that the side screws were only lightly adjusted. It would not do to have such a light adjustment for heavy cuts, and therefore he thought an apparatus could be made on the same lines which would give equally close results on heavy cuts. He did not think one could infer what the pressure would be on a heavy cut from what it was found to be on a light cut. On a light cut it was very much like cutting, but on a heavy cut it was more like punching, or tearing the metal off. In a heavy cut it might be found that it was not the edge of the tool which had been doing the work at all; after a certain amount of cutting had come off, the face of the tool was found to be quarried, and made rough behind the edge. While the edge remained as smooth as a razor, the part behind the edge had become pitted and made rough, and when that roughness had increased to the extent of producing such an amount of rubbing and friction upon the cutting as to create too much heat, then with the old steels the edge became ruined by the heat of the tool, although for a long time the edge resisted the heat, because it was constantly in contact with the cold metal of the stock which it was touching to clean up.

The author in the Appendix (page 19) had recorded some splendid work on a tool called D. According to his (Mr. Wicksteed's) calculation, about one ton of material had been removed at one setting of the tool. The tool had been working for 8 hours, and 269 lbs. of weight had been removed per hour, so that the one tool at one setting had removed a ton of material in shavings

$\frac{1}{10}$ inch thick, and at the end of that time its condition was still good. That was the class of tool required; endurance was wanted, not short runs. Makers did not want to know how much could be done in 20 minutes; they wanted to know the tool and the speed which would enable them to finish the job, without taking the tool out and without having to re-fix it. There was much more time lost in changing the tools than in running at such a moderate speed as would allow the tool to have the maximum amount of endurance.

There had been so much detailed criticism of the Paper that he would abbreviate any further remarks he had to make. He would, however, just like to ask the author to adapt a somewhat larger instrument to the slide-rest for holding the tool, one which would take off a moderate cut such as had been mentioned, in the case of D about $\frac{1}{10}$-inch feed and $\frac{1}{4}$ inch deep, and run it at 20 feet per minute, measuring the pressure, and then run it at 80 feet per minute, and again measure the pressure; and then inform them whether they must, in providing for quick cutting, make provision for a greater pressure upon the tool for removing the material quickly than was required for removing it slowly. He did not see why that result should be expected, but he did not think it could be known until the experiment was made. In many cases there was the flow of solids to consider. For instance, if one wanted to punch a nut which was 2 inches deep with a $\frac{3}{8}$-inch diameter hole, it could be done if the punch was given time, but if one tried to do it quickly the punch would fracture. There were many other cases, where time was an element in the resistance to the flow of solids, but he did not know that they were analogous to a lathe cutting, although there must be some flow of material in a shaving. A shaving which came off a lathe was by no means a strip cut off the material; all the particles were being pushed into fresh positions. It would be an advantage to the trade generally to be informed upon the point, whether, other things being equal, the fact of cutting rapidly put more pressure upon the tool than cutting slowly within the limits of, say, 20 feet and 80 feet per minute.

Dr. H. T. ASHTON said that, as he knew a good deal about the conduct of the experiments mentioned in the Paper, he might say

(Dr. H. T. Ashton.)

that amongst them one of the good results obtained was with a side-cutting tool, a knife tool, in which the angle was such as almost to draw the tool along the bed of the lathe by the dragging action which had been mentioned ; a great amount of metal was removed by that means. In regard to the criticism of the second paragraph on page 19, he thought it was evident from the context that what was meant was to compare one rapid-cutting tool-steel with another. The statement that the rapid-cutting tool-steel which was best for cutting mild material was by no means certainly the best for cutting very hard material seemed to him to be reasonable, because some rapid-cutting tool-steels resisted heat very much better than others, while not being necessarily so hard, and there was a much greater amount of heat involved in cutting a large amount of metal off a mild steel than there was in cutting a less amount of metal off a harder steel. In cutting very hard steels a very hard rapid-cutting steel was required, but not necessarily one which kept its temper at such a high temperature.

In regard to the question of gun-metals, he did not think the yellow metals of the Paper could be very well compared with the ordinary commercial yellow metals, because, for the most part, those used were very much harder than the trade brass, and, as he thought the author had pointed out, the experiments were not carried out specially to provide the information which was laid out in the Paper. He thought it was also fairly clear that the experiments detailed in the Paper were all carried out on lathes. From his own information he might supplement the knowledge given therein by saying that the tools were all on the centre of the work. Except as regards the position in which the shafting of the tool was held, provided they accommodated the angles accordingly, it was practically the same thing whether one placed the point of the tool slightly above or slightly below the centre. The same result could be obtained with the tool above or below centre, if the clearance angle and the top angle of the tool were both adjusted.

In regard to Mr. Adamson's query as to the diameter of the work affecting the clearance angle, he thought the statement in the Paper was literally correct, because one wanted to have as little rubbing surface of the tool and work in compression in contact below the cutting edge as possible. With a small radius of curvature the surface of the work was less in contact with the tool than with the large radius, that is to say, that it could stand a flatter and squarer angle.

A cutting tool was a sort of compromise. It had to do two things, first of all it had to part the material, for which purpose a razor edge would be best, but a razor edge could not be used on steel because the shaving had to be bent at right angles, and the razor edge would be broken away in the process of the bending. If the razor edge was being used, the line of sheer of the material being cut would be radial ; on the other hand, if the tool was absolutely square the line of sheer would be tangential; but in an actual cutting-tool there was a compromise between the two, which first of all parted the material and then bent the shaving over. It was in the bending over of the shaving that a large part of the work was done, as was pointed out by a previous speaker. It was no doubt owing to the greater pressure required to bend over the heavy shavings removed by the rapid-cutting tools, that these tools were found in the experiments to work better with a more obtuse cutting angle than that used with ordinary tools working the same material. In regard to the apparatus used for measuring the pressure on the tool, it was a very difficult matter to design any apparatus which would at once efficiently hold the tool and accurately register the pressure on the tool, because a very little movement would spoil the work of the tool.   With regard to the apparatus the author had mentioned in his remarks after the Paper was read, he thought perhaps it might make it a little clearer to the members if he stated that that apparatus measured directly, by means of a lever attached to the work, the torsional effort of the face plate of the lathe in turning the work round against the resistance of the tool, thus measuring not only the vertical pressure on the tool, but also the work of bending the shaving round.

Lieut.-Colonel R. E. CROMPTON, C.B., after thanking the author for his Paper, referred to the horizontal angle of the tool as shown

(Lieut.-Colonel R. E. Crompton, C.B.)

on Fig. 3 to be 30°. Here YY was the centre line of the tool viewed in plan. In the Paper nothing was said as to the advantage of increasing or reducing this angle, and he was anxious to hear what the author or other speakers might have to say on this point. It was clear that, as the tool was advanced in the direction of the arrow, the shavings cut from the work by the tool must have a rhomboidal cross-section as shown at C. It appeared to him that the horizontal angle of the tool determined the shape of this section, as, if the horizontal angle was small, the shaving was broad and thin. The question of this angle appeared to be closely connected with that of rounding the nose of the tool, and the question was one

FIG 3 —*Horizontal Angle of Lathe-Tool.*

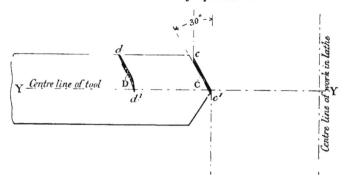

which had never been properly explained, and ought now to be looked into. He offered the suggestion that the advantages of a small angle and of rounding the nose of the tool, either taken together or combined, had the effect of making it easier to get rid of the heat, by spreading over a greater surface the scouring action of the shaving in impinging at a high speed on the top of the tool at a point somewhat behind the cutting edge, and to which Mr. Wicksteed had already called attention. The rounding action would have the effect of making the cross-section take the form D, and it would be seen that the length of the line $d\ d^1$ was considerably greater than the line $c\ c^1$. He hazarded this explanation, and would be glad of others' opinion on the point. Personally he had taken great interest in the cutting angles on machine-ground tools, but he found that the results obtained a few years ago had now to be greatly modified, owing to the advent of high-speed steel. In his own shops two distinct ideas were prevalent. In one shop the stiff straight tool recommended by the author was used, and in the other shop tools forged with a bend, and which were condemned by the author, were used. He believed that many tool foremen would disagree with the author in his condemnation of the bent forged tool.

Turning to another point, he had to thank Mr. Wicksteed for throwing light on a matter which had hitherto been a puzzle to him, that was, it had always been stated that, to get the full advantage from rapid-cutting steel, heavy stiff lathes ought to be used. This appeared to nullify the chief advantage which rapid-cutting steel offered to them, namely to increase the output by increasing the surface over which the cutting point ran in a given time. There was however a second reason, additional to that given by Mr. Wicksteed, why increased speed demanded heavier and stiffer lathes. This was in the case where the material to be removed was from steel castings. At high speeds the shocks due to irregularities in the castings would be considerably increased, and the tendency for all the supports to the cutting point to yield under the action of these shocks would be increased. For this reason only, it would be necessary to use heavier and stiffer lathes when high speeds were used on steel castings.

Mr. ROBERT DUMAS said he did not think the author, in giving the angles of the cutting tools in the Paper, had quite completely defined the tool, because it was clear that with a pointed tool, such as was dealt with in the Paper, the angle on the section line YY, Fig. 1 (page 8) was or might be dependent on the angle to which the non-active edge of the tool KB, Fig. 4 (page 42), was ground.

The diagrams of the tool (page 8) were reproduced in Fig. 4 (page 42), and lettered for reference. If the side clearance on face AK were uniform, the bottom edge of the tool would be shown by a dotted line parallel to the top edge AK, cutting the line YY at the point D. If the face BK was vertical, or if it was ground at

(Mr. Robert Dumas.)

any angle which was less than the side clearance angle, a constant angle of front clearance would always be obtained, which would be greater in amount than the side clearance. If the face BK were ground at a greater angle than the side clearance, it was evident that the front clearance was determined by the angle of clearance on the face BK rather than the angle of clearance on the face AK; so that, in order to define completely the grinding of the tool, some information was necessary about face BK, as well as about face AK. It was evident that if FG (*see* section on XX) was a straight line

FIG. 4.—*Cutting and Clearance Angles of Tools.*
(See also Fig. 1, page 8)

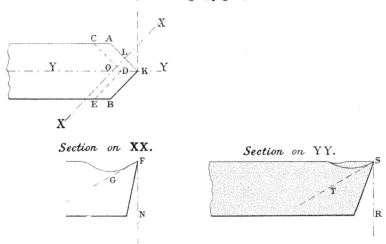

corresponding with LO in the plan of the tool, then the angle RST (*see* section on YY) must always be greater than the angle NFG, because the point O on the top face of the tool was farther away from K than L.

Referring to Table 3 (pages 14–15): in the case of tool No. II, the front clearance and the side clearance were both given as being 3°, whereas the front cutting angle was 59° and the side cutting angle 62½°· That seemed to him an impossible set of conditions. Either the angles had been incorrectly measured, or the nose of the tool had been rounded to such an extent that the distance OK was

less than OL, though even in that case a very curious form for the
top face of the tool would be necessary to get the angles given in the
Table.

He would like to make a few remarks upon the horizontal
angle referred to by Colonel Crompton (page 40). He was
considerably surprised to see the smallness of that angle as given in
Table 3 (page 14). He had always found that, with a small
horizontal angle, a considerable amount of power was taken in
removing the material. There was one reason for that which was
fairly clear, namely, that the material was being cut up into thinner
strips, and the smaller the horizontal angle was made the thinner
these strips became ; and under such conditions it seemed to him that
more power must be taken to remove a given bulk of material, for the
same reason that a milling machine required much more power than
a lathe removing the same amount of material by heavy cuts.

As an illustration of the importance of the horizontal angle being
large, he might mention that recently, when starting up a new boring
machine, it was found impossible to keep the belt on the pulleys, the
belt not being sufficiently powerful to drive it. Upon investigation,
it was found that the horizontal angle was about 45°; the tools were
replaced by others having a horizontal angle of about 90°, with the
result that there was an ample margin of power to drive the machine
on the same cut, the strain on the machine being very greatly
reduced. He thought there ought to be two or three shapes of tools
kept in the tool store for roughing work, in order to get the maximum
results. The shape of the tool, as had been pointed out by several
speakers, ought to depend on the depth of cut and on the feed. He had
drawn three shapes, Fig. 5 (page 44), of which shape (a) was suitable for
fair-sized feeds and deep cuts, shape (b) was suitable for medium feeds
and medium depth of cut, and shape (c) for very heavy feeds and a
small depth of cut; shapes (b) and (c) were in fact transition stages
between (a) and the square nosed tool used for scraping cuts at heavy
feeds. By using a tool which was really proportioned to the
character of the feed and cut employed, it was quite surprising how
much the output could be increased. The radius of the nose of the
tool should not be less than the feed employed, and it was preferable

(Mr. Robert Dumas.)

if it was slightly more. For general purposes he had found tool of shape (a) best, the angle being 90° or thereabouts. If the radius at the nose of the tool was varied with the feed, very heavy feeds and deep cuts could be successfully dealt with. This shape of tool had the additional advantage that it could be readily forged in a die, and it also lent itself to machine grinding. He was of opinion that, as far as consumption of power was concerned at the cutting tool itself,

FIG. 5.—Shapes of Roughing Tools.

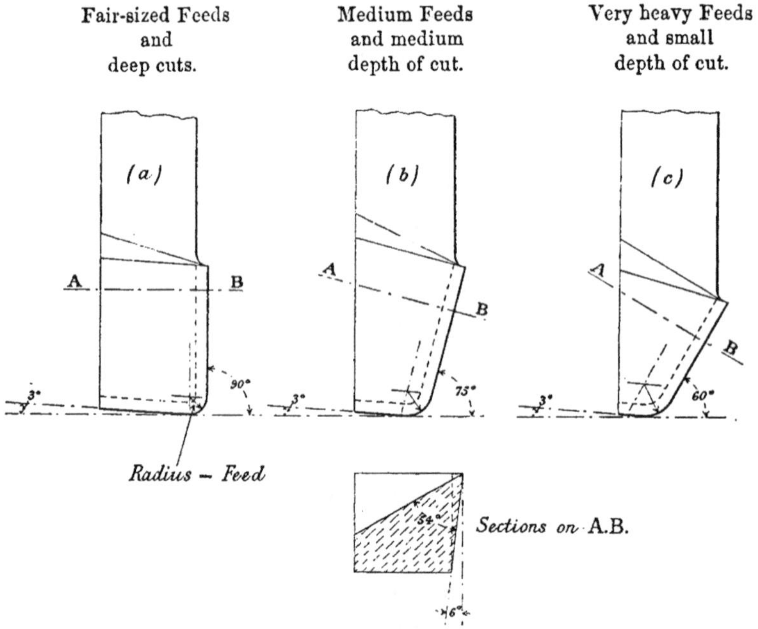

| Fair-sized Feeds and deep cuts. | Medium Feeds and medium depth of cut. | Very heavy Feeds and small depth of cut. |

Radius — Feed

Sections on A.B.

the most economical method was to have the radius of the nose of the tool, the depth of cut, and the feed all equal. The best results commercially might, however, be very different.

Mr. DONALDSON, in reply, said he had hoped he had made his position fairly clear when he said in the Paper (page 5) that the experiments detailed were not detailed as experiments upon that particular line, but were taken as applicable for use as a groundwork from which to start others in investigations. He had all along insisted that there was no finality about the experiments, and it was that want of finality which made him hesitate very much in writing the Paper at all. At the same time he could not but congratulate himself upon the result, in that the discussion had been most useful, which was his object in acceding to the request made to him to write the Paper.

Mr. Orcutt (page 28) referred to the question of whether the cuts mentioned were roughing cuts or finishing cuts. As a matter of fact they were both ; most, if not all, of them were roughing cuts, though some of them might be classed as finishing cuts, and he did not think he could say at the moment which was which. At the same time, one important point must not be overlooked, namely, that the experiments, as carried out, were made upon materials in general use, and in many cases the material was such that a heavy cut was an impossibility, and was often so hard that light cuts were essential. For heavy work the round tool had most distinct advantages in the amount of metal which could be taken off. With regard to Mr. Adamson's enquiries, he feared he had missed some of the particular points upon which that gentleman desired information. He could only say that, if he could afford Mr. Adamson and his society information or assistance in the experiments in which he was now embarked, he hoped his services would be commanded ; he would be only too happy to show him all the details of the experiments, for what they were worth. They were perhaps not experiments such as he would want, but might be found useful merely as indications.

A question had been asked with reference to the height of the tool above the centre. Practically in all cases the cutting point of the tool was level with the centre. Mr. Adamson had referred to the paragraph (page 7), " A heavy cut and slow speed will be more efficient in production than a light cut and quick speed." Mr. Adamson's idea of what he meant by those words was correct, that if the result was shown to be the same in figures, it did not necessarily follow that the one method would not in fact do more work than the other. At the bottom of the same page, reference had been made

(Mr. Donaldson.)

to the fact that the experiments were all made with pointed roughing tools, though in some cases, notably in specimens VI and VII, the point was somewhat rounded off, and upon that a criticism was based that the circumstances and conditions were not the same. He ventured to think they were sufficiently alike, taking the roundness of the pointed tool and the area of the rounded tool, to be able to obtain to a certain extent, he did not say accurately, some sort of comparison ; and as the experiments were not made for comparisons in the way they had been used in the Paper, he thought they must take what they could get in the matter of comparison. Mr. Adamson referred to the lubrication, and asked how it had been applied. So far as his recollection served him, in all cases it was simply dropped on, with a moistening, without such cooling effect as would be done with a pump.

With regard to the question of brass and yellow metal, he was afraid that he must plead guilty to the wording leaving something to be desired. As a matter of fact, what was referred to (page 16) as B.S., B.F.H., and B.F.S., and called Brass Soft, Brass Forged Hard, and Brass Forged Soft, were really the identical metals referred to as yellow metals in the G.M. hard alloy, and so forth. He apologised for calling them brass when they were perhaps something better. Mr. Adamson also referred to the clearance at 3°. He thought that perhaps it depended on which side of the tool the measurement was taken. He had been accustomed to measure it the reverse way to that in which he measured the front cutting angle, and perhaps that accounted for the discrepancy.

Mr. ADAMSON said that that was only for brass work.

Mr. DONALDSON said the brass was hard alloy ; he hoped he had made it perfectly clear. What was called brass on page 16 was the same sort of metal as was shown in the Table of analyses (page 12).

Mr. ADAMSON said the point he wished to make clear was that the front clearance for cutting brass ought to be greater than that for the steel.

Mr. DONALDSON said that Mr. Adamson had criticised the speed mentioned (page 23). The use of the words "speeds of work" there were not intended to convey the speed of the travel of the tool, but the speed at which the material was removed, and that he thought was clearly shown as ten times, as he had stated.

He was extremely obliged to Mr. Wicksteed for his suggestion in regard to carrying the tool-holder. Should it be advisable or possible to extend the experiments, of which at the present time he was afraid he did not see much possibility, he would be very pleased indeed to consider that particular suggestion. He was rather inclined to ask Mr. Wicksteed how he measured the 10 tons on the point of the tool of which he had spoken. Was it done electrically, or how?

Mr. WICKSTEED said it was a common thing to allow 150 feet of 4-inch belt to a foot of cut; and if one took the pull of a 4-inch belt as being 2 cwts., and multiplied it by 150 the figure of 10 tons he had mentioned was obtained.

Mr. DONALDSON said it came back to very much the same thing as if it was driven electrically, but there was a long distance in the travel of the power between the belt and the point of the tool.

Mr. WICKSTEED said the figures he had mentioned multiplied to 15 tons.

Mr. DONALDSON said that with reference to the remark about the increase of power if the speed was increased, he was unable to refer at that moment to any experiments, but he believed that if the speed was increased the necessary power of driving was also increased. Colonel Crompton's remark (page 40) on round-nosed tools had been partly dealt with by what Mr. Dumas had subsequently said. In the tool mentioned a longer cutting edge was obtained and the work was distributed over a better supported background. For heavy cuts he thought there was no doubt that the round-nosed tool was the best,

(Mr. Donaldson.)

of course for roughing cuts. As to Mr. Dumas's point in reference to the angle, he was sorry that a gentleman who had intended to be present was not able to do so, because he was in hopes that that gentleman would have described an angle which he had worked out some years ago for boring holes, where the angle so suited the machine, the feed, and the material worked upon that, having started the cut, he was practically able to throw off his feed, to keep the machine running and to have the boring bit feeding itself.

The PRESIDENT thought the members would all share his feeling, when he said he was extremely glad the author did not allow his views as to the want of finality of his Paper to prevent him from bringing it before the Institution. The Paper had elicited a most interesting discussion, and the members were extremely grateful to Mr. Donaldson for all the trouble he had taken.

Communication.

Professor ROBERT H. SMITH wrote that the cutting forces exerted by lathe tool-points had been measured in various series of investigations at various times since 1880, by an apparatus devised by him in that year. In the 1880 series, measurements had been taken under 240 different conditions. These were arranged systematically to show the effects (1) of kind of material cut, (2) of diameter of work cut, (3) of cutting speed, (4) of depth of cut, and (5) of breadth of traverse-feed. In each material all the combinations of four cutting-speeds, four depths of cut, and four traverse-feeds were tested. The effects of variations (1), (4), and (5) were shown clearly in the six diagrams of Figs. 6 and 7 (pages 50–51), which were borrowed from his book on "Cutting Tools," the first edition of which was published by Messrs. Cassell and Co. in 1882. The cutting speeds then experimented on were from 14 to $2\frac{1}{2}$ feet per minute on cast-iron ; $9\frac{1}{2}$ to $1\frac{1}{2}$ on wrought (forged) bar ; and 6 to 1·3 on forged bar steel. Slightly greater cutting force was needed for the higher speeds, but

the increase was unimportant.   Largely different diameters of work
were used only in the cast-iron series ; namely, $16\frac{1}{2}$, 7, and $3\frac{1}{2}$ inches.
The change in the diameter made but small difference in the
cutting force.   The steel and wrought-iron bars used were $2\frac{1}{2}$ inches
in diameter.

The forces measured on the cutting-tool point ran up to 920 lbs.
These were the vertical components of the whole pressure upon the
cutting-point.   This component alone was the active working force.
The two other components, which did not contribute to the horse-
power required for the work, were not measured.   The amount of
possible error in the measurements was determined, the character of
the apparatus making this determination easy and accurate.   It
varied from 1 per cent. on 200 lbs. to $\frac{1}{2}$ per cent. on 600–900 lbs.
The ratio of the cutting forces for different metals varied greatly
with the thickness of the shaving.   Thus wrought-iron took about
50 per cent. more than cast-iron for $\frac{1}{20}$-inch depth of cut, but for
half that depth it took only some 12 per cent. more, while for broad
very thin finishing cuts cast-iron took more force to cut it than did
wrought-iron.   At the greater depth steel took 45 per cent. more
than wrought-iron, and at the smaller depth over 100 per cent.
more.

In considering these variations, it should be remembered that
in most roughing work the traverse-feed was more nearly the
thickness of the shaving in the direction in which it was forced to
bend than was the "depth of cut."   This latter was really the
bending "breadth" of the shaving, or was at least proportional to
this breadth.   The force required to bend the shaving, and break it
by bending, must be nearly proportionate to the breadth, and might
be expected to be more nearly proportionate to the square of the
thickness.   But the force needed for this part of the cutting action
was by no means the whole.   Shear or semi-shear occurred at
regular small spacings, and the work spent on this shear was
probably proportionate to the first powers of both thickness and
width.

These results for what might be termed small work had
been confirmed by many subsequent repetitions of a less systematic

F

(Professor Robert H. Smith.)

FIG. 6.

*Cutting Forces exerted by Lathe Tool-points
measured by the apparatus shown in Fig. 8 (page 54).*

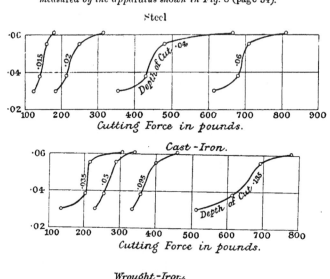

Steel

Cast -Iron.

Wrought -Iron.

Cutting Force in pounds.

FIG. 7.

Steel.

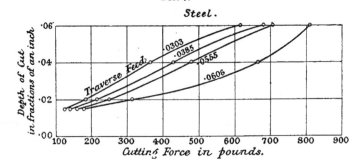

Cutting Force in pounds.

Fig. 7 (continued).

*Cutting Forces exerted by Lathe Tool-points
measured by the apparatus shown in Fig. 8 (page 54).*

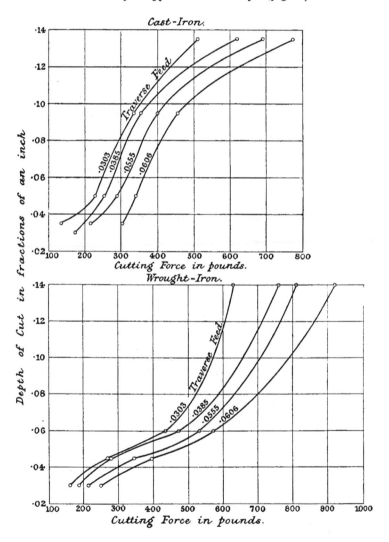

r 2

(Professor Robert H. Smith.)

character, carried out by the writer's college students. They had the advantage of a still better-made measuring apparatus, but their cutting tools were worse. In 1893 he had the good fortune to be allowed to make measurements on very heavy cuts in tearing down a steel ingot in the Sheffield works of Messrs. Jessop and Co. The design of the measuring apparatus was essentially the same as for small work; but, the tool being a solid bar 3 inches square, it assumed much larger proportions. In these tests the depth of cut went up to $1\frac{1}{2}$ inches with $\frac{3}{4}$-inch traverse-feed, and the cutting-pressure on the tool point went above 10 tons. In measuring the 10 tons, he ascertained that there was not so much as 2 cwts. error. The tool had three straight-edges with corners rounded to about $\frac{3}{8}$-inch radius, and with top surface level (that is, zero top-rake) and with ample bottom clearance—some 7° or 8°. This shape of tool was used for other reasons than that it was the best for diminishing the cutting horse-power required; it had not a long life under these very heavy cutting strains. He felt sure that a much longer life could be given to it, if the tool-box were differently designed. The strength of the tool and the expense of renewing its cutting-edge were here evidently considerations of paramount importance.

In trials preliminary to the 1880 series of tests, different forms of tool were tried, all of them round nosed, this being known to be the best for getting the metal cut away in quantity, and it was only for this kind of work that enquiry into the forces at action were of practical interest. The rake was practically all side rake, and amounted to 24° for cutting cast-iron and $46\frac{1}{2}$° for steel and wrought-iron. These were top-rakes and left the two cutting angles 62° and 41°. These cutting angles were much smaller than could be habitually used in ordinary workshop practice. Comparatively few workmen could use such tools without frequent breakages, and still fewer could temper them so as not to have the edges softened after very little work. The tools in question lasted admirably, but they were prepared by a specially clever tool-maker. The smaller the cutting-angle of the tool-edge the smaller was the horse-power required for cutting; but the angle actually required depended upon the necessity of avoiding (1) too frequent breakage, and (2)

overheating and consequent loss of temper in the edges. Thus the
limit of smallness of the cutting-angle which was practically useful
resulted from (a), the mechanical strength of the tool due to its form
and to the quality, especially to the toughness, of the steel, and (b)
the tempering quality of the steel. It seemed to the writer that
the merit of the new fast-cutting tool-steels, upon which he had
had no opportunity of experimenting, mainly depended upon their
peculiar power of standing much heating without losing temper.
Providing quality of steel was producible that would permit the
reduction of the cutting-edge angle with ordinary handling and a
reasonably long life, it seemed certain that such reduction of the
cutting-angle resulted in decrease of the horse-power needed for
driving. It must also not be forgotten that success in this direction
accumulated in a sort of geometric or compound ratio; because a
reduction of cutting-angle, by diminishing the work to be done, at
the same time decreased the amount of heat generated per second at
the tool point. In the above experiments no lubricant or cooling
water was used, and yet the 41° cutting-edge did two days' almost
constant cutting work without regrinding, and was in good condition
at the end of it.

It was well known that with a large amount of top-rake there
was risk of " digging in." If the depth of cut were small, the work
would push the tool outwards from the centre, a push which the
tool-clamp must resist. If the cut were deep, it would draw the tool
in towards the centre, to prevent which the tool-clamp must exert an
outward pull. For each top-rake and for each quality of material
cut—perhaps varying slightly with the cutting speed—there was one
definite depth of cut, to maintain which steadily the tool-clamp
needed to exert no radial force whatever. Similarly with the traverse-
feed. He had known turners able to adjust things so that the
transverse pressure of the shaving upon the tool-point in the
direction of traverse-feed was just able to draw the slide-rest along
the bed, giving a steady feed without assistance from either hand or
automatic feed-gear. It seemed that this was the ideally perfect
condition of things, and that it gave the correct relation between
depth of cut and top-rake angle. If so, it suggested the quickest

(Professor Robert H. Smith.)

and most accurate method of investigating this correct relation ; because it was evidently much easier to discover the depth of cut and amount of traverse-feed that would, with a tool of given angle cutting a given quality of material, produce this result than to proceed conversely to find the cutting-angle that would produce the same result for a given depth of cut in given material.

In Table 3 (pages 14–15) the angles, &c., were given which gave the " best results," but nowhere was it explained by what criteria the

FIG. 8.

Lathe-tool Cutting-Pressure Measuring Apparatus as at present used in King's College.

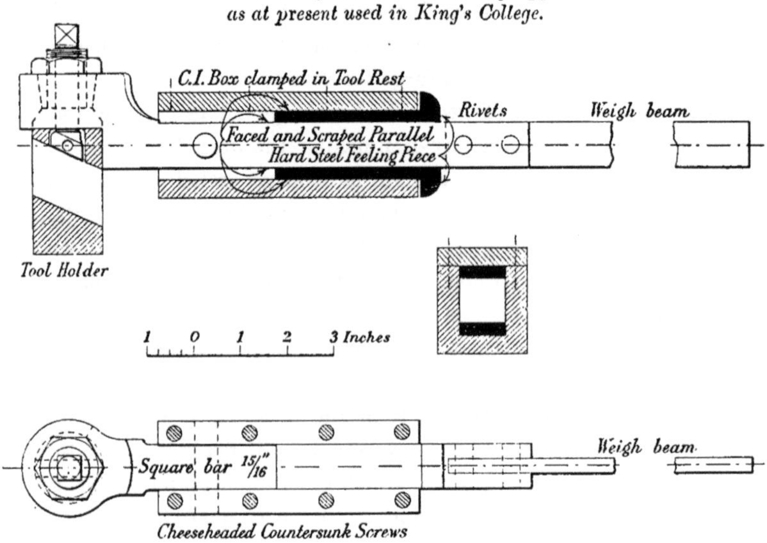

" best " results were judged. There were so many things desirable and to be aimed at, and what was deemed " best" was so invariably a compromise between losses and gains, that, without the criteria applied being communicated, no criticism was possible, and the given results lost much of their probable value.

In the measuring apparatus used by the writer, the form of which was illustrated in Fig. 8, the tool-bar or tool-holder was supported on a knife-edge as far forward as possible, and behind this its top and bottom surfaces were carefully trued up parallel by scraping. It lay in a special housing, whose top and bottom inside surfaces were similarly trued up parallel, the depth between them being about $\frac{3}{8}$ or $\frac{1}{2}$ inch greater than the thickness of the tool-bar. The upper and lower spaces between bar and housing were filled up by two hard-steel parallel strips, scraped true and smooth. When both were inserted, both could be moved easily when no upward or downward pressure came upon them, the weight of the tool-bar itself being exactly balanced by an adjustable spring sling arranged above the extension of the bar beyond the outside of the tool-box. These feeling-pieces filled up the tool-housing so completely and perfectly that they allowed no chatter in the tool. In the heavy-cut experiments at Sheffield a slight chatter became sometimes perceptible when the tool-pressure went above about 6 tons. Between 9 and 10 tons it once or twice became threatening, so that the turner withdrew the tool from the cut. When the hinder end of the tool-bar was lifted, the upper feeling-piece became quite tight, and could not be moved by the fingers, while the lower one could be moved with perceptibly greater ease than in the neutral position. Conversely, when the hinder end of the bar was depressed, the upper feeling-strip felt loose and the lower one could not be shifted. The tool-bar was extended outwards in the form of a scale-beam, and this was loaded so as to balance the working pressure on the tool-point. When the balance was perfect, both upper and lower feeling-strips could be moved to and fro by the fingers, and it was easy to detect any slight difference between the resistances to such shifting offered by the two. The balancing load was read from the scale on the beam and the weights in the scale-pan. As already mentioned, this arrangement provided very sensitive means of judging when the balance was correct, and moreover it avoided the introduction of any chattering, which in his opinion would almost destroy the value of any measurement, and maintained perfectly steady and quiet cutting. The special tool-housing was held in place by the ordinary studs of the tool-rest.

In the Paper, the author apparently admitted that the friction of the bearing points at the back end of his tool-bar prevented the free bearing of the bar upon his measuring piston with oil-abutment.

(Professor Robert H. Smith.)

Apparently his tool-bar, Plate 1, had a visible clearance between the upper and lower stops B and C, and he screws down his oil-plunger D until the tool-bar was visibly out of contact with both B and C. Seeing that D had a very appreciably elastic abutment, he failed to see how more or less violent chattering could be avoided, especially at the higher speeds. According to his own experience, the tool-holder must bear both top and bottom against sensibly inelastic and rigid abutments, in order to prevent chattering.

THE NATIONAL PHYSICAL LABORATORY AND ITS RELATION TO ENGINEERING.

A LECTURE

By Dr. R. T. GLAZEBROOK, M.A , F.R.S.,
Director of the National Physical Laboratory, Teddington,

AT A MEETING OF THE GRADUATES ON
Monday, 9th February 1903.

William H. Maw, Esq , *President*, in the Chair.

When I received your very gratifying invitation to lecture here to-night to the Association of Graduates of the Institution my first impulse was, as you know, to decline. Not that I failed to appreciate the offer and the opportunity, but I thought I could do more justice to my subject and bring before you results more worthy of the occasion, if, at a later date, I might be allowed to describe some of the researches which are now in progress at the Laboratory. The President reassured me by a most kind letter, in which he told me that I could do a service to the engineering profession by pointing out the directions in which the Laboratory might do work which would be practically and commercially useful to engineers. He invited me in particular to draw attention to the possibility of placing in our hands short enquiries on definite points to which an answer is required rapidly, as well as to those more long and costly

FIG. 2.—*Plan of Engineering Building.*

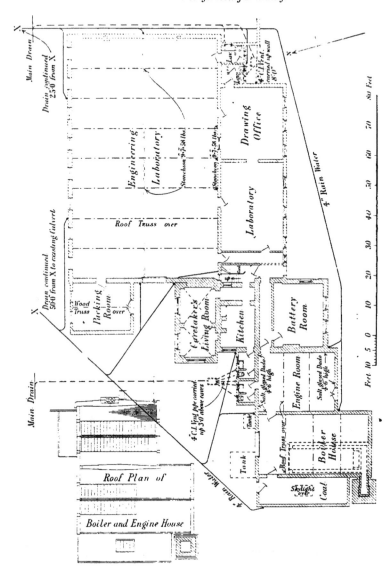

investigations which we hope to carry out. I was glad to be able
to accept that invitation, and I trust that what I have to say to-night
may really help members of your profession, and may lead them to
give us their support by sending us work to do and consulting us on
points on which it appears that we may help them.

Let me commence by a brief description of some of our buildings.
The Laboratory is at Bushy House, Teddington, Fig. 1, Plate 2, and
for many purposes the old house is admirably suited. The
engineering building, however, is almost entirely new, being
specially designed for the work we had to do. The engineering
laboratory itself, of which Fig. 2 (page 58), and Figs. 3 and 4,
Plate 2, give views, is a building 80 feet by 50 feet. It is divided
longitudinally into two bays, each of which is lighted from the north
by a weaving-shed roof. A shaft runs along one bay, and in it are
placed the machine tools comprising four lathes, a universal grinder,
a shaping machine, drilling machine, etc. The shafting is driven by
a Mather and Platt motor. The second bay is for experimental work ;
in it are the testing and part of the pressure-gauge testing apparatus,
Fig. 4, Plate 2. A boiler which can be worked at a pressure of 400 lbs.
per square inch has also been installed ; the apparatus given by Messrs.
Willans and Robinson for testing gauges and steam indicators under
steam will shortly be fitted, as well as a very complete alternate-
current outfit presented by Messrs. Siemens. A traversing crane
runs along this second bay, and has already proved of great assistance
in setting up the machinery and apparatus, and at one end is the
testing machine, Fig. 3, Plate 2.

The wind-pressure apparatus, about which I shall have more to
say later, is also set up in this bay. In sheds near the engineering
building are a small forge and a smelting-shop.

The engine-room contains a 75-kw. condensing turbine by Messrs.
Parsons. There is also an 18-H.P. Crossley gas-engine, driving a
dynamo by Messrs. T. Parker of Wolverhampton. The same firm
supplied a very convenient booster set for charging the storage
batteries. In the boiler-house is a 100-H.P. boiler, which also
serves for the heating both of the engineering laboratory and of
Bushy House.

Though most of the work specially interesting to engineers will be conducted in the engineering laboratory, several of the rooms in Bushy House are fitted for experiments which have a direct bearing on engineering practice. Fig. 6, Plate 3, shows the metallurgical room, in which it is hoped to carry on the work begun at the Mint by the distinguished Honorary Member of your Institution, the late Sir William Roberts-Austen, whose labours as Chairman of the Alloys Research Committee have helped so greatly to elucidate some of the changes which go on in metals.

In the centre is the Roberts-Austen pyrometer as fitted at the Mint. The apparatus was the gift of a staunch friend, Mr. George Beilby. To the right are the coke furnaces. The room was the old kitchen, and the chimney and fireplace have been thus utilized ; on the extreme left are the gas furnaces.

Fig. 7, Plate 3, shows the Thermometric Laboratory. This is arranged for the accurate measurement of temperature, specially of high temperatures. Here are placed the various baths and ovens by the aid of which steady temperatures up to 1,200° C. can be produced and measured. For temperatures up to 200° C. an oil bath is employed; from 200° C. to 500° C. a bath consisting of mixed nitrates of soda and potash is most convenient ; above 500° C. electric ovens afford by far the best means of obtaining a high temperature. Much of the apparatus in this room was the gift of Sir Andrew Noble.

Fig. 5, Plate 2, shows the basement room in which the measuring machine, dividing engine and standards of length are housed. These comprise a set of standard gauges, a Whitworth measuring machine, a special machine for measuring screws, also by Whitworth, a Pratt and Whitney measuring machine, and a dividing engine by the Geneva Society for the construction of Physical Instruments. Let me turn now from this enumeration of our rooms and apparatus to describe some part of our work ; but before doing so, I should like again to express my obligation to Mr. E. G. Rivers, the Chief Engineer of the Office of Works, and to the other Officials of the Office for the skill and care they showed in carrying out the work of erecting the Laboratory.

The Work of the Laboratory.—This I propose to treat of under three heads :—

(1) Routine Test Work.

(2) Original Investigations undertaken at the initiation of the Committee.

(3) Enquiries and Experiments made at the request of Engineers or others, and for which fees are received.

The first head need not take us long. Arrangements are complete for tests of length gauges of all kinds. The gauge room in a large shop is a very important part of the establishment, but it is not possible for every firm to maintain a set of standard gauges and the apparatus necessary to compare the shop gauges with these from time to time. This work we can do; I hope in a satisfactory manner and at a reasonable rate. The important Engineering Standards Committee is engaged in a great work at present in standardizing engineering practice. We hope to assist in that work by affording facilities to all to maintain the standards. What I have said as to length gauges applies even more to screw gauges, only here the need for standardization is more marked, because less has been done; with regard to the ordinary cylindrical gauges and length gauges, it must be remembered that the Board of Trade is the authorised guardian of the standards, and that gauges of the standard patterns are tested with great care and exactness in the Standards Department; but with regard to screw gauges this is not the case, while by the action of the War Office we are being placed in a very favourable position. With a view of securing uniformity of pitch in the leading screws of lathes in the various Government workshops, a lathe of special design, with a very accurate screw, is now being constructed by Messrs. Sir W. G. Armstrong, Whitworth and Co. This is to be placed at the Laboratory in a specially constructed chamber, and it will be our duty to issue both to Government Departments and to the public certified copies of this standard screw.

Mr. Donaldson referred to the machine at the last Meeting of the Institution.* Engineers will, I am confident, be interested when they read his Paper on the subject.

---

* Proceedings 1903, page 26.

A knowledge of the coefficient of expansion of the material he is using is often necessary to the engineer. A comparator of the Geneva Society enables us to measure this, at any rate in the case of materials which can be obtained in the form of rods and for moderate temperatures, while apparatus which will allow us to extend the measurements to a dull red heat is under construction.

Reference has already been made to the testing and standardization of pressure-gauges; for this the Laboratory is singularly well equipped. A mercury column nearly 50 feet in height has been erected in Bushy house; alongside this is a steel scale divided into millimètres, pounds per square inch, kilogrammes per square centimètre, and feet of water. Arrangements are made for applying pressure to the lower end of the column and the gauges in connection with it, from a bottle of compressed air. Thus pressures up to nearly 300 lbs. to the square inch can be read directly on the column, alongside of which a lift is fitted so that any point of the scale may be readily observed. In this case the gauges are read when cold; if it be desired to read them under steam, a direct weight-testing apparatus, given by Messrs. Willans and Robinson, will be employed. This same apparatus serves for testing indicators; steam can be supplied to this up to a pressure of 400 lbs.

For higher pressures a differential direct weight-testing apparatus, designed and constructed in the Laboratory, and described in *Engineering* (9 January, 1903), has been used very successfully. With this apparatus, pressures up to about 8 tons to the square inch are measured to an accuracy of about $0\cdot01$ ton. For comparing gauges among themselves, a set of pressure-pumps, supplied by Messrs. Schäffer and Budenberg, who also erected the mercury column, are employed.

In the Thermometric Laboratory high temperature measurements of real importance to engineers are being conducted. I propose to give one striking illustration of these later in the evening. Dr. Harker has very skilfully utilized the appliances put at his disposal, mainly by the generosity of Sir Andrew Noble. Our ultimate standard of reference up to temperatures of say 900° C. to 1,000° C. is the nitrogen thermometer. With this our secondary standards, the platinum thermometer and the platinum, platinum-rhodium or platinum, platinum-iridium thermo-couples, are being compared. These can be used up to perhaps 1,200° C. or 1,300° C.; a method of measuring still higher temperatures is still to be found, and for this some photometric method of measuring the radiation from a black body appears the most promising. To this we hope to go on; meanwhile we are prepared to standardise to a very high order of accuracy instruments for measuring temperature up to 1,300° C. The importance of the micro-photograph in the study of steel and its alloys is now generally recognised, and accordingly, urged by the advice of the late Sir William Roberts-Austen, a very complete outfit for photo-microscopic research has been installed. In setting this up we have been greatly assisted by Mr. Stead, who presented us with our polishing apparatus and helped Dr. Carpenter, the Assistant in charge, in getting to work. This has already led to some results, which I think will prove of interest on some of the alloys investigated by Professor Barrett and on some special magnet steels described by Mr. Ashworth; but I am speaking of it now as part of the regular testing outfit. We are prepared to examine sections of rails, girders, &c.

It must not be assumed that in this very hasty list I have completely enumerated all the tests we can undertake; there is one omission which will strike any engineer, I have said nothing as to the ordinary tests of material. It has been our object, at any rate in the first place, to fill up gaps; there are many places where the engineer can have access to a testing machine, and indeed such investigations are usually of necessity carried on at the steel works.

Still, it is felt, I know, by many prominent engineers, that we ought to have at the National Physical Laboratory a standard testing machine for ultimate reference in cases of dispute, and to carry out such tests as cannot conveniently be made on ordinary machines. Such a machine costs money and our funds are limited. However, if some generous benefactor will ask Messrs. Buckton and Co. to repeat for us the splendid 300-ton machine they are now making for the Conservatoire des Arts et Metiers in Paris, your President-Elect, Mr. Wicksteed, will, I know, be pleased, and the gift will be cordially welcomed

We already possess a small testing machine of his for our research work.

We have only been working nine months, it is true, and probably in that time ought not to expect much, still the number of engineering tests we have been asked to carry out has been few. Englishmen are conservative; the high-class maker does not want the tests, he knows his products are good; his cheap and nasty rival does not want the tests, they will expose his weaknesses. One firm of deservedly high reputation wrote to me, some short time back, "It would never do for us to admit that our goods stood in need of a certificate." Pardon me if I state plainly my opinion that, in many cases at any rate, the high-class maker is wrong. A certificate such as we can give can only help him and expose his inferiors; there was a discussion of great interest at the Institution of Electrical Engineers a short time since on photometry and the standardization of lamps.* One speaker, of considerable experience in the trade, stated that the want of a certificate in England enables continental manufacturers to dump down on our markets the goods they could not sell at home. Does the same criticism apply to other things besides lamps?

But now I will turn to more interesting matters, to some of the researches in progress. The lines of the first research were mapped out by a Committee of Engineers under the chairmanship of Sir A. Noble, and containing such men as Mr. Froude, Mr. Mansergh, Mr. Maw, Mr. Ferranti, Mr. Parsons, Sir William Preece, Captain Sankey, Sir John Thornycroft, Sir William White, Sir J. Wolfe Barry and Mr. Yarrow. This Committee placed first amongst its recommendations the continuance of the work of the Alloys Research Committee of the Institution of Mechanical Engineers, and indeed it was in great measure with this object in view that the outfit to which I have already referred was acquired. With this was connected very closely the question of the molecular changes in metals due to fatigue, motion, and the like; while the investigation of the pressure of wind on surfaces of various areas was mentioned as an important piece of research.

---

* Proceedings, Institution of Electrical Engineers, 11th Dec. 1902, page 119.

The Alloys Research was placed first by two other committees, that for Electricity and that for Chemistry. The Electricity Committee referred specially to the connection between the magnetic quality and the physical, chemical, and electrical properties of iron and its alloys, with a view specially to the determinations of the conditions for low hysteresis and non-ageing properties.

Accordingly it is to these two subjects—Alloys Research and Wind Pressure—that our energies are being primarily directed. But the alloys of iron are many, and it was not quite easy to select a starting point. Various enquiries, however, have tended to make us turn our attention to nickel-steel. Mr. Yarrow wrote with regard to some facts he had observed. Captain Longridge again was anxious for some information bearing on the motor industry. Nickel-steel is obviously a most interesting and a most important substance. Most of our knowledge with regard to it is due, I think, to Mr. Hadfield, whose Papers, read before the Institution of Civil Engineers, are a mine of wealth ; and to the researches of a special committee of the Berlin Society for the Promotion of Industry, under the presidency of Dr. Wedding. Reference should also be made to a Paper entitled " Nickel-Steel ; a Synopsis of Experiment and Opinion," by Mr. David H. Browne, of Cleveland, Ohio, and to Mr. Browne's remarks on the discussion on Mr. Hadfield's Paper and to M. Osmond's remarks on Mr. Hadfield's series of alloys.

Still there remains much to be done. Mr. Hadfield's results were all obtained with low carbon alloys; he realised that carbon exerted a strong disturbing influence, and wished to render this as small as possible. It is not probable, I take it, that much of the nickel-steel of commerce contains only the $0 \cdot 17$ to $0 \cdot 19$ per cent. found in his specimens with but one exception. In the German researches the percentage of carbon present has been varied, and still there is no doubt of its extreme importance. Mr. Hadfield discovered a remarkable alloy, having about 12 per cent. of nickel and $0 \cdot 18$ of carbon and $0 \cdot 9$ of manganese. This has a tensile strength of over 90 tons to the square inch, an elastic limit of 55 tons, but a low ductility. Professor Arnold has shown that, by

G

reducing the carbon to $0 \cdot 1$ per cent. and the manganese to $0 \cdot 15$, the ductility is greatly increased without seriously altering the breaking stress or the elastic limit. Experiments are, however, still wanting to show how a gradual increase in the carbon modifies the results ; it is known again from Professor Arnold's work that, when nickel steel containing a high carbon is rolled at a low temperature, nearly the whole of the carbon is separated as graphite. Under what conditions exactly does this take place, and what is the state of the carbon previously ? Again, what is the relation of the nickel to the iron and the carbon it contains when the carbon is low.

The researches of M. Osmond and others have revealed to us the various temperatures at which changes take place in the relation of the iron and the carbon as a mass of steel is cooled down, and the curves due to M. Osmond given in Mr. Hadfield's Paper show that these recalescence points remain on the whole, altered in position it is true to some extent, but still generally recognisable in Mr. Hadfield's series. It will be noted that the points Ar_3, Ar_2, have almost disappeared from the 15 per cent. nickel alloy ; a curve for the 12 per cent. alloy is not given. What, if any, is the connection between this and the properties of the alloy ? Will the microscope tell us anything further of its structure ? I gather that the sections of low-carbon nickel-steels are usually very homogeneous.

Mr. Ashton, however, communicated, in the discussion of Mr. Hadfield's Paper, some interesting facts, and M. Osmond has shown us that, at any rate for the larger percentages of nickel, by proper etching or other suitable treatment, something may be developed. Let me quote some sentences from Mr. Hadfield's Paper[*] :—" As already pointed out it is probable that the special advantage of the use of nickel is due not so much to the properties it confers on iron in carbonless or nearly carbonless iron alloys, but to its modifying influence upon iron in the presence of carbon when it probably forms a special carbide of nickel. In the latter case, it is

---

[*] Proceedings, The Institution of Civil Engineers, 1898–99, vol. cxxxviii, page 22.

hardly probable that high percentages of nickel will be required for ordinary uses to which steel is now applied. This theoretical opinion can only be accurately determined by the preparation of say ten specimens to each of the author's nickel percentages, the carbon varying say by tenths. This would mean the preparation of about 140 different specimen alloys. As each of these specimens would require about 500 to 600 separate tests of various kinds, it will be seen that the work will require close application to complete it even in a lifetime."

On the other hand, M. Osmond, in the correspondence which followed Mr. Hadfield's Paper, showed strong grounds for thinking that the direct effect of the nickel was very great in itself. Now without starting on a programme so ambitious as the one just outlined, we can take up some part of the work; we can prepare, for example, a series of alloys with a medium percentage of carbon, say about 0·45 per cent., and another with a high amount of carbon, say 0·9 per cent., varying, as Mr. Hadfield did, the amount of nickel, and then examine by means of tests such as his the properties of the alloys, the state of the carbon and its relation to the nickel. Though probably we shall not by this means find the exact point at which any critical change takes place, we may find limits within which it happens, and thus be able to narrow to its most important part the wider research.

And now I wish to point out the advantages the laboratory has for a research of the kind; it has its disadvantages too, I am aware, and for some of our work we shall have to go outside. We hope to prepare some, if not all the alloys, ourselves. We have a small smelting-house with an oil blast-furnace capable of melting 40 to 50 lbs. of steel. For rolling and working the specimens in large masses we shall have to trespass on our friends. I have a promise from Sir J. Thornycroft that he will help. For the various elastic and mechanical tests we shall have no difficulty. We can also trace the cooling curves and examine the sections under the microscope. Our chemical laboratory will enable us to analyse the specimens exactly and carefully. And, by the aid of the electric ovens of the thermometric laboratory, the effect of various forms of annealing and of exposure for stated intervals to definite high temperatures can be satisfactorily studied.

Mr. Hadfield's work gains added interest from the electrical measurements of Professor Fleming and the magnetic work of Professor Barrett; such measurements we can make readily and rapidly. The advantages of all working together under the same roof are immense. Speaking of the Berlin research Mr. Hadfield writes :—" The whole research and the complete manner in which it was carried out are but another proof of the great desirability of this country establishing a similar National Physical Laboratory to the Reichsanstalt in Berlin, where the above experiments were carried out." It seems fitting that we should attempt, at an early stage of our life, to perform a complementary series of experiments.

There is yet another point connected with the nickel-steels to which we hope to give special attention. The question of elastic fatigue in metals is one of great interest, and there is a general belief that nickel-steel, with its high elastic limit, stands repeated stress more easily and for longer periods than ordinary steel of similar tensile strength. Little is known as to how this comes about, or what the changes are which finally lead to fracture. A recent Paper by Mr. Smith, of the Owens College, has advanced our knowledge, and Professor Ewing has somewhat forestalled us by applying the microscope to the problem. We are building a machine by which stresses of known amount and of frequencies up to 1,200 or 1,500 per minute can be applied to our specimens, and hope to incorporate tests on this machine with our other results.

Before leaving this part of my subject, I should mention two other related questions which are engaging our attention. In his evidence before the Treasury Committee, to whose report the establishment of the laboratory is due, Mr. Hadfield gave the first place to a determination of the exact melting point of a series of iron, iron-carbon alloys commencing, say, with pure iron, then steel with $0 \cdot 1$ per cent. carbon, the latter element gradually increasing until white iron containing $3\frac{1}{2}$ to 4 per cent. of carbon is arrived at. For this work we think we have the means required and may extend it to some other high-melting-point alloys of iron.

Finally let me refer, though it is possibly of more interest to the electrician than to the engineer, to work on the important silicon and aluminium iron alloys investigated by Professor Barrett; we have already made some observations for him, and hope to be permitted to continue them.

Such is our programme of work on Alloys of Iron. I trust you will recognise in it an honest attempt to carry out the wishes of those appointed to direct our first researches. Such a programme demands funds, and we are deeply indebted to the three societies, the Institution of Civil Engineers, the Iron and Steel Institute and the Society of Chemical Industry, which by their generous confidence have made the work possible. I trust I am not abusing an opportunity kindly given, if I suggest that when the work of the Alloys Research Committee of this Institution has been closed by the issue of its final report, the Institution of Mechanical Engineers may also join the other three, and by its assistance claim some direct share in our results. The raw material for the research we owe mainly to the generosity of Mr. Hadfield's firm.

To turn now to the second large Research. The wind-pressure experiments are the outcome of those of Sir Benjamin Baker made at the time of the building of the Forth Bridge, and described to the British Association at Montreal in 1884. Sir Benjamin set himself to examine two points:—

(1) What is the maximum average pressure which a continuous surface of large area may be expected to have to withstand?

(2) In the case of complex structure such as a bridge composed of lattice girders, tubes, &c., what is the area on which the wind is effective and how far is the back surface of a girder or of the bridge shielded by the front?

A Committee of the Board of Trade, appointed shortly after the Tay Bridge disaster, advised the adoption of certain rules—I am quoting from Sir B. Baker—" Shortly stated these were (1) That a maximum pressure of 56 lbs. per square foot should be provided for. (2) That the effective surface upon which the wind takes effect should be assumed to be from once to twice the front surface according to the opening in the lattice girders. (3) That a factor of safety of four for the ironwork and of two for the whole bridge, overturning as a mass when gravity alone comes in, should be adopted. Sir Benjamin Baker's Paper contains results, from which it appears that the limiting presses of 56 lbs. is rarely, if ever, reached, and that when this does happen the great pressure is limited to a small area, the centre of a gust of great local intensity.

He conducted experiments for several years at Inchgarvie on the Forth with three pressure gauges. The first of these had a surface of $1\frac{1}{2}$ square foot, and was swivelled about a vertical axis so as always to face the wind. The second was of the same area, but was fixed with its plane north and south so as only to be exposed to easterly or westerly winds. The third was 200 times as large, 300 square feet, and was fixed with its plane also north and south.

TABLE 1.

Revolving Gauge.		Small Fixed Gauge.		Large Fixed Gauge.	
Mean Pressure.		Easterly.	Westerly.	Easterly.	Westerly.
lbs.	lbs.	lbs.	lbs.	lbs.	lbs.
0 to 5	3·09	3·47	2·92	2·04	1·9
5 to 10	7·58	4·8	7·7	3·54	4·75
10 to 15	12·4	6·27	13·2	4·55	8·26
15 to 20	17·06	7·4	17·9	5·5	12·66
20 to 25	21·0	12·25	22·75	8·6	19
25 to 30	27·0	—	28·5	—	18·25
30 to 35	32·5	—	38·5	—	21·5
Above	65	—	41·0	—	35·25
(One observation only above 32·5)					

The above Table, taken from Sir B. Baker's Paper, gives the results of two years' observations. The pressures recorded by the revolving gauges are set down in groups 0 to 5, 5 to 10, &c., and the mean of each group taken.

The mean of the corresponding readings at the same time of the small fixed gauge, and of the large gauge for easterly and westerly winds, are set down opposite. It will be observed that throughout the pressures recorded by the small gauge are in excess of those given by the large gauge. Moreover, on only one occasion was a pressure above 32·5 lbs. observed, and this it was shown was due without doubt to a sudden gust causing the needle of the gauge, in consequence of its inertia, to register an amount far in excess of the true pressure.

Sir Benjamin Baker has kindly placed at my disposal the results of further observations lasting up to 1900. The large gauge became unsafe in 1896 and was removed, and according to these, pressures of over 50 lbs. have been registered on three occasions on the top of the piers in November 1893, March 1898, and October 1898, when the gauge broke under a pressure of 60 lbs.

An instructive contribution to the question is contained in a Report on Anemometer Experiments by Mr. R. H. Curtis, printed as an appendix to the Report of the Meteorological Council for the year ending 31 March 1900. Each of the stations at Bidston and Holyhead is fitted with two anemometer gauges. One of these in each case is a Dines pressure-tube, the other is a spring plate. At Bidston the plate is free to oscillate in the usual manner. At Holyhead the plate is constrained. It can only move in the direction in which it is forced by the wind, and the position in which it is found at any moment indicates the maximum pressure to which it has been subject since it was last set. The two stations are about 60 miles apart and have approximately the same aspect. The observations recorded lasted, I believe, about a year. The maximum pressures observed on the two pressure-tube instruments during the period were much the same; that at Holyhead was between 20 lbs. and 21 lbs., while at Bidston it was 22·5 lbs. to the square foot. The plate observations at Holyhead agree closely with those given by the tube; they are uniformly somewhat less, but the difference decreases as the pressure rises. At Bidston, with the oscillating plate, the results are entirely different. At one time when the tube read 22·5 lbs. the plate gave 63 lbs. Records of from 30 lbs. to

FIG. 8.—*Comparison of Air-Pressure Records recorded by Plates and Pressure-Tube Anemometers.*

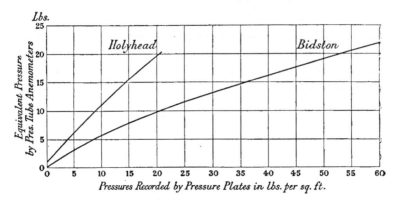

Pressures Recorded by Pressure Plates in lbs. per sq. ft.

50 lbs. per square foot were obtained on other occasions, in all of which the tube readings were below 17 lbs. On one occasion in 1871 the Bidston plate registered the huge pressure of 90 lbs. per square foot. The results are plotted in Fig. 8. In each of the two curves the ordinates give the pressure as read on the tube, the abscissae give the pressures on the plates. The Holyhead curve is nearly straight and inclined at 45° to the horizon, the readings of the two instruments are alike; the fact that the curve does not pass through the origin shows that there is a zero correction to apply to the plate readings. The coincidence of the values obtained from the two instruments proves that either of them is very approximately correct, and that during the period of observation the wind pressure did not exceed some 20 lbs. to the square foot. For Bidston the results are quite different; the plate readings start at low pressures by being nearly twice as great as those given by the tube; while at the maximum pressure recorded the plate pressures are three times as great as those given by the tube. The fact that the latter agree well with those obtained by both tube and constrained plate at Holyhead leads inevitably to the deduction that the plate results at Bidston are enormously too great; the high record is due in great measure to the inertia of the moving plate.

The diagram brings out clearly the uncertainty of the measurements and the importance of further experiments.

Now we do not get great storms at Bushy, and we can hardly hope to determine there the maximum wind pressure, but we may hope to determine (1) how far the average pressure over a large area differs from the maximum pressure at any one point, and (2) what are the screening effects due to the front member of a lattice girder or to the windward side of a bridge. This we are now doing on a Laboratory scale; but before describing our work I should like to call attention to two series of observations published also by Mr. Curtis of the Meteorological Office in the Quarterly Journal of the Royal Meteorological Society for 1882 and 1883.

Mr. Curtis made a large and careful series of observations by exposing perpendicularly to the wind four plates of areas 1 square foot, 2 square feet, $6\frac{1}{4}$ square feet, and 16 square feet: the second plate was round, the others square. A number of holes were bored in each plate, and the holes were connected by turns in pairs to the opposite arms of a water-gauge; the centre hole in each plate was taken as a standard for that plate and the difference in pressure at different points measured. Mr. Curtis was thus able to map out the distribution of pressure on his plates. Then the pressure at the centre of each plate in turn was compared with that at the centre of the round plate. As a result he finds (1) that the pressure at the centres of the various plates was practically the same, and (2) that the mean pressure per unit of area is greater on the large plate than on the small one. This result is apparently in contradiction to that found by Sir Benjamin Baker; it may well arise however from the difference in area of the plates experimented on, 16 square feet and 300 square feet respectively; or, as Mr. Curtis observes, from the fact that he measured the actual pressure on one side of a fixed plate, not the resultant force on the plate due to the difference in pressures on the two sides.

Our knowledge of the pressures on a girder or other engineering structure is due in large measure to Sir Benjamin Baker's experiments on models. A model of the girder was fixed to one end of a horizontal arm, at the other end was a flat plate or disc with its plane vertical and parallel to that of the girder. The area of the disc was variable.

The arm was supported at its centre by a vertical string, and allowed to swing as a pendulum in a plane perpendicular to the discs. When the resistances to the motion experienced by the disc and the model were the same, the plane of the disc continued at right angles to that of motion ; if these two resistances were different the system began to oscillate about the suspension. But by adjusting the area of the disc, a balance can be obtained. Thus a value can be found for an area which experiences the same resultant pressure as that acting on the girder.

Our own experiments now being conducted by Dr. Stanton combine the principles both of Mr. Curtis and of Sir B. Baker's observations, only in our case the disc or model to be subject to pressure is mounted on one end of a delicately suspended horizontal beam which can turn about a horizontal axis ; the plane of the disc is horizontal and the wind vertical. The other end of the beam carries a scale pan, and the resultant force on the disc or model is measured directly by the weights in the pan. These can be determined to the 1–100,000th of a pound. A number of holes in the disc can be placed in communication with one arm of a very delicate water-gauge, which allows a pressure difference due to 1–10,000th of an inch of water to be measured. The other arm of the gauge is connected to a fine tube placed in the incident stream of air, and thus the difference in pressure between a point on either side of the disc and one in the free air is measured, and the distribution of pressure over the disc determined. The diameters of the discs experimented on up to the present vary from 1 inch to 6 inches. These are placed in a current of air 24 inches in diameter, which has a practically uniform velocity across its section. The current is produced by an electrically driven fan, and the velocity can be varied between 5 feet and 30 feet per second. It is assumed, of course, that the wind pressures on structures can be calculated from experiments on models with the same kind of accuracy as is obtained in the calculation of the actual resistance of a ship from an experiment in a tank.

The first step is to determine the distribution of pressure on both sides of a flat plate as well as the resultant force on the plate, in view of the marked differences between the observed resultant force and that calculated from pressure measurements on the windward side alone. Concurrently with this, it is proposed to study the effect of the form and linear dimensions of the plate on the resultant force. Preliminary experiments have already indicated that the average intensity of pressure is very little affected by the form and size of the plates. The case of flat plates placed behind each other will then be taken, as there seems to be little agreement in the results of previous experiments as to the effect of the screening due to the windward plate. Then models of latticed girders of the usual type will be tried, and it is proposed to investigate the effect of wind pressure on solids, beginning with cylinders and rectangular prisms. These results may have some bearing on chimneys. We then hope to go on to the effect of wind pressure on oblique surfaces, first by trying a flat plate, and then approximating to the conditions of a roof by screening one side. Such is our programme; we have not advanced far with it, but we have learnt sufficient to feel fairly confident it will work, and I trust that when we have results to publish, it will be found that they have a real value to engineers. I should add, I think, that that part of the apparatus which has not been made in our own shop has been purchased by funds supplied by the Government Grant Committee of the Royal Society.

And now let me turn to the third section of our work—enquiries or experiments made at the request of engineers or others for which fees are received. It would not be difficult to make a list of enquiries of this kind, on which I think the results of our experiments on comparatively small points might be of service to the profession. I am glad, however, to be able to illustrate my meaning by a short account of two or three investigations which we have already carried out in response to such enquiries. Some few months since Mr. F. B. Behr brought us some specimens of cast steel which the maker had assured him would be found excellent as a tool steel, and which it was said could be produced at a less cost than that ordinarily paid for tool steel. He asked us to test the steel in any way we thought best, and I am indebted to his kindness for permission to use the results. Table 2 gives the result of the tests.

The specimens consisted of an ingot about 3 inches square in section and two rolled bars, one $\frac{7}{8}$ inch in diameter, the other $\frac{7}{8}$ inch square.

TABLE 2.—*Tensile Tests.*

No. 1.—Specimen turned from the round bar $\frac{7}{8}$ inch diameter in the state in which it was received.

No. 1A.—Specimen turned from the 1 inch × 1 inch bar forged from the ingot.

No. 2A —Specimen turned from the 1 inch × 1 inch bar forged from the ingot.

No. 3A.—Specimen turned from the 1 inch × 1 inch bar forged from the ingot.

No.	Initial Diameter.	Load in tons.		Stress in tons per sq. inch.		Elongation on total length.	Remarks.
		Yield.	Fracture.	Yield.	Fracture.		
	inch.					per cent	
1	0·425	6·19	8·93	43·4	62·7	6·5	Fracture shows remarkable fine uniform grain.
1A	0·460	—	9·66	—	58·2	5·0	Yield load not perceptible.
2A	0·341	—	6·00	—	65·9	5·7	Yield load not perceptible.
3A	0·380	—	6·55	—	57·9	4·0	Broke at collar of specimen. Yield load not perceptible.

General Remarks.—In all these tests the breaking stress is somewhat higher and the elongation lower than in an average sample of unhardened cast tool-steel. The appearance of the fracture in specimens 1A, 2A, and 3A was markedly different from that in No. 1, the grain being much coarser.

The mechanical tests consisted of—

(1) Tensile tests on specimens turned from the $\frac{7}{8}$ inch bar and from a bar 1 inch by 1 inch forged for us from the ingot by Sir J. Thornycroft.

(2) Tests of hardness on the round bar and also on the bar forged from the ingot.

(3) Cutting tests made on tools forged from the bars supplied and from the forged bar.

And in the first place, as to the treatment of the ingot at the Chiswick works. This was heated to a very bright red heat, and forged all over from 3 inches square to 2 inches square under a steam hammer. The bar was then raised to almost a white heat and drawn down to 1 inch square, when it was allowed to cool slowly in the air.

The breaking stress is somewhat higher and the elongation lower than in an average sample of unhardened tool-steel. It was clear from the fracture that the rolled bar supplied with the ingot had a finer grain than that forged from the ingot.

TABLE 3.—*Tests of Hardness.*

The Tests of Hardness were made by measuring the amount of indentation produced by a hardened steel knife-edge, ground to an angle of 90°.

Material.	Load per inch width of knife-edge.	Indentation.
	tons.	inch.
Normal cast steel	4·43	0·01465
Specimens of steel cut from ingot .	4·43	0·01585
Specimens of steel cut from bar forged from ingot	4·43	0·01405

These results show that, as might have been predicted from the tensile tests the forged bar is harder than normal cast steels and the material of the ingot less so, the variation in either case not being considerable.

In Table 3 (page 77) we have the tests of hardness. These were made by a method due, I believe, to Professor Unwin, by measuring the amount of indentation produced by a hardened steel knife-edge, ground to an angle of 90° and forced into the material.

The first line gives the result for some ordinary tool steel. The results show that the ingot is less hard than this steel, while the forged bar is harder. The variation in either case is not large.

TABLE 4.—*Cutting Tests of Tool Steel.*

Tools Nos. 1 and 2.—Cutters made from a round sample ⅞ inch diameter, hardened and tempered.

Tool No. 3.—Forged from a square sample 1 inch × 1 inch, hardened and tempered.

Tool No. 4.—Made from the 1 inch × 1 inch bar forged down from the ingot, hardened and tempered.

No. of Tool.	Duration of Test.	Mild Steel Bar on which Test was made.			Rate of cutting feet per minute.	Cubic inches of metal removed per min.	Remarks.
		Length.	Initial diameter.	Final diameter.			
	minutes.	ins.	ins.	ins.			
1	18·37	22	1·75	1·47	33·3	0·848	Cutting edge in good condition after test. Good supply of lubricant used.
2	10·77	22	1·47	1·28	49·8	0·835	Cutting edge blunt. Good supply of lubricant used.
3	40·32	48	2·00	1·70	38·7	1·04	Cutting edge blunt. Good supply of lubricant used.
4	18·75	39	1·69	1·49	57·5	1·04	Cutting edge in good condition. Good supply of lubricant used.

In Table 4 (page 78) we have the results of the cutting tests. The steel was hardened at a temperature of about 730° C. (1,350° F.), tempered in the usual way in the shop, and tested under ordinary workshop conditions in reducing a bar of mild steel in an 8-inch lathe. No attempt has been made to compare it with the effect of self-hardening high-speed tools working without lubricant.

It will be found, I think, from the figures that the depth of cut was from 0·10 to 0·15 of an inch, that the rate of cut was about

TABLE 5.

Analyses of Ingot and Bar.

	Ingot.	Bar.
	per cent	per cent.
Carbon	0·896	0·828
Sulphur	0·086	0·075
Silicon	0·253	0·200
Phosphorus	0·141	0·135
Arsenic	0·045	0·043
Manganese	0·313	0·310
Nickel	0·166	0·210
Chromium	trace	trace

Analysis of bar hammered from Ingot.

Carbon . . . . . = 0·890 per cent.

$\frac{1}{64}$ inch per revolution, and that the weight of steel removed was from 15 lbs. to 20 lbs. per hour. These numbers compare, I think favourably, with the results given in Mr. Donaldson's recent Paper (page 5).

These tests again show that the bar compares favourably with good commercial steels. They were carried out in the

Engineering Laboratory under Dr. Stanton's directions. Specimens of the steel were then analysed in the chemical laboratory by Dr. Carpenter. Table 5 (page 79) shows the results. In order to test if any serious change had taken place in the bar forged from the ingot, an analysis of its carbon was also made. The material was then subjected by Dr. Carpenter to a series of photo-micrographic tests.

The first photo-micrograph, Fig. 9, Plate 3, shows the structure of the original cast ingot magnified 102 diameters. With the exception of the black patches scattered here and there, and which are probably slag, the etched surface is coloured a uniform brown. This structure is characteristic of pearlite, the eutectic mixture of iron and carbide of iron, and is what might have been anticipated from a knowledge of the carbon content of the steel. The next photo-micrograph, Fig. 10, gives, under the same magnification, the structure of the bar hammered from the ingot, and shows very well the effect of hammering in removing slag. Only one patch is to be seen.

The character of the pearlite is shown in the next two photo-micrographs, Figs. 11 and 12, which are magnified 820 diameters. In the case of the ingot, Fig. 11, granular pearlite is found in the interior, lamellar pearlite near the edge. The banded structure seen in Fig. 11 is due to alternate lamellae of iron and carbide of iron. The white streaks are the carbide which is .not attacked by the nitric acid used for etching, and which therefore reflect the light back up the tube of the microscope, vertical illumination being used. The dark lines are the iron which has been pitted by the etching fluid, and which reflect the light in another direction. The effect of hammering in rendering the metal more homogeneous is seen in the photo-micrograph of the hammered bar, Fig. 12, where the lamellar and the granular pearlite have penetrated the one into the other. For the sake of comparison, Fig. 13 is reproduced from a photo-micrograph (magnified 850 times) of an unhardened spring steel, given on Plate 20 of the Fifth Report to the Alloys Research Committee.* It contains 0·78 per cent. carbon, and 0·17 per cent.

---

* Proceedings, 1899.

chromium, and its structure is quite similar to that of the hammered and unhardened bar, except there the pearlite is almost wholly granular.

The change of structure subsequent upon the hardenings of a soft tool-steel is seen in Fig. 14, Plate 4. A small section cut from the hammered bar was quenched at 741° C. in water at 15° C., reheated to straw temper and afterwards cooled in the air. It was then polished, and the structure of the surface developed by rubbing it on parchment with ammonium nitrate solution and calcium sulphate. Under this treatment the surface takes on a variety of colours—yellow, blue and black in parts, and when examined under high magnification, in this case 1,500 diameters, three types of structure are seen :—

(1) The white areas, which are ferrite, that is, iron.

(2) The light granular areas which are sorbite, that is, unsegregated or partially segregated pearlite.

(3) The dark areas which are troostite. As to the nature of this last-named structure, considerable doubt exists at the present time.

Thus a photo-micrograph of the hardened section gives some idea of the condition and distribution of the various constituents of the steel at the temperature of quenching; and, provided the chemical composition of the steel be known, it is possible to judge from the photo-micrograph within certain limits what the quenching temperature actually was.

With a view to examining the effects of hardening, small sections, 2 cm. × 1 cm. × $\frac{1}{5}$ cm., were prepared from the forged bar and heated in an electric furnace to 792° C., 741° C., 691° C. and 642° C. respectively. These were polished and examined for cracks. It had been found in tempering the tools that some care was required, and in one case a crack had developed. The specimens quenched after heating in the electric furnace were found to be free from flaws.

Thus finally we were able to conclude as follows :—

Mechanical tests.—The tensile tests show that the steel is of greater strength than average cast-steel, but that there is a corresponding reduction of elongation. The indentation tests confirm these results, and show further that the material of the ingot before working has a

H

degree of hardness very little less than a rolled bar of average cast steel. The cutting tests were satisfactory in every way. The results show that the steel is very suitable for use as a cutting tool, and bears favourable comparison with the best cast-steel used in lathes with a good supply of lubricant. It would seem that care is required in the forging and hardening of the steel to prevent the development of cracks, but with ordinary attention this can be avoided. These particulars seem to me to illustrate very clearly the point I wish to bring home—the commercial use of the Laboratory to the engineering profession. There are few places if any, I take it, except Bushy, where Mr. Behr could have had such a series of tests carried out; the value of the report is obvious and its utility clear. It should be noted that at least four departments of the Laboratory combined to obtain the results, and that all the work of preparation, except the forging of the ingot, was done on the spot.

The next case is a simpler one, but not less interesting. An engineer in practice in South Wales wrote to us with regard to the well-known method of determining the temperature of a furnace by the use of an iron ball and a calorimeter. The method is a bad one, it is obvious. Nowadays the platinum thermometer or the thermo-junction ought to supersede it in all permanent installations; still it has its uses. Our correspondent pointed out that there were considerable discrepancies in the values given for the specific heat of iron at high temperatures, or what comes to the same thing for the total heat emitted by a given mass of iron in cooling from a high temperature; he told us that he had made a number of experiments himself with great care, and always found a lower value than that given by Pionchon, whose work was generally looked upon as a standard, and he asked for assistance. He pointed out, moreover, that the value used by Pionchon for the melting point of silver, determined by Violle was 907° C., while the temperature is now known to be nearly 50° C. higher, or 962°, and suggested that this might be the cause of some of the difference. I asked Dr. Harker to look into the question. The electric oven gave us a ready method of heating the iron, while its temperature was easily found by the aid of a platinum thermometer, and a number of experiments were made. An oven was mounted with its axis vertical, and arranged so that it could be brought over the calorimeter and the iron dropped directly into the water. In some experiments, 42 (in others 26) grammes of Mr. Hadfield's Swedish charcoal iron were used, and the usual precautions necessary in calorimeter work taken. The diagram, Fig. 15, gives the results so far as the research has yet gone.

FIG. 15.—*Specific Heat of Iron.*

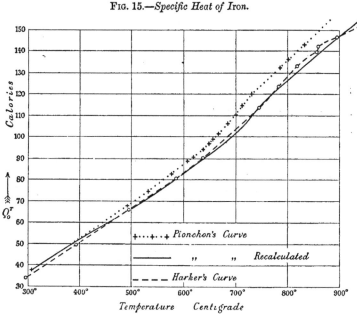

Temperature   Centigrade

The upper curve with the crosses gives Pionchon's results; the lower one with the circles shows the results obtained by Dr. Harker. He is, you will observe, below Pionchon, though not nearly so much below him as our Cardiff correspondent. But now Pionchon's temperature for the melting point of silver, 916° C., is too low by 50°. At lower temperatures, 200°, 300°, we may reasonably suppose his temperatures were accurate. Let us then modify his results by supposing that there was an error in his temperatures negligible in the lower part of the scale and increasing uniformly up to 50° at

H 2

900°; we obtain the full line, which agrees very closely with Harker's results. You will observe the break in Harker's curve at about 850°. M. Osmond determined the recalescent point Ar₃ of this iron, when the change from β iron to γ iron is supposed to occur, as 855° C. Dr. Harker's curve bears an independent testimony to a change at about this point, though of course the number of his observations is too small to afford very exact information. The work does not claim to be done in the same elaborate manner as Pionchon's. It was clear that when the hot metal was dropped into the water there was a certain amount of oxidization, and no attempt has been made to correct for this, but the error thus caused, which Pionchon avoided, must be very small, and certainly occurs in the practical application of the method.

Other examples of the work the Laboratory can do might easily be given. I was present here about three weeks ago at a very interesting discussion on the Cutting Angles of Tools introduced in a Paper by Mr. Donaldson, of Woolwich. Mr. Donaldson was the first to state that his results were not final, and it was clear from the discussion that much remained to be done. May I say without impertinence that what struck me most was the paucity of knowledge on a most important matter. I take it that the technical skill of the English workman has been such that it was best to leave the question to him, and allow him to grind his tools as experience led him to do. Is this a question on which we can render help, and on which our help will be of value? If so, perhaps the President would like to support a proposal at our General Board next month that we should take up the question. I am aware that our 8-inch lathe is not heavy enough, but that is a difficulty which can be overcome, and for the rest, with the kind assistance of our friends in providing material, we have all that we require to carry out the work. I cannot promise to measure forces of 10 tons on the point of the tool, as Mr. Wicksteed suggested might be required, but with Mr. Donaldson's help should hope to supplement the numbers given in his Paper.

I have already referred to the work on the aluminium and other iron-alloys used by Professor Barrett. Perhaps without going into the details of that work, I may show some photo-micrographs obtained; Plate 5 gives a photograph of three sections of the ingot from which the aluminium rod was made. You will notice in the first place the difference in structure in the different parts of the ingot. Figs. 20 and 21 are from the upper part, and show a series of small nodules forming a ring pattern on a ground mass of what appears from Fig. 21 magnified 1,260 times to be granular pearlite. It seems probable that these nodules are the oxide of aluminium. Fig. 22 and Fig. 23 are chiefly pearlite. There are still some nodules visible and a certain amount of ferrite, while in Figs. 24 and 25 from the bottom of the ingot we have chiefly ferrite. In Figs. 16 to 19, Plate 4, we have photos of the section of the rod 1167 H deeply etched to show grains in the high-power photographs. Figs. 16 and 17 are from near the edge. Note the ferrite on the extreme edge. The white patches in relief are probably aluminium. The small nodules of oxide are seen in the magnified photographs.

In Plate 6 we have photographs of the Swedish charcoal iron bar. Fig. 26 is the polished surface. Figs. 27 and 29 are different portions of the same after deep etching, showing a difference in the carbon present. This has to some extent been confirmed by chemical analysis. In Fig. 28 the ferrite grains are clearly seen; the one in the centre is about $\frac{1}{250}$ inch in diameter. Fig. 30 shows a ferrite grain bounded by pearlite.

In conclusion, may I again thank the President for the opportunity he has given me of placing some account of the National Physical Laboratory before the Graduates. I trust that my own zeal for what I believe to be a great and a greatly needed work has not led me to depict our efforts in too favourable colours, but that I may carry the opinion of the Meeting with me, if I express the hope that what we are doing is of real value to engineers, and if I urge on them the contention that the Laboratory is an institution which deserves their cordial support.

The Paper is illustrated by Plates 2 to 6 and 3 Figs. in tho letterpress.

The PRESIDENT, in proposing a hearty vote of thanks to Dr. Glazebrook for his interesting Lecture, emphasised the importance of the work which the National Physical Laboratory was capable of carrying out. There were, he remarked, very few engineering works in the country in which anything like an exhaustive enquiry into the physical properties of materials could be successfully carried out, as not only was the equipment wanting, but also the trained staff necessary to use effectively such an equipment. Engineers would therefore be well serving their own interests, if they referred to the National Physical Laboratory many questions which arose for settlement in the course of practical work.

The vote of thanks was carried with acclamation, and a brief discussion then ensued on some points raised by the Lecture.

Dr. GLAZEBROOK, in replying, thanked the Meeting for their kind reception.

Mr. J. S. WARNER, Chairman of the Graduates' Association, proposed a cordial vote of thanks to the President for taking the chair that evening.

The PRESIDENT suitably acknowledged the vote.

The Institution of Mechanical Engineers.

PROCEEDINGS.

FEBRUARY 1903.

The FIFTY-SIXTH ANNUAL GENERAL MEETING was held at the Institution on Friday, 20th February 1903, at Eight o'clock p.m. The chair was taken by the retiring President, WILLIAM H. MAW, Esq., who was succeeded by J. HARTLEY WICKSTEED, Esq., the President elected at the Meeting.

The Minutes of the previous Meeting were read and confirmed.

The PRESIDENT announced that the following two Transferences had been made by the Council :—

Associate Members to Members.

BRUCE, ROBERT ARTHUR, ... Leeds.
PLATTS, WILLIAM, ... Sheffield.

The following Annual Report of the Council was then read :—

ANNUAL REPORT OF THE COUNCIL
FOR THE YEAR 1902.

*For presentation to the Members at the Fifty-Sixth Annual General
Meeting on Friday the 20th February 1903, at Eight p.m.*

The Council have the pleasure of presenting to the Members the
following Report of the progress and work of the Institution during
the past year.

During the year the following honours have been conferred upon
Members of the Institution. His Majesty the King has appointed
Lord Kelvin a Privy Councillor, and has conferred upon Lord Kelvin
and Lord Rayleigh the new Order of Merit. His Majesty has also
conferred upon Sir Benjamin Baker, K.C.M.G., the honour of Knight
Commander of the Order of the Bath; the honour of Knighthood upon
Mr. George T. Livesey, Mr. William Mather, M.P., and Mr. John I.
Thornycroft; and has made Mr. H. W. Beatty and Mr. Maurice
Fitzmaurice Companions of the Order of St. Michael and St. George.

The Resolution passed at the last Annual General Meeting with
the object of increasing the number of Members of the Council to
twenty-eight, exclusive of Past-Presidents, was confirmed at a Special
Meeting held on Monday 10th March. The alteration now forms
part of the 23rd Article of Association.

The total number of all classes on the roll of the Institution at the end of 1902 was 3,892, as compared with 3,490 at the end of the previous year, showing a net gain of 402, compared with corresponding gains of 238, 243, and 325 respectively during the years which have elapsed since the Institution took possession of its new House. During the year 565 candidates were elected, of whom 43 were former Graduates elected as Associate Members ; and four elections became void, thus making 518 names added to the register. The total deductions have been 116, made up of 39 deceases during 1901 (see Report of 1901), 52 resignations which took effect on 1st January 1902, and 25 removals.

The rapid increase of the number of Members of all classes has thus been more than maintained, the increase of 30 per cent. during the three years 1899–1901 referred to in the last Annual Report having been succeeded by a net increase of 11½ per cent. during 1902.

The following forty-five Deceases of Members of the Institution were reported during the year 1902 :—

ABEL, Sir FREDERICK AUGUSTUS, Bart., G.C.V.O., K.C.B.,
D.C.L., D.Sc., F.R.S. (Hon. Life Member), . . . London.
AINLEY, HENRY, ... Oldham.
BOYD, JAMES TENNANT, ... Glasgow.
BROTHERHOOD, PETER, ... London.
BROWN, JAMES (Associate Member), ... Birmingham.
CHAPMAN, JAMES GREGSON, ... London.
COOPER, GEORGE, ... London.
COULMAN, JOHN, ... Hull.
DIXON, SAMUEL, ... Manchester.
DONKIN, BRYAN, ... London.
EDWARDS, WILLIAM BERNARD (Graduate), ... Derby.
GREGORY, GEORGE FRANCIS (Associate), ... Hawkhurst.
HOLDEN, ROBERT, ... London.
HOPKINSON, JOHN, ... Manchester.
HOUGHTON-BROWN, ERNEST, ... London.
HOW, WILLIAM FIELD, ... Sydney.
HUNTER, CHARLES LAFAYETTE, ... Cardiff.
HUTTON, WILLIAM (Associate), ... Johannesburg.

IMRAY, JOHN, .	London.
LEWIS, HERBERT WILLIAM,	Bombay.
LINN, ALEXANDER GRAINGER, .	Liverpool.
LOGAN, ROBERT PATRICK TREDENNICK,	Dundalk.
LONGBOTTOM, LUKE,	Stoke-on-Trent.
MACPHERSON, ALEXANDER SINCLAIR, .	Leeds.
MCTAGGART, JOHN (Associate Member),	Bradford.
MESSER, EDGAR HARRISSON (Associate Member),	Johannesburg.
MOBERLY, CHARLES HENRY,	London.
MUIR, ALFRED,	Manchester.
NEVILLE, EDWARD HERMANN (Associate), .	Madrid.
NEWBIGGING, JOHN (Associate Member),	Dublin.
PAJEKEN, JULIUS F.,	Berlin.
PARKER, HENRY ALBERT (Associate Member),	London.
PLATT, SAMUEL RADCLIFFE,	Oldham.
ROBERTS-AUSTEN, Sir WILLIAM CHANDLER, K.C.B., D.C.L., F.R.S. (Hon. Life Member),	London.
ROBINSON, JOHN,	Leek.
SHEPHERD, JOHN,	Leeds.
SOYRES, FRANCIS JOHNSTONE DE,	Bristol.
STEWART, ANDREW (deceased 1901), .	Glasgow.
SUTTON, THOMAS,	London.
TAYLOR, JOHN (Associate),	London.
THORBURN, WILLIAM,	Bilbao.
VINES, CHARLES GRANVILLE (Associate Member),	Kimberley.
WHEELOCK, JEROME,	Worcester, Mass.
WILSON, JAMES MICHAEL GRAHAM (Associate, deceased 1901),	Cape Town.
YOUNG, GEORGE SCHOLEY,	London.

Of these, Sir Frederick A. Abel and Sir William C. Roberts-Austen had been Honorary Life Members from 1883 and 1897 respectively; Mr. John Robinson was elected Member of Council in 1866 and President in 1878 and 1879; Mr. Bryan Donkin was elected Member of Council in 1895, and Vice-President in 1901; and Mr. Samuel R. Platt was elected Member of Council in 1900.

The Accounts for the year ended 31 December 1902 are now submitted (see pages 96–99 and 100), having been duly certified by Mr. Robert A. McLean, F.C.A., the Auditor appointed by the Members at the last Annual General Meeting.

The total revenue for the year 1902 was £9,884 11s. 10d., while the expenditure was £9,597 17s. 5d., leaving a balance of revenue over expenditure of £286 14s. 5d., exclusive of Entrance Fees £537 and Life Compositions £111 carried direct to Capital Account. The financial position of the Institution at the end of the year is shown by the balance sheet. The total investments and other assets amount to £70,405 18s. 3d., and, deducting therefrom the £25,000 of debentures and the total remaining liabilities, £2,598 17s. 0d., the capital of the Institution amounts to £42,807 1s. 3d., including the £4,351 L. and N. W. Railway 3 per cent. Debenture Stock set aside as the nucleus for a Leasehold and Debenture Redemption Fund. The remaining investments, consisting of £3,945 12s. Midland Railway $2\frac{1}{2}$ per cent. Debenture Stock and £1,000 Consols ($2\frac{3}{4}$ per cent.) are held by the Union of London and Smiths Bank as security for an overdraft. The certificates of the securities have been duly audited by the Finance Committee and the Auditor.

The completion of Sir William Roberts-Austen's Sixth Report to the Alloys Research Committee has been delayed by the lamented death of the Reporter, but a large amount of his experimental work, dealing with the tempering of steel, and also with alloys of the industrial metals, is available and is now being dealt with by the Committee.

No further Report will be made by the Gas-Engine Research Committee, until the large experimental engine has been put to work at the Birmingham University.

Professor T. Hudson Beare has been occupied at the University of Edinburgh in perfecting the apparatus for testing the Value of the Steam-Jacket.

Professor David S. Capper has now concluded his experiments at King's College upon jacketed and unjacketed steam cylinders, and a Report upon his comprehensive experiments is almost completed. Owing to the greatly regretted death of Mr. Bryan Donkin, this Committee will have to elect a new Chairman.

The question of the Standardization of Flanges has received the attention of the Council, and was dealt with at the April Meeting in a Paper by Mr. R. E. Atkinson. A considerable number of Members and others have since sent in contributions bearing on the best forms to be adopted as standards.

The Council desire to record their thanks to the executors of the late Mr. A. G. Linn, of Liverpool, and to the members and others who have given new books to the Library, also to the donors of various publications of societies and public authorities, and of technical periodicals, a list of which appears on pages 100 to 113.

The thanks of the Members are especially due to the 22 Members of the Council who presented to the Institution in May a fine portrait of the former President, Sir William H. White, K.C.B.; and to Mr. Henry Chapman for the gift of an electric lantern for use at the Meetings.

The Summer Meeting, held this year at Newcastle, was well attended. Interesting Papers were read dealing with the electrical developments in the city, and the marine, locomotive, mining, and steel industries of the North-East Coast, which were subsequently visited. Mr. H. I. Brackenbury acted as Honorary Local Secretary, and the members were hospitably received not only in Newcastle, but also in Sunderland, where Mr. Henry H. Wake and Mr. W. H. Dugdale organised a day of Visits, and in The Hartlepools, where Mr. J. R. Fothergill acted as Honorary Local Secretary. Sir Andrew Noble and Mr. Watson-Armstrong subsequently entertained a number of members at their country seats.

Monthly Meetings of the Institution were held throughout the year, with the exception of May to September, while on three occasions extra meetings were held to continue important discussions. These meetings, with the Newcastle Meeting, were occupied by the reading and discussion of the following Papers : —

Modern Machine Methods; by Mr. H. F. L. Orcutt.

Recent Developments in the Gas-Engine; Lecture by Professor T. Hudson Beare, F.R.S.E , at a Meeting of the Graduates.

Fencing of Steam- and Gas-Engines; by Mr. Henry D. Marshall.

Fencing or Guarding Machinery used in Textile Factories; by Mr. Samuel R. Platt.

Protection of Lift Shafts and Safety Devices in connection with Lift Doors and Controlling Gear; by Mr. Henry C. Walker.

Guarding Machine Tools; by Mr. W. H. Johnson.

Standardization of Pipe Flanges and Flanged Fittings; by Mr. R. E. Atkinson.

Liquid Fuel for Steamships; by Mr. Edwin L. Orde.

Pumping Plant for Condensing Water; by Mr. Charles Hopkinson.

Newcastle and District Electric Lighting Co.'s Power Station at Neptune Bank, Newcastle-upon-Tyne; by Mr. W. B. Woodhouse.

Some Experiments on Steam-Engine Economy; by Professor R. L. Weighton.

The Application of Steam Distributing Valves to Locomotives; by Mr. Walter M. Smith.

Mechanical Appliances in Mines (Coal Cutting and Drilling); by Mr. R. H. Wainford.

Oil Motor Cars of 1902; by Captain C. C. Longridge.

Recent Practice in the Design, Construction, and Operation of Raw Cane Sugar Factories in the Hawaiian Islands; by Mr. J. N. S. Williams.

The following Papers were accepted for publication in the Proceedings :—

Trial of a Triple-Expansion Vertical Condensing Steam-Engine for Hydraulic Pumping; by Mr. Charles J. Hobbs.

Notes on Turning Mirrors for Search Lights; by Mr. Arthur E. Pettit.

The Graduates have held monthly Meetings and have made Visits to Works during the year. The attendance has been fairly satisfactory, and each Meeting was presided over by a Member of Council or other prominent member of the Institution. The following Papers were read and discussed :—

Light Steam-Cars; by Mr. T. J. B. Drayton.

Twelve Months' Revision of a Drawing Office; by Mr. W. Stanley Bott.

The Construction of Modern Underground Electric Railways; by Mr. H. M. Rootham.

Coal-Gas Exhausting Plant; by Mr. O. Wans.

Weston-Twerton Bridge Undertaking; by Mr. H. H. Mogg.

Oil Engines; by Mr. E. A. Falconer.

At the February Meeting of Graduates, Professor T. Hudson Beare, F.R.S.E., delivered a lecture on "Recent Developments in the Gas-Engine," which was illustrated by lantern slides.

The Council have awarded a prize to Mr. W. Stanley Bott for his Paper on "Twelve Months' Revision of a Drawing Office," and one to Mr. H. H. Mogg for his Paper on "The Weston-Twerton Bridge Undertaking." Both of these Papers, read at Graduates' Meetings, will be published in the year's Proceedings.*

The Engineering Standards Committee, the constitution of which was explained in the last Annual Report, has held frequent meetings during the year, and its recommendations relating to Standard sizes for rolled sections will be published shortly.

The result of the Ballot for the election of President, two Vice-Presidents, and eight Members of Council, to fill the vacancies caused by retirement and decease, will be announced at the Annual General Meeting.

---

* Proceedings 1902, Part 5, pages 1003 and 1013.

Dr.   ACCOUNT OF REVENUE AND EXPENDITURE

Expenditure.

	£	s.	d.	£	s.	d.
To Expenses of Maintenance and Management—						
Salaries and Wages	2,547	2	0			
Postages, Telegrams, and Telephone	488	1	8			
Heating, Lighting, and Power	138	13	2			
Fittings and Repairs	200	16	5			
Housekeeping	137	7	3			
Incidental Expenses	48	8	4			
				3,560	8	10
„ Printing, Stationery, and Binding—						
Printing and Engraving Proceedings	1,393	17	5			
Stationery and General Printing	563	8	1			
Binding	23	15	0			
				1,981	0	6
„ Rent, Rates, Taxes, &c.—						
Ground Rent	875	17	2			
Rates and Taxes	859	7	8			
Insurance	33	4	0			
				1,768	8	10
„ Meeting Expenses—						
Printing	396	12	8			
Reporting	74	0	6			
Travelling and Incidental Expenses	221	12	4			
				692	5	6
„ Conversazione				166	11	7
„ Dinner Expenses				36	18	7
„ Graduates' Prizes				13	19	6
„ Books purchased				21	0	3
„ Engineering Standards Committee				250	0	0
„ Expenses in connection with Research Committees				13	5	0
„ Depreciation on Furniture and Fittings				65	4	8
„ Debenture Interest				1,000	0	0
„ Interest on Bank overdraft				28	14	2
Total Expenditure				9,597	17	5
„ Balance, being excess of Revenue over Expenditure (exclusive of Entrance Fees £537, and Life Compositions £111, carried to Capital Account), carried to Balance Sheet				286	14	5
				£9,884	11	10

FOR THE YEAR ENDED 31st DECEMBER 1902. *Cr.*

Revenue.

	£	s.	d.
By Subscriptions for 1902 9,273		0	0

	£	s.	d.		£	s.	d.
„ Estimated value of Subscriptions in arrear (*being equal to the amount received in* 1902) . . .	356	0	0				
Less difference between estimated value in last account and actual receipts in 1902	19	0	0		337	0	0
„ Interest, &c.—							
From Investments	120	12	4				
Income Tax refunded	14	4	9		134	17	1
„ Reports of Proceedings—							
Extra Copies sold					138	17	3
„ Debenture Transfer Fees					0	17	6

£9,884 11 10

Dr.   BALANCE SHEET

	£	s.	d.
To Debentures—			
250 *of £100 each at* 4%, *redeemable in* 1917, *or at par at any date after 1st Jan.* 1908, *on six months' notice to holder*	25,000	0	0

" Cash—

	£	s.	d.			
Union of London and Smiths Bank, overdraft .	992	17	4			
Less In the Secretary's hands	50	1	0	942	16	4

" Sundry Creditors—

	£	s.	d.			
Accounts owing at 31st Dec. 1902 *(since paid)* .	1,359	14	6			
Willans Premium Fund	9	10	8			
Unclaimed Debenture Interest (coupons not presented)	159	15	6	1,529	0	8

" Subscriptions paid in advance				127	0	0

" Capital of the Institution :—

	£	s.	d.			
Balance at 31st Dec. 1902	38,539	11	9			
Add :—						
Excess of Revenue over Expenditure for the year ended 31st Dec. 1902 . .	286	14	5			
Amount received from Life Compositions during 1902	111	0	0			
Amount received from Entrance Fees during 1902	537	0	0	39,474	6	2
Amount invested in £4,351 London and North Western Ry. 3% *Debenture Stock and interest thereon set aside for Redemption of Debentures and Institution's Leasehold Property, see contra*				3,332	15	1
(The Market Value of this Stock and interest at 31st Dec. 1902 *was about £4,478 3s. 4d.)*						

		£70,405	18	3

Signed by the following members of the Finance Committee :—

W. H. MAW,
B. BAKER,
E. B. ELLINGTON,
H. GRAHAM HARRIS.

AT 31st DECEMBER 1902.   *Cr.*

		£	s.	d.
By Investments	*Cost*	3,707	15	2

£
3,945 12s. *Midland Ry.* 2½% *Debenture Stock.**
1,000 2¾% *Consols.**
The Market Value of these investments at 31st Dec. 1902 *was
about* £4,203 11s. 11d.

„ Investment of Amount set aside for Redemption of Debentures
    and Institution's Leasehold Property, *see contra*   . . .   3,332 15 1
    £4,351 *London and North Western Ry.* 3% *Debenture Stock
    cost* £3,270 17s. 0d. *and* £61 18s. 1d. *balance of interest
    thereon to be invested.*
    *This stock with its accumulating interest is set aside for the
    above purpose.*

„ Subscriptions in Arrear, *estimated value (being equal to the
    amount received during* 1902) 356 0 0
„ Furniture and Fittings (*less depreciation*) 1,239 9 8
„ Books in Library, Drawings, Engravings, Models, Specimens,
    and Sculpture (*estimate of* 1893) 1,340 0 0
„ Amount in Union of London and Smiths Bank to meet
    unclaimed Debenture Interest (*coupons not presented*) . .   159 15 6
„ Proceedings—stock of back numbers, *not valued.*
„ Institution House (*see last Balance Sheet*) 60,270 2 10

£70,405 18 3

* See page 92.

I certify that all my requirements as Auditor have been complied with,
and I report to the Members that I have audited the above Balance Sheet, dated
the 31st December 1902, and in my opinion such Balance Sheet is properly
drawn up and exhibits a true and correct view of the state of the affairs of the
Institution as shown by its Books.

                                          ROBT. A. McLEAN, F.C.A.,
                                                    Auditor,
14th *January* 1903.   1 Queen Victoria Street, London, E C.

WILLANS PREMIUM FUND.

Investment £159 8s. 5d. of India 3% Stock . . . . cost £165 5s. 0d.

Dr.				*Cr.*		
	£	s.	d.		£	s.	d.
To Balance, held in trust .	9	10	8	By Interest, 1901 . . .	4	15	4
				„ „ 1902 . . .	4	15	4
	£9	10	8		£9	10	8

Audited, certified, and signed by the names on pages 98-99.

(For the Declaration of Trust, *see* Proceedings 1901, page 16.)

LIST OF DONATIONS TO THE LIBRARY.

BOOKS (*in order received*).

Memoir of John Elder, by W. J. M. Rankine; from Mr. Henry Chapman.

High Speed Steam Engines, by W. Norris and Ben. H. Morgan; from Mr. Ben. H. Morgan.

Congrès International des Méthodes d'Essai des Matériaux de Construction, Paris, 1900—Procès-verbaux in extenso des Séances du Congrès; from the publisher.

Manufacture of Paint, by J. Cruickshank Smith; from the author.

Seven Years in a Black Country Town, and a few Works designed by or upon which Mr. C. L. N. Wilson has been engaged; from Mrs. Wilson.

International Engineering Congress (Glasgow, 1901): Proceedings of Section VI, Mining and Metallurgy; from the Institution of Mining Engineers. Ditto; from the Mining Institute of Scotland.

International Engineering Congress (Glasgow, 1901): Proceedings of Section VIII, Gas; from Mr. J. W. Helps, Hon. Secretary of the Section.

Balancing of Engines, by Professor W. E. Dalby; from the publisher.

Papers by Hector MacColl—Strength of Cylindrical Boiler Shells; Shafting of Screw Steamers; Unusual Corrosion of Marine Machinery; from the author.

Mercantile Marine in War Time; from the publishers.

American Standard Specifications of Steel, by A. L. Colby; from the author.

Gas and Gas Fittings, by H. F. Hills; from the publishers.

The following from the Executive Committee of the International Engineering Congress (Glasgow, 1901):—Proceedings of Section I, Railways; Proceedings of Section II, Waterways and Maritime Works; Local Industries of Glasgow and the West of Scotland, edited by Augus McLean.

Jahrbuch für das Eisenhüttenwesen; from the publisher.

The Gas and Oil Engine, by Dugald Clerk; from the author.

Coal-cutting by Machinery in the United Kingdom, by S. F. Walker; from *The Colliery Guardian.*

Steam Engineering, by Professor W. W. F. Pullen; from the author.

Practical Treatise on Modern Gas and Oil Engines, by Frederick Grover; from the author.

Liverpool Trials of Motor Vehicles for heavy traffic 1899 and 1901, Judges' Reports; from Professor H. S. Hele-Shaw.

The Steam Turbine, by R. M. Neilson; from the author.

Manual del Tornero, by E. H. Neville; from Mr. J. Neville.

Grundlagen der Theorie und des Baues der Warmekraftmaschinen, by Professor Alfred Musil and Professor J. A. Ewing; from Professor Alfred Musil.

Inspection of Railway Materials, by G. R. Bodmer; Pipes and Tubes, by P. R. Björling; from the publishers.

Applied Mechanics for Beginners, by John Duncan; from the author.

Waterworks' Law as applied to small undertakings, by Percy Griffith; from the author.

Elements of Machine Design, Part II, by Professor W. Cawthorne Unwin; from the author.

International Engineering Congress (Glasgow, 1901): Report of the Proceedings and Abstracts of the Papers read; from Professor J. D. Cormack.

Motor Vehicles and Motors, by W. Worby Beaumont; from the author.

Mechanics Theoretical, Applied, and Experimental, by Professor W. W. F. Pullen; from the publishers.

Pumps and Pumping, by M. Powis Bale; from the publishers.

Gas and Petroleum Engines, by Professor William Robinson; from the author.

Elementary Photo-Micrography, by Walter Bagshaw; from the author.

Facts on Fire Prevention (2 vols.), edited by Edwin O. Sachs; from the Editor.
Report on the Engineering Trades of South Africa, by Ben. H. Morgan; from
the author.
Yacht Register, 1902; from Mr. H. Graham Harris.
Civil Engineer's Pocket-book, by John C. Trautwine; from John C. Trautwine,
Jun., and John C. Trautwine, 3rd.
The Life Story of the late Sir Charles Tilston Bright (2 vols.), by E. B. Bright
and Charles Bright; from Mr. Charles Bright.
The following from the Executors of the late Alexander Grainger Linn
(arranged alphabetically according to authors):—Steam Boilers, by Robert
Armstrong; Mensuration, by T. Baker; Hydraulic Tables, by Nathaniel
Beardmore; Orthographic Projection, by William Binns; Arches, Piers,
Buttresses, &c., by William Bland; Form of Ships and Boats, by W.
Bland; The Screw Propeller, by John Bourne; Bracing with its
Application to Bridges, by R. H. Bow; Railway Practice, First Series
(3rd ed.), Second Series (2nd ed.), Third and Fourth Series, by S. C.
Brees; Modern Marine Engineering, by N. P. Burgh; Modern Screw-
Propulsion, by N. P. Burgh; Blasting and Quarrying of Stone, by
Maj.-Gen. Sir John Burgoyne, K.C.B.; Limes, Cements, Mortars,
Concretes, Mastics, Plastering, &c. (2nd ed.), by G. R. Burnell;
Conservation and Improvement of Tidal Rivers, by E. K. Calver; Geology
of England and Wales, Part I, by W. D. Conybeare and William Phillips;
Universal Geography, Vol. I, by G. A. Cooke; Ships' Anchors, by George
Cotsell; Encyclopædia of Civil Engineering, by Edward Cresy;
Hydraulics, for the use of Engineers, by J. F. d'A. de Voisins, translated
by Joseph Bennett; Masonry and Stonecutting (1st and 2nd ed.), by
Edward Dobson; Historical account of the Great Level of the Fens, called
Bedford Level, by W. Elstobb; Ferguson's Lectures on Mechanics, by
C. F. Partington; Tertiary Fluvio-marine Formation of the Isle of Wight,
by Professor Edward Forbes; Geometry, Plane, Solid, and Spherical;
Mechanic's Calculator (6th ed.), by William Grier; Mechanic's Pocket
Dictionary (5th ed.), by William Grier; Elementary Course of Mathematics
5th ed.), by Harvey Goodwin; Estimates and Diagrams of Railway
Bridges, Roads, Culverts, &c., by J. W. Grover; Equilibrium of Arches
(3rd ed.), by Joseph Gwilt; Elements of Plane Trigonometry (2nd ed.), by
James Hann; Elements of Spherical Trigonometry, by James Hann; General
Theory of Bridge Construction, by Herman Haupt; Engineer's and
Mechanic's Encyclopædia, Vol. I, by Luke Hebert; Coal-Fields of Great
Britain (3rd ed.), by Edward Hull; Cast and Wrought Iron Bridge
Construction, Text and Plates (2nd ed.), by William Humber;
Mathematical Tables (new ed.), by Charles Hutton and Olinthus
Gregory; Abstracts of the Principal Lines of Spirit Levelling in England

and Wales, Text and Plates, by Colonel Sir Henry James, R.E. ; Dictionary of Geography, by A. K. Johnston; Masting, Mast-making, and Rigging of Ships (11th and 12th ed.), by Robert Kipping; First Mnemonical Lessons in Geometry, Algebra, and Trigonometry, by T. P. Kirkman; The Steam Engine (5th ed.), by Dionysius Lardner; Civil Engineering, Vol. III, Part I, by Henry Law; Memoir of the Thames Tunnel, Part I, by Henry Law; The Indicator and Dynamometer (2nd ed.), by T. J. Main and Thomas Brown; Mechanical Principles of Engineering and Architecture (2nd ed.), by Henry Moseley; Marine Engines and Steam Vessels (3rd and 4th ed.), by Robert Murray; Hydraulic Tables, Coefficients, and Formulæ (1st and 2nd ed.), by John Neville; The Oblique Arch (3rd ed.), by Peter Nicholson; Practical Builder and Workman's Companion, 3 vols., by Peter Nicholson; Civil Engineering (4th ed.), by W. J. M. Rankine; The Steam Engine (5th ed.), by W. J. M. Rankine; Report on the eligibility of Milford Haven for Ocean Steam Ships, and for a Naval Arsenal, by Thomas Page; Richards' improved Steam-Engine Indicator, by C. T. Porter; Rambles on Railways, by Sir Cusack P. Roney; The Electric Telegraph, by Robert Sabine; The Engineer and Machinist's Assistant; The Strains on Structures of Ironwork, by F. W. Sheilds; Practical Tunnelling, by F. W. Simms; Handbook of Iron Shipbuilding, by Thomas Smith; Nature and Properties of Algebraic Equations (2nd ed.), by R. Stevenson; Strength of Materials, by Thomas Tate; Art of Drawing in Perspective, by Elias Taylor; Reports of the late John Smeaton (2nd ed.), 2 vols.; Tracts on Vaults and Bridges; Elementary Principles of Carpentry, by Thomas Tredgold; Tyer's Block Telegraph and Electric Locking Signals (5th ed.); Shipbuilding, Theoretical and Practical, by Isaac Watts, F. K. Barnes, W. J. M. Rankine, and J. R. Napier; The Principles of Mechanics (8th ed.), by James Wood.

List of Sewers within the Borough of Liverpool, 1855; Ahn's remodelled German Grammar, by M. Meissner; Collection of the Public General Acts for the Regulation of Railways, 1830-1864 (11th ed.), by James Bigg; Bradshaw's Railway Manual, 1868; Engineer's and Contractor's Pocketbook, 1854 and 1859; Treatise on Arithmetic, by James Thomson; Virgil Delphini; Report from the Select Committee on Harbours of Refuge, 1858; Minutes of Evidence and of Proceedings on the Liverpool and Birkenhead Dock Bills, Sessions 1848, 1850, 1851, and 1852; Algerian Railway—Report on the Proposed Railway between Constantine and Philippeville, 1857; Minutes of Evidence and Proceedings on the Edgware, Highgate and London Railway Bill, 2 vols., 1866; Metropolitan Railway—Notting Hill and Brompton Extension, Additional Powers, New Works at King's Cross and Edgware Road, Extension to Trinity Square, Tower Hill, Plans and Sections, November 1863; Metropolitan District Railways—1, Railway complete an Inner Circle of Railways within London, North of the Thames ; 2, Railways to form an Outer Circle round the Metropolis North and South of the Thames, connected with Existing Railways, Plans and Sections, November 1863; Parliamentary Gazetteer of England and Wales, Parts 1-12 (wanting Part 2), with Supplement; Ditto of Scotland, Parts 1-6; Ditto of Ireland, Part 2.

In addition to the foregoing, a large number of duplicate volumes, Proceedings, &c., of Societies, Journals, Plans and Sections, Surveying Instruments, and Curves were also received.

The following Official Publications from the Government of New South Wales :— Annual Reports of the Department of Mines, 1900 and 1901; Report of the Department of Public Works for the year ended 30 June 1900; Mineral Resources, No. 9, 1901; Picturesque New South Wales: an Illustrated Guide for Settler and Tourist; Wealth and Progress of New South Wales, 1900-1901, by T. A. Coghlan; Year-book of New South Wales, 1902 and 1903.

The following Official Publications from the Government of Western Australia :— Supplement to Government Gazette of Western Australia; Report of the Department of Mines, 1900 and 1901; Report on the Working of the Government Railways and Tramways for the year ended 30 June 1901; Annual Progress Report of the Geological Survey, 1899.

Gold-Fields of Victoria, Monthly Returns; from the Chamber of Mines, Victoria.

The following from the Government of the United States of America :—Annual Report of the Chief of Ordnance, 1901; Notes on the Construction of Ordnance; Annual Report of the Chief of the Bureau of Steam Engineering, 1902.

The following from the U.S. Geological Survey :—Twenty-first Annual Report, Parts II, III, IV, V (with Maps) and VII, 1899-1900; Monograph XLI ; Mineral Resources of the United States, 1900 ; Bulletins, 134, 177-190, 192-194; Geology and Mineral Resources of the Copper River District, Alaska, by F. C. Schrader and A. C. Spencer; Reconnaissances in the Cape Nome and Norton Bay Regions, Alaska, in 1900, by A. H. Brooks, G. B. Richardson, A. J. Collier and W. C. Mendenhall.

PAMPHLETS, &c.

The Scarcity of Coal, by B. H. Brough; from the author.

Notes on Methods of Electric Traction, by Theodore Schontheil; from the author.

The Rapid Acting Vacuum Brake and its successes; from Mr. W. P. Walker.

Ventilateurs et Pompes Centrifuges for hautes pressions; Recherches Expérimentales sur l'Écoulement de la Vapeur d'Eau et sur l'Écoulement de l'Eau Chaude; from the author, M. Auguste Rateau.

Utilisation directe des Gaz de Hauts-Fourneaux pour la production de la force motrice, by H. Hubert; Notes on English and French Compound Locomotives, by Charles Rous-Marten; Presidential Address of Mr. James Mansergh to the Institution of Civil Engineers, 1900; Rank and Titles of the Engineer Branch, Royal Navy; Report on the Personnel of the United States Navy; Weak Points in Naval Administration, by C. M. Johnson; Statement: Engineer Officers, Royal Navy, 1898; from Mr. Edgar Worthington.

Presidential Address of Mr. Percy Griffith to the Society of Engineers, 1902; from the author.

The Kew Cowl Tests: Preface by Perry Fairfax Nursey; from the author.

The Hydraulics of the Resistance of Ships, by E. C. Thrupp; from the author.

Submarine Construction by Diving and other Compressed Air Methods, by Woodman Hill; from the Institute of Builders.

History of the London and North-Western Railway; History of the Chester and Holyhead Railway; History of the Chester and Crewe Railway; History of the Nottingham and Lincoln Line; History of the Loughborough and Nanpantan Edge-Rail-Way; Derbyshire Railway History; History of the Leicester and Swannington Railway; The Canals, Edge-Rail-Ways, Outram-Ways, and Railways in the County of Leicester; from the author, Mr. C. E. Stretton.

Calorimetry of Producer and Illuminating Gases, by J. F. Simmance; from the publisher.

Prevention of Strikes and Lock-outs, by B. H. Morgan; from the author.

On the Evolution of the Mental Faculties in relation to some Fundamental Principles of Motion, by Dr. Henry Wilde; from the author.

Naval Engineers: Their Position and Authority, by Ewing Matheson; from the author.

Purdue University Locomotive Testing Plant, by J. H. Smart and Professor W. F. M. Goss; from Professor W. F. M. Goss.

The Ironmaking and Shipbuilding Industries in Germany, by E. Schroedter; from the author.

Der Reguliervorgang bei Dampfmaschinen, by Benno Rülf; from the author.

Developpement des Associations d'Iugénieurs en Angleterre et en Allemagne, by M. Alby ; from the author.

Report on various Electric Tramway Systems in Europe, America, and Australia, 1901 (with Drawings), by Thomas Roberts; from the author.

The American Invasion, by B. H. Thwaite; from the author.

Economising on Steel, Tools and Fuel, by Francis Marsden; from the author.

El Gas Pobre y sus Aplicaciones en la Industria, by J. G. Neville ; from Mr. J. Neville.

Mechanical Road Traction, by A. E. A. Edwards; from the author.

Captain Wells' Improved Tractor ; from Mr. W. Lloyd Wise.

Prospectus of Sir Joseph Whitworth's Scholarships and Exhibitions for Mechanical Science; from the Board of Education.

Producer Gas and its Use in Engineering and Shipbuilding, by F. J. Rowan; from the author.

Tracé Graphique des Déformations Élastiques des Systèmes Triangulés, by J. Keelhoff; from l'Association des Ingénieurs de Gand.

Estudio y Proyecto del Puerto Comercial Saturnino E. Unzué, Entre Rios; from Mr. J. C. Calastremé.

Sulfure de Fer ses propriétés et son état dans le fer fondu, by Professor H. le Chatelier and Monsieur Ziegler; from Professor H. le Chatelier.

Recent Progress in Large Gas Engines, by H. A. Humphrey; from the author.

Technical Education in the University of London, by Professor J. D. Cormack ; from the author.

Rede zum Geburtsfeste Seiner Majestát des Kaisers und Königs Wilhelm II. in der Aula der Königlichen Technischen Hochschule zu Berlin, 26 Januar 1902; Festigkeitstheorien und die von ihnen Abhängigen Formeln des Maschinenbaues; from the Rector.

Report of the Hydraulic Engineer on the Water Supply of Queensland, 1902; from Mr. J. B. Henderson.

List of Chinese Lighthouses, Light-vessels, Buoys and Beacons, 1902; from the Inspector-General of Chinese Customs.

Twentieth Report of the Working of the Boiler Explosions Acts, during the year ending 30 June 1902 ; Board of Trade Reports on Boiler Explosions; from the Board of Trade.

Report to the Governors of the City and Guilds of London Institute, April 1902; from the Institute.

Crystal Palace Engineering School Magazine; from Mr. J. W. Wilson.

Seventh Annual Report, 1901, of the John Crerar Library, Chicago; from the Library.

Year-Book of the Royal Society, 1902 ; Sixth and Seventh Series of Reports to the Malaria Committee, 1901-1902 ; Reports to the Evolution Committee, Report I, 1901 ; from the Royal Society of London.

Classified Lists and Distribution Returns of Establishment, Indian Public Works Department, to 31 December 1901 and 30 June 1902; from the Registrar.

Calendars 1902-1903 from the following Colleges:—Royal Technical High School, Berlin; University of Birmingham; University College, Bristol (Calendar and Report); Glasgow and West of Scotland Technical College (Calendar and Report); Yorkshire College, Leeds (Calendar and Report); City of London College; King's College, London; South Western Polytechnic, London; University College, London; McGill College and University, Montreal; Redruth School of Mines.

The following from California University, U.S.America :—Register, 1900–1901; Secretary's Annual Report, 1900; Chronicle, Vol. 4, Nos. 1–6; Bulletin, Vol. 2, Nos. 8–12.

Universal Directory of Railway Officials, 1902; from the publishers.

Donaldson's Engineers' Annual, 1902; from Mr. P. R. Owens.

The following Abridgments of Specifications of Patents for Inventions, 1897–1900 :—Classes 1–40, 42, 47, 136; Subject List of works on Domestic Economy, Foods, and Beverages; Ditto of works on the Textile Industries and Wearing Apparel.

Technical Section Papers; from the Director of Railway Construction, Simla, India.

Bradford Municipal Technical College, Report 1900–1901; from the College.

University College of South Wales and Monmouthshire, Cardiff, Calendar 1901–1902; from the College.

University College, Dundee, Calendar 1901–1902; from the College.

Michigan College of Mines, Year-book 1901–1902; from the College.

Royal Technical High School, Munich, Report 1900–1901, Calendar 1901–1902; from the School.

CATALOGUES.

Ventilateurs et Appareils de Ventilation mécanique; from MM. E. Farcot Fils.

Surveying and Drawing Instruments, 1902; from Messrs. W. F. Stanley and Co.

Machine Tools; from Messrs. Cunliffe and Croom.

Labour Saving Appliances; from Messrs. Blake, Barclay and Co.

Farnley Iron Described ; from the Farnley Iron Co.

Diving Apparatus and Submarine Appliances; from Messrs. Siebe, Gorman and Co.

Railway Safety Appliances : Illustrations of Interlocking Installations; from Messrs. Saxby and Farmer.

Labour-saving Machinery; from West's Gas Improvement Co.

Steam Fittings, 1897 (with album); from Messrs. Thomas Noakes and Sons.

Limits and Limit Gauges: from The Newall Engineering Co.
American Machinery and Tools; from Messrs. Buck and Hickman.
Baldwin Locomotive Works, Records of Recent Construction, Nos. 30–38, 1902; from the Company.

The following PUBLICATIONS from the respective Societies and authorities :—

BRITISH ISLES.

Barrow-in-Furness Free Public Library; Nineteenth Annual Report.

Bradford Libraries, Art Gallery and Museums Committee; Thirty-second Annual Report.

British Association for the Advancement of Science; Report.

British Fire Prevention Committee.

Chemistry of Great Britain and Ireland, Institute of; Proceedings, and List of Fellows 1902-3.

Civil Engineers, The Institution of; Proceedings.

Civil and Mechanical Engineers' Society; Transactions.

Cleveland Institution of Engineers, Middlesbrough; Proceedings.

Cold Storage and Ice Association; Proceedings.

Electrical Engineers, Institution of; Journal.

Engine, Boiler, and Employers' Liability Insurance Company, Manchester Report (from Mr. Michael Longridge).

Engineers and Shipbuilders in Scotland, Institution of, Glasgow; Transactions.

Ipswich Engineering Society; Transactions.

Iron and Steel Institute; Journal.

Junior Engineers, Institution of; Transactions.

Literary and Philosophical Society of Manchester; Memoirs and Proceedings.

Liverpool Engineering Society; Transactions.

Liverpool Public Libraries, Museums and Art Gallery; Forty-ninth Annual Report.

London Mathematical Society; List of Members 1901.

Manchester Association of Engineers; Transactions.

Manchester Geological Society; Transactions.

Manchester Steam Users' Association; Report.

Marine Engineers, Institute of; Transactions.

Mining Engineers, Institution of, Newcastle-on-Tyne; Transactions.

Naval Architects, Institution of; Transactions.

North of England Institute of Mining and Mechanical Engineers, Newcastle-on-Tyne; Transactions.

North-East Coast Institution of Engineers and Shipbuilders, Newcastle-on-Tyne ;
    Transactions.
Patent Agents, Chartered Institute of ; Transactions.
Philosophical Society of Glasgow ; Proceedings.
Physical Society of London ; Proceedings.
Radcliffe Library, Oxford ; Catalogue of Additions during 1901.
Royal Agricultural Society of England ; Journal.
Royal Cornwall Polytechnic Society, Falmouth ; Report.
Royal Engineers' Institute, Chatham ; Professional Papers.
Royal Institute of British Architects ; Journal.
Royal Institution of Great Britain ; Proceedings.
Royal Irish Academy, Dublin ; Transactions and Proceedings.
Royal Scottish Society of Arts, Edinburgh ; Transactions.
Royal Society of London ; Philosophical Transactions (A), and Proceedings.
Royal United Service Institution ; Journal.
Science Abstracts—Physics and Electrical Engineering.
Society of Arts ; Journal.
Society of Chemical Industry ; Journal.
Society of Engineers ; Transactions.
South Wales Institute of Engineers, Cardiff ; Proceedings.
Staffordshire Iron and Steel Institute, Dudley ; Proceedings.
Surveyors' Institution ; Transactions and Professional Notes.
Waterworks Engineers, British Association of ; Transactions.
West of Scotland Iron and Steel Institute, Glasgow ; Journal.

Argentine Republic.

Engineers of River Plate, Institution of, Buenos Aires ; Journal.

Austria.

Zeitschrift des Oesterreichischen Ingenieur- und Architekten-Vereines, Vienna.
Zprávy spolku Architektův a Inženýrů v království českém, Prague.

Belgium.

Académie Royale de Belgique, Brussels : Bulletin.
Association des Ingénieurs sortis des Écoles spéciales de Gand ; Annales.
International Railway Congress (English edition), Brussels ; Bulletin.

Canada.

Canadian Society of Civil Engineers, Montreal ; Transactions.

China.

Shanghai Society of Engineers and Architects ; Proceedings.

France.

Académie des Sciences, Paris; Comptes Rendus des Séances.
Annales des Mines, Paris.
Association Technique Maritime, Paris; Bulletin.
Associations de Propriétaires d'Appareils à Vapeur, Paris; Compte Rendu des Séances.
Conservatoire des Arts et Métiers, Paris; Annales.
Génie Maritime, Paris; Mémorial.
Ponts et Chaussées, Paris; Annales.
Ports Maritimes de la France, Paris.
Revue Maritime, Paris.
Société Scientifique Industrielle, Marseilles; Bulletin.
Société d'Encouragement pour l'Industrie Nationale, Paris; Bulletin.
Société Industrielle de Mulhouse; Bulletin.
Société Industrielle du Nord de la France, Lille; Bulletin.
Société Industrielle de Rouen; Bulletin.
Société des Ingénieurs Civils de France, Paris; Mémoires.

Germany.

Zeitschrift für Architektur und Ingenieurwesen, Hannover.
Zeitschrift des Baverischen Dampfkessel-Revisions-Vereins, Munich.
Zeitschrift des Vereines deutscher Ingenieure, Berlin.

Holland.

Tijdschrift van het Koninklijk Instituut van Ingenieurs, 's Gravenhage

India.

Asiatic Society of Bengal, Calcutta; Proceedings and Journal.

Italy.

Associazione fra gli Utenti de caldaie a Vapore nelle Provincie Napolitane; Rapporto dell' Ingegnere Direttore.
Collegio degli Ingegneri Navali e Meccanici in Italia, Genoa; Atti.
Società degli Ingegneri e degli Architetti Italiani, Rome; Annali.
Real Istituto d'Incoraggiamento, Naples; Atti.

Japan.

Japan Society of Mechanical Engineers, Tokyo; Journal.

New South Wales.

Engineering Association of New South Wales, Sydney; Proceedings.
Royal Society of New South Wales, Sydney; Journal and Proceedings.

Norway.

Teknisk Ugeblad, Christiania.

Sweden.

Svenska Teknologföreningen, Stockholm.

United States.

American Academy of Arts and Sciences, Boston; Proceedings.
American Institute of Mining Engineers, New York; Transactions.
American Philosophical Society, Philadelphia; Transactions and Proceedings.
American Society of Civil Engineers, New York; Transactions and Proceedings.
American Society of Mechanical Engineers, New York; Transactions.
Association of Engineering Societies, Philadelphia; Journal.
Franklin Institute, Philadelphia; Journal.
School of Mines Quarterly, Columbia College, New York.
Smithsonian Institution, Washington; Annual Report.
United States Artillery, Fort Monroe; Journal.
United States Naval Institute, Annapolis; Proceedings.
United States Patent Office Gazette, Washington.
Western Society of Engineers, Chicago; Journal.

Victoria.

Australasian Institute of Mining Engineers, Melbourne; Transactions.

The following PERIODICALS from the respective Editors :—

BRITISH ISLES.

Appointments Gazette.
Arms and Explosives.
The Autocar.
Automobile Club Journal.
The Automotor Journal.
The British Architect.
The British Empire Review (from Mr. Henry Chapman).
British Refrigeration.
The Builder.
Camera Club Journal.
The Car.
Cassier's Magazine.
The Chamber of Commerce Journal (from Mr. Henry Chapman).
Cold Storage and Ice Trades Review.
The Colliery Guardian.

The Contract Journal.
The Cyclist.
Domestic Engineering.
The Electrical Engineer.
The Electrical Review.
The Electrical Times.
The Electrician.
The Engineer.
The Engineer and Iron Trades' Advertiser.
Engineering.
The Engineering Magazine.
Engineering Times.
Engineers' Gazette.
The Export Review.
Feilden's Magazine.
The Fireman.

The Fishing Gazette.
The Foundry Trade Journal.
The Journal of Gas Lighting.
Hardware, Metals and Machinery.
The Hardware Trade Journal.
Ice and Cold Storage.
Imperial Institute Journal (from Mr. Henry Chapman).
Invention.
The Iron and Coal Trades Review.
Iron Trade Circular, Ryland's.
The Ironmonger (and Diary 1903).
Locomotive Magazine.
London Technical Education Gazette.
London University Gazette.
The Machinery Market.
The Marine Engineer.
The Mariner.
The Mechanical Engineer.
The Mechanical World.
Midland Counties Herald.
The Mining Journal.

Model Engineer and Amateur Electrician.
Motor Car Journal.
Page's Magazine.
Phillips' Monthly Machinery Register.
The Plumber and Decorator.
The Practical Engineer.
The Public Health Engineer.
The Publishers' Circular.
The Quarry.
The Railway Engineer.
The Refrigerating Engineer.
The Shipping World.
The Steamship.
The Surveyor.
The Textile Recorder.
Traction and Transmission.
The Tramway and Railway World.
Transport.
Water.
The Woodworker.

Belgium.

Revue Universelle des Mines.

France.

L'Industrie.
Revue générale des Chemins de fer.

Revue Industrielle (from Mr. Henry Chapman).

Germany.

Glaser's Annalen.

Stahl und Eisen.

Holland.

De Ingenieur.

India.

The Indian and Eastern Engineer.
Railways.

Indian Textile Journal (English edition).

Italy.

Giornale del Genio Civile.

Spain.

El Ingeniero Español (London edition).

United States.

American Machinist.

American Manufacturer.

Architecture (from the B. and S. Folding Gate Co.).

Electrical Review.

Electrical World and Engineer.

Electricity.

The Engineering and Mining Journal.

Engineering News.

The Engineering Record.

The Iron Age (from Mr. W. H. Maw).

Marine Engineering.

Marine Review.

The Railway and Engineering Review.

Railway Master Mechanic.

The PRESIDENT, in formally moving the adoption of the Annual Report, thought it was a very satisfactory one in every way, both from the financial point of view and especially from the facts it gave as to the growth of the Institution. The diagram (page 118) graphically showed the progress made since 1890. It was particularly interesting to notice that the upper curve was a hollow one, and that the increase in the number of members was becoming year by year greater. The increase of the number of graduates was also a most gratifying feature, and he was sure that section would form a very valuable portion of the Institution.

No remarks being made, the Report was unanimously adopted.

The PRESIDENT next presented to Mr. W. Stanley Bott the Prize awarded by the Council for his Paper on " Twelve Months' Revision of a Drawing Office."

He announced that Mr. H. H. Mogg, who had also been awarded a Prize by the Council for his Paper on the " Weston-Twerton Bridge Undertaking," was unable to be present to receive his prize, which would be forwarded to him.

The PRESIDENT announced that the Ballot Lists for the election of Officers for the present year had been opened by a Committee of the Council, and that the following were found to be duly elected:—

PRESIDENT.

J. HARTLEY WICKSTEED, ... Leeds.

VICE-PRESIDENTS.

JOHN A. F. ASPINALL, ... Manchester.

A. TANNETT-WALKER, ... Leeds.

MEMBERS OF COUNCIL.

EDWARD B. ELLINGTON, ... London.

HENRY LEA, ... Birmingham.

MICHAEL LONGRIDGE, ... Manchester.

JAMES MANSERGH, F.R.S., . . . London.

JOHN F. ROBINSON, ... Glasgow.

JOHN W. SPENCER, ... Newcastle-on-Tyne.

JOHN TWEEDY, ... Newcastle-on-Tyne.

HENRY H. WEST, ... Liverpool.

For supplying the vacancy amongst the Vice-Presidents caused by the election of Mr. Wicksteed as President, the Council had appointed Mr. Edward B. Ellington as a Vice-President for the present year; and for supplying the consequent vacancy amongst the Members of Council, the Council had appointed Dr. Edward Hopkinson as a Member of Council for the present year, his name being the next highest in the voting for the election at this Meeting. Agreeably with the Articles of Association, both these gentlemen would retire at the next Annual General Meeting, and would be eligible for re-election.

The Council for the present year is therefore as follows :—

PRESIDENT.

J. Hartley Wicksteed, ...   Leeds.

PAST-PRESIDENTS.

Sir Lowthian Bell, Bart., LL.D., F.R.S.,   .   Northallerton.
Sir Frederick J. Bramwell, Bart., D.C.L.,
    LL.D., F.R.S., ...   London.
Sir Edward H. Carbutt, Bart.,   .   .   .   London.
Samuel Waite Johnson, ...   Derby.
Professor Alexander B. W. Kennedy, LL.D.,
    F.R.S., ...   London.
William H. Maw, ...   London.
E. Windsor Richards, ...   Caerleon.
Percy G. B. Westmacott, ...   Ascot.
Sir William H. White, K.C.B., LL.D., D.Sc.,
    F.R.S., ...   London.

VICE-PRESIDENTS.

John A. F. Aspinall, ...   Manchester.
Edward B. Ellington, ...   London.
Arthur Keen, ...   Birmingham.
Edward P. Martin, ...   Dowlais.
T. Hurry Riches, ...   Cardiff.
A. Tannett-Walker, ...   Leeds.

MEMBERS OF COUNCIL.

Sir Benjamin Baker, K.C.B., K.C.M.G., LL.D.,
    D.Sc., F.R.S., ...   London.
Sir J. Wolfe Barry, K.C.B., LL.D., F.R.S., .   London.
Henry Chapman, ...   London.
Henry Davey, ...   London.
William Dean, ...   Folkestone.
H. Graham Harris, ...   London.
Edward Hopkinson, D.Sc., ...   Manchester.

K 2

HENRY A. IVATT, ...	Doncaster.
HENRY LEA, ...	Birmingham.
Sir WILLIAM T. LEWIS, Bart., . . .	Aberdare.
MICHAEL LONGRIDGE, ...	Manchester.
JAMES MANSERGH, F.R.S., . . .	London.
HENRY D. MARSHALL,. . . .	Gainsborough.
The Rt. Hon. WILLIAM J. PIRRIE, LL.D., .	Belfast.
Sir THOMAS RICHARDSON, . . .	Hartlepool.
JOHN F. ROBINSON, ...	Glasgow.
MARK H. ROBINSON, ...	Rugby.
JOHN W. SPENCER, ...	Newcastle-on-Tyne.
Sir JOHN I. THORNYCROFT, LL.D., F.R.S., .	London.
JOHN TWEEDY, ...	Newcastle-on-Tyne.
HENRY H. WEST, ...	Liverpool.

The PRESIDENT remarked that the time had now come when he had to vacate the Chair in favour of his successor, Mr. Wicksteed. Before doing so, he desired to tender to his colleagues on the Council, to his fellow-members, and to the members of the staff of the Institution, his most sincere thanks for the hearty and loyal support accorded to him during his two years of office. There was a good deal of work connected with the office of President, but when one experienced such loyal support as he had received, and such a kind feeling as had been manifested by all with whom he had been associated, the labour became a pleasure. He had pleasure in vacating the Chair, and introducing his successor, Mr. Wicksteed, who was well known to them all as a staunch friend of the Institution.

The Chair was then vacated by Mr. MAW and taken, amid hearty cheers, by the new President, Mr. WICKSTEED.

Dr. ALEXANDER B. W. KENNEDY, Past-President, said it fell to his lot to undertake the extremely pleasant task, of which he was sure the members would heartily approve, of proposing a very cordial vote of thanks to the retiring President, Mr. Maw, for his conduct in the chair during the past two years. He knew from his own experience that the Presidency of the Institution was no light task, but it was easy in his time compared to what it had now become. The amount of work and the difficulty of the position were especially increased in Mr. Maw's case, by having to follow so brilliant a predecessor as Sir William White, and also by having to take the chair at a time when it might be supposed that the members of the Institution were perhaps a little weary after all the functions connected with the opening of the new House. Mr. Maw, however, had not been discouraged by those facts, and, in spite of these extra difficulties, had succeeded in still further steepening the upward angle of the curve of membership, as it appeared from the diagram shown (page 118). It appeared from this diagram that, during the two years Mr. Maw had been President, the increase in the membership averaged 11 or 12 per cent. per annum, and the increase in the Graduate class—a class which, he believed, in the future would become a feeder to the Institution in a similar manner to the Student class to the Civil Engineers—had been something like 40 per cent. per annum. If he might speak as freely as if Mr. Maw were not present, he wished to say that the late President had, in the first place, shown the members, what they knew before, that he was a thoroughly practical engineer, in touch with the whole of the practical details of the profession, and in fact a very mine of information on every subject connected with mechanical engineering. He had exhibited, as everybody knew who had had to do with him, a perfect readiness to give that information for the benefit of his friends at any time they wanted it. He had also shown that he had been in touch and in sympathy also with the scientific side of their work, which, for a President of the Institution, was as important as the practical side. Mr. Maw, in spite of being one of the busiest men among engineers, had somehow or other found time to devote himself most heartily whenever he was wanted, or

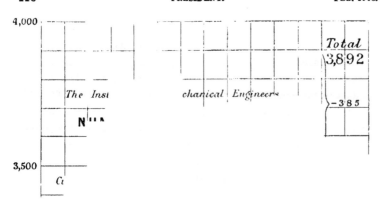

3,000

2,500

2,000

whenever he thought he was wanted, to the interests of the Institution; and he was quite sure that the personal interest taken by the late President in every matter connected with the Institution, the personal trouble he had given to it, and the time he had spent on its behalf, had largely had to do with the very great progress of the Institution during the last two years. He had very great pleasure in asking the members to accord a very hearty vote of thanks to Mr. Maw for his conduct in the chair of the Institution during the past two years.

Sir WILLIAM H. WHITE, K.C.B., Past-President, said he had very great pleasure in seconding the vote of thanks which had been so well proposed by Dr. Kennedy. When he vacated the chair two years ago and had the pleasure of introducing Mr. Maw as his successor, he ventured to say that up to that time there had been no more faithful servant or warmer supporter of the Institution than Mr. Maw had been, and that his past services were proof, if proof were wanted, that his presidential term would be one of great devotion to the interests of the Institution. He heartily endorsed every word which Dr. Kennedy had spoken in support of the work which had been done by the retiring President. It could not be known to the members of the Institution what an amount of time, thought, and care had been devoted by Mr. Maw to the work necessarily falling upon the President, and, in thanking him for all he had done in the past, the members might feel confident that he would continue, as a Past-President, to work as cordially and efficiently for the welfare of the Institution as he had done for so many years.

The PRESIDENT said that no duty could possibly be more welcome to him, as the first duty he had to perform from the chair, than that of putting the vote of thanks to the members. Dr. Kennedy's and Sir William White's remarks must be extremely gratifying to the retiring President, who had done so much work without, as it were, any words being used in recognition of it. He happened to know that Mr. Maw had made an extraordinary record as President. He

(The President.)

had never missed a meeting of the Institution, he had never missed a council meeting, and he had never missed a committee meeting during his two years of office, and he had never been late at any of those meetings. No wonder the line on the diagram had gone up ! He himself was much impressed to find that he had been elected President of an Institution containing nearly 4,000 members ; he was not aware, until he saw the diagram, that the Institution numbered so many members. It would be a wonder if the ascending aspect of the curve could be kept up ; he did not suppose it would be, but at any rate he was sure the members would agree with him that he must look upon Mr. Maw as an extremely difficult President to follow. None the less, and without any envy of Mr. Maw's great success, he most heartily put the resolution to the members.

The resolution was carried with acclamation.

Mr. MAW, in reply, said he felt it was quite impossible for him to reply adequately to a vote of thanks which had been proposed, seconded, supported, and carried in the manner in which that vote had been. He did not deny that there was a good deal of hard work connected with the Presidency of the Institution, but he thought the honour of being elected President was in itself an ample recompense for any trouble the holder of the office might have to take. Dr. Kennedy had referred to the manner in which the curve of the diagram exhibited on the wall had turned up during the last few years. That was not the result of his (Mr. Maw's) individual work ; it was the result, in the first place, of the work done at the time it was resolved to make the Institution a London Institution, and in the second place of the subsequent resolve to build a home of their own, and to take the position to which they were entitled among the important societies of London. Such sudden accessions of membership, as the diagram showed, were not obtained because any particular man was or was not President; they were obtained because the Institution was worthy of the support it had received, and as long as it was worthy of that support he was quite certain the Institution would continue to grow as it had been doing lately, and would take the place it ought to take amongst the Institutions of the world. He thought, perhaps, the country as a whole hardly realised how much it owed to engineers. The man-in-the-street did not quite understand what engineers had done for this country, but there was no profession which had done so much for England as engineers had done, and there was no branch of the profession which was more important than that which their Institution represented. He appealed to the members to work together to promote the interests of the profession and of the Institution, and so long as they did that he was quite certain that, whoever was President, the Institution would prosper. He thanked the members most heartily for the very kind way in which the motion had been received.

The PRESIDENT reminded the members that at the present meeting the appointment had to be made of an auditor for the current year.

Mr. DANIEL HORSNELL moved : " That Mr. Robert A. McLean, F.C.A,. 1 Queen Victoria Street, London, be re-appointed to audit the accounts of the Institution for the present year, at the same remuneration as last year, namely, twenty-five guineas."

Mr. C. J. SPENCER seconded the motion, which was carried unanimously.

The following Paper was read and discussed :—
" Hydraulic Experiments on a Plunger Pump " ; by Professor JOHN GOODMAN, *Member*, of Leeds.

The Meeting terminated shortly before Ten o'clock. The attendance was 163 Members and 75 Visitors.

HYDRAULIC EXPERIMENTS ON A PLUNGER PUMP.

BY PROFESSOR JOHN GOODMAN, *Member*, OF LEEDS.

A pump is sometimes regarded as a special form of ratchet by certain writers on the theory of machines, but if a pump always behaved in such a simple manner its behaviour would present no mechanical difficulties and its theoretical treatment would be a very simple matter; unfortunately, however, the simple theory is incomplete, and an actual pump behaves very differently indeed from the imaginary ideal pump, and moreover it presents some interesting problems which it is believed have never been fully solved. Those who have given attention to the subject, and have made careful experiments on hydraulic pumps and motors, could not fail to have observed some very curious actions which are not merely of theoretical interest but in many instances are of great practical importance.

In the early days of the petroleum industry in America the long pipe lines through which the oil was pumped gave constant trouble owing to frequent bursts, although the pipes had a large margin of safety over any static pressures they were likely to be subjected to. The problem of pumping through such great lengths of pipe was successfully solved by Henry R. Worthington, who minimised the fluid shock in the pipes by his well-known direct-acting or free piston pump, in which he caused the pump plungers to gradually come to rest and to pause at each end of the stroke, thus preventing any sudden change in the velocity of the fluid. When precautions are not taken to prevent sudden changes of velocity in the pipes and passages of fast running pumps, very serious "Shock" pressures may be set up even with short lengths of pipe. The following instance, which came before the author's notice some years ago, will suffice to show that "Shock" pressures may be very serious, and further that they are much greater than can be accounted for by the ordinary theory as laid down in Text Books or Treatises on the subject. The phosphor-bronze force pumps in use on the locomotives of an English railroad were constantly giving trouble through bursting. They were apparently well designed, and under the usual hydraulic test gave quite satisfactory results as regards the pressure at which they burst, that is, more than ten times the working boiler pressure.

A special hydraulic indicator was then fitted to the pump, also a hydraulic pressure gauge, and the following results were obtained, Table 1 :—

TABLE 1.

Diameter of Ram.	Stroke.	Speed of Ram.	Capacity of air vessel on delivery pipe.	Clacks in Pump.	Diameter of Suction Clack.	Diameter of Delivery Clack.	Diameter of Suction Pipe.	Diameter of Delivery Pipe.	Diameter of Boiler Clack.	Boiler Pressure.	Maximum pressure recorded in Pump and Delivery Pipe.	Remarks.
ins.	ins.	feet per min.	cubic ins.	no	ins.	ins.	ins.	ins	ins.	lbs. per sq. in.	lbs. per sq. in.	
2	26	931	73	6	2	2⅜	1¾	1½	1¼	140	3250	Pump with tortuous passages.
2	24	916	364	2	3¾	4½	2	1½	2½	140	850	After increasing the size of passages and otherwise easing the flow of water.
2	24	916	219	2	3¾	4½	2	1½	2½	140	950	

Further alterations were made in the way of increasing the valve openings and in making the passages more direct, but without further improvement as regards reducing the water pressure.

On calculating the pressure at the end of the stroke by the ordinary theory (*see* Appendix, page 168), it is found that one can only account for a pressure of about 200 lbs. per square inch in the first case and 150 lbs. per square inch in the others, apart from friction. There is no doubt that the water separated from the plunger at a speed of about one-third of that at which the pump was running.

Such results as those just mentioned, coupled with many similar mysterious actions that the author was constantly hearing from various sources, led him to carry out a series of experiments on a small plunger pump with the object of solving some of the mysteries. Possibly the smallness of the pump may have produced exaggerated results in some cases, but whether so or not the present series of experiments indicate the direction in which valuable data may be obtained by others who have larger pumps at their disposal for experimental purposes. Some of the hitherto mysterious actions have been satisfactorily cleared up, and light has been thrown on others, but some still remain as obscure as ever. The author has delayed the publication of this Paper with the hope that he might be able to offer more satisfactory solutions of some of the puzzles which have presented themselves; he has been urged, however, by one or two others who are working at the same subject to publish the results as far as they go, and he trusts that, although the theory is incomplete, the classified results of a very large number of carefully made experiments may prove of interest to other workers.

Many of the remarkable phenomena with which this Paper deals disappear when the pump is fitted with a suitable air or vacuum vessel on the suction pipe, but the use of such a vessel destroys the value of the experiments from the author's point of view, as he particularly wished to study the effects of water ram in pipes without any mitigation.

Description of the Pump.—The pump was made by Messrs. E. Green and Son, Wakefield, who very generously presented it to the

Engineering Department of the Yorkshire College, Leeds. From Fig. 1, Plate 7, it will be seen that it is an ordinary plunger donkey pump, such as is in common use for feeding boilers. The dimensions are :—

Diameter of Steam Cylinder . . . . . . 7 inches.
Diameter of Piston-Rod . . . . . . $1\frac{1}{4}$ „
Diameter of Plunger . . . . 4 „
Stroke . . . . 6 „
Length of Connecting-Rod . . . . . . 12 „
Diameter of Suction Pipe . . . . . . $2\frac{7}{8}$ „
Diameter of Fly-wheel . . . . 40 „
Moment of Inertia of Wheel in foot-pound units . . 1,200
Length of Suction Pipe . . . . . 63 feet and 36 feet.
Diameter of Delivery Pipe . . . . . . 2 inches.
Diameter of Suction and Delivery Valves . . . 4 „
Volume of Air Vessel on the Delivery Pipe . 0·6 cubic foot (approx)
No Vacuum Vessel on the Suction Pipe, except where specially mentioned.

The suction pipe was straight and horizontal ; it led direct to the sides of the measuring tank shortly to be described.

The delivery pipe, however, contained five bends, the outlet end of which was provided with a screw-down valve which was used to throttle the outlet and so regulate the delivery pressure. The pressure was measured by means of a pressure-gauge fitted to a Tee-piece near the end of the pipe ; the gauge was fitted with a plug-valve which was kept sufficiently closed to prevent the pressure gauge from fluctuating much on either side of the mean water pressure.

The extreme end of the pipe was provided with a loosely fitting bend, which could be readily turned to the right or left, and thereby to shunt the water into the upper weir basin of the main measuring tank or into another tank below, which stood on the platform of a weighing machine.

The pump itself was usually driven by steam in the ordinary way, but for certain experiments it was driven from the laboratory shafting by means of a belt running on the fly-wheel.

A Harding's Speed Indicator driven by a belt from the fly-wheel enabled the observer at the steam regulator valve to keep the pump running at approximately the desired speed. The method adopted for attaining the speed more exactly is explained later on.

Measuring Tank.—The general arrangement of the measuring tank is shown in Fig. 3, Plate 8.

The water is delivered into the chamber A, and in order to get still water over the weir it is passed through perforated zinc into the chamber B, which is filled with loose felt, and thence through a second sheet of perforated zinc to the upper weir basin C; it then passes over a bevelled edge V notch into the lower weir basin, from which the suction pipe draws its supply.

A hook gauge H is provided for measuring the head of water over the bottom of the notch, the scale provided enabled one to get readings within $\frac{1}{100}$th of an inch.

The valve for regulating the delivery pressure is shown at E and the pressure gauge at F. In order to check the accuracy of the weir coefficient the vertical pipe G is provided. After getting the pump to run perfectly steady, the head of water over the weir was measured accurately. The loosely fitting bend I is turned so as to deliver the water through the pipe G into the 200-gallon tank K on the weighing machine below. When the tank gets nearly full the water is shunted back into the top measuring tank as before. The exact time of starting and stopping the flow into K is taken by means of a stop watch; from the weight of water delivered in a given time the quantity of water delivered by the pump is readily arrived at, and should of course agree with that calculated from the flow over the weir. While the water is being delivered into K the lower weir basin is kept up to its normal level by means of a hose pipe.

This means of checking the delivery of the pump was adopted after [several experiments had been made. The first results obtained appeared to be so extraordinary that it was feared that some serious error was creeping in, but this direct measurement check completely corroborated the weir measurements.

A V notch was chosen rather than one of rectangular form on account of the greater constancy of its coefficient at various heads.

The experiments gave values from 0·606 to 0·611; a mean of 0·61 was taken for all the results quoted in this Paper.

Indicators.—Crosby indicators were fitted to the pump barrel, steam cylinder, delivery pipe, and suction pipe.

In the case of the indicator on the suction pipe, under certain conditions such a violent bang occurred at the end of the suction stroke that it was considered to be too risky for the indicator when used in the ordinary way, hence a special appliance had to be adopted. The reason for this violent bang is not altogether clear, as it very much exceeds in intensity the pressure that theory would indicate. It was feared at first that there was air in the pipes, but when it was proved to be otherwise, it was thought that possibly the indicator was not giving a true reading on account of the inertia of its moving parts. The indicator was therefore arranged with a screw "stop" gear, whereby the indicator spring could be initially compressed to any desired pressure. The water-pressure on the piston then simply lifted the piston off the stop if it exceeded the initial pressure on the spring. The " stop " gear is shown in Fig. 4, Plate 8. A stirrup A is fitted to the screwed head at the top of the indicator piston-rod; to the upper part of the stirrup a screwed link B is attached, which passes through a horizontal support C above the indicator; by tightening the thumb-nut D the indicator spring was compressed and the pencil thereby raised to any desired pressure.

When in use the thumb-nut was gradually tightened up until the instantaneous pressure (which occurred when the banging took place in the suction pipe) only just lifted the pencil, thereby indicating the maximum pressure that occurred in the pipe. Since the movement of the piston and the pencil lever was practically *nil*, the inertia effects of the instrument were eliminated. Much more regular results were obtained with this " stop " gear than when attempting to take ordinary indicator diagrams from the suction pipe.

In order to study carefully the action of the suction and delivery valves the special indicating gear, shown in Fig. 2, Plate 7, was fitted, a spindle A passed through a gland on the valve-box cover, a link B rigidly attached to one end of the spindle had its free end resting on the top of the mushroom valve at the other end of the spindle. A similar link C outside the valve box moved in a precisely similar manner to the link in contact with the valve. This link is provided with a pencil which draws a diagram on an indicator drum driven by a string from the crosshead reducing gear; thus it traces a curve whose height from the base line is the lift of the valve, and whose length at any point is proportional to the distance traversed by the plunger.

Object of the Experiments.—The experiments were carried out for the purpose of getting further information on the following points :—

(1) How the " Slip " * or the " Discharge Coefficient," of a pump, not fitted with a vacuum vessel, also the " Water Ram " pressure in the suction pipe are affected by—

(a) A change of outlet or delivery pressure when the speed remains constant.

(b) A change of speed when the delivery pressure remains constant.

(c) A change in the length of the suction pipe with the other conditions remaining unchanged.

(d) Running the pump without a suction valve.

(2) The exact behaviour as regards the opening and closing of the suction and delivery valves under various conditions of running.

---

* The author often has occasion to speak of the "Slip" and the "Discharge Coefficient" of a pump.

By the "Slip" of a pump he means the difference between the volume of the water actually pumped and the volume displaced by the pump plunger. In this Paper he adopts the expression "Discharge Coefficient" as representing the ratio.

$$\frac{\text{Volume of water actually pumped}}{\text{Volume displaced by the plunger}} = S$$

It is usually taken to be less than unity, and is ascribed to leakage past the valves due to their not opening and closing instantaneously. When the "Discharge Coefficient" is greater than unity, the "Slip" is said to be negative, *i.e*, when more water is pumped than can be accounted for by the displacement of the plunger.

In many of the experiments a pressure greatly above that of the atmosphere was recorded in the suction pipe towards the end of the suction stroke. Since this pressure is largely due to what is known as " water ram " in a pipe he will speak of it as " water-ram " pressure.

(3) The speed at which the plunger separates from the water during the early part of the stroke and catches it up later on, thereby producing a violent bang in the suction pipe.

(4) The loss of pressure due to the friction of the water passing through the valves and passages of the pump.

(5) The mechanical efficiency of the pump under various working pressures.

(6) The effect of fitting a vacuum or air vessel to the suction pipe.

Method of carrying out the Experiments.—In nearly all the experiments the speed of the pump was kept approximately constant by an observer stationed at the steam-supply valve, who was provided with a Harding's Speed Indicator driven by a belt from the fly-wheel. But in all the "Slip" experiments, the speed was obtained by a more exact method. A piece of white cardboard was attached to the pump crosshead at such a height that it could only be seen above a screen by the observer at the measuring tank, when the plunger was at the top of the stroke. Having got everything constant, the hook gauge H set, Fig. 3, Plate 8, and stop-watch in hand, the observer timed to the nearest one-fifth of a second, either 20, 40, or 60 revolutions of the pump according to the speed at which it was running, and the results were afterwards reduced to revolutions per minute. This method was considered to be more accurate than counting the number of revolutions made in one minute, because it is a very difficult matter to measure accurately fractions of a revolution. After getting accurately the time for the specified number of revolutions, the hook gauge was again examined, and if the head of water had materially altered, say, by more than $0 \cdot 02$ inch, the experiment was rejected, and fresh observations were taken; but if the alteration of head over the notch was within the limit stated, the mean of the head reading at the beginning and end of the experiment was taken. It is believed that the error in measuring the " Discharge Coefficient " is well within 1 per cent. By examining the curves in the diagrams it will be seen, however, that some of the observed points are situated more than 1 per cent. away from the mean curve. This is probably due to the great difficulty experienced in keeping the pump working at the exact specified speed and pressure.

Immediately after taking the "Discharge Coefficient" readings, indicator diagrams were taken from all the places enumerated above.

The results obtained were afterwards reduced from the measurements and the diagrams taken during the test.

Results of the Tests.

Discharge Coefficient when the Pump is running smoothly.—When making trials of pumping engines it is a very common practice to assume that the volume of water pumped is from 95 per cent. to 100 per cent. of the volume displaced by the plunger. The deficiency or " slip " is attributed to leakage and to small quantities of air being entrapped in the water. There are, however, a few isolated cases on record in which careful measurements have shown that the volume delivered was actually greater than the volume displaced by the plunger, but even to the present day many engineers regard such measurements with great suspicion and will not admit that such a result is possible.

It is, however, an easy matter to prove that not only are such results possible, but it is also possible to predict the probable amount of negative " slip " that will occur in a pump. An attempt to make such a prediction is set forth in the Appendix, but that it is only a partial solution of the question is evident from the fact that experiments show in some cases that more water actually passes than the expressions given would lead us to expect.

In addition to the treatment in the Appendix (page 168), it may be well to consider the question from a general standpoint. It is shown in text books dealing with the dynamics of the steam-engine that, in the case of an engine fitted with a very long connecting-rod, or its equivalent, the force required to accelerate or retard the reciprocating parts at either end of the stroke is equal in amount to the centrifugal force of a body concentrated as a ring round the crank-pin, the mass of the ring being equal to the mass of the reciprocating parts. It is further shown that if a short connecting-rod be used whose length is n times that of the crank radius, the pressure at the "in" end of the stroke is $\frac{1}{n}$th greater, and at the "out" end $\frac{1}{n}$th less than in the case of the very long rod. Now in the case of a pump having a suction pipe of the same diameter as the plunger, the water will, under normal conditions, move with the same speed as the plunger, and will be accelerated and retarded in the same manner. Hence the mass of the water in the suction pipe may be regarded as a part of the mass of the reciprocating parts, and therefore the suction or negative pressure required at the beginning of the stroke to accelerate the water, and the positive pressure set up at the end of the stroke when the water is retarded, can be readily calculated. If the area of the suction pipe be less than that of the plunger, the water in the pipe will move with a correspondingly higher speed, and the above-mentioned pressure (per unit area) will be increased in the direct proportion of the area of the plunger to the area of the pipe. The exact manner in which the pressure varies at all parts of the stroke is most easily followed by means of a diagram.

The line $a\,b$, Fig. 5, represents the distribution of pressure due to the reciprocating parts in the case of an engine having an infinitely

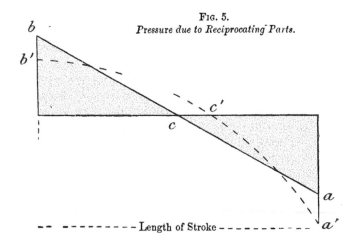

FIG. 5.
Pressure due to Reciprocating Parts.

long connecting-rod or a slotted crosshead, and the line a' b' for the case of an engine having a connecting-rod four times the length of the crank radius.

An indicator diagram taken from the barrel of a pump which is running very slowly and drawing its water from a source at its own level will be somewhat of the form e, f, g, h, Fig. 6 (page 134), or practically rectangular. As the plunger moves outwards the water runs smoothly into the barrel, and the suction line $h g$ practically coincides with the atmospheric line. But when the pump is running more rapidly the pressure required to accelerate and retard the water becomes more marked, and the suction line $h g$ becomes $a' b'$, which has the same form as $a' b'$ in Fig. 5. Underneath this ideal diagram an actual diagram taken from the pump is given, from which it will be seen that the suction line under the conditions of working is of the same general form as the calculated curve $a' b'$, the reason why the two curves are not of exactly the same form is explained later on (page 144).

In Fig. 6 the delivery line $e f$ is well above that of the maximum inertia pressure at b' reached by the water, but if the delivery pressure be lower than the maximum inertia pressure, as in Fig. 7 (page 134), the delivery valve will be forced open before the end of the suction stroke, and instead of getting the line $a' b'$ it will become $a' d' f$, and delivery will begin when the inertia pressure becomes equal to the delivery pressure, namely, at d'.

An actual diagram from the pump is also given in order to show a comparison between the theoretical and the actual results obtained.

That delivery does commence when these two pressures become equal is evident from the delivery-valve diagram given immediately above the actual pump diagram, from which it will be seen that the opening of the valve occurs at the instant the inertia pressure line cuts the delivery line.*

Since delivery commences before the completion of the suction stroke, and indeed while suction is still going on, it is clear that

---

* In reality slightly later than this, on account of friction (page 171).

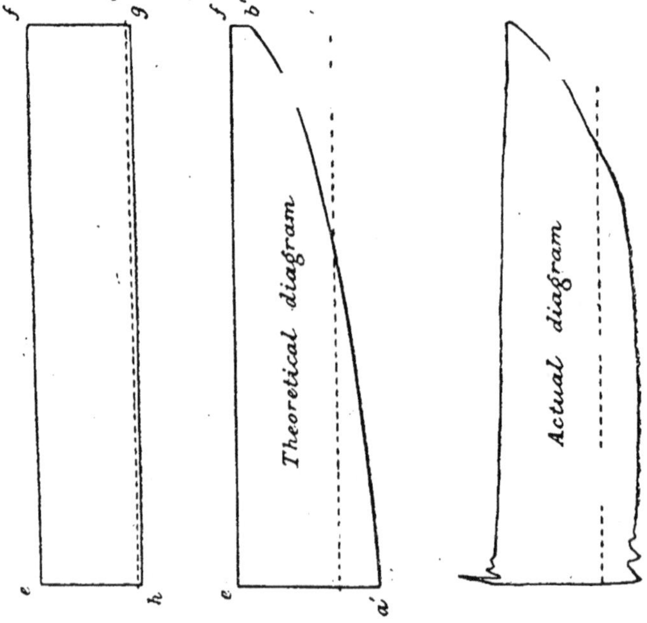

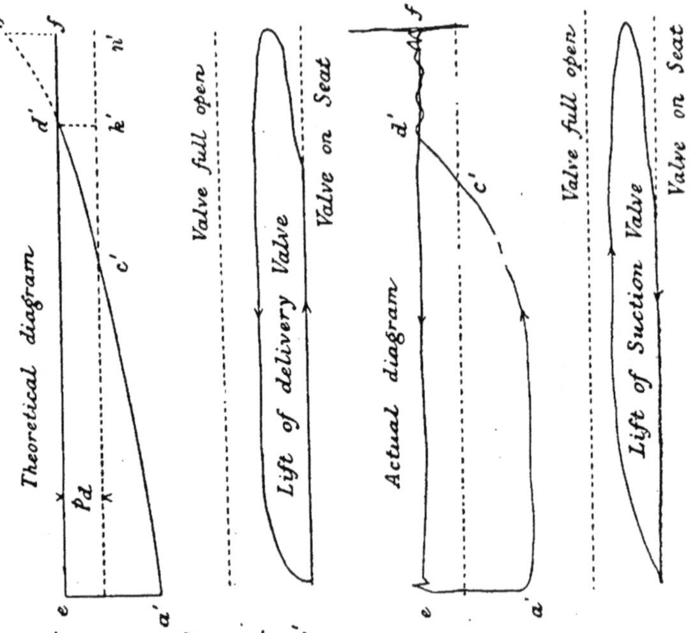

FIG. 7.—Pressure due to Inertia of water.

FIG 6.—Pump diagram at slow speed.

the pump will deliver a greater volume of water than the volume displaced by the plunger, or, in other words, that "negative slip" will occur. It is also evident that, for any given inertia pressure distribution or for any given speed of the pump, the delivery will commence earlier and earlier as the delivery pressure is reduced ; if this were the only disturbing factor in the pump the "negative slip" could be readily arrived at for all delivery pressures and speeds. It will be seen shortly that a much more serious disturbance is liable to occur.

·It should also be noticed that the suction valve does not close at the instant the plunger reaches the end of the stroke. The water is then moving at a high velocity along the suction pipe towards the pump and therefore possesses energy in virtue of that velocity. This energy is expended in forcing the water through the delivery valve and thus keeping both the suction and the delivery valves open at the same time. Later on we shall consider the extreme case in which the pump actually delivers water without any suction valve at all.

In Fig. 7 the line a' b' represents the distribution of pressure due to the inertia at all parts of the stroke. From a' to c' the pressure is negative because the pump is accelerating the water in the suction pipe.   At c' the inertia pressure becomes zero, and in consequence of the water being retarded after that point is reached it actually exerts a driving effort on the plunger.   When the plunger reaches d' the inertia pressure becomes equal to the delivery pressure and it immediately forces open the delivery valve, causing the water to pass up the delivery pipe before the completion of the suction stroke. Since the ordinates of the curve a' b' represent the water-pressure on the plunger, and abscissæ the horizontal distances the plunger has moved, the area c' d' f n' k' represents the work done by the retarded water in forcing the plunger forward, and the area d' b' f represents the work done in delivering the extra water during the suction stroke. Then, since the work done in lifting the water against a given resistance is the product of the volume of the water pumped and the pressure against which it is delivered, one has the extra volume of water delivered per stroke equal to the area d' b' f (for scale see Appendix)

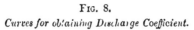

Fig. 8.

Curves for obtaining Discharge Coefficient.

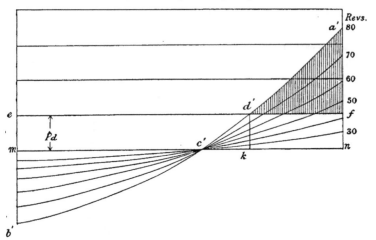

divided by the delivery pressure P_d. Let this be v and let V be the volume displaced by the plunger, then the discharge coefficient

$$S = 1 + \frac{v}{V}.$$

Curves necessary for obtaining this quantity are given in Fig. 8, from which the theoretical discharge coefficients have been obtained; an expression is given for S in the Appendix for the case in which the connecting-rod is very long.

In Figs. 9 to 25 (pages 137 to 141) a large number of discharge coefficient curves are shown; those obtained by experiments are shown in full and broken lines in Figs. 18–22 (pages 139 and 140), and those obtained by calculation are shown in dotted and dot stroke lines. It will be seen that the latter agree fairly well with the former. The small discrepancies are believed to be largely due to the friction of the water in the pump passages and pipe.

The discharge coefficients have been plotted to both an outlet pressure basis and a speed basis; in both cases it will be seen that over a wide range the coefficients remain practically constant. The reason for the increase in the coefficients with small outlet-

(*continued on page* 142.)

Figs. 9, 10 and 11.—*Discharge Coefficients obtained from Experiments.*

Suction pipe —— + 63 feet   Suction pipe - - - 36 feet.

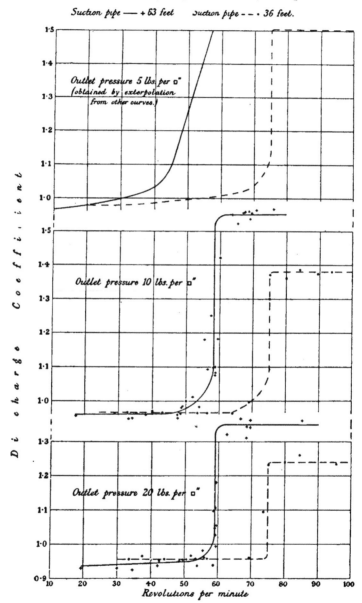

Outlet pressure 5 lbs. per □"
(obtained by extrapolation
from other curves.)

Outlet pressure 10 lbs. per □"

Outlet pressure 20 lbs. per □"

Discharge Coefficient

Revolutions per minute

FIGS. 12, 13, 14, 15, 16 AND 17.

Discharge Coefficients obtained from Experiments.

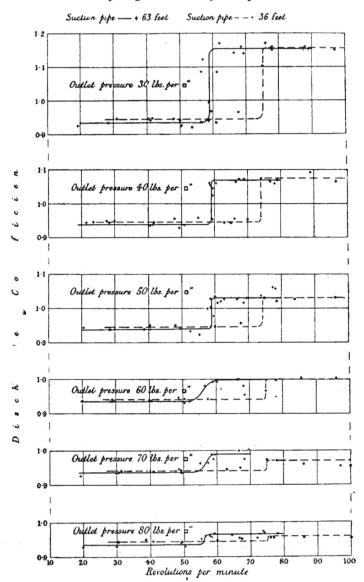

Suction pipe ——— ● 63 feet   Suction pipe – – · 36 feet

Outlet pressure 30 lbs. per ▫″

Outlet pressure 40 lbs. per ▫″

Outlet pressure 50 lbs. per ▫″

Outlet pressure 60 lbs. per ▫″

Outlet pressure 70 lbs. per ▫″

Outlet pressure 80 lbs. per ▫″

Discharge Co-efficient

Revolutions per minute

Figs. 18, 19 and 20.

Discharge Coefficients obtained from Experiments and Calculations.

Suction pipe ——• 63 feet   Suction pipe —— • 36 feet
Theoretical —·— „   „   Theoretical ...... „   „

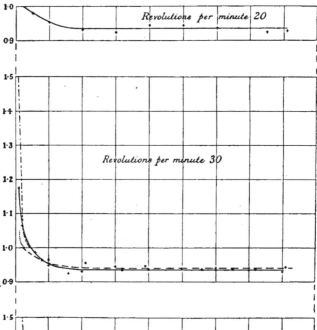
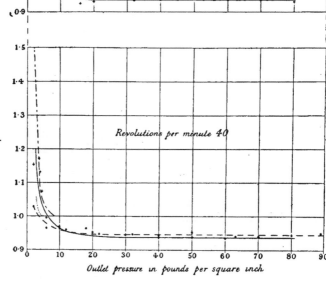

Outlet pressure in pounds per square inch

FIGS. 21, 22 AND 23.

Discharge Coefficients obtained from Experiments and Calculations.

Suction pipe ——● 63 feet   Suction pipe –––● 36 feet.
Theoretical – · – „ „   Theoretical ...... „ „

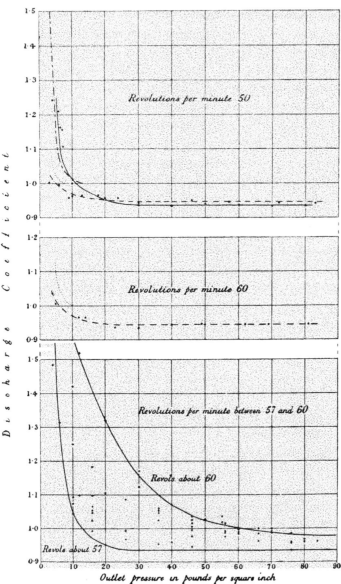

Outlet pressure in pounds per square inch

FIGS. 24 AND 25.

Discharge Coefficients obtained from Experiments and Calculations.

Suction pipe ⟶ 63 feet.   Suction pipe - -.36 feet.
Theoretical ...... „   „

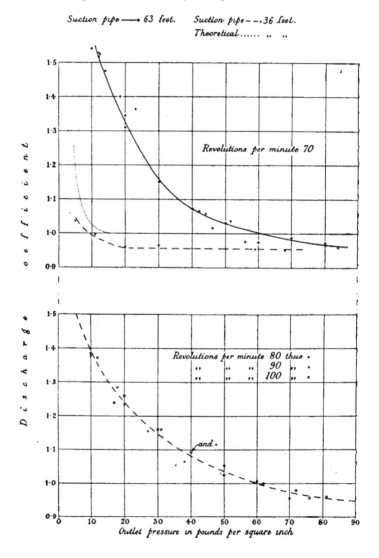

Revolutions per minute 70

Revolutions per minute 80 thus .
   „   „   „   90 „   .
   „   „   „   100 „   .

and .

Outlet pressure in pounds per square inch

pressures has already been considered, but it is a much more difficult problem to account fully for the sudden increase shown in the diagrams in which the pressure is kept constant and the speed is varied.

Discharge Coefficient when the Pump is not running smoothly.— It will be noticed that no calculated curves are given for the long pipe for any speeds above 50 revolutions per minute. The next set of observations were taken at or about 60 revolutions per minute. It will be shortly seen that at some intermediate speed an important change has taken place which quite upsets any calculations that have been made up to the present. During the suction stroke the water runs into the pump barrel, because the outside absolute pressure of the atmosphere acting on the free surface of the water in the suction supply is greater than that in the pump-barrel; but if the surface speed of the water entering the barrel be less than the speed of the plunger at any part of the stroke, the plunger will leave the water, and later on in the stroke, when the speed of the plunger is reduced, the water will catch it up and cause a more or less violent bang when they meet, the violence depending upon the velocity of impact. The speed at which this separation of the plunger and water occurs can be calculated (see Appendix, page 168). It has also been obtained experimentally by noticing the speed at which banging commences.

The following Table 2 shows that experiments and calculation agree quite closely in this instance:—

TABLE 2.

Suction Head.	Loss of Head due to Friction.	Length of Pipe.	Ratio of Cylinder Area to Pipe area.	Radius of Crank.	Speed at which Separation occurs.	
					By Calculation.	By Experiment.
feet.	feet.	feet.		feet.	Revolutions per minute.	
						between
0·07	8·5	63	1·83	0·25	60·5	56 and 62
0·10	8·8	36	1·83	0·25	77·9	73 and 78

Fig. 26.

Critical Point of Discharge Coefficient.

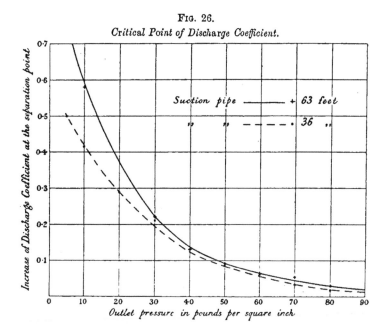

When the speed is near the critical point, that is, approximately that at which separation occurs, the discharge coefficients are very unstable, the slightest disturbance in the conditions causing the coefficient to rise or fall twenty or thirty per cent. ; in order to make this point clear, the diagram, Fig. 23 (page 140), has been plotted. As far as these experiments go, the coefficient appears to be constant at speeds above the critical for any given delivery pressure, Figs. 9 to 17 (pages 137–138), but when the speed is kept constant, it is found to vary with the outlet pressure.

This leap or sudden increase in the discharge coefficient at the critical point is much greater at low delivery pressures than at high pressures, Fig. 26, from which it will be seen that it practically disappears when the delivery pressure reaches about 100 lbs. per square inch. It should also be noticed that it is greater with a long than with a short length of suction pipe. The reason for this great and sudden increase in the discharge coefficient is not at all easy to see.

The diagrams given in Figs. 27 and 28 (page 145) may throw some light on the matter, but they do not by any means clear up the mystery. The full line a' b' is the inertia-pressure line already explained. The pressure required to force the water into the pump is e b' at the beginning of the stroke, and at any other point in the stroke it is given by the vertical height between the curve a' b' and the atmospheric line e f, if there were no friction in the pipes and passages of the pump. From a separate series of experiments the mean friction has, however, been ascertained (page 164), and since the friction varies as the square of the velocity of the water, and therefore of the plunger when the two are in contact, the magnitude of the friction can be calculated for any position of the plunger. Such a curve has been constructed and the vertical ordinates set down below the curve a' b', viz., a' m b'. Fig. 27 (page 145) is constructed for the pump when running at 50 revolutions per minute. For purposes of comparison an indicator diagram taken at 51·4 revolutions per minute has been superposed in broken lines to the same scale; it will be seen that the two agree fairly well.

But if the inertia pressure e b' be greater than the pressure available for forcing the water into the pump, Fig. 28 (page 145), the water will not flow in quickly enough to keep in contact with the plunger, and, consequently, the plunger will leave the water in the pump barrel as soon as it starts on its suction-stroke. The space between the plunger and the water continues to increase until a point m is reached, that is, where the pressure (e k) available for forcing the water into the pump is equal to that required to accelerate the water in the suction-pipe. After the point m has been reached, the water gains on the plunger and catches it up when the point n is reached, that is, when the area m n o is equal to k b' m. During the period k m n the pressure will on this hypothesis be constant, and the suction portion of the diagram will be as shown in thick lines. Fig. 28 (page 145) is constructed for the pump when running at 70 revolutions per minute. For purposes of comparison an indicator diagram, taken at 70·4 revolutions per minute, has been superposed in broken lines to the same scale, but unfortunately there is a serious discrepancy which the author is unable to account for.

Fig. 27.

Revs. 50.

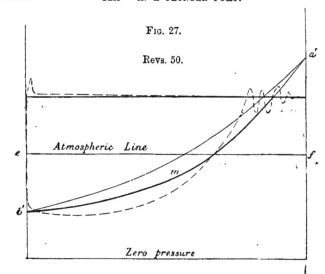

Fig. 28.

Revs. 70.

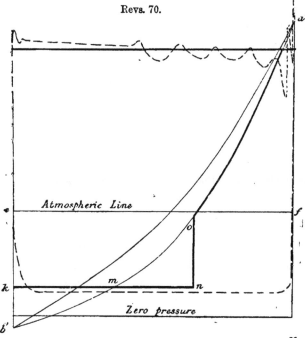

In Fig. 29 (pages 147 to 149) a series of pump and valve diagrams is given in order to show how the inertia effects develop as the speed increases.

After the completion of the experiments which are the subject of this Paper, the pump was fitted with an air or vacuum vessel on the suction-pipe coupled on at VV, Fig. 1, Plate 7; its capacity was 0·3 cubic foot. The effect of the vessel was very marked indeed, the violent banging which threatened to shatter the pump at high speeds was no longer experienced, and the pump now ran smoothly at speeds far above those considered safe before the vessel was added. The indicator diagrams, Fig. 30 (page 150 to 152), show a tolerably regular suction line with a slight quivering at the end of the suction-stroke, or rather, at the beginning of the delivery-stroke; the wavy lines are probably due to the inertia of the indicator.

Weir measurements of the discharge showed a practically constant coefficient varying only from 0·89 to 0·91 with 30 lbs. delivery pressure; there was no jump in the curve such as was found before the vessel was added.

The valve-lift diagrams, Fig. 30, also show that the valves opened and closed in quite normal fashion, although the lift was not quite as great with the vacuum vessel in use as it was before.

When no vacuum vessel was in use, the pump ran smoothly and silently until the critical speed was reached, but above that speed the bang was very violent, and made the whole pump and connecting pipes quiver; it is interesting, however, to note that if air were allowed to leak into the suction-pipe, or if the indicator cock (the indicator being in position) were opened, this bang was very much modified, and in fact almost disappeared.

The suddenness of the change at the separation point will be more readily followed by a study of the series of curves, Figs. 9 to 17 (pages 137 and 138), showing the relation between the speed and the discharge coefficient for any given delivery pressure. It will be seen that the discharge coefficient before the separation point occurs is practically constant over a wide range, and does not appear to vary with the speed, provided the inertia pressure is not greater than the delivery pressure.

(continued on page 153.)

Diagrams taken from the Pump at Various Speeds.

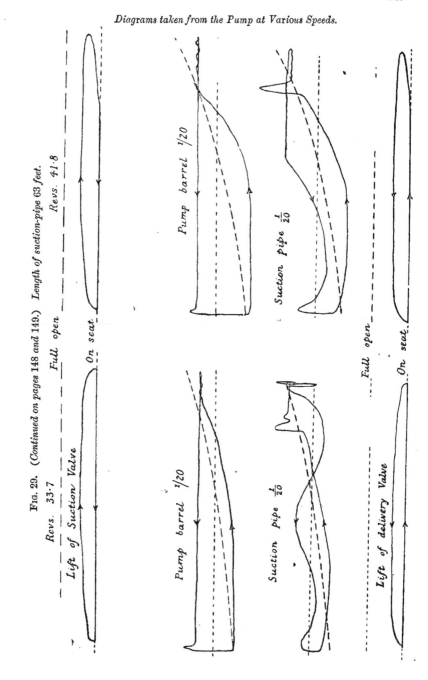

FIG. 29. (Continued on pages 148 and 149.) Length of suction-pipe 63 feet.

Diagrams taken from the Pump at Various Speeds.

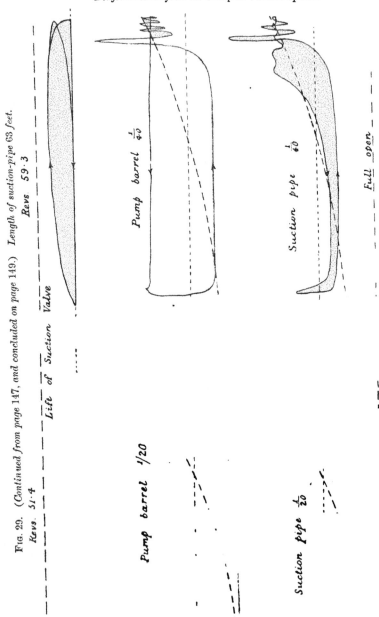

FIG. 29. (Continued from page 147, and concluded on page 149.)   Length of suction-pipe 63 feet.

Diagrams taken from the Pump at Various Speeds.

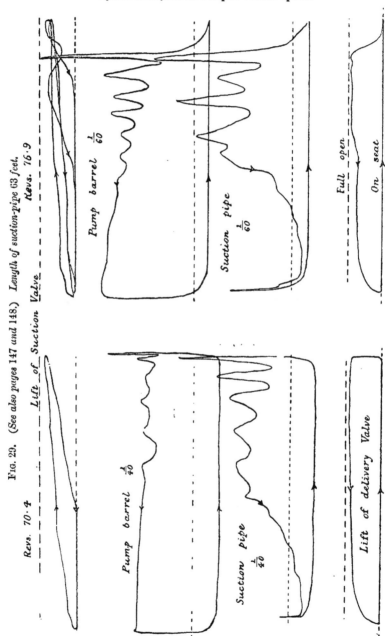

Fig. 29. (See also pages 147 and 148.) Length of suction-pipe 63 feet.

FIG. 30. (*Continued on opposite page.*)
Diagrams taken when the Vacuum Vessel was in use.
Suction-pipe 36 feet.

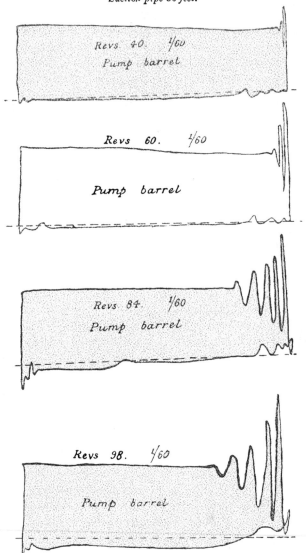

FIG. 30.   (*Continued on next page.*)

Diagrams taken when the Vacuum Vessel was in use.

Suction-pipe 36 feet.

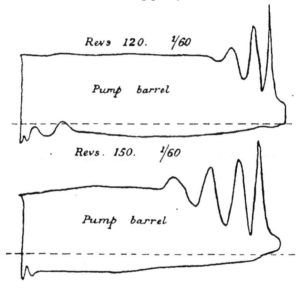

Revs 120.   1/60

Pump   barrel

Revs. 150.   1/60

Pump   barrel

FIG. 30.   (*Concluded from page 150.*)

Diagrams taken when the Vacuum Vessel was on the Suction Pipe.

Suction-pipe 36 feet.

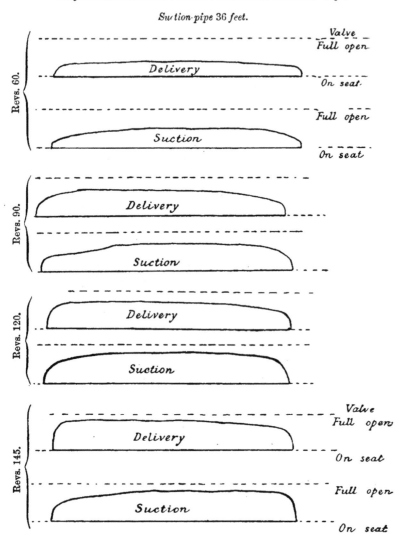

Many engineers have difficulty in believing that it is possible for a pump to deliver even 5 per cent. more volume of water than that displaced by the plunger. They will probably be astonished to hear that this pump actually delivered under some circumstances more than 50 per cent. greater volume of water than the displacement of the plunger. Of course the useful work done in delivering this quantity of water against the outlet pressure cannot be greater than the indicated work done in the pump barrel, in fact it will be less, on account of the friction in the pump passages. The following cases, Table 3, show that even with the highest discharge coefficient there is reasonable agreement between the indicated horse-power of the pump and that calculated from the measured discharge over the weir and the delivery pressure in the pump barrel :—

TABLE 3.

Revolutions per minute of pump.	Delivery pressure in lbs. per square inch.		Cubic feet of water over Weir per min.	Discharge Coefficient.	I.H.P. of Pump.	Horse-power calculated from pressure in pump-barrel and quantity passing over weir.
	In pump-barrel.	At end of delivery pipe.				
76·1	20	14	4·90	1·476	0·45	0·42
70·7	18	12	4·70	1·526	0·38	0·36
66·9	17	12	4·43	1·517	0·36	0 32
69·4	28	23	4·13	1·363	0·53	0·49
68·2	25	20	3·99	1·343	0·45	0·42
76·9	48	42	3·57	1·066	0·76	0·72
78·1	50	44	3·61	1·059	0·77	0·76

Perhaps the most remarkable of all the results obtained were the slip measurements taken when the pump was working without any suction-valve at all. From the diagram (Fig. 7, page 134) it will be readily seen how it is possible to get water delivered during the suction-stroke of the pump, simply by the inertia pressure in the suction-pipe forcing open the delivery-valve before the completion of the suction-stroke ; but the delivery of water during the short portion of the stroke represented by d' f, Fig. 7, is very small compared with the actual delivery. On examining the delivery-valve lift diagrams, Fig. 31 (page 155), which were obtained when the pump was running without a suction-valve, it is seen that only a very small portion of the delivery takes place through the inertia of the water during the latter part of the suction-stroke, but that the greater part of it occurs during the delivery-stroke. The question then naturally arises, why should the water be delivered against a very appreciable resistance in the shape of delivery head instead of returning along the suction-pipe which is perfectly free and open ? The reply is that the column of water which is flowing along the suction-pipe towards the pump during the suction-stroke either has to be suddenly reversed in direction when the delivery-stroke commences, or it must pass through the delivery-valve, but if the pressure necessary to reverse suddenly the direction of flow is greater than the resistance above the delivery-valve, this valve is lifted and water therefore passes up the delivery-pipe.

A study of the delivery-valve lift diagrams is very instructive. When the speed is kept constant and the pressure is varied, it will be seen that the period during which the delivery-valve is kept open gradually diminishes as the pressure rises, and at the same time the discharge coefficient also diminishes.

At every speed of the pump there was found to be a corresponding delivery pressure at which the flow ceased, which will be called the "stoppage pressure." It was anticipated that this pressure would correspond with the inertia pressure of the water (page 168); that it does so very nearly until the critical speed is reached, namely, 78 revolutions per minute, is clear from the following Table 4 (page 156). But after the critical speed a great change, similar to that already dealt with, occurs, and instead of the "stoppage pressure" being equal to the inertia pressure it is about six times as great.

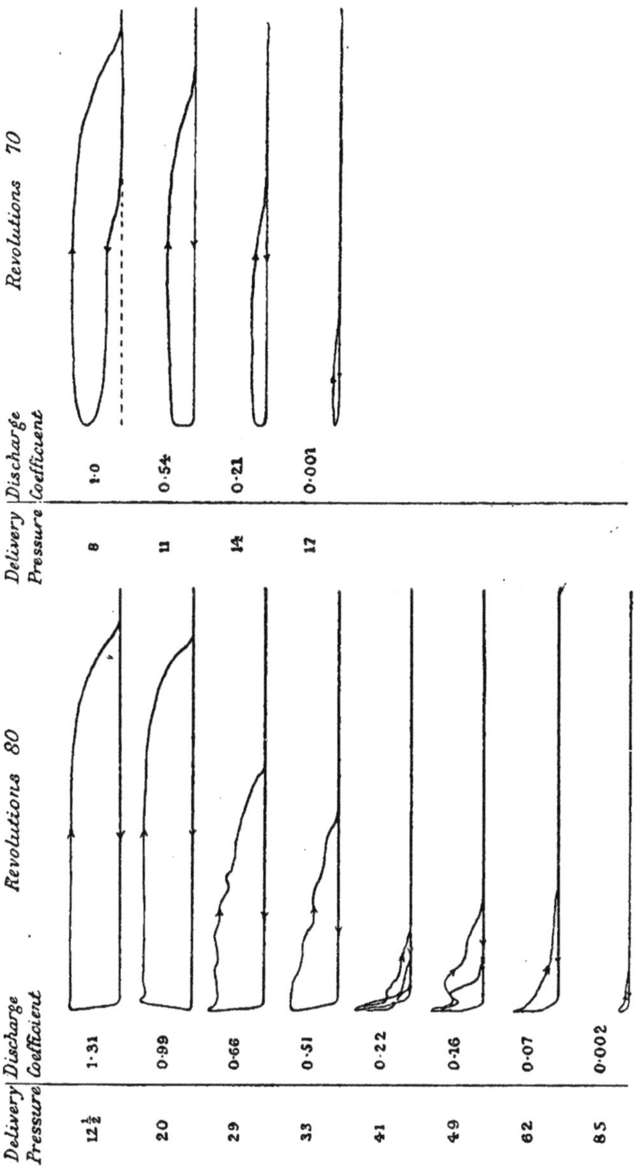

FIG. 31.—Delivery Valve Diagrams with no Suction Valve. Suction-pipe 36 feet.

TABLE 4.

Speed of Pump in revolutions per minute.	" Stoppage pressure " by experiment. lbs per sq inch.	Pressure due to the Inertia of the Water.
31·4	4·5	3·0
42·0	6 5	5·2
54·5	9·0	9·0
63·8	12·0	12·3
70·8	18·0	15·2
80·0	about 120·0	19·4

The relation between the speed, the outlet pressure, and the discharge will be seen from the curves in Fig. 32.

Fig. 32.

Relation between Speed, Outlet pressure, and Discharge.

Suction-pipe 36 feet.

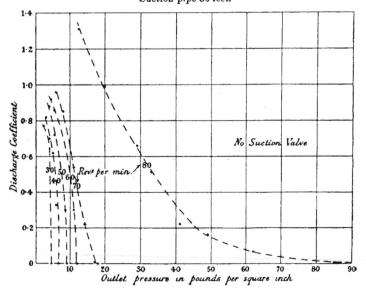

Pressure in the Suction-pipe due to the Inertia of the Water.—It has already been pointed out that, owing to a change in the velocity of the water at the beginning and end of the stroke, negative and positive pressures are set up in the suction-pipe. In the Appendix, it is shown how the magnitude of such pressures can be arrived at, and further that if this be the only action, the pressure due to this cause cannot exceed the delivery pressure beyond a small amount due to friction and to the resistance in passing the delivery-valve; but a series of experiments made to ascertain the maximum suction-pipe pressures clearly shows that the above-mentioned action is by no means the only one, for the pressures actually recorded far exceed those due to the simple change of velocity of the water treated as a part of the reciprocating parts of the pump. In the expression for this pressure the delivery pressure does not appear, and it is not easy to see how it can affect the water-ram pressure in the suction-pipe beyond limiting it to approximately that amount, but it is found that with the lowest inertia pressure, even when considerably below that of the delivery, the latter has a material effect upon the water-ram pressure.

The manner in which the delivery pressure affects the water-ram in the suction-pipe is shown in the curves in Figs. 33 and 34 (pages 158 and 159). The curves for both the long and the short pipe are of a similar character. In the case of the curves showing the relation between the speed of the pump and the water-ram pressure in the suction-pipe, Figs. 35 and 36 (pages 160 and 161), it will be noticed that there is a distinct change in the curvature at about 50 revolutions per minute in the long pipe and 60 in the short pipe. On referring to the curves showing the relation between the discharge coefficient and the speed, it will also be noticed that in those cases a change occurs at about 10 revolutions per minute higher than in these instances. The author has no reason to suggest for the apparent discrepancy.

It is worthy of note that in both instances the water-ram pressures appear to follow the same law, although greater in amount, as the theoretical pressures until the change point occurs. In the

(*continued on page* 162.)

FIG. 33.
Pressure in Suction pipe due to Water-ram.
Suction-pipe 63 feet.

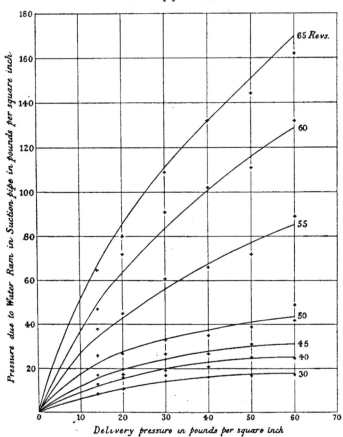

Fig. 34.

Pressure in Suction pipe due to Water-ram.

Suction-pipe 36 feet.

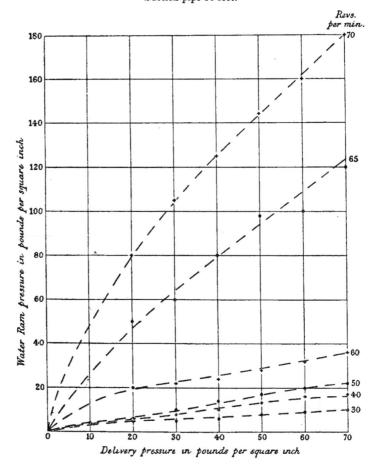

Fig. 35.
Pressure in Suction-pipe due to Water-ram.
Suction-pipe 63 feet.

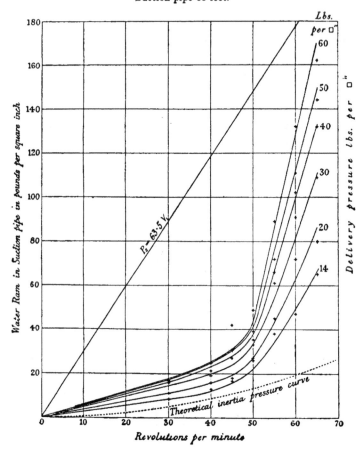

Fig. 36.

Pressure in Suction-pipe due to Water-ram.

Suction-pipe 36 feet.

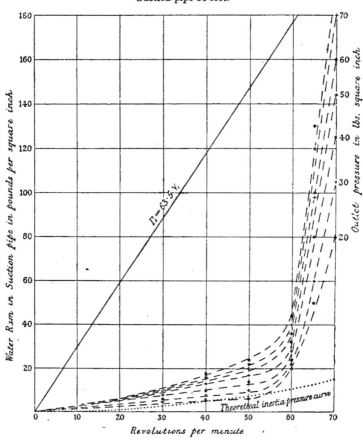

case of the 36-feet pipe the water-ram pressure, as found by experiment, is roughly one-half of that obtained from the 63-feet pipe, which is quite in accordance with theory, but after the change point the water-ram pressures appear to follow an entirely different law.

The violent banging action that occurs in the pump has already been referred to when describing the apparatus used for recording the pressures set up in the suction-pipe. If these pressures were due to the simple inertia action referred to (page 168), the pressure would rise gradually towards the end of the suction-stroke, but instead of doing so there is a violent bang at high speeds which threatens to shatter the indicator unless the adjusting screw D, Fig. 4, Plate 8, is tightened up. Hence one is bound to conclude that there is some other and more complex action going on. At first it was feared that the trouble was due to air in the pipes, but every care was taken to see that all joints were perfectly tight and that none leaked in through the indicator. To secure this the top of the indicator cylinder above the piston was partially filled with vaseline, and later a new well-fitting piston was used, but still with the same results.

After searching in many directions for a satisfactory solution of the water-ram pressures obtained, it was thought that possibly the cyclical variation in the angular speed of the pump might be responsible for a more rapid retardation of the plunger towards the end of the suction-stroke than would be given by a crank having a uniform speed of rotation. In order to overcome any possible cyclical variation of speed the pump was driven by a belt from a line of shafting both with and without the assistance or the steam-pressure in the cylinder of the pump. The results obtained were, however, practically identical with those originally got from the pump when driven by steam in the ordinary way, thus disposing of the possibility of the cyclical variation in speed producing any serious disturbance.

If the water in the suction-pipe were brought to rest absolutely suddenly at the end of the stroke, it is easy to show (see Appendix) that the resulting pressure would be $P_s = 63 \cdot 5 \, V_1$, where V_1 is the velocity of the water in the suction-pipe in feet per second. What the pressure would be with a valve that does not close suddenly is at present an unsolved problem, although more than one professes to have accomplished | it. But in every treatise that the author has come across the results obtained are very far from the truth. The manner, however, in which the pressure varies is shown by the curves in Fig. 37 ; the method of arriving at them is given in the Appendix for the case of a valve which closes at a uniform rate. These curves show how slowly the pressure rises at first and how rapidly it rises when the valve is nearly closed. This action gets more and more intensified as the length of the pipe increases. Possibly a complete solution of this problem might throw some light on the question. It is believed to be one involving complex wave motions. The author therefore fears that all he can do at present is to record the results of his experiments, and to hope than one more competent than himself will deem the matter to be of sufficient importance to lead him to investigate it still further.

It should be stated that the results given in Figs. 35 and 36 (pages 160 and 161) must be taken as approximate only. There is great difficulty in getting exact values, but they are probably not more than 10 per cent. from the truth.

FIG. 37.—*Pressure set up when closing a Valve.*

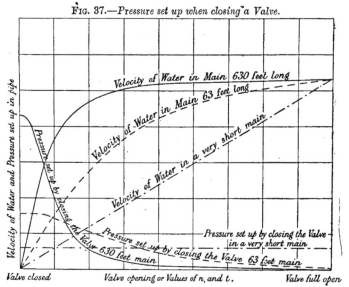

Velocity of Water through the Suction Valves.—The valve-lift diagrams taken at various speeds supply the data for finding the velocity of the water passing through the valves. At low speeds the valves, which were perfectly free as regards lifting, only opened partially, and as the speed increased the opening also increased; hence, knowing the lift of the valves and the quantity of water passing per second, by simple calculation the following results are obtained :—

Speed in revs. per min.	32·3	40·9	50·0	60·3	71·2	80	90	100
Lift of 4-inch valve in.	0·137	0·172	0·212	0·269	0·33	0·34	0·34	0·34
Velocity of flow . .	1·96	1·99	1·80	1·87	1·75	1·76	1·99	2·19

Frictional Resistance of Valves and Pump Passages.—The suction pressures as found from the pump-barrel diagrams are lower than those found from the suction-pipe diagrams. The difference between them is due to the friction of the water in passing through the passages and valve openings. From such measurements the friction curve in Fig. 38 has been obtained. It should be noted that it includes the resistance of three square-cornered elbows in the passages, as well as the resistance in passing the valve.

It will be seen from the curve that the resistance varies as the square of the speed, quite what theory would lead one to expect.

The Mechanical Efficiency of the Pump.—Although the mechanical efficiency of the pump is not of direct hydraulic importance, yet the results obtained when pumping at various heads may be of general interest. They were obtained from indicator diagrams taken from the steam and water cylinders when the glands were normally adjusted. The results were :—

Delivery Pressure in pounds sq. inch.	10	20	30	40	50	60	70	80
Mechanical Efficiency per cent. .	50	60	69	75	79	82	83	84

FIG. 38.

Loss of Pressure due to Friction in the Valves and Passages of the Suction side of Pump.

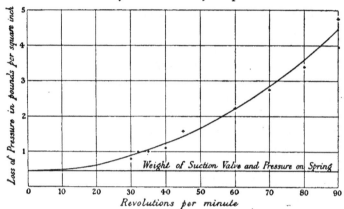

Revolutions per minute

General Conclusions.—The general conclusions that may be drawn from these experiments are :—

(1) That in a pump of this type without a vacuum vessel on the suction-pipe, the quantity of water delivered depends on the speed of the pump, and on a " coefficient of discharge " which depends on the action of the valves, and may under normal conditions of working be taken as about 0·94. This coefficient must only be used when the delivery pressure p_d is greater than

$$0·000148 \; L \; R \; N^2 \; \frac{A_p}{A_s}$$

and the speed N less than

$$54·5 \sqrt{\frac{(34 - h_s - h_f) \, A_s}{L \, R \left(1 - \frac{1}{n}\right) A_p}} \; ;$$

for the meaning of the symbols see page 168.

If the delivery pressure be less than that given by the above expression the coefficient of discharge may be approximately found by the method described on page 168.

Outside these limitations the quantity pumped may vary anywhere between 94 and 150 per cent. of the plunger displacement.

The results obtained in these experiments throw some light on the hitherto mysterious errors that occasionally occur in certain forms of piston water-meters.

(2) That the water-ram pressures in the suction-pipes of pumps which are not supplied with air or vacuum vessels may be very serious indeed, and are very much greater than one would be led to expect by the usual theory. The experience of others * shows that such vessels should be placed close to the pump in a continuous line with the suction-pipe and not on a branch at right angles to it.

(3) That the serious banging one often hears in the suction-pipes of pumps is due to the separation of the plunger from the water and their subsequent meeting.

Discontinuity between the plunger of a fly-wheel pump and the liquid column may be avoided—

(a) By using a vacuum vessel on the suction-pipe close to the pump.

(b) By having a suction-pipe of greater diameter and thereby reducing the velocity of flow.

(c) By using a shorter suction-pipe.

(d) By running the pump at a lower speed.

The requisite diameter and speed may be obtained from the equations given on page 171.

(4) That uncontrolled suction and delivery valves of pumps of this type do not open and close in the simple manner they are supposed to do, except when the pump is running at low speeds and when delivering against a moderate head. These experiments appear to show that the velocity of water through such valves should be about two feet per second.

The problem of running fly-wheel reciprocating pumps at high speeds has been very fully investigated by Reidler of Berlin, who very early recognised the fact that free, that is, uncontrolled, pump valves for both air and water act in a very erratic manner. He therefore designed a complete series of both air and water pumps in which the opening and closing of all the valves were mechanically

---

* Article by F. M. Wheeler in Cassier's Magazine, 1900, page 516.

controlled, with the result that his pumps run smoothly at far higher speeds than would otherwise be possible.

The author is fully aware that the results obtained on this small pump may not occur in so marked a manner in larger pumps, but still he trusts that the results may be suggestive, and of sufficient interest to lead other engineers to carry on similar experiments on a larger scale.   He, with many others in a similar position in England, often deplores the fact that a vast amount of labour and research is expended in engineering colleges in carrying out experiments on apparatus far too small to give really valuable data, and would suggest that engineers might, greatly to their own advantage and to that of students, lend plant to colleges for testing purposes, the results of the tests being given in return for the loan.

The Paper is illustrated by Plates 7 and 8, and by 35 Figs. in letterpress, and is accompanied by an Appendix.

APPENDIX.

Inertia Pressure and Discharge Coefficient with an Infinitely Long Connecting-Rod.

Let $R =$ the radius of the crank in feet.

$l =$ the length of connecting-rod in feet.

$n = \dfrac{l}{R}.$

$N =$ revolutions of the pump per minute.

$w =$ the weight of a unit column of water 1 foot high and 1 sq. inch section $= 0 \cdot 434$ lb.

$W =$ weight of the reciprocating parts per sq. inch of plunger in pounds; in this case the weight of a column of water of 1 sq. inch sectional area and whose length is equal to that of the suction-pipe, *i.e.*, $0 \cdot 434\ L = w\ L.$

$P =$ the "inertia pressure" at the end of the stroke, *i.e.*, the pressure required to accelerate and retard the column of water at the beginning and end of the stroke.

(See Goodman's "Mechanics Applied to Engineering," p. 145.)

Then, if the suction-pipe be of the same diameter as the pump-plunger, we have :—$P = 0 \cdot 00034\ W\ R\ N^2$, but if the area of the suction-pipe be A_s, and that of the plunger A_p, we have, substituting the value of W given above,

$$P = 0 \cdot 00034 \times 0 \cdot 434\ L\ R\ N^2\ \frac{A_p}{A_s}.$$

For the pump now under consideration $\dfrac{A_p}{A_s} = 1 \cdot 83$ and $R = 0 \cdot 25.$

Then $P = 0 \cdot 000068\ L\ N^2.$

In some of the experiments the length of the suction-pipe was 63 feet. Then $P = 0 \cdot 0043\ N^2$; for the 36-feet pipe $P = 0 \cdot 0025\ N^2.$

The manner in which the pressure varies from point to point in the stroke is given by the straight line ab, Fig. 39, and, as explained

FIG. 39.

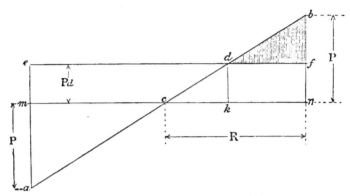

on page 156, the extra volume of water delivered is equal to the area $d\,b\,f$ divided by $144\,p_d$ (the 144 because the pressure is in lbs. per sq. inch, and the other dimensions are in foot-lbs.). We must now see what is the scale of the area $d\,b\,f$ or A.

Let the scale of the indicator diagram be s lbs. per inch, and let the stroke of the pump be m times the length of the diagram, then the area $d\,b\,f$ measured in square inches represents $\frac{s\,A\,m}{12}$ foot-lbs. of work per square inch of plunger, and the extra volume of water v (in cubic feet) delivered per square inch of plunger is $v = \frac{s\,A\,m}{12 \times 144\,p_d}$.

The volume V (in cubic feet) displaced per square inch by the plunger is $V = \frac{2\,R}{144}$, and the discharge coefficient is $S = 1 + \frac{v}{V} = 1 + \frac{s\,A\,m}{24\,p_d\,R}$. But the work represented by the shaded area may also be written $(P - p_d)\,\frac{\overline{d f}}{2} = \frac{R\,(P - p_d)^2}{2\,P}$ foot-lbs. of work per square inch of plunger, and the extra volume of water delivered v (in cubic feet) per square inch of plunger is $v = \frac{R\,(P - p_d)^2}{2\,P \times 144\,p_d}$, and the discharge coefficient is $S = 1 + \frac{v}{V} = 1 + \frac{(P - p_d)^2}{4\,P\,p_d}$.

The same result can, however, be arrived at by a more direct method without considering the question of scales. Since the ratio of the work done in delivering the extra volume of water to the work done under normal conditions is $\dfrac{\text{Area } d\,b\,f}{\text{Area } e\,f\,n\,m}$, and since the pressure scale is the same in both instances, the volume of water delivered will also be in the same proportion, whence the discharge coefficient $S = 1 + \dfrac{\text{Area } d\,b\,f}{\text{Area } e\,f\,n\,m}$.

Inertia Pressure and Discharge Coefficient with a Short Connecting-Rod.

It is shown in the above-quoted reference that the inertia pressure with a short rod is $P = 0 \cdot 00034 \ W\ R\ N^2 \left(1 \mp \dfrac{1}{n}\right)\dfrac{A_p}{A_s}$. The negative sign being for the "in" end of the stroke, and the positive sign for the "out" end. The line $a\,b$, Fig. 5 (page 132), then becomes the curved line $a'\,b'$. The area A then becomes (Fig. 8, page 136) $A_1 =$ area $d'\,a'\,f$ and the discharge coefficient becomes $S_1 = 1 + \frac{s}{24}\frac{A_p}{p_c}\frac{n}{P}$; an analytical solution of this case is more difficult than in the case given above. The discharge coefficient is most easily obtained, from the diagram, Fig. 8, by measuring the area A_1 with a planimeter, or it may be obtained as explained above, thus—

$$S_1 = 1 + \frac{\text{Area } d'\,a'\,f}{\text{Area } e\,f\,n\,m}.$$

Speed at which the Plunger leaves the Water in the Pump Barrel, when not fitted with a Vacuum Vessel.

Adhering to the notation given above, and further—

Let $h_s =$ the suction-head below the pump, *i.e.*, the height of the surface of the water in the sump below the bottom of the pump-barrel; if it be above, this quantity must be given the negative sign.

$h =$ the loss of head due to friction in the passages and pipes. Then the pressure required to accelerate the moving water at the beginning of the suction stroke, in this case when the plunger is at the bottom of the stroke, is as before

$$P = 0 \cdot 00034 \times 0 \cdot 434 \ L\ R\ N^2 \left(1 - \frac{1}{n}\right)\frac{A_p}{A_s}.$$

The height of the water-barometer may be taken as 34 feet, then the effective pressure driving the water into the pump-barrel is $(34 - h_s - h_f)\ w$. Separation occurs when this quantity is less than P; equating these two quantities we get the maximum speed N, at which the pump can run without separation taking place, and reducing we get

$$N = 54 \cdot 5 \sqrt{\frac{(34 - h_s - h_f)\ A_s}{L\,R\left(1 - \dfrac{1}{n}\right)A_p}}.$$

In cases in which friction plays an important part, the separation will not necessarily occur at the beginning of the stroke, because the water is then at rest and the friction zero, but the frictional resistance increases rapidly as the plunger moves; hence the separation will occur shortly after the beginning of the stroke.

Sudden Closing of a Valve.

Let P_s = the pressure set up when a valve is suddenly closed, in pounds per square inch.

V_i = the initial velocity of flow in the pipe, in feet per second.

then $P_s = 63 \cdot 5\ V_i$.

(See Goodman's " Mechanics Applied to Engineering," p. 501.)

Gradual Closing of a Valve.

Let V_1 = the velocity of flow in a pipe in feet per second.

V_v = „   „   „   through the valve.

A_1 = the sectional area of the pipe in square inches.

A_v = „   „   „   „   „ stream of water passing through the valve.

$n_1 = \dfrac{A_v}{A_1} = \dfrac{V_1}{V_v}.$

l = length of the pipe in feet.

D = diameter of the pipe in feet.

K = a coefficient depending on the roughness of the pipe (Darcy's Formula).

H = the total head of water above the valve.

Then, when the water is flowing, some of the head will be expended
in imparting kinetic energy to the water issuing from the valve and
some will be wasted in friction along the pipes. Thus we have per
pound of water in a horizontal pipe—

$$H = \frac{V_v^2}{2\,g} + \frac{l\,V_1^2}{K\,D} = \frac{V_1^2}{2\,g\,n_1^2} + \frac{l\,V_1^2}{K\,D},$$

whence

$$V_1 = \sqrt{\dfrac{H}{\dfrac{1}{2\,g\,n_1^2} + \dfrac{l}{K\,D}}}.$$

In Fig. 37 (page 163), various values of V_1 are plotted showing how
the velocity of flow in a pipe varies with the valve opening n_1 or with
the time t, if the valve be uniformly closed. The mass of water
moving along the pipe is constant, hence the manner in which the
pressure varies can be obtained by graphically differentiating the
curve, and thus getting values of $\frac{d\,v_1}{d\,t}$, which is the retardation of the
water at each instant, and since the mass is constant the pressure set
up is proportional to the retardation. Pressure curves are shown for
long and short pipes.

Discussion.

The PRESIDENT asked the members to show by acclamation their
appreciation of the Paper and their thanks to the author for bringing
it before them.

Professor GOODMAN, after thanking the President for giving him
an opportunity of making a few remarks upon the Paper before the
discussion commenced, said he would rather hear what the members
had to say before doing so. He had not given an account in the
Paper of the whole of the experiments he had carried out, but had
merely brought forward those that appeared to be the most important.
If, however, the discussion turned upon any point which he had not

fully dealt with, he would be only too glad to throw further light upon the matter, if he were able to do so. He thought he owed some apology to the Institution for bringing before it a Paper in such an incomplete state. He had very much wished that he could give a satisfactory theoretical treatment of all the strange phenomena which he had obtained with the pump, but he could honestly say that he had done his utmost, and trusted that some of the members, who were better able to treat upon the theoretical point than he was, would throw some more light upon it. If they did so, he would be amply repaid for any trouble he might have taken in making the experiments, and in bringing the Paper before the members.

Mr. GEORGE H. HUGHES said that the suction-pipe was mentioned as being straight and horizontal (page 126), but he could not see from the letterpress or the illustration whether, in the suction-pipe which led up to the pump, there was any real suction lift on the pump, or whether it was simply a gravitation pipe to the pump. He would also like to ask in connection with Fig. 2, Plate 7, what was the amount of lift of the valve until the valve reached the fixed stop. [See page 188.]

Mr. WILLIAM SCHÖNHEYDER remarked that he was very pleased the Paper had been read before the Institution, because it treated of a subject which, he was sorry to say, was not as much known as it ought to be, and very little was published in the text books about it. The Paper, no doubt, would do an immense amount of good in enabling many pump engineers to avoid the fatal shocks in suction pipes. At the same time, while he valued the information which the author had given, he thought in some instances he had given a little too much. For instance, he quite failed to see any utility in giving the total amount of pressure which might occur in a badly-designed pump. Pumps were not required with heavy shocks in them, and it was quite immaterial whether the pressure so called was 100 lbs. per square inch or 1,000 lbs. The point was not to make pumps strong enough to resist shocks, but to make pumps without shocks. In the same way the author calculated the exact

(Mr. William Schönheyder.)

position of the pump plunger, where the dilatory water again reached the plunger and caused collision. Engineers did not require to know that; they wanted to know how to make a pump without such collisions.

There were a few points to which he wished to refer. The author described (page 124) a locomotive pump, and gave the capacity of its air vessel as 73 cubic inches. He himself ventured to say that there was not any air in the air vessel. Thousands of air vessels were made, and the words " air vessel " were inserted on the drawings from which they were constructed, but that was all the air they ever had to do with. Unless there was a provision for continually pumping air into the air vessel, or for forcing it in by the main pump, the air would be gone in a very short time, and the pump would act as if there was no air vessel at all. He thought the makers had only themselves to blame, for making such an absurd locomotive pump to give the enormous pressure of 3,250 lbs. per square inch. What else but trouble of some kind could be expected from a 2-inch pump with a 26-inch stroke running at 215 revolutions per minute? In addition to the trouble from the smallness of the delivery vessel and valve, there was no doubt trouble from both the suction and delivery pipes. The author had not given the dimensions and length of these, otherwise he (the speaker) might have gone further into the subject.

With the author's statements about obscure actions in pumps he could not quite agree. There might be a few difficulties now and then, but it would be a great wonder if a pump could not now be made to fulfil certain conditions, if they were at all reasonable. The author gave an instance (page 128) in which trouble arose with a suction pipe, mentioning that at first there appeared to be air in the pipes. The probability was that, if there had been air in the pipes, they would not have had the shocks; it was the absence of air that caused the shocks. He did not know whether the author was aware that it was quite a common dodge of pump-makers, plumbers, &c., when they had a pump which caused trouble in the suction pipe, if the pipe was not too strong they simply took a chisel and hammer and knocked a hole in it. It very often cured the evil, and nothing more was said about it. The pipe then worked quietly.

The author referred to certain troubles which had occurred (page 166), almost as mysterious as the pump troubles in piston water-meters. He ventured to say that, in a well-designed piston water-meter, such difficulties would not occur. The second conclusion on the same page referred to a Paper by Mr. Wheeler.* He had read that article, and must say he was no wiser after reading it than before he read it. Very few dimensions were given, the pipes were evidently very crooked and small, and no conclusions could be drawn from the article. There was no such hard and fast rule as the dictum that a suction vessel must not be on a branch direct from the pipe. He had made a great many pumps of various kinds, and had not had trouble from that cause. He had been fortunate enough to experience only one difficulty with a suction pipe, namely, in the year 1869, since when no such difficulty had arisen. It occurred with a small vertical engine of 2 feet stroke, $8\frac{3}{4}$-inch bucket, intended to run at 40 revolutions per minute. The suction pipe had been made 7 inches in diameter, and was of a total length of 15 feet. The engine failed to make the 40 revolutions per minute without severe hammering, in fact the hammer was so great that a 4-inch crank-shaft was bent. Having investigated the matter, he discovered that the suction pipe was too small. Accordingly he enlarged it from 7 inches to $12\frac{1}{4}$ inches diameter, keeping the length the same. The pump was not altered, but it was simply joined to the new suction pipe by a conical pipe. With those alterations, and under otherwise precisely the same conditions as before, the pump ran freely at 50 revolutions per minute.

He had had to deal with many other suction and delivery pipes. For instance, in 1886, he had designed a suction pipe of a single-acting pump whose diameter was 42 inches with a 4-foot stroke, and making 20 revolutions per minute. The suction pipe was 48 inches in diameter and 850 feet long. For this he provided a closed-up suction vessel of the necessary dimensions. In another case in the southern part of America, a delivery pipe was required to furnish a gallon of water per second, which had to be pumped to a height of

* Cassier's Magazine, 1900, page 516.

(Mr. William Schönheyder.)

1,600 feet through 27½ miles of 4-inch main, causing an estimated friction of 600. The total pressure, therefore, would be about 1,000 lbs. per square inch. It was at first intended to use a Worthington engine for the purpose, and he was asked if he could design something more in accordance with the customs of this country. He accordingly designed a rotative engine with four

FIG. 40.—*Pump-delivery Diagrams, to discover size of Accumulator.*

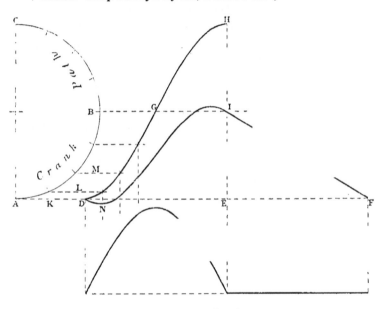

plungers, with four equal deliveries. As the pressure was so great, he dared not use an ordinary air vessel, because he thought the air would be absorbed by the water. He therefore adopted a kind of loaded plunger similar to that used in an accumulator, but he put the load on it by means of a steam-piston. That, of course, was not a novelty; but he added a self-acting valve, so that the pressure on the steam-piston was exactly right at all times, according to the pressure in the main. The plunger worked up and down most satisfactorily, and it took the place of an air vessel; since then there had not been any trouble with the arrangement. First of all he had to find out what size of plunger was required, and for that reason he made the simple diagram Fig. 40 (page 176).

In ordinary text books the curves of delivery of single-acting pumps, two pumps and so on, were shown ; for single-acting pumps the curve was similar to the lower one on the diagram. The first part of the lower diagram showed the volume of delivery of the pump for half a revolution, and there was no delivery after that. That was, however, only half the story, he considered, because it took no account of the standpipe, or the suction pipe, or the air vessel—the "equalising" vessel, to use a general term—which in nine cases out of ten must be connected with a pump of a fair size. On the upper diagram the half circle A B C represented half of the path of a crank-pin going at a uniform speed. The circumference was divided into sixteen parts. The line D E F was the time-base line, divided also into sixteen equal parts. When the crank-pin moved from A to the first division, the water would have been forced up into the stand-pipe (supposing it was of the same area as the plunger) by the amount K L. When the crank-pin had passed the two divisions, the water would have been forced up the stand-pipe to the level M, so that the successive levels of the water in the stand-pipe would be indicated by the S curve D G H. But at the same time in which the water was forced into the stand-pipe, it ran out of it and into the delivery pipe, so that the water, instead of being for the first division at the level L, it was down at N. At the second division, two volumes of water had run out of the stand-pipe, at the third division three volumes, and so on. . The actual levels in the stand-pipe were therefore shown by line D N I. Of course, after the half-stroke no water was delivered in the stand-pipe, but it kept on running out of it at a uniform rate which was indicated by line I F. From the curve D N I F one could estimate the necessary volume of vacuum vessel or air vessel or stand-pipe. This was a very simple construction, but he had not seen it given anywhere in the text books. The connecting-rod was assumed of infinite length, and to be vertical on the diagram. It was worth noting that for a single-acting pump the total variation of volume in the equalising vessel was about 0·56 of the delivery per revolution.

(Mr. William Schönheyder.)

Finally, he wished to say that it was pitiable to see the number of pumps of all kinds, met with everywhere, with an air-lock in them, that is, an air space in the upper part of the pump higher than the level of the delivery valve, so that air was constantly being trapped there, and the pump very frequently refused to act either on the suction or delivery stroke. It was still more deplorable to notice that kind of pump delineated in text books and catalogues.

Mr. WORBY BEAUMONT thought the members must all feel grateful to the author for bringing the Paper before them, because it was one which showed how very useful those in Professor Goodman's position might be to the practical engineer. Before he came to the meeting that evening he asked a friend whether he was coming to the meeting, and his friend replied, " Does not everyone know all about pumps already ? " He replied that he was not quite sure that they did, and after hearing the Paper he must admit that the pump, one of the eldest of the things to which mechanical engineers had turned their attention, was still a thing concerning which the indicator, as a mirror, might show them a very great deal. The experiments, amongst other things, directed one's attention to the way in which the inertia of motion of a fluid might do that which had escaped their attention, however carefully they might have considered the thing, such as a pump, which put that fluid in motion. The results of the author's experiments showed how a pump of a given size might cause motion in a fluid which would produce a greater output than the mere volume of the pump piston, and that one fact alone was of considerable interest. That action had been met with in other things, for instance, in gas-engines, where the inertia of the motion of a column, both of incoming air and of exhaust gases, had been made use of ; but it had not been put before the members previously in the manner given in the Paper, the figures showing not only how the inertia of motion effect, which was purposely set up, might be the cause of very great difficulties, but how, on the other hand, when properly understood, it might lead up to the production of a water-delivery machine, which would not necessarily be of greater efficiency than some of those already in existence, but might have a higher duty.

The author had referred to certain locomotive pumps. He was not sure, but he imagined that those pumps to which he referred were some that were in use at one time on the Brighton Railway; at all events, he happened to know of certain difficulties which were experienced with pumps on some locomotives of that railway. The Paper made it possible, or at all events helped, to interpret the causes of those difficulties. The proportions of the pumps, as well as the points which the author had mentioned with regard to speed, the sizes of suction pipes, delivery pipes, passage forms and dimensions, and so on, had a great deal to do with the matter. They all remembered the old long-stroke, small diameter pumps which were in use on the old London and South Western Railway engines, which, with the small air vessel, worked very comfortably at considerable speeds. Mr. Schönheyder had just suggested (page 175) that there was no hard-and-fast rule with regard to the position of the suction vessel, which had been experimented with in the Paper, and mentioned a Paper in Cassier's Magazine. Of course, there was no hard-and-fast rule, but one could see that one might soon arrive at a distance from the pump at which the suction vessel would begin to have a disappearing value. Once it was far enough away, the volume of water when it was put in motion would have the same effect at the pump and on its valves as though that vessel were not in the suction pipe at all.

He again thanked the author for having shown the members once more the real use of the indicator diagram, when employed by some one who was capable of putting it to practical uses, and interpreting the results in the way given in the Paper, namely, with relation to practical application, so that his remarks might be useful to those who had only time to do little more than adopt his conclusions.

Mr. Schönheyder said that Mr. Worby Beaumont had misunderstood his remarks. He did not mean to refer to the distance of the equalising vessel from the pump; it was its position in reference to the suction pipe to which he alluded, as mentioned in Mr. Wheeler's Paper in Cassier's Magazine. The nearer the air vessel could be placed to the pump, the better it would generally be.

Mr. W. S. LOCKHART said he would like to say a few words on the question of the air-vessel on the suction side. He noticed that the author seemed to use the terms air-vessel and vacuum-vessel alternatively. All through the Paper the expression occurred, air- or vacuum-vessels. He was not at all sure that he knew what a vacuum-vessel in this case was, but he knew what an air-vessel was. He would like to understand why the term vacuum-vessel was used in this way. In his opinion a plenum-vessel would be a very much better expression, because, in the matter under discussion, what had to be done was done not only by giving the water the opportunity of keeping moving while the suction valve was closed, and thus being ready to enter the pump immediately it opened, but also by storing up the momentum of the on-coming stream and returning it by the re-expansion of the air so that the water jumped, as it were, into the pump on the opening of the valve. To do this, and in fact to be of any use at all, the vessel must have air in it, for a vessel with a perfect vacuum would be useless, and would indeed produce exactly the trouble that took place in the pump. The blow in the pump was the result of its being for the moment a vacuum-vessel, and if the vessel on the pipe were really a vacuum-vessel the trouble would be increased, not diminished. There might not be much actual difference between a vessel in which there was an attenuated air-pressure and one in which there was an imperfect vacuum, but, as a question of terms, a vacuum-vessel was one in which, if possible, the vacuum should be perfect to give the best effect, and since in this connection it was clear that the difficulty sought to be remedied would be increased by a vacuum, the vessel on the suction side of a pump should, he thought, be called an *air*-vessel, and not a *vacuum*-vessel.

The question of the position of the vessel was important, and, to his mind, it could not be too close to the valve; if it could be actually incorporated in the valve, that would be its proper place. When the suction-valve closed, the water in the suction-pipe had either to stop dead or go somewhere, and, if it had the opportunity given it, would go into the air-vessel, creating a pressure there which, directly the valve opened, would cause the water to enter the barrel of the pump immediately and fill it up, instead of having to wait and be started again, and then come after the plunger and strike it a blow.

The point of the whole subject was very interestingly put in such an old publication as Bourne's Catechism of the Steam-Engine (1868), and he would like to quote the passage and make a sketch, Fig. 41 (page 182), to show the sort of thing described. The passage occurred on page 214, and was as follows:—

" All pumps lift much less water than is due to the size of their barrels and the number of their strokes. Moderately good pumps lose 50 per cent. of their theoretical effect, and bad pumps 80 per cent. This loss of effect is chiefly due to the inertia of the water, which, if the pump-piston be drawn up very rapidly, cannot follow it with sufficient rapidity ; so that there may be a vacant space between the piston and the water ; and at the return stroke the momentum of the water in the pipe expends itself in giving a reverse motion to the column of water approaching the pump. Messrs. Kirchweger and Prusman, of Hanover, have investigated this subject by applying a revolving cock at the end of a pipe leading from an elevated cistern containing water, and the water escaped at every revolution of the cock in the same manner as if a pump were drawing it. With a column of water of 17 feet they found that, at 80 revolutions of the cock per minute, the water delivered per minute by the cock was 9·45 gallons ; but, with 140 revolutions of the cock per minute, the water delivered per minute by the cock was only 5·42 gallons. They subsequently applied an air-vessel to the pipe beside the cock, when the discharge rose to 12·9 gallons per minute with 80 revolutions, and 18·28 gallons with 140 revolutions. Air-vessels should therefore be applied to the suction side of fast-moving pumps, and the suction-pipe should be made as short as possible."

The book was more than thirty years old, but the illustration given was, he thought, an interesting one. Another illustration which had come under his notice some years ago occurred in a pair of hydraulic engines under his care, Fig. 42 (page 182). They were a pair of horizontal engines of the ordinary type, pumping against an accumulator from an overhead feed-tank, and with suction-pipes as

(Mr. W. S. Lockhart.)

shown on the right side of the diagram. The engines had been carefully designed and were new, but they knocked very badly. The bearings and everything else that could possibly knock were examined, but without result, until suspicion fell on the suction pipes, which were then altered as shown on the left side of the diagram on both sides, and the trouble immediately disappeared. For pumps of this sort

FIG. 41.   FIG. 42.

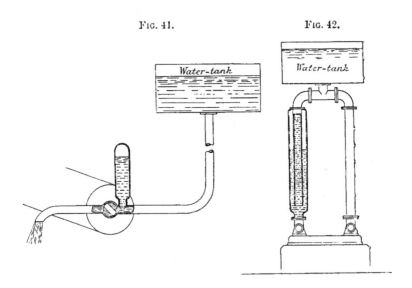

he thought the arrangement a good one, and the air-vessels as nearly in the right place as practicable.

In both these cases the source of supply was above the pumps, so that the vessels were plenum-vessels, but in the doubtless more usual case of the source of supply being below the pumps the principle involved was the same, and the difference was only one of degree in the actual tension of the air in the vessels. In any case he ventured to think that the vessel on the suction side of a pump must be an air-vessel with a pressure from anything upwards, but that directly it fell to nil it became a vacuum-vessel and ceased to be of use.

Professor DUNKERLEY said that there was one point to which he would like to refer, namely, the question of negative slip. In one experiment quoted in the Paper, the actual delivery was 50 per cent. in excess of the plunger displacement. The chief reason, very properly assigned by the author, was because the delivery valve opened prematurely before the end of the suction-stroke was reached, due to the fact that the water in the suction-pipe, in being retarded, exerted a force on the underside of the delivery valve sufficient to overcome the pressure on the top side. He would like to ask the author whether the suction-valve was found to open prematurely before the end of the delivery-stroke was reached. If so, it might possibly account for the observed great negative slip, because the pump would not only be delivering water during a portion of the suction-stroke, but it would be sucking water during a portion of the delivery-stroke.

Mr. Schönheyder (page 173) had deplored the lack of literature on the subject. As a matter of fact the literature was available, but we had to thank German scientists for it. The question of the theory of reciprocating water pumps and engines was very fully discussed in a book by Weisbach and Hermann, and an English translation, by Dahlstrom, was published by Messrs. Macmillan and Co. In that book all the points mentioned by the author were very ably and clearly treated ; and it appeared to him that the great value of Professor Goodman's Paper lay in the fact that it gave us detailed *quantitative* data of the various phenomena connected with the working of reciprocating pumps.

Mr. BRIDGMAN RUSSELL thought it might be of interest to the members if he stated that, during the next two months, there would be a series of trials of small pumps and wind engines, carried out under the direction of the Royal Agricultural Society at their Willesden Show Ground. As many as twenty-three wind engines and pumps of English, Canadian, and American manufacture would receive a thorough testing, the results of which he was sure would be of great interest to the members, some of whom might care to visit the grounds and see the various tests being carried out. The consulting

(Mr. Bridgman Russell.)

engineer, Mr. Frank S. Courtney, who would carry out the tests, was a Member of the Institution.

Professor GOODMAN, in reply, thanked the members for the kind remarks made in regard to the Paper. In reply to the questions asked, he wished to say that the suction-pipe was quite level. There was a small head of water, about $1\frac{1}{2}$ inches—the exact amount was given in Table 2 (page 142)—above the suction-pipe of the pump. The valves lifted $\frac{3}{8}$ inch ; the exact amount of lift was also shown in the diagrams. He had shown a zero line on all the pump-lift diagrams showing the valve on its seat, and an upper line showing the extreme range over which the valve could move. He had been much interested in what Mr. Schönheyder had said, but he differed from his statement (page 174) that there was no air in the air-vessel on the pumps referred to. Very great trouble was taken to see that air was there, and moreover a spring piston was also put on the pump, in order to make certain that there was no effect of that kind. In addition, the hydraulic indicator showed that the big pressures were set up, not on the delivery side, but on the suction side of the pump, again corroborating the result he had obtained with the pump under discussion.

Mr. SCHÖNHEYDER asked whether provision was made for renewing the air, and whether a continuous supply was available.

Professor GOODMAN replied in the negative. The air-vessel was looked to very carefully before starting out of the station, and could not very well, in a run of 50 miles, get completely water-logged. In addition, the spring piston fitted in the place of the air-vessel showed that there could not be any water-logging effect going on at all, and the fact that it occurred on the suction-stroke proved that there was no question of the action of the air-vessel at all. Whether the design was absurd or not, he was not prepared to say, but it was the work of an eminent engineer.

With regard to water-meters, he must admit there were many water-meters which did not give the mysterious effects described,

but years ago, when he had to test water-meters for the Colonies
for various purposes, he constantly came across such mysterious
actions.   Water-meters would be working tolerably well within
1 or 2 per cent. of their nominal value, and suddenly go wrong,
in some cases delivering far more water than they ought to, and
in other cases doing just the reverse.   He had looked into one or
two of the cases, and was quite sure that the reason of that was
the same as the reason why the pump he had described delivered
more water than the plunger displaced.

   With all respect to Mr. Schönheyder's opinion, he thought there
was a good deal of force in Mr. Wheeler's Paper.   It seemed to him
that, if water was travelling at a very high speed along a pipe, an
air-vessel on the end of the pipe, which did not interfere with the
continuity of flow, must be more efficient than an air-vessel placed at
right angles to the pipe.   A great deal of loss by shock was obtained
in the one case, which did not occur in the other.   In the case under
discussion the water flowed straight into the air-vessel, and returned
again in a smooth stream.   He had not made any actual experiments
on the point, but he was inclined to think that Mr. Wheeler's results
were reliable, and that there was much reason in what he had said on
the matter.

   He had been much interested in Mr. Schönheyder's curves
(page 176); he certainly had not come across them before in that
compact form.   He thanked Mr. Worby Beaumont for his kind
remarks, but he did not know that he had raised any special point
that required an answer.

   With regard to the question of air and vacuum vessels, he was
afraid he must plead guilty to having used the term rather loosely.
He used the word air or vacuum vessel, in order to make his meaning
perfectly clear.   He meant in all cases a vacuum vessel or partial-
vacuum vessel when referring to the vessel on the suction-pipe.   He
thought the expression " vacuum vessel " for the suction-pipe, and
" air-vessel " for the delivery pipe, was a desirable distinction.   Mr.
Lockhart's diagrams (page 182) were of very great interest, and he
should certainly look up Bourne's book on the subject.

   In reply to Professor Dunkerley's remarks (page 183), he wished

(Professor Goodman.)

to say that, when he obtained the very excessive slip, the suction-valve did remain open long after it ought to have done. Some of the diagrams showed that. He thought Professor Dunkerley would see that some of the suction delivery valve diagrams showed that point very clearly indeed. In one instance the suction-valve remained open for about three-quarters of the whole time ; it did not close until about half-way down the delivery stroke. That, however, was not the most interesting part of the whole Paper, and he looked forward with very great interest to the written communications on the subject. Of course, when cavitation occurred, which was really what occurred in the pump, the results were very extraordinary. He believed that practically no water, or a very slight amount of water, entered the pump during the suction-stroke, when it was cavitating, and yet one obtained 150 per cent. of water sucked in and delivered during the delivery stroke.

With regard to the great pressure of cavitation, he had heard of a case a few weeks ago in which some bolts were subject to the cavitation action. The top of the heads of the bolts were sunk about $\frac{1}{2}$-inch into the surface of the surrounding casting ; a small brass plate, somewhere about $\frac{1}{8}$-inch thick, was fitted over the top in order to make a smooth passage for the water to run over, but in every case the brass plate was flattened right down to the bolt heads, which were somewhere about $1\frac{1}{2}$ inches in diameter, and the amount of dishing was fully $\frac{3}{8}$-inch. That would show what an enormous pressure was set up, due to cavitation. He should follow the experiments of the Royal Agricultural Society, of which Mr. Russell had spoken, with great interest. In conclusion he thanked the Members for the kind way in which the Paper had been received, and trusted they would not let the subject drop, but would bring some of their mathematics to bear upon the subject, and try and clear up some of the mysterious points.

Communications.

Sir FREDERICK BRAMWELL, Bart., Past-President, wrote that he thought it might be of interest, in connection with the subject of the Paper, if he stated that which had occurred many years ago with a pair of Boulton and Watt's 6-column beam engines.

In the accompanying sketch, Fig. 43, XY denoted the plan of the two engines, SS their steam cylinders, AA air-pumps, CC cold-water or jack-pumps, and M the common suction-pipe of the jack-pumps, branching into *m*, *m'*, to each pump.

The engines were at a very slight elevation above the level of the river, from which the jack-pumps were supplied. The pipe M was small in relation to the pumps, and was of considerable length.

FIG. 43.

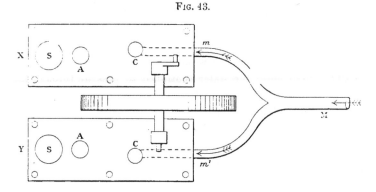

When the engines were put to work, the water spouted up from the two jack-pumps to the ceiling of the engine room, whereupon he (Sir Frederick) caused one of the pumps—say that of engine X—to be uncoupled from the beam, leaving the pump of the other engine Y to be worked. The result was that the working pump of Y behaved quite properly, and that the pump that was not worked by the engine X (having its bucket fixed), the foot-valve and the valve of that bucket were worked by the momentum of the water, and delivered apparently an equal amount with that of the pump that was worked by the engine Y. All the trouble, as regards the dashing of the water up to the ceiling, was at an end.

Mr. GEORGE H. HUGHES wrote, in continuation of his remarks at
the Meeting (page 173), that Professor Goodman had stated (page
184) in replying to the discussion, that the suction-pipe had been
laid with a fall of about 6 inches to the pump, and that the stops to
the valves had restricted the lift to about $\frac{3}{8}$-inch. A possible
solution of the abnormal pressure experienced in the suction-pipe
might lie in calculating the effect of the maximum velocity of flow
set up in the suction-pipe when the pump-plunger was on the centre
of the out stroke, taking the weight of water as a train on an incline
with a gradient due to the suction-pull of the pump.

The pipe, 36 feet in length, contained 48·816 lbs. of water, the
maximum velocity set up with a speed of 70 revolutions of the
pump he calculated was 183·26 feet per minute, amounting to a
maximum pressure of 8,946 lbs., and divided by the area of a 2-inch
pipe, the result was $\frac{8946}{3·14}$ or 2,848 lbs. per square inch as a possible
pressure by suddenly closing the valve. The reversal of the plunger
on its downward stroke was equivalent to closing the suction-pipe
partially, consequently the water continued to pass up through the
delivery valve, finding some relief, and setting up the pressure
ascertained by the indicator or gauge.

He would suggest that the experiments be continued with the
suction-pipe below the pump level, so that there might be a real
suction lift, and also with a suction-pipe quite vertical and not more
than 15 feet in length.

Mr. HENRY LEA, Member of Council, wrote that about forty-five
years ago he had been sent out to see to the fixing of some machinery
in a paper mill, and one of the machines was a small plunger pump,
driven direct from a crank on the shaft of a small steam-engine, with
a fly-wheel. This pump had been tested at the works before
despatch, and had worked very well. He did not remember the
revolutions per minute, but thought that they were somewhere about
100. The pump was fixed at the paper mill, with the suction-pipe
the same diameter as that which had answered very well at the works
test, but it was a very much longer pipe, probably 20 yards long, as
nearly as he could recollect. The engine was started, and very soon

the water came into the pump from the river, after being lifted about six or seven feet in vertical height. As soon as ever the pump became charged with water, and an attempt was made to run it at the intended speed, the crank-pin which worked the pump-plunger was promptly sheared off. A new crank-pin was fitted, and an air-vessel was put on the suction-pipe, as near as practicable to the pump itself, after which the pump worked at the required speed with great ease, and gave no more trouble.

This experience was an object lesson which fixed itself very firmly in his mind, and he had never since then put down pumps with long suction-pipes without capacious air-vessels thereupon. He did not think there was any danger of these air-vessels losing their air, because, being on the suction side of the pump, the tendency was to extract air from the water, and this air, naturally rising into the air-vessel, kept the same sufficiently supplied. Professor Goodman's experiments and conclusions were extremely interesting, but he appeared, in the writer's opinion, to have devoted a large amount of time to the investigation of a defect which no engineer ought ever to dream of allowing to occur in connection with pumping machinery.

Professor ROBERT H. SMITH wrote that Professor Goodman's Paper was a record of very interesting experiments, probably the first of their sort. In the Appendix (page 168) the ratio between the areas of plunger and suction-pipe sections was given as 1·83. According to the dimensions given, it was actually 1·94, but this 6 per cent. difference could have no important effect upon the subsequent calculations.

In referring to the oscillation at the end of the suction lines in Fig. 30 (page 150), the author attributed this (page 146) to inertia of the indicator parts. This could not be the real cause, as throughout the previous part of the line the pencil had remained practically motionless. As the oscillation was repeated in all the diagrams given, it seemed to indicate a real quiver in the pressure of the water reaching the indicator barrel. The same cause as produced this, perhaps a fluttering of the valve during closure, was no doubt, also influential in producing the very violent oscillation shown immediately

(Professor Robert H. Smith)

afterwards through the beginning of the delivery stroke, although here the momentum of the indicator masses very likely was responsible for much of the effect, and probably determined the rhythmic time-period of the oscillation.

It was also mentioned (page 146) that the opening of the indicator-cock had removed the "bang." One would infer from this that no indications could have been taken to show the character of this "bang," but the Paper did not explain whether this was so. It was not stated to which of the two lengths of suction-pipe Figs. 27 and 28 (page 145) corresponded. They could hardly be interpreted without this information.

The utility of the Paper would have been increased, if it had given some further data which were essential to the solution of the general problem. For instance, it was stated that an indicator had been fixed on the delivery, but no mention of cards obtained from this indicator was made. The positions of the indicators on the suction and delivery were not given, but were important, because at each instant the pressure varied largely along the lengths of both of these pipes. The position of the delivery air-vessel, and the volume of water between it and the delivery valve, were also very influential, but nothing was said about these. During the acceleration of the suction-water there was a downward pressure-gradient along the pipe from atmospheric pressure at the inlet to a minimum at the front end of the moving column. If this column were of uniform section, and if the water contained no air, this pressure-gradient would be uniform from end to end, and equal to $\frac{w}{g}$ a per unit area per unit length, if a be the acceleration and w the weight of unit volume of water. If there were changes of section in the length of the column, this expression was still correct for every part, but the velocity and, therefore, the acceleration and the pressure-intensity gradient changed in inverse proportion to the section. Thus the gradient of the total water-pressure over the whole section was the same in the large and the small parts of the column. The author took no account of the difference of pressure-gradient throughout the length of the suction-pipe and through the valve-passage and the pump-barrel; but as his suction-pipe was excessively long and the percentage of the accelerated water lying in these latter small, no material error arose from his not doing so. But superimposed upon this pressure-gradient due to the acceleration of the whole water-mass —which constituted an "unsteady" stream—there was the pressure-gradient between sections of different area due to change of velocity in the flow considered as a "steady" stream. This change of pressure, measured as a "head," was $\frac{1}{2g} (V_1{}^2 - V_2{}^2)$ between the sections 1 and 2. It would be a mistake to think that this had already been included in the "unsteady" acceleration gradient. The author omitted it from his calculation; and, as the velocities and their changes were both large in passing from the suction-pipe through the narrow valve-openings and into the larger-sized pump-barrel, this was an influential omission. On page 164 he attributed the difference between simultaneous pressures in pipe and barrel wholly to friction. There was really a considerable dynamic loss of head in passing the valve and a subsequent gain of head in entering the much larger barrel section. With the form of valve shown, the first loss was largely dissipated in eddy-making, and it could not be regained in any large measure in the following slowing down on entering the barrel; and, therefore, it was difficult to estimate how much the author's measurement of the friction deviated from the true frictional loss.

In considering the pressure effects of the suction-water being left behind by the plunger, it should not be forgotten that when pressure was suddenly withdrawn from a mass of water the space above the water surface filled with water-vapour so rapidly that the steam generation might be properly called instantaneous. The steam-pressure so generated corresponded with the temperature of the water and depended upon nothing else. In the present case therefore it was very small, but in the so-called air-pumps of condensers the water was hot and the steam-pressure was well above zero. But all undistilled water carried a considerable quantity of air mixed intimately with it, and this air was immediately evacuated when the surface-pressure was withdrawn. The pressure of air and

(Professor Robert H. Smith.)

steam together in the "empty" space below the plunger kept the plunger-pressure materially above absolute vacuum.

The air mixed with the water all along the suction-pipe acted in the same way as did the air-cushion in an air-vessel. Its effect seemed to the writer to be threefold. It softened the hammer blows due to sudden checks of flow. It made the whole column moderately compressible, and thus largely modified the laws of pressure-gradient obtained, as above, from considering the water-column as nearly incompressible. Thirdly, this permitted large-range waves of pressure to arise and run to and fro along the pipe, altering entirely the distribution of acceleration along this length from that assumed above. No doubt, in a "solid" water-column such pressure waves also travelled backwards and forwards; but the velocity of transmission of the waves was so extremely high that their passage must leave unaffected the mean pressures producing the comparatively slow accelerations here involved. The intermixture of a considerable quantity of air reduced in a very large ratio the speed of transmission of the pressure waves. With the air intermixed, the sudden closure or opening of a valve produced a rise or fall of pressure which at the valve face developed less quickly and travelled more slowly backwards towards the inlet end. Such to-and-fro surgings after suction-valve closure were testified to by Fig. 29 (page 147). In the design shown in Fig. 1, Plate 7, before the delivery-valve opened, the stream of water in passing from the suction-valve to the pump-barrel was deflected through about 1½ right angles, and the deflecting surface was for the most part the under-face of the delivery-valve or of dead water held in place by this valve. The pressure needed to produce this deflection was large, and was proportional to the square of the velocity of flow. This velocity was here probably proportionately large, owing to the contracted area of the suction-valve opening and to the volumes of dead water above this valve and in the corners of the box, which caused the same contraction of flow-section as if these volumes had been solidly filled with metal or other rigid material. Thus he himself would expect to find the lifting pressure per square inch on the under face of the delivery-valve to be much higher than that simultaneously indicated in the plunger-barrel. This, of course, conduced to early opening of the delivery-valve. When this valve opened, the water endeavouring to flow through it to the delivery met with a mass of water lying motionless between the valve and the air-vessel. This mass could not be instantaneously set in motion, so that the delivery flow through the newly-opened valve could not commence at once. It gained volume all the more slowly, because at the first instant there was little or no difference of pressure to act as a driving force to start the dead water between valve and air-vessel into motion. In fact, the relief of pressure under the valve caused by its opening might lead sometimes to the valve closing again. After commencement of the delivery flow, there was acceleration of the above previously dead water, and a down pressure gradient from the valve to the air-vessel. In this vessel the air-cushion prevented very large variation of delivery pressure, so that during this acceleration the pressure at the valve must be in excess of the mean delivery pressure. This excess was larger, and lasted longer, the greater the length between valve and air-vessel and the smaller the air-vessel, because a large air-vessel kept down the pressure at this point to little above the mean, while with a small vessel the pressure here rose quickly, and thereby diminished the pressure gradient available for quick acceleration. The conditions in the delivery, beyond the mere average delivery pressure, were therefore very influential in determining the " negative slip." These conditions affected the pressure on the plunger immediately after the delivery valve opened. The flow was now split after passing the suction-valve, and the work done by disappearance of kinetic energy during the remainder of the suction was divided between the plunger and the water driven along the delivery. The whole kinetic energy so given up was the area under the total length of the retardation curve, reckoning this curve from the velocities at the passage past the suction-valve where these velocities reached their maxima. Part so given up was wasted in eddy-making and consequent friction, and this part was unfortunately impossible to calculate. If from the remainder were subtracted the work done simultaneously on the plunger as shown by the barrel

P

(Professor Robert II. Smith.)

indicator-card, what was left was spent in driving water through the delivery. Part of this last was in the first place spent in producing kinetic energy in the water between valve and air-vessel, but, if the air-vessel worked correctly, this part should be recovered as useful water-delivery duty.

It was plain that the opening of the delivery-valve modified the retardation curve so as to lessen the retarding forces, and thus make the retardation slower and occupy a longer time. If such modification did not take place, the retardation would be finished—that is, the kinetic energy would be finally and completely used up—exactly at the end of the plunger stroke. It thus appeared that the effect of the premature delivery opening was to prolong the kinetic energy, or water-ram, effect into the first part of the delivery-stroke of the plunger.

Finally, it should be noted that with a small fly-wheel, such as was common on donkey-pumps, the curve of plunger velocity-accelerations was likely to be largely different from that calculated from uniform rotational speed of fly-wheel. Independently of variations of steam-driving torque, the violent variations of water-pressure on the plunger-face could not fail to modify largely the velocity-accelerations. It was too common an error to confuse accelerations with visible or measurable changes of velocity. An acceleration might be enormous, and yet act through so short a period of time as to produce no measurable change of velocity. Yet it was the acceleration alone, and not the velocity or its integral change, to which the pressures, sometimes rising to sharp hammer blows, were proportional. The accelerations of the plunger were at each instant proportional to the difference between the steam driving-torque and the water resistant-torque. These varied according to entirely different laws, each of the two being considerably erratic. Their difference thus varied rapidly and through large range. The acceleration of velocity, to which the ordinate of the important water-pressure curve here discussed was proportionate, equalled this erratic difference divided by the fly-wheel moment of inertia, and this inertia was so small in most pumps of this class that visibly large hitches in the rotational velocity of the wheel were more commonly present than absent.

Professor GOODMAN, in reply to the discussion and correspondence, observed that he feared some members did not quite realise that all research work necessarily brought out some points that might not be of immediate utility; but, in publishing the results of such work it would be folly to omit interesting matter simply because one could not see how it could be utilized. More than one Continental firm turned out high-speed reciprocating pumps which delivered very much larger quantities of water per ton of pump than was accomplished by any English maker, although some had unsuccessfully made the attempt. The author believed that a careful study of the results given in the Paper would throw much light on the reason why one maker succeeded and another failed to make satisfactory high-speed pumps.

Mr. Hughes' calculation (page 188) of the pressure in the suction pipe, namely 2,848 lbs. per square inch, was surely a slip. His method of arriving at it was, in the author's opinion, quite erroneous. If the water were brought to rest instantaneously, the pressure would only be $\frac{63 \cdot 5 \times 183 \cdot 26}{60} = 194$ lbs. per square inch (*see* Appendix), and it would be considerably less when brought gradually to rest, as in the present instance.

Professor Dunkerley (page 183) spoke of Weisbach and Hermann's book on "The Mechanics of Pumping Machinery" as giving a complete theoretical treatment of all the points mentioned in the Paper, but, since the meeting, the author had referred to it and found that they only treated of the simple case of a pump in which the obliquity of the connecting-rod was neglected, such as was given in Fig. 39 (page 169). The most interesting discovery of all, namely, the great increase in the discharge coefficient, also the enormous shock pressures set up after the separation or critical point was reached, was not even touched upon by them. The author much wished that Professor Dunkerley himself could spare time to look into the matter.

Professor Smith pointed out (page 189) that the ratio between the areas of plunger and suction-pipe section was given as 1·83, whereas the dimensions gave 1·94. The value given in the Paper, namely, 1·83, was the correct one. It was found by filling one

(Professor Goodman.)

length of the suction-pipe with water and afterwards weighing it,
the $2\frac{7}{8}$ inches given as the diameter was therefore only approximate.
He also referred to the " bang " not occurring when the indicator
cock was open. Of course no diagram could be obtained without
opening the indicator cock ; the author however fully agreed with
Professor Smith that the diagrams taken from the suction-pipe did
not represent a true state of affairs ; the fact that the water-ram
pressure as measured from the indicator diagrams was nearly always
found to be less than that obtained from the special indicator
arrangement shown in Fig. 4, Plate 8, was a proof of this. The

FIG. 44.

*Indicator Diagrams taken from the Delivery pipe at Various Speeds, the opening
of the Outlet Valve E, Fig 3, Plate 8, remaining the same in each case.*

Scale $\frac{1}{20}$.

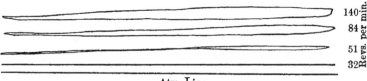

140 Revs. per min.
84 Revs. per min.
51 Revs. per min.
32 Revs. per min.

Atm. Line.

length of suction-pipe in Figs. 27 and 28 (page 145) was 63 feet. In
response to Professor Smith's request for some delivery diagrams the
author gave a few typical cases in Fig. 44.

The suction indicator was situated 9 inches from the face of the
pump suction flange, Fig. 2, Plate 7, and the delivery indicator
12 inches from the face of the pump delivery flange. The tubular
standards of the pump formed the delivery air-vessel, the volume of
which might be roughly obtained from Fig. 2, which was drawn to
scale.

The suction-pipe was of constant cross-section throughout its
length. With regard to the change of pressure due to the change
of section between the pump-barrel and the suction-pipe, it only
amounted to a small fraction of a pound per square inch, and

therefore appeared to be quite negligible when dealing with pressures up to 200 lbs. per square inch, and which were themselves very uncertain.

In conclusion, the author wished to express his indebtedness to his friend Mr. Andrew Forbes, and to Mr. J. W. Jukes, Mr. B. Humphrey and Mr. J. L. Wilson, students, who very kindly rendered him most valuable assistance in carrying out the experiments mentioned in the Paper.

Fig. 2. *Method of measuring the Cutting Resistance.*

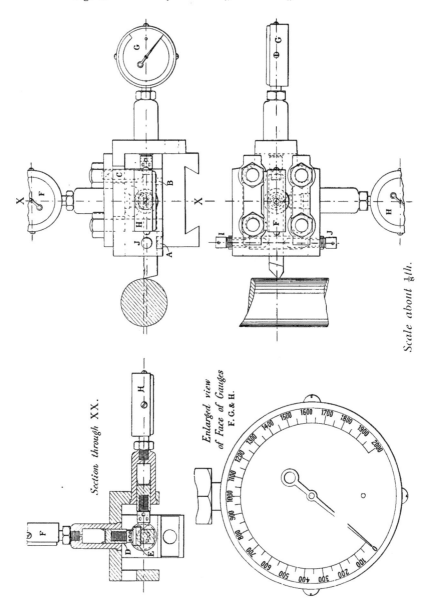

Scale about ⅛th.

Section through XX.

Enlarged view of Face of Gauges F. G. & H.

Fig. 5. *Physics Laboratory, Gauge Room.*

Fig. 4. *Engineering Laboratory, Pressure Tests.*

Mechanical Engineers 1903.

Fig. 9.
Steel Ingot × 102 diam:

F g. 1
d a   s.
ring
Fro...
.94

Fig. 7.   Thermometry, Outer Room.

Fig. 6.   Metallurgy.

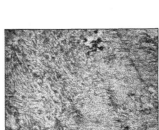

Fig. 12.
Hammered Bar × 820 d.

Fig. 11.
The Ingot × 820 diams.

Fig. 10.
Hammered Bar × 102 diams.

Mechanical Engineers 1903.

NATIONAL PHYSICAL LABORATORY.   *Plate 4.*

Fig. 14.   Hardened Hammered Steel Bar × 1500 diams.
Quenched at 741° C. and showing Ferrite, Sorbite, and Troostite.

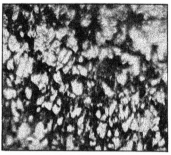

Photomicrographs of Tranverse Sections of Aluminium Rod 1167 H.
(Etched with 1°/₀ Nitric Acid in Alcohol for 3½ minutes.)

Fig 16.   × 60 diams.   *Fig. 17. The same × 500 diams.*

Edge (Ferrite)

Fig. 18.   × 60 diams.   *Fig. 19. The same × 500 diams.*

   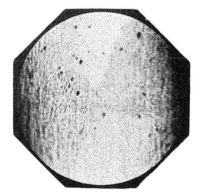

Mechanical Engineers 1903.

Photomicrographs of Transverse Sections of Ingot from which Aluminium Rod 1167 H. was obtained.

Fig. 20. *Portion A. × 60 diams.*   Fig. 21. *The same × 1260 diams.*

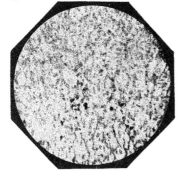
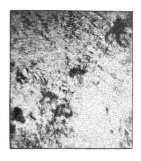

(Etched with 1°/₀ Nitric Acid in Alcohol for 50 seconds.)

Fig. 22. *Portion B. × 60 diams.*   Fig. 23. *The same × 1260 diams.*

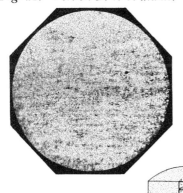

Positions of Portions

Fig. 24. *Portion C. × 60 diams.*   Fig. 25. *The same × 1260 diams.*

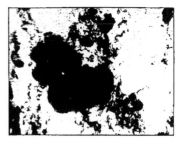

Edge (Ferrite)

Mechanical Engineers 1903.

Fig. 26.   *Plain Polished Surface* × *60 diams.*

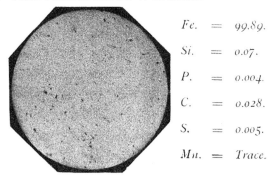

Fe. = 99.89.

Si. = 0.07.

P.   = 0.004.

C.   = 0.028.

S.   = 0.005.

Mn. = Trace.

(Etched with 1°/₀ Nitric Acid in Alcohol for 3 minutes.)

Fig. 27.   *A portion* × *60 diams.*

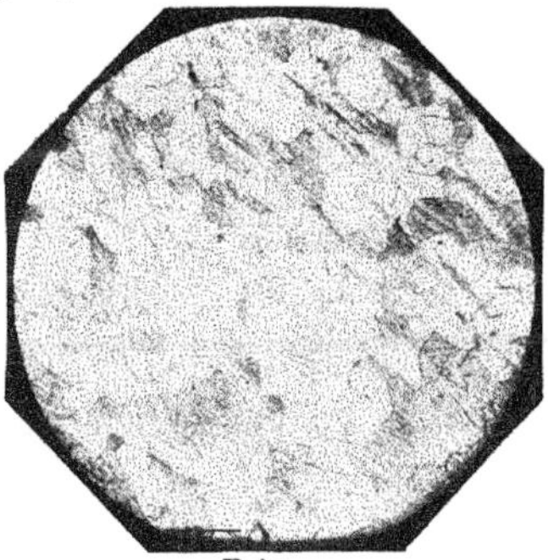

Fig. 28.   *The same* × *500 diams.*

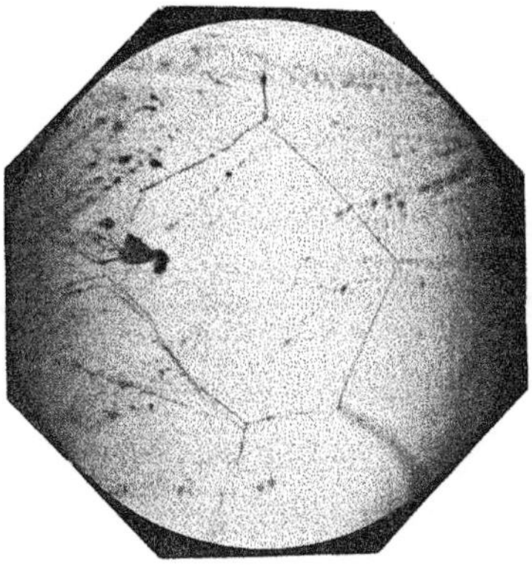

Edge

Fig. 29.   *Another portion* × *60 diams.*

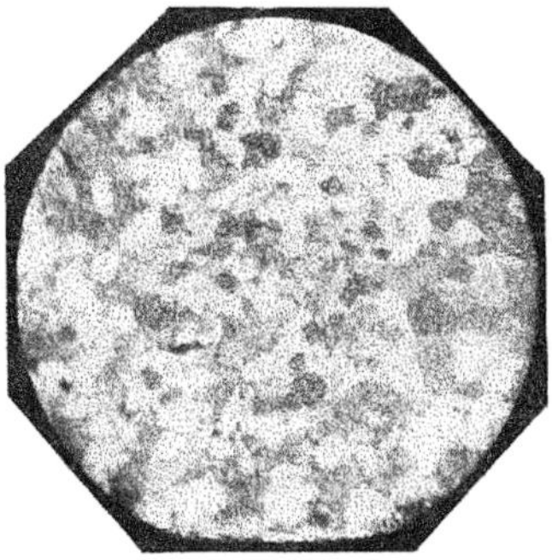

Fig. 30.   *Pearlite and Ferrite* × *500 diams.*

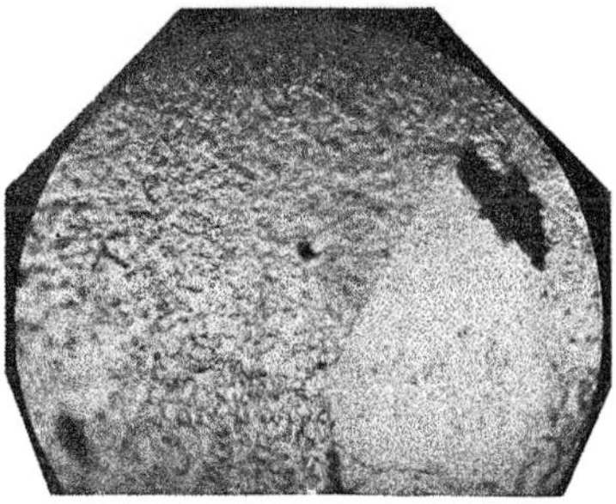

Edge

Mechanical Engineers 1903.

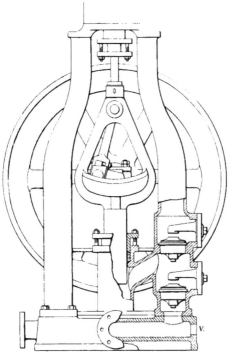

Fig. 2.   *Indicating Gear.*

Scale $^1/_{8th.}$

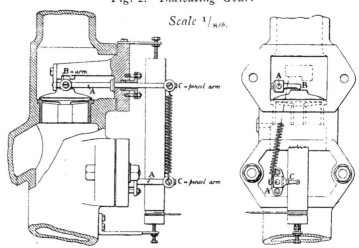

B = arm

C = pencil arm

A

A

C = pencil arm

A

B

C

A

Mechanical Engineers 1903.

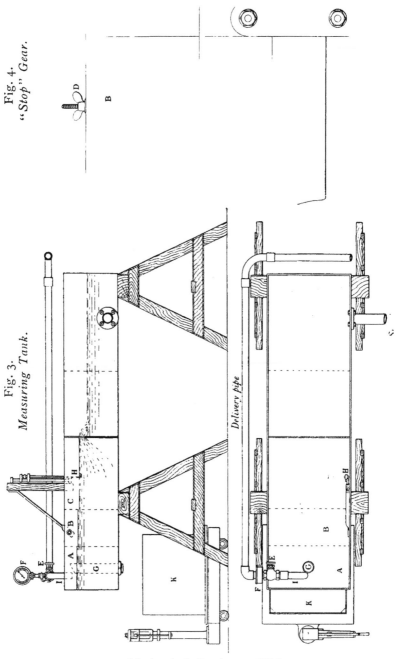

Fig. 4.
"Stop" Gear.

Fig. 3.
Measuring Tank.

Delivery pipe

Mechanical Engineers 1903.

J. H. Wicksteed.

PRESIDENT.

𝕿𝖍𝖊 𝕴𝖓𝖘𝖙𝖎𝖙𝖚𝖙𝖎𝖔𝖓 𝖔𝖋 𝕸𝖊𝖈𝖍𝖆𝖓𝖎𝖈𝖆𝖑 𝕰𝖓𝖌𝖎𝖓𝖊𝖊𝖗𝖘.

PROCEEDINGS.

MARCH 1903.

An ORDINARY GENERAL MEETING was held at the Institution on Friday, 20th March 1903, at Eight o'clock p.m.; J. HARTLEY WICKSTEED, Esq., President, in the chair.

The Minutes of the previous Meeting were read and confirmed.

The PRESIDENT announced that the Ballot Lists for the election of New Members had been opened by a committee of the Council, and that the following one hundred and one candidates were found to be duly elected :—

MEMBERS.

CAMPBELL, ALEXANDER,	Leeds.
CARR, WILSON STORY,	Newcastle-on-Tyne.
CARRUTHERS, ROBERT,	Bombay.
CHILD, HARRY SHAW,	Uitenhage, Cape Colony.
DEAKIN, WALTER,	Birmingham.
DICKEN, ROBERT CHRISTIAN JOHN,	London.
DOUGLAS, GEORGE CAMERON,	Dundee.
HAMILTON, GEORGE CLAUDE,	Leeds.
McLAREN, JOHN,	Leeds.
MELL, LEONARD PARRY,	Burton-on-Trent.
OSWALD, ROBERT,	Tientsin, North China.
OXLADE, HENRY JOHN WILSON,	London.
PEARCE, THOMAS, Engineer R.N.,	China Station.
PENDLEBURY, CHARLES,	Birmingham.

PLATT, FRED,	Wednesbury.
RAWORTH, JOHN SMITH,	London.
ROBINSON-EMBURY, RICHARD PERCY,	
Captain late R.E.,	Coventry.
SEVERN, WILLIAM HENRY,	London.
STURGESS, ARCHIBALD THOMAS,	Madrid.
SWIFT, JOHN CRANE,	London.
WAIGHT, JOHN,	Shanghai.
WATSON, JOHN,	Warrington.

ASSOCIATE MEMBERS.

BAKER, HENRY,	Manchester.
BAKER, HERBERT HENRY,	Newcastle-on-Tyne.
BARRETT, CHARLES EDGAR MATHEWS,	Kallai, Malabar.
BOULTON, JOHN HOWARD,	London.
BOWN, CHARLES HENRY,	London.
BROOKS, JOHN FREDERICK,	Southampton.
BURNAND, AUGUSTUS HENRY,	Poole, Dorset.
CHARNOCK, CLEMENT JOSEPH,	Sereda, Russia.
CHIVERS, FRANK EVANS,	Pernambuco, Brazil.
CLEMENCE, WALTER,	London.
COSTER, WILLIAM WALLACE,	Cape Town.
CROWTHER, HARRY,	Tunbridge Wells.
DAVIDSON, JAMES SAMUEL,	Belfast.
DAVIE, JAMES PATERSON,	Glasgow.
DOUGILL, ALFRED,	Leeds.
FERGUSON, RUSSELL FORRESTER,	Hastings.
GREEN, HARRY HEWLETT RICHARD,	Wolverhampton.
GROVES, THOMAS,	Selangor.
GUEST, GEORGE NEVILL,	Birmingham.
HAM, FREDERIC GEORGE SISON,	London.
HANSOM, OSWIN,	Fleetwood, Lancs.
HARTLEY, HERBERT ANDERSON,	Lincoln.
HERVEY, CHARLES,	London.
HIGGS, ARTHUR FRANKLIN,	Chicago.
HODGETTS, WALTER BEST,	Salford.
JEFF, WILLIAM,	Northfleet, Kent.

KAY, ERNEST EDWARD, ...	London.
LAMB, SIDNEY ERNEST, ...	Portsmouth.
LENOX, GEORGE DUNCAN, . . .	Pontypridd.
MACLEOD-CAREY, ARTHUR WILLIAM JOSEPH,	Middlesbrough.
MARSH, CHARLES FLEMING, . . .	London.
MOORE, NOËL STEPHEN,. . . .	Sheffield.
NORMAN, GEORGE FREDERICK ALEXANDER,	Germiston, S. Africa.
O'GORMAN, MERVYN, ...	London.
PASQUIER, ARTHUR EDMUND DU, . .	Cardiff.
REDMAN, SYDNEY GEORGE, . . .	Wallsend.
REYNOLDS, CHARLES HUBERT, . .	London.
ROBB, JOHN, ...	Edinburgh.
ROBERTSON, JOHN MURRAY, . . .	Lucknow, India.
ROBINSON, THOMAS HENRY, . . .	Widnes.
ROSE, ALEXANDER, ...	Ipoh, Perak.
SEAVILL, LEONARD CROWLEY, . . .	Rosario de Santa Fé.
SELZ, RUDOLPH, ...	London.
SIMCOCK, PHILIP, ...	Hong Kong.
SPROULE, GEORGE HUSTON RUSSELL, .	Johannesburg.
TAYLOR, ALFRED BERTRAM, . . .	London.
THOMSON, ALFRED MORRIS, . . .	Dundee.
WARNER, HAROLD METFORD, . . .	London.
WIDDICOMBE, ROBERT ALEXANDER, . .	Chicago.
WILKINSON, FRANK, ...	Manchester.
WILLIAMS, ERNEST THOMAS, . . .	Manchester.
WRIGHT, GEORGE ALFRED, . . .	Horsham.

GRADUATES.

BAKER, ARTHUR, ...	Dublin.
BRAND, WILLIAM DEANE, . . .	Thornton Heath.
BRITTON, EDWIN JOHN JAMES, . .	Gibraltar.
BROCK, DENIS TABOR, ...	London.
BURDOCK, ARTHUR, ...	London.
DEMPSEY, GERALD, . . . :	Dundee.
DUNELL, BASIL, ...	London.
FORSYTH, REGINALD JAMES, . . .	Salford.
FOSTER, EDGAR, ...	Sheffield.

GRANT, WILLIAM, ...	Aberdeen.
HILDITCH, GEORGE WALLS, . . .	London.
HUGHES, FREDERICK, ...	Cardiff.
JONES, HINTON JAMES, ...	Birkenhead.
LIDBETTER, CHARLES FREDERICK, . .	London.
MALDEN, HERBERT JOHN, . .	London.
MILLETT, CHARLES WALTER, . . .	London.
MINETT, ALBERT ERNEST SCULTHORPE, .	London.
MORRIS, WILLIAM EDWARD, . . .	London.
NEWILL, GEORGE ERNEST, . . .	Birmingham.
SHOOLBRED, ANDREW REID, . . .	London.
SOMMERVILLE, ALFRED, ...	Ayr.
STAVELEY, DOUGLAS DUNBAR, . . .	London.
SUTCLIFFE, INGHAM, ...	Ipswich.
TRESHAM, LANCELOT DAVID, . . .	London.
TRIMMER, GEORGE WILLIAM ARTHUR, .	Reading.
WELFORD, JAMES, ...	Glasgow.
WHITEHEAD, ALAN OCTAVIUS, . .	Edinburgh.

The PRESIDENT announced that the following Transference had been made by the Council since the last Meeting :—

Associate Member to Member.

LASH, HORATIO WILLIAM, ...   Sheffield.

The following Paper was read and discussed :—

"A Premium System applied to Engineering Workshops"; by Mr. JAMES ROWAN, *Member*, of Glasgow.

The Meeting terminated shortly after Ten o'clock. The attendance was 163 Members and 83 Visitors.

A PREMIUM SYSTEM
APPLIED TO ENGINEERING WORKSHOPS.

BY MR. JAMES ROWAN, *Member,* OF GLASGOW.

During the past few months a greater amount of interest has been shown by manufacturing engineers in the introduction of the premium, or bonus, system of paying wages than has ever existed before, and this may be an opportune time to lay before the members of this Institution some facts, gained from an experience of five years' working of the system, which may be of assistance to those who propose to introduce it into their workshops.

The system referred to throughout this Paper was fully described in a Paper entitled " A Premium System of Remunerating Labour," * which was read at the Mechanical Section of the International Engineering Congress held in Glasgow in September 1901, and can be adopted by any manufacturer, by engineers, by ironfounders, by brass-finishers, etc., manufacturing the heaviest articles, as also the lightest in any quantities. Already it is in use in workshops which are totally different from one another in the nature of their manufactures. It is in use in workshops where all

---

* Proceedings 1901, Part 4, page 865.

the machines are new, and in those where the machines are
anything but up-to-date. The author's experience of the premium
system has been gained in the manufacture of marine engines.

Generally speaking, those firms which have adopted the premium
system make visitors welcome, and before anyone introduces the
premium system into his own workshops, a visit should be made to
some of the workshops where it is in operation. This often turns
out to be of great value both to the visited and to the visitor.

Some firms keep their records in books; others on cards. The
author's firm keeps its records in books, but there are businesses where
the cards may be more economical and effective. This may be a very
difficult matter for a firm which has never used the card system to
decide, and the experience of others may be of great advantage to
them. He does not propose dealing with the advantages of the
premium system as compared with that of the time system or
piecework system of paying wages; many articles have been
published on the subject, and a list of the more interesting is given
in the Appendix (page 220). Some of these articles will be found
of great help to anyone who proposes introducing the system.

Before deciding to introduce the premium system, the author
would recommend that the following points be well considered,
namely, the amount of personal work involved, the prospective
expenditure of capital, the perseverance required on the part of the
management to maintain it, and the initial outlay on the Rate-fixing
Department. It is only by the most assiduous attention and the
utmost perseverance that it will be successfully carried through. It
should never be allowed to be a failure. It would be better not
to attempt it, if there is any chance of failure, and it certainly should
not be introduced into any workshop hurriedly or before arrangements
are thoroughly made.

The first thing to be done by a firm introducing a premium
system is to establish a Rate-fixing Department. It is of importance
that a separate department be started, and that the rate-fixing be not
placed as an additional duty upon the foremen. No one need be
alarmed at this as two men will do all the rate-fixing that is
required in an engineering workshop employing about 300 men.

In a very short time the Rate-fixing, Timekeeping, and Wages-costing Departments will merge into one, as the work done is so much allied.   Two men, with the assistance of two timekeepers and two boys, will do all the work that is required, until the wages are abstracted and invoiced against the different jobs.   The man in charge of the Rate-fixing Department should be a trained engineer, with a good deal of workshop experience and some experience in the drawing office.   An intelligent man, with this training, soon gains the confidence of the workmen, and in a short time is able to fix time allowances with wonderful accuracy, and his practical experience is of great assistance.

It is essential to have data as to the times taken by the men at the various jobs when working on time wages, before the premium system can be introduced into any workshop, and the following hints may be useful.   Let it be assumed that the workshop has been working on time, and that the times to do the various operations have been taken in detail by the timekeepers or recorded in some manner by the men themselves, as is done in many workshops, or where the work is large, such as marine engines, electric-light engines, etc., that the times to do all the operations have been recorded against the contract as a whole.   It may be considered by some employers that the times as thus taken by their timekeepers, or recorded by the men, afford them sufficient information to start upon.   Too much value, however, should not be attached to these times, and they should not be used as a basis for fixing time allowances.

Data should be gathered systematically and with great care. The gathering of data at the introduction of the premium system is in itself one of the most interesting, instructive, and valuable parts in the programme.   An intelligent, well-educated engineer should, at the outset, assist the rate-fixer when gathering data, by standing by a few machines from starting-time in the morning till stopping-time at night and noting the elements of each operation.   He should be given "Job Tickets" or "Lines,"* Fig. 1 (page 206), for the jobs the

* " Line " may be a local expression (in Glasgow) for " Job Ticket," but it is expressive, and will be often adopted throughout the Paper.

Fig. 1.

Fitting Shop Lines.

Machine Shop Lines.

Smith Shop Lines.

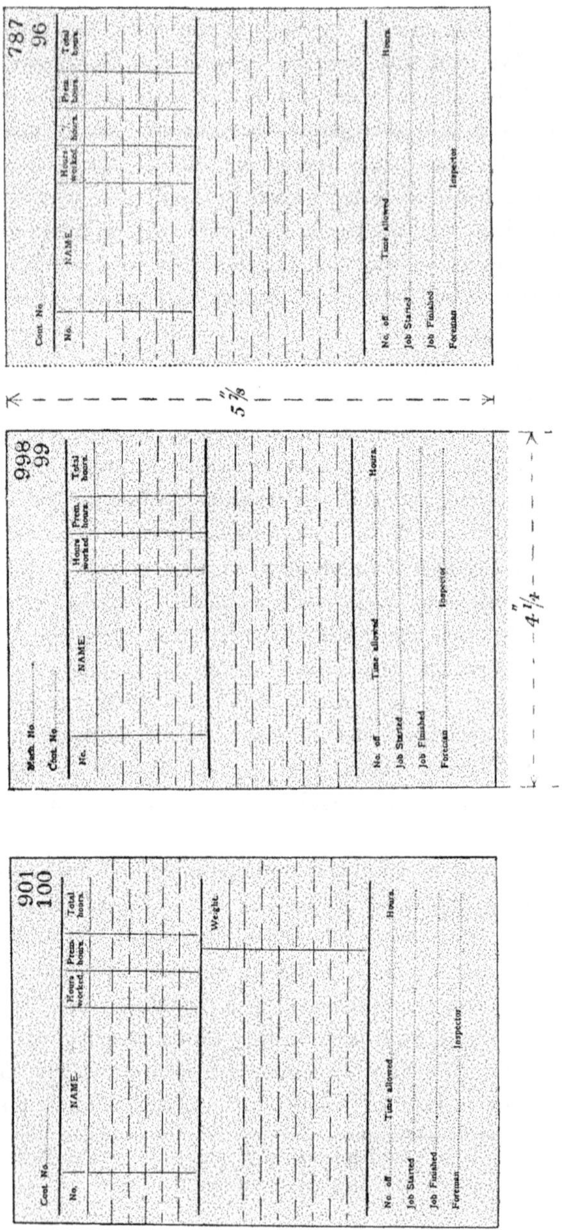

machines are employed upon, on each of which the rate-fixer should have written the time of starting and the details of the operation. When the jobs are finished, the engineer will initial and write on the lines the time finished, and return them to the rate-fixing department for recording.

The elements of an operation include the time for lifting an article on to the machine, for setting, for grinding tools, for machining, the speeds and feeds, the dimensions of the surfaces machined, the time taken to remove the job from the machine to the floor, etc., and also delays of every description and the reasons for them. This information, if properly gathered, is very useful, although later on it becomes valueless. These particulars, as well as the time each man at a machine takes to his job when working on an hourly rate of wages, should be carefully recorded by the rate-fixer in a book prepared for the purpose. A fair average time can thus be arrived at in which a man can do a piece of work, when working on time, and this can be fixed as the time to be allowed for duplicate jobs in the future, and used as a basis for fixing the times to be allowed for jobs of the same nature but larger or smaller.

Another method is to base the allowances on the best recorded or estimated performances, and add such percentage as may be considered proper.

In a workshop with about 150 machines, three men should gather all the data required, in from two to three months. Standard times may also, in the same manner, be arranged for work that is common in all workshops, such as drilling holes, tapping holes and inserting studs, machining flanges, etc. These times can only be accurately arrived at by personal and constant attention to the machines where this class of work is being done, but once the standard times are arranged, a time allowance can very quickly be arrived at for similar jobs, the times being arranged at so many holes per hour and so many minutes per diameter of flange, &c. It may be discovered by these observers that the time which is being taken to do a piece of work is excessive, and this should be carefully investigated before fixing a time allowance, as it may be caused in many ways apart altogether from the workmen. The usual method of the workshop may be a bad one. Feeds or speeds may be bad. It may be due to want of proper tools, or to the machine being out of repair, or too light. Very often it is due to the want of jigs, fixtures, or handy appliances or other causes. Were it done in a new and a good machine with proper appliances, it would be apparent to everyone that it could be done in much less time. The best way of fixing time allowances for jobs, under the above conditions, is to assume the work is being done in the best machine in the workshop, and then to add a certain percentage to the time allowed on account of its being done in an inferior machine.

It has been the author's experience that in most cases the time taken to do any piece of work can be reduced, although it may require capital expenditure together with the assistance of education and intelligence. This is, of course, common knowledge, but it may be questioned if the best use is made of this knowledge, except in workshops where a number of machines are employed in doing identically the same class of work. No doubt, in these workshops great attention is concentrated on the machines, with the result that their output is good, but where the machines are employed doing a variety of work, as in the manufacture of large engines, it is very hard to tell whether the time taken to do the work is good or bad. If, however, the time taken to do the work is down in black and white, and repeated again and again, the management should ultimately have no difficulty in deciding whether the work is done in good time or not. In estimating time allowances, after the system has been working for a period, there is nothing so useful as a comparison of the times allowed for similar jobs, and there is no question as to the value of this comparison, more especially in jobs which require considerable setting and some degree of skill and intelligence on the part of the workmen, and where the time the machine is cutting is only a minor part of the time actually required for the job. Data accumulate very quickly, and soon, no matter what job comes along, the rate-fixer can estimate a suitable time allowance for it which will be fair both to employer and workmen.

In the process of establishing time allowances, errors may be made, and it may be found necessary to either shorten or lengthen the times first tried, and every precaution should be taken to avoid the necessity for such changes by fixing time allowances as carefully as possible. It is also essential for success that anyone adopting the premium system should deal in all fairness when fixing time allowances.

In the Paper read at the Engineering Congress in September 1901, it is stated that the payment of premium does not take effect until 5 per cent. premium has been earned, and thereafter by increments of 5 per cent. This has been found by the author's firm to be a mistake, as the men watched that the time they took to their jobs came out at an even 5 per cent., that is 30, 35, 40 per cent. and so on. They might save 31, 36 or 41 per cent. of the time, but it was noticed that they never saved 34, 39 or 44 per cent. of the time, the reason being that by spending three or four hours more on their jobs they made a bigger premium. When they could not make 5 per cent. more, they spent three or four hours extra on their jobs until the time for the next even 5 per cent. was up.

Again, if the firm proposing to introduce the premium system has been working piecework, the time allowances should be fixed on such a basis that the men when put on the premium system will be able right away to earn a premium. It will often be found that the time under the premium system will be reduced as compared with the time taken when working piecework, and that the man's earnings will be increased. The author can quote only one case—and it is personally known to him—where the premium system has been instituted to take the place of piecework. This is a large firm which has been working on piecework for many years ; they stopped it and went on time for two months, when they introduced the premium system, and the work is now nearly all done in less time. The men are making better wages, and not a man has left the employment owing to the change from piecework to the premium system.

Every firm introducing the premium system should issue a pamphlet to its workmen. The statement, in the form of notes, which follows, will suit the requirements of most engineering establishments, but as one firm differs from another in manufacture, in district, and in many circumstances, it should be carefully studied by those issuing it. Naturally, with such a radical change in the payment of wages as is involved by the introduction of the premium system, the workmen are disposed to look upon it with suspicion and resent it. This is a common experience where it has been introduced, and when in America last summer the author found, even there, some firms which were anxious to introduce a premium system but could not do so, owing to the attitude taken up by the men. Explanations such as are given in the pamphlet will no doubt remove from the minds of reasonable workmen their principal objections to the premium system. There will always, however, remain a few who will argue that the introduction of the premium system will be the means of keeping a number of men idle, but it has been experienced from time immemorial, that the cheaper an article can be produced the greater will be the demand for it, and the number of men employed producing it will be increased. The following notes are based on the pamphlet which Messrs. Barr and Stroud, of Glasgow, (who were, as far as the author knew, the first to do so) issued to their workmen when they introduced the premium system into their works, and they have kindly given him permission to make use of it.

NOTES.

For the information of the Employees of Messrs. . . . .
regarding the Premium System of Wage Earning.

The Premium System may be briefly described thus :—

Each man is paid a regular hourly rate of wages. When a job is given out, a certain time is allowed for it. If the job is completed in less time than the time allowed, the workman becomes entitled to a premium, varying in amount with the time saved. If the job takes longer than the time allowed, the workman gets paid his regular hourly rate of wages, so that, no matter how short a time may be allowed for a job, the hourly rate of wages, at least, will be paid while engaged on that job, thereby preventing a premium on one job being lost through failure to make a premium on another. It will be evident from the above that, while the workman may increase his wages, he cannot lose money by the introduction of the system.

The system possesses two main advantages :—

(a) It enables a workman to increase his wages by his own individual effort, and the increase is immediately added to his wages, the premium being paid to the workman each pay-day along with his wages.

(b) The increased wages to the workman means also a reduced cost of production to the employer.

The details of the system as we propose to introduce it are as follows:—The amount of premium will bear the same relation to the ordinary wages due for the time taken to complete an operation, as the time saved bears to the time allowed. For instance, suppose a man is allowed 16 hours to do a job and does it in 12 hours, he saves $\frac{4}{16}$ of the time allowed, that is 25 per cent., and accordingly his wages will be increased by 25 per cent. for the time worked; that is, if the rate of pay is 8d. per hour the man will receive 8s. as time wages for the 12 hours worked, and the premium earned will be 2s., equal to 25 per cent. of the 8s. His time wages and premium will thus amount to 10s. for the 12 hours worked.

A convenient way for the workman to calculate his premium is to multiply the time taken by the time saved and divide the product by the time allowed. This will give him his premium in hours, which, multiplied by his ordinary time rate of wages per hour, will give his premium for the job.

Taking the ease already given as an example :—

$$\frac{\text{Time worked} \times \text{time saved}}{\text{Time allowed}} = \text{Premium in hours.}$$

$$\frac{12 \times 4}{16} = \frac{48}{16} = 3 \text{ hours premium.}$$

Multiplying this by his ordinary time rate he gets $3 \times 8d. = 24d. = 2s.$ as his premium for the 12 hours worked. Premium hours will be paid to the nearest quarter of an hour.

The following conditions will be observed :—

(1.) The time allowed for any job will be fixed by the management, and will be, as near as can be ascertained, the time which should be taken to the job when working on time.

(2.) The time allowed will include all the time necessary to procure tools, set up machine, and obtain material for doing the job.

(3.) For calculating the premium, the time taken on a job will include all working hours between the starting time of the job, and the starting time of the next job.

(4.) A time allowance, after it has been established, will only be changed if the method or means of manufacture are changed.

(5.) The hourly rate of wages will in all cases be paid for the hours worked. If a man takes longer to do a job than the time allowed, this will in no way affect the premium which he may have made or may make on any other job.

(6.) Overtime, nightshift, and other allowances will be paid to the men on the same conditions as already prevail.

(7.) In the case of overtime and nightshift, the premium will be calculated on the actual time worked, without taking the extra time due to overtime, etc., into account.

(8.) If an article turns out defective while being machined and is condemned, due to a flaw in the material, the workman will receive no premium on that article (of course, he gets his time wages); but if he has several articles on the one "line" and one of them is condemned, due to a flaw in the material, he will still get the premium, if earned, on the rest of the articles.

(9.) If a man's workmanship when finished does not pass inspection, he will receive no premium for that article unless he can make the work good within the time allowed, in which case he will still receive any premium earned.

(10.) In cases of dispute the matter will be referred to the management, whose decision shall be final.

(11.) Each workman on starting a job will receive a "Job Ticket," or "Line," on which he will find a description of his job, the time when started and the time allowed. When the job is finished he will return his "line" to his foreman, who, if satisfied with the work, will initial and write on it the time when finished, which will be the starting time of the man's next job.

(12.) In the case of a job requiring the services of a squad of men, a time allowance will be fixed for the complete job. If the total time taken by the squad is less than the time allowed, a premium will be paid to each man in the squad. This premium will have the same relation to his time wages for the job, as the time saved by the squad will have to the time allowed.

(13.) Fitting-shop apprentices in their first year will not receive "lines." Those in their second and third year will be considered junior apprentices, and 50 per cent. of the time they spend on a job will be calculated against it for premium. The percentage thus found will be paid on the whole time which they spend on the job.

For example, suppose a junior apprentice is allowed 16 hours to do a job and does it in 16 hours, 50 per cent. of this time which has been taken to the job, that is, 8 hours, is taken as a basis for calculating the premium, therefore he has made 50 per cent. on the job. If his rate is $1\frac{1}{2}d$. per hour he will receive 3s. instead of 2s. for the 16 hours worked.

Fitting-shop apprentices in their fourth and fifth year will be considered senior apprentices, and 75 per cent. of the time spent by them on a job will be calculated against it for premium. The percentage thus found will be paid on the whole time which they spend on the job. For example, suppose a senior apprentice is allowed 16 hours to do the job and does it in 16 hours, 75 per cent. of this time which has been taken to do the job, that is 12 hours, is taken as a basis for calculating the premium, therefore he will make 25 per cent. on the job. If his rate is 2d. per hour he will receive 3s. 4d. instead of 2s. 8d. for the 16 hours worked.

(14.) Apprentices at machines will be allowed 25 per cent. more time on a job than a journeyman.

The introduction of the premium system will, we believe, lead to the workmen suggesting improvements, devising better methods of doing many jobs and pointing out defects in machinery and tools, as they will at all times participate in the savings due to their suggestions.

The publication of such a pamphlet as the foregoing, in any workshop, will no doubt establish a bargain between the employer and the employed. It will enable the workmen to calculate their own premium and indicate how the system will benefit both the employer and the workman. A bargain of this description must be honourably adhered to. So soon as the requisite data have been gathered for a machine, the man working it should be put upon the premium system. A "line," as described Fig. 1 (page 206), is given to each man working a machine.

The "lines" are prepared in the Rate-fixing Department and issued to the foremen, who give them to the men. Extracts from them are entered on a "Daily Record Card," Table 1 (page 215).

There are five columns on this card, the first column is for the machine numbers, the second for the time allowed, the third for the number of articles on each "line," the fourth for the times the workmen have been working on the job up till 10.30 a.m. on the date on the card (any other convenient hour may be used), and the fifth for the record times—that is, the shortest time in which each job has been done previously. This card is made up daily, and although described in the Paper *—"A Premium System of Remunerating Labour '—it is repeated here, so much value does the author attach to it.

When a man is on the same job as on the previous day, the particulars are repeated; there is added, however, to the time taken the time worked by the man in the interval. This card keeps the Rate-fixing Department in touch with the lines that have been issued, and is also invaluable for letting employers and managers know how the work is progressing. They can learn how a job is progressing by a glance at the card, which can be folded in the centre and carried in the pocket.

When a man has finished his job, he returns the "line" to his foreman, who writes on it the hour when the job was finished, and initials it, if satisfied with the work. The foreman then hands it to the Rate-fixing Department, and at the same time intimates what the man's next job is to be. The working hours, between the starting time of the job and the finishing time of the job, are put on the

---

* Proceedings 1901, page 865.

TABLE 1.

Daily Record Card.

(Compare Proceedings 1901, page 874.)

5	—	—	—	—	70	21	42	3¾	10½					
6	225	75	22¼	111	164	57½	5	1	22½	Date 16. 12. 02.				
8	24¾	33	19¾	13	169	69	12	21½	44½					
19	35	28	3¾	17½	193	184	4	38¼	90	163	—	—	—	—
20	82	8	65½	55	204	36	36	18	18	208	14	2	9	10
23	30	1	2	20	236	13½	6	5¼	9	209	90	6	9¼	29¼
24	216	54	110	108										
25	48	8	33	24	56	15	51	1	×	160	77	15	20¼	50
30	12	6	3¾	9	63	35	5	7½	26	171	19½	9	12	10
31	13	18	9	×	69	36	15	17¾	18	191	22	1	12½	17
36	14	2	11½	9	82	50	1	19¾	38	222	2½	3	1¼	×
37	29½	2	20	21	142	7	20	3¾	×	224	11	1	7½	7¾
44	92	38	12½	55	153	11	1	5¼	×	231	—	—	—	—
61	16	1	8¼	8	165–6	88	4	17¼	29½					
71	35	70	4¼	17½	183	—	—	—	—	10	14¼	8	6¾	10
75	55	11	19½	27½	184	15	1	12¼	12½	66	39	3	9½	19½
167	—	—	—	—	206	29	1	12½	14½	214	30	2	1¾	18
168	30	2	10	20	225	28	1	20	21	215	26	8	13½	19
170	55½	3	25¼	24										
192	40	1	6¼	18	9	12	20	2	×	161	13	1	1	6½
210	78½	11	41¾	39¼	13	30	6	2¾	21	182	42	295	37¾	31
					18	12	1	6½	5½	194	60	3	43½	30
21	—	—	—	—	32	134	2	33	90					
22	35	1	19	12	33	20	28	10¼	×	64	37	728	18¼	27
26	11	9	7¼	8	62	44	13	31¾	×	201	31	302	20½	15½
27	6¾	9	6¼	4½	65	16	17	3½	×	239	57	114	7¼	29

TABLE 2.

Class :—Turners. Pay ending 4th December 1902.

No.	Name.		Time. F.	S.	M.	Tu.	W.	Th.	F.	S.	M.	Tu.	W.	Th.	Total Time.	Time Allowance.	Rate.	Amount.	Inf.	Total Wages.
																	d.	£ s. d.	d.	£ s. d.
61	John Mathieson	Lost Time															9			
		Overtime																		
		Daily Total	9¾	5¼	9¾	9¾	9¾	9¾	9¾	5¼	9¾	9¾	9¾	9¾	108			4 1 0	1	
		Premiums	9¼			9¼	1½					24	3½		47½			1 15 8		
		Money Allce.																		5 16 7
62	Wm. McGowan	Lost Time			3												9·25			
		Overtime																		
		Daily Total	9¾	5¼	6¾	9¾	9¾	9¾	9¾	5¼	9¾	9¾	9¾	9¾	105			4 0 11	1	
		Premiums	6			5						6	7¼		24¼			18 8		
		Money Allce.																		4 19 6
63	Wm. Riddell	Lost Time	¼		3					½							8·75			
		Overtime																		
		Daily Total	9½	5¼	6¾	9¾	9¾	9¾	9¾	4¾	9¾	9¾	9¾	9¾	104¼			3 16 0	1	
		Premiums			27	¼					21½				48¾			1 15 6		
		Money Allce.																		5 11 5
64	H. Abercrombie	Lost Time		¼	¼			¾	¼								8			
		Overtime																		
		Daily Total	9¾	5	9½	9¾	9¾	9	9½	5¼	9¾	9¾	9¾	9¾	106¼			3 11 0	1	
		Premiums							21½	2		3¼			26¾			17 10		
		Money Allce.																		4 8 9

Each page provides space for 10 men. Full working time per fortnight, 108 hours.

"line" (by the Rate-fixing Department) from the Workmen's Time and Wages Book, Table 2 (page 216), and the premium hours calculated, say, off a slide-rule, and put on the "line." The premium hours are then entered in the Time and Wages Book, under the hours worked on the day on which the job is finished. At the end of the pay * the hours worked are added together, and extended, giving the man's time wages for that pay; the premium hours are also added together, and extended, giving the man's premium for that pay; these two items—time wages and premium—are then added together, and this, with the addition of allowances for overtime, nightshift, etc., gives the man's total wages.

The men are paid their time wages up till Thursday night at stopping time, if the pay day is Saturday, and their premiums up till Wednesday at stopping time.

Most people would think that the workman who has been constantly working a machine should know the times the various jobs should take him, but, in a time shop, this, as a general rule, is not the case. He under-estimates his own ability, thinks he could not do the work much faster, and has simply no idea of the short time in which he can do a job, until he has the inducement of a substantial increase of pay to look forward to. When a workman first starts on the premium system, he is often dubious about the time allowed and looks on it with suspicion, but when he does the job and finds the time allowed is reasonable, if he has not made a premium the first time, he will likely do so the next time he gets a similar job. Although it is one of the features of the system, that a man need not earn a premium unless he chooses, it has been the author's experience that, if one gives the average workman the chance of honestly adding to his earnings, he will do so. Having once started to make an increase of wages, the workman naturally begins to educate himself as to the best methods of producing his work quicker. He devises new methods, and when he sees a man able to produce the work faster than himself, he adopts the methods of that

---

* "Pay" here means the fortnight or week over which a payment of wages is made.

other man. The employer and manager also take an interest in the
times the jobs take, and by means of the daily record card already
described soon find out the machines that take a long time to their
work. If it is the fault of the man, it can generally be improved by
telling him how the job can be done faster, educating him in fact.
If it is the fault of the machine, it may not be so easy to remedy; if
only requiring to be repaired, or to have some small improvement,
that is easily got over, but if the machine is not powerful enough—a
very common defect—or of bad design, often there is only one
remedy, and that is a new, accurate, and more powerful machine.
This is a drastic cure which would not suit every firm, but when
there is plenty of work for the new machine it should soon repay
itself. The chief point is, that the firm should be satisfied that they
are getting a reasonable output from their machines, considering
their condition.

As the workman gains confidence that his time allowance will
not be reduced, no matter how short a time he takes to a job, he
gradually reduces his time.

The Premium System was started in the author's works in
February 1898, and to illustrate the gradual improvement that has
resulted since its introduction, it may be mentioned that the times
taken by all the machinemen have, on an average, been reduced
during the four succeeding years by 20, 23, 31 and 37 per cent.
respectively. The earnings of the men have consequently increased
by these percentages.

   The author's firm has given every assistance in accomplishing
this reduction of time by general improvements in many directions.
Some would say that the men had formerly been loafing, or that the
works were indifferently managed ; there is, however, every reason to
believe that the men were honest enough and that the management
was fairly good.

A machine foreman should only have as many machines under
his charge as he can attend to, including careful inspection of the
work before it leaves the machine shop. Erectors, when working on
time, will take a job from the machines, and if not correct will make
it right without complaint, but if these same erectors are to actually

lose money by the carelessness of a machine man, as they would on premium, they rebel, with the result that great care is taken by the men and foreman that work when it leaves the machine is correct.

In his works the author has conferred at various times with many of the men, and they frankly admit that they are thoroughly satisfied with the premium system, and would not care to go back to the old time-system. As a matter of fact, there are now fewer changes amongst the workmen than before the premium system was started.

One advantage of the introduction of the premium system is, that it enables the management very readily to look personally into the progress of the work in the workshop, which is far more effectual than studying reports in the office on the progress of work.

Up to this point the system has been treated as applied to the machine shop, but it is equally applicable to the erecting and other departments. In the machine shop every man is given a "line" for his own job, that is to say, each man works on his own account, but there are a few exceptions to this; and it will always be found better, if possible, to arrange for each man to work on his own account. Departure from this rule should only be made when absolutely necessary.

The Paper is accompanied by an Appendix and one Figure in the letterpress.

APPENDIX.

(*a.*) "The Premium Plan of Paying for Labour," by F. A. Halsey, Vol. XII. of the Transactions of the American Society of Mechanical Engineers.

(*b.*) "A Partial Solution of the Labour Question," by Fred. W. Taylor, "Cassier's Magazine," February 1898.

(*c.*) "A Premium System of Remunerating Labour," by James Rowan, Proceedings, The Institution of Mechanical Engineers, September 1901, page 865.

(*d.*) * Articles in "The Engineer" of January 24th 1902.

,,	,,	,,	,,	,,	,,	31st	,,
,,	,,	,,	,,	,,	February	14th	,,
,,	,,	,,	,,	,,	March	7th	,,
,,	,,	,,	,,	,,	,,	21st	,,
,,	,,	,,	,,	,,	April	4th	,,
,,	,,	,,	,,	,,	,,	18th	,,
,,	,,	,,	,,	,,	May	9th	,,

(*e.*) Articles in "American Machinist" volumes for 1899, 1900, 1901, 1902.

---

* These articles have been reprinted in pamphlet form, under the title of "The Premium System of paying Wages."

Discussion.

The PRESIDENT said it was his privilege to ask the members to pass by acclamation a vote of thanks to Mr. Rowan for his Paper. There was a great deal more in the Paper than could possibly be assimilated, if they had not had the subject in mind before. But, although the author had given a Paper on the subject before the Engineering Congress * in Glasgow under the auspices of the Institution, yet the present Paper was quite fresh; it was nothing like a copy of the old Paper, although the preceding one was a help in understanding the beauty of the system. He thought the essential element, which underlay the newest and most complete system described, was that attention was compelled to every individual piece. The time that was wasted, or the time that was necessary upon each individual piece, was recorded, and attention was compelled to it; and either the design of the piece or the method of producing it might be changed, if it was costing more than it was worth. Again, the individual man who made the single piece had his distinct reward for what he did individually. Both those elements were very different from the old piece-work system, in which account was taken of the expense of completing the whole machine, and no account was taken of the individual parts nor of the share each individual man had taken. The recent system not only enabled one to follow up the improvement in detail, which was the only way in which improvement could be followed up, but it also secured the man his immediate reward when his wages were paid; whereas, under the other system it might be put off for one, two or four months, the man thereby losing interest in his work. It appeared to him that what had to be paid for those advantages was a very formidable organisation. The system, it seemed to him, could only be applied in works that were predominated by an extremely energetic intellect; the whole of the system from first to last was full of brain work and was opposed to all indolence, ease, or laisser faire.

---

* Proceedings, 1901, part 4, page 865.

(The President.)

In the course of the Paper, mention was made of the system carried out by Messrs. Barr and Stroud, which gave him a connecting link in selecting the speakers to open the discussion. Dr. Barr, who had works of his own, and was an employer of about 100 skilled mechanics, gave him some very interesting information, which he would like to hear him repeat, about the reforms which were made on the introduction of the card. This simply recorded the exact time spent on each individual job by each individual man, even before he offered any premium as an incentive to the man for extra effort. Another connecting link with the speakers in the discussion he found in the diagrams on the wall of which he was sure the members would be glad to have an explanation. The Institution was indebted for the diagrams to Mr. Pendred, Jun., whom he would call upon to open the discussion by giving an explanation of the diagrams.

He proposed that a very hearty vote of thanks be given to Mr. Rowan for his excellent Paper.

Carried by acclamation.

Mr. L. PENDRED remarked that when he sent in his name to speak on the Paper he had no notion there would be so many gentlemen present who were actually employing the premium system, otherwise he would have hesitated before putting the diagrams, Figs. 2 and 3 (page 223), on the wall. All that he (the speaker) had done was to combine two diagrams made by Dr. Barr. The lines of most interest in the diagrams were those which showed the wage cost per piece and the total cost per piece. With regard to the former, it would be noticed that whilst the time line fell quite regularly, the Rowan line declined in a curve; thus there would be a space between the two which exactly represented the amount spent in premium. Another pair of lines in Fig. 2 showed the total cost per piece made. This line was obtained by supposing that the establishment charges, the "on-cost," as they were called in the North, were equal to the wages without premium. It would be noticed that whilst this line was still above the time line, the distance between the two at the worst place was only half what it

Fig. 2 —*Rowan's System.*

$$\text{Premium rate} = \frac{\text{Time saved}}{\text{Time allowed}} \times \text{Wage rate.}$$

On cost at 100 per cent. of wage without premium.

Ratio of Time Saved to Time allowed

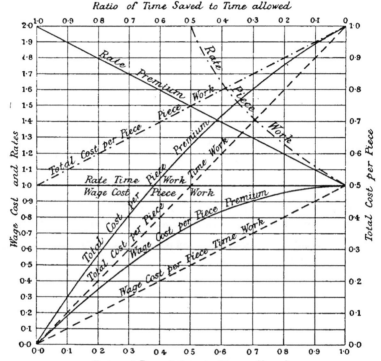

Ratio of Time taken to Time allowed

Fig. 3 —*Cost of Production per Piece by Premium Systems.*

Cost for 100 hours

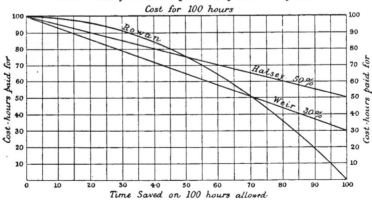

Time Saved on 100 hours allowed

(Mr. L. Pendred.)

was between the former pair of lines; and if the establishment charges were less than the wages, the two lines would still further approximate. Of course these lines showed that, as far as cost per piece was concerned, if only a man could, or would, do as much without the inducement of a premium as he did with it, the time system was the best. It was the fact that he would not do it that made the premium necessary. But although the cost per piece under the premium plan would be higher than under the time system, if the conditions were the same, the premium plan had the advantage, because the conditions never were the same. A man working under time never did more than he need, whilst under the premium plan he would do all he could. The consequence was that instead of having only one piece made in the time allowed for one, he might have many made. The middle point was never reached with the time system, but frequently with the premium plan. When the premium system was compared with piece-work, its advantages were much more clearly marked. In that case the top dotted line showed the total cost, and it would be observed that at every point it was higher than by the premium plan, and that, as more pieces were made in the time allowed for one, the comparison became worse and worse.

These points referred to the system as regarded from the employers' point of view; the advantage to the workmen was shown by the line marked "rate." It would be noted that this increased very rapidly; not so rapidly, indeed, as the rate under piecework would increase, but still in the initial stages, which were the important ones, very nearly as fast; and the fact that the rise became slower and slower than it did under piecework was an excellent thing, because it reduced the temptation to cut the rates; a circumstance which, as the author said, every precaution should be taken to avoid.

The second diagram, Fig. 3, was designed to bring out the difference between the three principal systems of payment under the premium plan. The three lines showed the Rowan system, the Weir system, and the Halsey system, the original system as it was used in America. Under the Halsey system the man received

30 per cent. of the time that he saved. Under the Weir system, if the man saved an hour he received in payment half an hour extra wages. Under the Rowan system the man received his payment in direct proportion to the amount of time he saved, that is, if he saved 50 per cent. of the time, his wages were increased by 50 per cent. Supposing a man saved 50 per cent. on a job occupying fifty hours instead of being paid for only fifty hours he was paid for fifty plus twenty-five hours, thus receiving seventy-five hours' wages. Under the Weir system the amount saved was shown on the diagram. Under the Halsey system it would be considerably lower; he would only receive 30 per cent. added. He asked the members to particularly observe that, for the first few hours saved, the Rowan system stood higher than either of the other systems, that is, a man received in payment more premium under the Rowan system for the first few hours which he saved, than he did under the Halsey or the Weir system. That, he thought, was very important indeed, because under the Halsey or the Weir system if the man succeeded in saving a very large number of hours he received a huge wage, in fact his wages could go up to such a rate that the master could not stand it with equanimity, the tendency to cut became so great that the master could not fight against it. The rate was cut, and when that happened all the troubles of the piece-work system came in. One of the great advantages of the Rowan system was that there was absolutely no tendency to cut the rate under any circumstances, because no matter how much time the man saved his rate could never be doubled. Supposing a man was allowed 100 hours to do the work, and he did it in one hour he saved only 99 per cent. of his wages; obviously he could never, under any circumstances, do the work in no time and so double his wages. It was very important indeed that that should be known.

The PRESIDENT asked whether, under the other systems, the wages could be doubled.

Mr. PENDRED replied in the affirmative; the wages under the Halsey or the Weir system could go up to an enormous amount.

(Mr. L. Pendred.)

He hoped the author would forgive him if he differed from him on certain points ; of course, as the author was an employer of labour, he naturally spoke with a great deal more weight than he (Mr. Pendred) did.   But he could not help thinking that Mr. Rowan made the premium system seem a little too difficult of inception.   The author recommended many excellent things, for example, the daily record cards which he agreed were very desirable, but they were not essential to the premium system, nor were they a part of it.   All card systems of keeping costs, rates, and so on were excellent in their way and entirely to be recommended, but they were not necessary to the premium system.   The premium system could be and had been, he knew for certain, introduced in certain works in a very simple way indeed.   All that was necessary was a card of a very simple description on which were written the time allowed for the work, the time taken, and the time saved.   By a simple rule, given in the Paper, a clerk—a boy in one of the works that he knew of— using a very simple form of slide-rule got out the premium in hours. That was sent over to the wages' clerk, who added it on to the actual number of hours the man had worked, thus ascertaining the total number of hours the man was to be paid, and then paying him in the ordinary way.   An employer did not want anything simpler than that ; it was so simple that even the management of the smallest and humblest works might start it without any difficulty.

Dr. ARCHIBALD BARR said he spoke with some diffidence on the Paper, because he felt that anything he said might be largely discounted from the fact that he was an employer of labour only upon a small scale, and also that the class of work with which he was specially concerned was different from that in which most engineers were engaged.   He would not have attempted to say anything on the Paper, were he not thoroughly convinced that the importance and practicability of the subject did not at all depend upon the nature of the work a shop was doing.   The author was right in saying that the premium system was applicable to the largest works as well as to the smallest ; it was also applicable to repetition work, and to work done to special designs.   He was afraid he could say little, except to emphasize the remarks the author had made, but he wished to say that he believed the importance of the Paper would not be fully realized by those who had not given the matter very careful study. The value of the premium system was not merely that it led to a saving being effected in the cost of work, nor that it increased the wages earned by the men ; its introduction into any workshop brought the management into touch at once with the methods employed in the workshop, and their results in detail.

If the premium system did nothing else but call the attention of the management, as it necessarily did, to the cost of every piece, and every operation upon that piece, he thought it was well worth introducing into any shop, large or small. In very badly managed engineering works, the costs of work were never known, and this was not only disastrous to the firms themselves, but to trade in general, because the result was a very unfair kind of competition, where the purchaser of the goods paid one part, and the creditors of the company paid another, and often a very large part of the cost of production. In many well managed works all that was known was the cost of a complete machine, and many engineers were quite content to know whether or not a product as a whole had cost more or less than they had estimated. But under the premium system one got to know the cost of every individual piece, and every individual operation upon that piece. The result was at once evident. He thought everyone who had adopted the premium system, no matter how great his previous experience of work had been, had frequently been astonished at the smallness of the necessary cost of some pieces and some operations, and the relatively large cost of others. An examination of the premium results would lead the engineer to see that some particular thing was costing much more than he had calculated, and that the design or method of production could be changed with great advantage. The introduction of the premium system, therefore, gave the management a great deal of information that was of value, independently altogether of the direct reduction of cost as the author had remarked upon (page 219). But perhaps the feature that would appeal most immediately to engineers was

(Dr. Archibald Barr.)

that it brought in the strong motive of self interest on the part of the men, to improve the productivity of machine-tool and handwork in the workshop. It thus led to the adoption of the best and most economical methods.

Another point, upon which the author had very properly laid stress, was the effect the system had in leading to the selection of the best machines for a particular class of work. The employer very soon got to know whether or not it would pay him better to retain a certain machine or to scrap it. Another point was that it brought the best men to the front. It was found in his own shop that there were certain employees who rarely made a premium, while there were others who seldom or never failed to make a premium. The management thus learned the values of different men in a way which they could not do, if the men were simply working upon time wages. Lastly, it greatly increased the general energy of a workshop, a fact which was known by everyone who had introduced the system. He believed the improvement in the results, as he had said before, could not be estimated by anyone who had not had experience of the working of the system.

Although the author had referred to the Paper he read before the Engineering Congress, he had not taken any credit for what he (the speaker) considered a very great step in the premium system, for which employers owed him their thanks. He knew that in speaking on that point he was trespassing upon ground which was more or less controversial ; but, from his own experience and study of the subject, he had not the slightest hesitation in saying that the author's system was much the best yet proposed. The Rowan system differed from Halsey's or Weir's system in two important respects, to which Mr. Pendred had already called attention. First of all, if by a miscalculation of the time required for a job, an excessive amount of time had been allowed, the remuneration of the workmen on the Halsey system went up to an unwarrantable amount. He thought he could challenge any engineer, whatever his experience of working on time-work or on piece-work had been, that he would sometimes greatly over-estimate the time a job would take, if done on the premium system. The first job that his firm gave

out on premium was one on which good experience had been
obtained, and yet the man did the work in from a quarter to a third
of the time that the management had estimated as a reasonable
time for him to do it in. The man had been doing the same work
before, but when he was put to the premium system he naturally tried
to reduce the time as much as possible. He arranged his work in a
better way and carried it out more systematically with the result
stated. If an excessive amount of time was allowed, Messrs. Weir's
system gave a very large premium to the workman, that is, the man
might get two or three times his weekly wage; and, even if no
mistake had been made, he did not think any man was entitled,
by extra exertion, to two or three times the wage for which he
undertook to give honest work upon the time system. On the
other hand Mr. Rowan's system gave a larger remuneration to
the workmen than Messrs. Weir's did, if the time had been well
estimated. Some people thought the Rowan system gave the man
too much under those circumstances, but a sufficient inducement
must be given to the men to reduce the time spent on a job, and
he asked employers to remember that though the wage-cost of the
work was the same as it had been before (as was the case on the
piece-work system), they still gained in on-costs or general
establishment charges upon the time saved. From the diagram,
Fig. 2 (page 223), it would be observed that the total cost per piece
—the wage-cost with on-costs added—varied to start with on
Mr. Rowan's system at the same rate, for very small savings of time,
as it did on the piece-work system, and nobody thought that a master
did not get a good enough remuneration under the piece-work
method, if the price was well calculated. The diagram was a
combination of some of the lines in the diagrams which he (the
speaker) had published in " The Engineer " in illustration of notes
which he had written on the subject, but it would take too long to go
over them. He would call the attention of members of the Institution
to the excellent pamphlet which the editor of " The Engineer " had
published, which included, among other things, the diagrams and
explanatory notes with regard to them.

(Dr. Archibald Barr.)

The President had said that he thought the premium system involved a very formidable organization. If he (Dr. Barr) had one thing to complain of in the Paper, it was that it rather gave that impression. In his opinion the author had put down the amount of extra work at a very much larger figure than was at all necessary. That introduced another large question, that of the keeping of time upon cards or in books. He was personally very strongly in favour of the Card system, and though he had with him data on the question, it would take too much time to enter into it. He, however, suggested to the Council that a Paper should be obtained on the Card system, which might be taken in connection with a discussion upon the whole subject. He believed his partner, Mr. Harold D. Jackson, an Associate Member of the Institution, would be prepared to write such a Paper. That gentleman had given the subject perhaps as much study as anybody else, and had devised a system which was used in their works, and which had received such an amount of recognition that it was copied almost exactly in other works. In their own works, which employed about eighty men, all the extra work necessary for getting out the premiums was done in about one hour per day by one boy. There was a very small amount of work involved.

A great many people said that the workmen would not understand the system. There was a sufficient answer to that in the remarks which the author made. That they at once understood the matter was evident from their treatment of the 5 per cent. method. The fact that workmen in a shop understood that at once, and knew their own interests, showed sufficiently, he thought, that they were alive to the importance of the question. But one point which he believed the workmen did not fully realise was this: If a man was given a single job to do on premium, under the Rowan system the man made most premium if he did it in half the time; he would make less premium if he did it in less than half the time. That was only the case if the man was on one isolated job, and the rest of his time was spent on time work; it was not so if he was on premium work all the time, but he thought that the men had the impression that it was. He had had experience of it in his own works, and the members had some evidence of it in Table 1 (page 215). It would be found that in the "record times" there quoted, there were only five cases in which the time was less than half, whereas there were nineteen out of the fifty-three which were exactly or almost exactly half time. He believed the men did not understand this aspect of the system; he thought it was not a coincidence but a misunderstanding on the part of the men which could, however, be quite easily explained to them.

The author had kindly referred to the pamphlet which his (Dr. Barr's) firm issued to their workmen. The pamphlet was reprinted in the booklet already referred to, issued by the Editor of " The Engineer." He had also referred to another point, that the firms who introduced the system made visitors welcome. He thought he might say that his firm had a good deal of experience on the latter point. They received very frequent visits from engineers connected with large works who wished to study the subject, whom they always made very welcome. They had been compelled to issue a third edition of their pamphlet, as so many applications were received for it from firms doing all kinds of engineering work, not only in this country, but on the Continent and in America. That, he thought, emphasised what the author had said, that the greatest interest was being taken by manufacturers in the subject. He firmly believed that there was nothing better calculated to improve workshop methods than the adoption of the Premium System.

Mr. J. F. ROBINSON, Member of Council, said that he had not seen his way hitherto to adopt the premium system in his locomotive building works for two reasons. In the first place, they had an enormous quantity of detailed work to do. The number of small parts in a locomotive of a given weight were, he believed, far greater than in any other machine of a like character. That meant that the number of operations required to produce a locomotive of a given tonnage were probably greater than they were in any other machine of that character. To introduce the system, it would mean that those operations must have their proper standard times fixed; and he confessed that the prospect of trying to fix such standard

(Mr. J. F. Robinson.)

times for all the operations required in the machine shop, in the boiler shop, and in the foundry, and so on, appalled him, and he had never been able to bring himself into a position to face the problem. Both the previous speakers had alluded to the point that, if it was found a certain piece could not be produced in a preper time, the design should be altered. But what would happen if the manufacturer was working to somebody else's design ? If he wont to the superintendent and said, "I cannot make it; it is too expensive ; I want you to alter it," the superintendent would reply "Certainly not ; I shall not alter it ; that is your affair not mine." It was all very well for gentlemen who were working to their own designs, and were doing more or less repetition work, reproducing the same thing over and over again, perhaps in different sizes, but where the character of the work was the same, to suggest such things ; but when an engineer had, as his firm had, to build work for other people, to other people's designs, who were constantly changing their types, that was a very great objection to the system. He did not say it was impossible to carry the system out, because their late rivals, and future colleagues, Messrs. Neilson Reid and Co., of Glasgow, were already introducing the system, and in a few months he hoped to be able to form an opinion as to whether the system really paid in locomotive building as compared with the piece-work system which had been in practice for so many years.

One point which had not been mentioned in the Paper, or if it had been mentioned it had escaped his notice, was that he understood one essential point of the premium system was the immutability of the standard allowance of time. He understood that when a time was fixed for a given article, it was bargained that that time should not be changed for twelve months. He would give an illustration of what occurred in his own shops. Supposing an order was received for twenty engines of all the same kind ; they were put down in the shops in fives. If at the end, or during the progress of the first five, it was found that the man could not make his balance, as it was called, or obtained too big a balance, that could be modified for the next set of five. But if the premium system was in use, as he understood it, the employer was tied hard and fast for a year. He did not quite see how that difficulty could be overcome. Otherwise, for operations in works where people built the same thing over and over again, of different sizes, and where one had the very great advantage of being able to design parts to suit one's self, the system seemed to him to be most excellent.

Mr. A. F. YARROW thought the subject under discussion had been very much neglected by engineers, the reason being, he presumed, because it was not very interesting. A large number of people ran to the other end of the world after a contract, but he thought it would very often pay them much better to stop at home and organise their works to the best advantage. No doubt it was very nice to take a trip to Japan or South America ; but he very much doubted whether it would not often be more profitable to stay at home. His firm had adopted a system which he copied from one of Mr. George Westinghouse's Works in Pittsburg. A certain length of time for a job was agreed upon which was the outside time it was estimated to take, and whatever time the work took was deducted from that figure, half the difference in time being registered in favour of the men, and half in favour of the firm. That system had been in operation five or six years ; certainly the men were quite satisfied with it, and, he believed, the firm was. Of course it required very great care in settling the time, because the success of the scheme, or any scheme of the kind depended upon never attempting to alter the time once it had been fixed, a point the author laid great stress upon. If a time were fixed, and the men made a considerable amount, and the time was then altered, the men felt that they must not hurry on and make a lot of money out of the job, otherwise the time would be altered in future, a feeling which took away the whole of the charm and benefit of the system. One could not take too much care in arranging a suitable time, which under no circumstances must be altered, provided the work had to be done in the same machine. He believed something like fifty of their men were working on the system. It was more important in regard to the men who were working machines, where an increased output on the machines, and on the capital embarked on those machines, was obtained.

s 2

(Mr. A. F. Yarrow.)

He would like to ask the author whether he found any difficulty through having day work and also the premium system in the same shop. His firm had adopted the two systems in the same shop ; he did not know they had found any difficulty, but he would like to know what the author's opinion was on the point. He would also like to know how the author dealt with the time that was lost. It sometimes happened that, if men made a lot of money, they did not come before breakfast, and possibly not on Monday morning. If a machine was thus idle for some time, it took away the benefit of the whole operation, because the machine being idle diminished the output, while the object of the scheme was to increase the output. He would like to know whether the author reckoned the time lost in any way as part of the operation, because it appeared to him that it was a very necessary thing. There was one point which appeared to be of very great importance, namely, that the time when a job finished should correspond exactly with the time when the next job began, otherwise anybody starting a scheme of the kind would find there was a big gap between the leaving off of the one job and the beginning of another. For the success of any system, it was essential that it was to the benefit both of the workman as well as of the firm.

Mr. HENRY LEA, Member of Council, said he wished to ask Dr. Barr one question. Dr. Barr had given an instance of the time allowance being fixed for a certain piece of work, which, when the premium system was introduced, was done by the workmen in a quarter of the time. He would like to ask whether the time allowance for that particular article had been preserved up to the present time, or whether some alteration was not found necessary.

Mr. R. PRICE-WILLIAMS considered that the Paper, although certainly on the confines of subjects which came within the province of the Institution to discuss, was a most interesting and valuable one, and well within the lines, involving as it did the all-important question of the economy of time in the workshops, which in fact was the essence of the Paper.

As regards the economy of time in the working of machinery—an important factor in the case—he might mention that he had recently visited some large works, the engineering manager of which, who had been at one time an assistant of his, and had served his time at the Crewe works, showed him very neat little diagram maps of every workshop recording every day the time the machines and engines had been at work, and those which had been lying idle. By this means he had been able to effect large economies, and he regretted that the gentleman, who was a Member of the Institution, was not present, as he would, he felt sure, heartily sympathise with and support this well-considered and thoroughly practical method of largely economising the time of the men in the workshops. The effect of this would, as the author had clearly shown, greatly benefit financially both the employer and employed, and at the same time teach both, as the author of the Paper has clearly shown, that they failed to realise at present the immense value which attached to economy of time in all commercial and more especially competitive trade transactions. This was strikingly illustrated recently in the case of a large time contract for the construction of a big viaduct in Egypt, which was secured on this very ground by an American firm and completed well within the contract time.

Dealing with the question of saving of time in the case of the locomotive, he quite differed from the view just expressed by Mr. Robinson as to the great difficulties attending the application of the premium system to it, owing to the great number of the parts which would have to be dealt with. He, however, saw no difficulty, and would remind Mr. Robinson that more than thirty years ago Mr. Webb and himself had been at the trouble of ascertaining the average life value and cost of each part of a locomotive and tender.

One other great advantage he recognised in the adoption of the premium system was its tendency to render closer and more friendly the relations of employers and employed, and it would enable the latter to realise, and in a direct way, that, by economising time at his work, he was not only benefiting himself pecuniarily, but his master also.

Mr. JOHN RICHARDSON remarked that he had used the piece-work system for about fifty years. Although he saw that there was an immense advantage in the bonus system over day-work, he was somewhat doubtful how it could be an advantage both to the employer and to the workman, formerly working on piece-work.

Supposing that the piece-work price for a particular job was 90s. and the workman's wages were 8d. per hour, if the man did that work in 60 hours he earned 1s. per hour, or a total of 60s. for 60 hours' work.

On the bonus system, if 90 hours was allowed for a certain job, and the man did it in 60 hours his wages being 8d. per hour, he would get 33 per cent. on his wages as bonus, or a total of 40s. + 13s. 3d. = 53s. 3d. Under these circumstances he could not see that the man could gain as well as the employer.

He understood the author to say that the bonus system was an advantage to the employer and employed. He, as an employer, wanted to gain something, especially if it could be done without making the conditions any worse for the men.

Mr. J. MACFARLANE GRAY said he was not an employer of labour, but some years ago he had paid considerable attention to the subject of the remuneration of the workmen. He pointed out what he considered to be an oversight in the construction of the Rowan premium scale. The statement in the Paper (page 218) that "the times taken by all the machinemen have, on an average, been reduced during the four succeeding years, by 20, 23, 31, and 37 per cent. respectively. The earnings of the men have consequently increased by these percentages," did not sufficiently illustrate the peculiarities of the scale, which at one end gave to the workman 59 minutes of the first hour saved on a 60-hour job, and reserved for the employer only the remaining one minute of the saving; while at the other end it provided that if the workman by supreme exertion completed the work in 10 per cent. of the set time, he would be rewarded with 10 per cent. of the time saved, the master pocketing 90 per cent. of it. There was, however, in this no concealed intention of favouring the master to the detriment of the workmen; it was merely an oversight, for up to a 50 per cent. reduction of time the workman got the larger share of the saving, and in the above quotation the maximum reduction stated was 37 per cent., of which 37 per cent., according to the scale, the workman was awarded 63 per cent., and the master received 37 per cent. The Rowan scale was that the saving effected was divided in the same proportion as the set time was divided, the workman getting, in addition to his work-time wages, the proportion of the saving representing the time taken to the job, and the master the proportion representing the time saved. This Rowan scale admirably provided, as the author told them was intended, "such a basis that the men when put on the premium system will be able right away to earn a premium." If the scope of the scale were limited to 50 per cent. saving, and beyond that, the saving were to be equally divided between the two parties, as at 50 per cent. it was now, the scale would better express what, no doubt, must have been the intention of the author. In the ideal case of the work being completed in 10 per cent. of the set time (100 hours), the workmen would then be paid 55 hours for 10 hours' work, and the master would be benefited to the extent of 45 hours. As the scale stood at present, in that case, the workman would be paid 19 hours for 10 hours' work, and the master would be benefited to the extent of 81 hours.

While he would like to congratulate the author of the Paper on his intelligent, energetic, practical endeavour to set things straight between capital and labour, he did not think that the premium system—or, indeed, any system of piece-work—could ever produce the final desired solution of this difficult problem. The time would come again when contracts could not be obtained, and the resolution never to reduce the once fixed set times must then be abandoned, or the equally disturbing proceeding of wages reduction adopted; and all the discontent, for the time happily obliterated, would reappear, and the task of Sisyphus would be just what it was before. He then broadly sketched a plan which he believed did contain all the essentials of an ultimate satisfactory solution. Let there be a new Act passed establishing a legal contract between employers and employed, providing that the capital of every

(Mr. J. Macfarlane Gray.)

firm working under this Act should be nominally declared, and the rate of usury to be paid for it stated ; the managing staff and the workmen to be hired at current market rates, subject to no restriction ; after the wages, the agreed usury, and the upkeep charges, the whole of the remainder would be profit to be equally divided between the proprietors and the employed, the moiety to the employed to be divided in proportion to wages earned. When trade declined, and contracts could not be obtained unless on reduced prices, the men could be told exactly how the matter stood, and in this way their honest co-operation would be secured when their present ideas of the magnitude of the profits of employers had been for ever dissipated. This, it might be thought, looked like giving too much to the men, but it was not so; and if it were tried, it would be found to be to the advantage of the employer as well as to the men. Everyone engaged in the works would be in spirit equivalent to a master, and would do his best for the advantage of the whole concern, while the saving in overlookers, inspectors, etc., would be a very important item.

Mr. C. D. ANDREW stated that he had seen the system in operation for about three years in the Westinghouse Co.'s Works in America, and came to the conclusion that it was a very good system ; in most cases it paid both the men and the employers. But it was generally necessary to increase the efficiency of the inspection. It was found that under the premium system it was necessary to provide a bigger outlay under that head as compared with piece-work. For instance, if a regular line of work was put through on the premium system, the first thing to do was to increase the staff of inspectors ; naturally at the outset the men tried to push the work with undue haste, and a close watch had to be kept in order to see that the gauges were adhered to. The expenses of inspection and shop-management were therefore made greater. Extra men also had to be put on to set the limits, which was an additional expense, although slight compared with the ultimate saving.

One important feature of the system was that, if an employer had a contract which he wished to push through very quickly, it could be done on the premium system, and even supposing the times were miscalculated he was sure to save very considerably compared with day-work, in addition to getting the work done much quicker, and in most cases better. Naturally the poor workmen showed up, and were put on to work better suited to them, while the men who were equal to the task got through the work well, and made a saving in time which was a benefit to themselves as well as to the employers. He had seen the system in operation for three years, and in most cases quite successfully. Of course there were cases where the time was miscalculated, but that could not be avoided. A man might make double time, or even more than that in a few instances, but it never paid to alter the rate. Once the rate was fixed, it ought to be adhered to for good; if it was altered, the men lost confidence in the system, and it would be difficult to re-establish it for that reason. He knew of several instances where the rate had been wrongly fixed, and the men had gone on earning an excess for possibly two or three years, and then perhaps the design or the operation was changed, when the price was of course revised. The men thoroughly understood that. Supposing a new job had to be put through, of which the employer only expected to get a few lines of the particular kind, he was always safer in applying the premium system than if he had a great many, because when the men had got through two or three jobs, prices could be based fairly on experience for a similar kind of work. But, in his experience, it was comparatively easy to fix the premium limit; one could generally arrive at a certain percentage, which could be obtained by calculation and observation, in comparison with the day-work. It was surprising what a big percentage was sometimes obtained; 50 per cent. was frequent on the operations a man generally did, because he put a lot of movements in which he did not need to perform at all. Just the same applied with regard to piece-work, but it seemed to be more apparent on the premium system, owing to its being progressive.

The system explained by Mr. Yarrow (page 233), whereby an allowance of half the saving was given to the firm and the other half to the man, was, he thought, a good system; it had been applied, as

(Mr. C. D. Andrew.)

explained by Mr. Yarrow, in the Westinghouse Works. Naturally the expenses of maintenance were much greater, especially in the way of providing jigs and special appliances. He had always found, and he believed it was the experience of everybody, that as soon as a man was put on the premium system he began to use his brains more than usual. Probably before, he had been using a wrongly-shaped tool, or had not been helping himself in some particular way, but he very soon found where he could improve on his work. The first thing he did was to go to the foreman and tell him what he wanted in order to get his work done quickly, and naturally the foreman gave it to him, because it was better for the firm and for the man. He had not had much experience of the system in England, but he saw it applied in America, where it worked very well.

The point of great importance was that the time must be adhered to. In any case, the saving was very marked even when a man made double time. Supposing a firm had been doing a certain job on day-work for two or three years, which took a man 10 hours ; and the firm then introduced the premium system and fixed a time of 8 hours, and the man did it in 5 hours, which was possible with improved appliances, the firm would save 5 hours' time on that particular job, as compared with the original day-work time, and the man would make 30 per cent. premium on his hourly rate of pay. In the long run the firm saved a great deal of time, and, as a consequence, a great deal on the wages sheet, in addition to getting quicker delivery. The scheme was, therefore, of great value.

Mr. ALFRED J. HILL remarked that he did not speak as one who had had any experience with the premium system, although it was a system in which he had been very greatly interested, and of which he would like to hear considerably more as compared with the piece-work system. The Paper imparted very valuable information, but the bulk of the comparison was with day-work and not with piece-work. The author said (page 209) "The author can quote only one case—and it is personally known to him—where the premium system has been instituted to take the place of piece-work. This is a large firm which has been working on piece-work for many years; they stopped it and went on time for two months, when they introduced the premium system, and the work is now nearly all done in less time." He could quite understand the author's reasons for not giving the name of the firm, but he would like to ask if there was any objection to his being privately informed of the name of the firm, so that he could make further inquiries as to the comparative results of piece-work and day-work.

He himself had had a very large experience of piece-work, and was at the present moment firmly convinced that it was an admirable system as compared with day-work, but he was quite ready to be convinced that the premium system was still better. It seemed to him that under the premium system there were one or two decided advantages as compared with piece-work. Under piece-work one had the disadvantage of the possibility of cutting the prices, which had the natural effect of the men trying to hold back their work. In the shops of which he was manager, where over 3,000 men were employed on piece-work, the men had been encouraged to earn large balances, and he had tried as far as possible to avoid cutting the prices. The average balance which those 3,000 men were earning was certainly over 30 per cent., and many of them were earning as much as 50 per cent. above their day wages. He would like to make the following comparison of the piece-work system and the premium system. A mechanic whose wages for a 9-hour day were 6s., equal to 8d. an hour, did a certain piece of work in 9 hours. Taking the piece-work price at 8s., the man earned his 8s. in 9 hours, thus earning 33 per cent. more than his day-work wages, which was roughly the average balance each man earned. For the man to earn the same amount on the author's premium system the time allowance would have to be $13\frac{1}{2}$ hours, the man would then save $4\frac{1}{2}$ hours, or one-third of the time allowed, and he would be paid 33 per cent. above his wages. Assuming that the man was able to perform the same work in 8 hours, under piece-work, the price remaining the same, he would receive 50 per cent. on his wages, but under the author's premium system the time

(Mr. Alfred J. Hill.)

saved would be 5½ hours or only 41 per cent., the cost of the work would be reduced and so would the man's wages. Looking at it from this point of view, he thought there might be a difficulty in introducing the system into a shop which had already been working on a really good piece-work basis, and that was why he was very anxious to get further particulars of a shop where the change had been made. In many shops it appeared to him there was work which it would be very difficult to fix a time allowance for, but for which a contract could be made.

He did not agree with Mr. Robinson's remarks as to the special difficulty of applying the system to locomotive shops, because it did not seem to him to be more difficult to fix a time for such work than it was to fix a piece-work price. Time and price were interchangeable, and as most of Mr. Robinson's men were on piece-work, he could not see how it would be difficult to fix a time, if it was possible to fix a price. Taking the case of a foundry, almost the whole of the men in his foundry were on piece-work. It would be interesting if the author could state how firms which used the premium system treated such men as trimmers, carriers, and general labourers in the foundry. How could they be placed on the premium system? In his shop such men were put on a contract or a tonnage rate. The moulders worked by piece, moulding a certain job for a certain price. The carrying up and general labouring of the shop was done on a tonnage basis, while the trimmers worked piece-work. The miscellaneous labour in a shop was a very important percentage of the staff in a large factory, and he had found a decided advantage and economy wherever possible to encourage that class of labour by putting them on piece-work or on a contract; he did not quite gather how any comparative system could be arranged under the premium system, and if the author could supply further information on that point he would be obliged.

He did not quite follow the author's statement (page 204), namely, "It is of importance that a separate department be started, and that the rate-fixing be not placed as an additional duty upon the foremen." In his shop the foremen in the main fixed the piece-work prices, subject, of course, in large contracts to supervision by the management, and he did not see why the foreman should not be able to fix his time-rate. He would like to ask to what extent, if any, the workmen were consulted as to the time allowance. In his own shop the piece-work price was fixed in consultation with the men, and was practically a bargain. Was that followed in the same way under the premium system, or did the management fix the time and tell the men they had to do it on that basis? He would also like the author to state the definite percentage of the men in his own works, or any other works, who were able to work on the premium system, or whether he was able to put the whole of the men on that system. He had already partly asked that question in connection with the labourers. He would also like to hear the answer to the question asked by Mr. Yarrow, as to whether it was found that in the same shop the premium system worked comfortably side by side with the day-work or with the piece-work system?

Mr. ROWAN, in reply, thanked Mr. Pendred for having explained the diagrams, Figs. 2 and 3 (page 223), which were exhibited. The premium system of paying wages was the subject of a pamphlet containing reprints of articles from "The Engineer" newspaper. Dr. Barr supplied some of the diagrams and Mr. Pendred others. He thought the diagrams were most useful to anyone introducing the premium system, who should also read the articles which appeared in "The Engineer." Dr. Barr brought out the point that the premium system was of greater value than the actual saving in money. As a matter of fact he did not think that in a large workshop spending thousands of pounds per annum a great saving in the wages would be found, but a very considerable saving ought to be effected in the on-cost, and the output ought to be increased. A gain was made in that direction, to which he could not put a figure, and he questioned whether anybody else could. Dr. Barr was quite correct in his remark that the introduction of the premium system brought the men and masters much more closely and intimately together, and also brought them into friendly relationship with one another, which was a very important point in a workshop. He should certainly urge the Council to induce Mr. Jackson to write a Paper on the

(Mr. Rowan.)

Card System as opposed to keeping records in books. As he mentioned in the Paper, his firm kept their records in books; they had used cards to a certain extent, and had looked into the card system, but they pinned their faith in books, although they were quite open to conviction on the other side, and after a long conversation he recently had with Dr. Barr he intended to go into the matter again. A Paper by Mr. Jackson would make it clear to a number of gentlemen where the saving was, because the saving obtained by those who used the card system was sometimes much more apparent than real. It was often said that the cards were kept by a boy who did the work in an hour per day, but on further inquiries being made it would be found that probably the manager and foremen devoted a great portion of their time to work which in his firm the rate-fixers did. In any case there was a certain amount of clerical work to be done, no matter who did it, and in his opinion the writing on cards, making up estimates of the cost of articles and the time they took to make, was not the work of managers and foremen. If they did that they were doing the wrong class of work; they ought to be looking after the men and machines and doing absolutely nothing else. The point brought out by Dr. Barr of there being so many records on the cards at exactly 50 per cent. was a feature which was disappearing. His firm had, as mentioned in the Paper, abolished the increments of 5 per cent.

He was afraid Mr. Robinson had not studied the premium system very carefully. In his opinion the application of the premium system to marine engineering was as hard, if not harder, than adapting it to locomotive works, although he did not know much about the latter. He failed to see the difference Mr. Robinson had pointed out (page 231). Mr. Robinson also made a great deal about the difficulties connected with the design. In marine engineering they had to work just as much to other people's design as engineers in locomotive works did. It was not actually in the doing of the work, but in the producing of it by means of jigs and fixtures, and one thing and another, where a great deal of economy could be effected. His notion had always been—of course it might be wrong

—that a locomotive workshop was an ideal place in which to introduce the premium system.

Mr. Yarrow (page 234) brought out a point which he had heard raised before, namely, how did the payment of premiums affect the men as to their time-keeping, etc. ?   He would reply most emphatically that it improved the men.   He had with him a list of the men and the premiums they earned per annum.   He was not quite sure how many hours were worked in his shops per annum, but on the list the maximum year was that of a man who had worked 2,654 hours, who, he believed, was the best time-keeper in the works. Other figures on the list were 2,635¾ hours, 2,604¼, 2,437, 2,654, 2,460, etc.   The names of the men on the list were put down from the time-book; they were all good time-keepers and highly respectable men.   In his opinion the men had been greatly benefited by the premiums they had received, and had taken full advantage of them. He had not been able to follow Mr. Macfarlane Gray's remarks. If a man did the work in 80 per cent. of the time allowed him he had 20 per cent. added to his rate of wages, while working on that job.

Mr. Macfarlane Gray thought the man received 80 per cent. of his saving.

Mr. Rowan replied that Mr. Macfarlane Gray was quite correct, but that the man receiving 80 per cent. of his saving, in the case quoted, was equivalent to 20 per cent. being added to his rate of wages while working on the job.

Mr. Macfarlane Gray said he wanted to know what the man did get.

Mr. Rowan replied that if a man was allowed 100 hours to do the work, and did it in 20 hours, he had 80 per cent. added to his rate of wages per hour or to the wages earned during that 20 hours ; he had 80 per cent. added to the wages which he earned on time in those 20 hours.

Mr. MACFARLANE GRAY thought that if the man saved 80 per cent. he received 80 per cent. of the 20.

Mr. ROWAN said that Mr. Hill made the remark that they allowed the men to earn large balances up to a certain point. It was just that point which defeated the piece-work system. In the premium system, as he mentioned in the Paper, they encouraged the men by all means to go beyond that point. A man when working on piece-work got more work out of his machine than when working on time, but he was not reducing the cost of the articles being turned out; under the premium system the employer encouraged the man to go beyond the point indicated by Mr. Hill, and helped him to make a bigger premium, because it reduced the cost of the work which came out of the machine.

In reply to Mr. Hill's further remarks (page 243), a workshop working on a piece-work basis was dealing with money, which was easily changed into time, and with a little careful study he fancied the piece-work system could be changed to a premium system. He would be very pleased to let Mr. Hill know the firm to whom he referred. Mr. Hill also asked whether they consulted the workmen. He replied in the negative. At first the workmen had been consulted, but that had been stopped, as it was soon discovered that none of them knew the length of time it would take to do work; the times had been fixed by the rate-fixers, and very shortly after the introduction of the premium system they knew far better how long it should take to do a piece of work than any of the workmen. Although that had been done, quite a number of the workmen had expressed themselves as thoroughly satisfied with the system as it was carried on in his works.

Mr. Hill had also asked the number of men employed on the premium system. Every man at a machine was on the premium system. On the heavy machines, the labourer assisting the machine-man was also on the premium system, but the men who swept up the floors, and the labourers who moved castings and forgings about the workshop, were on time. All the men who put engines on board the boats worked under the premium system, but there were a number of labourers on the boats who could not be treated in that manner, and were paid on time. The smiths and strikers were all on the premium system, with the exception of the forgers and others, who were on piece-work. So far as his firm was concerned, he did not think they would ever put the pattern makers on the premium system, which in the author's works consisted chiefly in altering old patterns. He thought that in the engine department about five-eighths of the men employed were on the premium system. He thanked the members very much for the kind manner in which they had received his Paper.

Communications.

Mr. DANIEL ADAMSON wrote to enquire of the author in what proportion the fixed charges (or working'expenses or shop charges or whatever they might be called in different works) bore to the wage-cost under the premium system when compared with the previous system (whether piece-work or day-work) that might be in operation before the adoption'of the premium system. The author had stated that in 1899 the workmen did 25 per cent. more work than was called for by the times fixed under the premium plan, and that in 1902 the men did; 58! per cent. more work than the premium limits call for. As, however, the author recommended fixing a generous time limit (pages; 207 and 209), one might assume that in 1899 the men were not really doing much more than they would have done under other systems, and he would therefore only compare 1899 with 1902. In 1902 the workmen did 26 per cent. more work in the same time than in 1899, and drew about 14 per cent. more wages, and he wished to know whether the item of fixed expenses (by which he meant the annual total of all charges on the cost of the finished articles except materials, net wages and profit) increased at all, and if so, whether in the ratio of the work done, that is 26 per cent., or of the extra wages paid, that is, 14 per cent. There must be some

(Mr. Daniel Adamson.)

increase in the fixed charges to pay for the increased supervision required when more work was being done, also to cover the extra wear and tear of machinery. Their own experience was that roughly five-sixths of the fixed charges would increase with the output and one-sixth (representing taxes, depreciation, &c.) would be stationary whether the output increased or was reduced, but he wished to know the author's experience on this point as it was very relevant to his subject.

Mr. PHILIP BRIGHT wrote that he was surprised that the " Rowan " system had given such good results as the author had stated. It was clear that the first reductions in time were the easiest to make, and for these first reductions the premiums were very high, whereas subsequent reductions in time, which required greater skill and efficiency on the part of the workman, brought a proportionately smaller reward. Under the "Weir" system, however (which was the one in use in the works with which the writer was connected), the proportion of the saving which went to the workman was the same, no matter whether the time saved was small or great. He would therefore have thought it probable that workmen, working under the "Rowan" system, would have relaxed their efforts directly their proportion of the saving began to decrease. In his opinion, wherever piece-work was worked, there was restriction of output, the quickest men holding back, for fear the rates would be cut, as this cutting would seriously affect the earnings of older and less efficient men. He had proved on many occasions that good premiums could be earned, if the standard times were fixed at the piece-work costs, but he thought it would pay an employer to add 5 per cent. or even 10 per cent. to piece-work rates when fixing standard times, in order to get the premium system introduced into works; thus, if a piece-work price were 9s., the standard time for a man rated at 9d. per hour would be about thirteen hours.

No one had any idea of the extraordinary results of premium work until operations had been repeated at least half-a-dozen times. The men reduced their time with each job until they finally reached a limit which was not capable of further reduction, and several cases had come under his observation where the net saving had been more than 50 per cent., notwithstanding which, the men had earned higher wages than they did when working piece-work.   He did not know of any other system which secured the co-operation of the workmen, as the premium system did, if it were fairly and justly administered, and he believed it would be universally adopted in progressive works before many years had passed.   This result would come about largely through the men themselves, the best of whom would prefer to work in premium shops.

Mr. P. T. J. ESTLER wrote that he had recently started a small engineering workshop, and had from the beginning fully recognised that some system of rewarding and encouraging labour was a necessity. It appeared to him that the chief merit of the author's system was its absolute fairness both to master and men, which, as far as he was aware, no other system possessed.   The piece-work system failed because it only led to constant friction, unless the men had their own way entirely.   Mr. Rowan's method was fair to both, whether the time specified was too liberal or too keen.   If too liberal, the men naturally did not obtain so large a percentage of the saving, and, on the other hand, if too keen, they obtained a large percentage of the saving.   It also induced the men to continue performing their work in a shorter time with each repetition of the same work, and did not necessitate the master modifying the time allowance when once fixed. This naturally prevented the mistrust which a piece-work system easily engendered, and which must be fatal to every system where mistrust was a possible contingency.   He could quite believe that it created a better feeling between master and man.

The greatest difficulty, he thought, lay in adopting this system after a piece-work system had been in force.   He would take an example which occurred in his own works.   The piece-work price for fitting together a certain piece of apparatus was one shilling each. The man engaged on the work did batches of fifty in about twenty-five hours, whilst formerly on day-work he required about fifty hours. The time had been gradually reduced by the man obtaining the conviction that the rate was not to be altered, and also because he

(Mr. P. T. J. Estler.)

now worked more systematically. The problem was what time allowance was to be made. It would of course have been quite easy if the man, who was paid 9d. per hour, had not been allowed to make more than 1s. per hour. In conclusion, he proposed to introduce the system at the very earliest opportunity and would be pleased to report later, if desirable, on its progress.

Mr. W. D. KILROY wrote that he thought the Paper was extremely interesting, and he felt sure that there were few who did not realise that the premium system possessed very many advantages over the other methods now in use.

There was a condition, however, which he would like to bring forward, as being one likely to cause serious trouble in the carrying out of this system. In the writer's works at present they were working piece-work and on time. It was essential that some of the men should be kept on time, no matter what system might be adopted for the remainder; because they had, besides purely experimental work (for which there was an experimental department), always a very fair amount of new work in hand, such as apparatus being made for the first time. It seemed impossible to put this work on the Premium System, and the men required for it must therefore be kept on time. Supposing he were to adopt the premium system for the remainder, there would naturally be a substantial increase in the amount paid to them on pay-day as wages. But what about the rate of pay to those men still working on time? He was sure that the human nature element in the British workman was over-developed rather than the reverse. They would certainly not be content to go home on pay-day with the same wages as heretofore and see their fellow-workmen drawing probably 25 per cent. or more pay. There would certainly be discontent, unless the wages of these men (working on time) were raised. How was this to be faced? Obviously, it would not do to raise their rate of wages until it compared with the average rate obtained by those working on the premium system. Generally speaking, it was the better workman who was kept on time, and how could he have similar chances of increasing his wages. He would very much like to know what the

author had to say to a state of affairs of this description. Did he
consider it impracticable to work some men on time and some men
on the premium system in the same works, without entirely separating
them by giving them separate shops and foremen ?

Mr. Macfarlane Gray had suggested the "ideal" method of paying
the workmen, namely, that of giving them as a bonus a certain
percentage of the yearly profits on a nominal capital (page 238).
This was certainly "ideal" in principle, but it would seem rather
an impracticable solution unless applied only to a few chosen men
whose honest work might be relied upon. This ideal method had
also the objection that it did away with "individual" effort of the
workman, which seemed to the writer to be the mainstay of the
premium system ?

Mr. HENRY A. MAVOR wrote that the work which Mr. Rowan had
done in bringing the premium system of remunerating labour under
the notice of the engineering trade in this country deserved hearty
recognition, and the writer had personally to express his indebtedness
to him for much valuable assistance in initiating the premium system
in the works with which he was connected. They had had it in use
there since May 1899, and a few details of their special experience
might be of service, the more so that the methods adopted differed to
some extent from those described by Mr. Rowan.

They adopted a card system of record, and found it very much
more convenient for their purposes than the keeping of records in
books. The Rate-Fixing Department was rendered very easy of
manipulation by means of the card system. The easy and regular
working of the records depended primarily upon the initiation of a
proper system of indexing. This was accomplished by means of a
specification. It would be seen that each article received descriptive
lettering, that each member of the article was indexed by the
numerator of a fraction, and each component by the denominator of
a fraction. The operations on each component were described by
means of letters as far as possible descriptive of the operation, for
example—

(Mr. Henry A. Mavor.).

P represented Planing.

M   „   Milling.

T   „   Turning.

and so on.

For example, the rough turning of a shaft for a Multipolar Enclosed Motor was indexed—

$$\text{ME. } 125 \tfrac{1}{2} \text{ T}_2,$$

and this enabled the duplicate job-ticket or line to which these symbols were attached to be filed under T_2, which brought it together with other turning operations, the record file for turning operations being sub-divided by the other symbols $\tfrac{1}{2}$, which meant an armature shaft, and ME. 125, which specified for which machine the shaft was made. It was thus easy to accumulate data before fixing the time allowances, and this data was correlated and grouped without clerical work other than the assistance of a boy in the cost room, who filed the cards according to the index letters and figures. The sub-division of the elements of each operation might be carried out to any required detail by the issue of job-tickets or lines for each detail required to be separately recorded.

It would be seen that all that was required to initiate this record was the preparation of a complete specification with a separate line for each piece or component, the lines of the specification being numbered, and that all instructions given in the workshop for carrying out operations upon these pieces be given by means of job-tickets. The job-tickets being made in duplicate, the man carrying out the operation kept one copy until completion of his job, the time of starting being placed upon it when it was handed to him by the premium clerk in the shops, and the time of completion and certification of accuracy of the work being inserted on the ticket by the foreman. This card was returned to the cost room, where already the duplicate had found its place on the cost file of the job, and was used as described above for keeping a record of the times, grouped in this case in the file set apart for recording the times taken on each operation. After a sufficient number of records had accumulated, they could be summarised on a permanent record card and the original individual lines destroyed.

He was able to confirm very fully Mr. Rowan's experience that the introduction of the premium system almost invariably resulted in a reduction of the time taken on the work. They had not issued any pamphlet to their workmen, but had by watching the men who were not earning premiums, and personally explaining to them that the object of introducing the system was not to " horse " or hurry the work at the expense of the workman, but to offer inducements to all departments and individuals in the factory to work to each other's hands; in other words, it was not for the purpose of chasing the industrial army to its work, but for the purpose of arranging that they should march in step and thus ensure steady even progress. It was of the first importance that the premiums earned should be paid to the men at the earliest possible moment, and it was found practicable to include the premiums of work finished during each fortnightly period with the pay at the end of the fortnight. A very important development of the use of the premium system was the payment to foremen of departments of a premium on their wages, calculated by taking the percentage of premiums earned to normal wages of the workmen under their charge and giving the foremen a premium at the same rate on their wages. This was a direct inducement to the foremen to encourage the men to earn premiums.

Another very important development was the effect upon apprentices. In making the time allowances to apprentices recognition was taken of the grading of apprentices. Thus, an apprentice in his first year had the time actually taken upon the job multiplied by $0 \cdot 2$ in calculating the premium earned. Thus, if ten hours were allowed for a job and ten hours taken by a first year's apprentice, he receives a premium of 80 per cent. This resulted in the men who had to work with apprentices selecting smart boys to work with them, and taking the trouble to teach them and develop their abilities, and the quality of the apprentices' work in the early years of their apprenticeship had thus been very materially raised, and the amount of work done by apprentices largely increased.

The second year's apprentice had his time taken multiplied by $0 \cdot 4$.

(Mr. Henry A. Mavor.)

The third year's apprentice had his time taken multiplied by 0·6.

The fourth year's apprentice had his time taken multiplied by 0·8, and the fifth year's apprentice had to work on the same time allowances as a journeyman.

The difficulty of inspecting the work after completion was not so serious as would appear, because, as Mr. Rowan pointed out, the fitters and erectors would not accept bad work which was to reduce their premium earnings.

In conclusion, he would emphasise Mr. Rowan's dictum that it would always be found better if possible to arrange for each man to work on his own account. Departures from this rule should only be made when absolutely necessary. It was part of the essence of the premium system that it individualised the workman and gave him a direct return for his skill and ability, and if the workmen were not so individualised, a large portion of the benefit of the premium system was lost to the employer and to the men.

Mr. WILLIAM POWRIE wrote that he thought the following Table 3 which he had prepared last year might assist the Members to understand the results of the different systems of payment on the premium system, which were likely to be in more general use in the future. The Table gave results with Rowan's, Halsey's and Weir's systems, for a workman paid at the rate of 9d. per hour on work estimated to take 48 hours to complete.

It would be noticed that Rowan's system gave the best results to the workmen for small reductions of time, and that both his and Weir's system gave similar results when the time allowance had been reduced one-half. Halsey's system did not give so much encouragement to the workman in the early stages of reduction, and would not be likely to find favour with men accustomed to piece-work where they had been allowed to make 25 per cent. to 50 per cent. in addition to the usual time wages, which was not unusual in the machine shops of this country. If the time allowance was fixed fairly, it was very unlikely that the work could be done in one-fourth of it, unless improved tools or methods were introduced. Either of

TABLE 3.

Rates of Premium.	Hours allowed.	Hours taken.	Premium earned per unit.	Premium earned in 48 hours.	Reduction in cost per unit.	Reduction in cost of 48 hours.
ROWAN.						
Per cent. of increase,	48	36	6/9	9/-	2/3	3/-
equal to per cent.	48	24	9/-	18/-	9/-	18/-
of saving.	48	12	6/9	27/-	20/3	81/-
WEIR.						
	48	36	4/6	6/-	4/6	6/-
One-half of saving.	48	24	9/-	18/-	9/-	18/-
	48	12	13/6	54/-	13/6	54/-
HALSEY.						
	48	36	3/-	4/-	6/-	8/-
One-third of saving.	48	24	6/-	12/-	12/-	24/-
	48	12	9/-	36/-	18/-	72/-

those systems was better than the usual piece-work plan of fixing a price in money or time per unit of product, inasmuch as the reduction in cost gave the employer a similar interest in increased production, which the increase in wages gave to the workman, and both might co-operate to their mutual advantage. Halsey's was probably better suited to machine departments, where the strain of increased production came to a large extent on the tools, and Rowan's was probably better for departments where the strain came on the workman, such as in fitting, forging, etc., but for general all round application the writer favoured Weir's, which was easily understood, and if the prices were properly fixed, gave a fair division of profits to both the parties to the agreement.

Professor ROBERT H. SMITH wrote that it might be interesting to mention, as a proof that such plans of wage-paying were becoming widely adopted, that Messrs. C. Griffin and Co. were about to publish an extended set of Tables to facilitate the reading off of the correct

(Professor Robert H. Smith.)

amounts of the premiums due, the Tables having been compiled by
Mr. Golding for use in Messrs. Bryan Donkin and Co.'s works.
Several speakers in the discussion had referred to these premiums
being calculated by the help of the slide-rule. The average workman
could not be expected to place implicit confidence in the accuracy of
a slide-rule calculation, and it seemed advisable to compute by means
of Tables which could be readily referred to in case of a mistake
being suspected.

Whether Mr. Rowan's or any other system were used, the
computation of premium seemed simplest and most readily
demonstrable to any doubting artizan if made in terms of extra time
paid for, and not in terms of extra hour-rate of wage. When the
extra time allowed as premium was obtained, then the ordinary
wages-tables could be referred to, to ascertain the corresponding
amount of money due. Thus, in Mr. Rowan's system, the extra time
allowance was $\frac{t(T-t)}{T}$. It might be noted that any "proportion
sum" like this could be readily solved by simple inspection of an
ordinary multiplication table extended to 100 times 100. If T and
$(T-t)$ be found in the first vertical column, and $100\,t$ be sought for in
the same horizontal line as T, then in the same vertical column as
$100\,t$ so found, and on same horizontal line as $(T-t)$, there would
be found 100 times the required quantity $\frac{t(T-t)}{T}$. This simple use
of the Table was so evident to everyone who had learnt elementary
multiplication, that both clerks and workmen would probably prefer
it to a slide-rule computation. It gave, of course, the result correct
for two-figure quantities, and for three-figure quantities by easy
interpolation.

In diagram form, Mr. Rowan's system was represented by a curve
of squares, the saving to the wage-payer being, in terms of time,
$\frac{(T-t)^2}{T}$; the proportion of which to the "standard time" was $\left(\frac{T-t}{T}\right)^2$.
It was difficult to see why such a curve plan should be preferred to a
straight-line arrangement. In the long run there must arise a
tendency to adjust standard times so that the saving $(T-t)$ was small,
and the curve offered the employer no inducement to effect very
small savings, because he was given no share of them, so far at least

as the saving appeared in the wages account. On the other hand, after a considerable percentage of time-saving had been effected, further saving became much more difficult, but by no means less desirable, while less and less inducement was offered to the artizan to assist in accomplishing this more difficult task. Not only did the share given to him of the whole time-saving go down as the whole became greater, but his share of the *increase* in this saving went down in a double ratio. Thus—

Proportion of Time Saving	0·1	0·2	0·3	0·4	0·5
Workman's Proportionate Share of Whole Saving .	0·9	0·8	0·7	0·6	0·5
Workman's Proportionate Share of Increase of Saving	0·8	0·6	0·4	0·2	0·0

Thus when the time-saving had reached near 50 per cent. no share of any further increase of it went to the workman. As the task proposed became more difficult, it would seem more advantageous to both parties to offer greater rather than less stimulus to the accomplishment of it. It was true that his wage reduced to per hour of actual work still went up; but the whole money received for the work done began to fall at a quickly accelerating rate. If the market always provided an infinite supply of work to be done at profitable wage-rates for the whole working-class population, no objection would probably ever be raised to this result of the system; but as this was not the case, it could hardly be expected that the system should remain permanently in favour in those conditions of the market when a plethora of supply of manufactured goods forced down the selling price and therefore also the standard wage-rate. It seemed that this danger would be better provided against by either a straight-line wage-premium system or else by one represented by a curve bent oppositely to that of Mr. Rowan's, that is, concave above and convex on its under side.

When it was remembered that every time-saving meant to the employer a great deal more in other ways than it did on the face of the wages-account, and that of this more important part of the saving no share at all fell to the workman under any of these systems, and probably could not rightly be made to fall to him except upon a co-operative-share system, there seemed all the more reason to be

(Professor Robert H. Smith.)

liberal to him in arranging that his share of the wage-saving should
not be scrimped under any probable or even possible circumstances.

Mr. ROWAN, in continuation of his remarks at the Meeting, wrote
in further reply to Mr. J. F. Robinson, who stated (page 232),
" One point which has not been mentioned in the Paper, or if it had
been mentioned it had escaped his notice, was that he understood one
essential point of the premium system was the immutability of the
standard allowance of time. He understood that when a time was
fixed for a given article, it was bargained that that time should not be
changed for twelve months." He thought it might have escaped
Mr. Robinson's notice that special prominence was given to the
above in the Paper (page 212, clause 4), where it was stated that
" A time allowance, after it has been established, will only be
changed if the method or means of manufacture are changed."

Further on Mr. Robinson stated that "he would give an illustration
of what had occurred in his own shops. Supposing an order was
received for twenty engines all of the same kind, they were put down
in the shops in fives. If, at the end or during the progress of the
first five, it was found that the man could not make his balance, as it
was called, or obtained too big a balance, that would be modified for
the next set of five." According to this statement of Mr. Robinson's,
there was a good deal of guesswork in fixing prices, which was quite
unnecessary if the time which was taken to do each piece of work
was properly recorded on cards, or in books, by a properly constituted
rate-fixing department. It was hardly conceivable that a man who
had five articles to do, and knowing there were fifteen to follow,
would exert himself to make a large balance, knowing well that the
balance might be modified for the next fifteen. It would be rather,
that, as the management apparently did not know the correct price of
a piece of work, that the man would not exert himself when doing
the first five, but would work in such a manner that for the next
fifteen he would get a better price.

In reply to Mr. Yarrow (page 234), in all workshops there would
usually be a certain number of men who received an hourly wage,
such as most men working in a tool room, men repairing machines

and keeping them in order, &c.. The rest of the men in the workshop might be working on the premium system or piecework, and it would always be found that the man who was receiving extra payment for extra exertion or extra skill, would make that exertion or use his skill, and do more work than the man working alongside of him who was working on simple time wages. The author's firm had never found any difficulty through having daywork and also the premium system in the same shop.

The author could not speak from his own experience on the point raised by Mr. Richardson (page 236), but the experience of others was that work at machines was after a time done quicker on the premium system than on the piecework system. The reason generally given for this was that the rates might be cut when the men began to make too great a balance when working on piecework, but when working on the premium system the same objection did not hold good, as it was to the employer's interest for the man to make as much as possible, consequently rates were not cut or should not be cut, and the man should be encouraged to make as much as he could. In the case quoted by Mr. Richardson, suppose a man were allowed on the premium system 120 hours for a job instead of 90, as suggested by Mr. Richardson, and he did it in 60 hours, his wages being 8d. per hour, he would get 50 per cent. of his wages as premium, and his rate would become 1s. per hour, the same as on piecework; but suppose the man did the work in 40 hours, which was not uncommon in workshops were piecework had given place to the premium system, he would now get 66 per cent. of his wages as premium. His rate of wage would be 13·3d. per hour, as compared with 1s. per hour when he was working on piecework, and the job would now cost the employer 44s. 4d. instead of 60s. when on piecework.

The case quoted by Mr. Macfarlane Gray (page 236) was one that the author had not yet encountered but hoped to do so in the future. If a man were given 100 hours to do a job and did it in 10 hours (supposing this were the first time this job had been done), it would be perfectly evident that a mistake had been made in the rate-fixing department, and would be corrected as provided for at the

(Mr. Rowan.)

bottom of page 208, and page 212, clause 4, of the Paper; but if, on the other hand, it was a job which the workman had gradually reduced to this time, the author's experience decidedly was that the employer had provided new methods or means of manufacture to enable the workman to do it in 10 hours, and the time should have been changed as provided for (page 212, clause 4).

The experience of the author's firm was that when the men were put on the premium system after they had been working on time, they would, through their own exertions or skill, and without any change in conditions, reduce the time on an average by about 15 per cent. If it was reduced to a greater extent, it was invariably due to the assistance which they had received from their employers, such as new methods or means of manufacture; in fact, one of the strong points of the premium system as introduced by the author was that it paid the men well at the earlier stages of reduction. At the later stages the men were paid to work hand in hand with the firm to reduce the time of production; but supposing that a man did a piece of work in a tenth part of the time allowed him, and he accomplished this through his own inventiveness or skill, the author would say, increase that man's wages, promote him if possible, and it might be incidentally mentioned that the premium system was of great help in discovering such men.

Mr. Daniel Adamson had raised a point (page 247) which the author regretted it was not in his power to answer. Since the author's firm introduced the premium system, their whole works had been re-organised, additions had been made, etc., so that no comparison could be made of the on-cost when working on time, with the on-cost when working on the premium system; the author's impression was that the on-cost was somewhat greater when the men were working on the premium system that when they were working on time, but that the on-cost was reduced per unit of output, due to increased output.

Mr. Philip Bright wrote (page 248) that "no one had any idea of the extraordinary results of premium work until operations had been repeated half-a-dozen times." The author's firm had had the same experience, but as before stated this had been due in a great measure to the management rather than to the men. It had often been stated by writers on the subject of the premium system and by the firms who had adopted the premium system, that it was a severer test upon the management than it was upon the men. Mr. Bright's experience was the experience of almost every firm which had introduced the premium system, where the firm had been working piecework or time, and many cases could be quoted in support of this. Tables with the same object in view as that given by Mr. Powrie (page 255) would be found on pages 21 and 22 in the pamphlet published by the " The Engineer," on the Premium System of Paying Wages.

𝕿𝖍𝖊 𝕴𝖓𝖘𝖙𝖎𝖙𝖚𝖙𝖎𝖔𝖓 𝖔𝖋 𝕸𝖊𝖈𝖍𝖆𝖓𝖎𝖈𝖆𝖑 𝕰𝖓𝖌𝖎𝖓𝖊𝖊𝖗𝖘.

PROCEEDINGS.

APRIL 1903.

An ORDINARY GENERAL MEETING was held at the Institution on Friday, 24th April 1903, at Eight o'clock p.m.; J. HARTLEY WICKSTEED, Esq., President, in the chair.

The Minutes of the previous Meeting were read and confirmed.

The PRESIDENT announced that the Ballot Lists for the election of New Members had been opened by a committee of the Council, and that the following fifty-four candidates were found to be duly elected :—

MEMBERS.

ASHLEY, HERBERT,	Portsmouth.
BOWER, WILLIAM NELSON,	Manchester.
CLARK, EDWIN KITSON,	Leeds.
CROFT, FREDERICK LISTER,	Bradford.
GREEN, EDWARD WILLIAM,	London.
HALL, JOSEPH,	Cheltenham.
HEALD, JAMES ARTHUR,	Cardiff.
HOMERSHAM, THOMAS HENRY COLLETT,	Bradford.
JONES, GEORGE PHILLIPS,	Loughborough.
McLAREN, JOHN,	London.
ROGERS, HERBERT MALCOLM,	London.
ROY, WILLIAM,	London.
SHANKS, ARTHUR THORNTON,	Colombo.
WILKINSON, HENRY DANIEL,	London.
WRIGHT, THOMAS,	Nottingham.

U

ASSOCIATE MEMBERS.

ADAMS, JORDAN DÁS, ...	London.
ANDERSON, EDWARD, ...	Warrington.
BATES, HERBERT, ...	Salford.
BERRY, HERBERT, ...	Bolton.
BULLEN, WILLIAM HENRY CHAMBERS, . .	London.
CARDEW, SYDNEY ROBERT, ...	London.
CURTIS, IVOR, R.N., ...	London.
CUTHELL, ERIC ANDREW, ...	London.
DAVIS, CHARLES RICHARD STANLEY, . .	Birmingham.
DENNIS, GEORGE STANLEY, ...	Manchester.
DEWS, JAMES WALTER, ...	Bradford.
HUDSON, ERNEST VICTOR, ...	Dursley.
JOHNSON, HILTON, ...	London.
LEAN, CLEMENT, ...	Leeds.
MACKENZIE, THOMAS ALEXANDER, . . .	London.
MARSDEN, FRED, ...	Bradford.
PAGE, WILLIAM CHRISTOPHER, . . .	London.
POWNALL, JOHN EDMUND, ...	London.
PRESTON, FREDERICK KENNERLEY, . . .	Manchester.
RENWICK, ROBERT, ...	Horsham.
REXWORTHY, HAROLD SIBREE, . . .	London.
ROBB, WILLIAM, ...	London.
ROBERTSON, GEORGE FINDLAY, . . .	Manchester.
SOPWITH, JOHN, ...	London.
STANGER, REGINALD HARRY HURSTHOUSE, .	London.
TWEEDIE, KENNETH DIGBY, ...	London.
WILLIAMS, JOHN ROBERT, ...	Sheffield.
WILSON, FRANK, ...	London.
WILSON, ORIN SALATHIAL, ...	London.

GRADUATES.

BEECHING, FRANK WILLIAM, ...	Birmingham.
HAGGARD, PERCY LESTER, ...	London.
HEADLEY, HALLOWELL, ...	London.

NESBITT, EDWARD ARTHUR, ... London.
ROBERTS, EDWIN GILBERT LLEWELLYN, . . London.
SAYNOR, LEONARD HENDERSON, . . . Loughborough.
SEAGER, FRANCIS PERCY, ... Sheffield.
SWARBRICK, FREDERICK, ... Loughborough.
TRAYTE, WALTER GEORGE, ... Woolwich.
TRENCH, WILLIAM CONNELL POWER, . . Shifnal.

The PRESIDENT announced that the following Transference had been made by the Council since the last Meeting :—

Associate Member to Member.

DONALD, DAVID BOSWELL, ... Redruth.

The PRESIDENT then delivered his Inaugural Address.

The following Paper was subsequently read and discussed :—
"The Education of Engineers in America, Germany, and Switzerland;" by Professor W. E. DALBY, *Member*, of London.

The Meeting terminated at Ten minutes past Ten o'clock. The attendance was 153 Members and 61 Visitors.

ANNIVERSARY DINNER.

The Anniversary Dinner of the Institution was held at the Hotel Cecil, Strand, London, on Thursday evening, 23rd April 1903. The President occupied the chair; and the following Guests accepted the invitations sent to them, although those to whom an asterisk * is prefixed were unavoidably prevented at the last from being present.

The Right Hon. the Earl of Selborne, P.C., First Lord of the Admiralty; * The Right Hon. Lord Strathcona and Mount Royal, G.C.M.G., Agent-General for the Dominion of Canada; The Right Hon. Lord Kelvin, G.C.V.O., O.M., D.C.L., F.R.S., Honorary Member; Sir Arthur Lawson, Bart.; * Sir A. John Durston, K.C.B., Engineer-in-Chief, Admiralty; Captain Sir G. R. Vyvyan, K.C.M.G., R.N.R., Deputy Master of the Trinity House; Sir James Williamson, C.B., Director of Dockyards; Sir John Gunn, President of the Institute of Marine Engineers; Sir Philip Magnus, Secretary, Technological Department, City and Guilds of London Institute; *Colonel E. Bainbridge, C.B., Chief Superintendent, Royal Ordnance Factories; Captain Richardson Clover, Naval Attaché, United States Embassy; Mr. H. F. Donaldson, Deputy Director General, Royal Ordnance Factories; Mr. Peter Samson, Marine Department, Board of Trade; Lieut.-Colonel Horatio A. Yorke, R.E., Chief Inspecting Officer of Railways, Board of Trade.

* The Hon. Henry Copeland, Agent-General for New South Wales; The Hon. H. Allerdale Grainger, Agent-General for South Australia; * The Hon. H. Bruce Lefroy, Agent-General for Western Australia; Mr. Thomas E. Fuller, Agent-General for Cape Colony.

Mr. J. C. Hawkshaw, President of the Institution of Civil Engineers; Mr. Aston Webb, President of the Royal Institute of British Architects; Mr. Arthur Vernon, President of the Surveyors' Institution; * Mr. J. Patten Barber, President of the Society of Engineers; Mr. E. G. Constantine, President of the Manchester Association of Engineers; * Mr. T. Hurry Riches, President of the South Wales Institute of Engineers; Mr. Philip M. Justice,

President of the Chartered Institute of Patent Agents; * Mr.
Silvanus Trevail, President of the Society of Architects; Mr.
Hardman A. Earle, Chairman, Manchester Section of the Institution
of Electrical Engineers; * Dr. J. H. T. Tudsbery, Secretary of the
Institution of Civil Engineers; Lieut.-Colonel R. M. Holden,
Secretary of the Royal United Service Institution.

Mr. T. Iliffe Weston, Master of the Salters' Company; * Mr.
John D. Gregory, Master of the Haberdashers' Company; * Mr. F. W.
Manson, Master of the Ironmongers' Company.

Mr. Arthur Greenwood, Chairman, Leeds Meeting Committee;
Mr. T. P. Reay, Treasurer, Leeds Meeting Committee; Mr.
Christopher W. James and Mr. E. Kitson Clark, Honorary
Secretaries, Leeds Meeting Committee.

Mr. Oliver R. H. Bury, General Manager, Great Northern
Railway; Professor W. E. Dalby; Dr. William Garnett, Secretary,
Technical Education Board, London County Council; Dr. R. T.
Glazebrook, F.R.S., Director, National Physical Laboratory; Captain
C. C. Longridge; * Mr. James Mansergh, F.R.S., Chairman,
Engineering Standards Committee; Mr. R. A. McLean, Auditor;
* Mr. H. L. Millar, Treasurer; Mr. James Rowan; * Fleet-Engineer
Thomas W. Traill; Mr. Charles J. Wilson, F.I.C.

The President was supported by the following officers of the
Institution :—*Past-Presidents*, Sir Edward H. Carbutt, Bart.; * Mr.
Samuel W. Johnson; Mr. William H. Maw; Mr. E. Windsor Richards,
and * Sir William H. White, K.C.B., LL.D., D.Sc., F.R.S. *Vice-
Presidents*, Mr. John A. F. Aspinall; Mr. E. B. Ellington; Mr.
Arthur Keen; and Mr. A. Tannett-Walker. *Members of Council*,
* Mr. Henry Chapman; * Mr. William Dean; Dr. Edward Hopkinson;
Mr. Henry Lea; The Right Hon. W. J. Pirrie, LL.D.; Mr. John
F. Robinson; Mr. Mark Robinson; and Mr. John W. Spencer.

After the President had proposed the loyal toasts, Sir Edward H.
Carbutt, Bart., Past-President, proposed that of " The Imperial
Forces."

The Right Hon. the EARL OF SELBORNE, First Lord of the Admiralty, in responding, said that within twelve months they had been fighting a civilised enemy in South Africa, under conditions which the best judges considered could never be exactly reproduced, and carrying on campaigns also in East and West Africa against totally different races and in countries with totally different physical features, the only similarity being the heat under which the operations had to be carried on. The war in Somaliland would be memorable as the first in which an organised attempt had been made to utilise the great invention of Marconi. At sea, wireless telegraphy had reached a point when the fact of any ship being unprovided with the apparatus might be thought to have a grievance. It was of the utmost possible convenience, and the certainty of its use never varied within a reasonable radius of action. It had been tried in peace, but how far it could be used in war time, when both combatants were provided with it, remained to be seen. This invention was an illustration of the fact that every application of science to modern warfare enlarged the field of engineering. Though the engineer brought into the science of war a continuous series of new inventions, he now and then left behind him problems only partially solved. Great as the difficulties of the Admiralty were in respect to boilers, they were greater and more troublesome in regard to condensers, and there awaited an enthusiastic welcome at the Admiralty for the man who first supplied a condenser that would cause no trouble or anxiety, just as a warm welcome also awaited the man who invented a powder and a gun that would be reasonable in the matter of erosion. On the subject of naval gunnery, comparisons were made as if all ships were alike, and the shooting took place with precisely similar guns and under exactly the same conditions. Nothing could be farther from the facts. A six-inch gun fired from a platform such as the "Powerful" was not quite the same thing as a six-inch gun even of the same mark fired from the deck of a second-class cruiser, and six-inch guns were not all alike. The difference in flatness of trajectory between the various marks was extraordinary. Therefore, comparisons might inflict a serious injustice. It had been said that a large number of naval officers were indifferent to the subject of gunnery, and cared only about the paint on their ships being bright. That was a ludicrous travesty of the truth. Any naval officer who did not place good gunnery as a first responsibility was unfitted for his position, and that was what naval men themselves felt. He had had opportunities of discussing gunnery with many flag officers and captains, and he had not yet found the officer who did not consider that if he had not done his utmost to bring the unit under his command to the highest possible perfection of shooting he had failed in his duty.

Major - General F. G. Slade, Inspector - General of Royal Garrison Artillery, also responded.

The toast of " Our Guests " was proposed by the President, and acknowledged by Lord Kelvin, G.C.V.O., O.M., Honorary Member, who expressed a hope that mechanical engineers would not consider inventions in small-instrument making beyond their province. The toast was also acknowledged by Captain Sir George R. Vyvyan, K.C.M.G., R.N.R., Deputy Master of the Trinity House.

Mr. E. Windsor Richards, Past-President, proposed the toast of "Kindred Societies," which was acknowledged by Mr. J. C. Hawkshaw, President of the Institution of Civil Engineers.

The toast of " The Institution of Mechanical Engineers " was proposed by Mr. Aston Webb, President of the Royal Institute of British Architects, and was acknowledged by the President, who said that the architect had something over and beyond what the engineer possessed, because besides having the science of construction, he had the sense of beauty, and beauty was something that transcended science. Sir Christopher Wren had had sense of beauty, yet in designing the splendid dome of St. Paul's, he did not scorn the aid of Vulcan, but introduced iron links to ensure the permanent stability of the structure. The President mentioned that the membership of the Institution of Mechanical Engineers now exceeded 4,000, and that the increase had been more than 10 per cent. last year. A baby institution could easily double its numbers in its second year, but when an institution was fifty-six years old, a 10 per cent. increase was a sign of strong growth.

ADDRESS BY THE PRESIDENT,

J. HARTLEY WICKSTEED, Esq.

Let me thank you for electing me to the honourable position of President of this great Institution, of which George Stephenson was the first President in 1847, and which now numbers nearly 4,000 members. We are associated as members of this Institution for the purpose of promoting " The Science and Practice of Mechanical Engineering." The history of Mechanical Engineering is co-existent with the history of Iron. As the production of Iron has developed, so have the works of the Mechanical Engineer. His monuments are in iron; with it he builds ships, guns, engines, and all sorts of machines, by which the labour of man is lightened and is increased in its productiveness, and the circulation of mankind through the arteries of the world is increased in volume and rapidity. A current of humanity is created through all the nations, and this is the very contrast to congestion and stagnation. It carries ideas and knowledge on its stream, and also the produce and specialities of distant climes and races. It stimulates production in all industries, and what industry is there to which the Mechanical Engineer has not lent his assistance?

Even in agriculture, where soil, water, and sunshine are the essential elements of production, the Mechanical Engineer has taken a hand. De Laval, with his iron bowl revolving at a speed which would disperse an earthen vessel by centrifugal force, has so intensified the force of gravity as to compel the creamy particles to separate from the rest of the milk in fewer seconds than it would take hours, if left to the action of gravity alone.

Let us trace the beginning of this iron, this great ally of the Mechanical Engineer. There is every reason to suppose that, in countries where iron ore existed on the surface, the knowledge of iron would be primeval. We only need to pre-suppose the possession of fire, and then imagine some one leaving a grid made of sticks daubed with red or brown hematite ore a few hours in a charcoal fire, and the result would be a lump of malleable steely iron in the embers. We can therefore assign no date to the first knowledge of iron, but in the earliest written records to which an accurate date can be fixed, namely, in the fourth millenium B.C., we find Pyramid texts which prove beyond question that iron was well known in Egypt at that time, and that it was forged into weapons, tools, and instruments. After an obscure existence of at least 3,000 years, iron became historically famous.

The time of Homer, 880 B.C., was notable for the attention that was given to iron. At that time it must have been cheaper than bronze, as is evidenced by the fact that objects have been dug from the mounds of Nineveh of about the time of Homer, and of which some consisted of cores of iron, round which bronze had been cast. Concurrently with this extended use of iron we shall find evidences of the primitive Science and Practice of Mechanical Engineering. For example, in the wall sculptures from Nineveh, with which Layard has enriched the collections of the British Museum, there is one in the Nimrûd gallery illustrating the campaigns of Assur-nazir-pal, 884 B.C. It shows a battering-ram and a chain to catch up the head of the ram to destroy its action. The beam of the battering-ram is hinged near the centre, and it is swayed up and down from the light end by men behind the protection of a barricade. The chain is hung down in a loop from the walls of the besieged city to catch the heavy end and break its force as it descends against the wall. Here we have primitive Mechanical Engineering contemporary with the Iron Epoch of the Mycenæan civilization. It is true that this marble sculpture does not inform us of what material the chain and ram-head were made ; the ram-head was probably a bronze casting, and the chain of forged and welded iron. A few centuries later, a chain for a similar purpose is specifically described by Thucydides, who says that the Plateans, during the siege of their city by the Thebans, 429 B.C., made use of long iron chains to suspend beams which they dropped, so as to break off the heads of the battering-rams brought up against their city. At this same period iron was known to make such formidable instruments of war, that, as is recounted by Pliny the Elder in the treaty which Porsena granted to the Roman people on the expulsion of the Kings (509 B.C.), there was a specific provision that iron was not to be used except in agriculture, and the most ancient authorities have preserved the fact that it was then that writing with a bone style came into practice. A classical friend has sent me a passage from Pliny's Natural History, Book 34, Cap. 39, which is prophetic of the importance of iron, and which is so quaint and poetical that I will now record it : "The first mines we should mention are those of iron ; iron is the best and worst implement of life, inasmuch as therewith we cleave the earth, saw trees, prune bushes, compel our vines to renew their growth yearly by shearing off their growth, build houses, shape stones, and for all other purposes use iron; but we likewise use it for wars, slaughters, robberies, not only at close quarters, but also as a swift missile, now hurled by engines and now by the arm and sometimes even winged, which latter I take to be the most infamous perversion of the human intellect, inasmuch as we manufacture this monstrous bird and give the iron wings that death may reach the man the quicker. Wherefore the blame of man must not be set down to the account of nature."

Mark the words "swift missiles hurled by engines." The Mechanical Engineer gave "the iron wings." He has increased the speed of its flight since then by putting gunpowder behind it. He has put steam into it and made it run and swim, and he has charged it with the unseen forces of magnetism and electricity. There is no writer of Latin of whom any considerable fragments are preserved earlier than Ennius, B.C. 239–160, and he used the word "ferrum" indiscriminately for war, sword, or iron. About his time iron began to emerge from its primitive application, to instruments of war and husbandry, and we shall find science beginning to be coupled with practice in developing its use, for in the 3rd century B.C. a great mechanical engineer came into the world, Archimedes, who, when a youth, went to Alexandria to the Royal

School of the Ptolemies, of which Euclid had been the ornament some half a century before. There he was educated in all the science of his day. On his return to his native city in Sicily he devoted himself to geometrical investigations, and by his great energy and inventiveness went far beyond what had been previously attained. Combined with his remarkable faculty of analysis was a power of practical application which enabled him to establish the science of engineering upon a solid mathematical basis. Among other achievements he made a burning glass; he discovered the principle that a body immersed in a liquid sustains an upward pressure equal to the weight of the liquid displaced. He measured the area of a circle by inscribing and circumscribing it by two polygons, each of ninety-six sides, and taking the area of the circle as lying between the measured areas of those two polygons. Thus he founded the fundamental principle of accurate measurement, viz., the establishment of upper and lower limits, the difference between the two being the gauge of the accuracy attainable or required. He also wrote many treatises, and one on arithmetical numeration applied to reckoning grains of sand, which contains the germ of the modern system of logarithms. He invented the Archimedian screw for elevating water, and he invented the steel-yard or Roman balance. By this he established the exact multiplying power of the lever mounted upon knife-edges and brought it into commercial use, a use which has never been superseded, but which has been so developed that in our day we measure loads up to 450 tons with steel-yards having a sensibility of 1 in 10,000.

During all the centuries which have been reviewed, and for many centuries afterwards, iron was only wrought by the hammer and was not run molten into moulds as bronze was; only half its properties were known, and its paramountcy among metals was not then established. The iron of antiquity was made direct from the ore, and was spongy malleable iron which could be made more or less steely; and it was only as reducing furnaces were enlarged and the blast increased that it came about within the last 400 years that cast-iron was produced on a commercial scale. Up to that time, bronze held the field for objects which could not be shaped by hand-hammering. The best authorities give the date 1490–1500 A.D. for the discovery of cast-iron, and it is remarkable that this discovery exactly corresponds with the Revival of Letters in England, when learning and art took fresh life, and simultaneously with which the world was enlarged by the discovery of an immense continent in the Western Hemisphere. From this time iron became as tractable as bronze, and the Iron Foundry was added to the Forge. The worker in iron of old days was typified by the Cyclops with the one glowing eye, as of a red-hot forging. The descendants of that race of giants have been gifted with a second eye with a glow as of the molten metal. With this extended outlook, the vocation of the Mechanical Engineer has gradually enlarged, slowly at first but with an increase like that of compound interest. In 1705 Newcomen made his steam-engine. In 1784 Henry Cort puddled pig-iron, made grooved rollers and rolled it into bars with a James Watt rotative steam-engine. Since this comparatively recent date, iron has spread into so many lines that I can only attempt slightly to follow one of them, which is a line that has entirely developed within the very last century.

Before 1808 iron had found its way into ships in the form of guns, boilers, and engines, but until that time all ships' cables and rigging were made of hemp. The substitution of iron cables for hempen ones directly led to testing the strength of materials, and it is this line of mechanical engineering which we will briefly follow.

When the first chain cables for ships were made by Robert Flinn and Captain Brown, they were found to be in every way superior to those of hemp, provided only that they were free from any single faulty link. From the first Captain Brown was of opinion that there was nothing more essential in completing an iron cable than the most rigid attention to proving. In the year 1812 chain cables were made by Brown at Millwall, and he had a machine made for proving them "which, by enabling him to put as great a strain upon the cables as was likely ever to be brought upon them when in use, thus detected any defective materials or insufficient workmanship" (see letter of John Knowles). The firm of Brown Lenox is still in possession of the original sketches made by Captain Brown of his machine. It was a 100-ton machine, consisting of a long bed, at one end of which was a powerful double-geared crab turning a chain barrel. Two side-link chains wound round the barrel, stapled to the two outer ends ; they converged as they coiled up towards the centre of the barrel, and thus united to give a central pull on the shackle to which the test-chain was attached. At the other end of the bed there were a pair of compound levers having a ratio of 200 to 1. The levers had knife-edged main fulcra, but instead of having knife-edged load fulcra they had rounded butts on which the side-link chains were placed, and from which they pulled at tangents, so that the fulcrum distances were the distances from the centre lines of these chains to the knife-edged fulcra. The upper lever carried a scale pan for weights.

Brown not only used his proving machine to prove his chains, but he tested the mechanical properties of the iron of which he wished to make the chains, and by so doing gave an impetus to testing which was much required, and which did much to place the country on a better footing with reference to the quality of iron and chain cables (see Chain Cables and Chains by Fleet Engineer Thos. Traill). This was the starting of testing the strength of materials in a commercial sense, and in 1829 a 130-ton vertical testing machine was designed and built at Cyfarthfa Works by Mr. Wm. Williams, engineer to the works. With this machine Mr. Williams made experiments on riveted joints with single butt straps, and also upon iron bolts up to 2 inches square. This machine had hydraulic cylinders for straining the specimens and knife-edged levers for measuring the stress. In the year 1831 the vessels of the Royal Navy were directed to be supplied with chain cables on a more extended scale, and the Admiralty ordered their first chain-testing machine from the firm of Messrs. Bramah, of Pimlico. The machine was finished in 1832, and was started in regular work at Woolwich Dockyard in April 1833. It was only as a result of all this careful testing that, in 1836, the underwriters ceased to charge a higher premium for vessels that had iron cables than for those that had hempen ones.

In the year 1846 the rules of Lloyd's Register were such that their surveyors were supposed to see that all new chains supplied to classed vessels were tested and that the test applied was marked on each length. In 1853 the rules of Lloyd's Register made it imperative that, before a vessel could be classed, a certificate should be produced as to the test of the chain cables, and in 1860 the rule was extended to anchors.   The use of testing machinery, originated for the purpose of proving chain cables, has been extended in many directions.   Researches in the strength of materials of construction have been pursued by many of the ablest mechanical engineers in this country and abroad, and there still remains a field open for experiments, especially on the strength of full-sized structures.   The French Government are now putting down a machine in Paris which will test a strut or column 90 feet long and 1 mètre square in cross section, and will impose a load upon it of 300 tons.   Experience has abundantly proved that testing iron conduces to improvements in its manufacture.   No doubt some people think tests an unnecessary expense, and others have thought that proving a chain makes it permanently weaker, like overstraining the muscles of a strong man ; and if the testing be not conducted rationally there is truth in these arguments.   But if, when you come to insure your ships or your boilers you save more on the premium than you have spent in satisfying the tests, you have a proof that the testing has been good economy.   So much has this proved to be the case that all the most successful shipbuilders buy and pay for tested material to put into their ships, whether those ships are to be classed at Lloyd's or not. It does not pay them to put work upon material which may afterwards have to be rejected and replaced ; with boiler-makers, gunmakers, and other mechanical engineers it is the same.   It is true that the Government step in with their Chain Cables and Anchors Act for the protection of life at sea, and make the testing of chain cables and anchors compulsory, which shows that enlightened economy is not alone sufficient to restrain unprincipled competition, but all chain-cable makers are glad of the Act on economic grounds, whether they be of a philanthropic turn of mind or not.

The leading railway companies also, for their goods department, without compulsion by any Act of Parliament, are now making a practice of proving all the sling chains and crane chains on their system in the most methodical manner, and are finding that the saving in casualties more than pays the whole expense of the testing department.

By testing the iron the mechanical engineer knows that his raw material, whether cast, malleable, or iron made into steel, has certain strengths and elasticities which he can count upon, and this knowledge prevents labour being wasted upon a quality of material unsuitable for the particular purpose he has in view.

May not the same principle of ascertaining the quality of the raw material applied to the case of his apprentices, before work is put upon them, may this not lead to similar economic results in producing a larger percentage of efficient mechanical engineers?

May we not regard our Technical Schools and Universities as the proving houses of the raw material for the engineering apprentice? May we not look for some exact knowledge of their mental and moral strength and elasticity before we assign a position for them in the structure of a mechanical engineer's business. The Government comes in with its Education Act, so that no one can evade their contribution to the general economy which is expected to result from education. The compulsory schools prepare the raw material for the proving houses, where they may be classed and certificated. The question of preparing and testing human capabilities is infinitely more complicated than that of testing iron, but it is quite as important for our welfare, and in proportion as it is more complicated it requires more consideration.

There are those who think scientific education an unnecessary expense, and that examinations overstrain the brain and disable it, and there is truth in these arguments if the education and testing be more severe than the physical raw material is able to bear; but without tests we are liable to encumber our works with apprentices, who do not possess the inherent qualifications for success and who turn out costly failures. There can be little doubt that the mechanical engineer of today needs more education than has sufficed to make him successful in the past. Formerly he rang the changes on the six mechanical powers, with heat, steam, the laws of motion, and hydraulics. Now electricity is added to the chimes, and the number of changes on his peal of bells is multiplied by a fresh factor. These considerations seem to make the most suitable education for a mechanical engineer of such pressing importance that I have thought a Paper and discussion on the subject this evening, which is the last of our present session, will be a much more profitable use of your time than a prolongation of general remarks in a Presidential Address.

The Paper which will follow has been expressly prepared for this occasion. It deals with facts and object lessons industriously gathered by travel and expert investigation. It is presented to us by a Member competent to analyse the facts ascertained. The Paper will be followed by your own contributions in discussion, and I hope you will now gladly devote your attention to that great question.

———

Sir Edward H. Carbutt, Bart., Past-President, said that, before the next Paper was read or the discussion commenced, he was sure that all present would wish that somebody on their behalf would propose a vote of thanks to the President. There were a great many disadvantages in growing old, but there was one advantage and one pleasure, namely, that if in one's youth or when one was a young man one had seen boys and become friendly with them, and found those boys grow up and take the high position which the President of the Institution did that day, it was a great pleasure added to life. He knew the President, he could not quite say how long ago, but he had followed his career through life and watched his doings, and he had always made for that which should be the object of everyone, to add to the sum of happiness of mankind, always endeavouring to leave the world better than he found it. He believed Mr. Wicksteed had done an immense amount of good. He had been elected to that proud position by his fellow-members, and that showed that his work in the world had been of some service. Mr. Wicksteed had given more attention than any man he knew to perfecting the testing machine, and by that means he had come across nearly all the Professors in the different technical schools and colleges, and that had given a bent to his mind the result of which had been seen in

x

(Sir Edward H. Carbutt, Bart.)

the Address just delivered. It showed that while he had been a
Mechanical Engineer he had kept up his thought for the improvement
of mankind in other ways, and had therefore thought it necessary to
see what could be done for education. He, Sir Edward, was
astonished to hear from Mr. Wicksteed that testing machines had
been in use so early, because he had a vivid recollection when serving
his apprenticeship with Palmers Shipbuilding Co., forty years ago,
that when the firm were using plates for boilers which would only
stand 15 lbs. per square inch steam pressure, he was required to test
some puddled steel plates. There was no testing machine anywhere
in the neighbourhood, and he had to test by dead weight. Since
that time, and since the colleges had taken up with the new testing
machines, as Mr. Wicksteed had said, and that in Paris a machine
was going to be erected which would test up to something like 3 feet
square, there was a great addition to knowledge of mechanical
engineering. The President had given his experience, and he was
sure it would be the wish of the members to pass a hearty vote of
thanks to him for his Address.

Dr. ALEXANDER B. W. KENNEDY, Past-President, said he need
hardly say it gave him very great pleasure to second the resolution,
not only because Mr. Wicksteed was a very old friend of his, but
also in recognition of the extremely ingenious way in which he had
worked out his ideas in the Address from the steel at the beginning
to the student at the end of it. He admired greatly the analogy which
he had found between unfortunate pieces of steel and iron which were
twisted and pulled and rent asunder, and the much more fortunate
young man who was to be treated, he understood, more or less in the
same way, but always within the limits of elasticity ! He had very
much pleasure in seconding the vote of thanks.

The vote of thanks was carried unanimously.

The PRESIDENT acknowledged the cordiality with which his
Address had been received.

THE EDUCATION OF ENGINEERS IN AMERICA, GERMANY, AND SWITZERLAND.

By Professor W. E. DALBY, *Member,* of London.

The President honoured the author by a request to prepare a Paper on this subject, because he thought of all the questions which might occupy their attention that of the Training of Engineers was one of the most important. The facts he was about to lay before the members were taken from reports on this subject which he had made for Mr. Yarrow.

The subject is one of increasing importance in view of the fact that they have now to meet in competition increasing numbers of scientifically-trained engineers of other countries.

Long before their competitors had any engineering industries at all, the British were fortunate in the possession of flourishing manufacturing concerns which practically gave them the monopoly of the world's trade, a monopoly partly due to their fortunate possession of coal and iron, but chiefly due to the genius, inventiveness, perseverance, shrewdness and business acumen of the great British engineers of the last century, many of whom may be found on the roll of Past-Presidents of this Institution. As the

x 2

industries developed, the method of training apprentices gradually took form and became a system, and probably most of those present have served an apprenticeship under some form of this system, namely a seven or five years' apprenticeship in the works. With scientific progress, changing methods of manufacture, and the advent of electricity as a necessary part of the engineering equipment of every mechanical engineer, there has been scarcely any change in the recognised method of training engineers. Sir Joseph Whitworth, President of this Institution in 1856-7, perceived that the engineers of the future must be trained in scientific principles as well as in workshop practice, and even so long ago as 1868 tried to find a remedy by the foundation of his scholarships—scholarships which have done a vast amount of good, but, he ventured to think, not entirely in the direction Sir Joseph Whitworth contemplated. In recent years, courses of instruction in the scientific principles of engineering have multiplied at the Universities and at Institutions of kindred types and equal standing from a purely engineering point of view. At the present time, therefore, there is no difficulty in obtaining a scientific training of a high character. The author thinks it will also be conceded that a youth can get a training in workshop practice in the first-class factories of this country second to none in the world. The weak point in the system is, however, the want of co-ordination between the workshops and the colleges. Many employers look askance at a college-bred youth, and there is no doubt that many college youths quite deserve it. But not more so than many who are trained entirely in the works. It is too often forgotten that a college cannot give a youth ability. All that can be done is to train what abilities he happens to bring to the college with him.

Presuming that the members all agree that the training of an engineer should be partly in scientific principle and partly in workshop practice, it is, he thinks, a fit subject for discussion in this Institution as to what course of training is the best adapted for carrying this principle into effect, so that future engineers of this country shall not be at a disadvantage in any one respect in comparison with the engineers of other countries.

As a basis of discussion, the author proposes to state a few facts in connection with the training of engineers in the United States of America, Germany, and Switzerland.

THE MASSACHUSETTS INSTITUTE OF TECHNOLOGY, BOSTON.

The foundation of this world-famed Institution was primarily due to the foresight and enthusiasm of William Barton Rogers, whose object was "to provide a complete system of industrial education supplementary to the general training of other Institutions, and fitted to equip the students with every scientific and technical principle applicable to the industrial pursuits of the age." Regular courses of instruction began in October 1865. The education he sought to provide, "although eminently practical in its aims, has no affinity with that instruction in mere empirical routine which has sometimes been vaunted as the proper education for the industrial classes. We believe, on the contrary, that the most truly practical education, even in an industrial point of view, is one founded in the thorough knowledge of scientific principles, and which unites with habits of close observation and exact reasoning a large general cultivation. We believe that the highest grade of scientific culture would not be too high a preparation for the labours of the mechanic and manufacturer, and we have in the history of social progress ample proofs that the abstract studies and researches of the philosopher are often the most beneficent sources of practical discovery and improvement."

This is the creed of the founder of the Institution, and is, the author ventures to say, that of those who so ably guide its fortunes at the present day. Small wonder that the Institution is such a feature in the educational life of the United States.

At the present time there are thirteen courses of undergraduate study. These and the numbers of students in each for the year 1901–2 are given in Table 1 (page 284).

TABLE 1.

Course.	Number of Students.
1. Civil Engineering	102
2. Mechanical Engineering	129
3. Mining Engineering and Metallurgy. . .	76
4. Architecture	40
5. Chemistry	35
6. Electrical Engineering	96
7. Biology	6
8. Physics	13
9. General Studies	9
10. Chemical Engineering	30
11. Sanitary Engineering	14
12. Geology	1
13. Naval Architecture	39
13A. Naval Architecture. A special course arranged for Naval Cadets (just started) .	3
	593

These numbers do not include first-year students, because choice of a course is not made until the second year. The total registration for the year 1901-2 was 1,415.

It will be seen that there is ample opportunity for specialising in different directions. The course lasts four years, and in the fourth year further options of specialising are offered, as will be seen from the course of study in mechanical engineering given below:—

Course of Instruction for Mechanical Engineers, (Course 2).

FIRST TERM.

(Common to all Courses.)

Algebra.	Freehand Drawing.
Plane Trigonometry.	French (or German).
Inorganic Chemistry; Chemical Laboratory.	Rhetoric and English Composition.
Mechanical Drawing.	Military Science.

SECOND TERM.

COURSES I. and XI.

Analytic Geometry.
Spherical Trigonometry.
Inorganic Chemistry; Qualitative Analysis; Chemical Laboratory.
Mechanical Drawing and Descriptive Geometry.
Freehand Drawing.
French (or German).
United States History.
Military Science.

COURSES II., III. (2), VI., VIII. (3), X., XIII.

Analytic Geometry.
Theory of Equations.
Inorganic Chemistry; Qualitative Analysis; Chemical Laboratory.
Mechanical Drawing and Descriptive Geometry.
Freehand Drawing.
French (or German).
United States History.
Military Science.

SECOND YEAR.

FIRST TERM.

Principles of Mechanism.
Drawing.
Carpentry and Wood-turning.
Differential Calculus.
Physics: Mechanics, Wave-motion, Electricity (lectures).
Descriptive Geometry.
German (or French).
English Literature.
European History.

SECOND TERM.

Mechanism: Gear-teeth; Machine-tools; Cotton Machinery.
Drawing.
Pattern Work.
Foundry (elective).
Integral Calculus.
Physics: Electricity, Optics (lectures).
German (or French).
English Literature and Composition.

THIRD YEAR.

FIRST TERM.

Steam Engineering: Valve-gears; Thermodynamics.
Drawing.
Industrial Electricity.
Dynamo-electric Measurements.
Forging.
Elements of Differential Equations.
Physics: Heat.
Physical Laboratory.
General Statics.
German (or French).
Political Economy.

SECOND TERM.

Steam Engineering; Boilers.
Drawing, Design, and use of Surveying Instruments.
Engineering Laboratory.
Forging; Chipping and Filing.
Physical Laboratory.
Strength of Materials: Kinematics and Dynamics.
German (or French).
Political Economy and Industrial History.
Business Law.

FOURTH YEAR.

FIRST TERM.	SECOND TERM.
Steam Engineering.	Hydraulic Motors.
Machine Design.	Engineering Laboratory.
Hydraulics.	Machine-tool Work.
Dynamics of Machines.	Strength and Stability of Struc-
Engineering Laboratory.	tures; Theory of Elasticity.
Chipping and Filing; Machine-	Foundations.
tool Work.	Industrial Management.
Strength of Materials; Friction.	Thesis.
Heating and Ventilation.	*Options.*
Metallurgy of Iron.	1. Marine Engineering.
Options.	2. Locomotive Construction.
1. Marine Engineering.	3. Mill Engineering.
2. Locomotive Construction.	4. Heating and Ventilation.
3. Mill Engineering.	
4. {Heating and Ventilation. Dynamo-electric Machinery. Hygiene of Ventilation.	

The students in this subject are bound to take this course, and examinations are held at regular intervals.

Laboratories belong to each department, and at the present time new buildings are in course of erection for the physical and electrical department, costing with their equipment £80,000.

All the courses given in Table 1 (page 284) lead to the degree of Bachelor of Science (S.B.).

There are advanced courses of study given which lead to the degree of Master of Science, after one year's work.

The average age of the undergraduates is $18\frac{3}{4}$.

CORNELL UNIVERSITY.

Cornell University, the existence of which is due to the combined bounty of the United States, the State of New York, and Ezra Cornell, was opened on the 7th October 1865. The object of the University is appropriately expressed in the words of the founder, Ezra Cornell : " I would found an Institution where any person can find instruction in any study." The object of the founder is realised to the extent indicated by the following departments, stated as being comprehended in the University in 1902. The number of students in each department is given also for the year 1900–01 :—

The Graduate Department	205
The Science and Arts Department	755
The College of Law	182
The College of Civil Engineering	183
The Sibley College of Mechanical Engineering and the Mechanic Arts	661
The College of Architecture	52
The College of Agriculture	193
The College of Medicine	347
The New York State Veterinary College	42
The New York State College of Forestry	23
	2,643
Summer Session	424
	3,067
Names counted twice 174	87
	2,980

There is a teaching staff of nearly 400 members, some non-resident.

Less than forty years ago the site on which the University stands was a bare hillside overlooking Lake Cayuga. Today one finds there one of the most picturesquely situated, one of the most beautifully grouped, set of University buildings in the world.

As it is the Education of Mechanical Engineers which is particularly under discussion, attention will be restricted to—

SIBLEY COLLEGE.

This college is divided into eight departments as follows:—

1. Mechanical Engineering.
2. Mechanical Laboratory Instruction.
3. Electrical Engineering.
4. Mechanic Arts (workshops).
5. Industrial Drawing and Art.
6. Machine Design.
7. The Graduate Schools of Marine Engineering.
8. The Graduate School of Railway Mechanical Engineering.

The staff consists of thirty-six teachers and instructors, and this number includes six professors and four assistant professors, and eight non-resident lecturers. The staff is inadequate at the present to deal with the numbers of students in the college.

The entries at Sibley College have increased from 63 in 1885 to 670 in 1901.

A great feature of Cornell is its workshops. Here instruction is given in pattern-making, moulding, forging, fitting and turning, and the work done in them is real. A student is advised to get as much academical education as he can before entering Sibley College, being told that "his success in the practice of his profession will be found to depend more and more in the future and always in a large degree upon the position he may be able to assume among men of education and culture." The courses at Sibley are organised solely to train men for the engineering profession.

All students in the college pass through the same course during the first three years. They may specialise in the fourth year in steam, marine, railway, or electrical engineering.

The course for mechanical engineering is shown in the following Table 2 (page 289).

The hours per week shown in the right-hand column of the Table are the hours for which a student must get credit. In the case of the workshops, drawing office and laboratories, credit is given for a fixed proportion of the hours actually worked. Thus to get credit for one hour in the schedule a student must work three hours in the workshops or drawing office, or two and a half hours in the laboratories. During the four years' course the student actually works 900 hours in the workshops.

The course given in railway mechanical engineering is given in Table 3 (page 290).

TABLE 2.
Regular Course.

	Hours per Week.	
	First Term.	Second Term
FRESHMAN YEAR.		
German or French	3	3
Analytic Geometry	4	—
Differential Calculus	1	2
Integral Calculus	—	3
Chemistry or Drawing	6	—
Drawing or Chemistry	—	6
Shopwork	3	3
In addition to the above the required Drill * must be taken.		
SOPHOMORE YEAR.		
Mechanics of Engineering	5	5
Descriptive Geometry	2	2
Physics	4	4
Chemistry or Drawing	6 or 5	—
Drawing or Chemistry	—	5 or 6
Shopwork	3	3
In addition to the above the required Drill must be taken.		
JUNIOR YEAR.		
Steam Machinery	—	4
Electrical Machinery	4	—
General Machine Design	1	4
Kinematics and Drawing	3	3
Materials of Engineering	2	—
Physical Laboratory	2	2
Mechanical Laboratory	3	3
Shopwork	3	3
SENIOR YEAR.		
Steam Engines and other Motors	5	2
Physical Laboratory	3	—
Mechanical Laboratory	2	1
Mechanical Laboratory	1	—
Engine Design (or 23)	3	2
Engine Design (or 22)	2	1
Shopwork	3	—
Thesis: Designing and Drawing, Mechanical Laboratory Investigations, Shopwork, time divided optionally, but subject to the approval of the Director	—	8
Elective	—	3-5
This course leads to the degree of Mechanical Engineer (M.E.)		

* Cornell received a grant of land under the Land Act of 1862, and must therefore give instruction in Military Tactics.

TABLE 3.

	Hours per Week.	
	First Term.	Second Term.
SENIOR YEAR.		
Rolling Stock (preceded by summer locomotive shopwork)	4	4
Designing of Railway Machinery . . .	3	(1)
Locomotive Testing (elective) 1 hour . .	—	(1)
Seminary	1	1
Electric Railways	1	1
Steam Engines and other Motors . . .	5	2
Mechanical Laboratory	3	1
Physical Laboratory	1	2
Shopwork)	2	—
Thesis	—	8
GRADUATE YEAR.		
Advanced Railway Mechanical Engineering .	5	5
Plant Designing	3	3
Seminary	1	1
Elective	6	6

A student does not enter on this course until he is in his fourth, that is, his senior year, and, as Table 3 shows, arrangements are made for continuing the instruction into the fifth year; so that a student anxious to take up this branch has first a three years' course in the fundamental principles of engineering, and then a two years' specialised course in railway work.

Admission to the course in the American college is by examination. To enter Cornell a student must be sixteen years of age, and to the Massachusetts Institute seventeen. The standard of examination is such that a youth from a good high school can pass.

There is no freedom left to the student regarding his course of studies when once he has chosen his department, except so far as is allowed in the options. He must go through a definite educational course of a definite number of hours' duration. Examinations are frequent, and promotion from one year to another depends upon the result of them. The courses are really a carefully-thought out and elaborately-organised species of educational drill. As a general rule a man must go through with it or fall out.

Königliche Technische Hochschule zu Berlin, Charlottenburg.
(Berlin Technical High School.)

The Berlin Technical High School is a State Institution, and is under the immediate direction of the "Cultur Minister," otherwise styled the "Minister for Ecclesiastical, Educational, and Medical Affairs."

The object of the school is to give a specialised training in industrial subjects founded on a preliminary scientific education. The course, lasting four years, begins with scientific subjects, and gradually becomes more technical until in the fourth year all the subjects are specialised.

Admission to the School.—German subjects are admitted to the school on the production of a Maturity Certificate from a German gymnasium or a Prussian real-gymnasium. The education given at the two kinds of schools corresponds somewhat with that given in the classical and modern courses of our public schools. The maturity certificate is obtained at the end of a nine-years' course. Those admitted by means of this certificate are styled *Students*.

Persons who cannot obtain or have not obtained this certificate can be admitted on school certificates of a lower value, but for the Departments of Architecture, Civil and Mechanical Engineering and Naval Architecture must, in addition, show that they have worked for at least one year in some works. Those entering in this way are styled *Hospitanten*.

Foreigners are admitted under special regulations only as *Students*.

The school has recently been given the status of a University, and *Students* may at the end of the course obtain a Diploma entitling them to be styled Diploma Engineer (written Dipl-Jng.), or the University Degree of Doctor Engineer (written Dr-Jng.).

Hospitanten are not allowed to proceed to a degree, but they may obtain a certificate of attendance.

The School is divided into six departments. These and the numbers of persons in each are given in Table 4 (page 292).

TABLE 4.

Students, Hospitanten, and Others (Hearers), in Attendance for the Winter Half-Year, 1902–3.

	Students.	Hospitanten.	Total.
1. Architecture	477	262	739
2. Civil Engineering	647	42	689
3. Mechanical Engineering :			
Specialising in Mechanical Engineering .	1319	180	1499
,, ,, Electrical Engineering .	270	51	321
4. Naval Architecture :			
Specialising in Naval Architecture . .	241	18	259
,, ,, Marine Engineering . .	106	17	123
5. Chemistry and Metallurgy :			
Specialising in Chemistry . . .	161	20	181
,, ,, Metallurgy . . .	169	11	180
6. General Science	6	—	6
Persons admitted under special regula- tions from affiliated Institutions. . }	—	—	80
Officers and Engineers from the Navy .	—	—	301
Total	3396	601	4378

The following Table 5 shows the teaching staff for the several departments and also the general staff. It should be understood that the students from all the departments follow courses of lectures given by the staff of the General Science Department in the earlier part of their work.

TABLE 5.

Teaching Staff.

	Architecture.	Civil Engineering.	Mechanical Engineering.	Naval Architecture.	Chemistry.	Science.
Professors	18	14	20	6	13	18
Priv. Dozenten	17	8	8	1	17	15
Construction Engineers . .	—	—	7	2	—	—
Lecturers	—	—	—	—	—	2
Assistants	1	1	13	3	15	4
Honorary Assistants . . .	53	33	67	9	10	27
Total	89	56	115	21	55	66

Total Teaching Staff 402
Office Staff 13
Library Staff 44
Attendants 8

Total 467

Academical Freedom.—A peculiar feature of this remarkable Institution is the jealous maintenance by the students of their right to choose their course of study as they please. This freedom is common to German Universities, and is similar to the academical freedom of Oxford and Cambridge Universities.

It follows that the educational authorities can only suggest courses of study in the respective departments, leaving the students free to follow their suggestions completely, or partially, or not at all. Nevertheless very complete and elaborate courses have been arranged, and as a rule are followed by the students.

The courses suggested in the departments of Mechanical Engineering are given in detail overleaf.

Course Recommended for Mechanical Engineering, Dept. 3.—For the first three years all students in this course follow a common plan. In the fourth year they may specialise in—

    (a) General Mechanical Engineering.
    (b) Railway and Locomotive Engineering.
    (c) Electrical Engineering.

Considering the lectures offered in detail, it may be remarked that the following list can be taken by any one student, the times being arranged to allow of this.

FIRST YEAR.

Experimental Physics.
Higher Mathematics.
Mechanics 1.
Descriptive Geometry.
Mechanical Technology 1.
Introduction to Machine Construction.
Experimental Chemistry.

SECOND YEAR.

Mechanics 2.
Mechanism.
Thermodynamics.
Strength of Materials.
Graphic Statics.
Cranes and Lifting Machinery.
Practical work in the Engine Laboratory.
Mechanical Technology 2.
Commercial Subjects.

THIRD YEAR.

Electricity.
Crane Design.
Steam-Engine Construction.
Hydraulic Machine.
Practical Work in the Engine Laboratory.
Masonry Construction.
Pumping, Compressing and Blowing Machinery
Steam Boilers.
Statics of Bridge Construction.
Finance and Banking.

FOURTH YEAR.

(a) *General Mechanical Engineering.*

Design of Hydraulic Machines.
Setting out Factories for various purposes.
Machine Construction.
Hydraulics.
Design of Boilers.
Practical work in the Engine Laboratory.
Ice-Making and Refrigerating Machines.
Practical work in the Electrical Engineering Laboratory.
Construction of Dynamos and Transformers.

(b) *Railway Engineering.*

Locomotive Carriage and Wagon Construction.   Permanent Way.
Railway working.
Setting out Factories, etc.
Iron Construction.
Iron Bridge Construction.
Electric Telegraph.
Practical work in the Engine Laboratory.
Practical work in the Electrical Engineering Laboratory.
Transport of Goods.
Construction of Dynamos and Transformers.

(c)

Practical work in the Engine Laboratory.
Construction of Dynamos and Transformers.
Practical work in the Electrical Engineering Laboratory.
Electrotechnics with relation to the Electric Telegraph.
Electric Measurements.
Electric Lighting.
Alternating Currents.
Electric Transmission.
Electric Railways.
The Physical Basis of Electrotechnics.
Electro-Chemistry.
Introduction to the Potential Theory.
Potential Theory and its Employment in the Study of Electricity.
Electric Waves.

In many of the foregoing subjects duplicate lectures are offered by different lecturers and professors, so that the student may select the particular lecturer who happens to treat that particular part of the subject in which he is most interested.

Remarks on the Engineering Course.—A feature to be noted is the absence of any attempt to teach workshop practice. Laboratory teaching is confined to the Engine Laboratory and the Electrical Laboratory, with a little practice in testing materials at the neighbouring Government testing establishment (Königliche mechanische-technische Versuchanstalt).

The Engineering Laboratory contains representative steam-, gas- and oil-engines, and hydraulic apparatus aggregating about 800-H.P., with the corresponding boiler plant. The shops for housing this plant are built away from the main school, and form a complete self-contained department under the direction of a Professor. The Electrical Laboratory is in the main building of the school, and has been formed by roofing over one of the courtyards.

The most striking feature of the course is the relatively large amount of time devoted to machine construction. (Under this head is included machine drawing, graphic statics, descriptive geometry, and the lectures connected with the various forms of machines, in which exercises in the drawing office are given.) This department is under the direction of Professor Riedler. Not less striking is the method of teaching the subject. Professor Riedler carries on a large engineering practice in the building, employing between twenty and thirty draughtsmen for the purpose. The majority of these men take part in teaching the subject, so that mechanical drawing and machine design are taught by practical draughtsmen engaged for the greater part of their time in actual designing. No better method than this could be devised, because to all intents and purposes the students are working under actual drawing-office conditions. There are a large number of rooms devoted to the teaching of mechanical drawing, and the engineering department in the main building of the school and in the recently-erected wing consists mainly of a succession of drawing offices.

There are various collections of apparatus forming small museums in connection with the various departments.   The most celebrated of these is the collection of Reuleaux's kinematic models.

A student passing through the course and graduating doctor-engineer has had a large amount of drawing-office practice of an advanced character, but has had very little practical work.   Whether this kind of training is the best is a matter of opinion, but the author thinks a course which makes less claim on the students' time for college work and allows more for practical work would in the long run produce better engineers on the average.   Special men it might suit admirably, but for the ordinary youth to arrive at the age of 23 or thereabouts, and still find himself with only one year's workshop experience is not, he ventures to think, a method of training suited to our English notions of what the training of a mechanical engineer should be.

L'Ecole Polytechnique Federale Suisse, Zurich.

This is a State Institution, managed by a Committee, of which Colonel Bleuler is the President, and Mr. Naville Vice-President.

The object of the school is to give a specialised training in industrial subjects.   The course lasts four years.

Admission to the School.—Admission cannot take place unless the applicant is 18 years of age.

Admission is by examination.   But a maturity certificate from a Swiss school is taken instead of an examination ; or a student may be excused part of the entrance examination by presenting certain school certificates.

Persons who desire to attend the courses, but not as regular students, must show that they possess the requisite knowledge to enable them to follow the work, and they may be examined to settle this point.   Those admitted in this way are styled "Hearers" (Auditeurs).

This school has not been given the academical rank corresponding to the Berlin Technical High School, and no degrees are given. Regular students may get a diploma.   The last half-year is taken up with the diploma examination, so that the actual time of instruction is seven half-years. It may be interesting to note in passing that, starting with a boy at six years old, his educational life would be somewhat as follows :—

Primary school	4 years.
Secondary ,, ...	2 ,,
Gymnasium	7 ,,
Polytechnic	4 ,,
Total	17 ,,

At the end of this long course the youth is 23 years old. The school is divided into six departments. These and the numbers in each department are stated in Table 6.

TABLE 6.

Students and Foreigners in Attendance for the Year 1901–2.

	Swiss.	Strangers.	Total.
1. Architecture	49	10	59
2. Civil Engineering . . .	181	53	234
3. Mechanical Engineering . .	230	186	416
4. Industrial Chemistry . .	124	92	216
5. Forestry and Agriculture . .	93	11	104
6. Mathematics and Physics . .	30	6	36
Total . . .	707	358	1,065

Academical Freedom.—The lectures and exercises as announced in the programme of the several departments are obligatory on the student. In each department, however, the students are allowed a choice in the third year. Once having chosen, they are obliged to follow the plan selected.

*Lectures and Exercises Announced for the Course in
Mechanical Engineering. Department 3.*

FIRST YEAR.

Differential Calculus.
Descriptive Geometry.
Analytical Geometry.
Machine Design. Chemistry.

SECOND YEAR.

Differential Equations.
Mechanics.
Physics.
Machine Construction.
Mechanical Technology.
Work in the Testing Laboratory, optional.

THIRD YEAR.

Thermodynamics.
Machine Design. Engine Design.
Hydraulics.
Practical Work in the Hydraulic Laboratory.
Electricity.
Mechanical Technology.
Hygiene.

FOURTH YEAR.

Construction in Iron.
Electrotechnics Laboratory. Experiments on Alternating Systems.
Engineering Laboratory. Boiler Construction and Engine Construction.
Dynamo Construction.
Hydraulic Laboratory. Construction of Hydraulic Machines.
Electric Installations.

Remarks on the Engineering Course.—The engineering laboratory
is combined with the hydraulic laboratory in one fine hall with an
arched girder roof. Drawing offices and lecture rooms are arranged
in the block of buildings forming the front of the hall. The total
cost of the building was about £32,000. The equipment and
machinery cost £20,000. Conspicuous amongst the steam plant is a
fine three-cylinder triple compound engine; each cylinder has a
different type of valve gear, each gear being controlled by a different
type of governor. The nominal H.P. of this engine is 150, its

maximum 200. The engine may be loaded in various ways, amongst them being a 3-phase generator capable of taking the full load, a direct current machine which can take the full load, and a line of shafting with rope and belt transmission to various machines and pumps. There is also a compound vertical engine of 100 II.P. maximum. There is a boiler house containing three different types of boilers, including a water-tube boiler which can be worked up to about 300 lbs. per square inch.

In the hydraulic part of the building there is a fine equipment of turbines, pumps and water-measuring apparatus. The cleanliness and order in the laboratory is most striking. The engines are not merely clean, they are glittering with brightness. Six assistants are employed in the laboratory whose business it is to keep the apparatus in order.

As at Berlin, no attempt is made to teach workshop practice ; but the bulk of the time is given to drawing-office work.

General remarks on the Preceding Courses.—A feature in all the courses quoted is the way they develop from a common scientific basis in the first two years into widely divergent and specialised branches in the remainder of the course. It should be understood that both in the States and on the Continent many of the specialised lectures are given by men in the full practice of their profession, and who are not regular members of the teaching staff. The best courses in this country are arranged on practically the same basis, but the longest being three years there is no time to develop the instruction into the specialised branches of engineering. At Boston, Cornell, Berlin, and Zurich, a student can get special instruction in the scientific part of locomotive and railway work. Not an Institution in this country can offer similar instruction. The locomotive comes in for scant notice in the general course on the steam-engine. The advantage to railway companies would have been great, if even one course of instruction in locomotive engineering existed where a laboratory was arranged in which an actual locomotive could be tested ; such an establishment, for instance, as exists at Purdue University, or in connection with some of the locomotive departments of the American railways. The course should be arranged for men who have already had an undergraduate course of study and have served two or three years at a locomotive establishment. In a laboratory of this kind the effect of varying the many quantities separately which are concerned in the working of a locomotive could be studied in a way utterly impossible on the road. Such tests taken in conjunction with dynamometer-car tests on the road would be of the greatest use to locomotive designers.

There is an essential difference in the method of training in America and Germany. In America the course of instruction is very exactly laid down, and the student is compelled to follow it step by step. Slight variations are permitted in the form of options, to use their term, in the later periods of the course. But whatever option is taken, the student must go through with it. He gets his degree from the gradually accumulating results of terminal and sessional examinations, ending finally with a thesis. He is in fact, as already mentioned, put through a thoroughly well-organised species of educational drill and must work or fall out.

In Germany the students of their great technical high schools enjoy the " Academische Freiheit " peculiar to the University system of that country. No student is compelled to take any special course. For his convenience definite courses are arranged and laid down in the school calendar, but the sequence of lectures therein stated are not binding. There is in no sense a prescribed course. The courses are only recommendations, and students may follow them or not as they please. Degrees are granted by examination, chiefly oral in character, by the professors of the schools.

At Zurich it will be noted the course is partly prescribed, partly selected.

Although a consideration of the lectures given, and the number of students, etc., give a general idea of what is being done in other countries, the peculiarities of the courses in comparison with what is usual in this country cannot be readily seen without having some idea of the weight attached to the different subjects of study. In order to facilitate this comparison, the author has estimated the percentage of the whole course devoted to different subjects, grouping them in the way familiar to members. The figures must only be considered as approximate, but they are sufficiently accurate to bring out the peculiarities of the several courses.

TABLE 7.

Table showing the percentage number of Hours' Instruction given in the Mechanical Engineering Courses under consideration.

	Massa- chusetts Institute.	Cornell.	Berlin Technical High School.	Zurich Poly- technic.
Mathematics . . .	8	5	14·5	19·2
Physics	5	8	6·8	6·0
Chemistry	7	7	1·7	3·0
Applied Mechanics . .	7	10*	22·5	19·5
Mechanism	4	—	8·0	—
Steam - Engine, including Thermodynamics . }	6	6	4·1	8·0
Mechanical Drawing† . .	26	20	31·0	39·3
Electrical Engineering . .	2	2	3·4	5·0
Commercial Subjects . .	2	—	8·0	—
Workshops	14	30	Nil	Nil
French	6	—	,,	,,
German	3	3	,,	,,
English	5	—	,,	,,
Engineering Laboratories .	5	9	‡	‡
	100	100	100·0	100·0
Approximate Hours . .	3,000	3,000	4,000	4,000
Distributed over . . ·{	Four Years.	Four Years.	Three Years.	Three Years.

* Includes Mechanism.

† Includes Freehand, Machine Drawing and the lectures connected with Machine Design.

‡ Laboratory courses are taken in addition, but it is difficult to estimate how much is recommended.

The Table shows that an American course may be taken as 3,000 hours' instruction distributed over four years. The Continental courses in the comparison are 4,000 hours distributed over three years, and this independently of laboratory work. The fourth year is not included, because it is so cut up with examination work. It must not be forgotten, however, that an American student actually receives about 3,000 hours' instruction ; a German or Swiss student is only recommended to attend courses aggregating 4,000 hours. Actually he may work just as many hours as he chooses.

The figures in Table 7 indicate a second essential difference between the American and Continental courses. In America a large proportion of the time is devoted to workshop practice. In Germany and Switzerland no time at all is given to this. In both Charlottenburg and Zurich there are fine engine laboratories, but no laboratories for the testing of materials belonging directly to the schools. What testing of materials is done takes place in the Government testing department, which happens to be adjacent to the school in each case. In America a testing laboratory is a feature of the course.

In brief, the American courses are more practical in character, they include more laboratory training than is recommended in the German course, and devote a proportion of the course to the teaching of handicraft skill.

In Charlottenburg and Zurich no attempt is made to teach handicraft skill, and the bulk of the training is given in the drawing office, though in addition a considerable amount of time may be given to engine testing.

In both Charlottenburg and Zurich a student finds himself at the end of his course with a degree or diploma, age 23, and no workshop training except a year, which is insisted upon as a preliminary to entry for some students at Charlottenburg.

In America a student finds himself with a degree or diploma, age 21, with what handicraft skill and workshop practice he has picked up in his college workshops. With these assets he has no difficulty in getting further practical training in the large works of the country. Employers take him, without premium, and pay a wage sufficient for maintenance straight away, recognising that his knowledge places him in a different position to ordinary apprentices. In this way they get highly trained men into their works, and by their own observation soon discover whether the youth possesses, in addition to intellectual acuteness, the qualities which go to make a successful business man or a good organiser, and recruit their staff accordingly.

After considering these facts, the question arises, is the British method of training engineers better or worse? Can they improve their methods in the light of what is being done abroad? Most people would consider the method at Charlottenburg and Zurich too academical. A youth must surely be handicapped afterwards, in whatever position he finds himself, if he is not familiar with the machine tools and appliances used in engineering works, from actual experience as a workman.

Many also, whilst admiring the American system of workshop instruction, think that it is better that this instruction should be obtained in a workshop under the actual conditions of practical work, for the reason that a youth who is to become a leader in the future requires to know as much as possible about the men, their habits of thought, and their point of view.

One thing is certain, however, the American, German, and Swiss student starts his course with a far better education on which to build than is the case with us. Much time is wasted at colleges here on teaching things which should have been taught at school.   ·

The great defect in the English method of training engineers, as the author has already stated, is the want of co-ordination between the colleges and the employers. If the employers will concern themselves with the question, he feels sure their attitude will speedily change.

The general opinion seemed to be that a course arranged so that the winter months are spent at college and the summer months in the works is a desirable one, and one from which good results may be expected. Such an arrangement obviously cannot be worked without the co-operation of the employers. This alternating system must not be regarded as experimental. The Admiralty have had something very similar in operation for forty years, and the system has produced a famous roll of chief constructors, and has been of benefit in the education of marine engineers. The Scottish Universities lend themselves to the system, and he believes Glasgow students in engineering consistently study in the winter and work in the summer. Mr. Yarrow has included this alternating system in the apprenticeship rules he has just made for his works. These rules were read during a recent discussion at The Institution of Naval

Architects, and the names of several firms who have agreed to admit apprentices under similar conditions were also given.

Leaving this point for further discussion, the author would merely remark that no system will be of any avail without the whole-hearted co-operation of the employers and manufacturers; with their co-operation almost any system may be made to work.

He wished to conclude with a few figures indicating the comparative appreciation in which professional education is held.

The figures concerning Germany and Great Britain are taken from a schedule published by the Association of Technical Institutions.* Those referring to the United States are quoted from the Report of the Commissioner of Education.

TABLE 8.

Number of Students during the Session 1901–2 *in the subjects stated.*

	German Technical High Schools.	United States. Institutions for Higher Education.	Great Britain. Universities and Technical Schools and Colleges.
Agriculture . . .	42	2,852	411
Architecture . . .	1,440	459	—
Civil Engineering . .	2,257	3,140	
Mechanical Engineering .	5,503	4,450	2,259
Electrical Engineering .		2,555	
Chemistry and Metallurgy	1,180	†	667
Naval Architecture . .	318	‡	33
	10,740*	13,465	3,370

* These figures do not include the students at Brunswick or Stuttgart.

† Chemistry not given separately, but probably included in "General Science Courses," the number attending being 10,925.

‡ Included in Mechanical Engineering.

* *See* also a pamphlet issued by The Association of Technical Institutions, which can be obtained from Professor Wertheimer, Merchant Venturers' College, Bristol.

The 10,740 students attending the German Technical High Schools would all be over eighteen years of age.

The students attending the United States Institutions are all probably over sixteen. The 3,370 students for Great Britain are only obtained by including in them all students over fifteen taking a day course of over twenty hours per week at the technical schools and polytechnics. Probably there are not more than 500 students over seventeen years of age in the whole of this country taking courses in the above subjects, which can be fairly compared with the 10,740 in Germany or the 13,000 in the States. These figures indicate that the bulk of engineer apprentices in the United Kingdom receive no scientific training of University rank at all.

It should be understood that the figures given in the Table refer only to students taking day courses of the type set forth in the various Schedules, so far as Germany and the United States are concerned. The excellent work done in the evening at Mechanics' Institutes and Technical Schools is not touched upon at all. This work should be looked upon, the author thinks, as a preparation for work of University character to be taken in a day course. In the United States there is a highly organised system of evening instruction, not only in technical subjects, but in classical and general scientific subjects as well.

The author would like to take this opportunity of thanking the many friends at the Institutions above mentioned for their patient help and kindly assistance.

Discussion.

The PRESIDENT said Mr. Yarrow's name had been mentioned in the Paper, and Tables 9 and 10 (pages 308–309) were contributed by him. He was sure the Meeting would be glad to hear some remarks from him on the subject.

Mr. A. F. YARROW thought the Institution was very fortunate in being favoured with such an interesting Paper as the one just read, which must have taken an enormous amount of work to prepare. Professor Dalby had placed before the meeting in a very complete and concise form the systems of training engineers adopted in two of the chief industrial centres on the Continent, and therefore, as he pointed out, it was a very fitting time to reflect whether what was done in this country might not be improved, based upon what was the practice elsewhere. Without doubt manufacturers in this country did not attach as much importance to scientific training as was done in the United States, Germany, and Switzerland, and it would be well to consider whether such training might not be introduced in the United Kingdom to a greater extent than at present with advantage. If a young man took the usual college course of three years, it must be borne in mind that, owing to the long holidays between the terms, he was practically only at work for about two years. Nearly one year of his life, so far as his scientific studies were concerned, was lost. That loss might, to a great extent, be avoided by what was termed the "sandwich" system, which Professor Dalby had referred to as being adopted to some extent in Glasgow. In consultation with both Professors and Employers, the scheme shown had been drawn up, so that those who were not familiar with the "sandwich" system might see how it could be practically carried out. The scheme represented what was possible for a lad studying marine engineering. It consisted of a six years' course, combining both practical and scientific training, and was divided up as follows, as shown on Tables 9 and 10 (pages 308–309).

(Mr. A. F. Yarrow.)

TABLE 9.

" Sandwich " System.

Suggested Six Years' Course for Pupils in Marine Engine Works.

Probable Age of Starting, 17 to 18.

1st YEAR.

6 months College.
1   „   Holidays.
5   „   Pattern Shop.

2nd YEAR.

6 months College.
1   „   Holidays.
5   „   Pattern Shop and Foundry.

3rd YEAR.

6 months College.
1   „   Holidays.
5   „   Machine Shop.

4th YEAR.

6 months College.
1   „   Holidays.
5   „   Erecting Shop.

5th YEAR.

7 months Erecting Shop and Fitting Machinery on board.
1   „   Holidays.
4   „   Drawing Office.

6th YEAR.

11 months Drawing Office.
1   „   Holidays.

Shop Practice   .   . 27 months
College Course   .   . 24   „
Holidays   .   .   .   6   „
Drawing Office   .   . 15   „
           Total   .   . 72 months, or 6 years.

TABLE 10.

College Course.	Workshop Practice.
October 1st	One Week's Holiday
November	April 14th
December	May
One Week's Holiday	June
January	July
February	Two Weeks' Holiday
March	August
April 7th	September 30th
Six Months' Study	Five Months' Workshop Practice
One Week's Holiday	Three Weeks' Holiday

He thought that the Tables might be of use to the meeting in promoting discussion, and to help in considering the problem which was practically raised by Professor Dalby, namely to ascertain what was the best system to adopt for the training of the rising generation of engineers. Of course, every branch of engineering would require a scheme adapted to its special class of work. It had nothing to do with premiums or salaries. The "sandwich" system was independent of any consideration of that kind. The education of young engineers who had in view to be leaders was mainly a matter in the interests of employers, and in coming in contact with many Professors of Engineering, he, Mr. Yarrow, found the general feeling was that there should be some expression of opinion on the part of manufacturers regarding the kind of scientific education which they considered would best fit students to be of value to them, and there could be no doubt that the persons who would be most benefited by such enlightened co-operation would be the manufacturers themselves. But the manufacturers must, as Professor Dalby had pointed out, show some interest in the rising generation. It must be admitted

(Mr. A. F. Yarrow.)

that the education for young engineers in this country was sadly
deficient, and if the present movement resulted in the subject being
carefully considered and the best system which could be devised
adopted, it would in the future be of undoubted advantage to the
manufacturing industry of this country; and it would be very
desirable, if some change were to be made, that it should not be
delayed, but that a commencement should be made after the
midsummer vacation of the present year.

So far as he had been able to ascertain the opinion of authorities
on such educational matters, there appeared to be a very great
majority in favour of the " sandwich " system. It certainly appeared
to have many advantages. Among others, it prevented a young man
who had gone through a technical course at a college forming a very
exalted opinion of himself when he entered the shops, because he
realised in the early part of his career that there were many other
qualities necessary to make a successful man besides that of being a
student. He believed that what was wanted was to blend a high
form of scientific education with the all-important accomplishments
of practical knowledge and the power to lead men, which latter two
qualifications were only to be obtained in the workshops.

He wished to refer to Mr. Maw's Presidential Address two years
ago.* His feeling in the matter was that that Presidential Address,
which was an exceedingly able one, had not received the attention it
deserved, and he would read one or two short extracts from it. That
Address dealt almost exclusively with the Education of Engineers,
and with regard to the scientific training in a college, Mr. Maw said
(page 437) :—" It is no uncommon thing for a young engineer after
he has been a few months in the workshop "—that is, after he has been
through a three-years' college course—" to begin to realise how many
opportunities he has wasted during his career at college." If one
had been working in a shop some part of the time and some part of
the time in a college, he soon began to appreciate the importance of
both, because one was so dependent on the other. Mr. Maw, in
another part of that Address, said (page 436), speaking of the student,

* Proceedings 1901, page 431.

" In fact he feels himself in the position that he has spent time acquiring knowledge which is of little immediate practical use to him, and that he has done this at the expense of not learning things which he badly wants to know." All that was through the want of having a certain amount of practical knowledge and seeing what information he required in the shop, and selecting in his studies those things which he found he most required. He thought anybody who was interested in educational matters for engineers would do well to study Mr. Maw's Address. It was certainly a very remarkable coincidence that two Presidents should both refer to the educational question, and it showed clearly the signs of the times.

Some years ago his firm, thinking over the education of pupils, proposed to have a class of their own, and to get a professor to take that class, but on looking into the matter it was found that it would in their opinion be impossible to institute successfully anything of the kind. There were only about 1,500 men, and in order to obtain the best teaching the best professor was required. It would also be necessary to have apparatus to illustrate and explain what the teacher wished to put before his students. His feeling in the matter at the time, and it had been considerably confirmed since, was that it was impossible for any moderate sized works to carry out in a satisfactory manner, though it might be possible for large locomotive shops with 4,000 or 5,000 men. He thought the leaders of the works must be highly educated, and the only way was to go to a University or a Technical College where the best opportunities for being taught existed. Whatever was done must be done on sound business lines, and for the interests of all. Workshops must not be looked upon as educational establishments or philanthropic institutions, but it must be apparent to all that the future prosperity of works was dependent in a very large measure upon the efficient training of those who would take the most active part in its future management. A scheme such as was shown enabled the heads of firms to form a good idea, towards the close of the six years' course, as to the comparative merits of young men, from whom they could select the best for the future leaders in their works. The college course, it would be seen, ended in the fourth year.

z

(Mr. A. F. Yarrow.)

He would read another extract from Mr. Maw's Address. In speaking of the secondary schools, Mr. Maw said :—" The secondary schools do not bring the education of the young man up to the standard that is necessary to go into a technical college, and the first year in a technical college is generally expended in teaching the lad what he ought already to know." He further said :—" The changes" —in secondary education—" which have so far been made in this direction are very limited compared with those really required, and there is still left to be done at the technical college much educational work which ought to have been done at school, the result being a waste of valuable time. This matter is one which merits the most careful attention of all interested in technical education." It was impossible to separate technical education from the general scheme. Whatever scheme was to be adopted, at any rate the artisan's son, when of exceptional ability, must be considered; this was secured to some extent by evening classes, scholarships, etc. Care must be taken that every facility to rise was given to the best man, irrespective of social position. Such men would be most frequently found amongst those not endowed with wealth, who had in consequence inherited the instinct for work, and by example and necessity would be more ready to go through the drudgery of learning their profession than those who were what was termed " socially above them."

Mr. DUGALD DRUMMOND said that the Institution was indebted to two members of the Institution for the position this question at present occupied with the engineers and the public of this country. The first was the late President, Mr. Maw, whose presidential Address was a valuable contribution to that important question. The second member was Mr. Yarrow. He did not believe any man in the profession had spent more time, labour, and money in bringing the question so prominently forward than that gentleman; and not the least of the services rendered by him was his selection of Professor Dalby to bring forward a clear and concise Report containing the information naturally required to place members in a position to understand the system adopted for teaching young engineers in other countries. After carefully reading over Professor Dalby's Paper, he did not think engineers would be wise in following either the system of education in force in America, or on the Continent. His opinion was that it was necessary to raise the majority of young men in the profession up to a higher standard than they had hitherto reached, so far as scientific knowledge was concerned. It seemed to him that in America and Germany the minority received that advantage. They were kept at College until they arrived at the age of twenty-two or twenty-three, the whole of their time from entering school until leaving College having been devoted entirely to attain scientific knowledge. His opinion of that was that it was the wrong way to go about it. He thought every lad who intended to follow the profession of an engineer ought to be well grounded in the primary National Schools of the country, and be prepared to enter on his profession at the age of sixteen. In his opinion they should then be allowed to receive their scientific training hand in hand with the practical training every year. After a lad had had six months' teaching of one hour per day during the first year, if he were unable to pass the standard which he was expected to do in the first year, he ought to retire and devote the rest of his energies to practical training. In the case of those who were able to pass the examinations of the first, second and third years the employer should do all that lay in his power to enable them to reach the highest standard of scientific training, because it was well known that it was not the gift of everybody to be able to attain to high scientific knowledge, and it would be a great mistake to press young men beyond their limit. After having passed the examinations held during the three years, one was able to judge exactly how far the limits of the boy would reach. Those whose abilities would enable them to attain the higher standard ought to be encouraged in every way to reach it. If their parents were unable to give them the education they required, it was for employers to help as far as lay in their power. That question had been discussed very fully at the Institution of Naval Architects, and it there received a reception worthy of it. He believed that every one who took part in the discussion on Professor Dalby's Paper at that Institution agreed thoroughly that

z 2

(Mr. Dugald Drummond.)

it was the right course to pursue. Therefore he thought that in the discussion on the present occasion the limit of any good which could be done by discussion or by letter writing in the papers had been reached, and he would therefore suggest for the approval of the Members of the Institution that they ought to ask the President, the late President (Mr. Maw), and Mr. Yarrow, to approach the incoming President of the Institution of Civil Engineers, who was a Past-President of this Institution, to form a Committee to invite kindred Institutions of the profession to elect suitable members from their Institutions, also Governors and Professors of Colleges, to form a strong united Committee to discuss this question in all its bearings in a broad and comprehensive spirit. There could be no doubt that talking would not be productive of any good in the matter, that is, it would not be productive of greater results than that already due to it. He was quite satisfied that the gentlemen he had mentioned, in conferring together, by this means would produce a scheme that would stand comparison with that of any other country in the world.

What was required was that engineers should put their houses in order ; it was necessary that their lads should be able to understand theoretically and practically all the work which they did. Employers were not to confine themselves entirely to the shops ; he thought the counting-house wanted overhauling as well. He was of opinion that every cost-clerk and every accountant connected with the workshops of this country should have a thorough training, practical as well as clerical, the office being the proper place to check leaks in the management of factories. Accountants and cost-clerks were the persons responsible for checking whether work was being executed at a profit or loss. It was very necessary to take the whole question out of the rut in which it had been running for so many years, and that the whole of the departments, clerical as well as workshops, should be overhauled and put on a safe and intelligent basis. He was perfectly satisfied that if that course were pursued British engineers would stand not only in the front rank, but the young men would have a better chance if educated on the lines which he had expressed than by the system either in America or on the Continent. He sincerely hoped that his proposition would be adopted with regard to the election of the President and late President and Mr. Yarrow as a Preliminary Committee, to bring this great question to a proper issue with other kindred Associations of the profession.

The following was a scheme which he, Mr. Drummond, had recently started at the London and South Western Railway Locomotive Works for the benefit of the apprentices working there.

London and South Western Railway.

Notice to Apprentices.

I am anxious that the apprentices in the London and South Western Railway Works at Nine Elms should have every possible opportunity afforded them of having a scientific education, arranged to go hand in hand with their practical every-day work, and so enable them to prepare, at the end of three years, to take up the higher scientific training to be obtained at the Technical Colleges during the last two years of their apprenticeship. The course will commence in October and end in February.

This has been arranged to enable you to have proper time for study in the evenings, and to compare the scientific teaching you receive at the classes with the practical work going on in the workshops, so as to cultivate the habit of thinking independently for yourselves.

I have arranged for a competent teacher to give one hour, from 8 A.M. to 9 A.M., on three mornings of the week for juniors, and one hour for two mornings of the weeks for seniors, which will form part of the day's work.

You will be expected to pass a preliminary examination in proportion, fractions, cubic and square root and mensuration, before being allowed to join the classes.

At the end of three months in each term an examination will take place to enable me to ascertain what progress has been made; the final examination to take place at the end of each term.

Those who pass the final examination will enter the higher class for the second year, and so on the third.

Those who fail to pass the first examination, but receive from their teacher a recommendation, will have the privilege of continuing in the same class another year to give them the opportunity of passing into the higher classes.

Those who do not receive a Certificate from the teacher, or who fail at the second opportunity of passing, will have to retire from the classes.

I will arrange for those who fail not to work overtime during the winter months, so as to enable them to attend evening classes if they so desire.

(Mr. Dugald Drummond.)

The apprentices who pass all their examinations satisfactorily will be allowed to attend the Engineering Colleges during the winter months to secure a higher education, and the time so occupied will be counted as part of their apprenticeship, and those who successfully pass the College examinations will have the privilege of entering the Drawing Office or the Chemical Laboratory during the summer months.

The privilege will continue for the last two years of their apprenticeship.

Those whose conduct is satisfactory and those who have shown ability both in workshops and in technical work shall have the first call for promotion. I therefore hope that every apprentice will do his utmost to improve his knowledge, and so become eligible for promotion.

Any lad whose parents have not had the means to keep him sufficiently long at school to give him an education such as would qualify him to pass the preliminary examination will call upon me, and I shall endeavour to make such arrangements as will enable him to acquire the necessary knowledge to do so.

The three subjects for the Session will be "Applied Mechanics," "Heat," and "Electricity." Only one subject will be dealt with, until the class is thoroughly capable of understanding it before the next is entered upon.

The Directors have kindly agreed to pay the teacher's fees for the first three years.

Sir ARTHUR RÜCKER, Principal of the University of London, said he did not expect to be called upon to speak, because he came as a learner, and had not much to say on the matter under discussion. He would begin by remarking how much pleasure it gave one who worked with the President years ago in Leeds on technical work, to find him in the distinguished position which he now occupied, on which he tendered him his hearty congratulations. With regard to the question under discussion, as to whether or not what had been called the sandwich or alternating system was the best, he did not propose to offer any very definite opinion; he was not an engineer, and probably the engineering professors and the great number of engineers who were gathered together would hardly look upon his opinion as being one of the most valuable on the subject. But if he might be allowed to speak as a comparative outsider in the matter, he would say that that system seemed to have very great advantages. There were two difficulties, a young man spending too much time in the college lost that deftness of hand and acquaintance with practical work, or did not reach it early enough in life; whereas if he went to the workshop for too long first, he lost the habit of sitting down to study in the college.   There was much to be said for the alternating system under discussion.   Assuming that it was a thoroughly good system, he strongly supported what Mr. Drummond said ; he believed that as much had now been said on the subject as was useful, and if the matter was to be carried through, some practical step should be taken.   It was not for him to suggest what steps the Committee should take, but he would definitely say that they would have to approach the teachers.   Whether they chose to approach them through the University, or to go to the colleges first and allow the colleges to move the University, was a matter for them to discuss ; but he was strongly of opinion that the University would do what it could to meet them in the way of making the courses of study for the degree to fit in with that system.   Although he was aware that in this country a degree did not carry any very great weight in the engineering world, yet surely it was desirable that a time should come when the fact that a man had gone through a thorough college course should be indicated by a degree, and that the degrees of the University should be brought within the grasp of men who were going to take their part in practical life.   The University of London authorities had been doing all they could to make that possible ; the University courses of study had been drawn up by the professors, and a certain number of practical engineers, who had given their help.   In addition, the University of London did not want now to make everything depend upon the examination system, with which the University of London had been so closely connected in the past.   He referred to an engineering course in which the student was not only to be tested by examination, but the work which he had done in the term and various things done in the workshop and laboratory would be considered by the examiners.   The candidate would be judged on his whole course, and not what he would do in a particular examination at the end of it.   The matriculation examination had been considerably modified, partly with a view to bringing it within reach of men who might be intending to go into practical life.   He wished to add that it appeared to him the scheme would very well fit with that shown in Table 9

(Sir Arthur Rücker.)

(page 308.)   He supposed that every one would agree that the earlier part of the course should be devoted to pure science, mathematics, physics and chemistry ; and that later on the student should go to the more practical subjects which would lead to the first part of the engineering course.   That drawn up by the University of London was devoted to pure science, and the student entering upon that course would pass those pure sciences first, and would then be at the right age to matriculate, i.e. sixteen or seventeen years of age. After another year's study he would be exactly at the age when Mr. Yarrow wished students to begin his course.   Therefore the time provided by his course was longer than that provided for by the University.   Provided the intermediate standard was reached first, there was a great deal more time in that course for engineering than was definitely required by the University.   It was expected at the University of London that the student would take more than the minimum the University required him ; in most cases the limit was below that which was insisted upon.   Thus there would be no difficulty, as far as the time requirements of the University were concerned, in fitting in with the proposed course.   The next point was one on which he did not think it was necessary for him to dwell at that moment, and he did not think it was quite easy for the college to manage.   One difficulty would be that the student would be spending six months over a course of physics when his colleagues would be going in for eight months, and it would be difficult to arrange the lectures to suit the two.   But if the lectures on pure science were accomplished first, there would be no difficulty, except in the engineering department itself.   But he would like someone present to answer one difficulty which occurred in the scheme proposed.   Practical men would remember that a certain amount of capital was sunk in starting a college, and in the provision of the necessary apparatus.   According to the proposed system it appeared to him that that laboratory would lie idle for six months.   That was a real difficulty, and it would be an advantage to know what would be done in the college itself during those six months.   If it were suggested that a certain grade of students was to go in for the summer six months, and another in the winter six months, it would be possible to get over that difficulty, but it was one which would have to be met. There was no difficulty at all to be dealt with from the University point of view. He only wished to repeat, in closing, that the University of London had been re-constituted, that is, it was determined to do its best to make itself useful; that it wanted to get into contact with practical men; and if the great technical associations, of which the Institution of Mechanical Engineers was one of the most distinguished, would be kind enough to take the University authorities into their confidence and say what they wanted, would come to them and talk over their difficulties, he could say without fear of contradiction that the University would try to be worthy of the trust placed in it.

The PRESIDENT said that, in consequence of the difficulty expressed by Sir Arthur Rücker, he thought it would be appropriate at that stage to ask a Professor from Scotland, where the system of six months' sessions had been in operation for a long time, to speak.

Professor ARCHIBALD BARR said that, in responding to the President's invitation to take part in the discussion, he would make his remarks as brief as possible. He might be allowed, in the first place, to answer Sir Arthur Rücker's question by saying that he believed the University laboratories could be used with great advantage for one of their chief purposes during the six months in which there were no ordinary students in attendance—a purpose to which they could not be so effectively put during the busy time of the session—he meant that they should be utilized for post-graduate study and research by holders of scholarships and the members of the teaching staffs.

He felt that so far the discussion that had taken place on the training of engineers had been left too much in the hands of the professors. He did not propose to say anything that evening from the point of view of a teacher of engineering science, except a word or two of comment upon the opinions of employers which he would quote. He might, however, as an employer himself on a small scale, say that his firm had always been ready to adopt the alternate six

(Professor Archibald Barr.)

months' system with their apprentices, and they were ready to do so in future.

He had in his hands contributions to the discussion which he had obtained (on the suggestion of Mr. Yarrow) from a few of the leading engineers in the Clyde district, and he would read one or two passages from each of these letters, dealing more particularly with what Professor Cormack had called the "sandwich" system of training.

The President had suggested, in the admirable Address which he had just delivered, that the Technical Schools and Universities might be regarded as proving houses of the raw material for the engineering profession. He (Professor Barr) feared that he could not quite agree with that suggestion as one for general application. He would much prefer that the youth who could do so should spend one year at least in the workshops to begin with before coming to the University. As the object in view was the discovery of suitable material of which to make practical engineers and not students, he thought that the most effective preliminary tests might be those founded upon practical work, and not upon ability to acquire a knowledge of the theoretical aspects of the subject-matter of engineering practice.

He did not propose to select the opinions in his possession; he would quote from each of the letters he had received. And first he would take one that expressed views quite against his own, but views worthy of every consideration, as coming from a man of great experience and responsibility, Mr. Hugh Reid, of the Hyde Park Locomotive Works, Vice-chairman of the new combine—The North British Locomotive Co. Mr. Reid, who, with his brother, went through the University course a good many years ago, wrote as follows :—

"As regards the University training, if it were possible to have a continuous course of instruction spread over the twelve months of the year, instead of six months only as at present, the course might be completed in two years, in place of the four years now required. The result of this would be that a much larger number of engineering apprentices than at present would avail themselves of the opportunity of attending our University classes.

"Every engineering firm, if working on sound commercial lines, must look at the education of its apprentices, more particularly from its bearing upon its own business, present and future. The more a business is removed from the purely manufacturing processes, and where new designs and developments are constantly required, so correspondingly will a larger number of highly trained engineers be required to undertake this work. Also, in Engineering Works situated far from or very close to large centres of population, highly trained engineers may in the one case be very difficult to obtain, and in the other, very difficult to retain.

"Under any of these circumstances, I could quite understand engineering firms being only too willing to put up with the undoubted inconveniences of the 'sandwich' system of training their apprentices.

"At the present time the Engineering Student attending our Colleges has no alternative to the 'sandwich' system of workshop and University training, other than to spend four years taking a University course (which might approximately be overtaken in two years with twelve months' sessions), and thereafter to undertake his workshop apprenticeship. It can scarcely therefore be maintained that the 'sandwich' system is the natural desire of the engineering student, but rather that it is the result of the circumstances in which he finds himself placed by the shortness of the present University Session.

"Looking at the whole question broadly, and from a national point of view, I am inclined to think that the best solution would be to effect a compromise that would result in giving our young engineers the best possible system of University and workshop training. By shortening both the University tuition and the workshop apprenticeship, the course of training could be completed within a period of five or six years, and much more satisfactorily than under the 'sandwich' system. The college course could be concentrated into two or three years (with a twelve months' session in each year) and the engineering apprenticeship, for those students who had satisfactorily passed their University engineering course beforehand, to a term of three years only. The University classes to be taken and completed before the apprenticeship.

"If my own son wished to be an engineer, the above is the system of training I would propose for him, if it were then possible.

"In the above suggestions I am not forgetting that our Professors and their staffs require holidays like other people, but this is a matter that could be easily arranged."

He (Professor Barr) would remark regarding these views that he entirely sympathised with the desire to have the total duration of the period of training shortened, but if the students worked as hard as the best students did with him during the six months' session, it would not be possible for them to keep it up for

(Professor Archibald Barr.)

two sessions of twelve months each. He, therefore, believed that the University work could not be well done in two complete years, whereas with four sessions of six months, with a complete change of employment in the intervals, it could be done well and thoroughly. Besides he considered it a very great advantage for students to have had some practical training in any case before they took the more advanced portions of the University course.

He was permitted to quote the views of Mr. John Inglis, LL.D., Past-President of the Institution of Engineers and Shipbuilders in Scotland, and of the Institute of Marine Engineers. Dr. Inglis studied for three years in Glasgow University, and served his apprenticeship partly on the " sandwich " system. He had sent two of his sons to the University, and both had graduated as Bachelors of Science, one on the side of naval architecture and one on the side of engineering. He had had other apprentices in his works on the " sandwich " system, who now occupied important positions in engineering and shipbuilding. He highly approved of the alternate six months' course of study and practice for promising youths, and considered that it was best to have at least one year of workshop or yard training before the University course was entered upon.

The next letter was from Mr. Robert Kennedy, B.Sc., of Messrs. Glenfield and Kennedy, Kilmarnock, who finished his University course ten years ago :—

"As you say, the consensus of opinion, as shown by the discussion, is strongly in favour of the ' sandwich ' system, which seems to me the most reasonable method of training ; though one is, of course, naturally biassed in favour of the system adopted in his own case. I know that I found my short previous experience in the shops, and at the elementary evening classes here, followed by further workshop training during the summer vacations, of great benefit in assisting me to derive fuller advantage from the Lectures delivered at the University.

"I remember that, among the Students of my time, those who took the highest places in the engineering classes almost all were men who had had some workshop training.

"As you know, a good many of our apprentices have adopted the same method as I did of going up to the University after two years or so of workshop training, and finishing their apprenticeship during the summer vacations, and it is intended to train my younger brother in this way. These men we, of course, take back during the summer; but we have so many applications for admission as apprentices to the works that we could not promise to take in outside students in this way.

*　　　*　　　..　　　*　　　*　　　*

"We do not accept premium apprentices under any conditions; and all must be under the complete control of their foreman and must attend regularly, keeping full working hours; otherwise they are liable to dismissal.

*　　　*　　　*　　　*　　　*　　　*

"I am afraid I cannot add anything of interest to what has already been said and written on the subject; but I am sure that this is unnecessary, as the advantages of the 'sandwich' system are so self-evident."

Mr. Andrew S. Biggart, of Sir William Arrol and Co., says :—

"I am, and always have been, strongly of opinion that the training should be a combined one, referred to by you as the 'sandwich' system, as I think a proper blend of both the theoretical and practical sides produces the best man, whether he ultimately follows the scientific or the practical side of the profession.

"You are aware that we have encouraged this idea, but owing to the peculiar position of our firm, we are subject to such strong influences in the choosing of pupils that sometimes we cannot avoid taking a class of young men and giving them a course which we do not wholly approve of. Because of this and other reasons, I cannot commit my firm to take in even a limited number of apprentices on the lines proposed by Mr. Yarrow.

"At the same time, I might mention we do so, and in the future, as cases are put before us, we will naturally do all in our power to follow that which we feel to be a proper course of training for the most promising young men.

*　　　*　　　*　　　*　　　*　　　*

"With regard to the premium apprentices, this is a matter we set our faces stedfastly against. We make it a rule that every one who enters our employ must accept a wage, and in addition to this conform to the rules of the Works in every particular."

Mr. James Hamilton, of Messrs. Beardmore and Co. (late Robert Napier and Sons), states :—

(Professor Archibald Barr)

" Personally I am in favour of the 'sandwich' system for training a few of
the young men in an establishment, on the chance of their turning out as
practical and more scientific than those who can only afford the constant
apprenticeship and night work. The opening must be kept for anyone who can
show he has the qualities required, no matter how attained "

Mr. James Howden, writes :—

" That a higher scientific education is necessary there can be no question,
but it is still as necessary as ever that it be combined with a good workshop
practical education. If a youth has had a fair ordinary education beforehand, I
would say that the workshop training is the more important of the two.

    *   *   *   *   *

" I have found that without a good practical experience, both in the
construction and the working of machinery, a proper knowledge of economic
production is not generally attained, so that those young men whose training is
chiefly scientific do not as a rule become so competent as those of equal ability
who have greater practical experience, though somewhat less theoretical
knowledge, in assisting to carry out in our workshops the production of
articles of the highest quality at a minimum cost. As it is really on these
two points that our success in competition ultimately rests, this greater
competency is of vital importance.

" In these remarks, I have had in view youths of average ability, and it is
after all by the average man that most of our work is done. There is always,
however, the gifted few to whom mechanical ideas come naturally. These with
a corresponding degree of energy, even with a very ordinary education, soon
take their proper position, and in these days when a scientific education can be
had in all mechanical centres, they would not fail to acquire, under almost any
conditions of apprenticeship, a sufficiency of such knowledge to keep them in
the forefront.

" What Mr. Yarrow has called the 'sandwich' system—the workshop in
summer, the college in winter—provides for a combination of practical training
with theoretical knowledge in a very admirable way, but it is only comparatively
few of the present class of apprentices who can take advantage of it. Employers
would, however, do well, and most of them I am sure would be willing, to give
the necessary facilities to apprentices who desire to carry out this combined
system of training. My firm has at present several apprentices working on this
system. I would recommend, however, all such apprentices to take a full year
in the shop before entering the college. This first year in the shop would give
them a grip of their work which they might not so well acquire afterwards. It is
also a good preparation for college training. I have taken this course with two
of my sons, and I think to their advantage. They took well to the practical
work, and it, I believe, made them better qualified for their University course.

Regular appearance at the works at 6 A.M, in all weather, is likewise a good physical and moral tonic.

"As regards the duration of apprenticeship under this scheme, I would consider five years sufficient if the apprentice began at sixteen years of age, or upwards, and had a fair ordinary education to start with. With a full year in the workshop before beginning his college training, three winter sessions at college should give any diligent student a good theoretical training, and as the last year of apprenticeship is generally in the drawing-office, where the working hours are shorter and the labour light, the evenings could be used for further study at the technical colleges for most of the year round if such is required."

Mr. Archibald Denny, President of the Institution of Engineers and Shipbuilders in Scotland, writes:—

"In regard to the education of an engineer, you will no doubt remember the remark of the Englishman, who said, 'There is no bad beer, but some beer is better than others'; well, that is what I think of the education of an engineer, any *system* is good, but my feeling is that the 'sandwich' system is the best.

"My own training consisted of three years' practical work in the yard and drawing-office, combined with evening study; then three years at the Royal Naval College; then I had about six months' practical surveying work in Liverpool under Lloyd's Registry, before finally joining my own firm.

"In the case of my younger brother, he served two years in the engine works as pattern maker and fitter, and then he attended the University for three years in winter, working in the yard drawing-office during the summer. It would have been better had he taken four years to his University course and served two periods of six months outside in the yard, on the boards and at iron fitting.

"What I do want to say, however, is, that we have practically tried the system which Mr. Yarrow advocates, and seemed to me to think a novel proposition in this country. I enclose you a copy of a letter we wrote to the Clerk of the School Board here in December 1890, more than twelve years ago."

The following is a passage from the letter referred to :—

"In order to encourage deserving pupils, who are successful in gaining a Denny Bursary in the Burgh Academy, we have decided to offer these students special facilities for taking up the study of engineering and shipbuilding. Any student who, having gained one of these bursaries, expresses the desire to become either a shipbuilder or a marine engineer, we will be willing to accept as an apprentice in our yard, or in our engine works, as the case may be, for a period of five years. He will be expected to attend during the winter the course prescribed for the B.Sc. Degree, in engineering or naval architecture,

(Professor Archibald Barr.)

according as he intends to be an engineer or shipbuilder, and during the summer he will have the privilege of entering our yard or engine works, and will be expected to follow out in his work the course we prescribe.

"He will have the same facilities given him for acquiring a thorough knowledge of his profession as are accorded to our other apprentices, and during the time he works in our yard or engine works he will be paid at the same rate as apprentices of the corresponding year. His time at college will be reckoned as part of his five years' apprenticeship, but he will not be paid any wages during his college terms."

Mr. Denny continues :—

"At a later date the privileges were extended to holders of bursaries which are open to the evening students in the Schools of Science and Art. As the result of this, eight young men have taken advantage of the privileges offered. Of these eight, one did not make sufficiently good progress, and we therefore declined to continue him further in our employ; one, after a trial of the work, was permitted to resign, as he desired to take up other work for which he thought he was more fitted, and one has not yet completed his course; that leaves five. Of these five one took his B.Sc. in Engineering; one took his B.Sc. in Naval Architecture; one took his B Sc. in Electrical Engineering (we have an electrical department in our yard); while two took their diplomas in Naval Architecture at the West of Scotland Technical College, and the man who is now a student is working for his B.Sc. in Engineering. All these men have done well in the drawing-office.

"Mr. Yarrow discusses the question of wages to be paid to those young men. I may say that we pay them the same rate of wages as to our ordinary apprentice draughtsmen in the yard—the wages start at £20 and rise to £60. I think that the wages question should not be introduced into this discussion; they must necessarily vary in different centres; but I think the benefit, at any rate during apprenticeship, is entirely on the side of the apprentice, and the employer has no hold upon the young man, whom he has trained, who may, at the completion of his apprenticeship, pack up his bag and walk, and in at least one instance this happened to us. It must not be forgotten, however, that our scheme deals exclusively with young men who have shown sufficient ability to gain bursaries either at school or in a technical class.

"I hold it is only really the best men who should have these exceptional privileges, though I quite admit that the best all-round men are not necessarily always the men who win bursaries ; but life is too short for a policy of perfection, and we have worked with the tools which were at our hand.

"This whole question of training is really in the hands of the employers, and it will only be taken up when employers themselves are sufficiently trained to know how to use fully trained men when they have them; it is not by any means at present the rule that employers have such knowledge."

In conclusion he (Professor Barr) would only remark that the opinion of engineers in the Glasgow district seemed to be almost unanimously in favour of one year or more being spent in the workshops before the University course was entered upon; and though he might be biassed in favour of such a course—as it happened to be the one he had himself followed—his twenty-seven years' experience as a teacher had confirmed his belief in the advantages of the " sandwich " system, and of the value of a year or more of practical work coming first in the course. Students who secured bursaries from school, and could not afford to relinquish them, would naturally enter the University at once; but where the choice was free, he advised students to get into a workshop for some time before entering the University.

He did not suggest that all apprentices should take a course in a University or at the day classes of a Technical College. It was only the more able men who would benefit by the combined course as against the ordinary apprenticeship and evening study, and such men were, perhaps, not so numerous as to make the " sandwich " system— if generally adopted for that limited class—impracticable from the point of view of the employers.

Mr. John A. F. Aspinall, Vice-President, remarked that Professor Dalby had said (page 282) that " the weak point in the system "—of dealing with students—" is the want of co-ordination between the workshops and the colleges. Many employers look askance at a college-bred youth, and there is no doubt that many college youths quite deserve it." There was a great deal of truth in that remark, and he read out that extract because he had always felt that young men who went to college first without having any workshop experience, had not, when they came to the workshop, done as well as they should do, and had been very apt, on account of their superior airs, to give a great deal of offence to those with whom they were working. He had, therefore, always advocated that a young man should start first in the workshop. When his company had been able to do it, as they had in many cases, they had said, " Come into the workshop for a year and then go for a year to the

2 A

(Mr. John A. F. Aspirall.)

college, and then come back again." If after that time the student
elected to go on working in the workshop for two years, he was
recommended—if he could afford it—to go again to the college and
complete his course. He was told that during the later period of his
time at the works it was possible, if he were an apprentice or pupil,
and had an opportunity of going in the drawing-office, to spend a
good deal of his time at evening classes, and thus prepare himself
for another year of college work. That system had certainly turned
out very well, and his company obtained some very good young men
in that way.

He thought, however, that in dealing with the question of the
education of engineers one ought not only to deal with those who
were sufficiently well off to enable them to go into works as
apprentices or as pupils, and who would ultimately become masters,
but some consideration ought to be given to the education of the
rank and file of the workshops, those who were not born with a
silver spoon in their mouths, and who ought to be able to get on
and advance themselves if they had sufficient brains, and ought to
have facilities for rising to the top. Professor Dalby had spoken
of what was done at Purdue University and Cornell University
and at the Technical Institute at Massachusetts. About two years
ago he sent an assistant of his, who was in America, out to the
Purdue University in order to investigate the system there; and he
had at present an assistant who was enquiring into the system in use
at Cornell. That had been done because the Lancashire and
Yorkshire Railway workshops at Horwich, or in connection with
them, there was a very large Mechanics' Institute at which there
were most excellent classes, where anybody in the works could
go for a mere trivial payment. Those classes were taught in the way
advocated by Professor Dalby, by men who were in actual work in
the shops. For instance, metallurgy was taught by the gentleman
having charge of steel melting and the steel foundry. Chemistry
was taught by the analytical chemists, whose duty it was to conduct
the whole of the analyses of everything connected with the railway.
The drawing-office work was dealt with by men who were practical
draughtsmen, and who were dealing with railway work every day

of their life. The result had been that that establishment—to which the young men were not in any way obliged to go, they could either go or stay away as they pleased—was competely filled during the whole of the winter evenings. A pleasing result from another point of view was that, now the railway was going into questions of electrical work and new departures of that kind, they would be able to have sufficient young men, quite young men, to take charge of that work.

Professor Dalby had spoken of the large number of students of engineering in the United States and in Germany, and mentioned the small number in comparison which were apparently studying the profession in England. But he thought Professor Dalby had probably not included such establishments as he, Mr. Aspinall, had indicated. He did not wish to say that the conditions obtaining at Horwich were singular; he had no doubt that similar Institutions were doing the same at Crewe, Swindon, Stratford, and elsewhere, and that the results were as good as at the Lancashire and Yorkshire Works. But he simply spoke of that which he knew. Probably a large number of those students whom Professor Dalby mentioned were workmen as distinguished from those who might be called pupils and premium apprentices, and that many at Swindon or Crewe or Horwich could be justifiably classed as students.

The great advantage, to his mind, of a young man going into the workshops at the commencement was, that he should get accustomed to the men with whom he had to work. He would thus find out that there were other people in the world besides himself, and would learn to appreciate the downright straightforwardness of the British workman who willingly imparted to him all that he knew himself about his trade, and it would obviate the student feeling when he came into the workshop that he was superior to the men whose knowledge he wished to acquire. A boy who went direct from school to one of the Engineering Colleges was too apt to spend valuable time at the trivialities of machine-tool work, whereas a year in the workshops taught him solid practical work which, when he went to the college in his second year, enabled him to know what would be really useful to him, and to lead his mind to consider what

2 A 2

(Mr. John A. F. Aspinall.)

were the theoretical points which he wanted to acquire in relation to his practical work. His third year commencing again in the shops saw him a much more thoughtful young man, considering as he took on each new class of work how he could apply the scientific knowledge—small though it might be—which he had acquired to enable him to grasp the knowledge of the profession which he hoped to complete.

Mr. W. H. ALLEN said that notwithstanding the able arguments which had been adduced in favour of the "sandwich" system as brought forward by Mr. Yarrow, he was sorry to say he could not agree with it at all. He was against any sort of see-saw education. He preferred continuity of thought rather than continuous change. In saying that, he did not think he was speaking out of the book ; he must have been one of the earliest practical manufacturers in the kingdom who welcomed the introduction of a pure college training, and that was brought about by a young Trinity College man twenty-two years ago who came to his office, and whom he, Mr. Allen, was unable to get rid of. Eventually he took that young man into his works, where he remained two years, jumping over the heads of all others. After a short space of time he took a high position in Lord Armstrong's works, at an excellent salary. That was the first time a college man had come before him; since then he had welcomed them in his works. There were now in those works twenty-five cap-and-gown men, and they could not do without them. Mr. Yarrow had intimated that, when he had that matter under consideration, he thought of starting a college within his works. That Mr. Allen thought was entirely wrong. Workshops and colleges should not be mixed ; one should be exclusively practice, the other principle.

He had been through Germany in 1892, and was brought face to face with the value of the college training going on there, a value which was evident in all phases of the workshop, in foremen, managers, and workmen. Accordingly he determined to build a lecture-room and have a lecturer who could attend to the pupils, and do more for them than he himself could. The old practice was for an apprentice to be in the stream of the men, and be very seldom seen again until the end of the term. But the plan he himself followed was to see that the pupils were taught in the principle of the work on which they were engaged. The demonstrator looked after them every day during the time they were in the works, and in the evening the lecturer attended in the lecture-room to discuss with them the points which they had encountered in the day. In the winter-time the lecturer gave lectures on the principles of the work, so that the pupils' ideas of principles were kept up during the time the students were in practical work. The advice he had given to parents was that their boys should have as high a school training as their income would allow, and that if the boy thought of being a student and was anxious to continue so he might go straight to college. If his desire was to start at practical work, he should determine what that work should be. At the end of his pupilage, by all means let him have a college education. In the future when choosing the superior men, that is, the staff, they should not be asked where they were apprenticed, but what college they belonged to, and what diploma they possessed. Upon the answer would probably rest an estimate of their value. Professor Fleming, with whom he had collaborated in the matter, had said that the difficulty he had found was that the workshop pupil, when he entered into the college, had lost much of the training he had at school, and therefore had to begin all over again. But if that knowledge were kept up in the way he had advocated, he thought the workshop practice did not altogether interfere with the regular student. Continuity of thought was most important. Those present had heard the method in which professors intended to deal with the colleges during the six months in the year when the pupils were not there, but he would also like to ask the professors whether they intended to keep up their own methods of lecturing. He imagined that such a long vacation as was now proposed would place the student out of the running of lecturing, because it was necessary to be in practice, whether for lecturing or for workshop purposes. The difficulty he would find in introducing students into the works in large numbers was how to deal with them; machines could not be kept idle during the six

(Mr. W. H. Allen.)

months of college work ; and it was a difficult thing to have young
fellows coming in with college thoughts in their head, without any
idea of what they were going to do. If the works were to be carried
on with a sense of business, the machines could not be allowed to be
idle during the six months of the college course.

There were three forms of education : the practical, the theoretical,
and a share of each. He believed that the proper course was that
students should take three continuous years in the works, and three
continuous years in the colleges, as was now done. As practical men,
what members ought to do was to induce students to take that
course which had not yet been taken up so readily in this country as
in the United States and Germany. If that could be done, the
apprentice would only have the ordinary six years' apprenticeship, as
they did in the old time, but the college training must be united, and
would immensely benefit the student.

The PRESIDENT said he wished at the present stage to learn the
wishes of the Meeting as to procedure. Should an adjourned
meeting be held—an extra meeting—for the purpose of pursuing the
discussion ? If so, it would have to be on the 6th May. He did not
wish to take it as a foregone conclusion that such extra meeting
should be held. There were no doubt many gentlemen present whom
the members would like to hear on the subject, especially Colonel
Crompton, Dr. Garnett, Mr. Stromeyer, Mr. Mark Robinson, and
others, but probably many people shared the views of Mr. Drummond,
that there was no great advantage in having many speeches upon the
subject, and that those who had anything solid to contribute could
do so in another form. He would first ask whether it was the wish
of the members to have an extraordinary meeting, and then, if it were
decided not to, he would ask whether it was the wish of the meeting
to carry Mr. Drummond's proposal. The effect of that would be
that the Council, who were the executive body of the Institution,
should organise a committee for the purpose of conducting what
might be called research into the subject, and to present a report of
their conclusions by way of a recommendation.

It was decided by vote to finish the meeting that evening.

Professor ROBERT H. SMITH said that the utility of the action of such a committee would be immensely enlarged, if it were to consist of representatives from the other great Engineering Institutions. He considered it almost futile for that Institution to act alone in such a matter. The other great Institutions were equally interested, and no great scheme could possibly be carried out without the co-operation of all. If the proposed committee were appointed, he thought it should be an instruction to the committee to endeavour to obtain the appointment of similar committees in the other Institutions, and to work along with them.

Sir ARTHUR RÜCKER said his name had been referred to as seconder of Mr. Drummond's proposition. He did not second the proposition, as he was not a member of the Institution. Personally he was in favour of the Council being left to consider what was best to be done.

The PRESIDENT then said that he thought the best way would be for him to place the wishes of the Meeting before the Council, without passing a formal resolution.

Professor DALBY, in replying to the discussion, proposed to confine his remarks to two points. First, as to the kind of education to which Table 8 (page 305) referred, the 3,370 students in Great Britain referred only to those taking a day course in a technical college of over twenty hours per week. The 13,465 students in the United States were returned, in the Commissioners' Report, as those receiving instruction in Institutions for higher education, that is, Institutions of University rank. The figures given for Germany referred only to day courses of University rank taken in the Technical High Schools. The figures therefore did not include any students receiving evening instruction only, either in Great Britain, the United States, or Germany.

Secondly, with regard to Mr. Allen's remarks (page 330), he was very pleased indeed to hear that a Trinity man had given such great satisfaction, but he might point out that at Cambridge the

(Professor Dalby.)

continuity of thought, or study, was interrupted from the end of May to the middle of October, unless a man came up specially for the Long Vacation term.

A cordial vote of thanks to the author of the Paper was, on the motion of the PRESIDENT, carried by acclamation.

Communications.

Dr. E. G. COKER (of McGill University) wrote that the timely Paper by Professor Dalby on the education of engineers was, in the main, limited to facts which were presented with a clearness that could not fail to impress the reader. The number of English engineers who had had the opportunity of an extended acquaintance with the system pursued on the American continent was probably not great, and the writer thought therefore that it was not out of place to offer a few remarks based on his experience of several years at the well-known engineering school of McGill University, which, as was well known, had been built and equipped by the wise liberality of Sir William Macdonald at a cost of about £200,000.

The professors and lecturers were mainly English by training, and therefore likely to be favourably disposed towards the methods of education in which they had been brought up, yet a comparison of the courses of instruction with those of the principal English and American schools showed a marked similarity to those of the latter. This was due to several causes, among which might be mentioned the strong desire on the part of the public that the higher education offered by Universities should have a practical bearing, which, in the case of engineering education, was intensified by the fact that the system of pupilage, as understood in England, was practically unknown. The college training was therefore charged with the duty of giving as much practical instruction as possible, and this had resulted in the equipment of fine workshops and drawing offices in addition to the laboratories, where students pursued a graduated course under suitable instructors. During the summer vacations, extending over four or five months, students were expected to enter engineering works, and as the " co-ordination between the workshops and the colleges " was extremely satisfactory, a large proportion did so, thereby obtaining further practical knowledge under actual working conditions.

Another difference between English and American schools was, that in the former a more general training was given, while in the latter the training became increasingly specialised as it advanced, until in the fourth and final year it was almost wholly devoted to one branch of engineering, civil, mechanical, electrical, mining, &c. That the system was a success was evidenced by the fact that for some years the demand for engineering graduates had been far greater than the supply, commencing salaries ranging from £100 to £250 per annum. English graduates who had difficulty in obtaining practical experience, and were unwilling to pay the heavy premiums demanded, might find it worth while to turn their attention to the United States and Canada, where practical experience might be gained, not only without cost, but at a salary commensurate with their ability.

Mr. James E. Darbishire wrote that he was heartily in accord with Mr. Drummond in believing that all prime-cost clerks should have passed through the workshop (page 314). He would, however, go farther, and say that all engineer students ought to pass through the prime-cost office. To become a successful mechanical engineer it was not alone necessary to know how to produce the best possible machine, but how to produce it at the lowest cost; yet he had never heard of a pupil or apprentice being instructed in cost of production and purchase of materials. This was not a College matter, but one to be learnt only in a manufacturing establishment. An acquaintance with prime cost and its book-keeping was the foundation of economical management—beyond this, an engineer should be instructed in correspondence, in estimating, and, if possible, in finance. This

(Mr. James E. Darbishire.)

might seem very prosaic, but it was common sense. He quite admitted that the difficulties in the way of the young engineer who wanted to know about his employer's costs, profits, etc., were almost insurmountable; but the matter was of such importance that he hoped it would not be lost sight of by the "Education Research" Committee, if that Committee should be appointed. After all, an employer must entrust these figures to some one—why not as readily to his pupils as to his clerks? There yet remained one subject to graduate in, and in that individualism was everything—the machine when made had to be sold, and a mechanical engineer to be successful must know how to sell; but as this could not be taught, in the educational sense of the word, it might be considered as beyond the present discussion.

Mr. JAMES HOLDEN wrote that he thought the thanks of the Institution were certainly due to Professor Dalby for the great care and trouble he had taken in preparing the Paper, and for the lucid way in which he had brought it before the members. While agreeing with Professor Dalby that the weak point in the English system was the want of co-ordination between workshop and college, he would like to point out that this co-ordination did exist to a certain extent in railway workshops and the Mechanics' Institutions which were, in almost all cases, connected with them. He was aware that Professor Dalby had taken the subject from the point of view of day Technical Institutions only, but he thought the system of having classes in connection with large factories, which extended throughout the country, must not entirely be lost sight of. He thought it would have added value to the Paper if Professor Dalby had been able to state how far this system was in existence in the United States and in Germany and Switzerland, so that one could form a more extended comparison between the technical education given in the various countries named.

He was of opinion that the "sandwich" system, alluded to by Professor Dalby and enlarged upon by Mr. Yarrow, formed the basis for the solution of the question. He thought, however, great care should be exercised that the amount of workshop experience was not

too limited. He agreed with Professor Dalby's remarks (page 304)
that the youth was handicapped who was placed in charge of works
and workmen without being familiar with the tools and appliances
used in the works, and who had not the opportunity of mixing and
working with the men, so that he might know their habits of thought
and general capabilities.

Having in view the fact that it was desirable for employers in
this country to encourage their pupils and apprentices to make a
greater effort to acquire a scientific as well as a practical training,
the Directors of the Great Eastern Railway had recently sanctioned
a scheme (page 338). This scheme had only just been adopted,
and it was therefore too early to speak of results. One important
point in the scheme was that only those students who showed
themselves capable and qualified to receive the higher technical
instruction were able to avail themselves of it. It gave such students
an opportunity which, so far as he was aware, had not yet been
conceded in any other direction, namely, one, and possibly more than
one, winter's course of instruction while receiving their pay, as
though they were working in the shops. This scheme was applicable
to all grades of young men who were serving their time with the
Great Eastern Railway Co. Of course, in order to qualify for the
benefit, it was necessary for the first two or three years of the students'
time that they made full use of the facilities afforded them to study
at the Company's Mechanics' Institute. He was aware that in the
opinion of some, the strain of working in the shops all day and
studying during the evening was too much, and in some cases partial
leave of absence was granted to these students, of course, in this case,
at their own expense. He thought that a lad who had served two or
three years of his time in a workshop was much more in a condition
to assimilate the instruction which he might receive at a technical
college than one who went straight from school, and if the
" sandwich " system was to be generally adopted, he would prefer
that at least the whole of the first year be spent in the workshops;
and assuming that a term of five years be considered a suitable one,
he would divide it somewhat in the following way :—

(Mr. James Holden.)

1st year.—In shops.

2nd year.—First six months in shops, second six months at Technical College.

3rd year.—Ditto.

4th year —Ditto.

5th year.—First six months in shops, second six months in Drawing Office.

GREAT EASTERN RAILWAY.

Notice to Employees in the Locomotive, Carriage and Wagon Department.

The Directors of the Great Eastern Railway Company desire to give increased facilities to their employees for the pursuit of Technical Study, and, subject to the following conditions, have agreed to grant to employee-students leave of absence with full pay for one or more winter sessions of about six months each.

Conditions.

1. An employee-student shall not be less than 18 or more than 20 years of age on July 1st of the year in which the session, for which leave of absence is granted, commences.

2. He shall have been at least three years in the service of the Great Eastern Railway Company, regular in attendance and industrious at work, and shall have given satisfactory evidence of his practical ability.

3. He shall have attended the Classes at the Company's Mechanics' Institute during at least two sessions, and shall be able to produce certificates from the Committee of the Institute that he has attended such classes regularly and worked in a satisfactory manner.

4. He shall be required to pass a local examination, which will embrace the following subjects, or produce certificates from the Board of Education (or in the case of Railway Carriage Building, the City and Guilds of London Institute), in the subjects and stages named, such certificates to have been obtained in the current or during the preceding year.

The local examination will be held in the month of May. Applications to sit thereat must be made to the Head of the Department not later than the 25th April.

a. { Machine Construction and Drawing . . . Advanced Stage.

         or

     Railway Carriage Building . . . . . Honours „

         or

     Building Construction . . . . . . Advanced „

b. Practical Plane and Solid Geometry. . . . Elementary „

c. Mathematics . . . . Second „

d. Steam . . . . Advanced „

e. Applied Mechanics . . . . „ „

f. Theoretical „ (Solids and Fluids) . . . Elementary „

g. Magnetism and Electricity . . . . . „ „

An employee-student who complies with these conditions must make application not later than August 1st of any year, and must submit for approval to the Head of his Department a scheme of the course of study he proposes to pursue. He will also be required to give satisfactory evidence by a monthly report, or by such other proof as may be required, that he is industriously working to the approved scheme.

In addition to the course of theoretical study, the employee-student will be required to devote a portion of his time to visiting manufactories, works in progress, etc., for which visits facilities will as far as possible be obtained. It is estimated that not less than one such visit can be made each month. A report of the visit must be submitted to the Head of the Department during the following week.

The leave of absence with pay will be cancelled in the case of any employee-student who fails to fulfil the necessary conditions or whose conduct is in other respects unsatisfactory.

JAMES HOLDEN,

Locomotive, Carriage, and Wagon Superintendent.

16th March 1903.

Mr. MICHAEL LONGRIDGE, Member of Council, feared that no entirely satisfactory scheme of education for engineers, and by engineers he meant engineers generally, could be evolved without the co-operation of the Public Schools and Universities. The time had gone by when the Church, the Navy, the Army, and the Law were considered the only professions fit for gentlemen, and no social inferiority now attached to the engineer because of his business or

(Mr. Michael Longridge.)

profession. The result was that an increasing number of engineers
were being drawn from the great Public Schools of England, and
they could not but rejoice that it was so, because whatever their
deficiencies in imparting utilitarian learning, these schools had
grasped the central ideal of education that character was a quality
higher than learning, and principle a leader safer than expediency,
and in these days when Acts of Parliament were needed to check
commercial corruption, he did not think that this kind of education
teaching men " to do the thing which lies before them and mean the
thing they say " could be too highly prized. But to gain the full
benefit of such a training, a lad should remain at school long enough
to experience the responsibility of ruling as well as the discipline of
obedience, and he could not do this unless he remained till he were
seventeen or eighteen years old, and at this age, if intending to be
an engineer, he should have acquired something more than the
knack of making bad Greek and Latin verses. He did not mean to
say that this was the extent of the curriculum of the great schools,
but he did think that the modern side should have the same prestige
as the classical side, and that the teaching of the modern side should
be co-ordinated as closely as possible with the science course of the
University, to which he held the engineering student should next
proceed. Part of this co-ordination should be the abolition of
compulsory Greek at the University.

He thought there were several good reasons for proceeding from
school to the University instead of to the shops, the office, or the
outdoor works of a contractor or civil engineer.

First, by spending two or three years in the shops much of the
knowledge, especially mathematical knowledge, which could only be
kept up by constant use, would be lost, and would have to be acquired
afresh.

Second, the habit of application to books would be lost, for he
did not believe in night classes for growing lads after a full day's
work.

Third, without the scientific training of the University, the
apprentice would not learn from his practical work one quarter of
what he would learn after having received this training.

Fourth, at seventeen a schoolboy was too young to be emancipated from all control, to live in lodgings and associate with workmen, with profit either to himself or them.

At the University the student should go through a course of study arranged to serve as a common foundation for every branch of engineering, so that on taking his degree he might be equally fit or unfit to enter either, as opportunity might occur. This course should be chiefly theoretical, following the German rather than the American system, and should include drawing and book-keeping. A certain amount of laboratory work in illustration of the subjects taught would be useful, but any attempt at making the student into a fiftieth rate handicraftsman should be discouraged. Specialising should not be compulsory, first, because few knew during their University course in what branch of engineering they would find an opening; and, secondly, because for a very large number special courses would, either for want of ability or lack of money, be useless or burdensome.

After leaving College the student should begin his practical working life as an apprentice, and it was then that engineers would be called upon to give their help. The duration of the apprenticeship would require consideration. He thought two or three years sufficient even for mechanical engineers, as he did not consider it necessary for men who were to earn their bread by their brains to waste time in attempting to become expert workers with their hands. He thought that to compel an engineer, as distinguished from a mechanic, to learn to file straight was as much an anachronism as to insist that he should learn Greek.

After finishing their apprenticeships of two or three years, those who were capable of it and minded so to do should have the opportunity of taking special courses in one or other of the various branches of engineering, either at one of the Universities or at a special central engineering college, where the coping-stone of the educational edifice could be put on by the best men and with the best apparatus money could procure. He felt very strongly that any course of engineering education which must be taken in its entirety by all would be a mistake, for it must give to some less than they

(Mr. Michael Longridge)

required, and to others more than they could digest, and above all he thought that while the education of all classes engaged in engineering, and indeed in all business, was desirable, the education of the leaders was incomparably more important; and if all the money wasted in engineering apparatus in technical schools had been spent on a central engineering college for the higher teaching, the finishing education of the leaders, the advantage to all classes would have been far greater.

Mr. T. Hurry Riches, Vice-President, wrote that he did not agree with the " sandwich " system suggested by Professor Dalby, but he quite thought that every mechanical engineer should be scientifically educated. The figures given of the numbers attached to the technical colleges of course must mean only those who were attending the day courses, as the school he happened to be chairman of had nearly 4,000 students, and from these evening classes they had trained many Whitworth scholars. Further, it must not be forgotten that, in many foreign countries, workshop practice was not looked upon as essential, whereas in England practice in the shops was deemed absolutely necessary.

It appeared to the writer that the remarks made at the meeting were too confined to the masters of the future and to the youths who could afford to pay college fees, and that the poorer financially did not get the consideration which they should. He was always proud of the fact that some of the best men he had turned out, including some of the Whitworth scholars, were sons of his workmen, and several of these were today holding responsible positions as locomotive superintendents, etc.

Mr. C. E. Stromeyer wrote that, when Professor Dalby read his first Paper before the Institution of Naval Architects, he himself touched on one or two points which deserved somewhat more detailed consideration. It was essential to the success of our future engineering managers that they should come in contact with the British workman at as early an age as possible, in order to acquire an instinctive knowledge of him; and if this plan was adopted, or

rather, if it were continued, for it is still the rule, then something
should be altered in his early education so as to remove the roughness
which seems to be one of the characteristics of many, if not of all,
really successful mechanical engineers. Bluntness of speech would
perhaps always remain as a distinctive engineering feature, for speech
reflected the inner working of the brain, and unless an engineer was
quite clear in his own mind about any particular subject, he was
likely to be pulled up with a round turn by the failure of one of his
creations ; he ought therefore not to be led to indulge in intellectual
flourishes. His reasoning must be true and to the point, and he would,
unless he expected to move in some limited paths of engineering, have
to refrain even from contemplating such possibilities as that black
might be white or, reaching back to the classics, that Achilles could
not overtake a tortoise.

To the writer's mind it seemed to be a matter of the utmost
importance that the teaching authorities of this country should fully
grasp the fact that there was a class difference between man and
man as regards brain power. One class, whose natural tendency
would be towards sciences, was endowed with correct abstract
reasoning powers; the other class, whose tendency would be towards
the learned professions, was endowed with an excellent memory.
One must, of course, concede that a good memory was a brilliant and
generally also a valuable gift, as its owner was ever ready with well-
tried arguments and numerous precedents to press forward his
objects ; but with engineers, abstract reasoning power must appear to
be the higher gift, even if for no other reason than that it came
later in life than memory.   Children were known to have most
marvellous memories, generally superior to that of their seniors,
though some men (as, for instance, Macaulay) not only retained but
improved this gift. On the other hand, abstract reasoning power
came to one relatively late in life. Some men, those who regulated
all their actions according to principles and precedents, had either
never acquired true reasoning power, or had weeded it out when still
a young and tender seedling. Such men sometimes appeared to
reason, but generally they only quoted what they or others did under
previous similar circumstances. Other men with scientific learnings

2 B

(Mr. C. E. Stromeyer.)

sometimes gained their reasoning power in early youth, when it seriously interfered with their lessons and examinations; but generally the public did not seem to concede the title of philosopher, that is, abstract reasoner, to even the most gifted individuals, until they had attained the age of thirty years.

It might appear that he intended to suggest that our boys should be trained to reason; that was not his wish. They were too young to think, but they could remember, and, therefore, their schooling must always largely consist in their remembering powers being taxed and developed by the assimilation of new facts and words and ideas. What he did intend to suggest, and he found himself in the best of company, was firstly that a young brain which might show signs of good reasoning powers, and which might, therefore, have only a medium memory, should not be overstrained to the breaking point, as was often done in schools, by being pitted against those infant prodigies of memory who headed all competitive examinations and carried off all scholarships. Scholarships, as they demanded cramming and bringing to the front the best memories, unless they killed these, were doubtless good for lawyers and similar professional men, but were bad for engineers; and Whitworth was never more disappointed than with the effect of his scholarship scheme.

His second suggestion was that young brains with medium memories should not be burdened with the classics: Latin, Greek and Euclid. The most progressive people the world had ever seen were the Greeks, yet they had no classics. The most stagnant of all nations were the Chinese, and their learning consisted of nothing but the classics. Why should one waste the time of thinking brains on remembering how the dead argued? What good were the classics to an Englishman? The stock answer was of course: one would not know English if one did not learn Latin, one could not think logically if one did not learn Euclid. To his mind this was a more crooked argument than any which the most sophistical of classics had ever produced. By all means learn the classics if one wanted to know them, but if one wanted to learn English, learn English but not Latin; if one could not think logically, study logic but not Euclid.

One's learning at school was now of course classic because one's schoolmasters must be classical scholars, but there were signs of a coming change, and he welcomed most heartily the bold innovation of the London University in making Latin an optional subject in the matriculation examinations. The recent Admiralty Education scheme was another step in the right direction, for it affirmed that one could be a gentleman without having passed through a school where the classics reigned supreme.

Engineers should take to heart these examples, and should encourage the creation of schools where the classics were dropped, but where all the general and linguistic accomplishments, which ought to be possessed by every gentleman, might be acquired, together with an insight into sciences; but such teaching, if good, would not be cheap. Sons of engineers might be favoured in the same way that sons of military men and of doctors and others were favoured at Wellington and at Epsom, &c.

He had already mentioned that to be a successful manager of men an engineer must enter the shop at an early age. What was the result? He would have learnt Latin, Greek and Euclid, and practically nothing else, perhaps not even English, and if he entered the shop say at fourteen, he would know nothing of chemistry, physics, or other sciences, nor of general or modern history, hardly any geography and nothing else. He would have come in contact with boys and their fathers who, if they could not express Newton's laws, knew what a falling weight or hammer would do, who knew the principle of a wedge, screw, &c., who could forge, drill, plane, saw and turn, who could, in fact, do things which the boy would like to do, and they knew things which every boy would like to know. Was it to be wondered at that, in a few months' time, he appreciated his fellow-workmen more highly than he ever did his schoolmasters? and was it surprising that after years of intimate contact with workmen, he should adopt their language and manners? and, finding himself at a social disadvantage in the company of his equals, should shape towards becoming a very rough diamond? It was this almost invariable effect, the result of sending young engineers early into the shop, which induced many rich manufacturers to send their sons to

2 B 2

(Mr. C. E. Stromeyer.)

those public schools were "they made gentlemen" of them. Of course, they then entered the shop when too old, and finding in after life that they were not so successful as their fathers, the factory was transferred to a company, and they retired.

He had already suggested as a remedy, that engineers should drop the classics out of their sons' school subjects, and should replace them by the rather more expensive general and scientific ones. Then, on entering a factory, even at an early age, a boy would at once feel that his fellow-workmen did not know very much more than their very limited trade, and he would neither look up to them nor copy their language and manners. Such a boy would grow up to be both thorough engineer and gentleman.

In conclusion, he would suggest that, should Mr. Yarrow's "sandwich" system be generally introduced, then it seemed to the writer doubly important that all unnecessary learning of dead languages and dead mathematics should be banished from those schools in which engineers received their elementary teaching; and as he wished every success to Mr. Yarrow's system, and yet feared that serious difficulties were opposing it, he would make the additional suggestion that the changes from workshop to college and back should take place not every six months as proposed, but every twelve months.

Professor HENRY SPOONER wrote that he was among those present at the meeting, and was sorry that time did not permit a larger number of representative employers to give expression to their views on the Paper, and to say how far they would be prepared to go in practically supporting such a scheme as Mr. Yarrow's. It would be too much to hope that any scheme of training that could be devised would give universal satisfaction, as opinions on such matters so widely differed, and the local dominant factors which practically decided any scheme, were rarely the same. This was well illustrated, so far as the college part of the training was concerned, in the Tables in the Paper (particularly the Paper read before the Institution of Naval Architects). They clearly showed that a pronounced difference of opinion and practice existed. For instance, it would be seen that the McGill University devoted 577·5 hours per year to the group of subjects, Applied Mechanics, Steam-Engine, Machine Design and Engineering Laboratory, while 225 hours, or 38·9 per cent. of that time, were considered sufficient at Yale University. Again, in mathematics, the McGill University devoted 135 instruction hours per year to 27·8 per cent. of that time, or 37·5 hours, at Cornell. Whilst as regards the college workshop, the diversity of practice was still more marked, as at one end of the scale there was Cornell giving 225 hours per year, and at the other end, the 10 hours per year which were apparently devoted by Yale to workshop visiting.

Thus, at Cornell, where the time devoted to the workshops was greater than in any of the other colleges, only 7½ hours per week (based on a college year of 30 working weeks) were spent in the workshops, compared with 6 hours, 3·375 hours and 3 hours per week at the McGill University, the Massachusetts Institute, and Harvard respectively. A still closer examination of the time-tables of the principal American colleges disclosed the fact that there was nothing approaching uniformity of training in that country, no more than there was in this, or probably any other country, and perhaps it was as well that it was so. It must be generally conceded that every system of training was a compromise to which some exception could be taken, and it seemed to the writer that of all the systems that were in operation or that had been proposed, the " sandwich " one was the best; but not the one that had received so much attention lately (the alternate slices in that being much too thick to be properly assimilated), but the one in which each day more or less represented a sandwich, the early part of the day, whilst the mind was fresh, being devoted to lectures, laboratories and class work, and two or three hours in the afternoon to the workshops. There was then no appreciable break in the continuity of the scientific training throughout the year (with three terms), and the student was kept in constant touch with the workshop. No one would seriously suggest that the college workshop could ever replace practice in the shops of a mechanical engineer, where the commercial factor was so important, but experience had proved that with a well-equipped

(Professor Henry Spooner.)

workshop, good practical instructors, and a succession of interesting jobs, students could be taught in a three years' course to do very good work at the bench and lathe, indeed, they were generally able, when entering works after their college course, soon to take the lead of those who had spent the whole of the same number of years in the works.

At the Polytechnic School of Engineering, London, the writer had had seventeen years' experience with the above system, and the results had been exceedingly satisfactory. It might be explained that students had from 9 to 11 hours per week in the workshops (in addition to workshop visiting in the summer) against the $7\frac{1}{2}$ hours at Cornell, but on the other hand their average hours per week were 28 against 24 at Cornell; and, again, at Cornell, the college year was apparently one of 30 weeks, whilst at the Polytechnic it ran for 40 weeks, in three terms.

Owing to the many objections to the "sandwich-year" system the writer did not believe that it would ever be generally adopted in this country. He considered that if, after all that had been said in favour of a youth commencing with a two or three years' college course, it were considered that he was, when older by those years, better able to stand the early hours of a factory and were less detrimentally influenced by associating with men so much older than himself, the balance would be in favour of such a course preceding the training in an engineer's works, always assuming that whilst at college sufficient time was spent in the workshops.

The writer feared that, when comparisons were made between the Technical Colleges of this country and the palatial ones in America, Germany and Switzerland, to the detriment of the former, it was often overlooked that our splendid system of Science, Art and Technical Evening Classes, was probably unsurpassed by any in the world. So long ago as 1825, before they were developed by the fostering care and public spirited support of the Board of Education, they excited the admiration of that famous educationalist, Baron Charles Dupin, who did all in his power (with great success for a time) to develop and emulate the system in France. Were it not that apprentices and young draughtsmen mechanics who availed themselves of these classes were handicapped by coming to them after a long day's work, and that they could not attend for more than three hours each evening, they would have in this system something approaching an ideal sandwich. And the writer ventured to think that in considering the broad question of the education of engineers, this most deserving class, a class from which so many distinguished men had sprung, and upon which the success of every works so much depended, must receive more than passing attention. He therefore heard with much pleasure the sympathetic remarks of Mr. Aspinall and Mr. Drummond (and of Sir William White and Mr. Yarrow at the Naval Architects), on the training of the young workers, whose only chance of getting a scientific education was by attending evening classes, and he trusted that it would be found practicable to do something to increase their opportunities of attending such classes. In the first place, this could be done by not calling upon any apprentice or young mechanic who was attending evening classes to work overtime, and better still, if the deserving ones could be allowed to leave off a little earlier on class nights, or start a little later the following morning, it would, doubtless, greatly increase their opportunities and efficiency. Many apprentices in the London district every year were compelled to give up their classes through having to work overtime ; indeed, some very bad cases have recently come under the writer's notice, where apprentices in some of the best-known works have been kept at work three or four nights a week till 9 o'clock, and even later.

If Mr. Yarrow could see his way to induce the masters to make such a concession he would be doing a great service for the country, and one that the writer had every reason to believe would be very much appreciated by a most worthy class.

CONVERSAZIONE.

A Conversazione was held at the Institution on Tuesday, 19th May 1903, when the Members and their friends were received by the President and Mrs. Wicksteed. During the evening the Band of His Majesty's Scots Guards performed a selection of music, and vocal music was rendered in the Library. A cinematograph exhibition, including pictures of the Delhi Durbar, was given in the large Hall, and a room was set apart for Electrophones. The number of Guests was about 850.

MEMOIRS.

HENRY BATES was born at Salford on 15th August 1846. He commenced his engineering career in 1858 as an apprentice at Tammerfors, Finland. Four years later he started at Messrs. Joseph Whitworth and Co., Manchester, where he finished his apprenticeship. His subsequent experience was gained at Messrs. Smith and Coventry's Works, passing from there a few years later to Messrs. Sharp, Stewart and Co., whose works at that time were in Manchester. He left the last named firm in 1870 to start business on his own account, but shortly afterwards, in 1872, joined Messrs. Crossley Brothers, gas-engine makers, where he remained a few years. He next became manager of the Breech Action Manufacturing Co., Birmingham. Some little time after this he joined the late Mr. W. W. Hulse as his chief in his business as consulting engineer at Manchester, and when that gentleman took over the Ordsal Works with the business of Messrs. Hulse and Co. in 1882, he went with him as manager. On the reconstruction of this firm, and its transformation into a private company in 1898, he was appointed managing director, which post he filled until the time of his death. During his twenty years' connection with the works of Messrs. Hulse and Co. he was closely identified with the design and construction of the great variety of machine tools for which that firm attained a high reputation. He had also for many years occupied the position of consulting engineer and valuer to several of the largest and most important firms in the country. His death occurred at Southport on 1st February 1903, at the age of fifty-six. He became a Member of this Institution in 1891.

JOHN WILLIAM BOOTH was born at Rodley, Leeds, on 17th September 1852, being the only son of Mr. John Booth, one of the founders of the firm of Joseph Booth and Brothers, Rodley, in whose works he served his apprenticeship from 1867 to 1873, being afterwards engaged as draughtsman from 1873 to 1883. From the latter date he became manager of the works until 1888, when he became a partner in the firm, and afterwards chairman on its conversion into a private company in July 1897. His death took place at his residence, Oaklands, Calverley, near Leeds on 19th May 1903, in his fifty-first year. He became a Member of this Institution in 1892.

JAMES TENNANT BOYD was born at Darvel, Ayrshire, on 12th January 1850. He served his apprenticeship with Messrs. M. Paul and Co., Levenford Works, Dumbarton, from 1866 to 1871, and on its expiration he was engaged as second engineer on the s.s. "Enterprise," of Londonderry. From 1873 to 1876 he served with the Peninsular and Oriental Co., as fifth, fourth, and third engineer. In 1876 he became chief engineer to the National Spinning and Weaving Co., at Bombay, and remained in that position until 1879. From that year until 1888 he was chief engineer and superintendent of the Bombay Ice Co., and during that period he re-erected the ice factory and machinery after a fire. In 1888 he founded and became sole proprietor of Boyd's Ice Factory in Bombay. He was also a director of the Mody Bay Iron Co., Bombay. In 1896 he returned to Scotland, and became one of the first directors of Stirling's Motor Carriage Co., of Glasgow and Hamilton. His death took place from paralysis at his residence at Lenzie, near Glasgow, on 28th December 1902, in his fifty-third year. He became a Member of this Institution in 1893.

ROBERT BARR CAMERON was born at Johnstone, near Glasgow, on 3rd May 1869. He received his elementary and scientific training at the public school, Johnstone, near Glasgow, the Glasgow and West of Scotland Technical College, and at the Royal College of Science, London. From 1884 to 1890 he served an apprenticeship in the engineering works of Messrs. Craig and Donald, of Johnstone, and was subsequently engaged as draughtsman for two years in the machine tool works of Messrs. William Robinson and Sons, of the same town. He was afterwards re-engaged by Messrs. Craig and Donald as leading draughtsman, and during this period he assisted with the designs of machinery for Messrs. J. and G. Thomson, Clydebank, and other shipbuilding firms. In 1895 he entered as draughtsman the torpedo department of Messrs. Armstrong, Whitworth and Co., and in 1897 was appointed assistant teacher of mathematics and mechanical science at the Municipal Technical School, Brighton. His death took place suddenly at Brighton on 20th March 1903, in his thirty-fourth year. He became an Associate Member of this Institution in 1898.

DAVID DAVY was born in Sheffield on 8th October 1837, and was the son of the late Mr. David Davy, of the firm of Davy Brothers, Sheffield. He was educated in Sheffield, with the intention of becoming a mechanical engineer, and turned his attention specially to colliery and iron undertakings, in which he was largely interested as chairman of the Manvers Main Colliery Co., Hickleton Main Colliery Co., the Midland Iron Co., and as a director of Messrs. Brown, Bayleys, and the Netherseal Colliery Co. He was a great lover of art and its application to Sheffield industries, and was one of the principal supporters of the Sheffield School of Art, being President in 1895. He took no part in municipal affairs, but was a member of the Ecclesall Board of Guardians. His death took place from typhoid fever at his residence at Parkhead, Sheffield, on 19th April 1903, at the age of sixty-five. He became a Member of this Institution in 1873.

ALFRED CHRISTIAN DOWNEY was born in Liverpool on 25th September 1831. He was educated at the Royal Institution in that city, and served his apprenticeship at the engineering works of Messrs. Fawcett and Preston. In 1852 he accepted an appointment with Messrs. Cochrane and Co., of the Woodside Iron Works, Dudley, in whose service he undertook his first responsible work of superintending the erection of the roof of the London and North Western Railway station in Birmingham, a work which to this day is admired for its gigantic and fine proportions. From Woodside his services were transferred in 1856 to the new Ormesby Iron Works of Messrs. Cochrane and Co. at Middlesbrough, where for several years he occupied the position as head of the drawing office and entire engineering department. In 1872 he entered into partnership with Mr. C. F. H. Bolckow with the style of Downey and Co., and erected the Coatham Blast Furnaces. He also reconstructed the Lackenby Iron Works, which were subsequently acquired by the firm, and opened the Stanghow mines for the supply of iron ore. For the last few years he was manager of Messrs. John Turner and Co.'s Works at Middlesbrough, and he also acted as consulting engineer. His death took place at his residence in Middlesbrough, on 19th April 1903, in his seventy-second year. He became a Member of this Institution in 1866, and was a member of the Ironmasters' Association, and a quondam president. He was among the founders of the Iron and Steel Institute, and was also a Member of the Institution of Civil Engineers, and other technical societies.

JOHN HENRY HARGREAVES was born at Hindley, near Wigan, on 4th May 1847. After being educated at Harrow School, he served his apprenticeship with Messrs. Hick Hargreaves and Co., of Bolton, of which establishment his father, the late Mr. William Hargreaves, was sole owner. He occupied a leading position until his father's death in 1889, and, on the business being converted into a company in 1892, he was appointed chairman and managing director, which position he occupied until his death. He took a great interest in the foreign business of the firm, spending some time on its account in Russia in 1898. He was also a director of the Bolton Iron and Steel Co., and of the Bridgewater Spinning Co. For some years he was an officer in the Loyal North Lancashire Regiment of Rifle Volunteers, and was also for some time a member of the Bolton Town Council. His death took place at Bolton on 8th January 1903, in his fifty-sixth year. He became a Member of this Institution in 1887.

WILLIAM FIELD HOW was born on 15th January 1856, and served his apprenticeship in the marine engineering works of Messrs. D. and W. Dudgeon, of Millwall, London, from 1871 to 1875. On the completion of his pupilage, he was engaged for three years, first in the locomotive works of the North London Railway at Bow, and afterwards with Messrs. Ruston, Proctor and Co., Lincoln, and Messrs. Richardson and Sons, at Hartlepool. In 1876 he obtained a Whitworth Scholarship, after which he gained further practical experience in various engineering works in England. In 1878 he obtained an appointment on the staff of the late Sir John Fowler, Bart., being occupied chiefly with the New South Wales Government business of that firm. In 1891 he went to New South Wales, where he acted up to the time of his death as the engineering representative of several English firms. His death took place from paralysis at Manly near Sydney, on 29th November 1902, in his forty-seventh year. He became a Member of this Institution in 1891.

MONTAGUE GEORGE ALFRED HUMPHREY-MOORE was born at Alton, Hants, on 10th November 1870, and was the youngest son of Mr. Arthur H. Moore of Alton. He was educated at King's College School, London, matriculating into the Engineering Department, where he completed a three years' training. From 1890 to 1893 he served a pupilage with Messrs. Bramwell and Harris, afterwards becoming one of that firm's assistants. In 1893 he was appointed fourth engineer and electrician to the Tombacci Société, and, shortly after, became the electrical engineer to the Regie Co., when he erected and set to work the first electric lights in Turkey. On his return to England he was again employed by Messrs. Bramwell and Harris as one of their chief assistants in supervising the erection of various electric light and other undertakings. In 1896 he was appointed assistant works manager to the British Plate Glass Co., St. Helens; after erecting new machinery, boilers, etc., he again returned as assistant to Messrs. Bramwell and Harris. In 1898 he was appointed an assistant engineer to the New River Co. His death took place from typhoid fever at his residence in Ealing, London, on 29th January 1903, at the age of thirty-two. He became an Associate Member of this Institution in 1900.

JAMES DICKERSON HUMPIDGE was born at Gloucester on 19th March 1861, and was educated at the Crypt Grammar School in that city. At the age of fifteen he was apprenticed to the firm of Messrs. Fielding and Platt for six years, during which time he passed through their shop and drawing office, being chiefly engaged on hydraulic machine-tool work, the leading speciality of the firm in question. After the termination of his apprenticeship he stayed on in the employ of the firm, about nine months later being appointed their chief draughtsman, and afterwards works manager. In 1885 he left Messrs. Fielding and Platt and went to Australia, where for a period of about twelve months he was engaged in the drawing office and works of Messrs. Hudson and Co., of Paramatta. Returning to England in 1886 he resumed his occupation of works manager with Messrs. Fielding and Platt, and continued in their service until 1889, when he commenced business on his own account in Gloucester as a mechanical engineer. In 1890 he disposed of this business to Messrs. H. Moffat and Co., and entered into partnership with Mr. John Platt (the eldest son of his late principal) as consulting engineers. In this connection he was engaged in laying out saw mills and other factories in various parts of the country, also acting as consulting engineer to the Gloucester Wagon Co., and Severnports Warehousing Co., of Sharpness, carrying out a good deal of trade for both firms. He also modernized the whole of the driving gear of Messrs. Marling and Co.'s large cloths mills near Stroud, putting in a new steam-engine, boilers, etc., and main driving.

In 1891, in partnership with his brother, Mr. H. Theo. Humpidge and Mr. G. E. Snoxell, he acquired an engineering business at Dudbridge, near Stroud, and commenced the manufacture of the "Dudbridge" gas-engine, doing at the same time a good general trade. In 1894 the firm purchased the business of their neighbours Messrs. Holborow and Co., steam-engine manufacturers, the title of the new Company being Humpidge, Holborow and Co., the late Mr. J. D. Humpidge being chairman of directors. In 1899 further changes were made in the constitution of the firm, the steam-engine business was sold, and the works devoted to the manufacture of gas and oil engines, the title of the Company being altered to the Dudbridge Iron Works, when he became co-managing director with Mr. H. Theo. Humpidge and Mr. T. G. Smith. From this period he was actively engaged in the production of gas and oil engines, making numerous alterations and additions to the plant and premises in order to facilitate the work. He was a most capable and energetic engineer, quick to grasp the requirements of a situation, fertile in design and means of executing work, withal combining great commercial capacity. In the formation of the Gloucestershire Engineering Society he took an active part, and was a member of the council from its inauguration, and in October 1902 was elected to the Presidency of the Society for the year 1903. He was fatally injured while watching the brake test of a large gas-engine when the flywheel broke into pieces, by one of which he was struck on the head. The accident occurred on 6th June 1903, and his death ensued the following day, at the age of forty-two. He was elected a Member of this Institution in 1894, and was also an Associate Member of the Institution of Civil Engineers.

WILLIAM HUTTON was born in Cape Town in the year 1857, and was educated at the Educational Institute in that city. From 1871 to 1876 he was engaged in the stores and engineers' department of the Cape Government Railway, after which he was employed as draughtsman on the Port Elizabeth harbour works for three years. In 1879 he set up in business on his own account as mining agent, and during 1889 and 1890 he was one of the managers for the Oceana Co. From 1897 up to the time of his death he was largely interested in the development of the Murchison Gold Fields (Selati district), Transvaal, and his opinions on properties in this district were highly valued. His death took place, after a short illness, at Johannesburg, on 28th September 1902, at the age of forty-five. He became an Associate of this Institution in 1896.

DAVID JOY was born in Leeds on 3rd March 1825, being one of the five sons of Edward Joy of the Oil Mills, Leeds. From his infancy he took the keenest interest in machinery of all kinds, and spent his leisure moments in making models of ships and engines. When he was about sixteen he began to study mechanical drawing at Wesley College, Leeds, filling up his spare time with mastering thoroughly a copy of Tredgold on the Steam-Engine. Having completed his education, he entered his father's works in 1841, and learnt trade routine and the business of seed crushing and oil refining. It soon became evident that the work was distasteful to him, and he was therefore apprenticed in the works of Messrs. Fenton, Murray, and Jackson, where he stayed until February 1843, when the works were closed. In June of that year he entered the Railway Foundry Works, Leeds, of Messrs. Shepherd and Todd, as a drawing office apprentice, where his first job was to prepare the plans of a "John Gray" engine for a steam pressure of 90 lbs. On Mr. E. B. Wilson taking over the Railway Foundry in 1844, Mr. Joy became manager of the drawing office, and it was in that capacity that he was so intimately associated with the designing of the "Jenny Lind" locomotive. In 1850 he was appointed superintendent of the Nottingham and Grantham Railway, then just opened. No engines were ready for the work, so that considerable ingenuity was necessary to get the engines in time, in order to comply with the conditions. His next appointment in 1853 was as the locomotive superintendent of the Oxford, Worcester, and Wolverhampton Railway, where he remained until the line was sold in 1856, when he returned to the Railway Foundry, Leeds. In 1855 he read a Paper * before this Institution on a "Spiral Coil Piston Packing." In 1857 he brought out a compound marine engine, in which a deep high-pressure piston acted as the distributing valve for the low-pressure cylinder. He also invented a steam reversing gear, and about this time he took out the first of three patents for hydraulic organ-blowers.† The first on a large scale was fitted to the organ at the Leeds Town Hall, and they are also in use at the Crystal Palace. In 1859 he accepted the position of manager of Mr. De Bergue's bridge-building yard in Manchester, and in the next year brought out a special form of steam hammer, for the manufacture of which he started in business for himself at the Cleveland Engine Works, Middlesbrough. For some years he was

* Proceedings 1855, page 171.
†   Do.   1857, page 184.

continuously engaged in inventing new machinery of various kinds. In 1871 his works were closed because the ground was required for the extension of one of the large shipbuilding yards there. During this and the succeeding year he organized the first serious effort to utilize slag as a residual product. A form of blast was used for pulverising the slag, and one of the results was silicate cotton. In 1874 he went to the Barrow Shipbuilding Co. as manager of the water-tube boiler department, the company having purchased the rights of the Howard boiler, and in June 1876 he also became secretary to the same company. During this period he worked out the details of his radial valve-gear, which was patented in 1879. It was taken up by the London and North Western Railway and by Messrs. Maudslay for marine work. Mr. Webb fitted a powerful six-coupled goods engine with it, and sent this engine to Barrow-in-Furness for the Summer Meeting * of this Institution in that town in August 1880. On that occasion Mr. Joy read a Paper † on the subject. In the same year he went to London to act as the London agent of the Barrow Shipbuilding Co., but only continued in this capacity for a little more than a year, the work in connection with his several inventions demanding all his attention. In 1882 he attended the meeting of the Master Mechanics' Association at Niagara, and there read a Paper on Mr. Webb's compound engine, and on his own valve-gear. Returning to London he continued, in partnership with his sons, to develop his various inventions, and he read Papers on the valve-gear and assistant cylinder before a number of Societies. The success of the valve gear is attested by the fact that it is now applied to engines aggregating one million horse-power, and a considerable number of his assistant cylinders are also in use. The latter device is successfully working on a large number of ships in the British and Foreign navies, mercantile marine, on private steam yachts, and on stationary engines, representing a total of over one and a half million horse-power. In 1894 he read a Paper before this Institution on a " Fluid-Pressure Reversing Gear for Locomotive Engines." ‡ He was a regular attendant at the meetings of this

---

* Proceedings 1880, page 432.

†   Do.   Do , page 413.

   Do.   1894, April, page 252.   2 c

Institution, to which he occasionally contributed remarks on the Papers. His death took place at his residence at Hampstead, London, from congestion of the lungs on 14th March 1903, at the age of seventy-eight. He was a Member of this Institution from 1853 to 1867, and re-joined in 1880. He was a Member of the Institution of Naval Architects of England and also of America, and of other Societies.

CUTHBERT RIDLEY LEE was born in Newcastle-on-Tyne on 26th May 1852. He served his apprenticeship from 1867 to 1873 in the works of Messrs. J. and G. Joicey and Co., of Newcastle-ou-Tyne, and on its completion remained with the same firm as draughtsman. From 1875 to 1876 he was engaged as draughtsman by Messrs. Bells, Lightfoot and Co., of Walker-on-Tyne, during which time he gained a local scholarship of £50 at the government science classes. Afterwards he superintended the erection of their machinery in the North. Subsequently and up to the year 1880 he acted as their manager in London, when he became engineer in the south for Messrs. John Abbot and Co., of Gateshead-on-Tyne. In this capacity he superintended the erection of their machinery at various docks and gas works. From 1880 to 1896 he was a partner in the firm of Messrs. J. Coates and Co., engineers of London and Melbourne, and from 1896 was in business as C. R. Lee and Co. His death took place at Dunmow, Essex, on 22nd April 1903, in his fifty-first year. He became a Member of this Institution in 1887.

GEORGE BRAITHWAITE LLOYD was born in Birmingham on 15th October 1824. His engineering training was received in Liverpool at the marine-engine works of Messrs. Bury, Curtis, and Kennedy, and at the Shildon Engineering Works of the Stockton and Darlington Railway, now part of the North Eastern Railway. About 1852 he started in Birmingham on his own account, under the title of G. B. Lloyd and Co., as manufacturer of iron boiler tubes for marine and locomotive purposes. After building up a successful business, his services were required upon the death of his father in 1859 in the family bank of Lloyds and Co. (the nucleus of Lloyds Bank), and he remained a director of the Bank until his death.   In 1870–71 he was Mayor of Birmingham.   For twelve years he was a director of the Midland Railway Co., and always took great interest in the locomotive and other engineering departments of that railway, for which his training had specially fitted him.   He was an active member of the Birmingham General Hospital Committee and was chairman for about thirty years of the Lunatic Asylums Committee.   His death took place at his residence at Edgbaston, Birmingham, on 8th February 1903, at the age of seventy-eight.   He became a Member of this Institution in 1854.

FRANCIS CARR MARSHALL was born at Bedlington, Northumberland, on 25th April 1831, and subsequently removed to Newcastle, where his father was employed at the locomotive works of Messrs. Hawthorn at Forth Banks.   On completing his education at Newcastle he was apprenticed to Messrs. R. and W. Hawthorn as an ordinary engineering apprentice, passing through the shops, and subsequently into the drawing office.   After being employed for some time as a draughtsman at Messrs. Hawthorn's, he became chief draughtsman to the firm of Messrs. Thompson and Boyd, of Newcastle, where he remained until the year 1860, when he took charge of the engine works of Messrs. Palmer Brothers at Jarrow.   While at Jarrow the whole of the engine works were remodelled and rebuilt under his régime, and a large amount of important marine engine work, including the engines for some of the largest Atlantic liners of that day were produced from the Jarrow works.   In 1870 he left Jarrow, and, in association with Mr. B. C. Browne (now Sir Benjamin Browne) and others, he purchased the Forth Banks Works of Messrs. R. and W. Hawthorn.   In the early " seventies " the firm commenced to do a considerable amount of government work, with the production of which it has in later years been so closely associated.   In addition to the government engines, he was responsible for the design of a considerable number of large mercantile engines in the early days of the new firm.   The Works continued to grow and develop until, finally, it was arranged that the whole of the Marine Department should be moved to St. Peter's (where

2 c 2

the Boiler Yard already was), and Mr. Marshall designed and constructed the present Marine Engine Works at St. Peter's. He was prominently associated with the first introduction of the high-speed marine engine for warship work, such as is now commonly in use. One of the earlier sets of machinery of this type, designed by him, was for the cruiser " Esmeralda," built by Messrs. Armstrong, Mitchell and Co., at Walker-on-Tyne. He also designed a number of other similar sets for various governments. He was associated with the first introduction of the torpedo type of engine into vessels of larger size, some vessels for the Russian Government and the Italian Government, engined by him, being the earliest representatives of the Torpedo Gun-Boat Class. He was largely instrumental in bringing about the introduction of forced draught on board ships, and designed some of the earlier installations which were effected. His name is also associated with the " Marshall " valve-gear—a single eccentric gear which has been fitted to a very large number of marine engines. When the firm of Messrs. R. and W. Hawthorn amalgamated with Messrs. Andrew Leslie and Co., Mr. Marshall became managing director of the Engine Works. After the amalgamation, he was responsible for a large amount of high-power engines for vessels built at the Shipyard, in addition to the warship machinery supplied to H.M. dockyards and also to Messrs. Armstrong, Whitworth and Co. He continued in his position as managing director at St. Peter's Works until 1897, when, owing to failing health, he was obliged to resign his active duties, although he remained upon the Board of Directors of the company for some years after. For the last few years he had been a confirmed invalid, and had not been able to take any interest in the business. In 1881 he contributed a Paper to this Institution on " The Progress and Development of the Marine Engine." * His death took place at his residence in Newcastle-on-Tyne on 24th February 1903, in his seventy-second year. He became a Member of this Institution in 1865, and was a Member of Council from 1882 to 1885, and 1892 to 1896, and a Vice-President in 1896–97.

---

* Proceedings 1881, page 449.

THOMAS NASH was born in Swindon on 1st March 1838. After receiving an elementary education he was apprenticed at the locomotive works of the Great Western Railway Co. He had barely attained his majority when, in 1858, he obtained an appointment as engineer in connection with the Bombay, Baroda, and Central India Railway. He remained in India for about eleven years, but the trying climate eventually affected his health, and he was obliged to come back to England. He then acted as inspecting engineer in Glasgow for the Bombay, Baroda, and Central India Railway until 1873, when he commenced business as a consulting and inspecting engineer in Sheffield. In 1880 he conceived the idea of establishing an independent testing and experimenting works in Sheffield. He took great pride in turning out accurate work, with the result that his business grew rapidly, and he found it advisable to add a chemical laboratory to his mechanical testing laboratory. In the course of his work he was frequently consulted regarding breakdowns of various kinds. He took no prominent part in politics or municipal work, but devoted himself unremittingly to his business, to which he was much attached. At the close of last year, being in failing health, he disposed of his business to his son, but only survived his retirement for a short time. He was a sufferer from bronchitis ever since his return from India, and he died at his residence at Nether Edge, Sheffield, on 24th February 1903, in his sixty-fifth year. He became a Member of this Institution in 1889, and was also a member of the Iron and Steel Institute.

WILLIAM HARRY STANGER was born on 24th September 1847, at Pietermaritzburg, Natal, which Colony had only then recently been added to the British Dominions. His father, the Hon. W. Stanger, M.D., F.G.S., was Surveyor-General of the district, and as such took part in the legislative and executive functions of the Government. In 1851 Dr. Stanger brought his family to England, when he came on two years' leave ; on his return to South Africa he only survived one year. The son—William Harry—received his early training at Norwich Grammar School, and completed it at King's College, London. He served his apprenticeship as an engineer at the

Hunslet Engine Works, Leeds, under Mr. Campbell, and after the expiration of his time, obtained employment in the locomotive department of the North Eastern Railway. His next appointment was as instructor of traction-engine-driving at Aldershot and Woolwich, and a few years later he went to South America as locomotive superintendent of a railway in Northern Brazil, but had to relinquish this position within two years owing to a severe attack of yellow fever. On his return to this country in 1873 he was appointed to the staff of the Crown Agents for the Colonies as engineering clerk, his duties consisting in advising the Crown Agents in engineering matters, and inspecting work generally. A few years later he resigned his position as a member of the official staff, and commenced business on his own account in Westminster, at the same time acting as inspecting engineer for the Crown Agents. In 1887 he established the Broadway Testing Works at Westminster for the mechanical examination of all constructional materials. Soon afterwards, recognising the importance of chemical as well as tensile tests, he took into partnership Mr. Bertram Blount, F.C.S. Since the commencement of this establishment he has continued to act as Inspector of Engineering Work for the Crown Agents, and for the Director of Works and the Chief Engineer of the Works Loan Department of the Admiralty; and in all these capacities he has tested cement for nearly all the great works carried out for naval defence, as well as for iron works for Gibraltar, Simon's Bay, Hong Kong, etc. At the request of the London Chamber of Commerce he carried out investigations during 1894–95 on the effect on Portland cement of the admixture with it of various foreign substances. One of the results was to create a standard which was accepted by the Chamber of Commerce, and a Paper giving the results of the investigations was contributed by him and Mr. Blount to the Society of Chemical Industry in 1897. For many years his health had suffered in consequence of the attack of yellow fever when in South America, and he was advised in December, 1902, to take the baths at Helouan, near Cairo, and to winter in the South of France. But as he was not well enough to land at Marseilles he proceeded to London, where his death took place on 13th February

1903, at the age of fifty-five. He became a Member of this Institution in 1875, and was also a Member of the Institution of Civil Engineers, and a Fellow of the Chemical Society.

JAMES MACINTYRE THOMSON was born in Glasgow on 5th December 1843. He commenced his apprenticeship in 1861 at the Clyde Bank Foundry of his father and uncle, Messrs. James and George Thomson. In 1864, when in the midst of his apprenticeship, the partnership was dissolved, his uncle, Mr. G. Thomson, acquiring the business. He then entered the drawing office of Messrs. Caird at Greenock. In 1868 his father bought some ground at Finnieston Street, Glasgow, and started him in business in conjunction with his elder brother—John, as the Finnieston Engine Works, and the title of the firm as John and James Thomson. The business progressed well, so that in 1882 the works were found too small for both engineering and boiler-making, and a large place was added for boiler-making at Kelvinhaugh. In 1891 the brothers resolved to retire from business, and let both the works in 1893 to Messrs. Barclay, Curle and Co., who had sold their own works to the Tunnel Co. (subway under the Clyde). Messrs. Barclay, Curle and Co. still continue as tenants in the two works referred to. In 1891 he entered the Town Council of Glasgow as one of the representatives of the Kelvinside Ward, and served as a magistrate for the usual period. He was also a member of the Clyde Trust, and convener of a section of the Sewage Committee. His death took place suddenly at his country residence, Montgomerie, Tarbolton, Ayrshire, on 4th February 1903, at the age of fifty-nine. He became a Member of this Institution in 1875.

JOHN TURNBULL, Jun., was born in Glasgow on 4th August 1841, and received his education at various schools in that city. He commenced his engineering apprenticeship in the Canal Basin Foundry under his father. On completing his time he went to London with two friends, and started a pattern-making business, which was very successful. In 1848 he went and fought under Garibaldi as one of the Glasgow Volunteers, and on his return to Glasgow, he was taken into partnership with Messrs. Turnbull,

Grant, and Jack of the Canal Basin Foundry. In 1877 he started in Glasgow as a consulting engineer, and made a speciality of steam-engine work and water motors. He brought out a compound engine and invented a simple cut-off valve gear, and was the author of "A Short Treatise on the Compound Engine," the "Engineer's Guide Book," "Arithmetical Questions," and "Water Wheels and Turbines." He was widely consulted with regard to water power and steam power, and was well known as an expert machinery valuer. His death took place at his residence in Glasgow, on 14th January 1903, in his sixty-second year. He became a Member of this Institution in 1885, and was also a Member of the Institution of Engineers and Shipbuilders in Scotland.

GEORGE UZZIAH WHEELER was born in Chiswick, London, on 25th February 1868. After being educated at Ealing, he served his time from 1884 to 1889 in the workshops and office of Messrs. R. Warner and Co., Walton and London. During this period he also studied at King's College and the Birkbeck Institute, London, and gained a Whitworth Scholarship in 1888. Shortly after completing his time he was engaged as a draughtsman in the office of Messrs. James Simpson and Co., of Grosvenor Road, London, until 1891 when he obtained a similar position at the Thames Iron Works and Shipbuilding Co., where he remained until 1895. During that time he was engaged on the design and erection at Crossness of the sewage pumping machinery for the Southern Outfall Main Drainage. He was next engaged in the drawing office of Messrs. Tangyes, of Birmingham, taking a leading part in the design of waterworks pumping machinery, triple expansion engines, etc. In 1899 he was compelled, by failing health, to relinquish active professional work, and in the autumn of 1902 went to South Africa in the hope of benefiting by the change of climate. He died at Cape Town on 5th January 1903, in his thirty-fifth year. At one time he was a contributor to the technical papers, and during the latter part of his illness was engaged on re-writing a series of articles for bringing out in book form. He became an Associate Member of this Institution in 1899.

John Mervyn Wrench, the third son of the Rev. T. W. Wrench of St. Michael's, Cornhill, London, was born on 14th November 1849, and received his early education at the Charterhouse School. In 1866 he entered the Great Western Railway Works at Swindon, as a pupil of Mr. Joseph Armstrong, locomotive superintendent, and in October, 1868, became a pupil of Mr. W. J. Kingsbury, of Westminster. He became assistant to Mr. C. F. de Kierzkowski Steuart in January, 1871, and in December of the same year was appointed Chief Assistant to the Public Works Construction Co. of Cannon Street, London, under Mr. de Kierzkowski Steuart, Chief Engineer to the Company, and remained with them for about two years, designing bridge, roof, and general ironwork, rolling stock, etc., and supervising the equipment in England of the East Argentine, Rio Cuarto, Andino, and other foreign railways. In October 1873, Mr. Wrench was appointed personal assistant to Mr. Charles Stone, chief engineer of the Scinde, Punjab, and Delhi Railway, at Lahore, India, and later acted as resident engineer on various railway extensions. In India he early distinguished himself by suggesting the utilization of worn-out rails from the permanent way to form girders for building small bridges, whereby a great saving was effected. He joined the staff of the Indian Midland Railway during the construction of that line in 1884 as assistant engineer, and was promoted to be chief engineer in 1887 shortly after the line was opened for traffic. Whilst with this company he carried out various extensions and improvements, and designed several long viaducts, in building which he made a further use of old rails as temporary scaffolding for the erection of the piers and long girders. On the amalgamation of the Midland Railway with the Great Indian Peninsular Railway in 1902, he was appointed chief engineer of the combined systems and moved his headquarters from Jhansi to Bombay. He brought out many inventions connected with his profession, among the best known being the system of interlocking points with signals known as " Wrench's Interlock." His death took place from bronchitis from which he had suffered for some time at his residence, Pali Hill, Bandra, Bombay, on 22nd January 1903, at the age of fifty-three. He became a Member of this Institution in 1881.

INDEX.

1903.

PARTS 1-2.

adjustment of screws for heavy cuts, 36; endurance of tool required rather than short runs, 37.—Ashton, H. T., Good results from side-cutting tool, 37; yellow metals; diameter of work and clearance angle, 38; apparatus for measuring pressure on tool, 39.—Crompton, Lt.-Col. R. E , Horizontal angle of tool, 39; rapid-cutting steel and heavy lathes, 41.— Dumas, R., Angles of cutting tools, 41; horizontal angle, 43.—Donaldson, H. F., Nature of experiments, 44; roughing and finishing cuts; height of tool above centre, 45; brass and yellow metal, 46.—Adamson, D., Front clearance for cutting brass greater than for steel, 46.—Donaldson, H. F., Measurement of 10 tons on point of tool; increase of speed requires increased power; round-nosed tool for heavy cuts, 47.—Maw, W. H., Thanks for Paper, 48.—Smith, R. H., Experiments on cutting tools in 1880, 48; ratio of cutting force for different metals, 49; measurements of very heavy cuts on steel ingot; forms of tools used, 52; measuring apparatus at King's College, 54; friction of bearing points at tool-bar, 55.

and Halsey systems. 254.—Smith, R. H., Tables for calculating premiums. 255 —Rowan. J., Immutability of allowance of time; day-work and premium system in same shop, 258; work quicker on premium than on piece-work system, 259 ; average reduction of standard time allowance. 260.

PRESIDENT'S ADDRESS, 271. *See* Address by the President.

PRESTON, F. K., elected Associate Member, 264.

PRICE-WILLIAMS, R., Remarks on Premium System, 234

PROSSER, R W. O., elected Associate Member, 3.

PRÜSMANN, C. A. L., elected Associate Member, 3.

PUMP EXPERIMENTS, 123 *See* "Slip" in a Plunger Pump.

RAKE, G. A., elected Member, 3.

RAWORTH, J. S., elected Member, 200.

REDMAN, S. G , elected Associate Member, 201.

REID, H , Remarks on Education of Engineers. 320

RENWICK, R., elected Associate Member, 264.

REPORT OF COUNCIL, ANNUAL, 89 *See* Council, Annual Report.

REXWORTHY, H. S., elected Associate Member, 264.

REYNOLDS, C H , elected Associate Member, 201.

RICHARDSON, J., Remarks on Premium System, 236.

RICHES, T. H., Remarks on Education of Engineers, 342

ROBB, J., elected Associate Member, 201

ROBB, W., elected Associate Member, 264.

ROBERTS, E. G. L., elected Graduate, 265.

ROBERTSON, G. F., elected Associate Member, 264.

ROBERTSON, J. M., elected Associate Member, 201.

ROBINSON, I. V., elected Graduate, 4.

ROBINSON, J. F., re-elected Member of Council, 114.—Remarks on Premium System, 231.

ROBINSON, T. H., elected Associate Member, 201.

ROBINSON-EMBURY, Capt. R. P., elected Member, 200.

ROGERS, H M., elected Member, 263.

ROSE, A , elected Associate Member, 201.

ROWAN, J., *Paper* on a Premium System applied to Engineering Workshops, 203.—Remarks on ditto, 243, 245, 246, 258.

ROY, W., elected Member, 263.

RÜCKER, Sir A.. Remarks on Education of Engineers, 316, 333

RUSSELL, B., Remarks on "Slip" in a Plunger Pump, 183.

"SANDWICH" SYSTEM OF EDUCATION, 304, 307. *See* Education of Engineers.

SAYNOR, L. H., elected Graduate, 265.

2 E

THE INSTITUTION

OF

MECHANICAL ENGINEERS.

ESTABLISHED 1847.

PROCEEDINGS.

1903.

PARTS 3-4.

PUBLISHED BY THE INSTITUTION,

STOREY'S GATE, ST. JAMES'S PARK, WESTMINSTER, S.W.

CONTENTS.

1903.

PARTS 3-4.

𝕿𝖍𝖊 𝕴𝖓𝖘𝖙𝖎𝖙𝖚𝖙𝖎𝖔𝖓 𝖔𝖋 𝕸𝖊𝖈𝖍𝖆𝖓𝖎𝖈𝖆𝖑 𝕰𝖓𝖌𝖎𝖓𝖊𝖊𝖗𝖘.

PAST-PRESIDENTS.

GEORGE STEPHENSON, 1847–48. (*Deceased* 1848.)

ROBERT STEPHENSON, F.R.S., 1849–53. (*Deceased* 1859.)

SIR WILLIAM FAIRBAIRN, BART., LL.D., F.R.S., 1854–55. (*Deceased* 1874.)

SIR JOSEPH WHITWORTH, BART., D.C.L., LL.D., F.R.S., 1856–57, 1866. (*Deceased* 1887.)

JOHN PENN, F.R.S., 1858–59, 1867–68. (*Deceased* 1878.)

JAMES KENNEDY, 1860. (*Deceased* 1886.)

THE RIGHT HON. LORD ARMSTRONG, C.B., D.C.L., LL.D., F.R.S., 1861–62, 1869. (*Deceased* 1900.)

ROBERT NAPIER, 1863–65. (*Deceased* 1876.)

JOHN RAMSBOTTOM, 1870–71. (*Deceased* 1897.)

SIR WILLIAM SIEMENS, D.C.L., LL.D., F.R.S., 1872–73. (*Deceased* 1883.)

SIR FREDERICK J. BRAMWELL, BART., D.C.L., LL.D., F.R.S., 1874–75. (*Deceased* 1903.)

THOMAS HAWKSLEY, F.R.S., 1876–77. (*Deceased* 1893.)

JOHN ROBINSON, 1878–79. (*Deceased* 1902.)

EDWARD A. COWPER, 1880–81. (*Deceased* 1893.)

PERCY G. B. WESTMACOTT, 1882–83.

SIR LOWTHIAN BELL, BART., LL.D., F.R.S., 1884.

JEREMIAH HEAD, 1885–86. (*Deceased* 1899.)

SIR EDWARD H. CARBUTT, BART., 1887–88.

CHARLES COCHRANE, 1889. (*Deceased* 1898.)

JOSEPH TOMLINSON, 1890–91. (*Deceased* 1894.)

SIR WILLIAM ANDERSON, K.C.B, D.C.L., F.R.S., 1892–93. (*Deceased* 1898.)

PROFESSOR ALEXANDER B. W. KENNEDY, LL.D., F.R.S., 1894–95.

E. WINDSOR RICHARDS, 1896–97.

SAMUEL WAITE JOHNSON, 1898.

SIR WILLIAM H. WHITE, K.C.B., LL.D., D.Sc., F.R.S., 1899–1900.

WILLIAM H. MAW, 1901–02.

The Institution of Mechanical Engineers.

OFFICERS.

1904.

PRESIDENT.
J. HARTLEY WICKSTEED, ... Leeds.

PAST-PRESIDENTS.
SIR LOWTHIAN BELL, BART., LL.D., F.R.S., ... Northallerton.
SIR FREDERICK J. BRAMWELL, BART., D C.L., LL D., F.R.S., London.
SIR EDWARD H. CARBUTT, BART., ... London.
SAMUEL WAITE JOHNSON, ... Nottingham.
PROFESSOR ALEXANDER B. W. KENNEDY, LL.D., F.R.S., .. London.
WILLIAM H. MAW, ... London.
E. WINDSOR RICHARDS, ... Caerleon.
PERCY G. B. WESTMACOTT, ... Ascot.
SIR WILLIAM H. WHITE, K.C.B., LL.D., D.Sc., F.R.S., .. London.

VICE-PRESIDENTS.
JOHN A. F. ASPINALL, ... Manchester.
EDWARD B. ELLINGTON, ... London.
ARTHUR KEEN, ... Birmingham.
EDWARD P. MARTIN, ... Dowlais.
T. HURRY RICHES, ... Cardiff.
A. TANNETT-WALKER, ... Leeds.

MEMBERS OF COUNCIL.
SIR BENJAMIN BAKER, K.C.B., K.C.M.G., LL.D., D.Sc.,F.R.S., London.
SIR J. WOLFE BARRY, K.C.B., LL.D., F.R.S., ... London.
HENRY CHAPMAN, ... London.
HENRY DAVEY, ... London.
WILLIAM DEAN, ... Folkestone.
H. GRAHAM HARRIS, ... London.
EDWARD HOPKINSON, D.Sc., ... Manchester.
HENRY A. IVATT, ... Doncaster.
HENRY LEA, ... Birmingham.
SIR WILLIAM T. LEWIS, BART., ... Aberdare.
MICHAEL LONGRIDGE, ... Manchester.
JAMES MANSERGH, F.R.S., ... London.
HENRY D. MARSHALL, ... Gainsborough.
THE RIGHT HON. WILLIAM J. PIRRIE, LL.D., ... Belfast.
SIR THOMAS RICHARDSON, ... Hartlepool.
JOHN F. ROBINSON, ... London.
MARK H. ROBINSON, ... Rugby.
JOHN W. SPENCER, ... Newcastle-on-Tyne.
SIR JOHN I. THORNYCROFT, LL.D., F.R.S., ... London.
JOHN TWEEDY, ... Newcastle-on-Tyne.
HENRY H. WEST, ... Liverpool.

HON. TREASURER.	AUDITOR.
HARRY LEE MILLAR.	ROBERT A. McLEAN, F.C.A.

SECRETARY.
EDGAR WORTHINGTON,
The Institution of Mechanical Engineers,
Storey's Gate, St. James's Park, Westminster, S.W.
Telegraphic address:—*Mech, London.* Telephone :—*Westminster,* 264.

PRESIDENT, 1890–91.

(Deceased 1894.)

The Institution of Mechanical Engineers.

PROCEEDINGS.

July 1903.

The SUMMER MEETING of the Institution was held in Leeds, commencing on Tuesday, 28th July 1903, at Ten o'clock a.m.; J. HARTLEY WICKSTEED, Esq., President, in the chair.

The President, Council, and Members were received in the Rooms of the Philosophical and Literary Society, Park Row, Leeds, by the Right Honourable the Lord Mayor of Leeds, JOHN WARD, Esq., J.P., and by the other Chairmen and the Members of the Reception Committee.

The LORD MAYOR of LEEDS, in according a hearty welcome to the Members, remarked that the Institution was an important body, numbering nearly 4,000 members, who were somewhat widely scattered throughout the world. The Institution also had a long list of illustrious Past-Presidents, the first of whom, he noticed, bore the honoured name of George Stephenson. The present meeting at Leeds would be presided over by a Leeds gentleman, and he was told that a Leeds man had rarely, if ever, before had that honour. Leeds was an industrial and, therefore, a manufacturing centre; it was somewhat smoky at times, but some of the members would not perhaps consider smoke as an unmixed evil. The Corporation had a very good electric lighting installation to show the members; they had also one of the best systems of tramway traction, which would be of interest. He noticed in addition that the members were to visit a number of places in and about Leeds, which would naturally

2 F

(The Lord Mayor of Leeds.)

be of interest to gentlemen of their occupation and craft. Leeds had something to show the members in the way of public buildings, notably, first and foremost, its magnificent Town Hall. The Corporation also had been trying in recent years to do something in the way of beautifying the City. Streets had been widened, and shortly there would be one or two large open spaces in the centre of the City, which would tend to its improvement. Generally speaking, Leeds had been built more on the principle of utility than grandeur; but the members would observe, in their comings and goings through the streets of Leeds, that the Corporation was doing something in the way of improving them. He trusted that the visit of the members of the Institution to Leeds would be one of interest, pleasure, and profit, and that they would carry away with them from the meeting pleasant recollections.

The PRESIDENT, on behalf of himself, the Council, and the Members of the Institution, thanked the Lord Mayor most sincerely for his presence, and for the words of welcome with which he had greeted them. The Institution met in Leeds twenty-one years ago, but did not then have such a civic reception as they had had on the present occasion; in fact, Leeds was at that time a town and not a City. The Chief Magistrate of the City had risen to the occasion, and had met the members with a reception worthy of the change from a town to a City. The Lord Mayor had been twice Mayor of the town of Leeds, and was now Lord Mayor of the City of Leeds; he had been tested twice, and the present was the " third time of asking " that he had taken the position. Although it was twenty-one years since a Summer Meeting of the Institution was held in Leeds, the Institution had visited Leeds since that time, because a special Autumn meeting was held in 1886, by the special desire and invitation of the inhabitants of Leeds, to celebrate the opening of the new buildings of the Engineering Department of the Yorkshire College, on which occasion a very large gathering took place. When this took place the Yorkshire College was put in possession of the best equipped Engineering Laboratory at that time in any of the Colleges and Universities in England. It had done extremely good work ever since. It had never looked behind it since that inauguration, and it was one of the most successful departments of the College. He was sure many present would be acquainted with men who had been educated at that College, and as he moved about he heard from time to time of the very good reputation of the men who had been trained in the Engineering Department of the College.

He reminded the Lord Mayor that in Leeds in 1812 the first locomotive that was put on rails, and made to draw usefully a mineral train, was put down to draw coal from the Middleton Colliery to Hunslet, near where the present Midland goods station was situated. The conception of that steam locomotive was due to Mr. Blenkinsop, and it was built by an engineer of great ability named Matthew Murray, an eccentric man of genius who built works in Leeds after the model of a cyclopean steam cylinder. The main body of the works was circular, and had a projection on it exactly like the steam-chest of a cylinder. Up the steam-chest the staircases went; the doors were like the ports of the steam cylinder, the floor was like the piston of the steam cylinder, and the pillar that supported the roof was like the piston-rod of the engine; and the men, in doing their work in the huge cylinder, went in and out just like particles of steam. The fame of Matthew Murray became so great that the firm of Boulton and Watt were jealous of what he might do in the way of steam-engines, and they bought up the land all round the building in order to prevent him extending his borders very much. The rails on which this mineral train ran were cast-iron, about 3 feet long, which stretched from chair to chair. Since that time rails, instead of being 3 feet long, had become 30 feet long, and then 60 feet long; but in Leeds, under the auspices of the Corporation, and, if he was rightly informed, under a considerable amount of initiative from one of the Members of the Institution, Mr. Alderman Frederick W. Lawson, the rails in Leeds had become indefinite in length. If there was a track 6 miles long in Leeds, the rails also were 6 miles long. It was notable, he thought, that that was the first introduction into England of the continuous welding together of the rails. He would not enlarge at the present

(The President.)

moment upon the advantages which would be derived from the welding of the joints, but he thought it was notable that as Leeds was the first town to put down old cast-iron rails 3 feet long, it was now the first place, in England at least, to put down continuous rails of miles and miles in length.

Another interesting point that struck him as being worthy of notice was with regard to Mr. Blenkinsop and his amateur efforts in the direction of steam locomotion. Before Blenkinsop had the locomotive built by Matthew Murray, he and other gentlemen in Leeds turned their attention to the steam locomotive for common roads. They had a great many meetings, of which they kept minutes, which minutes he had seen, and the bulk of their discussion related to the form of brake which would enable them to prevent the road locomotive from running away. They built the locomotive and built the brake, but could not make the locomotive go at all, so the brake was never proved. It was a curious fact that one of the most prominent Papers to be discussed at the present meeting was upon the subject of a brake, and a brake for the very purpose of arresting the motion of a road locomotive, although he believed it might be chiefly valuable for preventing it going backwards, as that had been a danger experienced when an autocar was brought to rest on a steep gradient. It would be seen from the history he had given that, while for nearly one hundred years the subject of a good brake had been in the minds of mechanical engineers, the thing was not so simple—although it was said that mechanics had no mystery about them—but that it could be developed and improved by fresh invention even now. He could not express more than he had done the gratitude of the Institution for the very kind and noble reception that the Corporation and its Chief Magistrate had given to the Institution.

The Minutes of the previous Meeting were read and confirmed.

The PRESIDENT announced that the Ballot Lists for the election of New Members had been opened by a committee of the Council, and that the following one hundred and thirty-four candidates were found to be duly elected:—

MEMBERS.

ARKWRIGHT, BERNARD GEORGE,	Newcastle-on-Tyne.
BALDWIN, JAMES,	Keighley.
BARCLAY, ANDREW WALKER,	Birmingham.
BARLOW, ARTHUR BERNOULLI,	Manchester.
BENTON, WALFORD JOHN,	Leeds.
BICKERTON, HENRY NIELD,	Ashton-under-Lyne.
BJÖRNSTAD, JÖRGEN,	Erith.
CAMPS, HAROLD EDWARD JOSCELYN,	London.
DANIELS, GEORGE,	Stockport.
FLOCKTON, BENJAMIN PERCY,	Manchester.
HALL, JOSHUA,	Hyde.
HARDISTY, JOHN,	Chesterfield.
JENKINSON, EDWARD HENRY,	Sydney.
KNIGHT, WILLIAM EDWARD,	Havana, Cuba.
LAMBERT, HAROLD,	Leeds.
LINDLEY, HERBERT,	Manchester.
LOW, WILLIAM,	Dundee.
MACDONALD, JAMES TURRIFF SINCLAIR,	Bristol.
MASTERTON, WALTER NEILL,	Singapore.
MOORE, WILLIAM HENRY,	Manchester.
PECKETT, THOMAS,	Bristol.
PLATT, ISAIAH,	Wednesbury.
PUDDEPHATT, ERNEST ORLANDO,	Bradford.
REDMAN, JOSEPH FELL,	London.
ROBB, ALEXANDER,	Hull.
ROBERTS, HARRY,	Manchester.
RÖPER, ANTON,	Düsseldorf.
SAMPSON, EDWARD DUDLEY,	Hubli, India.
STOREY, PERCIVAL,	Torquay.
THOMPSON, JOHN,	Dundee.
WESTON, THOMAS WILSON,	York.
WILKS, EDWARD CHARLES,	London.

ASSOCIATE MEMBERS.

ADAMS, DAWSON, ...	Preston.
ALLBUTT, JONATHAN ENOCH HILL, .	Chesterfield.
ALLKIN, HENRY WESTON, . .	Wynberg, Cape Colony.
AVELINO, RALPH CLIFFORD, . .	London.
BINNS, WALTER, ...	Newcastle-on-Tyne.
BIRD, FRANK NOEL, . .	Yarloop, W. Australia.
BODEN, ARTHUR LIONEL, . .	Stockton-on-Tees.
BRITTON, SYDNEY ERNEST, . .	Motherwell.
COLTMAN, WALTER WILLIAM, . .	Loughborough.
COOMBES, THOMAS, . . .	Torquay.
COOPER, ARTHUR THOMAS, . .	Reading.
COWCHER, GEORGE, . . .	Dudbridge.
DAYSON, STEPHEN FREDERICK, .	London.
DEAKIN, GEORGE WELSBY, . .	Birkenhead.
DEWHIRST, JAMES, . . .	Chelmsford.
DOUGLAS, FRANK, . . .	Delagoa Bay.
DUKE, REGINALD FRANCIS, . .	Littlehampton.
DYSON, JOHN DANIEL, . . .	Bradford.
ELLIS, BERNARD PEMBERTON, . .	London.
FARNSWORTH, ALFRED WILLIAM, .	Derby.
FELLOWS, THOMAS EDGAR, . .	Wolverhampton.
FROST, HENRY MAURICE, . .	London.
HAMILTON, ALEXANDER, JUN., .	Almeria.
HARDIE, EDWARD PETER, . .	London.
HEDLEY, WILLIAM SCOTT, . .	Newcastle-on-Tyne.
HINES, ARTHUR HENRY COBDEN, .	Cambridge.
HODSON, HORACE MALCOLM, . .	London.
HOLLIS, SYDNEY AINSLIE, . .	Bloemfontein.
HOLMES, RICHARD JOHN MONTAGUE, .	Newcastle-on-Tyne.
HORTON, LEONARD WILSON, . .	Wednesbury.
INGLE, HERBERT EDWARD, . .	Dagenham.
INNES, AUGUSTUS MONTAGUE, .	Hitchin.
JONES, CHARLES, ...	London.
KIDD, JOHN WILLIAM, . . .	Hadley, Salop.
LEWIS-DALE, HENRY ANGLEY, .	London.

Longland, Walter, ...	London.
Lucas, Ralph, ...	London.
Maitland, Dascon James, . . .	Madrid.
McDonald, William, ...	Sheffield.
Miller, William Thomas Ward, . .	Sheffield.
Moffet, James Scott Duncan, . .	Rochdale.
Monkhouse, Frank Lionel, . . .	Newcastle-on-Tyne.
Monro, John Duncan, Lieut. R.E. . .	Chatham.
Morgan, Edward George, . . .	Dublin.
Naylor, John Hardcastle, . . .	Woolwich.
Neachell, Edward John, . . .	Liverpool.
Pickering, William Sellers, . .	Exeter.
Pocock, James Herbert, . . .	Glasgow.
Reeve, Walter, ...	Bradford.
Rew, Reginald John, ...	Exeter.
Richards, Walter Granger, . .	Middlesbrough.
Roberts, Charles Drury, . . .	Pretoria.
Roberts, Joseph, ...	Melbourne.
Roberts, Trevor, ...	Swindon.
Roe, Harrison, ...	London.
Roseveare, Leslie, ...	Plymouth.
Salter, John Ruffell, . . .	Liverpool.
Sanders, George Ernest, . . .	King William's Town.
Saunders, Albert Frank, . . .	Fremantle, W. Australia.
Schmidt, Frederick Albert, . .	Tunbridge Wells.
Shanan, Charles Henry, . . .	Liverpool.
Shiner, Albert Edward, . . .	London.
Smart, Lewis Anderson, . . .	London.
Smith, Jonah Walker, . . .	Barrow-in-Furness.
Somers, Richard Mousley, . . .	Leeds.
Stevenson, John Le Che, . . .	Birmingham.
Stewart, Charles, ...	London.
Sunderland, Wallace, . . .	Wakefield.
Thunder, Cyril Joseph, . . .	Pernambuco.
Turner, John Alexander, . . .	Peterborough.
Usherwood, Thomas Scriven, . .	London.

WALTON, DENYS, ... London.
WARNER, JAMES SUTHERLAND, . . London.
WESTLEY, CHARLES BARNARD, . . Bristol.
WIGRAM, WALTER GOTT, . . . Leeds.
WILLCOCKS, JOHN, ... Buckfastleigh, Devon.
WILLSTEED, CHARLES WESLEY, . . London.
WILSON, ROBERT JAMES, . . . London.
WINTOUR, ERNEST RICHARD, . . . Clevedon.
WOODS, HARRY, ... Cachar, India.

GRADUATES.

ASTON, REGINALD GODFREY, . . . Stafford.
BULKELEY, GEORGE VICARY OWEN, . . Wolverhampton.
BURGE, RODON LUDFORD, . . . London.
CARR-GOMM, MARK CULLING, . . . Erith.
CHEESEMAN, ALFRED BRYSON EAST, . Rochester.
CUTLER, GEORGE BENJAMIN, JUN., . . London.
DUFF, PHILIP JOHN, ... Erith.
EVANS, CYRIL AUSTIN, ... London.
GRAVES, LIONEL PERCIVAL, . . . London.
HODSON, REGINALD CECIL, . . . London.
HUGHES, ARTHUR MUMFORD, . . . London.
MACDONALD, IAN, ... Edinburgh.
MOBBS, HERBERT, ... Northampton.
MOORE, FRANCIS CHARLES, . . . London.
PICKETT, HENRY ARTHUR, . . . Wolverhampton.
PRICE-WILLIAMS, DOUGLASS, . . . London.
ROOSE, FITZROY OWEN JONATHAN, . . Bournemouth.
SHIRREFF, JAMES ARTHUR, . . . London.
SMART, DAVID HOWARD ST. BARBE, . . London.
STANFORD, ERIC SPENCER, . . . London.
WOOD, WILLIAM BERTRAM, . . . London.
WORSSAM, LESLIE HENRY, . . . London.

The PRESIDENT announced that the following nine Transferences had been made by the Council since the last Meeting :—

Associate Members to Members.

BAMBER, HERBERT WILLIAM,   .   .   .   London.
BARKER, THOMAS PERRONET,   .   .   .   Birmingham.
LIVESEY, ROBERT MARTYN,   .   .   .   Gibraltar.
MASSEY, LEONARD FLETCHER,   .   .   .   Manchester.
PICKERING, FRANK, ...   Cape Town.
SEGUNDO, EDWARD CARSTENSEN DE,   .   .   London.
WISEMAN, ALFRED, ...   Birmingham.

Associate to Member.

HARVEY, JULIUS, ...   London.

Graduate to Associate Member.

WILSON, ALEXANDER COWAN,   .   .   .   Birkenhead.

The following Papers were then read and discussed :—

" The Diesel Engine " ; by Mr. H. ADE CLARK, of Leeds.
" Notes on High-Speed Tool Steels " ; by Mr. HENRY H. SUPLEE, of New York.

At a Quarter to One o'clock p.m. the Meeting was adjourned to the following morning.

The ADJOURNED MEETING was held in the Rooms of the Philosophical and Literary Society, Leeds, on Wednesday, 29th July 1903, at Ten o'clock a.m.; J. HARTLEY WICKSTEED, Esq., President, in the chair.

The following Papers were read and discussed :—

" A new form of Friction Clutch "; by Professor H. S. HELE-SHAW, LL.D., F.R.S., *Member*, of Liverpool.

" Economy of Fuel in Electric Generating Stations "; by Mr. HENRY McLAREN, *Member*, of Leeds.

The PRESIDENT proposed the following Votes of Thanks, which were passed with applause :—

To the Right Honourable the Lord Mayor of Leeds, to Sir James Kitson, Bart., M.P., Sir Arthur T. Lawson, Bart., and to the Reception Committee of that City, especially Mr. Arthur Greenwood, Chairman, and Mr. T. P. Reay, Honorary Treasurer, for the welcome they have extended to the Members, and for their hospitality in entertaining the Members to Luncheon in Leeds on three occasions, and also at York, and at Fountains Abbey.

To the Chairman and Directors and Officers of the Great Northern Railway for conveying a large party of Members from Leeds to Doncaster, to visit the Locomotive Works, and for entertaining them in the Works at Luncheon. To the same Company, and the Great Central Railway, for conveying the party to Frodingham, and thence to Leeds. Also to the Frodingham Iron and Steel Company for showing the Members over their interesting Works.

To the Right Honourable the Lord Mayor for so hospitably entertaining the Members, and Ladies, at a Reception in the City Art Gallery.

To the Municipal Authorities and Owners of Works in Leeds and Neighbourhood for their kindness in throwing open their Works to the visit of the Members, and for making arrangements to convey Members by special tramcars.

To Messrs. Pope and Pearson for inviting a party of Members to their Colliery, and entertaining them at Luncheon; and to the Owners of Works in Wakefield for inviting the Members to that City.

To the Most Honourable the Marquis of Ripon, K.G., for kindly throwing open the Grounds of Fountains Abbey to Members and Ladies; and to the Very Reverend the Dean of Ripon for expounding to the Members the History of Ripon Cathedral.

To Sir Edward Green, Bart., and Mr. Frank Green, for inviting the Members to Treasurer's House, York; to the Very Rev. the Dean of York, and the Organist, for showing the Minster to Members, and arranging for a recital on the New Organ; and to Messrs. Rowntree for showing the Members over their Chocolate Works.

To the Honorary Local Secretaries, Mr. E. Kitson Clark and Mr. Christopher W. James, for the admirable arrangements which their forethought and energy have provided for each day of the Meeting, and for so kindly conducting the large parties on the enjoyable excursions to Fountains Abbey and to York.

The Meeting then terminated at One o'clock. The attendance was 412 Members and 41 Visitors.

THE DIESEL ENGINE.

BY MR. H. ADE CLARK, OF THE YORKSHIRE COLLEGE, LEEDS.

Introduction.—The Diesel Engine is an internal combustion engine, intended to work with gaseous, liquid, or solid fuel. At present developed as an oil engine, it works on the four-stroke cycle, and is vertical. The author of this Paper, while being fully aware that the actual performance and the whole cost of the engine must be the ultimate basis of comparison with other engines, is also of the opinion that the scientific inception and development of the] engine is not without interest and value. To confine the Paper to performance and cost would be to make it very incomplete, and not nearly so satisfactory as if it contained a preparatory statement of principles and a brief statement of development.

Principles and Development.—Herr Rudolph Diesel, after careful analytical and experimental study of the present forms of Heat Engine, and after comparing, their cycles of operation with the ideal Carnot cycle, concluded that all existing engines were worked on defective principles, and that no radical improvement in their efficiencies could result until the principles of working were altered.

Rationally following up this destructive criticism, he then proceeded to construct a new process for the utilization of the heat of a combustible in the internal combustion engine, and finally laid down the following conditions as being essential for the maximum utilization of the heat in the combustible :—

(1.) The highest temperature in the cycle of operations must be produced by isothermal and adiabatic compression of the air only, previous to the introduction of the combustible.

(2.) The combustible must be gradually introduced into the highly-heated air in such a fine state of division that it is instantaneously burned, thereby supplying the heat necessary to keep the gases at a constant temperature during the period of combustion equal to that attained during the compression of the air.

(3.) The use of a large and definite excess of air proportioned to the heat value of the fuel is necessary, so that the temperature of combustion may be kept at the highest temperature of compression, and so determined that the engine may be worked and lubricated without using a water-jacket.

The complete cycle for such an engine would be :—

(1.) Isothermal and adiabatic compression to the maximum pressure and temperature.

(2.) Isothermal combustion and finally adiabatic expansion to atmospheric pressure and temperature.

This is, of course, the Carnot cycle, and Herr Diesel has here laid down the conditions which are necessary to obtain the most perfect combustion compatible with maximum utilization of the heat in a possible engine.

The cycle is illustrated in Fig. 1, Plate 9. These original analyses, proposals and experimental work will be found in detail in Herr Diesel's book entitled " A Rational Heat Motor." *

Development.—In the original attempts to carry out this cycle of operations in 1893, the high pressures and high ratios of expansion used required a three-cylinder engine. One designed to burn dust-coal was constructed and experimented with. The experience then obtained brought to light many points of practical importance, and indicated how far materials and methods of construction at present allow the ideal to be carried into practice.

* English Translation by the late Bryan Donkin. E. & F. Spon.

The next step was to consider where and how far the ideal cycle could be modified, without unduly diminishing the actual and the theoretically possible thermal efficiencies. After two years' experiment and discussion of results, the following modifications were adopted :—

(1) The attempt to obtain an isothermal step in the compression period was abandoned, adiabatic compression alone being employed. This allows the high temperature necessary at the end of the compression to be obtained without using an excessively high pressure; but it results in more heat being thrown away in the exhaust.

(2) The expansion was not continued to atmospheric pressure, in order to reduce the bulk, weight, and cost of the cylinder per horse-power.

The effects of these changes are indicated on the diagram in, Fig. 2, Plate 9, the black shaded ends being eliminated and the compression curve being an adiabatic. An engine was built and worked on this modified cycle. It had a water-jacket, the upper temperature limit was 700° C. with a maximum pressure of 64 atmospheres, or 910 lbs. per square inch, and a theoretically possible thermal efficiency of 64 per cent. This experimental engine gave 12 H.P., 71 per cent. mechanical efficiency, and 26 per cent. thermal efficiency.

Having solved the initial difficulties of making an engine work successfully on the proposed cycle, it became necessary to make a powerful engine without using an immoderately large cylinder. The solution was obtained by carrying on the combustion at the pressure of 30 to 40 atmospheres (430 to 570 lbs. per square inch) instead of isothermally; the result was of course an increase in temperature of such character that the water-jacket became necessary, and this, in the final form of the engine, was adopted.

The final result of these compromises between the theoretical requirements and the practical necessities is the cycle of operations shown in Fig. 2, Plate 9, and is there compared with an actual indicator card. The cycle is :—

Adiabatic compression $d\ h\ f$, practically attainable.

Isothermal combustion $f\ f_1\ a$, theoretically desirable.

Adiabatic expansion $a\ b$, practically attainable (broken line).

The actual card from the 80 B.H.P. engine being d e h a c, the curve a c indicates the higher temperatures of the actual gases over what would be the case if the curve f a could be attained ; at the same time the maximum pressure attained is lower and the mean pressure higher than the theoretical card gives. The thermal efficiency is of course lower, but still both the theoretically possible and the actually obtained thermal efficiencies are much higher than for any other type of engine.

The actual Diesel engine has the following cycle of operations :—

1st stroke takes in air alone at atmospheric pressure and temperature.

2nd stroke compresses this air to a high pressure (35 atmospheres = 500 lbs. per square inch) and to a temperature of about 1,000° F. This compression is neither isothermal nor adiabatic, since the operations are conducted in a water-jacketed conducting cylinder.

3rd stroke is the working stroke, during the first part of which the combustion of the fuel is carried on at constant pressure for a period which is determined by the amount of oil to be sprayed in, which quantity is controlled by the governor. The second part of this stroke is approximately an adiabatic expansion.

4th stroke exhausts the gases.

80-B.H.P. ENGINE.

This size of engine is made by the Vereinigte Machinenfabrik, Augsburg, and Carels Frères, Ghent. On Plate 10 is illustrated a longitudinal section and a transverse section of the engine, with standard arrangements of piping for petroleum, lubricating oil, cooling water, air-blast, starting air and exhaust, in fact, all the engine connections, except the water-cooling tanks when such are used.

Description.—The engine is of the vertical type, with a strong cast-iron A frame, the upper part of which forms the outer wall of the water-jacket ; into this upper part is fitted a cylinder of special close-grained cast-iron. The cylinder cover is deep and hollow, being thoroughly water-jacketed ; it carries five valves, two being shown in Fig. 4, Plate 10, the central valve being the oil sprayer, the other being the starting-valve, which may be made to act as the suction-valve for the air-pump. In the longitudinal section three valves are seen in the cover: in the centre is the oil-sprayer (for detail see Fig. 6, Plate 11), on the right is the air-inlet, and on the left is the exhaust-valve. The fifth valve is the air-pump suction valve. The oil-spraying valve opens upwards or outwards, and the others open downwards or inwards; all three are spring closed, the air- and exhaust-valves being kept closed by pressure inside the cylinder. All valves are opened by the action of the bent rocking-levers seen upon the right in the transverse section; the movements of the levers are determined by the cams placed upon the horizontal cam-shaft. The cam-shaft is driven at half the speed of the crank-shaft by means of the bevel gearing and the vertical shaft seen on the right in the longitudinal section.

The governor is of the loaded centrifugal type, and is placed at the top of the vertical shaft; its action is explained in connection with the oil-pump, Fig. 5, Plate 11.

The piston is of the usual open trunk type, directly connected to a connecting-rod of the marine type. There are seven piston-rings of the Ramsbottom type, six near the top and one much lower, in order that it may pass the lubricating channels.

The crank-shaft is solid, and has three bearings fitted with ring lubricators. The fly-wheel is built in halves, and on the inner edge of the rim is a toothed ring into which work two ratchet-pawls, actuated by a rocking-lever; this device is for bringing the engine into the starting position, that is, with the crank just beyond its top dead-centre.

The petroleum pump is shown on the left of the longitudinal section, Fig. 3, Plate 10. This pump is connected by a pipe to the petroleum-filtering tanks, and by a pipe of small bore to the oil-spraying valve. The plunger of this pump is driven by a crank-pin placed in a disc at the end of the cam-shaft, and so has a constant stroke.

The air-pump is shown on the right of the transverse section, Fig. 4; its cylinder is thoroughly water-jacketed, and the plunger is driven by connecting-rod, rocking levers and connecting links from the small

2 G

end of the connecting-rod.   This pump takes its air from the engine
cylinder just before the end of the compression stroke, still further
compresses this air, and delivers it to the air-blast reservoir.   In the
earlier designs the pump took its air direct from the atmosphere, and
was then much more bulky and less efficient.

The air-blast reservoir is the smaller one on the right of the
cross section, and is connected to the oil-spraying valve for injecting
the petroleum into the cylinder against the high pressure of
35 atmospheres already existing there; for this purpose the pump
maintains a steady pressure in the reservoir of 5 to 15 atmospheres
higher, that is from 40 to 55 atmospheres.   It is also connected
by an overflow valve to the air-starting reservoir, this valve allowing
air to pass from the blast to the starting reservoir when the pressure
in the former exceeds a predetermined value.   There may be a similar
overflow valve connecting the engine cylinder to the starting
reservoir, in order that the maximum compression pressure may
be controlled.

The cylinder lubrication is forced by means of the pumps seen on
the left of the engine in the longitudinal section, the lubricant
entering at five or six points in a horizontal plane, below which one
piston-ring passes.   The crank is lubricated by the ring and oil ways
seen in the longitudinal section.   The lubricant is forced to the
small end of the rod inside the piston.

The water-jacket is very complete, as the sections show, entirely
enveloping the engine cylinder walls and end, and also the air-
pump.   The water enters at the bottom of the engine jacket, and
passes upwards through this jacket and the air-pump jacket, from
the top to the cylinder cover, the outlet being close to the exhaust pipe.

The oil-spraying or pulverising valve is illustrated in detail in
Fig. 6, Plate 11.   The horizontal section shows the petroleum and
air-blast passages to the central valve, and the vertical section
shows the body of cast-iron, with the petroleum passage and overflow
or test valve, the central needle valve with its guiding sheath and
at the base the pulverising device, consisting of a set of four metal
rings of special form perforated by small holes and separated by
four metal bands; the terminating nose piece has narrow channels

cut in it, through which the pulverised oil passes to the expanding orifice, and is sprayed into the cylinder when the needle valve is raised. It will be seen that the petroleum pump delivers petroleum to the nozzle by the narrow passage, and that the nozzle is in direct communication with the air-blast reservoir, the pressure in which is kept steadily about 150 lbs. higher than the maximum pressure of compression. This device works well.

The petroleum pump is shown in Fig. 5, Plate 11. The plunger has a constant stroke, being driven from the end of the valve cam-shaft. Passing through the head of the plunger is a lever working upon an eccentric fulcrum; to this lever is attached the valve-rod which opens the inlet valve against the action of a spring. The fulcrum of the rocking lever is eccentric to the shaft upon which it is placed, and this shaft is caused to rotate by the action of the governor, thus altering the stroke of the valve-rod and through it the opening of the inlet valve, thus controlling the amount of oil passing to the pump chamber to be forced past the outlet valve to the injecting valve.

TABLE 1.—*Leading Dimensions of 80-B.H.P. Engine.*

	Mm.	Feet.	Inches.
Diameter of cylinder	400	1	3·75
Stroke of piston	600	1	11·62
Length of piston	905	2	11·65
Length of connecting-rod	1,610	5	3·4
Distance between crank-bearings	920	2	7·5
Distance between fly-wheel bearings	1,440	4	8·7
Diameter of fly-wheel	3,400	11	1·8
Air-pump diameter	60	0	2·36
,, stroke	140	0	5·5
Blast reservoir diameter	204	0	8·03
,, ,, length	900	2	11·42
Starting reservoir diameter	340	1	1·38
,, ,, length	1,785	5	10·27
Petroleum filters diameter	330	1	1
,, ,, length	600	1	11·62
Over all length of engine, including railing	3,900	13	2·6
,, ,, width ,, ,, ,, ,,	3,900	,,	
,, ,, height ,,· ,, ,, ,,	3,900	,,	
Depth below floor of engine	1,500	4	11·1
Depth of foundations	2,400	7	10·6
Height required for erection	5,900	19	4

There is a controlling device shown, by which the inlet valve can be held open, thus allowing the plunger to pump back the oil to the pump reservoir tank instead of passing it on to the engine by way of the outlet valve. The tappet of this device encircles the valve-rod foot tappet.

Starting and Running of the Engine.—The engine is started by compressed air, which is stored in the starting reservoir under a pressure of about 55 atmospheres or 800 lbs. per square inch, by the air-pump during the previous run of the engine, or, in the case of a new engine, the makers send out these reservoirs ready charged; there is little or no danger of these vessels losing a charge. An engine of this type at the Sewerage Farm at Harrogate was started by air stored at Augsburg eight weeks previously, and in the case of an engine sent to India the vessels were charged four months before the engine was started from them.

Assuming the engine to be left standing from the previous day's run, the procedure for starting would be:—

(1) See that all lubricators are in order.

(2) See that the petroleum tanks are charged.

(3) Open the test cock on the oil-spraying valve, using the oil pump as a hand pump, and if necessary charge the oil-spraying valve with petroleum.

(4) By means of the hand-lever, rack the engine over till the crank is just past its top dead-centre.

(5) Pull over the lever which puts the starting lever into connection with the starting cam and relieves the oil spray. (*See* starting position on transverse section, Fig. 4, Plate 10.)

(6) Open the screw-down valves of the blast reservoir and the starting reservoir.

The engine now starts off as an air-driven engine. Allow it to make five or six revolutions under this condition, and then throw back the lever; the engine will then start on its normal cycle as an oil engine. Fig. 8 (page 414) shows the Starting Card.

The actual starting of the engine is very simple and quite certain, and the author has seen this operation successfully performed on all occasions, whether the engine has been stopped for a few minutes or for days; in fact, he has seen an engine in course of erection by the makers, Messrs. Carels Frères of Ghent, started right away.

The normal running of this engine is on the ordinary four-stroke cycle with the following distinguishing characteristics :—

(1) Very high compression of the air, to about 500 lbs. per square inch, and a temperature 1,000° F., so that the fuel burns at once on being injected, needing no igniting device whatever.

(2) The gradual injection of the fuel into this volume of highly heated air, by means of a blast of air at about 100 to 150 lbs. per square inch higher pressure than is already in the cylinder.

(3) The gradual and complete combustion of the fuel as distinguished from the explosive combustion of the ordinary type of gas- or oil-engine.

Performance of the Engine.—Until April 1902 all published trials of Diesel engines were either of Continental or American origin. So the author, upon learning that the Harrogate Corporation were putting in a Diesel engine, at once applied to the Borough Engineer, to be allowed to test it; this permission was most courteously granted, and the author, assisted by Mr. R. Dudley, proceeded on 9th April 1902, to test the engine on the brake. The results of this test are given in full in Table 2 (page 404). The oil used was a crude petroleum from Texas, having a specific gravity of 0·922 and a calorific value of 19,150 B.Th.U. per lb. The result of the test, less than half a pound of this oil per B.H.P. hour, was very gratifying to the author, after usually getting nearly twice this consumption from the ordinary type of oil engine.

This result impelled the author to get a special indicator and make another trial. The indicator was readily obtained from the Diesel Engine Co., of London, and on 10th May 1902, a second trial of this 35-B.H.P. engine was made; the results are given in Table 2 (page 404).

As a complete report of these trials has been published,* no further comment on them is necessary, except to say they are the first trials in England of the first Diesel engine put to work in this country.

---

* "The Engineer," 22 August 1902.

TABLE 2.

35-*B.H.P. Diesel Engine.*

Test of 35-B.H.P. Diesel Engine at Harrogate, 9 April 1902.

Summary of Results.

1. Load		0·25	0·5	0·75	Full
2. Duration . . . mins.		60	60	60	40
3. Revolutions per minute . .		185·4	184·6	183	181·4
4. Load on the brake . . lbs.		74·3	138	213	299·6
5. Pull on spring balance . lbs.		3·14	3·36	14	27·8
6. Effective load . . lbs.		71·16	134·66	199	271·8
7. Brake horse-power . .		10·87	20·44	30·07	40·26
8. Total oil . . lbs.		8·95	10·95	14·35	12·39
9. Total oil per hour . . lbs.		8·95	10·95	14·35	18·45
10. Oil per B.H.P. hour . lbs.		0·823	0·536	0·478	0·459

Summary of Results of a second trial made on 10 May 1902.

1. Load		0·40	0·57	0·89	0·98
2. Duration . . . mins.		30	58	30	60
3. Revolutions per minute .		182·7	181·7	182·7	182·5
4. Load on the brake . . lbs.		111·3	159·4	246·6	281·7
5. Pull on spring balance . lbs.		3·0	4·94	21·6	16·1
6. Effective load . . lbs.		108·3	154·4	225·0	264·6
7. Brake horse-power . .		16·29	23·03	33·1	39·21
8. Total oil . . . lbs.		4·75	12·25	7·875	18·06
9. ,, ,, ,, per hour .		9·5	12·67	15·75	18·06
10. Oil per B.H P. hour . lbs.		0·583	0·550	0·476	0·461
11. Mean effective pressure lbs. per sq. in.		69·7	93	103·2	117·6
12. { Indicated horse-power . .		31·45	42·2	46·88	52·27
{ I.H.P. of air-pump		—	—	—	3·0
13. { Oil per I.H.P. hour . lb.		0·302	0·301	0·336	0·346
{ Oil per net I.H P. hour . lb.		—	—	—	0·366
14. Cooling water per I.H P. hour in lbs.		—	37·6	—	33·6
15. Rise of temperatures of water		44° F.	44° F.	—	55·3°F.
16. B.Th.U. per net I.H.P. minute		—	—	—	117·5
17. ,, converted to work .		—	—	—	42·4
18. ,, rejected in cooling water		—	—	—	32·9
19. B.Th.U. in the exhaust gases .		—	—	—	23·4
20. { Thermal efficiency on net I.H.P . . per cent.		—	—	—	36·3
{ Thermal efficiency on B.H.P. per cent.		—	—	—	28·7
21. Mechanical efficiency per cent.		51·8	54·7	70·5	75

By the time this Paper is read the author hopes to repeat these trials on the engine, after its twelve months' work of 16 hours a day and 7 days per week. (*See* Table 6, page 422 and page 419.)

80-B.H.P. ENGINE.

For the purposes of this Paper, the author, at the invitation of the Diesel Engine Co., of London, visited the works of Messrs. Carels Frères, Ghent, Belgium, in company with Mr. A. J. Lawson, Engineer to the British Electric Traction Co., and Mr. H. W. Anderson, of London, to see the building of the Diesel engine and the running of an 80-B.H.P. engine. In compliance with the wish of the two gentlemen named, he made four trials on the 80-B.H.P. engine at that time on the erecting bed; the data and results of these trials are given in Table 4 (page 412). As no preparations for a trial had been made, these results are not so complete as for the other two engines. The engine was connected by a belt to an overhead shaft, which was belted to a four-pole dynamo and to a motor running light. The output of the dynamo was absorbed by a water resistance. The petroleum tank supplying the engine was weighed by the author immediately before and after each trial. Indicator cards were taken every five minutes and the revolutions by a counter attached to the air-pump rocking-lever.

The effective H.P. of the engine is obtained from the I.H.P. by deducting 22 H.P. This is warranted from the results of repeated tests on the engine when running without load, the indicated H.P. being then 21 to 22. This result may be found on reference to Professor Eugen Meyer's tests in June 1902, and Messrs. Carels assured the author that was also the result of running the engine under compression from the motor. The author has, however, since corroborated this statement for himself.*

There was not an opportunity during these tests to run the engine light, owing to the time required to remove the heavy belts and the limited time the author could then devote to the work.

---

* *See* result of running the double engine quite light on the 10th March 1903. *See* result of test on 80-B.H P. engine on 16th May 1903, Table 7 (page 423).

The oil used on this occasion was a Texas crude petroleum, a sample of which the author has since tested at the Yorkshire College with the following results :—

Specific gravity . . . . . . 0·922
Gross calorific value . . . . . 19,300 B.Th.U. per lb.

This oil was completely burned by the engine at all but the lightest load, the exhaust being clear and free from smell of unburnt oil, except when running under very light load, a little smoke being then visible.

TABLE 3.

Crude Petroleum used in the Test of 80-*B.H.P. Diesel at Ghent.*

Weight of Vessel.	Weight of Vessel and Oil.	Weight of Oil.	Initial Temperature.	Final Temperature.	Rise of Temperature.	Calorimeter constant $1\cdot75 \times 1830 \times t$ $\frac{}{w}$	
Grammes.			F.°	F.°			
48·78	50·48	1·70	49·0	59·2	10·2	19,420	} 19,300
48·70	50·23	1·53	49·5	58·6	9·1	19,260	

This result is in close agreement with the result obtained in May 1902 for the same kind of oil.

160 B.H.P. ENGINE.

On 7th March 1903, by the kindness of Messrs. Carels Frères, the author made complete trials on a double cylinder 160-B.H.P. Diesel engine, at the works in Ghent. Measurements for the following quantities were taken :—

    a. Indicated horse-power of each cylinder.

    b. Total oil used during any run.

    c. Total cooling water used during a run.

    d. Inlet and outlet temperatures of the cooling water.

    e. Temperature of the exhaust gases.

    f. Total revolutions of the engine.

g. Speed at any instant by tachometer.

h. Analysis of the exhaust gases.

i. Effective horse-power.

The indicated horse-power was obtained by taking indicator cards from each cylinder every five minutes during one hour, using calibrated indicator springs which gave a scale of 372 lbs. per square inch of piston per inch of height.

The total oil was determined by weighing the oil and tank before and after each run. Two oil tanks were used, each connected to the pump supply-pipe, one being placed upon the weighing machine. During preparation the engine was run from the stand-by tank, and when running a trial the engine was run from the one placed upon the weighing machine; the cutting out of one tank and putting in of the other was done at the same instant, the operation being always performed by the author at the instant he gave the signal to commence or to cease a trial. The weighing of the tank was also done by the author, other quantities being taken by assistants, check readings only being taken from time to time by the author. A gauge glass was attached to each oil-pump reservoir to give the level in each case, and a mark placed to allow exact measuring. The cooling water was measured by a small Siemens water-meter graduated on the metric system.

Temperatures were taken by thermometers graduated in centigrade degrees, that of the exhaust gases being a mean of the temperature existing quite close to the exhaust valve inside the exhaust pipe. The revolutions were recorded by a counter attached to the air-pump rods. The gases for analysis were collected over water from the exhaust pipe close to the exhaust valve, and analysis by the Orsat apparatus was made on the spot.

The effective horse-power was determined by subtracting from the indicated power the indicated power of the engine when running quite light, the driving belt being removed. This was necessary owing to the engine being belted to a line shaft and then to a dynamo and to a motor. The load trials were each of one hour's duration, and of half-an-hour when running light. The data and results are given in Table 5 (page 413).

On Fig. 7 three pairs of curves are shown, one pair for each of the engines tested by the author, the lower curves (straight lines) showing the oil-consumption of these engines as the load varied. These curves of total consumption show well the action of the governor in cutting down the oil-supply to the requirements of the

FIG. 7.—*Oil Fuel for Diesel Engines.*

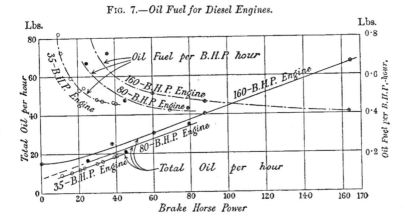

engine, a result not so nearly approached by other types of oil engine; the curves show that the oil-supply is directly proportional to the horse-power, except at very low loads, that is, loads below 25 per cent. of the normal. In the 35-H.P. engine the total oil-consumption is fairly well given by the equation —

Total oil in lbs. per hour = $4 + 0 \cdot 357 \times$ B.H.P.

for powers over 10 B.H.P.

In the 80-B.H.P. engine the equation is :—

Total oil in lbs. per hour = $7 \cdot 5 + 0 \cdot 336 \times$ B.H.P.

In the case of the 160-B.H.P. engine the equation becomes :—

Total oil per hour = $10 + 0 \cdot 345 \times$ B.H.P.

for B.H.P. is larger than 50 when both cylinders are at work.

The three upper curves show the oil-consumption per B.H.P.-hour; it will be observed that those curves descend to nearly the same value at top loads, and that the rate of increase of consumption is not high, even down to half load. Both these facts point to the economy of this engine when distributed in small units, making it unnecessary to concentrate the power in large units and then use expensive transmission to the points where the power is required.

The performance of the 160-B.H.P. engine under variation of load was as follows :—

Load on the dynamo 700 ampères at 115 volts, that is, 108 H.P., tachometer reading 89.

The whole load thrown off at once, the tachometer ran up to 96·2 and settled back to 88·7 in about a minute.

The engine running light, tachometer 91, a load of 700 ampères at 115 volts was put on at once, when the lowest reading of the tachometer was 87·8 and the engine settled back to 90 in about half-a-minute.

The calculations from average values of the results for trial No. 2 of 7th March on the 160-B.H.P. engine (page 413) are as follows :—

$$\text{Average revolutions} = \frac{9{,}424}{60} = 157 \text{ per minute.}$$

$$\begin{aligned} \text{I.H.P. L Cylinder} &= 64\cdot7 \times 157 \times 0\cdot005806 \\ &= 59 \\ \text{,, R Cylinder} &= 74\cdot1 \times 157 \times 0\cdot005806 \\ &= 67\cdot6 \\ \text{Total I.H.P.} &= 126\cdot6 \end{aligned}$$

$$\begin{aligned} \text{Air Pumps H.P.} &= Pe \times n \times 0\cdot0000358 \times 2 \\ &= 291\cdot5 \times 157 \times 0\cdot0000716 \\ &= 3\cdot26 \end{aligned}$$

$$\begin{aligned} \text{Net I.H.P., or the I.H.P. developed by the oil,} &= 126\cdot6 - 3\cdot26 \\ &= 123\cdot34 \end{aligned}$$

This assumes all the Pump H.P. returned to the working cylinder.

$$\begin{aligned} \text{Effective H.P.} &= \text{I.H.P.} - 39\cdot6 \\ &= 126\cdot6 - 39\cdot6 \\ &= 87 \end{aligned}$$

$$\text{Mechanical Efficiency} = \frac{\text{Eff. H.P.}}{\text{I.H.P.}} = \frac{87}{126\cdot6} = 68\cdot7 \text{ per cent.}$$

Pounds of Oil per net I.H.P. hour $= \dfrac{40 \cdot 58}{123 \cdot 34} = 0 \cdot 329$

,,   ,,   ,,   Eff. H.P. hour $= \dfrac{40 \cdot 58}{87} = 0 \cdot 466$

Jacket Water per minute $= 43$ lbs.

$$\left.\begin{array}{l}\text{Thermal units per I.H.P.} \\ \text{minute in Jacket Water}\end{array}\right\} = \dfrac{43 \times 43}{126 \cdot 6} \times \dfrac{9}{5} = 26 \cdot 3$$

Amount of exhaust gases. From the gas analysis one obtains:—
CO_2 $4 \cdot 1$ per cent., O_2 15 per cent., and by difference N_2 $80 \cdot 9$.

From the well-known formula :—

$$\frac{\text{Total air}}{\text{Consumed air}} = \frac{N}{N - \dfrac{79}{21} \cdot O} = \frac{21}{21 - 79 \cdot \dfrac{O}{N}}$$

where N and O are the percentage volumes of nitrogen and oxygen in the exhaust. One gets for this case—

$$\frac{\text{Total air}}{\text{Consumed air}} = \frac{21}{21 - 79 \times \dfrac{15}{81}} = \frac{21}{21 - 14 \cdot 6} = 3 \cdot 31$$

The approximate composition of the petroleum is—

Carbon	$85 \cdot 0$ per cent.
Hydrogen	$13 \cdot 5$,, ,,
Incombustible	$1 \cdot 5$,, ,,

1 lb. of such oil would use $14 \cdot 56$ lbs. of air for complete combustion, calculated as below :—

Carbon. . . .	$0 \cdot 85 \times 11 \cdot 6 =$	$9 \cdot 86$ lbs.
Hydrogen . . .	$0 \cdot 135 \times 34 \cdot 8 =$	$4 \cdot 70$ lbs.
	Total	$14 \cdot 56$ lbs.

The resultant products being $15 \cdot 56$ lbs. per lb. of oil.

From the analysis $3 \cdot 31$ times this air is used.

Therefore air per lb. of oil in this case gives—
$$3 \cdot 31 \times 14 \cdot 56 = 48 \cdot 2$$

The resulting products of combustion—
$$49 \cdot 2 \text{ lbs. per lb. of oil.}$$

Therefore weight of exhaust per I.H.P. minute—
$$= \frac{49 \cdot 2}{60} \times \frac{40 \cdot 58}{126 \cdot 6} = \frac{15 \cdot 75}{60} \text{ lbs.}$$

The specific heat at constant pressure of these gases may, with sufficient accuracy, be taken at $0 \cdot 25$.

Consequently the British Thermal Units per I.H.P. minute carried away by the exhaust approximates to—

$$\frac{15 \cdot 75}{60} \times 0 \cdot 25 \times 239 \times \frac{9}{5}$$
$$= 28 \cdot 2 \text{ B.Th.U.}$$

Thermal Units per I.H.P. minute supplied by the oil fuel
$$= 0 \cdot 329 \times 19{,}300 \div 60$$
$$= 106 \text{ B.Th.U.'s.}$$

Thermal units supplied per I.H.P. minute by the air
$$= \frac{15 \cdot 75 - 0 \cdot 329}{60} \times 12 \times \frac{9}{5} \times 0 \cdot 25 = 1 \cdot 4$$

Thermal equivalent of I.H.P. minute $= 42 \cdot 4$.

Heat Balance.

Supply $106 + 1 \cdot 4$.

I.H.P. minute .	.	.	$42 \cdot 4$	or	$39 \cdot 5$ per cent.
Cooling Water	.	.	$26 \cdot 3$	,,	$24 \cdot 5$,, ,,
Exhaust Gas .	.	.	$28 \cdot 2$	,,	$26 \cdot 3$,, ,,
Unaccounted .	.	.	$9 \cdot 1$	,,	$9 \cdot 7$,, ,,
			$106 \cdot 0$		$100 \cdot 0$,, ,,

Thermal efficiency on the net I.H.P. $= \frac{42 \cdot 4}{106} = 40$ per cent.

Thermal efficiency on the B.H.P. would be—
$$40 \times \frac{87}{123 \cdot 4} = 28 \cdot 3 \text{ per cent.}$$

As a check on the gas analysis results, the volume of air per lb. of oil may be approximately calculated as follows :—

The working volume of the cylinder is $2 \cdot 663$ cubic feet; assuming this to be filled with air at atmospheric pressure, at each stroke one will have for this two-cylinder engine weight of air per revolution—

$$= 2 \cdot 663 \times 0 \cdot 0809$$
$$= 0 \cdot 2163 \text{ lbs.}$$

(*Continued on page* 418.)

TABLE 4.

80-*B.H.P. Diesel Engine.*

Test of Single-Cylinder Diesel Engine at Messrs. Carels Frères, Ghent,
on 7 Feb. 1903.

This trial was run in response to the request of Mr. A. J. Lawson, Engineer to the British Electric Traction Co., Mr. H. W. Anderson, London, and the author, when on a visit of inspection to Messrs. Carels Frères' Works at Ghent. Preparations for running this trial not having been made, it is not so complete as the others.

Summary of Results.				
1. Load	0·25	0·5	0·75	Full
2. Duration . . . n.ins.	30·5	30·1	32	30·75
3. Total revolutions . . .	4,890	4,863	4,800	4,797
4. Revolutions per minute . .	163	161	160	160
5. Mean effective pressures. .	48·5	68·4	89·1	109·4
6. Indicated horse-power . .	46·8	67·0	82·8	101·5
7. Effective ,, ,, . .	24·8	45	60·8	79·5
8. Electrical ,, ,, . .	17·3	35·6	52·1	68·8
9. Mechanical efficiency . .	53	67·0	73·5	78·3
10. Total oil . . . lbs.	8·38	10·73	Lost	17·63
11. Oil per I H.P. hour . lb.	0·352	0·320	—	0·339
12. ,, ,, effective H.P. hour lb.	0·664	0·477	—	0·434
13. ,, ,, electrical ,, ,, lb.	0·952	0·601	—	0·500
14. B.Th.U.'s per I.H.P. minute .	114	104	—	111
15. {Thermal efficiency on the} {I.H.P. . . per cent }	37·2	40·8	—	39·9
16. {Thermal efficiency on the} {B H.P. . . per cent.}	19·7	27·3	—	31·2

The transmission to the dynamo consisted of a belt to a line shaft and a belt from the shaft to the dynamo; there was also belted to the shaft a motor which was running light.

TABLE 5.

160-*B.H.P. Diesel Engine.*

Test of Two-Cylinder Diesel Engine at Messrs. Carels Frères, Ghent,
on 7 March 1903.

Summary of Results.

		Full	0·53	0·375	0·25	None
1.	Load	Full	0·53	0·375	0·25	None
2.	Duration . . . mins.	60	60	61	60	30
3.	Total revolutions . .	9,258	9,424	9,639	9,476	4,770
4.	Revolutions per minute .	154·5	157	158	158	159
5.	{ Mean effective pressures L	113·4	64·7	51·9	38·2	46·2
	{ ,, ,, ,, R	114·5	74·1	60·7	46·2	39·7
6.	{ I.H.P. cylinder 4L . .	101·0	59·1	46·6	34·4	42·6
	{ ,, ,, 5R . .	103·4	67·7	53·4	40·5	36·7
7.	,, total or mean . .	204·4	126·8	100·0	74·9	39·64
8.	,, of air pump . .	3·3	3·25	3·0	2·97	2·98
9.	Net I H.P. from the oil .	201·1	123·6	97·0	71·93	36·66
10.	Effective horse-power . .	164·8	87·2	60·4	35·4	—
11.	Mechanical efficiency . .	80·7	68·8	60·4	49·2	—
12.	Total oil . . . lbs.	67·0	40·58	30·45	25·6	7·6
13.	Oil per net I H.P. hour lb.	0·333	0·329	0·309	0·356	0·415
14.	,, ,, effective H.P. hour lb.	0·408	0·465	0·505	0·724	—
15.	{ Cooling water per minute	49·6	43·0	26·25	16·1	—
	{ lbs. . . . Inlet	9·5	9·5	9·5	9·5	10·0
16.	{ Temperatures of water					
	{ C. . . . Outlet	62·0	52·5	60·5	65·5	67·0
17.	Change of temperature .	52·5	43·0	51·0	56·0	57·0
18.	Temperature of exhaust C.°.	384·0	239·0	197·0	158·0	—
19.	B.Th U. per net I H.P.-minute	108·0	106·0	102·0	114·0	154·5
20.	,, converted to work .	42·4	42·4	42·4	42·4	42·4
21.	{ ,, rejected in cooling	22·9	26·3	24·3	21·7	—
	{ water					
22.	{ B.Th.U. rejected in exhaust	32·1	28·3	14·4	13·5	—
	{ gases					
23.	{ Thermal efficiency on net	39·25	40·0	41·5	37·2	—
	{ I.H P. . . per cent.					
24.	{ Thermal efficiency on Eff	32·3	28·3	26·1	18·3	—
	{ H.P. . . per cent.					
	Exhaust gas analysis:					
	{ Carbon dioxide	7·0	4·1	3·1	2·9	—
	{ vol. per cent.					
25.	{ Carbon monoxide	—	—	Trace?	0·1?	—
	{ vol. per cent.					
	{ Oxygen . . per cent.	11·3	15·0	16·6	17·6	—
	{ Nitrogen by difference	81·7	80·9	80·3	80·3	—
	{ per cent					

Note.—The tap of the carbon dioxide bottle was not quite gas-tight, but was
well smeared with vaseline to prevent serious leakage.

Full-size Indicator Diagrams from 160-B.H P. Diesel Engine.

FIG. 8.—*Starting-Card.*

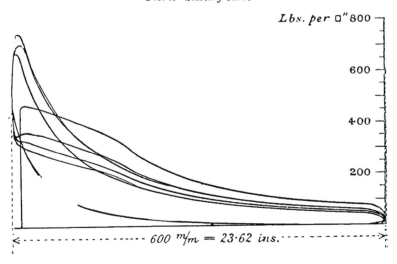

FIG. 9.—*Air-Pump Card.*

Taken at 160 revs. per min. with a pressure in blast reservoir of 800 lbs. per sq. inch.

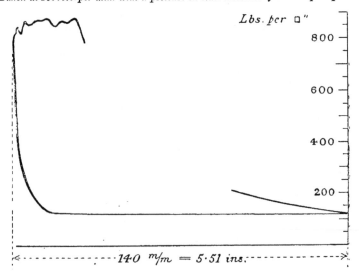

Full-size Indicator Diagrams from 160-B.H.P. Diesel Engine.
FIGS. 10, 11, 12 AND 13.

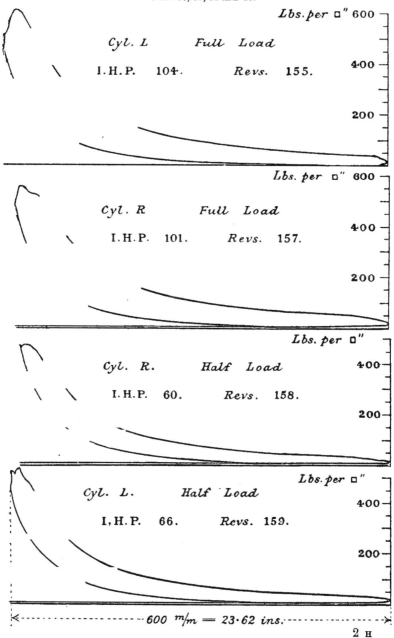

Full-size Indicator Diagrams from 160-B.H.P. Diesel Engine.
FIGS. 14, 15, 16 AND 17.
Engine running light.   Cyl. L Driving.   Cyl. R Compressing.
Revs. per min. 159.

Lbs. per □".

400

L.

200

Lbs. per □"

400

200

Lbs. per □"

400

200

Lbs. per □"

400

200

Full-size Indicator Diagrams from 160-B.H.P. Diesel Engine.
FIGS. 18, 19, 20 AND 21.
Engine running light.   Cyl. R Driving.   Cyl. L Compressing.
Revs. per min. 159.

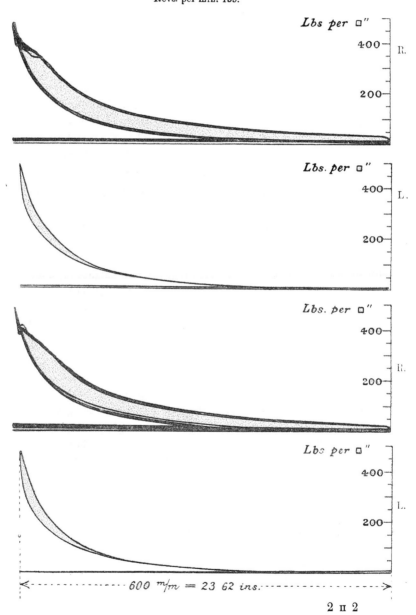

2 II 2

The weight of oil used per revolution in Trial 2 (page 413)—

$$= \frac{40 \cdot 58}{9,424} = 0 \cdot 00431 \text{ lbs.}$$

Therefore ratio of air to oil $= \frac{0 \cdot 2163}{0 \cdot 00431} = 50 \cdot 3$, as against $48 \cdot 2$ by the gas analysis—a difference of 4 per cent. Nothing is allowed for rise of temperature, as the air enters the cylinder.

The indicator diagrams shown in Figs. 8–21 (pages 414–417) were taken during the trial of 160-B.H.P. engine on 7th March 1903. Fig. 8 (page 414) shows the starting of the engine under air-pressure; the card shows four air-pressure strokes, followed by the first three oil-fired strokes; following Fig. 8 come a series of cards showing the development in the body of the card under various loads, the action of the governor in giving more oil for more work being well shown.

In the no-load run only one cylinder is at work, the other is running under compression, and, as will be seen by the accompanying card, is really working negatively; this negative work has, however, been neglected in getting out the power required to drive the engine without load, and it has been assumed that this will balance anything in the nature of an increase in friction under the heavier loads.

The whole of the cards taken in a given run are very regular, any single card being very little different from the mean; there are of course no misfires.

In Fig. 22 (page 419) is shown a comparison of the pressures, powers, and clearances of the three types of engine, namely, the compound condensing engine using high-pressure steam with jacketed cylinders, the scavenging type of gas-engine using gas of 600 B.Th.U. per cubic foot, and the Diesel engine using crude petroleum. The indicator cards there shown are drawn to the same scales and for the same total volume of cylinder.

The clearances are as follows :—

The steam-engine $3 \cdot 5$ per cent. of the volume of the H.P. cylinder; the gas-engine 27 per cent. of whole volume; the Diesel engine 7 per cent. of the whole volume.

The I.H.P. at 160 revolutions work out at :—

Diesel engine	. 105
Gas-engine | . . · .	. 66
Steam-engine	. $24 \times 4 = 96$

The multiplier 4 being due to the steam-engine having a working stroke every stroke.

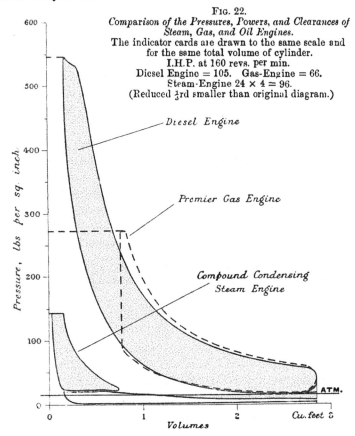

FIG. 22.

Comparison of the Pressures, Powers, and Clearances of Steam, Gas, and Oil Engines.
The indicator cards are drawn to the same scale and for the same total volume of cylinder.
I.H.P. at 160 revs. per min.
Diesel Engine = 105. Gas-Engine = 66.
Steam-Engine 24 × 4 = 96.
(Reduced ⅓rd smaller than original diagram.)

Further trials were made in England to confirm those on the makers' testing beds. Table 6 (page 422) gives the results of a trial made on the 35 B.H.P. Diesel engine at Harrogate on 8th May 1903, that is, twelve months after the trials given in Table 3. Taking trials at 33 B.H.P. the results are as follows:—

Year.	B.H.P.	I.H.P.	Mechanical Efficiency.	Oil per hour I.H.P.	B.H.P.
1902	33·1	46·88	70·7	0·336	0·476
1903	33·2	48·80	68·1	0·344	0·5u5

A month previous to this trial a new exhaust valve and a new piston ring were fitted. The result is quite satisfactory, as the higher I.H.P. shown in the 1903 trial was probably due to a little slackness in the indicator pencil gear.

Table 7 (page 423) gives the results of a trial made in London on 16th May 1903, at Messrs. Stafford Allen & Sons' Drug Mills in Cowper Street, Finsbury. This engine is of 80 B.H.P. at 160 revolutions, and was made by Vereinigte Maschinenfabrik, Augsburg, and there is therefore additional interest in comparing it with the trial given in Table 4 on the engine of the same size made by Messrs. Carels Frères, Ghent.

The engine was run on the works' load which consisted largely of grinding mills, and in order to keep the load steadier, a plank brake was applied to the fly-wheel, the load being adjusted from time to time as the indicator cards showed a slackening off in the works. In the trial at half load the maximum I.H.P. was 73, the minimum I.H.P. 44, the mean being 62 I.H.P., taken from 52 indicator diagrams. In the three-quarter load the highest I.H.P. was 99, the lowest I.H.P. being 69·2, and the mean I.H.P. was 81·5 from 36 diagrams. In the no-load trial the highest I.H.P. was 31·2. The lowest I.H.P. 16·35, and the mean from 26 cards, each containing five or six diagrams, was 23·8; the governor "hunted" a little when working without load.

This trial was made when the engine was running under load for the first time since its erection at Messrs. Stafford Allen and Sons' Works, and was not then out of the makers' hands, consequently the no-load I.H.P. is probably a little high, and a month's running of the engine would probably reduce this.

Cost of Power.

Comparison of total cost of power-production when using—

    (a) Diesel engine.
    (b) Gas-engine.
    (c) Steam-engine.

In the following Table 8 (pages 424 and 425) a comparative and fairly accurate statement of costs is drawn up for three sizes of engine, namely 35, 80, and 160 B.H.P.

Initial capital cost is stated for the three items, Land, Buildings, and Plant. Annual charges are stated under the heads, Interest on Capital, Maintenance and Depreciation for Buildings, and also for Plant, Fuel, Oil and Water, and Wages.

> Line 11 gives the total Capital Cost in £.
> „   12   „   „   Capital Cost per B.H.P. in £.
> „   21   „   „   the total Annual Expense in £.
> „   22   „   „   Annual Cost per B.H.P. in £.
> „   23   „   „   Cost per B.H.P. hour in pence.

The gas-engine taken for comparison is the Crossley using a Dowson gas-producer. The steam-engine compared with is the high-speed compound condensing engine.

The fuels (in all cases delivered) are :—

> For the Diesel .   .   . Crude petroleum at 45s. 0d. per ton.
> For the gas-producer . . Anthracite cobbles at 24s. 0d. per ton.
> For the steam-engine . . Coal at 12s. 6d. per ton.

The oil-consumption for the Diesel has been taken, in each case, as the mean between the full and half-power rates of consumption.

The fuel-consumption in the case of the gas-producer has been taken at 1·5 to 1·25 lbs. per B.HP.-hour plus an allowance for stand-by losses.

The fuel-consumption for the steam-engine has been taken at 4 lbs. and 3·5 lbs. per B.H.P.-hour for the lowest and highest powers; assuming an evaporation of 8 lbs. of water per lb. of fuel.

The final result, which is much in favour of the Diesel engine, is as follows :—

Total cost per B.H.P.-hour.			
Engines.	Diesel.	Gas.	Steam.
	d.	d.	d.
35 B.H.P.	0·59	0·69	0·89
80 B.H.P.	0·39	0·52	0·63
160 B.H.P.	0·32	0·40	0·49

TABLE 6.

35-*B.H.P.* *Diesel Engine.*

Test at Harrogate after running 16 hours a day for 12 months,

8 May 1903.

Summary of Results.		
1. Load	0·91	0·82
2. Duration . . . . . mins.	50	29
3. Revolutions per minute	179·9	179·8
4. Load on the brake . . . . lbs.	236	204
5. Pull on spring balance . . . lbs.	12	8·4
6. Effective load. . . . . lbs.	224	195·6
7. Brake horse-power	33·2	28·9
8. Total oil . . . . . lbs.	13·97	6·875
9. Oil per hour . . . . . lbs.	16·76	14·24
10. Oil per B.H.P. hour. . . . lb.	0·505	0·493
11. Mean effective pressure . lbs. per sq. in.	106·5	99·4
12. Indicated horse-power	48·8	45·5
13. Oil per I.H.P. hour . . . . lb.	0·344	0·313
Temperature of Jacket Water Inlet . .	50° F.	50° F.
„ Outlet .	166° F.	155° F.
15. „ „ „ Rise . .	116°	105°
21. Mechanical Efficiency . . per cent.	68	63·6

TABLE 7.

80-*B.H.P. Diesel Engine.*

Test at Messrs. Stafford Allen & Sons, London, 16 May 1903.

Summary of Results.				
Load	0	0·5	0·75	Full
Duration . . . . mins.	49	145	78	—
Total revolutions . . .	—	—	—	—
Revolutions per minute . .	158	157	156	—
Mean effective pressure . .	25·6	69	91·2	—
Indicated horse-power . .	23·8	61·9	81·5	—
Effective horse-power . . .	0	38·1	57·7	80
Mechanical efficiency . . .	0	61·6	70·4	—
Total oil . . . . lbs.	6·52	48·22	35·69	—
Oil per hour . . . lbs.	7·98	19·91	27·48	—
Oil per I.H.P.-hour . . lb.	0·374	0·322	0·337	—
Oil per effective H.P.-hour lb.	—	0·523	0·479	—
Cooling water per minute · lbs.	—	24·1	25·7	—
Temperature of water inlet . .	—	—	—	—
,, ,, outlet .	66° C.	60° C.	64° C.	—
Temperature of room . . .	25° C.	24° C.	25° C.	—
,, ,, oil . . .	21° C.	21° C.	18° C.	—

TABLE 8 (*continued on opposite page*).

Total Costs in Power Productions.

	Diesel Engine.	Gas Engine.	Steam Engine.
1. Means of production . .{			
2. Brake horse-power . . .	35	35	35
3. Size of engine house . feet	12×20×15	12×20×15	10×20×15
4. Ground for gas-holder . feet	—	10×20	—
5.{Size of boiler house and gas-producer . . . feet}	—	8×16×12	14×40×15
	£	£	£
6.{Cost of land at 7s. per square yard . . . . .}	10	22	30
7. Cost of buildings . . .	90	130	230
8.{Engine foundations, etc., boiler settings . . . .}	10	15	60
9.{Engine, tanks, piping and condenser, etc. . . .}	585	280	240
10.{Boiler or gas-producer and accessories . . . .}	—	200	290
11. Total capital cost . . .	695	647	850
12. Capital cost per B.H.P. . £	19·8	18·5	24·3
Annual Charges.	£	£	£
13. Interest on capital, 4 per cent	27·8	26	34
14.{Building maintenance and depreciation, 5 per cent. .}	4·5	6·5	11·5
15.{Engines and machinery. Maintenance and depreciation, 10 per cent. . . . .}	58·5	28	24
16.{Boilers. Maintenance and depreciation, 12 per cent. .}	—	—	35
17.{Gas-producer and holder. Depreciation, 5 per cent. .}	—	10	—
18. Fuel	47·25	90	130
19. Lubricant, waste, etc. . .	20	20	20
20. Wages, removal of ash, etc. .	75	90	95
21. .	233·05	270·5	350
22. Annual cost per B.H.P. . £	6·66	7·73	10
23. Cost per B.H.P. hour penny	0·59	0·69	0·89

(*concluded 'from opposite page*) **TABLE 8.**

Total Costs in Power Productions.

Diesel Engine. 80	Gas Engine. 80	Steam Engine. 80	Diesel Engine. 160	Gas Engine. 160	Steam Engine. 160	} 1. 2.
$16 \times 24 \times 20$	$16 \times 24 \times 20$	$14 \times 24 \times 20$	$20 \times 27 \times 20$	$20 \times 27 \times 20$	$17 \times 27 \times 20$	3.
—	11×24	—	—	12×26	—	4.
—	$10 \times 18 \times 15$	$15 \times 46 \times 20$	—	$10 \times 20 \times 16$	$20 \times 48 \times 20$	5.
£	£	£	£	£	£	
15	29	40	21	41	56	6.
140	190	360	180	235	420	7.
20	28	80	35	40	100	8.
900	635	500	1680	1140	800	9.
—	260	350	—	380	600	10.
1075	1142	1330	1916	1836	1976	11.
13·44	14·25	16·7	14·28	11·98	12·4	12.
£	£	£	£	£	£	
43	45·7	53·2	76·64	73·4	79	13.
7	9	18	9·0	11·75	21	14.
90	63·5	50	168	114	80	15.
—	—	42	—	—	72	16.
—	13·0	—	—	19	—	17.
98	205	260	186	343	440	18.
35	35	35	55	55	55	19.
75	95	110	87·5	110	140	20.
348	466	568·2	582	726	887	21.
4·35	5·82	7·1	3·63	4·45	5·5	22.
0·39	0·52	0·63	0·32	0·40	0·49	23.

The author wishes to thank those gentlemen who have rendered help to him whilst making these investigations, and in particular Professor John Goodman, M.I. Mech. E., of the Yorkshire College; Mr. G. Carels, and the firm of Carels Frères, of Ghent, Belgium; Mr. F. Bagshaw, C.E., Borough Engineer, Harrogate; Mr. G. Wilkinson, M.I. Mech. E., Borough Electrical Engineer, Harrogate; Messrs. Stafford Allen and Sons, Cowper Street, Finsbury, London; and The Diesel Engine Co., London.

The Paper is illustrated by Plates 9 to 11 and 16 Figs. in the letterpress.

Discussion.

Mr. H. ADE CLARK said that, since the printing of the Paper, he had made a few more tests of the oils and the engine. The oils had been tested in an ordinary Thompson type of calorimeter, and had since been subjected to a test in a bomb calorimeter which gave more complete results, shown in Table 9 :—

TABLE 9.

No. of Test.	Liquid Fuel used at	Total Calorific Value.	Available Calorific Value.	Specific Gravity at 60° F.
		B.Th.U.	B.Th.U.	
1	Harrogate, 1902 Trials .	20,000	18,800	0·922
2	Harrogate, 1903 Trials .	19,880	18,680	0·932
3	{Stafford, Allen & Sons,} London, May 1903 .	19,970	18,770	0·932
4	{Ghent, Belgium,} 7th Feb. 1903 . .	20,200	19,000	0·920
5	{Ghent, Belgium,} 7th March 1903 . .	20,220	19,020	0·918

Each of these values was the average of several tests agreeing very closely with each other. The calorimeter was calibrated with pure sugar, having a calorific value of 3,955 calories per gramme, and giving a water equivalent to the calorimeter of 1,360 grammes, of which 1,000 grammes were actually water.

The average specific gravity of the oils was 0·925 at 60° F., and the average total calorific value as given by the bomb calorimeter worked out at 20,050 thermal units. The available calorific value, allowing 1,200 thermal units for the latent heat of the steam produced, was 18,850 thermal units, which was rather a lower figure than he had used in working out the thermal efficiency of the engine. In connection with thermal efficiency, he wished to draw attention to a Table in the middle of page 411. The I.H.P. minute absorbed 42·4 thermal units out of a total of 106 supplied by the oil, which gave a thermal efficiency on the I.H.P. of 39·5 per cent.; the cooling water carried away 24·5 per cent.; the exhaust gases accounted for 26·3 per cent. of the total heat supplied by the oil, and there was 9·7 per cent. of heat to be accounted for by radiation and errors of observation.

The four oils that he had used were slightly different in their densities. Of the samples which were exhibited, No. 1 was that which he had used at Harrogate in 1902; No. 2, which was a much denser oil, was the one used in 1903, and accounted to some extent for the fact that the engine results were not quite so good as in 1902. The same remarks applied to the other two oils. No. 4 was the one he used at Ghent, and No. 3 he used at Messrs. Stafford, Allen and Co.'s Works in London; the latter was a denser oil than No. 4, the densities being 0·920 against 0·932. Not only that, but the denser oils contained a great deal more sediment.

In testing the Harrogate engine it stopped, and on examining why it had stopped—because that was the first time he had an experience of the engine stopping under work without provocation—it was found that dirt had penetrated into the inlet valve of the oil pump and held it open. If the inlet valve was held open, it would be seen from the construction of the pump, Fig. 5, Plate 11, that the engine could not get any oil, because the pump would simply force the oil back again into the reservoir. The oil was fed to the oil drums from the supply tank and taken off near the bottom, the filtering being downwards; that arrangement had been improved upon. The oil was now fed to the oil tanks at the bottom, filtered upwards through the filtering mesh, and taken from about the

(Mr. H. Ade Clark.)

middle of the tank. That method obviated the difficulty which arose in this particular case of a great deal of dirt passing through with the oil into the engine.

Since those trials, he had attempted to run another one to get over the difficulty, mentioned in the Paper, of having to determine the effective H.P. by subtracting from the total I.H.P. of the loaded engine the I.H.P. of the engine when running quite light. The result obtained was not far off the actual truth, but it would be much better to get a direct output; with the 80 or 160-H.P. engines a difficulty arose, unless a special brake was available. The engine he tested on 7th March in Ghent was now being put up for the Temple Press in London, and ten days ago he tested it with a dynamo on its shaft direct. It only ran at 160 revolutions. A four-pole dynamo, capable of giving an output of 75 kilowatts at 530 volts, was being used, which was really a bigger load than the engine was capable of doing continuously, so that the dynamo had to run under load. The tests were made on the engine after its first erection, before it had been handed over to the buyer. The following results were obtained after a run of one hour:—I.H.P. 82·5, which was 80 per cent. of the full load of the engine, electrical H.P. 51·4, which only gave a gross efficiency of 62·3 per cent. The dynamo was only running at about 52 per cent. of its full load. He did not know the dynamo efficiency at half-load. The oil per I.H.P.-hour amounted to 0·365 lb., which was higher than the figure he had obtained in Ghent, when the engine was tested on the bed. In fact the figures were all a little higher than were obtained at Ghent, but they were so close that they confirmed the original result satisfactorily. One could not get absolutely identical results, when in one case an engine was tested on a constant load and in another on a varying load. When he tested Messrs. Stafford, Allen's engine it was on a varying load, and he attempted to keep it steady by applying a brake to the wheel. In the case of the Temple Press engine, the oil used per electrical H.P. was 0·572 lb. per hour, which was fairly good, but evidently there was something wrong. The speed only came out at 154 revolutions, but the engine was designed to run at 160, which indicated that the governor was not properly set. He pushed the load up to about 90 per cent. of its full load and started another trial, and then a bearing became hot, which was probably the primary cause of the unsatisfactory trial. He had not been able to complete the trials, but he thought the first one, where he obtained about 50 per cent. of full load, was reasonably satisfactory. The engine would have to be made better undoubtedly before the buyer took it over.

He had mentioned nothing in the Paper about the weight of the engines, but he had made a comparison to show the powers for a given size of cylinder. A 20-H.P. engine at the present time weighed about 4·72 tons; in 1899 it weighed 6·4 tons. A 30-H.P. engine now weighed 7·68 tons; in 1899 it weighed 9·35 tons; a 40-H.P. engine now weighed about 10 tons, whereas in 1899 it weighed 11·8 tons; so that the weight had been reduced on an average of about 25 per cent. in the last four years, and in addition the price had fallen about 30 per cent.

With regard to mechanical efficiency the following figures would be interesting. In 1897 it was 74·8 per cent. on a 35-H.P. engine; in 1902 it was 80 per cent. on a 40-H.P. engine, and in 1903 it was 82 per cent. on an 80-H.P. engine. At Stockholm the short-stroke high-speed type of the Diesel engine had been developed; a much lighter engine was thus obtained, which, it was claimed, had a mechanical efficiency of 85 per cent. on an engine developing 120 useful H.P. The weight had been very much reduced by the development of the short-stroke high-speed type for the 30-H.P. engine, the weight being from 4·85 to 6·15 tons, and for the 60-H.P., 7·04 to 8·38 tons, depending upon the use of a light or heavy fly-wheel. The members would no doubt be interested to know to what extent the engine was being adopted. Within the last three years 60 engines had been sold in England and the Colonies, chiefly in England, giving a total B.H.P. of 3,220. Germany had taken 85 engines, giving 2,910 B.H.P., and Russia 224 engines, giving 16,200 H.P. At the present moment a large municipal tramway station at Kieff was being equipped entirely with four groups of Diesel engines, giving 400 to 500 H.P., which were to work sixteen hours a day, with a guaranteed consumption of crude oil of 0·679 lb. per kw.-hour. He had not

(Mr. H. Ade Clark.)

said anything about lubricating oil in the Paper, because he had not much experience of it, but the guaranteed consumption in the case of that particular station worked out at about 3·18 gallons per 54-hour week.

With regard to the temperatures of the cycle, he had already mentioned (page 398) that the temperature in the engine at the end of compression was about 1,000° F. Taking the indicator cards for the 80-H.P. engine—about 10 to form a fair average card—and assuming 30° C. for the temperature of the air, when the piston started on its compression stroke the temperature at the end of the compression worked out at 474° C., or 890° F. At the end of combustion a temperature of 1,079° C. or 1,940° F. was obtained; it should have been 890° F., because the idea was to get the combustion without a rise of temperature; and at the end of the expansion curve the temperature worked out at about 483° C. or 870° F., almost the same temperature as was obtained at the end of compression. The actually observed temperature, for that particular trial from which he took the cards, at the end of expansion was 384° C. or 722° F. against the calculated temperature of 483° C. or 901° F. He wished to call attention to the fact that the figures he had obtained in England confirmed his results obtained at the maker's works, Fig. 23, Plate 12.

There was one point he would like to draw attention to in connection with the comparison of costs he had made. He did not think he had stated the fact that they were made for 50 weeks of 54 hours. The fuel allowance for starting and stand-by losses, in the case of a steam-engine, was estimated at the rate of two hours a day; for the gas-engine for 50 weeks of 54 hours, 10 hours stand-by and getting-ready loss was allowed for per week. In the Diesel engine during the same period he had not allowed any stand-by losses at all, because there was no fuel spent in getting ready or in slacking off.

The PRESIDENT was sure the members would desire to thank Mr. Ade Clark for his capital Paper.

Mr. E. R. DOLBY said that the Paper was extremely valuable, and placed before the members practically all the data that were obtainable. The figures that had previously been put before the world had been so rosy that he, as a consulting engineer, thought it was his absolute duty to look into them carefully, so that, for electric light installations and similar work, he might be in a position to advise his clients whether or not they should use a Diesel engine in preference to some other type. About November last year he had an opportunity of doing that. He required to put down two engines to develop about 20 B.H.P. each, and went somewhat carefully into the question of the initial cost, and the cost of maintenance. He put himself into communication with the Diesel Engine Co., and also wrote to various people who had used the engine. The Diesel Engine Co. were extremely kind, and offered him every facility. He visited a place in London where a 10-H.P. engine was in use, and was very much struck by the workmanship of the engine, and the ease with which it was started. The next thing to do was to find out what was the cost of the engine, and what would be the exact cost of the fuel to his client. That gave somewhat surprising results. He was sorry he did not make a careful note of the figures before he came to the meeting, but if he were allowed to give the figures from memory he would correct them afterwards. The cost was given (page 421) as very favourable to the Diesel engine; but after his enquiries he found that in order to get a 20-B.H.P. engine he would have to wait four or five months—that was in November 1902—and that the cost would be £500 for the engine delivered on the bed. Another serious point was that when he came to enquire about the cost of the oil, he found that the costs given in the Paper, and elsewhere, referred to what was called " oil naked on the wharf." On enquiring from the Shell Transport Co., it was found that, in order to deliver it to his client's country-house, which was in Kent, the cost of the oil would be practically doubled; and in order to get it there at any cheap rate it would be necessary to have a special carriage constructed, with a sort of boiler-shaped tank, such as one frequently saw in the streets. The points he had mentioned, he thought, were worthy of attention if anyone was considering the desirability of putting down a Diesel engine.

Mr. GEORGE WILKINSON thought the author had brought forward his interesting Paper at a particularly opportune time, because he (Mr. Wilkinson) thought that in the near future very considerable developments would be seen in the application of internal-combustion engines for the production probably of electric power and other purposes. The Diesel engine, while it was well known on the Continent, was not sufficiently known in this country, and he congratulated the author on producing such a comprehensive Paper on the subject. He had shown very clearly in the illustrations and descriptive matter the character of the engine ; he had also furnished a number of tests, and had treated the theory of the subject very exhaustively. While the Paper covered much of the ground from a theoretical stand-point, and gave full information as to tests on various-sized engines, it did not contain any practical suggestions such as would be useful to a man having an engine in daily work. As he (the speaker) had put down the first engine in this country for the Corporation of Harrogate about 16 months ago, which was referred to on page 403, and had since had it under his supervision and control, besides being connected with the installation of other engines more recently, he might be permitted to make a few observations based on daily experience of the running of these engines.

It was some seven or eight years since an experimental Diesel engine of about 30-H.P. was built in Scotland and tested by Professor W. H. Watkinson of Glasgow, and it was found to have a thermal efficiency not greatly inferior to the present-day engines ; it was found, however, that certain mechanical difficulties prevented this engine from running more than, say, thirty hours without internal cleaning ; another difficulty and danger was due to premature admission and combustion of oil, causing extremely high pressures which, in some cases, burst off the cylinder cover ; this difficulty was also likely to occur owing to oil collecting in the cylinder and burning prematurely.

Another attempt to build the Diesel engine was made near Stockport only about three years ago, but the engines then produced (of which one was exhibited at the Glasgow Exhibition) failed owing to the builders attempting to make them give an impulse every revolution before they had built engines having the ordinary "Otto" cycle. These engines were also wrongly proportioned in their bearings, which caused the crank-pins to run hot before the engines had been at work many minutes. English experience and practice furnished him with no encouragement in adopting the Diesel engine, and it was only after a careful investigation and study of the engine in Germany that he felt justified in recommending the engine of German manufacture for driving the sewage pumping plant which he was engaged in putting down for the Corporation of Harrogate.

An experience of sixteen months, during the greater part of which time the engine had been at work fifteen or sixteen hours per day, had confirmed the wisdom of what was at the time considered a bold step to take. It would be seen (page 419) that the consumption of oil per B.H.P. per hour was 0·476 lb. in 1902, and 0·505 lb. twelve months later at the same B.H.P.

It was of the first importance that a clear exhaust be maintained, as a black exhaust meant a deposit of carbon in the passages and around the oil-spray nozzle, and the trouble rapidly became worse, finally resulting in the blocking of the oil-spraying passages, a good deal of local heat, and stoppage of the engine. The exhaust could readily be cleared by turning paraffin through the engine for a few minutes in place of the crude oil for the purpose of washing out the passages, and by altering the air-injection pressure to suit the particular crude oil one might be using. Another good plan, which he adopted for washing out the passages clean, was to work with paraffin instead of crude oil for the last ten minutes before closing down at night. To show the losses represented by a black exhaust, he might mention an 80-H.P. engine which ran with 0·46 lb. of oil per B.H.P. hour, which he tested on the brake when giving a black exhaust, the consumption per B.H.P. was then increased to 0·59 lb.

Care should be taken to obtain crude oil which contained a minimum of sulphur, otherwise trouble would be found from the burning of the exhaust valve, which, with a sulphurous oil, would be found to burn and pit in a curious way, generally at only one or two places, causing leakage on the seating; also the engine would require internal cleaning much oftener.   2 I 2

(Mr. George Wilkinson.)

Another important point was to ascertain that the various consignments of oil had the same viscosity, otherwise the spraying nozzle and the air injection would require re-adjustment to suit the variation in the oil. Of nine samples of oil supplied in bulk for various Diesel engines he had found only small variations in specific gravity ranging between $0 \cdot 889$ and $0 \cdot 936$, but there was wide variation in viscosity, and the engines had in each case required ome re-adjustment to suit each sample. It was sufficient to clean the valves every four or six weeks when using the thinner crude oils, but the thicker oils required heavier injection air-pressure, and the valves and passages required more frequent cleaning.

The engines were admirably governed, as explained on page 401, but, in his opinion, it was very desirable to have regulation on the injection air as well as upon the oil, and this could easily be obtained. At light loads the power cylinder received a small injection of oil but an excessive injection of air, which tended to cool the gases and prevent perfect combustion, besides which it imposed an excessive amount of work on the air-compressing pump. The engine as made at present was necessarily complicated, and in the valve gear there were several parts, to remove any one of which it was first necessary to remove several others.

It was important to clean the oil filter pipes and small oil pipe to the engine occasionally—say once in six months—or they became blocked up. He had known two engines brought to a stand-still without any apparent reason, and in each case it had been discovered that the small pipes were quite made up with sediment from the oil. In conclusion he must express regret that they were compelled to purchase these engines abroad, and at the present time there did not appear to be a firm in England who were sufficiently enterprising to take up the manufacture.

Mr. MICHAEL LONGRIDGE, Member of Council, said he ventured to intervene in the discussion because he had had an opportunity of testing a Diesel engine. The engine was called a 30-B.H.P. engine, and had a cylinder $11 \cdot 8$ inches diameter by $18 \cdot 12$ inches stroke, the nominal speed being 180 revolutions per minute. It was intended to make a series of progressive trials with loads varying from about 40 B.H.P. to the friction of the engine, but the intention was only partly fulfilled owing to want of time and to imperfect combustion during the first three trials.   This was due to no fault of the engine, but to ignorance on his part and on the part of the attendant.   He had never seen a Diesel engine of the type before, and the driver had only been in charge a few days, so that they failed to regulate the air-pressure in the receiver properly at first. The oil was crude Texas oil, costing 50s. per ton at the works, and having a calorific value determined in a Berthelot-Mahler calorimeter of 17,900 B.Th.U. per lb.   In all, seven trials were made, giving the figures shown on Table 10 (page 436).

The first three of these trials were unsatisfactory, because the smoky exhaust indicated imperfect combustion in the cylinder.   The sixth failed because, with no load on the brake, the governor hunted so much that the indicated horse-power could not be determined. The remaining three trials, Nos. IV, V, and VII, were satisfactory, and showed that the difference in the consumption per B.H.P. was lowest at the normal load (trial IV), increased slightly at 20 per cent. overload (trial V), and still more at half load (trial VII), but in neither case was the increase very serious. The increase at half load was due to the fall in the mechanical efficiency, and not to loss of thermal efficiency, indeed both the oil per I.H.P. per hour and the thermal efficiency were highest at half load.   It occurred to him whether it would not be possible in future, with the Diesel engine, so to increase the quantity of air that the thermal efficiency would be raised high enough to compensate for the loss of the mechanical efficiency at light load.   It might be possible—theoretically it was possible—to so increase the weight of air per lb. of oil as to do away with the water-jacket altogether.   If that were done, an enormous source of loss would be taken away; and it might be possible so to raise the thermal efficiency as to bring up the real commercial efficiency, somewhat to what it was at full load now, when the engine was running at half load.   If that could be done, the engine, even as it was now constructed, presented enormous advantages in some cases, because it could be run in small units and at small loads

(Mr. Michael Longridge.)

TABLE 10.

		I.	II.	III.	IV.	V.	VI.	VII.
No. of Trial								
Duration of trial	minutes	55	60	60	60	30	30	60
Revolutions per minute		164	173	171	173·6	171·2	178·8	177·7
Mean effective pressure on piston	lbs. per sq. in.	97·6	87·4	85·2	81·7	107·4	?	53·4
Indicated horse-power		40·2	37·95	36·5	36·5	46·1	?	23·75
Brake horse-power		32·65	28·75	28·35	28·6	36·3	?	14·76
Mechanical efficiency		0·813	0·76	0·777	0·805	0·788	—	0·622
Oil used per hour	lbs.	24·2	16·4	18·25	14·03	18·14	6·44	8·087
Oil used per I.H.P. per hour	lb.	0·602	0·432	0·500	0·386	0·394	—	0·366
Oil used per B.H.P. per hour	lb.	0·741	0·569	0·633	0·491	0·500	—	0·585
Thermal efficiency		0·238	0·331	0·283	0·361	0·364	—	0·391
Pressure of air in receiver	lbs. per sq. in.	675	675	700	830	810	675	645
Temperature of exhaust	F.	809°	678°	735°	611°	725°	—	397°
Jacket water per minute	lbs.	?	28·8	34·4	30·9	33·1	9·06	27·2
Initial temperature of water	F.	—	40°	48°	48°	48°	48°	48°
Final temperature of water	F.	—	107°	108°	102°	108°	101°	86°

almost as economically as at full load. It was also entirely self-contained; it required nothing except a small water-supply for cooling the jacket, and it could be put down in any place. Those, to his mind, were very great advantages.

He had carefully enquired into the working of the engine with regard to the difficulties which one might have expected with the high pressures in use, and from the failures which had occurred to those who had tried to make the engines in England; and he had found that, so far, there really were no difficulties at all worth speaking of. The difficulties of the varying viscosity of the oil and the cleaning of the exhaust valve were easily got over, and those were really the only two difficulties about the engine; and from his nine months' experience he most heartily recommended it to the members.

Mr. Henry McLaren said that a comparison was made (page 421) between the Diesel oil-engine, the gas-engine, and the steam-engine. He protested against $0 \cdot 89d.$ being taken as a fair figure for the steam-engine.

The President asked whether it was quite clear that the author and Mr. McLaren were both speaking about the cost of coal.

Mr. Ade Clark said it was total cost, not cost of fuel only.

Professor Arnold Lupton, although thinking the Paper was exceedingly interesting and instructive, was somewhat surprised, after waiting for three years, to find that the Diesel engine was only the oil-engine, whereas the principle of the Diesel engine, being the slow combustion of gas in the cylinder, was applicable to engines that were driven by means of a gas-producer. However excellent the oil-engine might be when it was applied to a moderate extent, it could never be greatly used as a substitution for the steam-engine, whereas the gas-engine might. The gas-engine might be used all over the world in substitution for the steam-engine, because gas could always be produced from coal; but if any great number of people started to put down oil-engines, the price of oil would be

(Professor Arnold Lupton.)

doubled and trebled before long. The production of coal in the world was something like 800 million tons per annum, and that of petroleum was only about 20 million tons, a comparatively trifling amount. In his opinion the real interest in the Diesel engine consisted in regarding it as a slow-combustion engine, and he should very much like to hear from the author if he had any experiments or information to give upon the Diesel engine as a gas-engine.

Professor JOHN GOODMAN said the author had very kindly referred to him in the Paper; but he felt that he (Professor Goodman) owed the author thanks for the very kind way in which he had brought the figures given in the Paper before him from time to time. They had thoroughly discussed them together, and in that way he felt that he had learned a great deal more about the Diesel engine than would have been possible from simply reading the Paper. He thought they must all look upon the Diesel engine as a great triumph for scientific men. The oil-engine and the gas-engine, as far as he was aware, had previously never given a thermal efficiency of over 20 per cent., but at one stride the Diesel engine is brought up to over 40 per cent. He believed he was right in saying that Herr Diesel in his very first attempt, made an engine based on purely theoretical reasoning, which gave an efficiency of approximately twice that which had ever been attained before. A firm not far from Leeds, makers of the old type of oil-engines, told one of their students a few weeks ago, that " There was no need to go to the Yorkshire College to study Theory; Molesworth would teach you all you want to know." It would be interesting to know why this firm did not find out from Molesworth how to design a Diesel engine, and how to get such high results as Diesel obtained.

He had prepared a diagram, Fig. 24, Plate 12, showing the effect of compression on the efficiency of an oil-engine. He believed Mr. Dugald Clerk first pointed out some 20 years ago, in a Paper before the Institution of Civil Engineers, the reason why a higher efficiency was obtained from a gas-engine by increasing the compression. In his treatment he assumed that the explosive

mixture was adiabatically compressed, heated by the combustion of
the gas, then expanded, the expansion curve being a hyperbola.
Although such a treatment showed distinctly that there was a
material gain by compressing the mixture before ignition, yet it did
not do justice to the value of high compressions. In an actual oil-
engine, however, both the compression and expansion curves much
more nearly approximated to a curve of the nature $PV^{1.3} =$ a
constant. Taking it as such, the increase of efficiency due to high
compressions was much greater than in the approximate hyperbolic
treatment—a comparison of the two was given in the figure. In all
cases it was assumed that the same weight of combustible was used,
and that there was a rise of temperature of 2,000° F. due to the
combustion of the charge, which fairly represented actual facts. It
was readily shown that the useful work done per impulse cycle was
given by the expression—

$$\frac{V_1\ (P_1 - P_3) - V_2\ (P_2 - P_a)}{1}$$

In which $V_1 =$ the clearance volume of the cylinder.

$V_2 =$ the total volume of the cylinder up to the end
of the stroke.

$P_1 =$ the pressure after the burning of the mixture,
sometimes termed the maximum or the
explosion pressure.

$P_2 =$ the release pressure.

$P_a =$ the pressure at the beginning of the compression
process.

$P_3 =$ the pressure at the end of the compression
process.

$N = 1.3$ in this case.

In the Diesel engine the compression was carried to about
35 atmospheres, and in the ordinary engine to about 5 or 6
atmospheres ; if the work done in the two cases were compared, one
reason would be apparent why the Diesel was so much more
economical, but at the same time from this diagram one would not

(Professor John Goodman.)

expect the Diesel to be twice as good as an ordinary oil-engine, since the compression effect only partially accounted for the excellent results obtained. After discussing the matter with the author they have formed the opinion that a much more complete combustion of the oil ensued at the high-compression pressures and temperatures than at those ordinarily reached in the oil-engine. Some of the members were aware that great strides were now being made by English engineers in producing economical oil-engines, not exactly on the Diesel lines, but by introducing water with the oil into the cylinder. It had the effect of reducing the temperature of the explosion. If the oil were raised to a very high temperature indeed, it was liable to become dissociated, that is, it was split up ; and, as far as was known at present, it did not combine again at the temperatures obtained in the ordinary oil-engine. Consequently some of the heat of the oil was not usefully developed in the cylinder, but by injecting water with the oil, as some English makers were now doing, the explosion temperature was reduced, thus keeping the oil always in the state in which it could be completely burned ; and moreover the expansion line of the indicator card was kept up by the use of water, that is, the value of N was still further increased, with a corresponding rise in the curve shown in the accompanying diagram. He thought it would be worth while for oil-engine makers to look carefully into this question. In conclusion, he wished to say that he thought English engineers might do a little more than they had done in the way of careful trials with oil-engines. If they would only go as carefully into every detail of the losses in their oil-engines as Mr. Clark had done in the case of the Diesel, they would then see why their engines were not giving as good results as the Diesel. Why should English makers be behind the times ? If Germany could produce an engine giving such an excellent result, English makers ought to see to it that their engines at least approached or even surpassed it.

Mr. MARK ROBINSON, Member of Council, said he wished to straighten a little misunderstanding which had arisen in connection with Mr. McLaren's remarks in regard to the comparison of the steam-engine with the Diesel engine and with the gas-engine. The Paper stated (page 421) that the coal consumption for the steam-engine had been taken at 4 lbs. and $3\frac{1}{2}$ lbs. per B.H.P. for the lowest and highest powers. He believed Mr. McLaren had not noticed those figures when he spoke, and supposed the consumptions given were still higher; but he wished to say, as a steam-engine builder, that Mr. McLaren's criticism was, in principle, perfectly well founded. The figures given were absurdly high for any good high-speed condensing steam-engine, and comparison on such a basis could only mislead. The Diesel engine came into the field as an engine in which economy ranked before everything, and therefore comparison should be made, not with bad steam-engines, or even with average steam-engines, but with engines of some admittedly economical type, such as a well-informed user would be likely to purchase at the present time.

Mr. Henry H. Suplee said he had had the opportunity of visiting Germany, and of examining a number of Diesel engines built by the Vereinigte Maschinenfabrik at Augsburg. He had had the privilege of the personal acquaintance of Herr Diesel at the time when he was first beginning his work, and he knew how very thoroughly Herr Diesel worked out the details theoretically, as Professor Goodman had stated, and how he had made a number of experiments bearing on some of the questions which had been raised. For instance, his first engine was made without any water-jacket, with the idea of utilizing the heat entirely in power, and of using only air to prevent the temperature rising beyond working limits · but he found it was not altogether satisfactory, and, although the engine ran and gave good results, the engines were now made and run with water-jackets.

Again, Herr Diesel had worked out not only a vertical, but a horizontal form of marine engine, which was being used in some canal boats. Engines of 50 H.P. to 100 H.P. had been made, and he believed that some of the engines were now being tried for marine purposes in connection with submarine boats instead of gas-engines. The works at Augsburg at present were overcrowded

(Mr. Henry H. Suplee.)

with orders. He saw on the floor of the works an engine of
500 H.P., which was to be sent to the power-station at Kieff, and
also engines down as small as 8 H.P. Every engine was thoroughly
tested with a great variety of oils. They had in the laboratory
between 150 and 200 varieties of oils from different parts of the
world, each of which had been tested in the engines, and also
tested for its calorific power, and the results recorded. The whole
engine had been worked out to such an extent as to make it
undoubtedly a scientific instrument. The business had been
eminently prosperous, so that there seemed to be some sound
commercial basis in the matter. In New York, about two years
ago, he had had the opportunity of examining a 20-H.P. engine,
and the combustion was undoubtedly perfect. The engine ran at
full power for two weeks; the cylinder was then taken off, and he put
a clean handkerchief in and wiped it, and could hardly discolour it.

Mr. ALFRED SAXON thought that, as engineers, they ought to
welcome all new forms of power, because in a country like England,
where natural water-power was not available, all other means of
conserving the coal-supply should be entertained as far as possible.
What engineers desired was to have a fair comparison. In
connection with the steam-engine, the price of coal was taken at
12s. 6d. per ton. He did not know whether that applied to London
and other towns remote from the coal-fields, but, in the manufacturing
districts of Lancashire and Yorkshire, coal for engine purposes could
be purchased at 8s. a ton delivered at the mills; and although that
was only one item, still perhaps that would have some little effect on
the final figure. Secondly, he wished to say that all new powers had
their separate spheres of application. Hardly any of them were
universal or would entirely displace other forms of power, but each
seemed to have its best sphere of application. Upon examining
Table 8 (page 425), it would be found that in the 160-B.H.P. engine
the percentage was very slight between the three forms of power. The
Diesel engine, which was probably shown to its best advantage, was
put down at $0\cdot32d.$, the gas-engine at $0\cdot40d.$, and the steam-engine at
$0\cdot49d.$ Those figures could be brought down still nearer, provided

the price of coal was reduced to what he would call the average amount for a manufacturing district. He had no doubt that the price for the crude petroleum had been put as low as possible.

Mr. A. J. LAWSON said he was in the happy position of being able to speak from practical experience of the use of such oil-engines on traction work. The author of the Paper had taken coal at 12s. 6d. per ton. He (the speaker) was paying exactly the same rate for coal in one station, where he was, in addition, using a Diesel engine. The coal cost worked out at $0 \cdot 637d$. per unit, and the Diesel engine, with oil costing £4 12s. 6d. per ton, instead of 45s.—because they had to take it in barrels, and pay the heavy freight charges to an out-of-the-way place—worked out at $0 \cdot 375d$. That was the practical result of ordinary working at a traction station. So satisfactorily did the engine work that he had put in two more, one of 160 H.P. to run a 100-kw. dynamo. There, taking the same consumption, and with oil costing a little over 45s. per ton, as it would in that district, the cost per unit would be about one-sixth of a penny. Taking England as a basis, and not the favoured Yorkshire and Lancashire districts, where coal was to be had at 8s. per ton, 12s. 6d. per ton was rather under than over the average price of coal. In the South of England he paid 18s. for the same quality coal delivered into the bunkers, and therefore the cost would be all the more in favour of the Diesel engine in that district. There was no particular skilled labour required for working the engine. He had employed a man who was a stoker a year ago; he was afterwards put on to running a steam-engine, and, having done very well at it, had now been put on to running the Diesel engine, and did it excellently. There was therefore no extra charge to be put against the Diesel engine on account of complication, or higher wages for the working.

He observed that the author referred to the stopping of the engine. He had seen such an incident once. At an exhibition of the Diesel engine—such things always occurred at exhibitions—they forgot to fill the fuel reservoir; the fuel gave out, and the engine came gradually to a standstill. With regard to the capital expenditure

(Mr. A. J. Lawson.)

for the 160-H.P. engine already mentioned, the total cost of land, building, engine, dynamo, storage-tanks, water cisterns, etc., worked out at £28 10s. per kw., erected in complete working order.

Mr. CHARLES DAY stated that Mr. Wilkinson and Professor Goodman had said in the course of the discussion that in England there had not been sufficient enterprise to take up the Diesel engine. The word "England" had been used, he took it, to mean, as was customary, Great Britain. If that was the case, those remarks were not quite true. The Diesel engine had been manufactured in Scotland and in England, but unfortunately the manufacture was taken up before the engine was sufficiently developed by its patentees. It was simply, then, a case of nursing somebody else's baby; such things did not pay, and both firms stopped manufacturing. However, the Mirrlees Watson Co., of Glasgow, of which firm he had the management, had again taken up the Diesel engine, and during the latter portion of last year completed a 35-H.P. engine. They had had that engine running now for some time driving a portion of their works. He thought that was the only engine running in Great Britain which had been manufactured in the country. The results, so far as they had as yet been determined, were just about equal to those mentioned in the Paper, and by various other speakers. The troubles in regard to the engine getting choked up with dirt occurred in the old engines, but had been quite got over; in fact, his firm were so satisfied that the troubles had been overcome, and the engine was now a commercial one, that they were preparing to build the engines on a large scale, and commence their sale.

Mr. F. HOWARD LIVENS wished to make a few remarks in regard to the difference in the cost, as shown by the author, between the 35-H.P. Diesel engine and an ordinary oil-engine of 35 H.P. The author stated (page 403) that in the ordinary oil-engine the consumption was nearly twice that of the Diesel engine. If one turned to page 424 and glanced down the first column, one would see that £585 was set down as the price of the Diesel engine. He ventured to say that

that was quite double the price of an ordinary oil-engine. On the other hand, if one turned to the figure for maintenance and depreciation, 10 per cent. on the price was put down, and 4 per cent. for the interest, so that the maintenance of the engine and interest on capital came altogether to about £82. If an engine costing half that money was bought, that figure would be divided by 2, which would allow £4 to be added to the fuel bill. The fuel for the Diesel engine was put down at £47, £41 added to this would be £88, or nearly double, which would practically repay the extra cost of the oil which the ordinary oil-engine might use. He was referring particularly to engines used for small powers, which came into the hands, as he knew by experience, of men who were not engineers and very often unintelligent, and did not know how to manage a high-class machine, and he wished to emphasise that for such powers the ordinary oil-engine was not so difficult to manage. It did not require so much care as the Diesel engine, because the maximum pressure in the Diesel engine was something like 750 lbs. per square inch, whereas in the oil-engine it was not more than about 200 lbs. So far as small powers were concerned, he did not see that the Diesel engine offered the advantages which, at first sight, it might appear to have.

In the usual way a user of an engine had to work with any oil of the required kind that came to hand, and it was difficult to get him to adjust his engine for the specific gravity; it would be still more difficult for him to have to take into account the varying viscosity of the oil, as Mr. Wilkinson had pointed out as necessary.

Professor W. H. WATKINSON said that Mr. Wilkinson had stated that in the engine which he tested in Glasgow, made by Messrs. Mirrless, Watson and Co. some years ago, difficulty was found with premature ignition and blowing off of the cylinder cover. Those difficulties were not encountered in Glasgow, but in Germany. At that time the engine was quite as defective, as constructed in Germany, as the one constructed in Glasgow, and the German engines did not work any better than the Glasgow ones.

Mr. H. ADE CLARK, in reply, said that several of the points which had been raised had been already answered by subsequent speakers. His price of 12s. 6d. a ton for coal had been rather severely criticised. It was the average cost taken from the returns of the electric lighting stations of this country for the last year. With regard to the figures of 4 and $3\frac{1}{2}$ lbs. of fuel per B.H.P. hour which had been taken for the steam-engine, he did not dispute them, although the best boiler and the best engines and the best workmen could get very much less, but on the average they did not. He took the fuel oil consumption figure purposely at about $\frac{3}{4}$ load for the Diesel oil-engine, so as not to give it its best consumption. He thought that for the sizes of engine with which he had compared it the figures were fairly correct, and agreed that better figures could be attained.

The question had been raised as to the Diesel engine being a slow combustion gas-engine. It had been run with gas, but it had been developed as an oil-engine, and therefore he treated it from that point of view. It was a commercial engine, and he treated it from a commercial standpoint, not solely from the scientific. As would be seen from the Paper, he had devoted most attention to practical points. With regard to the criticism that, compared with an ordinary oil-engine, the cost of the Diesel was higher, naturally a great deal of money had been spent in developing it, but the price would come down. The cost of the fuel was low, whereas in the ordinary oil-engine it would cost at least three times what it cost in the Diesel engine, when first cost of oil and quantity used were considered.

Communications.

Mr. JOHN HARDISTY wrote that the author of the Paper appeared to consider that the Table (page 421) conclusively proved that the Diesel engine was a cheaper source of power than the Crossley gas-engine using Dowson gas, and that this in turn was cheaper than a high-speed compound condensing engine. In fact, it was distinctly stated that the result "is much in favour of the Diesel engine." Now, he was of opinion that nothing of the kind had been proved in the Paper. It was possible to run a gas-engine with fuel much cheaper than 24s. per ton ; in fact, the Power Gas Corporation was running gas-engines with fuel at 6s. per ton, and no more of it was used than was stated by the author, namely, $1\frac{1}{4}$ to $1\frac{1}{2}$ lbs. per B.H.P. per hour to be the consumption of anthracite cobbles in the Dowson gas-producer. The locality of the gas-engine had a considerable influence on the cost of fuel, stated in the Paper to be 24s. per ton ; but the same engine worked in Swansea could have anthracite coal at 17s. per ton, which in Chesterfield would cost 27s. per ton. The substitution of these figures for the 24s. named in the Paper would make a material difference in the position of the gas-engine as compared with the Diesel engine using crude petroleum at 45s. per ton.

Illuminating gas used for power cost 2s. 6d. at Chesterfield, 5s. at Baslow (8 miles away) and 5s. 10d. at Hayfield (25 miles away) per 1,000 feet consumed. Therefore, while it might be desirable to put down a gas-engine at Chesterfield, it might be the wrong thing to do at Baslow and Hayfield. And to take a mean of these three prices, namely 4s. 5d. per 1,000 feet, as the cost of gas in Derbyshire, would be misleading for all the places. But this was exactly the way in which the author had treated gas and steam-engines in his Paper.

It was to be presumed, in the absence of information to the contrary, that 45s. was the price of oil at Harrogate, but Mr. Dolby pointed out in the discussion (page 431) that the same oil delivered in Kent was double the price "naked on wharf"; therefore if the price of oil delivered in Kent had been used in calculating the cost of power as produced by the Diesel engine, and the price of anthracite at Swansea taken for the gas-engine, the gas-engine would have been shown to be a more economical power producer than the Diesel engine. Similarly, the coal consumption for the steam-engine was given as $3\frac{1}{2}$ to 4 lbs. per B.H.P. per hour, assuming 8 lbs. of water to be evaporated per pound of coal. Did this mean

2 K

(Mr. John Hardisty.)

8 lbs. from and at 212° F. or from the temperature of feed, and at the boiler pressure, which was a materially different thing? but whichever was meant, the water consumption equalled 28 to 32 lbs. per B.H.P. per hour, which was about double the consumption of the best modern engines of the type described. If this correction were made in the calculation for cost of power for steam-engines, the result would be very different to what was stated in the Paper.

The cost of coal was given as 12s. 6d. per ton, and while coal might cost more than this in some places, it was certainly much cheaper in many other localities. For instance, at the Southgate pit near Clowne, Shireoaks slack could be got for 5s. 2½d. per ton, and Shireoaks unscreened coal for 7s. 2½d. per ton, while the prices delivered at Chesterfield, about 10 miles away, were 6s. 9d. and 8s. 9d. per ton respectively. Therefore if for the author's price of coal and consumption per B.H.P. one substituted Shireoaks coal or slack, and assumed a first-class modern engine of the type described, working say at Chesterfield, the cost of power would be materially less than for the Diesel engine working at Harrogate.

He thought the author had erred in taking averages of gas and steam-engines and prices of coal to compare with an oil-engine favourably situated. It was, in fact, a mistake to deal with averages in this manner at all. Anyone desiring to get his power at the cheapest possible rate must first decide where he wanted the power produced and applied; he could then consider :—

(I) The proper type of steam-engine, its steam consumption, and cost of coal delivered into his power-house.

(II) The proper type of gas-engine, its consumption, the various processes of gas manufacture, and cost of coal delivered into his power-house.

(III) Electric power from Corporation generating stations.

(IV) And finally, the proper type of oil-engine and the cost of oil delivered into his power-house.

After this he would be in a position to select the kind of engine suitable to the locality, and he ventured to say that for some localities the steam-engine would be more economical than either gas or oil-engines. For other localities the gas-engine would be better than either steam or oil-engines, and there might be places where the oil-engine would be preferable to gas or steam-engines; but every locality and amount of power required must be considered on their merits. It was quite impossible to accept the doctrine laid down that a Diesel engine was more economical than a Crossley gas-engine using Dowson gas, and that this in turn was better than a high-speed compound condensing steam-engine in all situations.

Mr. LEWIS A. SMART wrote that he had inspected Diesel engines at work, and he wished to be assured by the author that for wide variations of load he did not find it necessary for the attendant to alter the air-supply. He was informed by the makers that such was not the case, but those engines he had seen at work emitted black sooty smoke at the exhaust pipe when the load varied to any considerable extent.

In saw-mill engines the load might vary at any minute 50 per cent., due to, say, heavy frame saws and a 5-foot circular saw ceasing work simultaneously.

In marine engines it might be necessary to have clutch reversing gear, and here again the load might vary considerably. The question of the air-supply being entirely automatic appeared to the writer, therefore, to be of great importance. He was quite aware that the makers of the Diesel engine were experimenting with a four-cylinder engine for marine propulsion which would not require a clutch reversing gear. He had seen evidence in these engines that at light loads there was a difficulty in maintaining the high-pressure air-supply, due, he believed, to the lower pressure in the engine cylinder allowing the high-pressure air to flow more freely. He would like to know if this had been overcome?

As regards a point raised in the discussion by Mr. Lupton, namely, that extended use of these engines would send up the price of crude oil, he had approached a firm supplying this oil and found that they were prepared to contract at the prices quoted in the Paper for five years, and to give a guarantee that in the next twenty years the price would not exceed 50s. per ton f.o.r. at their stores in this country.

2 K 2

Mr. H. ADE CLARK wrote, in further reply to the discussion, that he would first deal with the criticisms of the engines. Mr. Wilkinson (page 432) had referred to the early troubles of pre-ignition and incomplete combustion during the experimental period. These troubles had been entirely overcome by the oil-injection valve, which, when properly set, made pre-ignition impossible, because there was nothing to ignite until the correct instant; if all valves were tight, so that the proper compression was reached, the exhaust would remain quite clear from full to a quarter load, and at very light loads or no load there would be a slight blue smoke. If the engine were started quite cold to run on no load, there would be more smoke for a longer period than if the engine was immediately put under load until hot and then run at no load.

Mr. Wilkinson's reference to the use of paraffin for the last few minutes of the day's run for the purpose of "washing out the passages," must be understood to refer to the pipes from the oil-tank to pump and from pump to spray, and that the precaution was only necessary when using the thickest and most viscous of crude oil, or when one wished to clear out accidental deposits of sediment.

Mr. Livens (page 445) seemed to think that the specific gravity of the oil demanded some adjustment to the engine; that was not the case for any ordinary oil ranging between "Royal Daylight" and "Texas Liquid Fuel," or the range from $0·8$ to $0·95$ specific gravity. The difficulty of the viscosity of the oil varying in different supplies reduced itself to the matter of there being ample passage provided for the most viscous samples. Usually the adjustment of air-pressure would meet any such variation as would occur between one batch of oil and another. In very exposed places where the temperature variation of the air was considerable, the oil might be kept sufficiently fluid by passing the whole or part of the hot jacket-water through a coil of pipe inside the oil-drum.

Mr. Smart (page 449) referred to the question of adjusting the air-blast pressure to suit different loads. For ordinary temporary changes there was no necessity for adjustment; for the most economical working at any given load there was no doubt a pressure of injecting air most suitable, and the author's experience indicated that the blast-pressure suitable for full-load working was not suitable for continuously running at a very light load. The adjustment, however, was merely that of screwing down a valve until the pressure was maintained steady.

Mr. Michael Longridge (page 435) suggested the possibility of using larger quantities of air with the object of getting higher hermal efficiency and eliminating the water-jacket. This was Herr Diesel's original intention, and had been actually done ; but the engine was far too bulky for the power obtained, and the mechanical efficiency fell off. If the water-jacket was eliminated, the ratio of expansion must be very much increased and the period of combustion decreased ; otherwise there would be more exhaust gas at a higher temperature than now obtained, throwing away the heat, or much of it that was now carried away in the jacket-water. The water-jacket was not such an unmitigated source of loss as was sometimes thought, for when all facts were fully recognised, one was bound to admit that the transformation of a quantity of heat to work involved a penalty, the amount of which depended upon the temperature range as evidenced by the Carnot efficiency formula—

$$\frac{T_2 - T_1}{T_2}$$

The possible ways in which the Diesel type of engine could have its efficiency increased were :—

(I) Increase in mechanical efficiency ;

(II) Increased range of temperature in the cycle by using higher pressures and greater expansion ratios, but this must not be accompanied by a corresponding fall in mechanical efficiency ;

(III) The more perfect attainment of slow or rather gradual combustion of the fuel. In the direction of isothermal combustion he thought undoubtedly lay the greatest possibility of increasing the efficiency of internal-combustion engines.

The comparison of costs, and particularly the fuel quantities and prices, had called forth a certain amount of criticism. The author had hoped that the speakers would have asked for and given the

(Mr. H. Ade Clark.)

definite statements of costs in the case of the gas- and steam-engines from both makers and users of them, culled from actual experience and accompanied by all the data and conditions pertinent to the particular cases. He was fully aware of the possible fallacy of comparisons drawn from averages where care was not exercised in seeing that the averages were drawn from like conditions. He was careful in this matter to give the price of fuel as "in all cases delivered," and where the author's prices did not hold, the comparison must be altered to suit. On the question of costs, and fuel, &c., Mr. E. R. Dolby (page 431) said that the oil prices in the Paper were those for oil "naked on wharf"; that was not the case however. The oil price quoted was 10s. a ton higher than the price of the oil in bulk at the wharf at the time of writing. The price per ton delivered into railway tank wagons at London, Barrow, and Birkenhead, was 37s. 6d. London and Barrow charged 1s. 6d. per ton for the wagons; at Birkenhead this charge was included. Four-ton lots were supplied in road tank wagons at a charge of 2s. 6d. a ton in London. Railway carriage from Birkenhead to Manchester would be about 5s. 6d. per ton, to Leeds or Birmingham about 8s. per ton. To take the oil in barrels was expensive, the charge being 25s. per ton, the ton requiring six barrels, each of which would be worth about 3s.; the railway rates on the barrels of oil was, however, higher than for tank wagons. As the cost of a road tank wagon to carry 4 tons would be about £10 to £15, it would certainly pay the consumer to invest in a wagon rather than have the oil in barrels. The 20-B.H.P. engine fixed ready for work in England would cost £180.

Mr. Henry McLaren (page 437) and Mr. Mark Robinson (page 441) both condemned the figure he gave for coal as compared with that for oil. He would, however, meet their criticism by taking what must be considered very good experimental results instead of probable practical results, namely:—1·5 lbs. of best Welsh coal per I.H.P. hour, 90 per cent. mechanical efficiency. Therefore coal per B.H.P. hour $= 1·5 \div 0·9 = 1·66$ lbs. Such coal would cost on the average in England 20s. per ton. The annual coal bill for the 160 B.H.P. engine working 54 hours per week with a 2-hour per day for stand-by and getting ready, loss, and working 50 weeks per year would be

$$\frac{1 \cdot 66 \times 3300 \times 160}{2240} \times 1 = £393$$

against £440 as given in Table 8 (page 425). Having taken an excellent result for the coal in the steam-engine, he would do the same for the oil-fuel in the Diesel engine, namely :—0·408 lb. of oil per B.H.P. hour at £2·5 per ton ; the fuel cost was then for 50 weeks at 54 hours per week

$$\frac{0 \cdot 408 \times 2700 \times 160}{2240} \times 2 \cdot 5 = £198,$$

against £186 as given in Table 8 (page 425). These changes would alter the final total costs per B.H.P. hour to

160 B.H.P. Diesel Engine = 0·326 penny.
   ,,   ,,   Steam   ,,   = 0·464 penny.

The figures produced by Mr. A. J. Lawson (page 443) more than supported the average figures given in the Paper ; moreover, costs of fuel were exceptionally high on account of the locality where the plant was situated.

Mr. Livens (page 445) drew a comparison of first cost and upkeep in the cases of an ordinary oil-engine and a Diesel oil-engine. The author thought that costs for fuel would be more like the following than those given by Mr. Livens, because the ordinary type would use quite 1·5 as much oil, and that oil would cost twice as much per ton as the crude oil :—

Diesel engine, 35 B.H.P., using crude oil, say, £50 a year.
Ordinary engine, 35 B.H.P., using Russoline, £150 a year.

Again, comparing speed with speed, and power with power, the Diesel would be the smaller engine. The ordinary type would require more internal cleaning on account of incomplete combustion. The Diesel engine, in the author's opinion, demanded more careful attention, because the results of mismanagement would be more serious in the case of the more perfect and more expensive engine.

(Mr. H. Ade Clark.)

Mr. Hardisty (page 446) commented on what he termed the general conclusion that the Diesel engine was the cheapest form of power under all conditions. There was no need to examine his figures, as there was not enough data as to conditions. Further, the author disclaimed such general conclusion being in general agreement with Mr. Hardisty's line of argument on averages. When evaporative power of fuel was stated without qualification, it could only mean under standard conditions, namely from and at 212° F.

That any steam-engine of the size and class used in these comparisons was producing a B.H.P. for 15 lbs. of steam per hour in practice was to be seriously questioned, unless very highly superheated steam was used and coupled with the very best types of auxiliary apparatus; this would materially increase first cost and upkeep.

TABLE 11.

	Longridge	Clark	Longridge	Clark
I.H.P.	23·75	31·45	46·1	46·9
B.H.P.	14·76	16·29	36·3	33·1
Mechanical Efficiency . .	62·2	51·8	78·8	70·5
Oil per I.H.P. hour . lbs.	0·366	0·302	0·394	0·336
„ „ B.H.P. „ . ..	0·585	0·583	0·500	0·476
Calorific Value of Oil . .	17,900	18,800	17,900	18,800

Mr. Michael Longridge (page 435) gave the result of a set of trials on an engine having the same dimensions as the one at Harrogate. The figures were interesting; for instance, trials II and IV showed the effects of working with and without a smoky exhaust, 9 per cent. extra fuel being used in trial II than in IV. If a comparison of similar trials by Mr. Longridge and the author were instituted as in Table 11, it would be seen that Mr. Longridge obtained much lower indicated horse-powers than those by the author, and consequently higher mechanical efficiencies. The oil-consumption per B.H.P.-hour was alike at half load in the two, but at the higher load the author's result was about 5 per cent. lower than Mr. Longridge's; on the other hand, Mr. Longridge gave the calorific value of his oil 5 per cent. lower than that mentioned by the author.

In conclusion, the author would like to express his thanks for the way in which his Paper had been received.

NOTES ON HIGH-SPEED TOOL STEELS.

By Mr. HENRY H. SUPLEE, of New York.

The following notes represent officially verified data as to the use of High-Speed Tool Steels in the works of the Union Pacific Railroad at Omaha, Nebraska, and as such are offered as a brief contribution to the subject. As is now well known, these steels are similar in constitution to the Mushet air-hardening steel, the principal difference being that a much higher temperature is used in the tempering process. The steels contain both chromium and tungsten in varying proportions as well as molybdenum. The method of treatment consists in heating the tool up to about 2,000° F., then cooling rapidly down to about 1,700° F. in a lead bath, and then slowly in air, lime, or oil.

These steels, of which the Taylor-White is the best known and earliest example, are able to maintain a cutting edge even when operated at speeds producing a red heat, and in fact unless such speeds and temperatures are maintained they do not give satisfactory results. These tools should be used only for roughing purposes, and the great economy resulting from their use appears when it is found that the forgings can be made with less care as to size, the roughing down to finishing dimensions being made more rapidly and economically in the machining processes than in forging.

In the samples of chips shown on Plate 13, the small chips, Fig. 1, were turned from car-wheel tyres at lineal speeds of 5 to 8 feet per minute, the weight of metal removed being about 8 lbs. per hour; this was with ordinary tool steel. The spiral turning, Fig. 2, from a locomotive tyre, was made with high-speed steel at a speed of 24 feet per minute, removing 100 to 120 lbs. per hour; while the heavy chip, Fig. 3, was taken at 18 feet per minute, removing 450 lbs. per hour. This latter cut was too heavy for the powering of the lathe, however, and the rate could be maintained for only a short time, but the tool showed no signs of distress.

The accompanying Tables (pages 459–466) give further interesting data from the Union Pacific shops at Omaha, for which the author is indebted to Mr. R. Emerson, Secretary of the Union Pacific Railroad Board of Tests, and the photographs, Figs. 4 and 5, Plate 13, show the wheel lathe and the planer referred to in some of the Tables.

The Paper is illustrated by Plate 13, and is accompanied by an Appendix.

APPENDIX.

TABLE 1.

DATA OF NOVO TOOL STEEL WORK.

Machine tool	Lathe.
Maker	Pond.
Model	32 inches.
Spindle	20 revolutions per minute.
Circumference of cone pulleys . . .	$23\frac{5}{8}$, $34\frac{1}{4}$, $45\frac{1}{4}$, $55\frac{3}{4}$, 67 inches.
Width of belt	4 inches.
Thickness of belt	$\frac{5}{16}$ inch.
Speed of belt, feet per minute . . .	735 (smallest pulley used in test).
Condition of belt	
Material machined	Soft cast-iron.
Form machined	Piston-valve bushing, outside for compound cylinder.
Speed of cut	74 feet per minute.
Depth of cut	$\frac{1}{2}$ inch.
Feed	$\frac{3}{32}$ inch.

Hardening, etc., of tool : Air Novo steel, forged at high lemon colour, cooled slowly in air, re-heated to white heat, till nose of tool begins to run, then cooled in steady air blast or in oil, the latter giving better results.

Remarks—Authentic records in shops of Union Pacific Railroad, at Omaha, Nebraska, United States.

R. EMERSON,

Secretary of the Board of Tests and Methods.

TABLE 2.

DATA OF NOVO TOOL STEEL WORK.

Machine tool . . . . Lathe.

Maker . . . . Pond.

Model . . . . 32 inches.

Spindle . . . . 18 revolutions per minute.

Circumference of cone pulleys . . . $23\frac{5}{8}$, $34\frac{1}{4}$, $45\frac{1}{4}$, $55\frac{3}{4}$, 67 inches.

Width of belt . . . . . . 4 inches.

Thickness of belt . . . . . $\frac{5}{16}$ inch.

Speed of belt, feet per minute . . . 661 (smallest pulley used in test).

Condition of belt . . . . .

Material machined . . . . . No. 1 scrap iron.

Form machined . . . . . Piston-rod, 4 inches.

Speed of cut . . . . . . 18 feet per minute.

Depth of cut . . . . . . $\frac{3}{4}$ inch.

Feed . . . . $\frac{1}{16}$ inch.

Hardening, etc., of tool: Air Novo steel, forged at high lemon colour, cooled slowly in air, re-heated to white heat, till nose of tool begins to run, then cooled in steady air blast or in oil, the latter giving better results.

Remarks.—Authentic record in shops of Union Pacific Railroad, at Omaha, Nebraska, United States.

R. EMERSON,

Secretary of the Board of Tests and Methods.

TABLE 3.

DATA OF NOVO TOOL STEEL WORK.

Machine tool	Lathe.
Maker	Pond.
Model	$27\frac{1}{2}$ inches.
Spindle	
Circumference of cone pulleys . . .	$20\frac{1}{2}$, $30\frac{1}{4}$, 39, 48, $56\frac{3}{4}$ inches.
Width of belt	$3\frac{1}{2}$ inches.
Thickness of belt	$\frac{5}{16}$ inch.
Speed of belt, feet per minute . . .	580 (smallest pulley used in test).
Condition of belt	New.
Material machined	No. 1 scrap iron.
Form machined	Crank-pin.
Speed of cut	26 feet per minute.
Depth of cut	$\frac{1}{2}$ inch.
Feed	$\frac{1}{8}$ inch.

Hardening, etc., of tool: Air Novo steel, forged at high lemon colour, cooled slowly in air, re-heated to white heat, till nose of tool begins to run, then cooled in steady air blast or in oil, the latter giving better results.

Remarks.—Authentic record in shops of Union Pacific Railroad, at Omaha, Nebraska, United States.

R. EMERSON,

Secretary of the Board of Tests and Methods.

TABLE 4.

DATA OF AIR NOVO TOOL STEEL WORK.

Machine . . . . Vertical boring mill.

Maker . . . . Niles.

Model . . . .

Spindle . . . .

Circumference of cone pulleys . . . 25, 34¾, 44½, 54¼, 64, 73¾ inches.

Width of belt . . . . . . 4 inches.

Thickness of belt . . . . . $\frac{5}{16}$ inch.

Speed of belt, feet per minute . . . 815 (smallest pulley used in test).

Condition of belt . . . . . Old overhead belt.

Material machined . . . . . Tyre-steel.

Form machined . . . . . Locomotive driver tyre.

Speed of cut . . . . . . 40 feet per minute.

Depth of cut . . . . . . ⅛ inch. Roughing cut.

Feed . . . . ⅛ inch.   ,,   ,,

[Hardening, etc, of tool: forged at high lemon colour, cooled slowly in air, re-heated to white, almost running heat, then cooled in steady air blast or in oil, the latter giving better results.

Remarks.—Belts would not stand faster speed. Roughing cut and finishing cut taken at same time, tools on two boring heads Authentic record in shops of Union Pacific Railroad, at Omaha, Nebraska, United States.

R. EMERSON,

Secretary of the Board of Tests and Methods.

TABLE 5.

DATA OF AIR NOVO TOOL STEEL WORK.

Machine . . . . Vertical boring mill.

Maker . . . . Bullard.

Model . . . .

Spindle . . . . . .

Circumference of cone pulleys . . . $27\frac{1}{2}$, $34\frac{3}{4}$, 42, $49\frac{1}{2}$, $56\frac{1}{2}$ inches.

Width of belt . . . . . . $2\frac{1}{2}$ inches.

Thickness of belt . . . . . $\frac{5}{16}$ inch.

Speed of belt, feet per minute . . .

Condition of belt . . . . . Good.

Material machined . . . . . Cast-iron.

Form machined . . . . . Piston-head.

Speed of cut . . . . . . 20 feet per minute.

Depth of cut . . . . . . $1\frac{3}{2}$ inch.

Feed . . . . $\frac{1}{8}$ inch.

Hardening, etc., of tool: forged at high lemon colour, cooled slowly in air, re-heated to white, almost running heat, then cooled in steady air blast or in oil, the latter giving better results.

Remarks.—Above belting on machine. Belt to counter-shaft 4 inches by $\frac{3}{16}$ inch. This machine was built for electric drive. The variable speed obtained by cone pulleys as above. Instead of motor, power is obtained from counter-shaft. Authentic record in shops of Union Pacific Railroad at Omaha, Nebraska, United States.

R. EMERSON,

Secretary of the Board of Tests and Methods.

TABLE 6.

DATA OF AIR NOVO TOOL STEEL WORK.

Machine tool	Horizontal cylinder boring mill.
Maker	Bement Miles.
Model	
Spindle	
Electric drive, motor	Westinghouse direct.
Horse-Power	10
Volts	230
Ampères, highest speed	38
Motor	650–900 revolutions per minute.
Material machined	Low-pressure 19-inch cylinder.
Form machined	Cast-iron (very hard).
Speed of cut	18 feet per minute.
Depth of cut	$\frac{3}{8}$ inch.
Feed	$\frac{1}{8}$ inch.
Weight material removed per hour . .	

Hardening, etc., of tool : forged at high lemon colour, cooled slowly in air, re-heated to white, almost running heat, then cooled in steady air blast or in oil, the latter giving better results.

Remarks.—Four tools used on boring head. This test was up to limit of motor, as controller rheostat gave signs of burning out. Cutting tools remained in perfect shape. Authentic record in shops of Union Pacific Railroad at Omaha, Nebraska, United States.

R. EMERSON,

Secretary of the Board of Tests and Methods.

TABLE 7.

DATA OF AIR NOVO TOOL STEEL WORK.

Machine tool	Driver lathe.
Maker	Pond Machine Tool Co.
Model	88 inches.
Spindle	$1\frac{1}{2}$ revolutions per minute.
Electric drive, motor	Westinghouse direct.
Horse-Power	15
Volts	230
Ampères, highest speed	56
Motor	650-925 revolutions per minute.
Material machined	Driver tyre hardened by sliding on sand.
Form machined	
Speed of cut	24 feet per minute.
Depth of cut	$\frac{3}{8}$ inch.
Feed	$\frac{3}{32}$ inch.
Weight material removed per hour . .	90 lbs.

Hardening, etc., of tool: Forged at high lemon colour, cooled slowly in air, re-heated to white, almost running heat, then cooled in steady air blast or in oil, the latter giving better results.

Remarks.—Limit of machine. Authentic record in shops of Union Pacific Railroad, at Omaha, Nebraska, United States.

R. EMERSON,

Secretary of the Board of Tests and Methods.

TABLE 8.

DATA OF AIR NOVO TOOL STEEL WORK.

Machine tool	Planer.
Maker	Pond.
Model	30 feet table.
Spindle	
Electric drive, motor	Westinghouse direct.
Horse-Power	20
Volts	230
Ampères highest speed	74
Motor	875 revolutions per minute.
Material machined	No. 1 scrap iron.
Form machined	Connecting or side rod.
Speed of cut	15 feet per minute.
Depth of cut	$\frac{9}{16}$ inch.
Feed	$\frac{1}{4}$ inch.
Weight of material removed per hour .	247–458 lbs.

Hardening, etc., of tool: forged at high lemon colour, cooled slowly in air, re-heated to white, almost running heat, then cooled in steady air blast or in oil, the latter giving better results.

Remarks.—Authentic record in shops of Union Pacific Railroad at Omaha, Nebraska, United States.

R. EMERSON,

Secretary of the Board of Tests and Methods.

Discussion.

Mr. Suplee stated that the Tables and the other matter concerning them simply represented the results obtained at the Union Pacific Railroad shops, and he had no means of verifying them ; he knew little if any more about the Tables which had been sent to him by the testing manager of the works than the members did. With regard to the use of high-speed steels, Messrs. J. A. Maffei, in their locomotive works at Munich, had made some interesting experiments with them ; but they found, as many shops in the United States had found, that the best results could not be obtained with the present tools ; in other words, the power of the machines was not sufficient to stand the heavy cuts which the tools were capable of resisting. At the works of Messrs. Escher, Wiess and Co. at Zürich they had had a small experience ; but Mr. Naville, the engineer there, informed him that he had made some interesting experiments with the ordinary Mushet steel treated in the same manner as the new steel, and that he had obtained very good results by taking the ordinary Mushet steel and tempering it at the higher temperatures mentioned in the Paper. Some complete reports had been published of the tests made, particularly on the Taylor-White steel, by the Franklin Institute, that steel having been entered as a candidate for award. It was tried at various speeds, and a most thorough and exhaustive report was given, which could be obtained from the Secretary of that Institute, Philadelphia. At the present time he believed comparative tests were being made with a view to a second report on the Taylor-White steel, which report would appear in the Journal of the Institute when it was completed.

The principal point, on which a misunderstanding had arisen in connection with the steels, had been the attempt to use them for finishing tools. They were not suited for that purpose. They were rather brittle and would not stand a light cut, because they wanted the heat of the high speed to attain their proper toughness. As a matter of fact the tools did not do well unless they were used on a heavy cut at a high speed and practically at a temperature which

(Mr. Suplee.)

should cause the chips to come off a blue colour. At the Bethlehem Steel Works, where those steels were first used, the test adopted was to observe whether blue chips were taken off. If the chips came off blue, the workman was running his tools at the proper capacity.

The PRESIDENT was sure it would be the wish of the members that a hearty vote of thanks should be conveyed to Mr. Suplee for his interesting Paper. The author had filled a distinct gap by bringing forward a Paper on such a subject, and in such a modest and unassuming way. The subject itself was a very great one; it was the greatest revolution that had happened in his own particular trade in his own lifetime, and it was not yet half understood nor a quarter developed.

He had the pleasure of seeing among the audience an old friend of his, Mr. Joseph Barrow, who of all men knew what could be done with the old steels; and he thought it was extremely useful, before anything new was adopted, to know exactly what could be done with that which went before. Mr. Barrow with great perseverance so perfected the action of the best steel he could get in his time—long before anything was known of the Novo steel or the steels that were hardened at a very high temperature—and so developed the supporting of the tools, the proper speeding of them, and the proper lubricating of them, that he was able to remove a thicker shaving from a parting groove than had ever been removed before, and was able to form a large square thread at one passing up of the dies in an automatic screwing machine. With that slight introduction, he would call upon Mr. Barrow to give his experience upon the new development in steel.

Mr. JOSEPH BARROW, after thanking the President for the kind manner in which he had introduced him to the meeting, said that in the spring of this year a well-known engineer, very largely interested in the manufacture of engines in America, who had just been through one of the large works in Glasgow and had seen some of the large cuttings taken there at moderate speeds, visited him and said he had been greatly impressed. The Bethlehem Company, who had taken up the subject of high-speed cutting more enthusiastically than anybody, had, he thought, to use an American phrase, converted the whole of their machinery into an "unsaleable wreck." He wished to say first that the use of high-speed cutting steel was of great value. Where one could satisfactorily increase the speed from 18 or 20 feet up to 60 feet a minute that was undoubtedly a very great advantage—such as, for instance, in dealing with rolled bars and converting them into finished shafting and screws, also using high speeds for the cutting of annular grooves in marine plates, for tube stays and things of that kind. But this advantage must be taken in a very limited sense, and not in the enthusiastic manner in which the President spoke of it, as a great revolution. No doubt they would "muddle through" somehow, and would ultimately come to the decision that, wherever a small amount of material had to be removed, high-speed cutting could be used with advantage, but where a great amount of material had to be removed it was necessary to go back to the low-speed cutting. He personally thought that theory could best be deduced from fact; it was easy to understand that a 12-inch length of cut of material done at 60 feet a minute took three times the power necessary than if it were done at 20 feet a minute. If it were done at 60 feet a minute and there was 20 feet of belt used for each foot of cut, then at 20 feet a minute 60 feet of belt would be used to remove each foot of cut and power to take very heavy sections of cut obtained. One could not take more out of a tool than was put into it, and therefore one must arrive at the fact that the power of taking the cut was the power put in to produce it.

Another thing which struck him was the total ignorance of such people in regard to shavings. If one used a round-nosed tool it cut a crescent shape, which when impinged upon the flat of the tool became flat and was converted into a shape with serrated edges. It required a very considerable amount of power to do this, because if one took a rolled bar of crescent section, and by pressure converted it to a flat section, he would not only require power, but would get what he called the "mark of the beast," the splitting into serrated edges of the cut, just as the lathe did. In the use of round steel in a tool-holder, the surface that was taken off was really a very small

(Mr. Joseph Barrow.)

portion of the circle, and was nearly a flat surface; in light work these tool-holders were good, but in heavy cutting were, in his opinion, practically useless. The leading principle therefore in the shape of tool was to make it flat and not round. He said therefore that, subject to the above facts, a great advantage could be obtained from high-speed steel cutting, but its limit was soon reached. A well-constructed lathe, that was balanced in all its parts, was well able to take all the value that could be obtained by high-speed cutting; it was of no use having a driving headstock, with an enormous amount of power in it, unless the loose head and the tool rests and saddles were capable of resisting this increased pressure put upon them. There were makers, not only in Yorkshire but elsewhere, making lathes which he called a sort of mechanical hydrocephalus, that is, lathes with big heads and little bodies. The buyer paid 50 per cent. more for that lathe for having a big head on it, and very soon it was in the hands of the doctor for the purpose of repairs. The shape of the tool, as the author had truly said, had a great deal to do with the amount of cut that could be taken off. In the works in which he was a partner he experimented with ordinary lathes, and he had been able to take cuts at 60 feet a minute, without stress and without any lathe trouble or disease of the kind of which he had spoken. He thought the members would come to the conclusion—and if they did they would save money—that where they turned work in the lathe up to about one-third of the height of centres in diameters, and where there was not much material to come off, they would greatly benefit themselves by using high-speed cutting; beyond that there was a gradual ratio of reduced cutting speeds in inverse ratio to the increased diameter.

Mr. CHRISTOPHER W. JAMES disagreed in almost every respect with the last speaker. He gathered from what he had heard of his remarks, that Mr. Barrow considered if a large amount of material had to be removed from the work it did not pay to use high-speed steel. Mr. Barrow appeared to think it was only advantageous to use high-speed steel in an ordinary lathe, and when a small amount of material had to be removed. He had made some experiments in a atbe built for the express purpose of using high-speed steel. He quite agreed with the last speaker in one respect, that the hydrocephalus lathe was not a thing to be desired, but he thought Mr. Barrow's remarks indicated by negative evidence the proper type of machine to use, and that a lathe must not only have a very powerful head if high-speed steel was used, but all its parts must be equally strong. If one tried in an ordinary lathe to speed it up to use high-speed steel, it could be done to a certain extent ; it could be run at high speed and a cut taken off, but not a heavy one. If one tried to take only a moderately heavy cut, it would probably refuse ; the belt would slip and vibration would be set up, and instead of turning out a round article one would turn out a polygon.

Referring to the Paper, he was somewhat surprised at the results that were given of work done with the particular brand of steel mentioned. He was not interested in any special brand of steel, as he was using a good many, there not being much difference between them, provided each one was appropriately treated. He found in turning ordinary mild steel of a tensile strength of about 25 tons, in a 12-inch lathe with a cutting speed of 45 feet per minute—which was not the maximum speed at which it was possible to work the lathe —it would easily remove, provided it was suitably designed, 14 lbs. a minute. It would be noticed that the weight of material given in the Paper was usually weight removed per hour, and the figures were far below the true capacity of high-speed tool steel in nearly all the cases quoted, the reason obviously being the inadequacy of the machine. It was interesting to find that with a stronger and harder steel, a steel containing $0 \cdot 65$ carbon, and about 55 tons per square inch tensile strength, the same special 12-inch lathe, with the same tool steel, would cut at about 24 feet a minute and remove from $2\frac{1}{2}$ to 3 lbs. a minute. The amount of material removed depended very largely upon the carbon content and the tensile strength. If any members had made experiments upon lathes of the ordinary type, which had been the standard up to now, he thought they would find that on a 12-inch lathe it was impossible to remove as much as 14 lbs. a minute ; he was unable to do it on any of his ordinary lathes, and it was only with lathes specially constructed for the purpose that he

(Mr Christopher W. James.)

was able to do so. He did not think it paid to use high-priced high-speed steel in the ordinary lathes.

Another point in the construction of machines for using steel of the kind to which great attention required to be paid was that not only the parts must be very rigid, but the feed motion required to be very powerful. In " turning " work, which was the most usual, it was often not necessary to reduce the article very much in diameter, but it was desirable to remove the surplus material in a minimum time, and it was very important therefore to have a powerful feed. He found as a matter of experience that if a cut $\frac{1}{4}$ inch deep was being taken the machine worked more easily, and more material would be removed in a given time, if there was a feed of $\frac{1}{4}$ inch per revolution, taking off approximately a rectangular section, than if an equivalent cut of say $\frac{1}{2}$ inch by $\frac{1}{8}$ inch feed was taken.

In regard to planing machines, dealing more particularly with castings rather than with forged work, he found a great difficulty in getting any of the high-speed steels to stand the high speeds possible in a lathe From 35 to 40 feet a minute was about the limit to which one could work. One could drive the planing machines faster if properly designed ; but whether it was the alternate heating and cooling of the steel, or whatever the reason might be, it was very difficult to get the tools to stand more than 35 feet a minute on cast-iron of normal English character, fairly hard. The point that struck him about the Paper was that the experiments were not tests to any real extent of the tool steel. It appeared to him that the limits of the steel had not been reached by a long way, but that the amount of cut, and the speed at which it was taken, and the material removed per hour or minute, were limited by the machines themselves, because they had not been built for the purpose. It was a delusion to imagine that by merely speeding up a machine the economical advantages of the new high-speed steels could be attained. Once vibration was set up it was hopeless to continue ; nothing like true work could be produced. A machine far more rigid than usual must be obtained, and more power must be put into it, if it was intended to cut at 60 instead of 20 feet per minute. He did not think the power used was an item worth considering. The cost of power in an average engineering workshop was a very small item indeed upon the cost of the work that was turned out.

Mr. E. WINDSOR RICHARDS, Past-President, asked Mr. James whether he had any experience of silicon steels for cutting tools.

Mr. JAMES replied in the negative.

Mr. WINDSOR RICHARDS replied that Mr. James evidently did not know anything of the material being used in America for cutting tools. Mr. Keen and himself were in the Bethlehem Works in January last year, and saw most remarkable results obtained from those particular tool steels. They saw forgings, and ingots in the rough, which must have been at least 40 or 50 feet long, in the lathes, and three or four tools working them, taking off immense cuts. It was not carbon steel of which the last speaker spoke, but was probably silicon steel. The results were most marvellous, and he hoped the members would give the matter the most careful consideration, and obtain more information as to what was being done in America before arriving at the decision not to adopt these high-speed cutting tools.

The PRESIDENT said the Institution was much indebted to Mr. Windsor Richards for bringing the matter forward, it was a most important element in the discussion; but he would like to ask him whether the tensile strength of the material operated upon was a good gauge of its resistance to cutting, ignoring whether its tensile strength was produced by means of carbon and silicon or anything else.

Mr. WINDSOR RICHARDS replied that he thought it was quite misleading to talk of tensile strength in connection with that particular class of tool steel.

Mr. JOSEPH BARROW said Mr. Windsor Richards knew more about armour plates, which were called nickel steel. The speed at

(Mr. Joseph Barrow.)

which they could be cut was 7 feet a minute. If one had a planing machine which was speeded to do 30 feet a minute, and put a piece of nickel steel upon it, what would happen? Or, if a steel casting that was insufficiently annealed was put on such a machine, where would they be? The fact also had to be taken into consideration that in a planing machine it was not the cutting speed which dominated the machine, but the return of the table. If the return of the table was 30 or 40 feet a minute, one might go up to 30, 40, or 50 feet with less trouble and a great saving of tools by cutting up to the high speed. On the contrary, where a cut had to be taken at 7 feet per minute, then mechanical difficulties came in—converting the 7 feet of motion into a quick return, say of 50 feet. It became almost impossible to do it. Therefore there were other points connected with the steel which had to be considered. Mr. Windsor Richards had spoken about silicon steel, which he presumed must be a soft material.

Mr. WINDSOR RICHARDS said it was quite the contrary.

The PRESIDENT said he was afraid the discussion had got into rather a muddle. Mr. Barrow thought that Mr. Windsor Richards alluded to something very soft; Mr. Windsor Richards intended to allude to something very hard; but none of the members knew whether the article Mr. Windsor Richards referred to was the cutting tool or the material which was cut.

Mr. WINDSOR RICHARDS replied that he meant the cutting tool.

The PRESIDENT thought Mr. James's observations were directed to the materials which were cut.

Mr. CHRISTOPHER JAMES stated that he had nothing further to add to his remarks except that he advocated in the most emphatic manner the use of high-speed cutting steel, particularly where a large amount of material had to be removed in a given time.

Mr. WILSON HARTNELL stated that some years ago he had read before the Leeds Engineers Society a Paper which was afterwards published in "Engineering," 21 and 28 November 1879, the point of which was that it was desirable that the manager of a Works should ascertain how many square feet of surface he could tool per hour on different classes of work.* A very large amount of the work which the tool did in engineering works was to face off a thin surface, not to take off a great weight of metal as in marine work. The rate of surfacing depended on the speed and feed; with the new steel, although the feed was the same (being limited usually by the rigidity of the article being tooled), the speed of the cutting edge could be doubled or trebled in roughing cuts, and therefore the rate of surfacing be doubled or trebled as compared with standard examples given in the Paper referred to. This he could confirm from actual observation. No academical discussion as to whether the steel could be used was needed; the members could be shown what it was doing. At an engineering works in Sheffield, where they were making engines of 10,000 H.P., he saw a crank-shaft being turned at the rate of 80 feet a minute. Such results were not published, nor attention drawn to them. There were other large steel works—he was scarcely at liberty to mention their names—where with the new steel they were running at very high speeds, and had experimented with higher until the tools ran red hot. At one of the largest firms, covering 60 acres of ground, they were breaking down the gearing of their tools right and left, saying that they would drive them at the high speeds now practicable. This speedy cutting was being done every day in England. He had himself turned cast-iron, taking light cuts off, at 80 feet a minute. That work used to be done at 17 or 18 feet a minute. He had tried the high-speed steels of two or three makers, and all of them gave excellent results.

With regard to large headstocks, if a shallow cut had to be taken off, it did not follow that because the speed was trebled the

---

* The square feet surfaced per hour was shown to be "the speed of the cutting edge × 5 and divided by the cuts per inch." Many examples were given.

(Mr. Wilson Hartnell.)

strain on the lathe bed was greater. It was found that the belt power usually provided was insufficient. As the diameter of the coned pulleys was limited it was necessary to make them much wider—usually by reducing their number and also lengthening the headstock. The smallest cone pulley was much enlarged. This gave an unusually heavy appearance in the headstock. The range of speed was maintained by two sets of speed pulleys on the countershaft and at least double the belt power on each set.

When the lathe was driven electrically (and the speed varied by electrical adjustment) the maximum of belt-power could be obtained on an ordinary size of headstock by using one broad pulley only. He hoped the members present who were using quick-speed steels would give their experience.

Mr. JOHN HARDISTY said that he was constantly turning dead soft Swedish steel at from 90 to 130 feet per minute, light finishing cuts only being required, and he used ordinary tool steel for that purpose. He had tried several of the special high-speed tool steels, and found them all much inferior to ordinary tool steel for this kind of work. They did not retain their cutting edges, and did not produce such a good finish as ordinary tool steels.

Mr. R. W. ALLEN agreed with previous speakers that the production of the high-speed steel was a most wonderful invention. Three years ago he visited the Taylor-White Works at Bethlehem, and, as Mr. Windsor Richards had stated, it was one of the most wonderful sights he had ever seen. He endeavoured to obtain a licence from the Company, but as the price was so high he was unable to do so. On his return he put himself in the hands of Messrs. Seebohm and Dieckstahl, who brought out a steel which he considered was almost equal to the Taylor-White process, namely, the "Capital" and the "New Capital." He was certain that the high-speed steel was the right thing, because, speaking from his own experience with men, it gave them a great incentive to work. Really the proof of its worth was in the cost of the articles machined. Since he had adopted the high-speed steel throughout his works, the cost of machining had been reduced at least 50 per cent. With regard to cast-iron, they found Messrs. Armstrong Whitworth's steel the best, and now used it exclusively on the planing machines, and found it superior to " Capital " steel for this particular purpose.

Mr. JAMES BUTLER, of Halifax, said that, as a lathe maker of thirty years' standing, he had tried a good many experiments in his time, especially lately with high-speed steel. There seemed to be an erroneous impression before the meeting in regard to its uses. On a lathe of the old style, if one wished to take off a small amount of material, say to reduce a shaft not more than $\frac{1}{8}$ inch in diameter, with the new steels the machine could run at least four times the speed as with the old steels. Again, if one wished to reduce a shaft end or an axle from 4 inches down to $2\frac{1}{2}$ inches, one could run with the new steels at five times the speed as with the old, thus reducing the shaft and turning it out in less time than it would take to heat it in a forge and get it ready for forging. Those instances were sufficient to call attention to the uses of high-speed steel. There were five or six high-speed steels in use at the present time, and his own experience was that the higher priced ones did more work than the lower priced and cost less for grinding. They stood longer in the lathe without having to be taken out. Only the previous day he tried to see what it was possible to do with the new steels in reducing a shaft. He put on a 4-inch shaft, and endeavoured to reduce it down to $2\frac{3}{4}$ inches at one cut. He found that practicable at the end of the shaft nearest the centre ; but the springing of the shaft—he had no steady rest on—would not allow that to be taken off at a great distance. In order to get over that he put on first one, two, three and four tools side by side, took a cut with it and found one could reduce a shaft in a good modern lathe from 4 inches down to $2\frac{1}{2}$ inches at a cutting speed of 50 feet a minute. That, he thought, showed the use of the high-speed steel.

Mr. ARTHUR D. ELLIS stated that the Taylor-White steel was made in the American branch of the firm of Sanderson Brothers, in the Syracuse Works in America. The high-speed steel was

(Mr. Arthur D. Ellis.)

actually a Sheffield product. It was not necessary to go to America
for it, several brands being made in Sheffield itself. As an example
of what was being done with high-speed steel in Sheffield, not long
ago he saw a large lathe which was cutting off no less than 3 tons of
turnings every twelve hours, or 560 lbs. an hour, from crank-shafts.

Mr. LACEY R. JOHNSON, of Montreal, said he was sorry that before
leaving Canada he did not know the Paper was to be read, because
he could have brought with him figures and chips corroborating
every figure given by the author. In the Canadian Pacific Railway
shops in Montreal, and in several shops along that system, the
" Novo " steel had been used for some time past, the effect being
nearly to double the output of the machine shops ; in fact, it had
had the effect of pulling up the output of the machine shops, which
had lagged behind the erecting shops, until now in the erecting shops
they were obliged to work overtime in order to keep pace with the
machine shops. So that it would never do, to use an Americanism,
to " buck the high-speed steel." It had come to stop, and the sooner
engineers used it and learnt its possibilities the better it would be
for the credit of the machine shops of the country.

Mr. JOSEPH HILL said a great deal had been said in the discussion
about lathe and planing machine tools, but nothing about rotary
tools. The firm with which he was connected manufactured cold
sawing machines. He found from his experience of the last fifteen or
sixteen years with these special tools that for soft and ductile steels a
much less feed was required, but a quicker periphery speed. Steels
were now being used in saw-blades a great deal harder than were
previously used. It was found in the operations upon hard steels
that a slower periphery speed was required, yet a strong feed. He
had just lately adopted an increased periphery speed by introducing
special steels for cold-saw blades, which were giving excellent
results. This somewhat corroborated Mr. James's remarks that in
machine tools a strong feed was necessary. The method employed
in holding the tool was of great importance, and it was also a very
prominent feature in cold-saw blades. The form of the tooth was
likewise an essential matter.

With regard to the Taylor-White steel, he believed that as good, if not better steel, could be found in England. When he was an apprentice, one of his greatest difficulties was in forming a tool for parting a large shaft, but it was a very easy matter now to do such work by the aid of the cold saw. The observation of Mr. A. D. Ellis, who mentioned the steels used in America, coincided with his own views.

Mr. DANIEL ADAMSON asked the author to supplement the valuable statistics in the Paper by obtaining information as to the length of time each of the tools was running. It was an important point, in estimating the capacity of the tool, to know whether it was in use for half an hour or half a day. It would also be valuable to know the size of tool steel used.

Mr. W. WORRY BEAUMONT thought the support which Mr. Barrow gave to the President's opening remarks seemed to be a little defective. The defect related chiefly to his remarks in regard to economy. He referred to the apt phraseology of an American remark and he might, in connection with his remarks as to economy, repeat another. He was speaking to an American in New York some time ago, and saying to him that in a certain matter they were using a great deal of power to do certain things, and the American replied, " Well, I reckon power is about the cheapest thing you can want, and if you can use it you had better have it." He thought that was the case with the high-speed tool steels and the extra power required to do more work in less time.

The PRESIDENT thought Mr. Beaumont's remarks suitably closed the discussion on the question of power. The question was how much power was used per pound of metal removed. About 1 H.P. was used per pound of metal removed per minute. Of course, if it were being done quickly, more power was being used during that time; but if more metal in proportion was being removed, they were not consuming more power to do the job; so that it was quite ridiculous to say that the thing was of no service because it required more power. It did

2 M

(The President.)

not require more power per pound of metal removed. The other point which he had strongly in his mind was that the same characteristics of the steel which enabled it to take a light cut quickly where it suited the work also seemed to enable it to take an exceptionally good heavy cut. Although it took the heavy cut not quite so quickly, it took it quicker than the older steels did ; in fact, roughly speaking, the speeds might be doubled without risk of failure to the tool. If the old speed was 5 feet a minute for very hard material, it could be made 10; if the old speed was 30 feet a minute it could be made 60 ; the softer the material the faster it could be cut by old or new steel in the proportion of about 2 to 1. He did not now think the enquiry as to endurance was so important as he thought at one time, because he found that if the tool were not right it soon gave out, whereas if it were right and did not give out soon, it would usually endure and finish the job. He was surprised that his friend, Mr. Hardisty, had not had a fortunate experience in high-speed cutting, because he was so well up in the use of steel. He thought still more could be made of the tools if pyrometers were used in the gas-furnaces, and they were able to tune up the furnace to the exact temperature at which any particular steel would harden best.

Mr. SUPLEE, in reply, said Mr. Barrow's remarks were very interesting, but they referred more to the shape of the tool and the method of doing the work rather than to the material. The experience which had come from the older steels had come also to the new ones with equally good results plus the speed. The mark of the beast was caused because the tool was driven to its heaviest capacity, and naturally the material broke down. It showed what would happen. He wished to add that the figures in the Paper were not tests; they were reports of daily work. They were not offered as maximum tests of what the steel would do in any particular case. The tests made at the Franklin Institute were done on a special lathe with a variable-speed gear. The tools were tested to destruction, so that the Tables given in the Report would be more decisive.

In the case of the Bethlehem works, the steel was devised to meet a great emergency. The works had orders which they were unable to fulfil with their existing capacity, and the time required to erect additional buildings and build new tools made the completion of the work out of the question. By putting in the new steel they were able to carry out the contracts, which showed what could be done by the change of tools. Apart from that the interest and depreciation charges on buildings and plant naturally must be taken into account. Of course new machine tools must be made, and heavier and steadier lathes with variable-speed gear. As far as the working was concerned, it had been found that the introduction of the new steels went hand in hand with the introduction of piece-work and the premium system. The workmen wanted the new steel because they could earn more when they were paid on that system. There were plenty of good steels made, both in England and America, so that there was no difficulty in getting them.

In a communication from the makers of the steel used in the tests mentioned in the Paper, subsequent to its presentation at the meeting, it was stated that a superior method of tempering the steel was to employ a blast of compressed-air, after the steel had been raised to white heat; this gave even better results than when oil was used, although the latter was very good when the compressed-air was not readily available.

A NEW FORM OF FRICTION CLUTCH.

By PROFESSOR H. S. HELE-SHAW, LL.D., F.R.S., *Member*, OF LIVERPOOL.

The Friction Clutch is a very old invention, and is probably as familiar to every one in its simplest form, as any other machine-detail. It is however only in recent years that inventors have—in any great numbers—seriously attempted to overcome the defects of the ordinary cone clutch, and their attempts have been attended with such success as to lead to the introduction of a very large number of clutches of moderate power into machinery. This is proved by the fact that in recent years a good many engineering workshops have been devoted, either wholly or in part, to the manufacture of clutches of certain well-known forms. A number of such makers have supplied clutches for many thousands of horse-power for purposes in which they were little used a few years ago. A familiar illustration is the extensive introduction of friction clutches to countershafts in place of the ordinary fast and loose pulley. It might therefore be thought at first that, in such an apparently simple matter, there was very little more to be said and nothing more to be discovered. The importance of the friction clutch, however, justifies a very careful consideration of the problem, and as to whether, in view of the modern developments of motive power, particularly in internal-combustion engines and of high-speed machinery of great power, there are not yet improvements possible in this subject.

The usual object of a friction clutch is to impart motion from a piece of machinery to another piece of machinery at rest, and thus the friction clutch differs in its object, in a very essential manner, from any ordinary form of coupling, or from the well-known jaw clutch. In the latter there is not, or should not be, any relative motion between the parts connected. Hence the strains involved are of a statical nature, and however great, do not involve a mutual action of the nature of a shock, whereas where two pieces of machinery, not having the same speed, have to be connected with each other, there would be a shock due to inertia, unless the connection was made by means of a friction clutch, or some equivalent, such as a slipping belt, allowing the two pieces of machinery to gradually come to the same speed. This gradual effect is therefore the primary object of the friction clutch, and how important this object is will be realised when it is remembered that it is impracticable to start any machine instantaneously, for the magnitude of the stresses and strain caused when a machine is set in motion by another, increases with the suddenness of the operation, and all appreciable shock can be obviated by allowing the action to take place gradually.

During the whole time that the two pieces of machinery are coming to one speed—which in the case of the friction clutch is caused by one starting another into motion, while in the case of a brake it is caused by one bringing the other to rest—during this time work is being lost at the surfaces where the slipping is taking place, and consequently, heat is generated. The more gradual the action therefore the greater the amount of heat, which reaches its greatest amount in the case of the friction dynamometer, where slipping takes place the whole time, and all the work of the moving machine is converted by the friction into heat. Further, it may be said—speaking practically—that wherever there is friction between two surfaces, there is a certain amount of wear taking place, which may be diminished by the use of suitable lubricants. The better however the surface is lubricated, the less the effect of friction, and consequently the less the gripping power of the clutch.

Stated briefly, the four conditions which seem to be involved in the problem of the friction clutch are :—

(1) It must have sufficient gripping power.

(2) Undue wearing of the surfaces must be avoided.

(3) Provision must be made for conveying away the heat where there is much slipping contact in the clutch.

(4) Motion should be imparted to the driven shaft without shock.

One is thus met at the outset with the contradictory conditions which have made the problem of the friction clutch such a difficult one. The author does not remember seeing that in any previous writings on the subject, or in the statement of inventors themselves, the important matter mentioned in condition (3) has been provided for. This may account for the large number of instances in which friction clutches have failed to give satisfactory results for any but the smallest powers.

Looking at the various clutches in use, they may be classified and represented diagramatically as follows :—

(1) The cone clutch (Fig. 1), where considerable pressure between the surfaces is obtained by the wedge action of the cone.

(2) Various forms of rim clutch (Fig. 2), in which the action is obtained by means of levers.

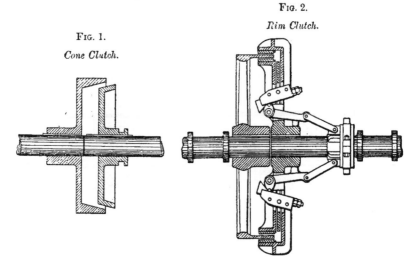

Fig. 2.

Rim Clutch.

Fig. 1.

Cone Clutch.

(3) Clutches with rings or segments expanding within a drum or
   annulus (Fig. 3).

(4) The brush clutch (Fig. 4) in which brushes of wire are
   thrust into a finely serrated or grooved plate.

(5) The coil clutch (Fig. 5), in which a coil of metal or wire rope
   with blocks is employed to give great gripping power.

(6) The " Weston " clutch (Fig. 6), in which the friction effect is
   produced by a number of circular discs connected alternately
   with the driving and driven machine.

FIG. 3.

Drum Expanding-Clutch.

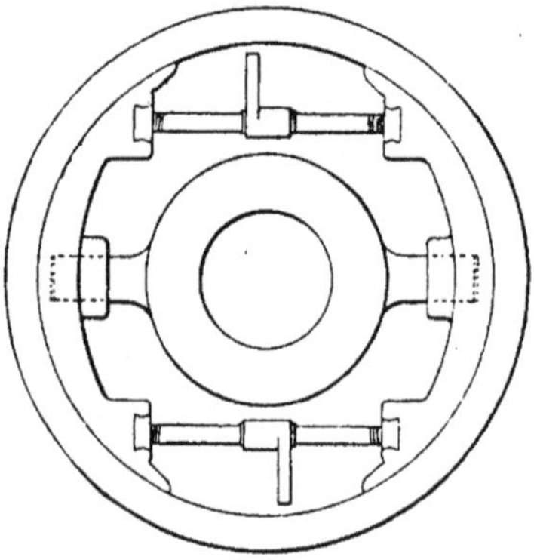

FIG. 4.

Brush Clutch.

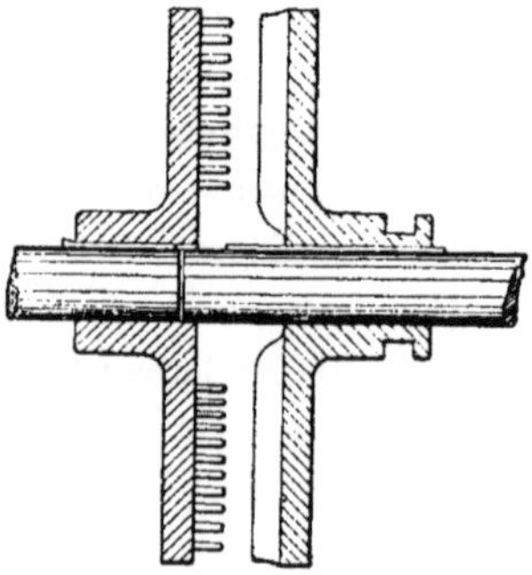

FIG. 5.

Coil Clutch.

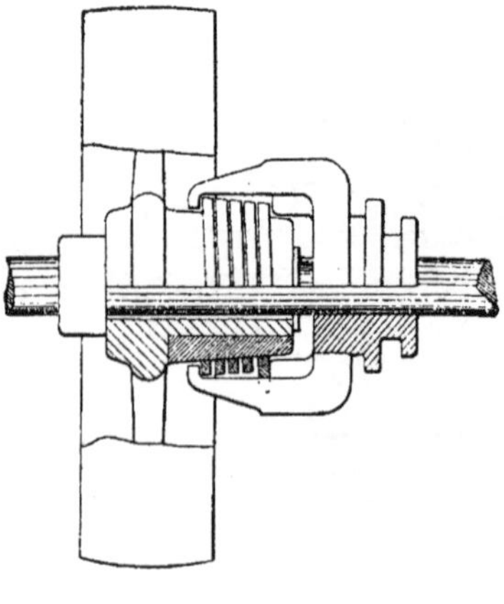

FIG. 6.

" Weston" Clutch.

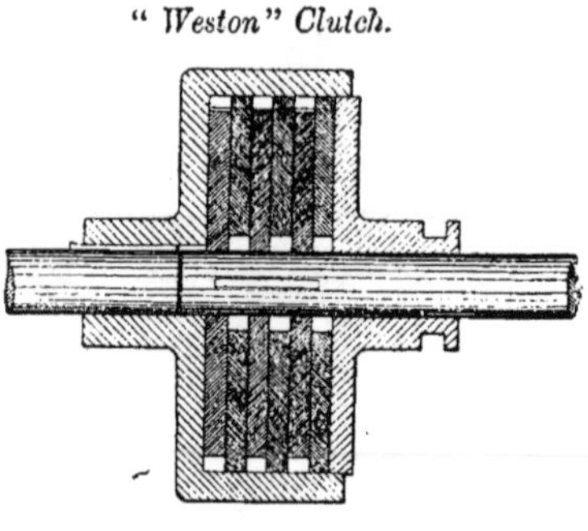

There are numerous modifications in detail of all these clutches, but the inventions in connection with them—which are very numerous —relate principally to the mode of obtaining the requisite pressure between the friction surfaces. Thus taking the case of the expanding ring or segment, which is one of the most successful forms, wedges, right- and left-handed screws and toggle joints, have all been used in different ways for expanding the rings or segments : but no clutches appear to have been designed for the purpose of allowing slipping to take place to any considerable extent, so as to prevent, at the same time, undue wearing and heating of the friction surfaces. These clutches, if not transmitting considerable power and not required to slip, serve their purpose very well. Some, such as the coil clutch, the "Weston" clutch, and the expanding ring clutch, may be made to transmit great powers, but it may be safely said that not one of the foregoing clutches has yet been designed so as to be capable of slipping for more than a very short time, without being seriously injured, even if the surfaces in contact were not actually destroyed.

There are plenty of illustrations in mechanical science where it has hitherto been impossible to reconcile conflicting conditions, such for instance as the variable change speed gear, especially when required to be of high power and high efficiency, and it is an important question whether the present case forms another example or not. The author believes there is a way out of the difficulty, and this he preposes to bring forward in the present Paper. It is quite possible that there are other ways, and if engineers consider the matter of sufficient importance, it may put them on the track of overcoming the difficulty by a way other than that which will be now described.

Suppose a circular disc of metal to be corrugated, the section of the corrugation being the frustum of a cone, and that the disc is placed upon another one similarly corrugated. It will be observed that not only do portions of the frusta not make contact with each other, but there is also a space left between the flat portions of the discs.

By placing these discs together as in Fig. 7, and turning one alternately to the other, an amount of friction is produced which depends on the acuteness of the angle of the frusta. If a

Fig 7

Pair of Plates, showing Clearance due to Corrugations.

number of these plates are now placed in a box of the type of the " Weston" coupling, so that the plates alternately engage with two sleeves, one connected with the driver and the other with the follower, as in Fig. 8, it will be found : first, there is very considerable gripping power ; second, there is a tendency to part rapidly with heat, owing to the separation of the discs of metal.

Fig. 8.

Ordinary Arrangement of Corrugated Plates.

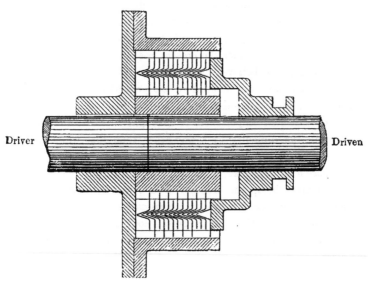

Driver   Driven

By the ordinary theory of friction, the former effect—in a box of given length, if the gripping power is compared by actual measurement with the "Weston" clutch—will be found to be much greater, and yet this cannot be accounted for by the ordinary laws of friction, since the increased effect of the wedge action (which

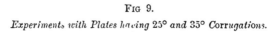

FIG 9.

Experiments with Plates having 25° and 35° Corrugations.

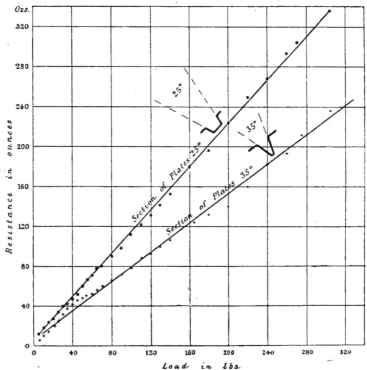

varies with the cosecant of the angle of the cone) is directly in proportion to the diminution in the number of plates which it is possible to put in a box, which diminution varies according to the same law. There is however a still more striking difference between the gripping action of the flat and corrugated plates when a lubricant is introduced.

The results of experiments which have been made with corrugated plates of different angles are graphically recorded in Fig. 9 (page 489). Some explanation is needed of the increase in effect over and above

FIG. 10.

Experiments with Dynamometer (Fig. 20) using No. 3 Plates (Fig. 21).

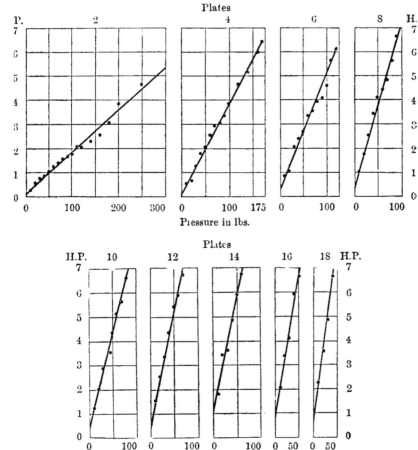

that to be expected from calculations based on the ordinary laws of friction. The author has arrived at the conclusion that it may be due—in part at any rate—to the necessary deviation from the truly circular

form of the corrugations in the plates.   It may be interesting to give
a series of results which have been obtained by actual experiments,
with the dynamometer attached to the reversing clutch illustrated
in Fig. 20 (page 498), on the effect of multiplying a number of
plates, and gradually increasing the pressure, and these results
have been graphically plotted and are given in Fig. 10 (page 490).
Fig. 11 shows the results of this series of experiments in Fig. 10,
(page 490) represented so as to give the H.P. for any number of
plates with varying pressures.

FIG. 11.

Power Transmitted, using various pressures.

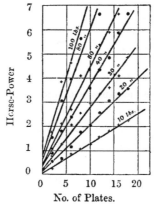

No. of Plates.

In order to ensure efficient lubrication of the surfaces in
contact, the faces of the plates are drilled, as shown in Fig. 12.

FIG. 12.

Holes in Corrugations, for Circulation of Liquid.

This method is also indicated in the sectional view, Fig. 13.
It will be noticed from this illustration that the number of plates in
a given space depends upon the angle of the corrugation thus:
four plates with 30° occupy the same space as six plates with 50°.

FIG. 13.

Circulation of Liquid with varying Angles of Plates.

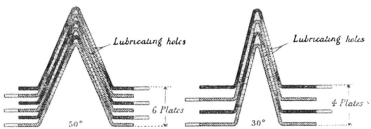

These four plates however give a better grip than the six plates with
50°, and have the great advantage of allowing a freer circulation of
liquid, as may be easily seen from the figure. The plates are also
much more rigid with the more acute angle, and indeed this
increased rigidity appears to be closely associated with their greater
gripping power.

Two views of a standard type of clutch for shafts up to 2 inches
diameter are shown in Fig. 14.

FIG. 14.

Type of Open Clutch with air cooling, showing triggers for holding off pressure.

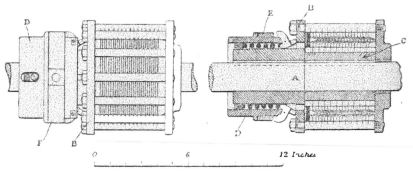

The shaft is divided at A, the outside case B being keyed to the left-hand piece of shafting, and driving the set of plates having external driving teeth, the core C keyed to the right-hand shaft driving the plates with internal driving teeth. Pressure is applied to the plates as follows :—The sliding sleeve D containing a coil spring is fitted with pins which project through the outside case of the clutch ; these pins press against a flat disc, which in turn presses against the plates causing the clutch to drive.

When the operating lever is worked so as to release the plates, the ring E encircling the sleeve withdraws the trigger pins from the

FIG. 15.
Closed Clutch (medium size).

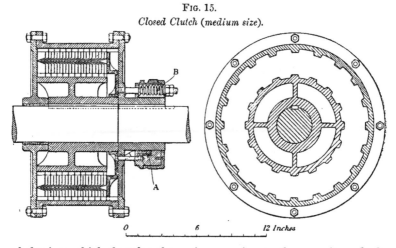

0   6   12 Inches

holes into which they fit; the spring pressing on the opposite end of the trigger pin causes the trigger to fly up and the clutch is thereby kept out of operation. By moving the lever so as to force the ring E against the trigger, the pin end falls into the hole opposite to it, and the coil spring is then allowed to transmit its pressure to the plates.

Fig. 15 shows a section of a larger type of clutch as fitted to a 3-inch shaft. The action of this clutch is similar to that described in Fig. 14. The triggers, however, are worked by separate coil springs A, the pressure is also applied to the plates by means of separate springs B instead of a single one, as in the type illustrated in Fig. 14.

A considerable amount of lubricant can be contained in this clutch owing to the construction of the casing. The right hand view indicates in section the form of driving teeth or notches adopted for preventing the plates turning when transmitting power.

A view of a clutch transmitting 80 H.P. at 60 revolutions per minute is given in Fig. 16. One of the conditions of working of this

FIG. 16.

Clutch at Plate-Glass Works, transmitting 80 H.P. at 60 revolutions per minute.

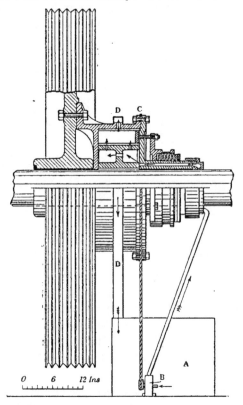

clutch is that it should run for at least two hours daily transmitting its full force when slipping 50 per cent., that is, the shaft shown in the figure running at 30 revolutions with the main shafting running at its full speed of 60 revolutions. It must pick up its load gradually and without shock. The cooling and lubricating fluid is contained in the tank A placed on the ground. The fluid is raised to the clutch case by means of the small rotary pump B, driven by a band from the clutch case at C, the path of the fluid being indicated by arrows. It leaves the casing through a series of holes in the periphery, being discharged through these by centrifugal action. It is then collected in the annular trough D, which, fitting loosely over the revolving casing, directs the fluid back into the

FIG. 17.

20-H.P. Reversing Clutch for Petrol Motor Launch.

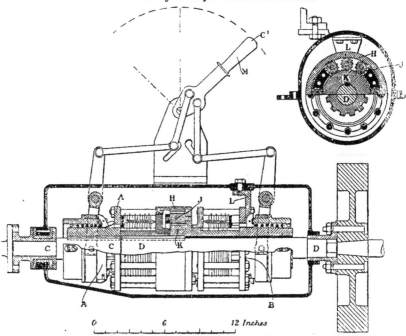

supply tank A. This clutch has been at work at Messrs. Pilkington's Plate-Glass Works, St. Helens, for a period of six months, working day and night, and fulfilling every requirement satisfactorily.

The clutch is the chief factor in a wheel train reversing gear, and the author has recently designed such a reversing gear in which the new clutch is employed. This reversing gear, shown in Fig. 17, consists of two clutches A and B. The outer case of the clutch A

2 N

is keyed to the reversing shaft C, which may be a propeller shaft, the core of A being keyed to the engine shaft D. Inside the rim of the outer case of A teeth are cut in the direction of the shaft, forming an annular wheel H.

To the internal core of clutch B a series of pinions are fixed as shown at J, these pinions gear with the annular wheel H. They also gear with the wheel K, keyed to the engine shaft, the three together forming an epicyclic train. The outer case of B is held stationary, being fixed to the frame of the machine by a bracket L.

The gear operates as follows :—When the lever M is in position C^1, clutch A is made free and clutch B comes into action. The core of B carrying the pinions J is now fixed to the outside case of clutch B (which is permanently at rest), and, as the core of B is free on the engine shaft, the toothed wheel keyed to this shaft at K transmits motion to the pinions, which, being also in gear with the outside case of A, causes the shaft C, to which A is fixed, to rotate in a reverse direction and at a slower speed than that of the engine.

When the operating lever M is in mid position both clutches are inoperative and the reversing shaft is at rest.

When the lever M is in position A^1 clutch B is free, at the same time the plates in clutch A are caused to grip, and as the outside case of A is keyed to the reversing shaft and the core of A to the engine shaft the two rotate together in the same direction. The toothed gearing at H, J, and K, now rotate as a fixed mass, the teeth themselves not being in operation. The presser rings of clutches A and B on this gear are fitted with the spring and trigger arrangement described previously in connection with Fig. 14 (page 492).

The locking system of operating links and levers renders it impossible for both clutches to grip simultaneously, no matter how suddenly the operating lever M may be moved from one extreme position to the other. The enclosing case of the whole gear acts as an oil bath, and thereby ensures an efficient system of lubrication.

This reversing gear has been successfully applied in the new 20-H.P. motor launch made by Messrs. Thornycroft for His Majesty, and is being applied by the same firm to a motor boat of 120 H.P.

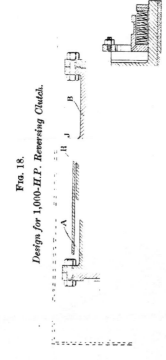

Fig. 18.

Design for 1,000-H.P. Reversing Clutch.

FIG 19.

Epicyclic Gear for Reversing Clutch shown in Fig. 18.

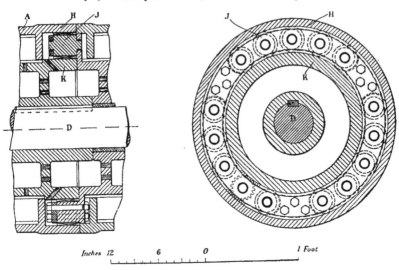

Inches 12   6   0   1 Foot

FIG. 20.

Experimental Reversing Clutch with Fly-wheel and Dynamometer
(See also Fig. 22.)

FIG. 21.

Sections of Plates for various Speeds and Powers.
The Angle of the V being 35° in each case.

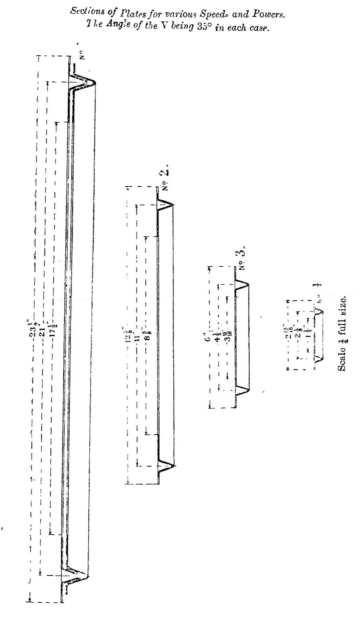

Figs. 18 and 19 (pages 497 and 498) show the design of a reversing gear of this type for a steam turbine of 1,000 H.P., the number of revolutions per minute being 700 to 800. The action of this type of reversing gear may be made as quick or slow as desired, but with the fairly heavy flywheel attached to the gear shown on Fig. 20 (page 498), it has been found possible to change from full speed in one direction to full speed in the opposite direction in 5 seconds.

FIG. 22.

Front and Sectional Elevation of 150 H P. Dynamometer at the Walker Engineering Laboratories, Liverpool

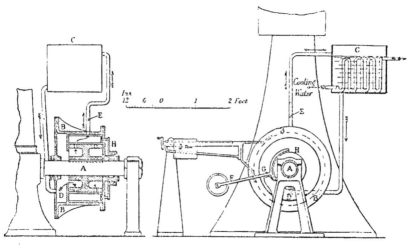

Applied as a dynamometer the corrugated surfaces are quite successful. Fig. 22 shows one fitted to the 150-H.P. triple-expansion engine at the Engineering Department, University College, Liverpool. This dynamometer, which replaced one of the "Froude" type previously used, consists of forty plates, the corrugations including an angle of 35°. A is the engine shaft on which is keyed a slotted core, this core driving one set of plates; the outer casing B is stationary; it is slotted inside its periphery, and carries the stationary plates; bolted to this casing is a wrought-iron arm, this arm being connected at its free end to a calibrated steelyard on which the torque is read off in foot-pounds. Above the dynamometer proper a tank C is fixed containing the cooling and lubricating solution; circulation through the dynamometer is maintained by gravity, the cool liquid enters the casing at D, and, becoming heated, returns to the tank above by means of the pipe E fixed to the top of the casing. Inside the cooling tank a coil of piping is arranged through which water from the town mains is allowed to flow by pipes; by this means the arrangement is kept cool for tests of any duration. Varying pressures can be applied to the plates by means of the hand-wheel and screw fixed to the end of the lever F, the fulcrum being at G; the other end of this lever carries a claw which transmits the pressure to the ring H, carrying the pins which press against a flat disc, this disc transmitting its pressure to the corrugated plates. As each pair of plates have flat springs inserted between them the engine is enabled to run quite free when starting. A friction dynamometer of this type of 600 H.P. has been constructed for Messrs. Vickers, Sons and Maxim, Barrow. It may be pointed out in connection with these dynamometers that the speed can be reduced to a very low number of revolutions under the conditions of pressure at the engine without impairing the accuracy of the observations, while the running remains perfectly steady for very great loads. Although this dynamometer has been at work for many months the plates have not worn to any appreciable extent.

The author has mentioned a number of applications of the clutch which have been already made, but the application which really led to its invention was in connection with the motor car.

A satisfactory clutch is one of the most difficult things to secure for this purpose. The ordinary type of cone clutch, which is generally employed, the cone of which is covered with leather, can rarely be maintained in a normal condition for the following reasons:—A great command over the car, especially in driving through traffic, is secured by allowing the clutch to slip. When slipping has been going on for some time the surfaces are generally so altered in their condition that either the clutch will not grip at all or it grips violently and harshly. The result in driving is not only most unpleasant but it is very injurious to the car itself.

The author originally fitted one of his new clutches to a fairly heavy car, on which the engines had been changed from 6 to 12 horse-power. This clutch, although the plates were only 6 inches in diameter, drove the car so satisfactorily that the 6-H.P. gear has, under very trying conditions of the British Association tests on road resistance, been found quite strong enough for the purpose. Since then he has changed the cone clutch of a 24-II.P. Darracq car for one of the type described in this Paper with remarkable improvement in its action. It has been found capable of driving the car so altered, although a fairly heavy one, for long distances, even in hilly country and without ever going off top speed, which is a direct gear to the driving wheels. The actual speed at the normal revolutions of the engine on this car is 70 kilomètres (*i.e.*, about 45 miles) per hour. This example shows the possibilities of the clutch, with which the car can be driven for any required distance at the speed of 3 or 4 miles per hour, or even less, and after running for an hour or two under these conditions the clutch does not get hot. Of course the main object of the change gear, which is to increase, if necessary, the turning force at the tyres, is not attained, since the slipping of the clutch can only involve loss of power, but this is a case in which, with an engine of sufficient power, variation in the range of speed is really a more important object rather than the saving in power.

There are many other instances where this property of variation in speed without using the change gear is required, but this example suffices to illustrate the operation of a slipping clutch. There are many cases, such as the use of electricity with petrol motors on light railway wagons and also with what is called the Petrol Electrical Car, where expensive means have been employed to effect what is done in a perfectly simple manner by a slipping clutch, and members will themselves be familiar with numerous other illustrations of this in engineering work.

The Paper is illustrated by 22 Figs. in the letterpress.

Discussion.

The PRESIDENT, on behalf of the members, thanked the author for the most interesting Paper he had given, and especially for his unusually attractive exhibition. He congratulated Professor Hele-Shaw on the perfection with which the apparatus demonstrated his Paper. It was unusual in ordinary practice to have such a perfect apparatus for indicating the speed and the variation in speed between the two things, and the whole subject had been extremely well placed before the members. Although he had expressed such words of admiration they were not a mutual-admiration Society, and it would be quite suitable for anyone to offer criticism, even if it might be more or less of a destructive character, upon the invention.

Mr. S. DAWSON JONES asked if there was a spring between each of the plates, to cause them to separate when the clutch was thrown out of gear. He also enquired whether the author found that the plates, on account of their small thickness, did not get out of the vertical and so cause the cones to wear on the sides irregularly. Finally, did the amount of wear which was taking place, in consequence of the plates being so very thin, cause the distortion of the plates and take away from their durability and efficient gripping action.

Professor ARNOLD LUPTON thought the friction clutch described by the author was precisely what was required in the mining world. The driving of hauling apparatus and other machines in mines by friction clutches was nearly universal, and such a beautifully arranged clutch as that shown, which enabled one to vary the speed at difficult turns in underground roads and to either reduce the speed or put it on without altering the speed of the electric motor, which was now so commonly used as a means of driving hauling apparatus where the work was at a long distance from the source of power, it seemed to him, would be fraught with great advantages. It would also be very suitable for starting electrically-driven underground pumps, where it often happened that there was a long column of

(Professor Arnold Lupton.)

water to be gradually raised to a high speed against a great pressure. He therefore congratulated the author upon the work he had done, and hoped to hear more of it in the future.

Mr. R. C. JACKSON asked whether any application of Professor Hele-Shaw's interesting development had been made to planing work, or to lathes for variable speeds and for heavy turning.

The PRESIDENT wished to contribute to the discussion a criticism which occurred to him upon the arrangement of the dynamo. There was an illustration, Fig. 22 (page 500), of the application of the brake to a 150-H.P. triple-expansion engine, in the Engineering Department of the University College, Liverpool. One portion of the brake was kept from revolving by a steel-yard with an adjustable poise-weight, and by the adjustment of that poise-weight the driving force could be measured. The pressure was brought upon the plates, which constituted the clutch so ably described, by a claw working between collars upon the sleeve that carried the central plates. The claw swivelled upon a fixed bracket, and was described as being forced into position by a hand-wheel and a screw. There was, therefore, a force brought to bear upon those collars equal to the force with which the plates were pressed against each other. The force upon the collar would produce friction ; that friction was not intensified by the wedge shape of the clutch itself, nor did it act at so long an arm from the centre. But it appeared to him that the engine, the power of which was being measured by the dynamometer, had also to overcome the friction of the collars, and that therefore the steelyard would not measure the whole of the power that was transmitted by the engine, but that there would have to be a deduction made for the friction upon the collars.

Mr. WILSON HARTNELL said that for a long time he had been looking for a durable clutch for obtaining variable speeds on cranes driven by polyphase motors (which run at one speed), especially for foundry purposes, where one wished to lift slowly at first but afterwards at a higher rate of speed. The amount of power

wasted in slipping need not be considered.   In the contrivance shown
by the author they had presented for the first time a clutch which
could be allowed to slip more or less, under heavy pressure, for a
considerable time without injury or want of regularity in action.
These qualities enabled one to lift weights at a slow and steady
speed, and accelerate the revolutions as desired until the full speed
was obtained.   A point mentioned by the author which had not been
commented upon was that, with the same end-pressure, the amount of
friction was greater than with flat plates, to a higher amount than
anticipated from the increased pressure due to the angle of the
V channels.   He would like to know whether the author had taken
into consideration that when flat discs were being used the average
radial distance of the surface might be considerably less than with
the V discs shown.   Mr. Barnes had suggested to him—if the members
would excuse the homely analogy—that if a lid was put on a tobacco
box it turned round freely when it was true; but if the lid was
slightly distorted there was great friction.   Would distortion and
buckling of the discs account for the increase of friction which the
author had measured.   There was a certain amount of distortion—
it might be slight—which took place at every revolution.   The
Paper was not brought forward as a speculation; it was the
description of an invention which had been successfully worked.
That fact rather disarmed scepticism, otherwise he would have said
in regard to the wear and tear, that the plates should wear through
and go as thin as paper, in the course of a few years, and the
clutch would then be useless.   That anticipation, however, was said
to be contrary to experience, so that really there was very little to
criticise in the clutch itself.

Mr. J. E. LLOYD BARNES, referring to a complaint sometimes
made that Professors' Papers read at their meetings were not always
of a very practical nature, pointed out that the present Paper, at any
rate, was not open to this criticism, for Professor Hele-Shaw had
brought before them an exceedingly practical development in a very
important branch of engineering.

(Mr. J. E. Lloyd Barnes.)

Friction clutches and reversing gears were required in almost all engineering work, and he was afraid that for large powers the progress of invention in this particular branch, up to the present, had hardly kept pace with general engineering progress. Weston clutches, with flat plates, were very well known, but one could not help remarking, if one had had the advantage, as he had, of seeing the author's clutch compared with the flat-plate clutch, upon the enormous increase in the frictional resistance and smoothness of action which resulted from the corrugations. He thought, as Mr. Wilson Hartnell had mentioned (page 505), that the friction might not be quite the ordinary kind of friction to which engineers had been accustomed in this connection; it might not be a surface friction, but more related to a kind of molecular friction due to infinitely small distortions, which gave a steadier result. He thought that those who had to deal with friction clutches would be most interested in the diagram (page 490). If any members having experience with friction clutches such as were for sale to-day, plotted the torque or turning effect of such clutches with the normal pressure exerted to get that torque, they would find that a curve was obtained which rose very gradually at the beginning, but which, at the point at which seizure took place, ran up very steeply; and this had been the great objection, so far as he had been able to gather, to the ordinary friction clutch when applied, say, to motor-cars, to single-phase, or to polyphase motors, or to large gas-engines and the like. There was a slight increase in the turning effect up to certain extent; then the full increase took place suddenly with all the evil effects of suddenness.

Referring again to the diagram (page 490), it would be seen that on plotting the pressure and the power transmitted, the resulting curve was very nearly a straight line, and it was this property which enabled one, with the author's clutch, to produce the graded and regular results which, so far as his experience went, one did not seem to be able to get with any other clutches. Another remarkable feature which occurred to him was that the author should go out of his way to lubricate the clutch; one did not do that usually. The author seemed to say that the more the clutch was lubricated the

better it worked, which seemed to suggest that there was something
in the nature of internal molecular friction, the wearing friction
being reduced or eliminated by the lubrication, and that steady even
effect obtained, which made the clutch so very unlike other clutches
with which he had had to deal. There seemed to be a profitable
market for the clutches in connection with electro-motors, such as
the one exhibited which was used to drive the author's reversing
gear. In this single-phase motor there was a centrifugal clutch in
the pulley, to allow the motor to get up its full speed before putting
on the load; such an arrangement was not the most desirable one
because it often failed when it was most wanted, but a clutch like
the author's would allow one to start up a single-phase motor
without load and to bring the load gradually on. The other
important point that seemed to him of considerable advantage was,
that since the plates were loosely packed in the box one might have
a considerable want of alignment between the two shafts. Professor
Lupton, who referred (page 503) to the mining application, would
appreciate that point. Looking at some of the bearings that one had
to deal with in practice, the shafts would be found very wanting in
alignment. If there was not proper alignment of the shafts the
ordinary clutch could not be got to work, but in the clutch exhibited,
owing to the elasticity of the plates, one could do with a considerable
want of alignment, and therefore have this considerable advantage
with the clutch. He believed that the clutch could be operated with
a " Bowden " wire, because the triggers simply had to be released,
and this could be done with the expenditure of very little work. He
thought the author must be gratified with the reception his Paper
had received, and that the Institution was to be congratulated on
having such a Paper brought forward at Leeds, which was the centre
of an industry with great potentialities for a really good friction
clutch.

Professor HELE-SHAW, in reply to Mr. Jones (page 503), said
that springs were placed on alternate plates which did not revolve
relatively, as the plates were in pairs. He had with him three plates
which would enable members to understand how this was done,

(Professor Hele-Shaw.)

the springs being riveted on. (Model produced.) The spring
pressed against, in the same groove, the projecting parts of the next
plates of the same kind, that is, they missed the inner plate which
ran round inside, and which was just beneath the projecting pieces.
Therefore the spring did not rub in any way upon the plate next to
it, but pressed along the groove against the next plate, which did
not revolve relatively. This simple expedient had effected a
wonderful change in the clutch, particularly in its use as a
dynamometer. The dynamometer, previously to this being done
would not free itself properly when the load was taken off. It was
also found that even when it was freed there were a few pounds on
the steelyard, which was due to the end plates coming away and the
other plates remaining too closely together, so that these had to be
more or less shaken apart by any oscillation in the machine. But
directly the springs were added, the plates freed themselves
instantaneously, even if the engine was at rest. It would be seen
from the working model that when he threw the handle over they
were tied together, and when he reversed the handle they were all
opened simultaneously. The springs did that; with them there was
no sticking between the plates.

He had been exceedingly pleased to hear what Professor Lupton
had to say. The action of the new clutch when there was want of
alignment in the shafting was a great point. The plates were
dropped into the boxes quite as a loose fit, and the moment they got
up any speed they tended to adjust themselves, just as a top did, and
spun round truly in the box the moment the clutch was taken off or
on. Even if at first they were not in alignment, they became so at
the first chance they got, and then they ran true.

With regard to the use of the friction clutch on a planing
machine, he had not the slightest doubt that the clutch, or some other
slipping clutch—his might not be the only slipping clutch capable
of doing it, although he did not know of another—would enable the
introduction of chain driving for this purpose. Speaking of
transmission in the workshop generally, he said advisedly that the
chain drive was impossible where there was not some way or other
for the motor and the driving machine to be in elastic or yielding

connection with each other. Therefore with chain driving some form of slipping clutch was necessary. It was the want of this which caused the use of belts to be retained more than anything else. It was often supposed that the forest of belts in the workshop was, like the horse upon the road, a sort of arrangement of nature which could not be obviated, but he did not believe that the one was more necessary than the other. He believed it was possible to have chain driving if some slipping clutch was used. In reversing a planing machine by belts there was an enormous squeaking directly the heavy table was reversed. His contention was that this squeaking indicated wear, and that if that wear could be put into a liquid, which did not wear appreciably, then they were putting the wear into something prepared for wear, and that a strap, which was not prepared to slip or wear, only did so because it could not help itself.

With regard to the President's objection (page 504), the turning moment, as indicated on the steelyard, was 5,000 lbs. Taking 220 revolutions, at the distance of a circle, 3, divided by 33,000, 100 H.P. was obtained. The pressure on the lever was about 200 lbs. at 6 inches; it was 12 inches diameter. It was kept well lubricated, and therefore the pressure on the wheel and lever was somewhere about 500 lbs. But the friction was as about 1 per cent. of that; therefore they had, roughly speaking, 5 lbs. at 6 inches, which gave $2\frac{1}{2}$ foot-lbs. as against 5,000. That was about $\frac{1}{20}$th per cent. Professors were often told by practical men that they went too closely into decimals; and it was fair to retort upon such an able practical man as the President that if students measured accurately within $\frac{1}{20}$th per cent. they were doing very well. With regard to the lid of the tobacco-box theory, he did not like to advance theories unless sufficient proof were forthcoming, because it was not pleasant, if an explanation turned out to be not the true one, to have advanced it in too positive a manner, but as he had intimated in the Paper he believed that the theory advanced was correct. He believed in it for the reason chiefly that he found the web of the plate played such an important part in the power transmitted. If he applied the same force with a thin web he could not obtain nearly so much power out of the dynamometer or clutch; if he made the web deep it seemed to be able

(Professor Hele-Shaw.)

to increase the power capable of being transmitted. He believed that was due to the rigidity of the two rings. The plates were not circular or uniform; the grooves were not quite circular, and as a proof he found that, after a good deal of wear, the black in many places was not worn out. He believed that when the plates were pressed together there was a want of concentricity, and that it was like trying to drive one flexible eccentric round another fixed one, thereby doing work, and that the work and resistance, to his mind, might account for a part of the action in the new clutch.

The PRESIDENT said he had been asked to express thanks to the Leeds Corporation Electricity Department, who had lent the motor, and also to one of their staff, who had assisted at the experiment.

ECONOMY OF FUEL IN
ELECTRIC GENERATING STATIONS.

By Mr. HENRY McLAREN, *Member*, of Leeds.

All Public or Municipal Lighting Authorities are required to furnish " Yearly Returns " to the Board of Trade, giving their costs of generating current. From these returns " The Electrical Times " has compiled a Table of Costs, which is kept up to date, and published weekly. The following remarks and figures are based on this Table, published on 1st January 1903.

Most of the Municipal Bodies end their financial year on 31st March; therefore the figures given in Tables 1 to 11 (pages 523–536) mainly refer to the financial year ending March 1902. It must be borne in mind that this was a period of exceptionally dear coal.

The first point that attracts attention is the great difference in cost per unit generated in these stations. They are located in various parts of the country ; some are condensing, others partly condensing, or non-condensing stations, the price paid for coal varies largely, and different systems of supply are used. All these considerations affect the costs.

In the returns for the Lighting Stations, the figures are based on units sold to consumers, and therefore include the large leakage or loss in the mains, which is debited to the Generating Plant. This may be as much as 20 per cent., and is a very important factor to be kept in mind when comparing purely Lighting Stations with Tramway or Railway Plants, that "meter" the current as it leaves the Station.

2 o

Engineers would naturally expect that condensing stations would show a better fuel economy than non-condensing, both for tram and lighting purposes, but so far as lighting only is concerned, it is just the reverse ; the average, both in fuel and works costs, is considerably in favour of non-condensing stations. Where trams are added to lighting, the condensing plants show a better economy than non-condensing, but in all cases the partly condensing station is the most economical both in fuel and works costs.

Works costs include fuel, oil, water, waste, stores, wages of workmen, repairs and maintenance. It is manifestly unfair to compare purely Lighting Stations, which have only a poor load factor, with stations that combine Lighting and Tramways, and generate large quantities of current all day long, and meter it as it leaves the station.

The author has therefore subdivided the stations given in the Table of Costs published by " The Electrical Times," and arranged them as follows :—

22 METROPOLITAN STATIONS.

Lighting only.

			Units sold.
Non-Condensing	.	. Table 1 (page 523) .	24,014,141
Partly Condensing	.	. „ 2 („ 524) .	11,305,581
Condensing	. .	. „ 3 („ 525) .	41,318,891

These stations are fairly comparable ; they are all generating current for lighting purposes, and pay London prices for coal and labour. In comparing the costs of these three types, it is found that the non-condensing stations produce current at 18 per cent. and the partly condensing 21 per cent. less cost than the condensing stations. For fuel only, the saving in favour of non-condensing and partly condensing plants is about 14 per cent. and 17 per cent. respectively.

116 PROVINCIAL STATIONS.

Lighting only.

			Units sold.
Non-Condensing	.	Table 4 (pages 526–7) .	18,430,119
Partly Condensing	.	„ 5 (page 528) .	20,873,145
Condensing	. .	„ 6 (pages 529-531).	35,933,215

These stations, as a rule, are small, and situated in various parts of the country; on the whole, they will have the advantage of cheaper coal and labour, as compared with the Metropolitan stations, and they show that their average cost of producing current is 18 per cent. less than the Metropolitan stations. Here again the non-condensing and partly condensing stations are more economical than the condensing, the saving in works costs being 14 per cent. and 16 per cent. respectively, and the saving in fuel $12\frac{1}{2}$ per cent. and $20\frac{1}{2}$ per cent.

32 PROVINCIAL STATIONS.

Supplying Current both for Lighting and Tramways.

			Units sold or generated.
Non-Condensing	.	Table 7 (page 532) .	7,968,426
Partly Condensing	.	,, 8 (,, 533) .	30,040,162
Condensing . . .	.	,, 9 (,, 534) .	39,893,928

In these stations the condensing plants are found to be more economical than the non-condensing by about 9 per cent. in works costs, and 19 per cent. in coal, but the partly condensing stations are still better than the condensing, showing an economy of 23 per cent. in works costs, and nearly 4 per cent. in fuel.

4 PROVINCIAL STATIONS OR POWER-HOUSES.

Generating Current for Electric Tramways.

All Condensing.

Total units generated 20,066,089

Table 10 (page 535) gives particulars of these Stations.

It is surprising to find that at least two of the Lighting Stations, Leeds (Table 6, page 530) and Edinburgh (Table 4, page 526), deliver current at the consumers' meters at 9 per cent. lower average cost than these four power-stations averaged at their switchboards ; moreover Edinburgh is a non-condensing station.

Glasgow, the most economical of the four power-houses, with an annual output of over 10,000,000 units and a tramway load-factor, delivers current at the switchboard at a cost of $0 \cdot 6$ penny per unit works costs ; whereas Leeds Lighting Station, with only 3,000,000 output and a lighting load-factor, shows a works costs of $0 \cdot 62$ penny per unit, including the heavy loss in the mains.

2 o 2

If the Leeds current was metered at the switchboard the works costs would , read considerably lower than Glasgow, the most economical power-house in the country.

2 METROPOLITAN AND 1 PROVINCIAL POWER-HOUSES.

Generating Current for Electric Railways.

Total units generated 26,513,997
Particulars of these Stations are given in Table 11 (page 536).

These are beaten in economy by the tramway power-houses, already mentioned, by over 20 per cent. in works costs, and $5\frac{3}{4}$ per cent. in fuel costs. The extra for coal in the three railway power-houses may be accounted for by their geographical position, but the 20 per cent. excess in works costs appears unduly heavy.

The Diagram on the opposite page shows in graphic form the average costs of the various classes of stations or power-houses. The results from any station given in the various Tables, may be compared with the average in its own, or any other class of station by means of this Table. Take Leeds Lighting Station, with fuel costs of $0 \cdot 27$ penny per unit, or works costs of $0 \cdot 62$ penny, it is found to be considerably lower than the average of the tramway power-stations, the lowest average on the diagram.

Taking coal at 10 lbs. for a penny, or 18s. 8d. per ton, then each decimal point of a penny on the diagram represents 1 lb. of coal. If the Metropolitan (Condensing) Lighting Stations, Table 3 (page 525), purchased their coal at 18s. 8d. per ton, they would appear to be using on an average over $11\frac{1}{2}$ lbs. of coal per unit sold, or $8 \cdot 6$ lbs. per electrical H.P., or say $7\frac{1}{2}$ lbs. per I.H.P. per hour, taking a combined efficiency of 87 per cent. During the period under review these stations were no doubt paying more than 18s. 8d. per ton. This would modify the above figures, but it is impossible that coal could have been sufficiently dear to bring them within reasonable limits. An economical condensing engine and boiler should not require more than $1\frac{1}{2}$ lbs. of good coal per I.H.P. per hour, or $2\frac{1}{4}$ lbs. per unit generated. In everyday work, $2\frac{1}{2}$ lbs. per unit at the switchboard may be considered very good economy.

From these figures it would appear that the Metropolitan Condensing Stations, Table 3 (page 525), after allowing 30 per cent.

Diagram of Average Costs.

	Metropolitan Stations.						Provincial Stations.												Electric Tramways.	Electric Railways.	
	Lighting only.						Lighting only.						Lighting & Tramways.									
	Fuel per unit.			Works-Costs per unit.			Fuel per unit.			Works-Costs per unit.			Fuel per unit.			Works-Costs per unit.			Fuel per unit.	Works-Costs per unit.	Fuel per unit.	Works-Costs per unit.	
Table.	1	2	3	1	2	3	4	5	6	4	5	6	7	8	9	7	8	9	10		11		Table.
Cost per unit in pence.	Non-Condensing.	Partly Condensing.	Condensing.	Non-Condensing.	Partly Condensing.	Condensing.	Non-Condensing.	Partly Condensing.	Condensing.	Non-Condensing.	Partly Condensing.	Condensing.	Non-Condensing.	Partly Condensing.	Condensing.	Non-Condensing.	Partly Condensing.	Condensing.	Condensing.		Condensing.		Cost per unit in pence.

Works-Costs include Fuel, Oil, Waste, Water, Stores, Wages of Workmen, Repairs, for outside losses and banking fires, &c., were using at least twice the amount of fuel that economical plants ought to require. They generated during the year over 41,000,000 units, at a cost for fuel of over £200,000. A very large percentage of this amount could have been saved by the use of well-managed, economical generating plant. It must be remembered that most of these stations were pioneers in the electric lighting industry of this country. Each extension was in a measure a large experiment. They have had to contend with difficulties too numerous to mention, and, like "Topsy," they have just "growed" to what they are. Could their engineers start afresh, with a "clean slate," there is no doubt that these stations, with their enormous output, would take their place amongst the most economical in the country. There is a good deal to be said in favour of the "clean slate." Taking the average station throughout the country there is great room for improvement; on the other hand, one finds a few stations that leave nothing to be desired; some remarkably economical results are recorded even in small provincial stations.

Condensing Plants.—Separate condensing plants are used in most lighting stations. The vacuum in the main engine low-pressure cylinder is usually poor, owing to the long lengths of exhaust pipe, often overcharged, and having numerous sharp bends; 11 inches vacuum in the cylinder has been recorded with 23 inches in the condenser. Station engineers as a rule do not pay sufficient attention to this. The engines of separate condensing plants, and other auxiliary machinery, are often fitted with piston valves. Such engines may be fairly economical when new, but soon develop a great appetite for steam. They are usually fixed in inaccessible places out of sight, and out of mind; their inflated exhaust passes direct to the condenser unobserved. It is quite common to find this type using 10 per cent. of the steam required by the main engines, when working at their full load, this percentage increasing rapidly as the load falls off. A recent test of a separate condensing plant revealed the fact that it used 4 lbs. of steam per unit generated by the main engines.

Engines fitted with their own condensers have a great advantage over those exhausting into one main serving several engines; usually a saving of 5 to 10 per cent., due to the better vacuum, is effected. Add to this the 10 per cent. of steam used by the separate plant; the loss in the auxiliary steam main with its traps; the loss in starting up separate plants for short runs; the loss at light loads; the loss of vacuum through leaky atmospheric exhaust valves; and it is not so surprising that under these conditions, the economy due to condensing reaches the vanishing point, or even falls 13 per cent. behind well-equipped non-condensing stations, as shown by the Tables. The author's firm usually guarantee $7\frac{1}{2}$ per cent. less steam per kilowatt-hour if the engines are to be fitted with their own condensers, and drive their own pumps. The total cost of these engines is less than those fitted with separate condensers, and the steam required by the auxiliary machinery is altogether saved.

Partly-Condensing Stations.—It is rather remarkable that the partly-condensing stations show the best average economy. Partly condensing is a very vague term; it may mean one engine in ten is non-condensing, or *vice versâ*. No doubt, most of them might be classed as condensing stations. Without more exact data it is impossible to come to any definite conclusion as to the reason for this good average economy.

Non-Condensing Stations.—In "The Engineer" of 15th May, 1903, the compound engines for the Tooting Electric Tramways are illustrated, and a few particulars are given as to guarantees, &c. The lowest guarantee, when working condensing, is $13\frac{1}{4}$ lbs. of steam per I.H.P. per hour; the lowest when working non-condensing is 16 lbs. per I.H.P. per hour. If the latter figures can be obtained, the money sunk in condensers, &c., will only yield a poor return, after the steam used by the condensing plant has been debited to the main engines. It is well to note that the figures here given are those guaranteed; the figures actually obtained have not been published.

About fifteen years ago, the author's firm devoted considerable attention to the manufacture of small compound non-condensing engines and boilers combined. Professor Barr and Mr. Druitt Halpin carefully tested one of the smallest sizes of these engines, with the following results, when using an exhaust steam feed-heater:—

Size of Cylinders :—5¾ inches and 9 inches by 15 inches stroke, working
   at 24 I.H P.

Steam per I H.P. :—18·8 lbs. per hour.

Welsh coal per I.H P. :—1·8 lbs per hour.

Water evaporated per lb. of coal, from a temperature of 203° F. :—
   10·44 lbs. Boiler pressure, 140 lbs.

Mechanical efficiency of engine, 86 per cent.

The above results in fuel and steam may be considered very good
for non-condensing compound engines even of much larger sizes.

The results obtained from the smallest size of triple-expansion
condensing engines, made by the same firm, and tested by Mr.
Wilson Hartnell and Professor Goodman, were as follows :—

Sizes of Cylinders :—9 inches, 14¼ inches, and 22½ inches by 24 inches
   stroke, working at 120 I H.P.

Steam per I.H.P. per hour :—12·54 lbs.

Welsh coal per I.H.P. per hour :—1·25 lbs.

Water evaporated per lb. of coal from a temperature of 110° F. :—
   10·11 lbs.

Water evaporated per lb. of coal from and at 212° F. :—11·34 lbs.

Boiler pressure :—157 lbs. per square inch.

Mechanical efficiency of engine, including the driving of all the pumps,
   and brake-wheel friction :—93·5 per cent.

These triple engines were fitted with surface condensers, and
drove their own air, circulating, and feed-pumps. The feed-water
was pumped direct from the hot-well to the boiler. Taking
18·8 lbs. of steam per I.H.P. for the non-condensing, and 12·54 lbs.
of steam for the triple-condensing engine, the economy in steam, due
to condensing, is 33 per cent., the economy in coal 30 per cent.
The triple-condensing engines had the advantage of being larger in
size, and working at 17 lbs. higher boiler pressure, but as regards
economy of fuel, the non-condensing engine had the advantage of
the feed-heater. If 25 per cent. be taken as the saving in fuel due
to condensing, when the best engine of each type, working at the
same steam-pressures, is used, it will be very near the mark.

The Tables show that in Lighting Stations the non-condensing
engines beat the condensing by 13 per cent. in fuel economy; this
is the average over the whole country, 13 + 25 = 38 per cent. to be
accounted for to bring matters as they ought to be. Uneconomical,

three-crank compound, tandem, condensing engines, largely used in Lighting Stations, no doubt account for some of this, and, as already shown, separate condensing plants are also steam wasters.

All non-condensing stations should be fitted with efficient exhaust steam feed-heaters; their cost is comparatively small, and the saving in fuel is a clear 13 or 14 per cent. over cold feed. Feed-heaters or economisers fixed in the boiler flues cool the waste gases, and to a certain extent spoil the draught. If this has to be made good by using steam-jets under the grates, the actual saving in fuel may be small. The best results are to be got by employing both types of heater, the water entering the economiser at nearly boiling point. It is interesting to note that Edinburgh and Motherwell, the two most economical non-condensing stations given in Table 4 (pages 526–7), are both fitted with exhaust steam feed-heaters. About fifteen years ago, the author fitted this type of heater on a condensing engine with good results. It is now the common practice in America to fit these heaters on condensing engines.

Engines in General.—The steam losses in electric generating engines are mostly due to valve or piston leakage; mainly the former. This applies more especially to condensing stations using engines fitted with piston valves; unless these are very carefully looked after, heavy leakage is likely to occur, the steam passing direct to the condenser. In non-condensing engines the exhaust pipe usually gives warning of leaky valves. Many station engineers do not appear to realise how serious this loss may be, and allow their piston-valves to run in a very leaky condition. Others give this matter most careful attention, and are well rewarded for their trouble; in fact, some of the most economical non-condensing stations are fitted throughout with piston-valve engines, carefully looked after and kept steam-tight. Balanced slide-valves, and valves of the Corliss type, having some pressure on the back to keep them up to the port faces, require much less attention, and will run for many years practically steam-tight.

Two recent tests of condensing engines (in different stations), fitted with piston-valves, disclosed the fact that they required over 45 lbs. and 50 lbs. of steam per kilowatt-hour respectively. When previously tested the consumption was about 30 lbs. per kilowatt in each case. Outwardly they appeared to be running as well as ever.

Coal to the value of £839,613 was used in the 177 stations given in the Tables, during the year under review. It is safe to assume that £100,000 was lost in engine leakages alone. There is no doubt it would pay to employ an engineer in each of the larger stations to do nothing else but make tests and report on fuel losses with a view to their remedy.

Boiler-House Plant.—There are great differences in the qualities of steam generated by boilers. Engine-builders find much less difficulty in getting their guarantee figures from the honest steam of the Lancashire or other cylindrical boiler than they do from the high-pressure "Scotch Mist" given off by the now fashionable box of water-pipes. In lighting stations there are great fluctuations in the demand for steam. Only those who really understand the " art " of firing, can appreciate the difficulty of working mechanical stokers with economy under such conditions. Extra grate area may be the salvation of the " tyro " at the higher loads, but with such men, it is fatal to economy at the lighter loads, especially when steam-jets to force the draught are used.

An efficient method of forced draught is a decided advantage in lighting stations, if it is properly used, for it enables steam to be maintained over the peak of the load. The latest system of Mr. Halpin's thermal storage is the ideal method of obtaining the same end. At light loads the steam-jet method of forcing draught is often greatly mismanaged. A peep through the small side door, at the back of the grates, in a Babcock boiler, fitted with moving grates and forced draught, and carrying, say, half load, is often a revelation of how not to do it. One-third of the grate may be covered by a spouting volcano of fire, the remainder employed in carrying away dead ashes and clinker, allowing large quantities of cold air to pass, all of which has to be heated up to the temperature of the escaping gases. Under these conditions, a low flue temperature is no criterion of economy.

Some makers of steam-jet draught-producers only require 3 per cent. of the steam generated to supply their blowers ; others state that they require 3 per cent., but their fittings are found to be large enough to pass four times that amount, even when the full rating of the boiler is taken. The average fireman, as a rule, goes for all he can get under any conditions ; hence it comes that he is able to maintain steam under the conditions already described, and waste large quantities of fuel by heating up the great excess of air passing through the uncovered bars.

Some years ago the author made a steam-economy test of a large pumping plant in the Colonies. The boilers were fired with wood, and the draught was got by means of steam-jets. It was found that they used 19 per cent. of the total steam generated. In a recent test of a modern stoker the numerous steam-jets used 12 per cent. of the total steam generated. By the use of fans, especially when heated air is used, good economy is obtainable. Feed-pumps and other boiler-house auxiliary engines require a considerable percentage of the steam generated. In condensing stations, better economy is obtained by utilizing the exhaust of the auxiliary engines in heating the feed than working them as condensing engines. In lighting stations the losses in the ring steam main, with its numerous traps, is a serious consideration ; it is seldom less than 5 per cent., and may be very much more if the traps are not kept in perfect order and properly regulated.

Superheating, especially when piston-valve engines are used, shows a good steam economy, but the saving in fuel is not so marked. In lighting stations its proper regulation is rather difficult. The extra wear and tear on the main engines, and on the superheaters themselves, are items that have to be considered.

The Station Engineer.—He should note that the average works costs for the non-condensing lighting stations, Table 1 (page 523) and Table 4 (pages 526–7), are $16\frac{1}{2}$ per cent. less than the condensing stations, given in Tables 3 and 6. One explanation of this may be that less plant requires less looking after and less repairs ; there is also less capital invested. But a well-managed condensing station with economical engines, fitted with their own condensers, &c., is certainly the most economical type of station, especially where trams are coupled with lighting. The figures already given prove that the great economy due to condensing may be frittered away by using unsuitable plant, and by careless management. In his specifications, he should insist that the total steam used by the main engine and its condensing plant be included in the guaranteed consumption; and that the evaporation of the boilers be taken in dry steam, delivered to the engine, and exclusive of steam used for various purposes in the working of the boiler. He would then be much better able to decide on the best type of plant to use.

Steam Turbines.—There are only four stations given in the Tables wholly using steam turbines:—Newcastle District, Table 5 (page 528), Scarborough, Cambridge, and Melton Mowbray, Table 6 (pages 529–530). Newcastle Station is the lowest, both in coal, and works costs; yet it is easily beaten by South Shields, a station with vastly less output, situated on the same river, and using triple-expansion condensing engines, which comes out 35 per cent. less in fuel, and 25 per cent. lower in works costs than the turbine station.

Gas-Engine Stations.—In the costs Table of the "Electrical Times," particulars are given of only two gas-engine stations." Northwich, using "Mond" producer-gas, has an output of 104,000 units. Its fuel cost is 0·48 penny per unit; works costs are 1·48 per unit. This is handsomely beaten by Leigh, Table 4 (page 526), a station with a larger output, using non-condensing steam-engines, which comes out at 28 per cent. less in fuel, and 15 per cent. less in works costs.

Redditch, the other gas-engine station, using "Dowson" producer-gas, has an output of over 172,000 units; coal costs are 1·3 pence per unit, and works costs are 2·61 pence per unit. There are over a dozen smaller sized steam-engine stations given in the Tables that beat these results.

It is encouraging to note from recently published returns that the costs in various stations throughout the country show a decided tendency to improvement. This is partly owing to cheaper coal, but there is also a healthy rivalry between station engineers, stimulated no doubt, by the "Electrical Times" publishing their results. This should in time bring the electric generating stations into the front rank as regards economy of fuel.

If any remarks in this Paper assist the station engineer towards obtaining this better economy, the author will feel amply rewarded.

TABLE 1.

Metropolitan Stations.   Non-Condensing.

LIGHTING ONLY.

	Coal per Unit sold.	Units sold.	Units × Coal Cost.	Works Costs per Unit sold.	Units × Works Costs cost.
	d.		*d.*	*d.*	*d.*
Blackheath . . .	1·22	359,554	438,655	2·18	783,827
Chelsea	0·81	2,011,150	1,629,031	1·70	3,418,955
Islington	1·16	2,186,044	2,535,811	1·98	4,328,367
Notting Hill . . .	0·84	947,491	795,892	1·58	1,497,035
St. James's . . .	0·86	5,842,496	5,024,546	1·37	8,004,219
Shoreditch . . .	*1·48	2,247,055	3,325,641	2·25	5,055,873
Southwark . . .	1·35	484,431	653,981	2·29	1,109,346
Stepney	0·5	1,008,037	504,018	0·81	816,509
Westminster . . .	1·04	8,927,883	9,284,998	1·59	14,195,333
TOTAL . . .	—	24,014,141	24,192,573	—	39,209,464

$$\frac{24,192,573}{24,014,141} = 1\cdot007 \text{ pence per unit sold for Coal.}$$

$$\frac{39,209,464}{24,014,141} = 1\cdot632 \text{ pence per unit sold for Works Costs.}$$

* Steam partly supplied by Refuse Destructor.

TABLE 2.

Metropolitan Stations.   Partly Condensing.

LIGHTING ONLY.

	Coal per Unit sold.	Units sold.	Units × Coal Cost.	Works Costs per Unit sold.	Units × Works Costs cost.
	d.		*d.*	*d.*	*d.*
Brompton . . .	0·77	1,702,749	1,311,116	1·27	2,162,491
Charing Cross .	0·85	6,553,792	5,570,723	1·39	9,109,770
Hampstead . .	1·21	2,127,173	2,573,879	1·89	4,020,356
Lambeth . . .	1·67	921,867	1,539,517	2·74	2,525,915
TOTAL . . .	—	11,305,581	10,995,235	—	17,818,532

$$\frac{10,995,235}{11,305,581} = 0\cdot972 \text{ pence per unit sold for Coal.}$$

$$\frac{17,818,532}{11,305,581} = 1\cdot576 \text{ pence per unit sold for Works Costs.}$$

TABLE 3.

Metropolitan Stations.   Condensing.

LIGHTING ONLY.

	Coal per Unit sold.	Units sold.	Units × Coal Cost.	Works Costs per Unit sold.	Units × Works Costs cost.
	d.		*d.*	*d.*	*d.*
City of London . .	1·06	12,719,872	13,483,064	1·81	23,022,968
Clerkenwell . . .	0·81	2,839,713	2,300,167	1·53	4,344,760
Hammersmith . .	0·79	1,461,427	1,154,527	1·47	2,148,297
Kensington . . .	0·81	3,058,680	2,477,753	1·54	4,710,367
London Electric .	1·28	3,546,461	4,539,470	2·36	8,369,647
Metropolitan . .	1·57	11,122,022	17,461,574	2·42	26,915,293
St. Pancras . . .	1·04	4,729,840	4,919,033	1·90	8,986,696
Wandsworth . . .	1·07	1,511,392	1,617,189	2·32	3,506,429
Woolwich District .	1·12	329,484	370,022	1·91	629,314
TOTAL. . . .	—	41,318,891	48,322,799	—	82,633,771

$$\frac{48,322,799}{41,318,891} = 1\cdot169 \text{ pence per unit sold for Coal.}$$

$$\frac{82,633,771}{41,318,891} = 1\cdot999 \text{ pence per unit sold for Works Costs.}$$

TABLE 4 (*continued on opposite page*).

Provincial Non-Condensing Stations.

LIGHTING ONLY.

	Coal per Unit sold.	Units sold.	Units × Coal cost.	Works Costs perUnit sold.	Units × Works Costs cost.
	d.		*d.*	*d.*	*d.*
Aberystwyth . .	1·72	78,815	135,561	2·95	232,504
Alderley . . .	1·03	29,853	30,748	2·23	66,583
Bangor. . . .	2·05	86,820	177,981	3·3	286,506
Barnsley . . .	0·79	200,112	158,088	1·41	882,157
Birkenhead . .	0·61	391,516	238,824	1·34	524,631
Bury St. Edmunds	1·21	73,042	88,380	2·08	151,927
Chelmsford . .	1·79	82,906	148,401	2·72	225,504
Crewe	0·59	310,505	183,197	1·01	313,610
Darlington. . .	0·57	162,563	92,660	1·16	188,573
Dewsbury . . .	0·60	263,076	157,845	1·61	423,552
Dublin	1·85	625,580	1,157,323	3·24	2,026,879
Edinburgh. . .	0·34	7,760,307	2,638,504	0·65	5,044,199
Exeter	1·49	311,160	463,628	2·7	840,132
Folkestone . .	1·25	464,392	580,490	1·9	882,344
Govan	0·42	409,397	171,946	1·22	499,464
Greenock . . .	1·00	157,681	157,681	1·75	275,941
Harrow. . . .	1·53	183,908	281,379	3·02	555,402
Hove	1·15	629,427	723,741	1·8	1,132,968
Ilford	0·88	403,144	354,766	1·33	536,181
Isle of Wight . .	1·56	182,384	284,519	2·43	443,193
Leigh	0·35	151,048	52,866	1·01	152,558
Leith	0·84	511,765	429,882	1·36	696,000
Leyton	0·68	811,141	551,575	1·36	1,103,151
Liverpool District.	1·13	126,689	143,158	2·4	304,053
Llandrindod . .	1·35	79,420	107,217	2·8	222,376
Morley	1·03	99,205	102,181	2·28	226,187
Carried over	—	14,585,856	9,612,541	—	18,236,575

TABLE 4 (*concluded from opposite page*).

Provincial Non-Condensing Stations.

LIGHTING ONLY.

	Coal per Unit sold.	Units sold.	Units × Coal cost.	Works Costs per Unit sold.	Units × Works Costs cost.
	d.		*d.*	*d.*	*d.*
Brought over		14,585,856	9,612,541		18,236,575
Motherwell . .	0·46	166,860	76,755	0·98	163,522
Nelson	0·68	106,189	72,208	2·15	228,306
Newmarket . .	1·19	60,676	72,204	2·1	127,419
Northampton . .	0·80	286,890	229,512	1·7	487,713
Oswestry . . .	1·37	50,649	69,389	2·87	145,362
Pontypool . . .	0·75	70,491	52,868	1·88	132,523
Richmond . . .	1·19	217,789	259,168	1·96	426,866
Shrewsbury . .	0·99	249,263	246,770	1·65	411,283
St. Annes' . . .	0·57	98,369	56,070	1·39	136,732
Stirling. . . .	0·90	111,369	110,255	1·98	220,510
Stockton . . .	0·84	205,322	172,470	2·2	451,708
Watford . . .	1·94	290,947	564,437	3·05	887,388
West Hartlepool .	0·61	268,523	163,799	1·69	453,803
Wimbledon . .	*1·18	806,280	951,410	1·7	1,370,676
Winchester . .	1·36	192,385	261,643	1·69	325,130
Windsor . . .	1·24	366,534	454,502	2·01	736,733
Wycombe . . .	1·17	295,727	346,000	1·66	195,179
TOTAL . . .	—	18,430,119	13,772,001	—	25,137,428

$$\frac{13,772,001}{18,430,119} = 0·747 \text{ pence per units sold for Coal.}$$

$$\frac{25,137,428}{18,430,119} = 1·363 \text{ pence per unit sold for Works Costs.}$$

* Steam partly supplied by Refuse Destructor.

2 P

TABLE 5.

Provincial Partly-Condensing Stations.

LIGHTING ONLY.

	Coal per Unit sold.	Units sold.	Units × Coal Cost.	Works Costs per Unit sold.	Units × Works Costs cost.
	d.		*d.*	*d.*	*d.*
Altrincham . .	1·83	354,615	648,945	3·76	1,333,352
Belfast . . .	0·61	1,206,699	736,086	1·27	1,532,507
Birmingham . .	1·00	3,391,099	3,391,099	1·82	6,171,800
Bournemouth .	1·04	1,003,401	1,043,537	1·77	1,776,019
Cardiff . . .	0·67	1,005,763	673,861	1·46	1,468,413
Coventry . . .	0·80	404,968	323,974	1·8	728,942
Eastbourne . .	1·82	447,295	814,076	2·74	1,225,588
Glasgow . . .	0·36	9,282,043	3,341,535	0·95	8,817,940
Guildford . .	1·97	99,150	195,325	2·78	275,637
Middlesbrough .	0·60	284,332	170,599	1·36	386,691
Morecambe . .	1·28	202,967	259,809	2·61	529,743
Newcastle District . }	0·69	1,619,671	1,117,572	1·22	1,975,998
Preston . . .	0·51	912,142	465,192	0·92	839,170
Tunbridge Wells	1·5	659,000	988,500	1·98	1,304,820
TOTAL. . .	—	20,873,145	14,170,110	—	28,366,620

$$\frac{14,170,110}{20,873,145} = 0·678 \text{ pence per unit sold for Coal.}$$

$$\frac{28,366,620}{20,873,145} = 1·359 \text{ pence per unit sold for Works Costs.}$$

TABLE 6 (*continued on next page*).

Provincial Condensing Stations.

LIGHTING ONLY.

	Coal per Unit sold.	Units sold.	Units × Coal cost.	Works Costs per Unit sold.	Units × Works Costs cost.
	d.		*d.*	*d.*	*d.*
Accrington . .	*0·16	97,869	15,659	1·54	150,718
Ayr	0·97	490,496	475,781	1·53	750,458
Barking Town .	1·16	285,916	331,662	2·07	591,846
Barnes . . .	1·34	44,370	59,455	2·27	100,719
Barrow . . .	0·96	284,709	273,320	1·93	549,488
Bath	0·68	743,607	505,652	1·59	1,182,335
Bedford . . .	1·49	655,574	976,805	2·22	1,455,374
Bexhill . . .	1·22	231,734	282,715	1·7	393,947
Bristol . . .	0·91	2,756,624	2,508,527	1·55	4,272,767
Bromley . . .	1·01	306,323	309,386	1·54	471,737
Burton . . .	0·36	241,534	86,952	1·76	425,099
Bury	0·49	323,337	158,435	1·26	407,404
Cambridge . .	0·77	382,224	294,312	1·67	638,314
Canterbury . .	*0·80	288,455	230,764	1·32	380,760
Cheltenham . .	1·25	653,364	816,705	2·15	1,404,732
Chester . . .	0·56	840,960	470,937	1·22	1,025,971
Colchester . .	1·29	180,166	232,414	2·24	403,571
Derby. . . .	1·43	754,952	1,079,581	2·28	1,721,290
Ealing . . .	1·71	705,786	1,206,894	2·21	1,559,787
Eccles. . . .	1·35	155,247	209,583	2·63	408,299
Fareham . . .	1·12	64,115	71,808	2·39	153,234
Gloucester . .	0·58	270,713	157,013	1·32	357,341
Hanley . . .	1·05	697,602	732,482	2·16	1,506,820
Harrogate . .	0·79	648,677	512,454	1·54	998,962
Hastings . . .	1·55	703,814	1,090,911	2·31	1,625,810
Hereford . . .	0·64	96,012	61,447	1·71	164,180
Carried over	—	12,904,180	13,151,654	—	23,100,963

* Steam partly supplied by Refuse Destructor.

TABLE 6 (*continued on opposite page*).

Provincial Condensing Stations.

LIGHTING ONLY.

	Coal per Unit sold.	Units sold.	Units × Coal Cost.	Works Costs per Unit sold.	Units × Works Costs cost.
	d.		*d.*	*d.*	*d.*
Brought over		12,904,180	13,151,654		23,100,963
Huddersfield .	0·51	1,179,849	691,722	1·14	1,345,027
Hull	0·84	1,490,099	1,251,683	1·74	2,592,772
King's Lynn . .	0·59	286,679	169,140	1·26	361,215
Kingston. . .	2·46	346,332	851,976	3·35	1,160,212
Lancaster. . .	*0·82	341,536	280,059	1·79	611,349
Leeds. . . .	0·27	3,055,165	824,894	0·62	1,894,202
Leicester. . .	0·55	1,042,302	573,266	1·34	1,396,684
Lincoln . . .	0·58	263,216	152,665	1·35	355,341
Llandudno . .	*0·69	279,292	192,711	1·35	377,044
Melton Mowbray	0·93	43,742	40,680	2·19	95,794
Newport . . .	0·50	578,275	289,137	1·39	803,802
Norwich . . .	0·85	1,332,109	1,132,292	1·59	2,118,053
Oxford . . .	0·89	569,561	506,909	2·21	1,258,729
Paisley . . .	1·41	370,521	522,421	2·92	1,081,921
Portsmouth . .	1·11	1,847,790	2,051,046	1·7	3,141,243
Prescot . . .	0·99	215,231	213,078	1·7	365,892
Rathmines . .	0·85	362,552	308,169	1·34	485,819
Reading . . .	0·89	464,580	413,476	1·42	659,703
Rochdale. . .	0·89	91,632	81,552	2·24	205,255
Salford . . .	1·35	716,355	967,079	2·87	2,055,938
Salisbury. . .	0·46	150,222	69,106	1·21	181,768
Scarborough. .	1·31	379,743	497,463	2·1	797,460
Sheffield . . .	0·46	2,487,584	1,144,428	1·0	2,487,584
South Shields .	0·44	985,646	433,684	0·92	906,794
Stafford . . .	0·62	245,803	152,397	1·77	435,171
Carried over	—	32,029,996	26,872,687	—	50,275,735

* Steam partly supplied by Refuse Destructor.

TABLE 6 (*concluded from page* 529).

Provincial Condensing Stations.

LIGHTING ONLY.

	Coal per Unit sold.	Units sold.	Units × Coal Cost.	Works Costs per Unit sold.	Units × Works Costs cost.
	d.		d.	d.	d.
Brought over		32,029,996	26,872,687		50,275,735
Torquay . . .	2·41	199,369	480,479	3·47	691,810
Wakefield . .	*0·32	331,183	105,978	1·05	347,742
Wallasey . . .	0·69	269,424	185,902	1·42	382,582
Walsall . . .	0·80	340,232	272,185	1·83	622,624
West Ham . .	0·95	1,583,421	1,504,241	1·7	2,691,815
Wolverhampton .	0·67	746,318	500,033	1·44	1,074,697
Wrexham . .	*1·01	76,390	77,153	2·32	177,224
Yarmouth . .	1·85	356,882	660,231	2·63	938,599
TOTAL . .	—	35,933,215	30,658,889	—	57,202,828

$$\frac{30,658,889}{35,933,215} = 0\cdot853 \text{ pence per unit sold for Coal.}$$

$$\frac{57,202,828}{35,933,215} = 1\cdot591 \text{ pence per unit sold for Works Costs.}$$

---

* Steam partly supplied by Refuse Destructor.

TABLE 7.

Lighting and Tramways.

NON-CONDENSING.

	Coal per Unit sold or generated.	Units generated or sold.	Units × Coal cost.	Works Costs per Unit sold or generated.	Units × Works Costs cost.
	d.		*d.*	*d.*	*d.*
Blackpool . .	1·01	2,018,132	2,038,313	1·82	3,673,000
Nottingham. .	0·46	4,094,897	1,883,652	0·81	3,316,866
Southampton .	0·94	1,430,222	1,344,408	1·67	2,388,470
Stockport . .	0·62	425,175	263,608	1·43	608,000
TOTAL . .	—	7,968,426	5,529,981	—	9,986,336

$$\frac{5,529,981}{7,968,426} = 0 \cdot 693 \text{ pence per unit sold or generated for Coal.}$$

$$\frac{9,986,336}{7,968,426} = 1 \cdot 253 \text{ pence per unit sold or generated for Works Costs.}$$

TABLE 8.

Lighting and Tramways.

PARTLY CONDENSING.

	Coal per Unit sold or generated.	Units generated or sold.	Units × Coal cost.	Works Costs per Unit generated or sold.	Units × Works Costs cost.
	d.		*d.*	*d.*	*d.*
Aberdeen . .	0·66	1,546,569	1,020,735	1·1	1,701,225
Ashton . . .	0·69	977,044	674,160	1·18	1,152,911
Blackburn . .	0·71	2,002,141	1,421,520	1·09	2,182,333
Bolton . . .	0·44	3,120,709	1,373,111	0·80	2,496,567
Liverpool . .	0·50	20,018,142	10,009,071	0·81	16,214,695
Sunderland . .	0·73	2,375,557	1,734,156	1·11	2,636,868
TOTAL . .	—	30,040,162	16,232,753	—	26,384,599

$$\frac{16,232,753}{30,040,162} = 0\cdot540 \text{ pence per unit sold or generated for Coal.}$$

$$\frac{26,384,599}{30,040,162} = 0\cdot878 \text{ pence per unit sold or generated for Works Cost.}$$

TABLE 9.

Lighting and Tramways.

CONDENSING.

	Coal per Unit sold or energated.	Units generated or sold.	Units × Coal cost.	Works Costs per Unit sold or generated.	Units × Works Costs cost.
	d.		*d.*	*d.*	*d.*
Bootle . . .	0·59	1,327,432	783,184	0·90	1,194,688
Bradford. . .	0·39	4,901,172	1,911,457	0·76	3,724,890
Brighton . .	0·74	4,860,430	3,596,718	1·43	6,950,414
Burnley . . .	0·40	505,736	202,294	1·07	541,137
Carlisle . . .	0·71	494,475	351,077	1·61	796,104
Cork	0·32	1,701,996	546,638	0·61	1,038,217
Croydon . . .	0·92	1,810,387	1,665,556	1·34	2,425,918
Dover . . .	1·24	781,809	969,443	2·13	1,665,253
Dundee . . .	0·82	1,041,800	854,276	1·33	1,385,594
Grimsby. . .	0·84	292,094	245,358	1·46	426,457
Halifax . . .	0·48	2,557,548	1,227,623	1·09	2,787,727
Manchester .	0·41	10,502,299	4,305,942	1·15	12,077,643
Newcastle . .	0·48	2,562,383	1,229,943	0·94	2,408,610
Oldham . . .	0·71	1,042,055	739,859	1·33	1,385,933
Plymouth . .	1·17	832,435	973,948	1·72	1,431,788
Southport . .	0·49	1,462,407	716,579	0·97	1,418,534
St. Helens . .	0·35	910,736	318,757	0·82	746,803
Taunton . .	1·11	385,907	428,356	1·63	629,028
Tynemouth . .	0·74	449,467	332,605	1·41	633,748
Whitehaven .	0·7	228,887	160,220	1·44	329,597
Wigan . . .	0·45	579,904	260,956	1·13	655,291
Worcester . .	0·85	662,569	563,183	1·45	960,725
TOTAL . .	—	39,893,928	22,383,972	—	45,614,129

$$\frac{22,383,972}{39,893,928} = 0·561 \text{ pence per unit sold or generated for Coal.}$$

$$\frac{45,614,129}{39,893,928} = 1·143 \text{ pence per unit sold or generated for Works Costs.}$$

TABLE 10.

Electric Tramways.

CONDENSING.

	Coal per Unit generated.	Units generated.	Units × Coal cost.	Works Costs per Unit generated.	Units × Works Costs cost.
	d.		*d.*	*d.*	*d.*
Blackpool and Fleetwood	0·46	917,930	422,247	1·16	1,064,798
Dublin . . .	0·43	6,651,113	2,859,978	0·79	5,254,379
Glasgow . . .	0·19	10,527,489	2,000,222	0·6	6,316,493
Hull	0·54	1,969,557	1,063,560	0·99	1,949,861
TOTAL . .	—	20,066,089	6,346,007	—	14,585,531

$$\frac{6,346,007}{20,066,089} = 0\cdot316 \text{ per unit generated for Coal.}$$

$$\frac{14,585,531}{20,066,089} = 0\cdot721 \text{ per unit generated for Works Costs.}$$

TABLE 11.

Electric Railways.

CONDENSING.

	Coal per Unit generated.	Units generated.	Units × Coal cost.	Works Costs per Unit generated.	Units × Works Costs cost.
	d.		d.	d.	d.
Central London	0·38	16,753,292	6,366,250	0·92	15,413,028
City and South London . .	0·28	4,850,420	1,358,117	0·94	4,559,394
Liverpool Over-head . . .	0·24	4,910,285	1,178,468	0·83	4,075,536
TOTAL . .	—	26,513,997	8,902,835	—	24,047,958

$$\frac{8,902,835}{26,513,997} = 0·335 \text{ per unit generated for Coal.}$$

$$\frac{24,047,958}{26,513,997} = 0·907 \text{ per unit generated for Works Costs.}$$

The Paper is illustrated by 1 Figure in the letterpress.

Discussion.

Mr. McLaren said the Paper had been written deliberately to promote discussion, so he hoped the members would not be sparing in their criticisms. Before the discussion commenced he wished to emphasise the last three lines in paragraph 3 (page 511), namely, " The price paid for coal varies largely, and different systems of supply are used. All these considerations affect the costs." He had taken the whole of the stations in the country, so that the average cost of fuel might be got throughout the country. Metropolitan stations were kept separate, and compared with metropolitan, as he presumed they paid the same for coal and labour in all these stations. The different systems of supply was a very important point, and he hoped that some of the electrical members present would throw some light on their effect, so far as works costs were concerned. The losses in the mains were debited to the generating plant in all lighting stations that measured the current at the consumers' meters. Trouble would have been saved if the current had been metered at the stations, and comparisons of the merits of the various generating plants could have been made much easier. Metering the current at the consumers' premises, however, emphasised the fact that there must be a great reduction in the costs of generating and distributing current, if central power stations were to compete for mill and factory driving.

Mr. Ade Clark in his Paper had compared the most economical oil-engine with a steam-engine that no self-respecting power-user would keep for long on his premises, and he considered that advocates of central power-stations were doing very much the same thing. They compared the best they could put down with old antiquated engines, thus making a good comparison ; but he would like some of the mill-engine builders to give those engineers a few figures of what they would have to compete with, if they wished to get the driving of mills or factories of any size. The wages of the engine men was a large factor in a small power plant; it was not so important when one got to 100 H.P. and upwards ; in one or two cases that he had investigated, the mill engines were generating

(Mr. McLaren.)

their power on the premises at three-eighths of a penny per I.H.P., including interest and depreciation, etc. He considered that the best oil-engine should be compared with the best steam-engine; a gentleman should be compared with a gentleman, not with a rogue. With regard to gas-engines in central stations, he hoped that some of the manufacturers of gas-engines would shed some light on how it was that, with their highly perfected engines, they were not getting any better results, if as good, as the steam-engine. Costs per unit generated in central stations fitted with turbines did not show up better than the steam-engine stations. The figures he had given were taken from actual work. The steam-engine makers, gas-engine makers, and turbine makers were all pouring their most economical engines and prime movers into the capacious maw of the central station, and how was it that they did not get any better results out of the central station? He thought, perhaps, if they were to try two or three highly economical Diesel engines they might give out a better economy, but as yet they had not been tried. So far as gas-engines, steam-turbines, and steam-engines were concerned, the cost per unit at the consumers' meters were high out of all reason.

The PRESIDENT was sure it would be the pleasure of the members to convey their thanks to Mr. McLaren for his most interesting Paper; he had thrown down the gauntlet most handsomely.

Mr. DRUITT HALPIN thanked the author for his kindly reference to himself in connection with boilers and Thermal Storage. Before proceeding to deal with the subject of the Paper, he wished to give one or two figures of comparison between gas and steam-engines which might be of use to the members, which came under his notice within the last fortnight. The gas plant was quite new, having been put down last year with all the latest improvements. The engines were of 250 H.P., and very accurate accounts were kept of what they were indicating, braking, and burning. Close by them in the same works there was a steam-engine, and the owners noted carefully for a month what that steam-engine was doing. The engine was a tandem compound with a jet condenser, was of about 80 H.P., working at 60 lbs. pressure, which it had been doing during the last 28 years. The boilers that were driving it had been working for 32 years at 60 lbs. pressure. The gas-engine used 0·89 lb. of coal; the steam-engine used 3 lbs. of coal. But that was not the end of the story. The coal for the gas-engine cost 23s. a ton; the coal for the steam-engine cost 6s. 6d. a ton, so that, if it were taken out in money, the gas-engine was using 5 per cent. more money than the engine 28 years old. This charge was for fuel only, and was exclusive of repairs, which were very heavy in the case of the gas-engine plant.

Coming to the subject of the Paper, it might possibly be interesting to give some particulars of what was being done at the present moment. For a good many of the figures he intended to give he had to thank Mr. Miller, the engineer of the joint station of the Kensington and Notting Hill Co. at Shepherd's Bush. That Company was making polyphase current at 5,000 volts, and sending it down to town where it was transformed. The full line, Plate 14, showed the load curve for the maximum load in January 1903, reaching about 2,000 kw., and the maximum load for January 1904, reaching 2,750 kw. The stations were increasing their load very fast, and the engineers found they would have to provide for an additional 1,200 I.H.P. in the coming winter. That 1,200 H.P. was shown by the difference in the height between the top and bottom load curve lines, and was built up from the result of their previous experience in their extensions. The dotted lines showed either coal or thermal units put into the boilers. When they were working in the old ordinary way the coal curve would be the top one for 1904, and with Thermal Storage in use it would be the lower one, the hatched area showing the amount of heat stored in the Thermal Storage vessels. The increase in steam-producing power when the boilers were being fed from the Thermal Storage vessels was 22 per cent., which was taken off the top of the curve; that was to say, the work stored in the bottom cross-hatched areas was given out on the peak of the load. If anyone wanted to say that was incorrect he would not be wrong, because the heat curve was given taking the varying feed temperature into account. The feed temperature was 150° F. with

(Mr. Druitt Halpin.)

a large volume of water, and also a large volume of gas going through the economisers, and it was higher at light load, coming up to as much as 220° F.

Fig. 2, Plate 15, showed the combined general arrangement of one of the existing boilers and the Thermal Storage vessels now being fitted, two to each boiler. The existing boilers were marked A, and the boundary line below the top girders which carried the condensing tanks limited the available room. In the remaining space the vessels were placed. The only alteration that was made was that a hole was cut in the top of the boiler shell, a saddle piece riveted on, and a pipe put up which acted as a support to the vessel and had a connection for the steam and feed-water into the boiler, so that the pressure and the temperature in the Thermal Storage vessels and in the boiler drums were always constant. The Thermal Storage vessel was only rigidly connected at one point, so that it was perfectly free to expand in all directions, being carried at the other end by a suitable cast-iron supporting saddle-piece. The only other change that was made was in the feed-pipes, which were disconnected from the boiler steam drums and carried up into the Thermal Storage vessel, and the whole of the feed passed direct into the vessels, and then passed through the feed regulating cock and pipe into the boiler, as shown on the diagram. Supposing in the morning with a light load the water was down to the lowest water-level in the Thermal Storage vessel: if 10,000 lbs. an hour was being used in the engines, 20,000 lbs. would be pumped in; 10,000 went into the engines and the remaining 10,000 gradually accumulated until 3 or 4 o'clock, when the heavy load came on, so that when the vessel was full there was a charge of 20,000 lbs. available in each vessel at the same temperatures and pressures as existed in the boilers; thus at the top of the load, instead of having to evaporate from 150° F., they were only putting in the latent heat, as the temperature of the feed from the vessels was already 388° F., equal to the temperature in the boiler. In order to give some idea of what that meant, he might say, that as 20,350 lbs. were stored in each vessel, and 4,843,300 British thermal units were therefore contained in each vessel (this amount being the product of the available amount of feed-water into the difference between the temperature at which the feed was received at top load, viz., 150° F. and the temperature at which it entered the boiler, viz., 388° F.), the total working pressure was 215 lbs. per square inch, this increase temperature being, of course, 238° F., and there were ten of those vessels. The steam used for the engines in that particular station was 23 lbs. per kw.-hour.

Each of the ten vessels stored up 250 kw.-hours, which gave 2,500 kw.-hours to help them over the top of the load. The reason that system was adopted in connection with the new plant was very simple. The engineer went very closely into the question of the figures, and obtained tenders for the boilers and buildings, and without making any allowance for the value of land, which was very high in London, it was found that Thermal Storage only cost 70 per cent. of the price of new boilers without contingencies. In connection with the question of price, Mr. Miller was good enough to give him the figures that actual electric storage batteries power had cost him. Four batteries were put in of 1,200 kw.-hours, which, without buildings, cost £7,500, so that the electric battery power cost exactly 6¼ times as much as the Storage. Another point in connection with the battery was what would be got out of it. Fortunately on that head he was able to give very exact figures. In one of the Thermal Storage vessels Messrs. Willans and Robinson had put up at their Rugby Works they kept the vessel filled with steam for eleven hours. They put a thermometer in the top, in the middle, and in the bottom of the vessel, and calibrated them carefully with a very accurate steam-gauge. They worked with the thermometer to see what result was being obtained, the final planimetered that in the test the drop in the temperature of the feed-water entering the boiler was 3° F. below the steam temperature. A drop of 3° F. represented a loss of 0·8 of 1 per cent. He did not know whether any of the electric accumulator people would get a less loss than 25 per cent. in watts delivered.

Referring to what the actual losses by radiation were on the vessels, the results obtained at Rugby on a vessel 4 feet in diameter by 16 feet 8 inches long, showed that they were condensing

(Mr. Druitt Halpin.)

something like 24 lbs. or 25 lbs. of steam per hour, representing 3·2 lbs. of coal. In another vessel 6 feet in diameter by 19 feet long, in another place, the measured condensation over ten hours was 1·125 lbs. of coal per hour. That showed that at Rugby 1 lb. of coal was sufficient to deal with the radiation from 73 square feet of heating surface, whereas in the other case 1 lb. of coal would deal with the radiation of 234 feet of radiating surface, a proportion of 3 to 1 merely in the efficiency of the two different non-conducting materials. Messrs. Willans and Robinson used a most beautiful and costly material, but unfortunately it had not the same effect. He could give some other figures which might be of interest. Last April he tested with an assistant a plant of the old type, which had been at work for ten or twelve years. For one day they ran nine-and-half hours without storage, and the next day ten hours with storage. The water was very carefully measured by a Schönheyder water-meter, which was calibrated before and after the test; the coal was also weighed, its value on both days being tested by a calorimeter, so that they knew exactly what they were doing. The result was an increase of evaporative efficiency of 21·7, when Storage was being used, without giving credit for the difference in the thermal value of the coal.

A test much more interesting to the meeting would be the last one he would give, which took place practically without his knowledge; the tests were finished before he knew they were being made, his clients determining to test the apparatus for themselves. They ran from the 10th May last to the 7th June, day and night, Sundays included, with varying loads, which were very light during eight hours of the night, carefully measuring the whole of the coal and water. The result was that 334,560 lbs. of coal were burned, and 2,786,720 lbs. of water evaporated, showing that with exceedingly low class Midland slack, costing 6s. 6d. per ton delivered in the bunkers in Birmingham, an evaporation of 8·12 was being done from and at 212° F. They then ran from the 21st June to the 19th July under the same conditions, evaporating 2,223,270 lbs. of water, and burning 352,464 lbs. of coal, and in that case the evaporation from and at 212° F. as before was 6·7. That test confirmed the figures he obtained in his own test, his figure for two months being 21·7, while his clients' was 21·35. In addition to the advantage of an increase of economy during the whole time of evaporation, there was a further advantage in getting the greater part of the dirt and lime in the feed-water deposited in the vessel, and in some cases, the whole of the dirt in the feed-water was precipitated; whereas, if they passed the feed-water through an economiser in order to get the benefit of the increased temperature, the lime in the water was deposited inside the pipes, where, owing to the relatively high temperature outside, it was baked on in a very hard condition, and was very difficult to remove.

On page 23 of Mr. Stromeyer's report for 1902 to the Manchester Steam Users' Association, he referred very fully to the undesirable amount of heat stored up in the brickwork of water-tube boilers, which became most inconveniently evident when the boilers had suddenly to be taken off the load. In this case, in order to get rid of this accumulated heat, it was necessary to continue pumping in feed-water, and allow the steam produced to blow off through the safety valves. Where Thermal Storage was used this waste did not occur, as the heat remaining in the brickwork was all usefully applied in raising the temperature of the feed-water pumped into the Thermal Storage vessels for use during the time the next heavy load came on.

The apparatus had also been applied to locomotives, the boiler being fed from the Thermal Storage vessel when going up hill, the vessel having been previously filled while the engine was running down hill. By this means an increase in boiler steaming power was produced amounting to 30 per cent.

Mr. GEORGE WILKINSON thought the Paper was full of interest to all engineers connected with the generation of power; it simply bristled with points upon which profitable and interesting discussion might be raised. The one statement which impressed him most in reading the Paper was that on page 512, namely, that, taking the average results from the tabulated statements in 1902, condensing stations were not, by some considerable extent, so efficient, in point of works costs and fuel costs, as non-condensing

(Mr. George Wilkinson.)

stations. Having read that, he looked through the Paper to find out if the author gave any reasons for such a startling conclusion. He found one (page 521) under the heading "The Station Engineer," where the author said "One explanation of this"—*i.e.*, the economy of non-condensing plants—"may be that less plant requires less looking after and less repairs; there is also less capital invested." The capital invested did not affect the works costs, but the other did. The author continued, "But a well-managed condensing station with economical engines, fitted with their own condensers, &c., is certainly the most economical type of station." He had emphasized the "well-managed" because he thought that probably condensing stations, taking them all round, were as well-managed as non-condensing stations. He was thoroughly in accord with the author when he advocated the use of complete units, such as he described (page 516). The author pointed out there a general instance of a station which was fitted with a separate condensing plant common to the whole of the generators in the station, and it was not surprising to him, after having seen a number of such stations, that the economy came out so low; because when one came to consider that the range of output of those stations, especially lighting stations, varied from 100 per cent. for a short time down to 10 per cent. for a very much longer period, and that in such stations one often found the condensing-plant engine of larger size than the day-load engine, and of a less economical type, it was a fertile source of loss; whereas if each generating unit were fitted with its own condenser, and made to drive its own air and circulating pumps, very much better results might be obtained.

He would like to ask the author whether he could give more in detail the black sheep amongst the condensing stations which lowered the averages so much that they compared unfavourably with non-condensing plants. The author mentioned (page 521) a fertile source of loss in station plants, namely, the capacious appetite of the boiler feed-pumps for steam. That was a loss which was common to both non-condensing and condensing stations, but he did not hesitate to say that in stations where the output varied very considerably, as it did in most electricity works, feed-pumps were altogether a mistake for boiler-feeding purposes. He had designed and successfully introduced into several electric-supply stations in the country a new arrangement of injectors for this purpose, shortly described as follows:—Injectors in various sizes (so that a wide range of feed could be obtained) were " banked " in a row; the bank was divided in the centre by valves inserted in the steam, water, and delivery pipes. The steam to the injectors was taken in at each end of the bank, thus providing a duplicate service; a dual water-supply was also furnished by connecting at each end of the bank, and in like manner duplicate deliveries to the boilers were provided. Instead of using live steam from the steam ring, all the water separators and all the steam-jackets of the engine cylinders were connected to a small steam-pipe, which in its turn fed the bank of injectors. By this method, although the duty of each individual injector was curtailed, the enormous advantage was obtained of doing away with steam-traps; and the injectors were able to pass the whole of the condensed water from the separators and cylinder jackets through again into the boilers. It was found also that the injectors were quite capable of taking hot water from any ordinary hot-well, providing the injectors had no high lift on the suction. This arrangement was simple, efficient, and very economical, as proved by a number of years of continuous working.

The author drew certain conclusions (page 522) between steam-turbines and reciprocating plant, and he rather joined issue with him on that matter. The Newcastle station was quoted as being the most economical steam-turbine station, but that came out rather at a disadvantage when compared with the South Shields station mentioned also in the same paragraph. He thought it was only fair to say that many of the steam-turbines in the Newcastle station were of an old type; much of the pioneer work had been done there, and their economy could not be equal in any sense to the present-day turbines. Another point was that South Shields, which had triple-condensing engines, had a load factor of 14·6 per cent., whereas the Newcastle station had a load factor of only 13·67 per cent. Of the other steam-turbine stations mentioned, Scarborough had a load factor as low as 9·81 per cent., Cambridge only 8·59 per

2 Q 2

(Mr. George Wilkinson)

cent., and Melton Mowbray only 6·16 per cent. In connection with
this matter he had a few months ago the advantage of testing a
300-kw. triple condensing engine, direct-coupled to a generator, and a
300-kw. turbo-generator working under precisely the same conditions,
excepting that in the case of the vertical triple-expansion engine
the superheat was somewhat greater than in the turbine, but in the
turbine test the vacuum was a little better. At full load the
vertical triple engine came out at 23·9 lbs. per kw. generated against
the turbine 22·76 ; at half-load the vertical triple engine came out
at 29·4 against the turbine 24·1. He did not mention these figures
in any spirit of cavil, but to show that there was little difference
between a modern steam-turbine plant and a high-class triple-
expansion reciprocating engine plant in point of steam consumption.

Likewise in connection with gas-engine stations, the author
quoted Leigh against Northwich. Leigh had a load factor of
12·32 per cent., and also had the advantage of being on a coal-field
where coal was cheap, whereas Northwich had a load factor of only
9·34 per cent., and purchased their gas from the Brunner-Mond Co.,
which company, no doubt, made a profit on the transaction. With
regard to losses in mains, the author allowed 20 per cent. (page 511).
He thought he had erred on the safe side there, because at any rate
in alternating systems 30 per cent. or 33 per cent. would be much
nearer the mark.

Mr. CHARLES FORGAN thought the Paper would have been of
more value to central station engineers, if the author had investigated
his figures before publishing them, not merely taking them from the
" Electrical Times," because he did not think they went to the point.
For instance, the author stated (page 514) that the works costs of the
three railway power-stations were heavy. He would like to know
whether the other power-station results were obtained on a similar
basis as the railway generating stations. Being connected with one
of those railways, the Central London, he thought the figures were
very misleading ; more details might have been given. For instance,
on page 536 the Central London Railway's works costs per unit was
given as 0·92d. That included power-house costs, the wages of

locomotive drivers, costs of repairs to locomotives, and the wages of the men employed in that department. The total cost which the $0 \cdot 92d$. referred to was £31,537 for the half-year. The power-house costs were £18,049, while £13,488 represented the cost of the locomotives and car department. Taking the power-house costs and works costs alone, the figure obtained was just a little over $0 \cdot 5d$. per unit generated, that was taking the wages of the office staff, working expenses, water, gas, waste, store repairs and maintenance of the power station. He thought $1\frac{1}{2}$ lbs. of coal per I.H.P. was a very good figure if it was maintained in most stations.

He thought the author was in the same boat as the speaker referred to on the previous day in comparing badly equipped condensing stations with well-equipped non-condensing stations; if the author had taken modern well-equipped condensing stations, he would have found the comparisons come out better. For instance, the Central London Railway had not the best condensing plant, and the engineers very soon knew it in the matter of boiler-power if they had trouble with the condensers. With reference to the Scotch mist, there was no doubt that Lancashire boilers had good staying powers; and he thought it was an advantage to power stations to have them in parallel with water-tube boilers, because the Lancashire boiler was very good for steady loads, but if one was in a hurry a water-tube boiler was wanted. He had had a good experience with water-tube boilers at the London Electric Supply Corporation and the Central London Railway, and in an extension which had just been carried out four Lancashire boilers had been put in, and a saving resulted when the boilers were run with the water-tube boilers.

He thought it would have been more to the advantage of station engineers had the author dealt more with the different costs of coal. The mechanical stoker had a good deal to do with the question. A lot of engineers thought the mechanical stoker a boon: he thought otherwise. At the Central London Railway two years ago they changed from using the mechanical stoker to hand-firing—he would not mention the particular type—because he did not think there was much to choose between them. Difficulty was experienced in getting

(Mr. Charles Forgan)

the men to do exactly as one wanted, but after six months' running
the coal costs were £4,000 a year less than they were previously
with mechanical stokers. He could not say that that saving was all
due to the stokers, because he had not much to do with the company
while the mechanical stokers were in use ; but, under instructions
from his chief, he made the change he had referred to. The units
generated were practically the same in both cases, while the engines
had only undergone the usual overhaul and adjustment; therefore
the stoking had a good deal to do with the saving in fuel.

Mr. J. D. BAILIE thought the author was not quite fair to
station engineers in general in regard to condensing. They were not
blind to the advantages of a good vacuum, and the case cited (page 516)
of 23 inches at the condenser and 11 inches at the engine exhaust
was quite an exception. Driving from the main engine had one
disadvantage, in that if anything went wrong with the air and
circulating pumps it disabled the main engine temporarily, until
the pumps were disconnected. Probably, in general, the best
method for central-station working would be to drive the pumps
by motors, and to have intercommunication pipes and valves
between the condensers, so that, in the event of any hitch occurring
on the one condenser, a reasonably good vacuum could be obtained
by coupling on to another one. He did not think that the last
sentence of the first paragraph on page 517 was quite correct,
for the author stated that the steam required by the auxiliary
machinery was altogether saved. That, of course, was not so ; for
the air-pumps and the circulating pumps had work to do in moving
the water against a certain head, which entailed an expenditure of
power and steam, whether by the main engine or by an auxiliary
engine.

The author quoted (page 518) a test of a 120-H.P. triple-expansion
condensing engine, the exceptionally good result of 12·54 lbs. of
steam per I.H.P. per hour being obtained, with a boiler pressure
of 157 lbs. per square inch, and no superheat. The mechanical
efficiency of this engine, including the driving of all the pumps and
the brake-wheel friction, was given as 93·5 per cent. This certainly

seemed a high mechanical efficiency for a triple-expansion engine, and 12·54 lbs. per I.H.P. was an exceedingly low steam consumption under the conditions stated. Working these figures out they came to 13·4 lbs. per B.H.P., and, if coupled to a dynamo with only 90 per cent. efficiency, which was low for a dynamo, an electrical H.P. consumption of about 15 lbs. would be obtained, or a consumption per kilowatt of say 20 lbs. With all respect, he did not think the author's firm would care to guarantee a consumption of 20 lbs. per kilowatt per hour under those conditions.

Mr. McLaren interjected that he had given the names of the men who had carried out the tests.

Mr. Bailie said he would like to know, in connection with these tests, what the vacuum and the temperature of the steam were. It did not state the amount of superheat, if any, and what allowances were made for leakage, etc.

Mr. McLaren replied that there was no superheat.

Mr. Bailie, continuing, said that the author advocated running the auxiliary engines non-condensing and using the exhaust steam for heating the feed. Why not combine the two, by running both the main engines and the auxiliary engines condensing, and, as was done at West Bromwich, pass the exhaust steam through feed-heating tubes first and then through the condenser, imparting say 60° F. to the feed, and maintaining a vacuum of say 26 inches on the condensers? He thought perhaps the greatest economy was obtained in that way.

The author referred in a somewhat disparaging manner to steam-turbine stations. It was rather a pity he had done so, unless, as he said, it was for the purpose of provoking discussion, because it was quite unfair to compare an old station like the Newcastle and District with a practically new station such as South Shields. The Newcastle and District station was partly condensing; the South Shields

(Mr. J. D. Bailie)

station was wholly condensing. The Newcastle and District station had to pay for the carting of coal and also for the carting of ashes, whereas in the South Shields station the coal was practically taken from the boats to the furnaces. The heading " Fuel Costs " was quite misleading, in that it did not take into account the different items which were included. The author also gave the stations mentioned by him as the only four stations wholly equipped with turbines, but of course there were others, including West Bromwich, Shipley, etc. Three of the four cited, namely, Newcastle, Cambridge, and Scarborough, were the oldest. The fourth, Melton Mowbray, was a very small station, with a total capacity of only 150 kilowatts, so could not fairly be compared with a large station. For comparison with turbine stations, the author selected South Shields, a comparatively new station with every facility for economical working, but he omitted to take a station such as Oxford. At Cambridge the total cost of fuel per unit sold was $0 \cdot 66d.$, whilst at Oxford it was $0 \cdot 74d.$ The stations were started in the same year, and the conditions as to irregularity of load were probably much alike, both being University towns, yet the cost of coal, the total works cost, and everything else came out higher at Oxford than at Cambridge. A comparison between Cambridge and Oxford would have been fairer than one between Cambridge and South Shields, though probably the author would dispute this, and would say that the reason for the higher fuel cost at Oxford was the system of distribution in vogue; if so, his whole Paper fell to the ground at once, for it showed the fallacy of abstracting isolated columns from tables such as those published in the "Electrical Times," and assuming the figures in them to represent the economy of the engines. Melton Mowbray was, as previously stated, a very small station, with 150 kw. capacity and an output of 43,000 odd units per annum, as against 1,600 kw. capacity and a million odd units output at South Shields.

After reading the author's Paper, he wrote to the Melton Mowbray engineer, who informed him in reply that, since the addition of a larger turbo, the cost of coal had been reduced to $0 \cdot 388d.$ per unit sold, and that it was coming down every week.

He also wrote to the engineer of West Bromwich, who had very kindly furnished him with a copy of their official returns as prepared for the Board of Trade. The total works cost there, although it was comparatively a small station, came out at 0·65d. per unit sold, and the cost of fuel at 0·35d.—they were below that at the present time. That was a better result by far than South Shields, considering the difference in size of the stations, etc.

In regard to the question of superheat, he presumed the author's remarks referred to separately fired superheaters. If they referred to flue superheaters, he did not agree with him; there was very little upkeep or attention required with the flue type of superheater, except in cleaning out the soot deposited from time to time—a simple matter. With superheaters of the ordinary Babcock and Wilcox type, about 1,500° F. superheat was regularly maintained at a number of stations which he knew—West Bromwich, etc. The economy of superheat to that extent was about 15 per cent. of the steam consumption, so that he could not quite see that superheaters were a disadvantage in a station.

Mr. T. L. MILLER thought the point in the Paper which probably interested engineers most was that referring to the superior economy of the non-condensing electric lighting stations over the condensing. He did not think, however, that if all the facts bearing upon the question were taken into consideration, the difference shown would be found to be due to whether the engine were non-condensing or condensing. In the remarks he made after the reading of the Paper, the author, he thought, gave one reason for the figures, and Mr. Wilkinson in his remarks gave another which he proposed to deal with very shortly. Dealing first with the relative economy of non-condensing and condensing engines, it was a well-known fact that working under similar conditions the condensing engine, at all loads, gave a greater economy in water consumption per horse-power hour than the non-condensing engine. Provided, then, that the conditions of loading were the same in the non-condensing and condensing stations, it appeared to him that there could be no

(Mr. T. L. Miller.)

increased economy of the non-condensing engine over the condensing. This being so, it was necessary to look in some other direction for an explanation of the very startling figures given by the author. Now, in addition to the type of plant employed, the efficiency of a generating station, stated in terms of cost of coal per Board of Trade unit sold, depended, amongst other things, upon:—

1. The system of supply.
2. The "load factor."
3. The cost of fuel.
4. The personal factor in the management.

All these factors had a considerable bearing upon the costs of production. Dealing with the first factor, namely, the system of supply, it might be pointed out that the stations in this country could be divided roughly into two classes, one the high-tension and the other the low-tension stations. The high-tension stations, for the most part, were put down in districts where the demand was sparse, when one had to go for considerable distances in order to get consumers. On the other hand, low-tension stations were put down in more densely populated districts. Leeds was an example to the contrary, and there were other such exceptions throughout the country. In the high-tension stations the difference between the units sold and the units generated was very much greater than in the low-tension stations; for example, from an examination of the records of fourteen low-tension stations and twenty high-tension stations, he (the speaker) found that the average percentage of units sold to units generated was 86·6 in the low tension, and but 67·3 in the high-tension stations. Allowing for these losses, it might reasonably be expected that the coal costs in the high-tension stations would be higher than those in the low-tension stations. In addition to this, however, the low-tension stations had the advantage of batteries, which, in the smaller stations, tended to further reduce the costs; as well as a considerable motor load during the day, giving an increased "load factor." Bearing in mind these facts, and dividing the stations into two classes, high and low tension, it would be seen what a great effect the system of supply had on the costs, and how fully their ideas of the economy, in terms of coal consumption per unit sold, of the condensing engine over the non-condensing were borne out. Leaving out of consideration the Metropolitan Company's stations, part of which were condensing and part non-condensing, and also the eight stations where steam was partly supplied by refuse destructors, the stations might be grouped as follows :—

	Metropolitan.		County.	
	H. T.	L. T.	H. T.	L. T.
Non-condensing . . .	2	6	6	36
Partly condensing . .	3	1	8	6
Condensing	5	3	36	17

It would be noticed from these figures that in the non-condensing stations the low-tension systems of supply were greatly in excess of the high-tension, and from the data already given regarding the better efficiency of the low-tension systems, it would be readily understood that the non-condensing results would show a high average efficiency. In the condensing stations, on the contrary, the high-tension systems of supply were in excess of the low-tension, and as a consequence a lower average efficiency would be obtained for the condensing stations. This disproportion between the high and low-tension systems of supply in the non-condensing and condensing stations appeared to him to account largely for the supposed superior economy of the non-condensing stations referred to by the author.

Dealing, then, with the author's figures of coal costs per unit sold, and omitting as before the figures for the Metropolitan stations, and the eight "destructor" stations, he (the speaker) subdivided the purely "lighting" stations into two groups from which he obtained the following figures for the average coal costs per unit sold :—

(Mr. T. L. Miller.)

	Metropolitan.		Country.	
	II. T.	L. T.	II. T.	L. T.
Non-condensing . . .	1·17	0·93	1·72	0·63
Partly condensing . .	1·14	0·85	1·03	0·54
Condensing	1·04	0·95	0·90	0·64

Referring to the high-tension stations, he pointed out that in both the metropolitan and country stations the condensing stations showed a greater economy than either the partly condensing or non-condensing, and that the partly condensing stations in like manner showed a better economy than the non-condensing. In the low-tension stations, however, the coal costs in both non-condensing and condensing stations were sensibly the same, while the partly-condensing stations appeared to better advantage than either the condensing or non-condensing stations. This difference in the metropolitan low-tension stations, he thought, might be accounted for by the better load factor of the non-condensing stations, which, according to the "Electrical Times," varied from 11·38 per cent. to 22·74 per cent. in the non-condensing stations, was 18·6 per cent. in the case of the partly condensing, and varied from 7·87 per cent. to 16·55 per cent. in the case of the condensing stations. In the country stations again, the load factor and the price of coal would very largely affect the results, some stations being in the neighbourhood of the coal-fields, and having good motor loads, while others had to pay a heavy price for coal and at the same time had practically no day-load. In addition to these points the personal factor in management had a very considerable influence on the reduction of costs, and while he (the speaker) thought that the system of supply had a great influence on the question of costs, particularly in the smaller towns, he was fully aware that it could not account for all the variations noted. The "policy" of putting condensing plants down in purely lighting stations was not under discussion, and he did not therefore touch on it.

The next point to which he wished to refer was with regard to the driving of auxiliaries. The author had given some very high figures of steam consumption for driving auxiliaries, but he thought that in most modern stations the auxiliaries, and especially the air and circulating pumps, would be electrically driven. In that connection he would refer to a Paper read before the Manchester Section of the Institution of Electrical Engineers* by Messrs. Taite and Downe, in which the results of the running of the plants at the Salford and Southport stations were given. In those stations the auxiliaries were electrically driven and the energy metered. Regarding the metered units, it was found that the condensing plants taken throughout the year accounted for 4·67 per cent. of the total units generated, and the boiler feed-pumps 1·18 per cent. The electrically driven plant cost about three and a half times the steam plant; but even allowing for the increased capital cost there was still a considerable saving in electrical driving. Mr. Druitt Halpin earlier in the discussion gave some figures with regard to gas-driven plants; he (the speaker) would like to give some figures with regard to another gas-driven plant, a 500-H.P., in which Mond Gas was used, without ammonia recovery. The load factor of the plant was 55 per cent., and the cost of fuel 11s. 8d. per ton. Under these conditions the costs were as follows:—

Fuel	0·138 penny per unit	
Stores	0·055 ,,	,,
Labour	0·151 ,,	,,
Total . .	0·344 ,,	,,

The only other point to which he would refer was the question of the gas-driven stations. Attention had already been drawn to the fact that the Northwich station, referred to by the author, bought its gas from Messrs. Brunner Mond and Co., the price paid being 2d. per 1,000 cubic feet. The author also gave some figures for another gas station, Redditch, where the fuel costs were 1·3d. per unit. The most recent gas station was that at Walthamstow, where Dowson gas-producers and anthracite coal were used. From the return of the past

---

* Journal 1903, vol. 32, page 1050.

(Mr. T. L. Miller)

year's running, which had just been issued, he found that the fuel
costs at Walthamstow amounted to $0 \cdot 42d$. per unit, and the works
costs to $1 \cdot 18d$. per unit, figures that were very much below those
given for Redditch.

Mr. MARK ROBINSON, Member of Council, thought the members
were very much indebted to the author for deducing so many
instructive and valuable conclusions from statistics which indeed
were always .before them, but which without such guidance they too
often found dry and uninteresting.   No doubt what would attract
attention chiefly were the remarks upon the comparative economy
of condensing and non-condensing engines—a very important subject,
to which, in his opinion, not nearly sufficient attention had been
given by engineers.   He could not help thinking that the last
speaker had chiefly been showing how figures might be made to prove
anything, because there was nothing to be surprised at in the fact
that the non-condensing lighting stations gave the best results.
The old idea of a non-condensing engine was that it was a sort
of poor relation, a sort of deformed example, of the only proper
kind of steam-engine, the condensing engine—a view which might
have been reasonable when all steam-engines worked at very low
pressures.   If one cut off the part of the indicator diagram which
was below the atmospheric line, in a case where the initial pressure
was not very far above that line, there was of course very little
diagram left, but when one worked with high pressure, the part of
the diagram above the atmospheric line was very much the largest
part.   In order to retain the smaller part, below the line, one had to
spend much money upon condensing plant, and if the engine
ran only for a few hours in the week, it was quite natural that it did
not pay to do so.   And not only would they have to pay interest
upon the cost of the condensing plant, but the engine itself must
be enlarged.   It was sometimes thought that if one had a
non-condensing engine and then added a condenser to it, more
power should be obtained from it.   That was a wrong idea
altogether; the use of condensing, that is of the artificial removal
of atmospheric back-pressure, was to enable expansion to be carried

further, but for that purpose the cylinder volume must be increased to correspond. The non-condensing engine therefore was much cheaper and smaller than a condensing engine giving the same power, altogether apart from the outlay on air-pumps. When these considerations were taken into account, it was not at all surprising that in stations not running with good load curves the non-condensing plants came out the best. In a power station, with a fairly even load, of course condensing paid, but for almost any other class of station it appeared to him that the system of partial condensation was likely to be the best. When, on the normal average load, condensation could be used, the occasional higher load, the peak of the curve, would be best met by starting an extra non-condensing engine or engines.

It was with satisfaction that he noticed in Mr. McLaren's tribute to the non-condensing engine, what was really a recognition of the work of another Leeds man, the late Mr. Willans, who nearly 20 years ago made non-condensing engines which used less than $18\frac{1}{2}$ lbs. of steam per I.H.P., the same engines using about $12\frac{1}{2}$ lbs. when working condensing. Between those figures there was no doubt a great difference, but in an engine which ran only intermittently, and for a very short time in each day, it quite possibly might not be worth while to save it. The author had endeavoured to account for the comparative non-success of the London condensing stations by speaking of them as pioneer stations, but he did not think that was correct, except with regard to the big station at Deptford, which fortunately was a law unto itself, and exceptional. The old pioneer plants of the London stations had long since been done away with; the real pioneer stations in London were the non-condensing stations, such as those of the St. James's and Pall Mall, and of the Westminster Companies; those stations used Willans non-condensing high-speed engines, and gave excellent results. Only one power station, the great tramway station at Glasgow, seemed to do better in works costs per unit than the best of the lighting stations, and then it was only by the tiniest decimal, the figures for Glasgow being $0 \cdot 6d.$ against $0 \cdot 62d.$ for Leeds and $0 \cdot 65d.$ for Edinburgh. He did not know the comparative cost of

(Mr. Mark Robinson.)

coal in Edinburgh and Leeds, but he imagined Leeds had the advantage in that respect, perhaps sufficient to alter the order in which they stood. It was very remarkable that Edinburgh, which took this high position in the entire list, was another example of a non-condensing station. In view of this he felt justified in entering a protest against the author's condemnation (page 519), where he mentioned " uneconomical three-crank compound, tandem, condensing engines largely used in lighting stations," because these words seemed to refer to the very engines, the Willans high-speed three-crank engines, which, even without the advantage of condensation, gave such splendid results at Edinburgh. Further, in connection with the London stations, it would be found in Table 3 (page 525) that the Kensington condensing station, which had those very three-crank engines, tied with another London condensing station for the second place on the list, and was only just beaten by one other station, which had engines of another kind.

Mr. McLAREN, interposing, stated that he specially mentioned that the most economical stations were fitted with non-condensing engines.

Mr. MARK ROBINSON, continuing, said that piston-valves were somewhat strongly condemned in the Paper. He could say, from an experience of many years, based upon the use of many hundreds of Willans engines fitted with piston-valves, that these valves, when properly made (which was often not the case) remained tight and efficient. His firm had the strongest reason for discovering the best valve, and their continued faith in the piston-valve was simply the outcome of long and oft-repeated experiments ; they could simply find no sort of valve which in actual use was, and remained, so tight. The Edinburgh engines were, of course, piston-valve engines, and they had for the most part been running for years.

Mr. MICHAEL LONGRIDGE, Member of Council, said the most striking fact brought out by the Paper was the superior economy of non-condensing over condensing engines. One reason had been given by a previous speaker, but there was another which he thought must have suggested itself to many of the members, that was the exceeding badness of many of the condensing plants put down by electrical engineers. It was true that electrical engineers had taught steam engineers much, but it was equally true that they had much to learn themselves from the steam engineers. The surface of evaporative condensers, the capacities of air pumps, and the water cooling arrangements were often utterly inadequate, and the idea of the independent central condensing plant which seemed to find favour among electrical engineers was, when reduced to practice, generally a fraud. He strongly advised a separate condenser for every engine. The objection generally urged against this arrangement was the high speed of the engines, which made the driving of the air-pump difficult. But he would remind the members that there was a condenser which required no air-pump, and was therefore independent of the speed of the engine, a condenser which moreover occupied very little floor space, and that was the barometric 33-feet pipe condenser. The water could be supplied either by a centrifugal pump electrically driven, or by a small steam-pump such as the Worthington pump. If, however, a steam-pump was used, he thought it well to caution the members that the stop-valve for admitting the steam to the cylinder must be made exceedingly small. The ordinary stop-valve supplied with such pumps was well adapted for starting the run when the water had to be lifted to the top of the condenser; but the moment the vacuum was established, and the steam-pressure in the pump had only to overcome the friction of the water in the rising main, the quantity of steam taken by the pump became so small that the ordinary stop-valve had only to be opened a fraction of a turn, and he had found that when working in this way the vibration of the pump was very liable to shut the valve and stop the action of the condenser.

Professor ARNOLD LUPTON said that he had several barometrical condensers working successfully. There could be no doubt at all, as stated by Mr. Mark Robinson (page 557), that in considering the kind of engine required for a central power station it was necessary to take into account the load factor. One kind of engine was very

2 R

(Professor Arnold Lupton.)

suitable for 5 per cent. load factor, and another kind more suitable for 50 per cent. load factor. Most of the stations were electric-lighting stations. But in future they would have to deal more with electric power distributing stations. He had had some share in getting passed through Parliament two Acts providing for two large electric power generating stations involving two millions of capital each, so that naturally he had given a good deal of consideration to the question. He quite agreed with the author's remark that it was rather absurd to think that one could get an engine, turn its power into electricity, take it ten miles and drive a factory, and compete with another factory which bought its fuel as cheap and could buy an equally good engine. He and his colleagues never proposed such a thing. They thought they might supersede existing steam-engines of a bad type, using from 5 to 10 lbs. or even 20 lbs. of fuel per horse-power. Where extensions of power were required, and it was not convenient to put down a big engine, then electrical transmission would come in useful. There were economies. He had made a few calculations to show what could be done theoretically. Taking the cost of fuel at a central generating plant distributing power over, say, a radius of ten miles, and taking the cost of fuel in a town for one of the bad engines, which someone said was iniquitous, he found that £1 spent in fuel at an electrical power central generating station would do as much work upon the material that had to be worked upon at the factory as £10 spent on fuel at the factory. Taking the cost of 4s. a ton at the central generating station near the coal mines—which he knew to be a reasonable price in some districts—and taking the cost at 10s. a ton in the town at the small mill of 20 or 30 H.P., at the generating station, with every known device for economy, the I.H.P. would require only 1½ lbs. of fuel per hour ; and at the suggested small factory with second-class engine, it would require 6 lbs., thus making a ratio of 4 × 2½ = 10 to 1 in fuel cost. In the electrically-driven factory the motor would be on the machine or on the countershaft, and in each case the power actually utilised in the machine was taken at one-half of the I.H.P. This calculation took no notice of any costs but the fuel cost.

Mr. W. PRICE ADELL thought that not the least important part of Mr. McLaren's Paper was the pertinent diagnosis of the work done at central stations, enabling him to point out with such clearness the weak spots, some of which were a surprise to engineers—namely, that non-condensing plants on an average acted more economically than condensing plants. He (the speaker) could not pass over Mr. Forgan's condemnation of mechanical stokers; some certainly failed, but the recognised good makes of mechanical stokers got over the three great causes of loss in hand-firing in steam generating, namely, first, the entrance of cold air whilst charging the furnace by hand; secondly, the entrance of cold air whilst cleaning the bars; and, thirdly, the entrance of an excess of air though parts of the furnace charged too thinly, or where combustion had been quicker than in the remaining parts. In most of the stokers used in the Leeds electric light and power stations, members could see how the third loss was prevented by the utilization of a movable bar taking its own restricted air; so that in case of one part of the fire burning faster than another, it did not allow excess of air to pass through the thin part and rob the thicker layer of fuel of its allotted share. In several factories with which he had been connected, attention to these three points had reduced the fuel consumption very considerably, even in cases where expensively watched scientific hand-firing had been practised, and much more so when the stoking had been done with efficient mechanical stokers.

Mr. ALFRED SAXON said there was one aspect of the triple-expansion condensing engine which had not been referred to, and which, he thought, would account for its apparent loss of economy, namely, that in a short run a triple-expansion engine would not be seen at its best advantage nor even at light loads. A triple-expansion engine of 1,400 I.H.P. made by his firm was tested in November last year. In the first test of three hours in the morning the steam consumption worked out, with saturated steam, to 12·6 lbs. per I.H.P. In the afternoon run the steam consumption worked out

(Mr. Alfred Saxon.)

at 12·21 lbs., with practically the same load.* The following day the tests were continued with almost identical results. That served to show that in a long run, up to a given point, the triple-expansion engine became more efficient the longer it ran. The author had asked for some cotton mill engineer to give the cost of running a cotton mill for a week. A cotton mill in Lancashire, with an engine developing 1,000 H.P. could be run for £20 worth of coal per week of fifty-six hours, including the heating of the rooms and the stand-by losses. Taking fifty working weeks to the year and £20 per week, this would equal £1,000, and thus an engine of 1,000 I.H.P. could run for fifty-six hours in the week at a cost of £1 per I.H.P. for a year.

Mr. McLAREN, in reply, said that Mr. Wilkinson asked for the names of the black sheep. He had only to look down the Table of the costs per unit to find them. In connection with auxiliary machinery Mr. Wilkinson mentioned his injectors. That was a very interesting system. He had seen the method at work at Harrogate, and had adopted it with great success, using the injectors to feed the boilers, and taking the steam for the injectors from all the steam separators. Some engineers might think that injectors would not work with water in the steam supply, but they did ; as a matter of fact they worked splendidly.

He was not at one with Mr. Wilkinson with regard to the turbine and the triple engine. Mr. Wilkinson's triple engine was practically a compound engine, the cylinders being proportioned for a low steam-pressure—it was only 120 or 125 at the time of the test ; he had now got up to 130 by squeezing the Boiler Insurance Co. The condenser coupled to the engine was of the ejector type, and only gave a very poor vacuum indeed, something about 15 inches. The turbine worked with a splendid vacuum. It was well known that

---

* Corroborating the figures given by the author, a similar new triple-expansion engine to the one described above was indicated by the Engine and Boiler Insurance Co., on 2nd April 1903, and showed an efficiency of 91 per cent.

about $1\frac{1}{2}$ per cent. in economy was gained for every extra inch of vacuum obtained; if one intended to compare engines with engines, this must be borne in mind.   Mr. Forgan explained (page 547) how it was that the works costs on the Central London Railway were apparently high, through outside expenses, that were not included in other stations, being added to them.   He, the author, specially mentioned in the Paper that that was a point which required explanation, and the explanation given was satisfactory.   Mr. Bailie mentioned (page 548) that the Metropolitan was not all condensing. As he already explained, he took the figures from the Tables published by the " Electrical Times."   Mr. Bailie also doubted the possibility of the $12\cdot5$ and the $93\cdot5$ per cent. efficiencies being correct.   There was no need for doubt in regard to the $12\cdot5$; it was e sily obtained with good engines, as Mr. Robinson and Mr. Saxon had explained; but the $93\cdot5$ per cent. certainly required explanation. The efficiency came out rather over 91 for engine and brake combined. The brake wheel was 10 feet in diameter, and the trough-shaped rim was for water cooling.   Their contract stated that they finished at the shaft-coupling.   They had nothing to do with the alternator, so he demurred to driving this large brake and including its friction as engine losses.   Mr. Wilson Hartnell settled the point; he sent for a motor, obtained a light strap, and drove the wheel uncoupled, in its own bearings without brake straps on, but with equivalent weight added so as to give it the same pressure on the bearings; it was rather surprising to see how long it took to get the meter pointer quite steady; he believed it was found that $2\frac{1}{4}$ E.H.P. was required to run the wheel to proper speed.   That was how the figures were got out, and he knew no man whom he would sooner trust to make a brake test than Mr. Hartnell.

Mr. Bailie also made some remarks with regard to the turbines and engines.   He took the Newcastle station because it was the best. He thought it was an error in the " Electrical Times " tables to say that the district station at Newcastle was partly condensing; it was a fully condensing station.   The turbines were not antiquated and out-of-date; and there were some of the best and latest types of them in that station.   There was very little difference between Newcastle and

(Mr. McLaren.)

South Shields as regards coal supply. He believed that South Shields at that time was paying very much more for coals than Newcastle. Mr. Bailie also mentioned Oxford and Cambridge; but there was no parallel between those two towns. Oxford was an isolated case practically, in regard to the system of distribution. The current left the station at 4 lbs. of coal per kw.; and the members could easily reckon out from the Tables what it amounted to when it was distributed. The price of coal was much higher at Oxford than at Cambridge, Cambridge being on one of the main coal lines to London. Mr. Miller gave some interesting details in connection with the mains and their effect on the costs; but he would reply to his remarks more fully in writing (page 573).

There was one point he would like to mention in connection with the method of distribution of power by electric energy. Ten years ago in his works they took out their steam-engines and boiler used for driving the works and put in a new one, thereby saving over 60 per cent. in fuel. If there had been any electricity employed, all that saving would have been put down to the electricity, but luckily there was no electricity connected with it. It would thus be seen that a great saving could be made, by simply putting in an efficient steam-engine in place of a wasteful one.

Communications.

Mr. E. G. CONSTANTINE wrote that Mr. McLaren was to be heartily congratulated on bringing before the Institution such an important subject as " Fuel Economy in Generating Stations " and the anomalies which existed in the generation of steam in this enormous and rapidly extending industry. It might come as a surprise to the author that, although, as he stated (page 520), "Engine-builders (may) find much less difficulty in getting their guarantee figures from the honest steam of the Lancashire or other cylindrical boiler than they do from the high-pressure ' Scotch mist ' given off by the now fashionable box of water-pipes," an analysis of the figures supplied showed that the boxes of water-pipes gave better results in working than either Lancashire or other shell types of boilers. Of the more economical stations (lighting and traction) referred to in the Tables, 29 generated or sold electricity at a coal cost per unit of 0·5d. or under. Of these stations 10 were equipped exclusively with Lancashire boilers, 3 with shell boilers other than the Lancashire type, 10 with a mixture of water-tube and shell boilers, and 6 stations with water-tube boilers exclusively.

The average coal cost per unit sold or generated at the 29 stations worked out as below :—

				d.
10 stations with Lancashire boilers only	.	.	.	0·435 per unit.
3 ,, ,, shell boilers other than Lancashire	.	.	0·393 ,, ,,
10 ,, ,, mixed water-tube and shell boilers	.	.	0·390 ,, ,,
6 ,, ,, water-tube boilers only	.	.	.	0·385 ,, ,,

It would thus be seen that the coal cost per unit was highest in those stations in which only Lancashire boilers were employed, and that those exclusively equipped with water-tube boilers generated their electricity at the lowest cost for fuel. In passing, it might be noted that in the very best example given in the Tables, that of Glasgow Electric Tramways, the cost of 0·19d. per unit was obtained with the author's "boxes of water-pipes." Had the 10 stations fitted with Lancashire boilers been equipped with "boxes of water-pipes" working at the average cost for the 6 stations given in the Tables, the saving in cost of fuel would have amounted to over £7,000 per annum.

Carrying the comparison farther, it was interesting to observe that the works costs per unit generated were also the lowest in the stations furnished exclusively with water-tube boilers, and highest in those having Lancashire boilers exclusively—the relative figures being 0·86d. in the water-tube boiler stations against 1·024d. in the Lancashire boiler stations. It might be contended that the better results shown by the water-tube boiler stations were due to cheaper fuel, more economical engines and better management, but the first reason was disposed of by the fact that of the Lancashire boiler stations six were actually on the coalfields, and four were within a

(Mr. E. G. Constantine.)

distance of 30 miles of collieries producing steam-coal; while of those stations equipped with water-tube boilers—one was in London, two were in Ireland, two in the Glasgow district, and only one on the coalfields. Doubtless economical engines and good management were responsible to some extent for the results, but these advantages were certainly not peculiar to exclusively water-tube boiler stations. The Lancashire boiler was a good old faithful servant, but where economy, space, and high pressures were in question it was not the best boiler, and must give place to the "box of water-pipes."

Although the "Scotch mist" box of water-pipes (illustrated in connection with Mr. Halpin's remarks, Plate 15) was by no means the most efficient water-tube boiler in use, even it has been proved more efficient than the Lancashire boiler; and one might almost be tempted to infer that, when once the engine-builders' guarantees were fulfilled, engines worked more economically on "Scotch mist" than on the honest steam referred to by the author.

Lt.-Colonel R. E. CROMPTON, C.B., wrote that Mr. McLaren dealt with matters of supreme interest not only to electrical engineers but to all users of power. They were indebted to him for calling attention to certain matters which were far too much neglected by all users of steam power. He thoroughly agreed with his remarks on page 519, in which the author attributed a very large share of the want of economy of steam-engines of all classes to valve and piston leakage, especially the former. He also agreed with him that huge losses occurred through leaky piston-valves, and that Corliss valves, plain slide-valves, and properly balanced flat slide-valves were much easier to keep steam-tight. The author's remarks on this matter must be read in connection with Callendar and Nicolson's valuable Paper* on "The Law of Condensation of Steam." Their Paper showed conclusively what a large proportion of the so-called condensation losses were really due to valve leakage losses and subsequent re-evaporation of the steam, and that this leakage was great when saturated steam was used, but became negligible when

---

* Proceedings Institution of Civil Engineers, 1897-98, vol. cxxxi., p. 147.

the steam was sufficiently superheated; in fact that the reduction of this leakage was the chief cause of the economy shown when superheated steam was used. Experiments made since the date of their Paper showed that certain steam-engines driving dynamos, and which had been designed to give high economy, entirely failed to do so when using saturated steam, on account of the heavy leakages past the piston-valves, and that this leakage had been so reduced by superheating the steam that the consumption had been reduced in the proportion of 100 to 45, a reduction which could not be explained thermodynamically.

He thought that Mr. McLaren was aware of this two years ago from remarks which he had made to the writer in conversation which took place at Newcastle ; and since that time he himself had found that by carefully considered superheating he had been able to obtain economies from ordinary engines, which had enabled him to obtain such economy in electric light and power stations that in two important cases he had reduced the works costs to figures far below any of those given in Mr. McLaren's Tables for condensing stations ; for instance, the whole of the power at his own works had been for some years past supplied by a separate power station. In this case he had reduced the fuel in pounds and pence per unit delivered into the works from 5 lbs. and $0 \cdot 42d$. in 1900 to $3 \cdot 8$ lbs. and $0 \cdot 32d$. in 1903. It would have added greatly to the value of the Paper if the author had attempted to do what he did in former cost-papers addressed to the Institution of Electrical Engineers, namely, attempted in each case to find the actual weight of coal used per unit as well as its cost in pence.

Although they were greatly indebted to Mr. McLaren for bringing such interesting facts together, he was sorry to say that he could not agree with some of the conclusions the author had arrived at. For instance, on page 520, under the heading of boiler-house plant, he stated that engine-builders found much less difficulty in getting their guarantee figures from the honest steam of the Lancashire or other cylindrical boiler than they did from the high pressure " Scotch mist" given off by the now fashionable box of water-pipes. He considered such a statement to be not only erroneous but mischievous,

(Lt.-Colonel R. E. Crompton, C.B)

and it was entirely contrary to his own experience. It was no doubt made in good faith, but was in no way based on the results of average practice. No doubt Mr. McLaren and any other engine-builder putting down plant and taking his supply of saturated steam from new and clean Lancashire boilers would find it somewhat easier under those conditions to get sufficiently dry steam, than he would by taking the supply from equally new water-tube boilers ; but water-tube boilers were now almost invariably fitted with superheaters, and with reasonable management it was found in practice that, under varying conditions of load, it was far easier to obtain regularly superheated steam from these latter boilers than from Lancashire boilers with the separate superheater usually employed with them. At any rate the writer's best results, which were certainly as high as any that had ever been obtained in practice, had been obtained with water-tube boilers using chain-grate stokers, but, and here followed a condition of the greatest importance, the setting of these boilers must be as carefully considered as the design of the boilers themselves. It was only of late years that Mr. Miller had shown at the Wood Lane joint station of the Kensington and Notting Hill Companies, that the space allowed in the combustion chambers of this class of boiler had been altogether too small when bituminous or smoky coals were used. If this matter was attended to, not only was the production of steam from this kind of boiler quite as regular and as free from moisture as from the Lancashire boilers, but the superheating could be carried on in the same flues with extreme regularity.

Another point in which he criticised the Paper was that little notice was taken of the question of lubrication. Whenever condensation was used, it was most desirable that the quantity of oily matter which passed into the return feed-water should be minimised, and one of the problems of the modern station engineer was to effect this. Of late years it had been found that the sight-feed lubrication was unsatisfactory. Far more satisfactory lubrication had been obtained by positive lubricators, in which either the oil or a paste of oil and graphite was forced into the steam-pipe with perfect regularity. Callendar and Nicolson drew attention to the important part that lubrication played in the steam-tightness of sliding surfaces. His own experience agreed with theirs. He found that if they desired to obtain satisfactory and sufficient lubrication, so as to minimise the amount of greasy matter that went forward with the steam and still kept the sliding surfaces in the most steam-tight and perfect condition, they did so best by keeping the steam dry. All water passing forward with the steam rapidly removed the lubrication from the sliding surfaces, but, on the other hand, he found that with supply at the higher temperature involved with regular superheat even up to a total temperature of 800° F., on account of the facts just mentioned, surfaces could be kept in better order with a small use of oily matter than was possible with engines using saturated steam said to be sufficiently dry for economy.

Mr. CHARLES DAY wrote that from the various Tables given by Mr. McLaren it would appear that no advantage was to be derived from the adoption of condensing at electric-lighting stations; such a conclusion was, however, quite contrary to facts, as condensing plant properly applied was of very material benefit even in a generating station where current was supplied for lighting only, and this fact was best appreciated when the condensing plant at a station was for any reason stopped for a time.

It was to be noted that the figures from which Mr. McLaren made his deductions in regard to condensing might equally be used to show whether water-tube boilers gave more economical results than Lancashire boilers: whether slow-speed engines were more economical than high-speed, or for almost any comparison in regard to the equipment of electrical stations that one might wish to make. Conclusions based on figures covering such a wide area, and including so many variables, proved nothing beyond the truth of the old saying that " Statistics can prove anything."

Apart from the question of economy in coal, the Tables appeared to show that the works cost, exclusive of coal, was lower in stations which had condensing plant in addition to their engines, than it was in stations which had not got the additional condensing plant;

(Mr. Charles Day.)

thus, in the Metropolitan non-condensing lighting stations the total works cost was $1 \cdot 632d$. per unit, and the fuel cost $1 \cdot 007d$. per unit, the difference being $0 \cdot 625d$. per unit. Turning to those Metropolitan stations which were partly condensing, one found that instead of the difference being $0 \cdot 625d$. it was $0 \cdot 604d$. Again, in Provincial lighting and tramway non-condensing stations the works cost, exclusive of coal, was $0 \cdot 56d$. per unit, whereas the corresponding cost in the partly condensing stations was $0 \cdot 338d$., and in the condensing stations $0 \cdot 405d$.; thus, the works cost, exclusive of coal, in the non-condensing lighting and tramway stations was about 70 per cent. higher than in those stations which had some condensing plant also.

Excluding coal, the works cost included oil, waste, water, stores, wages of workmen, repairs and maintenance. The most enthusiastic advocate of condensing plant would, he thought, hardly be prepared to say that the addition of condensing plant would reduce any of these items except water. Generally speaking, therefore, the Tables showed that condensing plant, instead of reducing fuel cost, reduced the other items forming the works cost; this deduction was, however, so absurd that he believed the sounder conclusion was that the Tables proved nothing.

Mr. McLaren instanced (page 516) a case of 11 inches vacuum being recorded in the cylinder, whilst 23 inches were recorded in the condenser; this case only showed that the pipes between the cylinder, or the ports into the cylinder, or both, were absurdly small, or that there was a distinct air-leakage near the cylinder. If independent condensing plant was placed within a reasonable distance of the engine with which it was connected, there was no difficulty whatever in getting a vacuum at the cylinder outlet within $\frac{1}{2}$ inch of the vacuum in the condenser. The author instanced a case where the steam used by a separate condensing plant was excessive. There was no reason why such a state of affairs should exist any more than that the steam used by the main engines should be excessive. Each could be kept satisfactory by proper attention, as had been shown in another part of the Paper.

It was stated (page 517) that when condensers were attached direct to engines the steam required by the auxiliary machinery was altogether saved. That statement was clearly wrong, as in either case the same amount of work had to be done in driving the pumps; the only difference in steam consumption was that arising from the possibility of the larger engine using less steam per H.P. than the smaller condensing engine. On the other hand, when an engine was but lightly loaded, the independent condensing plant might be run at half-speed, in which case the steam consumption would be less than if the pumps were driven direct from the engine. Several disadvantages were brought in when pumps were coupled to the main engine. First, the design of the main engine usually required modification to permit of direct connected pumps. Second, it was generally now preferred to run engines driving dynamos at higher speeds than it was desirable to run air-pumps. Consequently if the pumps were so connected, they themselves were not so satisfactory as if worked independently and at lower speed ; or the speed of the main engine had to be reduced to suit the pumps, with consequent increase in its cost, and in that of the dynamo. In designing engines, vibration troubles required careful consideration, and these troubles were very likely to be increased if the pumps were connected to the engine direct.

The author's suggestions as to the various ways in which improvements in economy could be obtained were excellent, but in almost every instance they applied with equal force to stations which had condensing plant as to those which had not. Mr. McLaren stated (page 518) that 25 per cent. might be taken as the saving in fuel due to condensing, other conditions being equal. The writer's own experience quite confirmed this statement, but he would modify it by saying that, after taking into account the capital invested in condensing plant, and the slight increase of maintenance and works costs, excepting fuel and water, the safe estimate of the net gain obtained by the adoption of condensing was 20 per cent.

Mr. McLaren wrote, in further reply to those who took part in the discussion, and to the written communications since received, that it was evident from Mr. Druitt Halpin's remarks (page 538), that there was a manifest saving in fuel, by simply passing the feed-water through the storage vessel before it reached the boiler proper,

(Mr. McLaren.)

quite apart from the advantage derived from storage. Trapping the
dirt and lime at the storage vessel instead of allowing it to pass into
the boiler was also a great advantage, especially where water-tube
boilers were used.

Mr. George Wilkinson stated (page 546) that the losses in the
mains, between the station and the consumer, might be as much as
30 to 33 per cent. where alternating current was employed. This
was a very important point to be borne in mind when settling the
system of supply from a central station.

Mr. Charles Forgan remarked (page 547) that if well-equipped
condensing stations had been compared with the non-condensing, the
former would have shown up better, which was no doubt true, but
any result could have been obtained had stations been selected in
each class. As already stated, the Paper was based on the broad
fact that all the stations throughout the country were taken. It was
impossible to compare station with station, unless the price and
calorific value of the fuels used in the various stations were known,
but by taking all stations, using fuel at all prices, it was possible to
compare one class with another in a fair manner.

Mr. J. D. Bailie considered (page 548) that the author was
rather unfair to the engineers of condensing stations. This was
disclaimed by the author, who was personally a strong advocate of
condensing stations, and in making suggestions where the efficiency
of these stations could be improved, he did so in the most friendly
spirit. The figures given (page 516), namely 23 inches at the
condenser and 11 inches at the engine, were certainly the worst
the author had found, and were not intended to be taken as normal
conditions in condensing stations. Respecting Mr. Bailie's remarks
concerning steam used by auxiliary condenser engines not being all
saved, the author would call his attention to the whole of the
paragraph (page 517) from which he quoted. It read that the author's
firm usually guaranteed $7\frac{1}{2}$ per cent. less steam per kw.-hour if
the engines had their own condensers and drove their own pumps.
The $7\frac{1}{2}$ per cent. was the net saving in the main engines, after
deducting the power required to drive the pumps. It was found
that the better vacuum obtained at the engine cylinders more than

compensated for the power required to drive the pumps. It was under these circumstances that the whole of the steam that would have been used by the auxiliaries was saved.

It was the practice in this country for engineers who made engines with separate condensers not to include the steam used by the condenser engine in their guarantee for the main engines. The author pointed out that this steam ought to be included in that used by the main engines.

Mr. Bailie also considered that the steam turbines had been treated in a disparaging manner, and objected to Newcastle District Station being compared with that of South Shields. The author admitted that there were objections to comparing individual stations, unless the whole of the conditions and cost of fuel, etc. were known. He had therefore got out the average cost per unit generated in the four turbine stations, and found that it was 5 to 6 per cent. higher than the average in the non-condensing steam-engine stations, both for coal and works costs. Great steam economy was claimed for turbines when tested individually, but the figures given in the " Electrical Times " Table of Costs did not substantiate that claim for turbine stations as a whole. Mr. Bailie added some particulars concerning Melton Mowbray and West Bromwich, to show that the costs in these stations had come down rapidly during the present year. This applied equally to steam-engine stations, but as the Paper only dealt with the complete year 1901 to 1902, a period of dear coal, these additional figures could not be discussed.

Mr. Bailie in his remarks concerning superheaters said that he could not quite see that they were a disadvantage. The author agreed that superheating was a great aid to economy, and only mentioned the fact that the economy of fuel was not so marked as the economy of steam. This was especially the case where separately-fired superheaters were used.

Mr. T. L. Miller added a valuable contribution to the discussion (page 551). His remarks went to prove that the system of supply had an important bearing on costs. He selected fourteen low-tension stations, and showed that the loss between the switch-board and the consumer was 13·4 per cent. and that in the high-tension stations it

(Mr. McLaren.)

was as much as 32·7 per cent.   The latter figures closely agreed
with Mr. Wilkinson's estimate of the loss in high-tension alternating
systems.   Had Mr. Miller taken the whole of the stations in the
country as a basis, his figures would have been more valuable,
though perhaps less startling.   In dealing with the coal costs of the
lighting stations using low tension, he found that the coal costs in
the condensing and non-condensing stations were practically
identical, whereas the condensing stations ought to have shown a
large saving.   In the high-tension stations in London, he showed
the condensing stations to come out better than the non-condensing,
but to obtain this result, he omitted the Metropolitan Companies'
Stations from the condensing list.   They generated 11 million
units out of a total of 41 millions, and their cost per unit was about
50 per cent. higher than the remaining stations which he took ;
whereas on the non-condensing side, he omitted Shoreditch, which
only generated 2¼ million units, at about the same cost per unit.
Had he included all stations, the non-condensing side would have
certainly come out the best.   In the Country Lighting Stations
there were only seven working non-condensing, and using the high-
tension system of supply.   From these he selected six, the seventh
(Wimbledon) having a destructor.   The six selected stations were
Aberystwyth, Chelmsford, Dublin, Exeter, Morley, and Watford.
The total output of these stations was under 1½ million units, out of a
total of over 75 millions generated in the Country Lighting Stations.
Dublin accounted for over 42 per cent. of the total units generated
in the six stations, at a cost for coal of 1·85d. per unit. which was
abnormally high for a large station.   The cost of coal in the other
five stations was also high.   It would therefore appear that high-
tension non-condensing stations were not economical.   Mr. Miller
somehow managed to divide the whole of the current generated, as
coming from separate high- or low-tension stations, but about
20 per cent. of the whole came from stations using both systems of
supply.   It would be interesting to know how he ascertained the
amount of current supplied through each system at these stations.
On the whole, Mr. Miller's figures went to prove that the system of
supply had a very important bearing on the cost per unit of
electricity supplied to the consumer.

Mr. Mark Robinson mentioned (page 558) that Edinburgh and Kensington, both economical stations, had three-crank compound engines. The author ventured to assert that had these stations been fitted with three-crank engines, having only three cylinders, side by side (triple-expansion engines), they would have come out still better in economy. A 300-H.P. three-crank tandem compound engine was simply three engines of 100 H.P. each, and therefore had a larger proportion of loss due to clearance, initial condensation, and other losses inherent in small engines. The great economy in the stations mentioned by Mr. Robinson was no doubt largely due to good and careful management. The author would remind him that quite a number of firms made, or had made, three-crank tandem compounds.

Mr. E. G. Constantine wrote (page 564) giving some particulars concerning water-tube boilers, which he said might come as a surprise to the author, but, strange to say, he gave no figures to disprove the only thing the author inferred, namely, that the quality or dryness of the steam given off by the water-tube boiler was not so good as that given off by the cylindrical boiler. To an engine builder who guaranteed the amount of saturated steam his engine would use, the quality of the steam was of vital importance; 10 lbs. of "Scotch Mist" per lb. of fuel might in the long run be cheaper to the station engineer than 8 lbs. of honest steam, but under the circumstances already mentioned, the "Scotch Mist" penalised the engine builder, yet added to the apparent evaporation of the boiler, a point to be borne in mind by the station engineer when making engine and boiler tests. Mr. Constantine's remarks and figures were certainly interesting, but they did not bear on anything said by the author. When superheaters were added to water-tube boilers, the steam of course was dry, and good economy was obtained from its use. Mr. Constantine did not state how many of the water-tube boilers mentioned by him were fitted with superheaters. The cost of cleaning and scaling a water-tube boiler, as compared with a Lancashire, was an important item; repairs and upkeep had also to be considered; these came in the works costs, and were important items to the station engineer. As

(Mr. McLaren.)

Mr. Constantine had shown, works costs were in some cases lower where water-tube boilers were in use, and no doubt many cases could be cited where high works-costs and water-tube boilers went together.

Lt.-Colonel Crompton considered (page 567) the "Scotch Mist" statement not only erroneous, but mischievous; then he proceeded to state that it might be easier to get drier steam from a new cylindrical than a new water-tube boiler, but a water-tube boiler fitted with a superheater he considered gave good dry steam. With this statement most people would agree. The author was dealing with the quality of the steam given off by the boiler, not by the superheater.

Mr. Charles Day wrote (page 569) that he considered the conclusions drawn from the figures given in the Tables quite contrary to facts. The author would remind him that the figures given were actual facts; conclusions drawn from them might be influenced by what Mr. Day termed variables, and they might be drawn on too wide an area for comparisons to be made between individual stations. This was the very reason that induced the author to include the whole of the stations throughout the country. A wide base demanded a proportionate structure; there was no getting away from the fact that the average cost per unit of electricity generated in the whole of the non-condensing stations was 16 per cent. less than that generated in the whole of the condensing stations throughout the country. His remarks on condensers and bad vacuum &c., had already been answered by the author in his reply to other contributors to the discussion. Mr. Day's remarks on works costs exclusive of coal were ingenious but specious. If he took the wide base, he must also take the whole structure, not select a portion of it.

In conclusion, the author desired to thank all those who had taken part in the discussion. Many points had turned up well worthy of consideration by those interested in the Economy of Fuel in Electric Generating Stations.

EXCURSIONS.*

On TUESDAY AFTERNOON, 28th July, after luncheon in Powolny's Rooms, by invitation of the Reception Committee, the Members visited the following Works, by means of special free tramcars provided by the Corporation :—

Leeds City Refuse Destructor, Armley Road.
Leeds City Electric Lighting and Power Station, Whitehall Road.
Leeds City Tramway Power Station, Crown Point Road.
Leeds Corporation Subway, for Cables, Gas, Water, and Sewer, York Street.
Joshua Buckton and Co , Well House Foundry, Meadow Road.
Deighton's Patent Flue and Tube Co., (Motor Wagons), Vulcan Works, Pepper Road, Hunslet.
Fairbairn, Macpherson and Co , Wellington Foundry.
John Fowler and Co., Steam Plough Works.
Greenwood and Batley, Albion Works.
Hathorn, Davey and Co., Sun Foundry.
Hudswell, Clarke and Co., Railway Foundry.
Hunslet Engine Co., Jack Lane, Hunslet.
J. Kaye and Sons, Lock Works, South Accommodation Road.
Kitson and Co., Airedale Foundry.
Lawsons, Hope Foundry.
Manning, Wardle and Co , Boyne Engine Works, Jack Lane.
Mann's Patent Steam Cart and Wagon Co , Pepper Road Works, Hunslet.
J. and H. McLaren, Midland Engine Works, Jack Lane.
Monk Bridge Iron and Steel Co., Whitehall Road.
Walter Scott, Leeds Steel Works, Hunslet.
Tannett Walker and Co , Goodman Street, Hunslet.
Joshua Tetley and Son, The Brewery, Hunslet Lane.
Thermit Welding of Tram Rails, Wellington Street.

---

* The notices here given of the various Works, &c., visited in connection with the Meeting were kindly supplied for the information of the Members by the respective authorities or proprietors.

The following Works were open in the afternoon, and during the Meeting:—

Leeds Corporation Gas Producing Plant, Meadow Lane.
Leeds Corporation Sanitary Depôt, Dock Street.
J. Booth and Brothers, Rodley.
Campbells and Hunter, Dolphin Foundry.
Chorley and Pickersgill, The Electric Press, Cookridge Street.
Clayton, Son and Co , Moor End Iron Works.
Coghlan Steel and Iron Co.
Alf. Cooke (Executors of the late), Crown Point Works, Hunslet.
Graham, Morton and Co , Pepper Road, Hunslet.
Wilson, Hartnell and Co., Volt Works.
W. Johnson and Sons, Castleton Foundry, Armley.
Robert Middleton, Sheepscar Foundry.
S. T. Midgley and Sons, Boot Manufacturers, Crown Works.
Scriven and Co., Leeds Old Foundry, Marsh Lane.
Taylor, Wordsworth and Co., Water Lane.
Wilson, Walker and Co , Leather and Glue Works, Sheepscar Street.

In the evening the Institution Dinner was held in the Town Hall. The President occupied the chair; and the following Guests accepted the invitations sent to them, although those to whose name an asterisk (*) is prefixed were unavoidably prevented at the last from being present.

The Right Honourable the Lord Mayor of Leeds, John Ward, Esq., J.P., Honorary Chairman of Reception Committee; Sir James Kitson, Bart., M.P., Honorary Chairman of Reception Committee; *Sir Arthur T. Lawson, Bart., M.P., Honorary Chairman of Reception Committee; Alderman Edmund Butler, Deputy Lord Mayor; The Rev. Dr. Gibson, Vicar of Leeds; The Rev. Charles Hargrove; Mr. Roland H. Barran, M.P.; *Mr. Henry S. Cantley, M.P.; Mr. A. G. Lupton, Chairman of the Council, Yorkshire College; Dr. N. Bodington, Principal, Yorkshire College; Mr. W. J. Jeeves, Town Clerk; Alderman John H. Wurtzburg, President, Leeds Chamber of Commerce; Professor A. Smithells, F.R.S., President, Leeds Philosophical and Literary Society; Mr. George R. Goldsack, President, Leeds Association of Engineers; Mr. F. Herbert Marshall, President, Cleveland Institution of Engineers; Dr. A. G.

Barrs, F.R.C.P.; Mr. E. E. Lawson, Chairman, Leeds Corporation Sanitary Committee; *Mr. W. J. Grinling, Chief Traffic Manager, Great Northern Railway; *Mr. E. L. Davis, Chief Passenger Agent, North Eastern Railway; Mr. John B. Hamilton, General Manager, Leeds City Tramways; Mr. Ben Day, Secretary, Leeds and District Engineering Employers' Association.

Leeds Executive Committee.—Vice-Chairmen, Colonel T. W. Harding; Alderman F. W. Lawson; Alderman A. Tannett-Walker; *Treasurer,* Mr. T. P. Reay; *Honorary Secretaries,* Mr. E. Kitson Clark and Mr. Christopher W. James.   Mr. Henry Berry; Mr. William Booth; Mr. A. Currer Briggs; Mr. James Campbell; Mr. Harold Dickinson; Mr. Arthur D. Ellis; *Mr. W. E. Garforth; Mr. Ernest G. Gearing; Mr. G. C. Hamilton; Mr. Wilson Hartnell; Mr. Francis L. Lane; Mr. Ewing Matheson; Mr. Henry McLaren; Mr. John McLaren; Mr. A. H. Meysey-Thompson; Mr. Robert Middleton; Colonel F. W. Tannett-Walker, V.D.; Mr. Edwin Wardle.

Professor Roberts Beaumont; Mr. T. A. Carpenter; Mr. H. Ade Clark; *Mr. R. H. Fowler; Professor H. S. Hele-Shaw, LL.D., F.R.S.; Mr. F. J. Kitson; Mr. W. W. Macpherson; Mr. E. Richards; Mr. Henry H. Suplee; and Mr. J. F. Walker.

The President was supported by the following Officers of the Institution :—*Past-Presidents,* Sir Edward H. Carbutt, Bart., *Mr. Samuel W. Johnson, Mr. William H. Maw, and Mr. E. Windsor Richards; *Vice-Presidents,* Mr. John A. F. Aspinall, Mr. Arthur Keen, Mr. Edward P. Martin, and Mr. A. Tannett-Walker; *Members of Council,* Mr. Michael Longridge, Mr. John F. Robinson, Mr. Mark Robinson, Mr. John W. Spencer, and Mr. Henry H. West.

After the President had proposed the loyal toasts, Sir Edward H. Carbutt, Bart., proposed the toast of " The Houses of Parliament," remarking that in the elevation of Mr. W. L. Jackson to the peerage as Lord Allerton a great honour had been conferred upon Leeds. Mr. Rowland H. Barran, M.P., suitably acknowledged the toast.

In proposing " The City and Trade of Leeds," Mr. William H. Maw said that the position which Leeds held in the engineering

world had been thoroughly earned by the quality of its products and
the sterling character of its citizens. The LORD MAYOR, in
responding, observed that Leeds had made phenomenal progress
At the beginning of last century it had a population of about
53,000, but today its inhabitants numbered 440,000, and its increase
in this respect had during the last decade been greater than that of
any other city in England. The trades of Leeds being numerous
contributed to its progress, and he believed it was destined to occupy
a very high position among the cities of the kingdom.

Mr. E. WINDSOR RICHARDS, in proposing the toast of " The
Reception Committee and Our Guests," remarked that the list of the
Committee showed a large and influential body representing the
district. The Members were in the home of the " Best Yorkshire
Iron," a manufacture which had been carried on in this district for
over one hundred years, and he maintained that there was no better
iron made anywhere.

Sir JAMES KITSON, Bart., M.P., in response, observed that all
sections of society and politics were engaged in examination and
enquiry, and the Institution was also engaged in an examination into
the mechanical position of Leeds. The Members would probably
find that Leeds was not as actively occupied at the present time as
they were accustomed to be, but he believed there were brighter
times to come. Colonel F. W. TANNETT-WALKER, V.D., also replied,
and remarked that one heard a great deal about high-speed tools,
but he maintained that their best mechanical tool was the workman
himself. He urged the necessity for the better training not only of
the men who were to become foremen and managers, but of the boys
who entered their employment to learn the trade.

The concluding toast of " The Institution of Mechanical
Engineers" was proposed by Mr. A. G. LUPTON, Chairman of
Council of the Yorkshire College, who attributed a great deal
of the success of the engineering department of the Yorkshire
College to the great interest which had been taken in it by the
Engineering Committee. In acknowledging the toast, the PRESIDENT
said that the importance of the Institution was reflected in its
membership of over 4,000. One language was common to them

all, they could all read a mechanical drawing, and the value of their published Proceedings consisted largely in the excellence of the mechanical drawings, for which this Institution had been pre-eminently distinguished through all its history.

On WEDNESDAY AFTERNOON, 29th July, after luncheon in Powolny's Rooms, by invitation of the Reception Committee, the Members visited some of the Works that were open on the previous day. The following were also open :—Messrs. J. Barran and Sons, Wholesale Clothing Manufacturers, Hanover Lane ; the Leeds Copper Works, Hunslet; and the Yorkshire College.

In the evening a Reception—to which Ladies were also invited— was held by the Right Hon. the Lord Mayor and the Lady Mayoress, in the Fine Art Gallery of the Municipal Buildings.

On THURSDAY, 30th July, three alternative Excursions were made.

One was by a special free train to Doncaster, where a large party of Members visited the Wagon Works, Running Sheds, and Locomotive Works of the Great Northern Railway, under the guidance of Mr. Henry A. Ivatt, Locomotive Engineer. After luncheon in the Mess Room, by invitation of the Directors, the Members proceeded by special free train on the Great Central Railway to Frodingham. There they were conducted over the Works of the Frodingham Iron and Steel Co. by Mr. M. Mannaberg and other officials, and the 100-ton furnace, which is worked on the Talbot continuous principle, was tapped during the visit. The return journey was made by special train.

Another visit mide was to the Farnley Iron Works, by special tram to Wortley, and thence by wagonettes provided by the Farnley Iron Co. The Members were conducted over the Works by Mr. Ewing Matheson, managing director. Subsequently they returned to Leeds, and were invited by the Reception Committee to luncheon at Powolny's Rooms.

A third Excursion was made by ordinary train to Altofts, where the colliery of Messrs. Pope and Pearson was visited under the guidance of Mr. W. E. Garforth, and the Members subsequently partook of luncheon, by invitation of the Directors.

In the Afternoon, the Members proceeded by ordinary train to Wakefield. On arrival at Kirkgate Station, they were conveyed by brakes to visit the Works of the Seamless Steel Boat Co., thence to Messrs. Green's Economiser Works, and Messrs. Cradock's Steel Rope Works. The return journey was made by ordinary train from the Westgate Station.

On FRIDAY, 31st July, two alternative excursions were made.

One was made by special train to Ripon, where the Members and Ladies visited the Minster under the guidance of the Dean of Ripon They then proceeded by brakes to Studley Park, across which they walked to Fountains Abbey. After luncheon at the Abbey in the Dormitory Undercroft, by invitation of the Reception Committee, Sir Edward H. Carbutt, Bart., proposed the health of the Marquis of Ripon, K.G., who had honoured them with his presence, and the toast was suitably acknowledged by his Lordship. The Members were then shown over the Abbey by the Honorary Local Secretary, Mr. E. Kitson Clark, F.S.A., who related its history. Tea was subsequently provided in the King's Café, Ripon, after which the return journey was made to Leeds by special train.

The other Excursion was made by special train to York, under the guidance of the Honorary Local Secretary, Mr. Christopher W. James. Proceeding along the Walls and over the Bridge, the Members and Ladies visited the ruins of St. Mary's Abbey, where Mr. S. H. Davis of Messrs. Rowntree related its history. The Minster was subsequently visited, and a recital on the new Organ was given by the Organist, by permission of the Dean. After luncheon in the Station Hotel, by invitation of the Reception Committee, the party visited Messrs. Rowntree's Cocoa and Chocolate Works, where they were shown the processes of packing chocolate and other sweetmeats in boxes. The excursion terminated with a visit to the ancient Treasurer's House at York, where the Members were hospitably received by Mr. Frank Green, the owner of the house, who explained the many historical features of that interesting building. The return journey was made by special train.

CITY ELECTRIC LIGHTING
AND POWER STATION, LEEDS.

This Station is situated in Whitehall Road, and has been enlarged from time to time. In 1891 a Provisional Order was granted to the Yorkshire House-to-House Electricity Co., and in 1893 the supply of electricity to the public was commenced. The undertaking was however transferred in December 1898 to the Leeds Corporation, who found it necessary to make a large extension of the works to provide for future increase. The works comprise an area of 290 feet by 181 feet, and consist of engine-house, switchroom, cable cellar, boiler-house, stores and workshops.

The engine-house is 220 feet in length by 65 feet in width, and 33 feet in height from floor line to crane rail. The plant consists of one McLaren engine and one Belliss engine, both coupled to alternators made by the Electric Construction Co. The capacity of the dynamos is 1,500 kilowatts each. The engines are of 2,400 horse-power, and are of the triple-expansion enclosed type with forced lubrication. The dimensions of the engines are as follows :—

	McLaren Engine.	Belliss Engine.
	Inches.	Inches.
High-pressure cylinder, diam.	27½	22½
Intermediate ,, ,,	39½	35
Low-pressure ,, ,,	64½	55
Stroke	24	30

The speed of both engines is 200 revolutions per minute. The valves of the McLaren engine are trick-ported slide-valves, balanced horizontally and vertically, and the governor is of Messrs. McLaren's automatic expansion type. The Belliss engine has piston valves, and the governor is of the throttle type.

The dynamos are made to supply two-phase energy at 50 periods per second, and generate 2,000 volts on each phase. The alternator shaft is bolted up direct to the engine coupling, and runs therefore at 200 revolutions per minute. The diameter over the field magnets is 13 feet 6 inches, and the weight of the magnets and wheel shaft is 28 tons. The exciter for each machine is mounted on the alternator shaft, and is of 8 kilowatts capacity.

In the cellar below the engines are placed two surface-condensing plants, made by Messrs. Cole, Marchent and Morley, of Bradford, one condenser for each engine, in connection with which are two centrifugal pumps made by Messrs. Belliss and Morcom, of Birmingham. The condensers have each 3,400 square feet of heating, and the number of tubes in each is 2,245 of $\frac{3}{4}$-inch internal bore, and 7 feet 9 inches in length.

The steam, feed, exhaust, suction, and air-pump discharge pipes, oil filters, &c., were supplied by Messrs. John Spencer, of Wednesbury. The steam-pipes are lap-welded steel, with flanges welded on. The exhaust pipes are wrought-iron, riveted, and the remainder of the pipes are cast-iron. There is a 30-ton overhead travelling crane, supplied by Messrs. J. Booth and Brothers, of Rodley. This crane is equipped with a two-phase motor.

The switchboard was made by Messrs. Ferranti, and consists of six sets of their two-phase oil break switches on the dynamo board with necessary bus bars, synchronising apparatus, &c. The cable cellar running under the floor of the switchroom is so arranged that the cables leaving the board are carried down direct on to shelves supported on brackets carried on iron pillars, and the cables then pass away into the cable subway, which runs as far as Queen Street. The subway is capable of holding 200 cables.

The boiler-house is arranged to hold twenty-four water-tube boilers with 4,300 square feet of heating surface each. At present there are installed eight Babcock and Wilcox water-tube boilers fitted with superheaters and purifiers. The tubes are arranged in seventeen sections, with twelve solid-drawn steel tubes to each section. The boilers, fitted with Meldrum's coking stoker and forced draught, are arranged with twelve on either side of the boiler-house. Immediately above is the coal store, from which the coal gravitates by means of inclined shoots into the mechanical stokers. A Green's economiser, consisting of 960 tubes, is placed over the boilers between the coal store and the western wall of the engine-house. Provision is also made for three other economisers.

The pump-house is situated at the south end of the boiler-house, and is fitted with two feed-tanks, and is capable of holding seven Weir pumps. At present there are provided two pumps with a capacity of 4,000 gallons per hour each, and one with a capacity of 8,000 gallons per hour. There are also installed two Worthington pumps, one with a capacity of 12,500 gallons per hour and the other with 25,000 gallons. The coal-conveying plant is supplied by the New Conveyor Co., and will deal with 40 tons of coal per hour.

The capacity of the generating plant is 8,740 kilowatts, and the number of 35-watt lamps (or their equivalent) connected to the mains is over 260,000. The price of current for lighting purposes is 4d. per unit, less 5 per cent. discount, and for power is $1\frac{1}{8}d$. to 2d. per unit, less 5 per cent.

The engineering and electrical work was arranged by Mr. Harold Dickinson, M.I. Mech. E., the Corporation's Engineer.

CITY REFUSE DESTRUCTOR,
ARMLEY ROAD, LEEDS.

This Destructor Depôt, built in 1879 from the designs of Mr. Morant, Borough Engineer, was one of the first destructors erected in this country, and up to the present time has cost about £13,000. It consists of twelve furnaces, constructed by Messrs. Manlove, Alliott, and Fryer, of Nottingham, and of four larger cells of the Horsfall type designed and erected by Mr. ¡Thomas Hewson, City Engineer. The boiler is multitubular, 12 feet long, 8 feet diameter, containing sixty-four 4-inch tubes, and has been continuously at work since September 1879. The engine is 20 horse-power horizontal, made by Messrs. Manlove, Alliott, and Fryer, and drives two mortar-mills. The, sale of mortar for year ending 25th March 1903 amounted to about 2,250 tons. The chimney is 40 yards high by 5 feet 3 inches square.

The amount of refuse destroyed during the year ending 25th March 1903 was 28,046 loads at a net cost of £2,262 8s. 8d., or 1s. 7d. per load.

The works are quite near to the Armley side of Wellington Bridge ; the Corporation have also destructors at Beckett Street, Meanwood Road, and at Kidacre Street, near Hunslet Goods Station.

CORPORATION SUBWAY,
YORK STREET, LEEDS.

The Corporation of Leeds, having in common with other large towns, experienced continual breaking up of the surface of streets for necessary extensions to tramway, water, gas and lighting undertakings, decided, that in the construction of various main roads in their Unhealthy Areas Improvements Schemes, to build subways in which most of these mains might be placed. Deputations visited and inspected the subways at London, Nottingham, and St. Helens, and eventually adopted the arrangement shown on the accompanying sketch, Fig. 1. The subway, which is 7 feet high and 8 feet wide, is

FIG. 1.

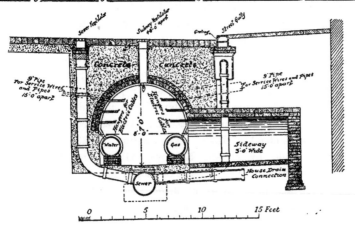

Subway now being constructed by the Leeds Corporation.

constructed to contain gas- and water-pipes, telegraph, telephone, electric light and tramway cables.

The sewer is laid under the floor and is accessible by means of manholes sealed with air-tight covers. The sewer is ventilated in the usual manner. The subway itself is ventilated by openings in the roof 48 feet apart. At distances of 40 feet, sideways are constructed (4 feet high and 3 feet wide), under the floor of which will be placed house-drain connections. The floors of the sideways are formed of bricks on edge laid dry so that ready access is given to the house-drains. Nine-inch pipes are inserted every 15 feet in the walls of the subway to take the surface wires and pipes to abutting houses.

THE YORKSHIRE COLLEGE,
LEEDS.

The Yorkshire College, which is one of the three constituent Colleges of the Victoria University, is widely known for the range of its technological work, which has received a wider development here than in any other university college. The buildings appropriated to its arts, science, and technological departments are situated in College Road, Leeds, about a mile from the railway stations, and occupy part of a site of about five acres in extent. The land, buildings, and equipment for these departments have cost upwards of £180,000. The Medical School, situated near the General Infirmary, has cost in site, building, and equipment about £50,000 additional.

The College Road block includes lecture and class-rooms for the departments of Latin, Greek, English, French and German languages and literature, Romance philology, history, philosophy, law, economics, education, mathematics, physics, chemistry, biology, geology, civil, mechanical, electrical and sanitary engineering, mining, textile industries, art, dyeing, leather industries and agriculture, as well as fully equipped laboratories in all the above subjects requiring experimental treatment. It also contains an examination hall, which on the occasion of important functions is used for general meetings, and in which about 800 persons can be seated, and a library for the use of students. In a separate block is the refectory, and at a little distance from the College a hall of residence has been built for the accommodation of such students as desire the conditions of a common life under the direction of a warden.

The medical school includes all the usual departments requisite for the complete medical and surgical curricula.

The practical work of the agricultural department is carried on at the Manor Farm, at Garforth, which comprises about 300 acres, half being grass and half arable land.

The number of students in the day classes last year was 819, and in the evening classes 346, making a total of 1,165, who paid £13,225 for their year's tuition. The engineering department, which is one of the most successful branches of the college, had 98 day students in its classes and laboratories last session and 80 evening students, these numbers being included in the totals given above.

MESSRS. JOSHUA BUCKTON AND CO., WELL HOUSE FOUNDRY, MEADOW ROAD, LEEDS.

Self-acting machine tools have been the speciality of these Works since they were founded by the late Mr. Joshua Buckton in 1842. The present heads of the firm are Mr. J. Hartley Wicksteed, chairman and managing director, Mr. Christopher W. James, joint managing director, Mr. Norman D. Lupton, secretary and director, and Mr. R. A. Bince, works manager and director.

The firm has turned out considerably more than 10,000 machine tools, varying in weight from 5 cwts. to 120 tons and composed of individual pieces up to 30 tons weight cast in their own foundry. The heavier tools are such as armour-plate finishing machines, large lathes, shearing machines for hot steel blooms and cold steel-plate, plate shears, and billet shears; and the lighter tools are such as medium-sized lathes, drilling, boring machines, &c. A speciality of the firm is a planing machine with double-cutting tool-boxes, and with power feed and traverse to the tool-boxes.

Amongst the machines finished or in progress which were open to the inspection of the members, the following represented recent developments:—Double-cutting table planing machine, 8 feet by 8 feet by 24 feet, with power feed and tool-box traversing motion; standard 12-inch lathe for high-speed cutting; two-spindle radial drill; locomotive crank-axle profile slotting machine; four-spindle rail drilling machine; vertical and horizontal planer with fixed table, and moving upright; chain and anchor testing plant; 300-ton universal testing machine; vertical testing machines; and $4\frac{1}{2}$-inch square cold billet shears.

FARNLEY IRON WORKS,
LEEDS.

These Works were established in 1844 by the Messrs. Armitage Brothers, of Farnley Hall, to utilize the minerals—coal, iron, and fireclay—on their estate. At first their operations were confined to the manufacture of pig-iron in cold-blast furnaces, but in a few years the works were extended to include all the operations for making finished iron. The minerals were conterminous with those of the Lowmoor Iron Co. and the Bowling Iron Co., both of Bradford, whose works are near, and the processes for producing what became generally known as "Best Yorkshire Iron" were the same. At the present time about 1,500 workmen are employed at Farnley.

The ore is that known as the Black bed, and in the raw state as mined contains about 33 per cent. of iron, and when calcined about 40 per cent. Amongst the Farnley Collieries the seam known as the "Better-bed" gives a coal remarkably free from sulphur, and which consequently can be brought into contact with the iron without the risks which always attend fuel of a less pure kind. Sulphur in iron, unless reduced to the merest trace, renders it "hot short," and is fatal to the highest qualities of ductility and safe

2 T

welding. The special process of manufacture at Farnley commences
with the use of cold-blast furnaces for melting the ore. When the
works were first started, the hot-blast system, invented and introduced
by Neilson about 1830, was only coming gradually into common use ;
while in the year 1900, when about 400 blast-furnaces were working
in Great Britain, only about 10 still used the cold-blast. The
Farnley furnaces are among the survivors, and it is of this cold-
blast pig, refined, puddled, hammered and rolled, that Farnley
" Best Yorkshire " iron is made. The waste gases from the blast-
furnace are utilized as fuel for the boilers of the blowing engine, as
the usual purpose to which such gases are applied, namely, in
stoves for heating the blast, is of course absent. The blowing
engine is an old one, but with up-to-date valve arrangements for
economising steam. The pressure of the blast ranges from $2\frac{1}{2}$ lbs.
to $3\frac{1}{2}$ lbs. per square inch. As it has become desirable to have a
stand-by blowing engine, the opportunity was taken of adopting
one of Parsons' steam turbines for this purpose. This turbo-engine
works at about 5,800 revolutions per minute, and has proved
efficient.

As is well known, the decarbonising of iron so as to render it plastic
is generally effected by boiling and stirring pig-iron in puddling
furnaces, but at Farnley the pig-iron is first re-melted in open
fires, a process which eliminates the silica and sulphur. It is the
resultant refined castings that are taken to the puddling furnaces.
The after processes of hammering and rolling resemble those used
in the manufacture of ordinary wrought iron, but are more elaborate.

From the earliest days of the Farnley works their best Yorkshire
boiler plates were in great demand as being more durable and more
safely worked than any other kind, and it became usual among
engineers to specify " Farnley, Low-Moor, or Bowling " as the only
acceptable brands, when a high quality was required. The Bowling
works were closed in the year 1896. The introduction of steel and
the facility and cheapness with which cast ingots can be rolled
down into plates of almost unlimited sizes and weight have reduced
the manufacture of iron boiler-plates to a very low point. They

are still used by those who are acquainted with their merits for special purposes where the service is a severe one, as in contractor's locomotives ; or where the water is impure ; or where repairs may have to be effected in remote places, and where durability is of importance. Farnley plates cost more than double the price of steel plates, and it is only if the life of a boiler and the expenses of renewal are taken fully into account that the ultimate advantages afforded by Best Yorkshire Iron are realised.   Last year the boilers of thirty main-line engines built in England for an important British Colony were made entirely of Farnley plates.

Farnley bar-iron is as much used as ever, and it appears that there will always be occasions when the finest wrought-iron, made regardless of cost, can be employed with advantage.   Safety in welding is the first consideration in comparing iron with steel, and durability against oft-repeated percussion is the other.   The range of temperature within which smithing and welding can be effectually done is much narrower with steel than with iron.   Steel chains are unknown.   Railway shackles, hooks, couplings, piston rods and connecting-rods subject to rapid alternations of tensile and compressive stresses demand the highest quality of iron, as also do the suspended cages and other vital parts of mining machinery. In the highest class of engines and carriages these parts are always made of Best Yorkshire Iron.

A complete plant for the manufacture and distribution of Mond gas has been established at Farnley, partly for supplying fuel to the widely distributed furnaces, and partly also for obtaining the convenience of a central station from which power can be distributed by electricity.   This plant is now successfully working, and many of the separate boilers and engines which supplied power in the works have been superseded.   The Mond gas is found quite suitable for gas-engines, two of which, each 250 H.P., made by the Premier Gas Engine Co., of Sandiacre, are now working, and a third engine of 650 H.P., by the same makers, is nearly ready for fixing.   A complete plant for the recovery of the by-product sulphate of ammonia has been established with success.

The manufacture of Fireclay goods has become as large a branch of manufacture at Farnley as that of Best Yorkshire Iron. The fireclay found under the works and under the neighbouring hills is of a highly refractory nature, and affords material not only for fire-bricks and retorts, but for glazed bricks and ware to which very hard glazes can be applied, needing very great heat to melt them—a heat far in excess of what can be endured by less refractory clay. This circumstance and the long experience in the details of manufacture are the main circumstances that have caused Leeds glazed bricks and ware to have the pre-eminence they enjoy in London and elsewhere.

MESSRS. JOHN FOWLER AND CO. (LEEDS), STEAM PLOUGH AND LOCOMOTIVE WORKS, HUNSLET, LEEDS.

These Works are situated in Hunslet about a mile from the railway station, and cover from 10 to 11 acres. The principal manufactures are steam ploughing engines, traction engines, road roller light locomotives, semi-portable engines, winding and pumping engines, and all descriptions of colliery plant, alternators, dynamos, switchboards, and electrical plant.

The principal work in progress at the time of the meeting was the manufacture of ploughing and traction engines, steam road-rollers, large winding engines and general electrical work. The number employed is about 1,500 men.

MESSRS. GREENWOOD AND BATLEY, ALBION WORKS, LEEDS.

This business was originally established in the year 1856 at the Albion Foundry, in premises still belonging to the Company, and occupying about an acre of land in East Street, Leeds, with a frontage and wharfage on the River Aire (Aire and Calder

Navigation), the buildings on which are now used as warehouses and storage for the main portion of the large and valuable collection of models, the property of the Company.

In 1859, these premises being found too small, new and more commodious workshops were built on a site containing upwards of ten acres of land fronting on Armley Road, Leeds, and having a wharfage behind on the Leeds and Liverpool canal; and these, the Albion Works, have since been very largely extended, so that at present the total shop area is 31,594 square yards, and the establishment gives employment in fairly brisk times to some 2,000 workpeople, and is equipped with upwards of 1,500 machine tools.

The works are connected to the Great Northern Railway system by a small branch line, and railway trucks can be taken into the principal workshops.

The business is worked by departments, the main being:—

Special and general machine-tool making, including the construction of machinery used in the manufacture of war material of all kinds, testing machines, forging machinery, minting machinery, special wood-working machinery, cloth-cutting machines, etc.

Oil mill and general millwright work department, including machinery for warehousing and crushing every kind of seed, and refining oil.

Textile machinery department, embracing all machinery for the preparation and spinning of silk, silk waste, China grass, and other fibres.

Electrical department, for the manufacture of every kind of electrical plant.

Turbine department, for the manufacture of De Laval's patent Steam Turbine Motors, Turbine Dynamos, Turbine Pumps and Fans (for Great Britain and Colonies, China, Japan and Egypt).

Ordnance department, for the supply of the Whitehead fish torpedo, etc., the manufacture of small arms cartridges, projectiles, horse-shoes, etc., etc.

Sewing machine department, viz.:— patent boot and shoe sewing machines, and machinery for making and sewing leather belts, etc.

MESSRS. HATHORN, DAVEY AND CO.,
SUN FOUNDRY, LEEDS.

This company was formed in 1872, and took over the business of Messrs. Carrett, Marshall & Co., who for many years previously had manufactured steam and hydraulic engines. The new firm from time to time enlarged the works and installed modern machinery for the manufacture of steam, hydraulic and electric pumping machinery, which has been supplied to water works, sewage works and mines in all parts of the globe. The differential gear (Davey's Patent) has been largely manufactured here, both for new pumping engines and also for application to existing ones in substitution for the old-fashioned tappet gear.

Members visiting the Sun Foundry would see in an advanced stage of construction (besides other sets of pumping machinery for water works and mines) :—

Two sets of vertical, triple-expansion pumping engines, for the British Government, and the Rosario de Santa Fé Waterworks respectively ; (an engine of this type, recently tested by Professor Unwin, gave a duty of 151,670,000 foot-pounds per 1,000 lbs. of saturated steam) :

A compound engine, actuating a centrifugal pump with horizontal axis, to raise 140 tons of water per minute a vertical height of 13 feet, for draining the Fen District :

Two vertical engines with overhead beams, to raise water out of a well and deliver it into a tank 100 feet above ground level, for the London County Council.

MESSRS. HUDSWELL, CLARKE AND CO.,
RAILWAY FOUNDRY, LEEDS.

These Works were established in 1860 for locomotive engine building and general engineering. They are lighted and driven by electricity. The number of hands employed is from 400 to 500.

HUNSLET ENGINE CO.,
HUNSLET, LEEDS.

These Works are situated in Jack Lane, Hunslet, covering an area of about three acres, and employing about 400 men. The business was established in 1864 especially for the manufacture of locomotive tank-engines, and their productions are well known not only at home, but abroad, and particularly in most British Colonies. Approximately, 75 per cent. of their yearly turn-out is exported, and engines, for no less than thirty-six different gauges, have been designed and executed.

The works are well equipped with all kinds of tools, many of them of a special character suitable for the work generally produced ; within the last twelve months steam-engines for motive power have been abolished, and the entire works driven by electric power supplied by the Leeds Corporation. At present considerable extensions are in progress, the most important item being a new boiler shop, about 200 feet long by 53 feet wide, which will be equipped with all the most efficient and up-to-date tools.

MESSRS. JOSEPH KAYE AND SONS,
LOCK WORKS, HUNSLET, LEEDS.

These Works were established by Mr. Joseph Kaye 1865, and are situated close to the new Suspension Bridge, which spans the River Aire, and joins Hunslet to East Leeds. Although this firm turns out about 1,000 dozen of the seamless " K " oil-cans per month, the works are chiefly occupied in the manufacture of their patent safety railway carriage door locks ; also their patent locks for the doors of institutions such as asylums, hospitals, workhouses, hotels, and where a large number of locks are made to differ, with sub-master keys, master keys, grand-master keys, and superior grand-master keys.

MESSRS. KITSON AND CO.,
AIREDALE FOUNDRY, LEEDS.

The Airedale Foundry was started in 1839 by Mr. James Kitson, the father of the present chairman, Sir James Kitson, Bart., M.P., with whom Mr. T. P. Reay is associated as managing director, and Mr. E. Kitson Clark and Mr. E. C. Kitson as members of the Board. The Works were founded for the purpose of locomotive engine building, and the records of the firm represent the history of the development of locomotive engineering from its commencement up to the present time. Portions of the works are also laid out for dealing with large castings and pieces of machinery suitable for blowing engines, rolling mills, pumping engines, and steam hammers, of the largest size, but at the present time only locomotive engines are in hand.

In the erecting shop an interesting contrast will be found of eight-wheels-coupled engines designed for heavy goods traffic on two widely different gauges. For the Cape Government there is an engine made for the 3 feet 6 inches gauge, with $18\frac{1}{2}$ by 24-inch cylinders, 31 square feet grate area, and a total weight (engine and tender loaded) of 98 tons, which will haul 2,700 tons train-load, on a straight and level road, at a speed of 10 miles per hour. The curves round which this large engine has to go are as small as 330 feet radius, and the difficulty of putting into a narrow-gauge engine a fire-box and grate area sufficient to raise steam for such a powerful machine has been met in an interesting manner. Instead of making a very long and narrow fire-grate, as was the practice until a few years ago, the fire-box has been shortened in length, reduced in height, and spread boldly out over the tops of the wheels and the

framos.   The result arrived at is a high boiler centre, but a
thoroughly efficient boiler.   A passenger engine close by for the
Cape has a similarly large fire-box, space for which is obtained
between the frames and connecting the hind portion to a widely-
spread steel-casting.   It will be noted that the Colonies have accepted
for the time the American bar frames, and members may be interested
to observe the complication of the attachments involved.   Besides
these large engines a tank engine is being built, also for South
Africa, with a loaded weight of 25 tons, the latter being a
reproduction of an engine designed by this firm and sent out to
Natal about 30 years ago.

In the same shop, engines of a thoroughly English type are
under construction for the Great Central Railway, 8 wheels coupled,
cylinders $19\frac{1}{2}$ by 26 inches, specially designed for hauling the coal
traffic from the centre of England to the East Coast.   The fire-box
is "Belpaire" in accordance with modern practice, by which special
facilities are gained for washing out, and a proper flexibility given
to the fire-box top.   The engine and tender when fully loaded
weigh about 104 tons, and will haul on the level 2,830 tons train-
load on a straight road, at a speed of 10 miles per hour.

A large portion of the machinery in the works is specially
designed to deal with locomotive details, and the shops are
generally equipped with the ordinary engineers' machinery, which
does not require special comment.   The large boiler shop however
may be noted, which has been found necessary, owing to the increase
in the size of the boilers, and the more rapid renewals consequent
upon higher pressure of steam and the more difficult conditions of
modern locomotive service.   This shop is built with two bays, each
of which is 68 feet wide, and crossed by a high bay of 48 feet wide.
An extension has been lately added, so that the total length of the
shop is now 426 feet by 136 feet wide, making a floor area of 6,437
square yards.

It may be remembered that in the programme for the last
meeting in Leeds, in 1882, a tool by Messrs. Joshua Buckton and
Co. was described "as made in 1848 and is still working." *   The

---

* Proceedings 1882, page 453.

same tool, which came from the firm of which the present President of The Institution of Mechanical Engineers is Chairman, is working satisfactorily after 55 years' service.

The number of men employed is about 1,500.

LEEDS COPPER WORKS,
STOURTON, LEEDS.

The Works of this Company are situated at Stourton, on the site formerly occupied by the Elmore Works. They consist of boiler and economiser house, engine and dynamo station, copper-refining furnaces, depositing-room, draw bench and machine room, &c. These have been built and equipped on the pattern of the kindred works in France, which have been in operation for some years past, at Dives, in Normandy. The various shops at present cover 9 acres; they are connected both with the Aire and Calder Canal and with the Midland Railway.

The engine-room measures 215 feet in length by 65 feet in width, and contains :—

(a) The depositing set of engines, consisting of four pairs of 250-horse-power Bollinckx cross-compound condensing Corliss engines. The high-pressure cylinder exhausts into a receiver placed underneath the engine-room; the receiver supplies the necessary steam to the low-pressure cylinder, which exhausts into the condenser. Each pair of engines drives by belt transmission a 1,700-kilowatt continuous-current Brown-Boveri dynamo, for depositing.

(b) The power set, formed of three pairs of 400-horse-power Bollinckx cross-compound condensing Corliss engines, similar in design to the above, each pair of which drives a Brown-Boveri dynamo of 425 horse-power, of 525 volts and at 325 revolutions.

(c) The lighting set, which consists of two 210-horse-power Willans' high-speed engines, direct-coupled to Elwell-Parker dynamos.

(*d*) The main switchboard, in two parts—one for depositing and one for power—and the lighting switchboard.

The engines are supplied with steam at 140 lbs. per square inch, from seven Galloway 300-horse-power boilers, 32 feet long and 8 feet 6 inches in diameter, connected to a chimney 200 feet high and 10 feet inside diameter at top. They are fed by two Worthington feed-pumps. The feed-water is taken from the canal; but previous to being used in the boilers it runs through a pulsometer " torrent " filter. The boilers work in conjunction with seven economisers, located in a space between the boiler-house and the engine-room. All the steam pipes, valves, and connections are in duplicate.

The copper used in the depositing process is supplied to the works in the shape of 96 per cent. Chili bars or blister copper; this is treated for the removal of arsenic and other impurities, in three refining furnaces of 12 tons each. The refined copper is then cast into trough-shaped moulds, 13 feet in length, for the production of the copper anodes. These are triangular in section, with slightly concave sides and vertices rounded off. The anodes, as obtained in the refining-shop, are of practically pure copper. They are cleaned in sulphuric acid and water baths, previous to being placed in the depositing tanks.

The depositing department forms a special building, 265 feet long and 200 feet wide, with six bays ; it contains 216 acid sulphate-depositing tanks, in 24 lines of nine tanks placed end to end, each bay covering four sets of nine baths each, connected in pairs side by side for the mechanical action. The plant is capable of a total weekly output of 75 tons. Two electric motors are placed one on each side of the depositing room, in a special compartment ; these motors work alternately for driving an underground main shaft, placed underneath the front end of the sets of tanks. Link gearing, which starts from the main shaft, gives the rotary motion to the mandrels (cathodes) inside the tanks, around which the copper tubes are deposited ; the gearing ensures also the backward and forward travelling of the burnisher frame. For tubes up to 4 inches inside diameter the mandrels are of brass ; for those above 4 inches inside diameter the mandrels are of cast-iron, with a brass neck for

taking in the current. The latter mandrels, previous to being used
in the acid sulphate-depositing tanks for the production of copper
tubes, are treated in an alkaline electrolytic bath, containing sheet
copper anodes, in which they are coated with a thin covering of
copper. The alkaline bath is slightly heated in order to forward
the depositing action. All the mandrels, before being placed in the
acid sulphate-depositing tanks, are carefully covered with black lead,
to facilitate the removal of the finished tube.

When starting an operation, the pure copper anodes and the
mandrels are placed in the tank, the mandrels rest at both ends on
insulated supports, and are nested alongside the anodes. Acid
sulphate is then led into the tank by gravitation from five cylindrical
reservoirs built on the side of the depositing-room ; the positive
current is supplied through suitable lead conductors to the copper
anodes, the negative current being led to the mandrel through copper
conductors and flexible brushes. The mandrels are made to revolve,
and agate burnishers, held in wood supports and fitted to a frame
placed transversely over the tank, travel automatically up and down
the whole length of the tubes during the time the depositing
operation is in progress. The frame travels on paths on the sides of
the tank, and is driven, as above mentioned, by gearing from the
main shaft. The burnishing gives the tubes a uniform density and a
perfectly smooth surface. The works can manufacture tubes up to
4 feet in diameter and 13 feet in length, though practically there is
no limit to the size. One of the current sizes is 12 feet by $2\frac{1}{2}$ inches
outside diameter by about 8 W.G. in thickness.

The depositing process being carried out with practically pure
copper, the sediment in the cisterns is very easily dealt with ; it has,
at times, been found to contain as much as 250 ounces of gold to the ton
of residue. When the copper-coated mandrels are removed from the
depositing tanks, they are carried to an adjoining machine-shop and
placed in a tube-expanding machine, in which they are made to
revolve under two friction rollers, which travel along the whole
length of the tube. They are then placed in a power draw-bench in
which they are held down ; one end of the mandrel is fastened to a
grip, on being hooked to a flat-link chain drawn by a sprocket

wheel, and removes the mandrel from inside the tube. A number of the tubes manufactured are drawn in the draw-benches through dies, in order to give them the required dimensions to meet certain specifications. This has the effect of hardening the copper; the tubes after drawing are placed in a reverberatory furnace, in which they are heated to a low red heat; they are then quenched in water, and cleaned in acid and water baths. Besides the tubes, the specialities of the works are the following : calico-drying cylinders, calico-printers' rolls, paper-makers' cylinders, seamless copper cylinders for hydraulic machinery, pump liners, and hydraulic ram covers. The copper tubes made by this process are able to stand, without showing any defects, the following mechanical tests: doubling close cold and then doubling over; doubling close cold and opening out; creasing up on end, expanding, flanging (diameter across the flange being equal to three times the diameter of the tube), reducing the doubled edge of the tube to a knife-edge, and turning back flat in its hard state, without breaking.

MESSRS. MANNING, WARDLE AND CO., BOYNE ENGINE WORKS, HUNSLET, LEEDS.

The Boyne Engine Works, of which Mr. Edwin Wardle is sole owner, cover nearly six acres, and are situated in Jack Lane, Hunslet, Leeds. They were founded in 1858, and take their name from the land, it having been purchased from Lord Boyne. The premises run alongside the Midland Railway, a siding from which connects direct with the yard. The front, where the offices are situated, is in Jack Lane, adjoining the railway bridge. The buildings for the works department are behind, and include all the usual and necessary workshops for the manufacture of locomotive engines of all descriptions.

MESSRS. J. AND H. McLAREN,
MIDLAND ENGINE WORKS, LEEDS.

These Works were established in 1876 by the present proprietors, Messrs. John and Henry McLaren; they are situated in Jack Lane, Hunslet, about one mile from the centre of the city. The ground occupied is about two acres, bounded on the south by the main line of the Midland Railway, from which there is a siding into the works. The principal specialities are traction engines and wagons; steam ploughing engines, and electric generating engines.

The boiler shop is equipped with modern hydraulic machinery for flanging and riveting; the machine shops are supplied with the latest type of labour-saving machinery, and are fitted up with electric and steam overhead travelling cranes. In addition to the home trade, the works are largely engaged in the production of steam-engines for shipment, and there are several progressive colonial agencies run in connection with the firm. The largest engine constructed by the firm was supplied to the Leeds Corporation in 1902 for generating electricity at the Whitehall Road Station. It is a high-speed enclosed triple-expansion engine, of 3,000 I.H.P. The number of men employed is about 230.

MESSRS. S. T. MIDGLEY AND SONS,
BOOT MANUFACTURERS, CROWN WORKS, LEEDS.

These Works are situated to the east of Roundhay Road, about a mile from the centre of the city. The business was established by the late Mr. S. T. Midgley at Halifax in 1851, being transferred to Swinegate, Leeds, in 1886. The sole proprietor is now Mr. A. W. Midgley, the son of the founder. In 1897 the present factory was built at Harehills, Leeds, and a large amount of new machinery was introduced, Mr. A. W. Midgley having decided to make a great change in the character of the business, which had hitherto been confined to the production of what was known as the standard Leeds goods, namely, boots of a heavy class. Recognising that the demand was increasing every year for boots of a lighter character, he determined to undertake the production of light or better class goods, such as had previously only been made in Stafford and Northampton, although it was impossible to obtain in Leeds the necessary labour requisite for such a change. In spite of great difficulties he succeeded, and large quantities of light as well as strong boots are now manufactured here every year.

Three years ago another branch of boot manufacturing was undertaken, that of Army boots, these having previously been made in Northamptonshire only. At the request of the Government, the firm offered to supply the War Office, and now there is a special department in which nothing but hand-sewn boots for the Government are produced, turning out 1,500 to 2,000 weekly. All the hands in this department have had to be specially trained, or imported from other districts.

The new works erected in 1897 soon proved too small, and an addition of almost the same size was made in 1901, the factory now being regarded as one of the finest in the north of England. The machinery includes all the latest in the trade, consisting of sewing machines, presses, sole rounder, heel builders and attachers, lasting, riveting, screwing, and stitching machines, also Goodyear welting (hand-sewn method) plant, finishing machines, etc.

In the closing room the bulk of the machines have been supplied by Messrs. Wheeler and Wilson, and Messrs. Singers. Of the latter make, the firm's new high-speed machine has just been put down; it has the working parts of gun-metal, and runs on ball-bearings. The accessory machines in this department are many and important, including a new addition in a Lufkin folding machine, punching and eyeletting machines, power eyeletter, button-hole machine, wax-thread machines, etc. In the rough-stuff room the operations of rolling, ranging, cutting, sorting, stamping, fitting up, skiving, moulding, etc., proceed in their proper consecutive order; and where machinery can be used to advantage a machine is used. A new Gunton press and Julian sole rounder have recently been put down in this department.

In the Goodyear welted work department, the British United Shoe Machinery Co. have supplied a number of machines. The first processes these machines deal with is in the preparation of the Gem inner sole, which is done by three small machines—a channelling machine, a lip-turning machine, and the re-enforcing machine, which covers the inner sole with canvas in such a way that it makes it damp-proof and prevents squeaking. The inner sole is now tacked upon the last, and the upper pulled over ready for the Consolidated lasting machine, and lasted in precisely the same manner as by hand labour. The toes are lasted by means of a wiper, and a small soft wire is laid round the feather, which leaves the toes the perfect shape of the last. The attachment of the welt to the upper and the inner sole is accomplished by a new hand method lockstitch welt and turnshoe machine, which is the first and only machine of its kind placed upon the market. This machine has a steam generator for heating the necessary parts, and the wax for lubricating the thread. The clever mechanical movement of the shuttle, the thread measuring-off device to suit the various substances of the material, and the locking of each stitch in the inner sole, differ from anything yet seen in welt-sewing machines, and make a seam equal to the best hand-sewn. The following operations are then proceeded with—trimming the welt-seam, hammering out the welt by the Universal welt beating-out machine, and filling the bottom with rubber solution. The outsole receives a coating of solution, and, together with the boot, is placed upon a jack of the sole-laying machine, which moulds the outsole to the bottom shape of the last. This method of attaching the outsole appears to be a great improvement over the old method of attaching soles by rivets.

The next operations are performed by the Universal rounding and channelling machine, and the Universal shank skiving machine. The channels are opened, and the work then stitched together by the Goodyear lockstitch machine. The boot is now levelled on the Goodyear automatic levelling machine, which is operated by a boy, who places the boot on a jack and starts the machine, whilst the machine is working automatically, rolling from side to side, changing its motion from toe to heel, thoroughly rolling out the bottom of the boot. During this operation the boy is preparing another boot on jack No. 2, which is ready to be rolled as soon as the first boot is taken away. The last operation is the attaching of the heel to the boot by the new lightning heeler, which attaches the heel while the last is in the boot. The firm has recently laid down a plant for a high-pressure system of gas lighting, which has been found to work most satisfactorily.

MONK BRIDGE IRON WORKS, LEEDS.

The Works of the Monk Bridge Iron and Steel Co. occupy an area of about 10 acres, and are divided into two parts by the Whitehall Road. The principal manufactures of the Company are :—Best Yorkshire iron bars, plates, forgings, &c.; Special Steel tyres, crank-axles, straight-axles, forgings, &c.; and cast-steel locomotive wheel centres. The Company owns blast-furnaces in another part of Leeds, where they make pure cold-blast pig-iron for the manufacture of " Best Yorkshire Iron." The pig-iron from which " Best Yorkshire Iron " is made is treated in two open-hearth refineries, which are supplied with cold blast by a horizontal blowing-engine with a 48-inch air cylinder.

Puddling Furnaces.—There are 22 puddling furnaces, the iron from which is shingled and hammered at 4 hammers before being taken to the mills and blooming hammers. There are altogether 12 steam hammers for the manufacture of " Best Yorkshire Iron " into blooms for finishing into plates and bars, steel blooms for tyres and axles, and finished crank-axles, forgings, &c., in iron or steel.

Hydraulic Press.—The hydraulic press of 1,200 tons capacity is used for making large forgings and for pressing crank-axle ingots, the hydraulic pressure of $2\frac{3}{4}$ tons per square inch being supplied by an

2 u

engine having 2 cylinders of 20 inches diameter. In the same shed are 2 powerful hammers employed in the manufacture of steel tyres and axles, &c. Steel is manufactured by the Siemens-Martin process, and the melting furnaces are supplied with gas from Wilson's producers. The Steel Foundry is under the same roof as the steel furnaces, and is fitted with three 30-ton overhead electric cranes.

Rolling Mills.—These consist of 19-inch, 14-inch and 10-inch bar mills ; the 19-inch mill is driven by a 36-inch cylinder horizontal engine, reversed by a patent clutch ; the 14-inch mill is driven by a compound beam-engine with 40-inch and 48-inch cylinders. This engine also works a train of forge rolls. The 10-inch mill is worked by a 30-inch cylinder horizontal engine. The plate mill consists of 2 pairs of plate rolls, driven by a 36-inch cylinder horizontal engine, and reversed by a patent clutch. The tyre mill is driven by the same engine that works the large bar mill.

Machine and Turning Shops.—These are electrically driven. Steam is raised in 13 mechanically-fired boilers, in 2 groups of 5 and 8 boilers, and also in 17 boilers arranged over the puddling furnaces. The feed-water for the boilers is heated by 2 sets of Green's economisers, and 3 Berryman's heaters.

MESSRS. WALTER SCOTT,
LEEDS STEEL WORKS, HUNSLET.

The Iron and Steel Works at Hunslet on the south side of Leeds were acquired by the Leeds Steel Works Co., in the year 1888, and commenced manufacturing operations under the direction of Mr. Walter Scott, of Newcastle-on-Tyne, as Chairman of the Company in the following year. The Works, together with collieries in Durham, and other properties belonging to Mr. Scott, were amalgamated as a company in July 1900, under the name of Walter Scott, Limited.

The Leeds Steel Works comprise a blast-furnace plant, basic Bessemer steel department, rolling mills, girder constructional and tram-rail finishing yards; also a basic slag artificial manure plant. They occupy an area of 25 acres of land in one of the busiest parts of Leeds, and are favourably situated for railway communication, being close to the Midland Railway sidings.

The blast-furnace plant consists of three furnaces of equal size, measuring 65 feet in height with a diameter of 18 feet inside the fire-brick lining at the bosh. They are closed at the top with the usual hopper and conical bell arrangement. There are eight tuyeres at each furnace. The ores used are chiefly from Lincolnshire and Northampton, together with puddling furnace cinder. The coke used is part Durham and part Yorkshire, and the total quantity consumed per week is more than 2,000 tons. The limestone used as flux is the ordinary blue mountain limestone from Ribblesdale. The materials are filled into the charging barrows, which are weighed, and are lifted to the top of the furnaces by a vertical steam-hoist. The quantity of materials dealt with is about 9,000 tons per week; the hoist engine lifting the material at the rate of nearly a ton per minute.

There are eight " Cowper " stoves of uniform size, measuring 75 feet high and 23 feet diameter. These serve for heating the blast to an average temperature of 1,350° F. before it reaches the furnaces. The air is driven into the furnaces by three vertical blowing-engines, the air-cylinders of which are each 100 inches in diameter. These engines were built by Messrs. Kitson and Co., of Leeds, and the air-pressure is maintained at about 6 lbs. per square inch. An electrical pyrometer is used for testing and registering the heat of the blast. This is connected with the furnace manager's office, and an observation can be made of the temperature of the blast at a moment's notice. The waste gas from the top of the furnaces is utilized for heating the blast, and also for generating steam by means of Lancashire boilers, no coal being needed for boiler firing. The quality of the iron produced is what is known as basic Bessemer, and the make for the three furnaces averages 2,200 tons per week, the record being 2,396 tons.

In addition to the iron there is about an equal weight of slag produced in the furnace from the fusion of earthy matters in the ores, etc. The liquid slag floats on the surface of the molten iron as it collects in the hearth, and the molten slag is run off into iron tubs, and, after cooling down, is broken in stone-breaking machines, and screened in revolving screens to various sizes suitable for road metal and concreting purposes. About one-half the production of blast-furnace slag is utilized in this way, the remainder being tipped as refuse on land belonging to the company at Woodlesford.

The Bessemer Steel Department is about a quarter of a mile distant from the blast-furnaces. The plant consists of two Bessemer converters, each producing 7 tons of steel at a cast, and two large converters of a capacity of 10 tons each. There are two semi-circular casting pits provided with top supported hydraulic cranes for supporting the casting ladle, and other smaller hydraulic cranes for arranging ingot moulds in the casting pit, and for removing ingots. There are three cupolas, each 8 feet in diameter by 36 feet high, for remelting the iron cast on the pig beds at the blast furnaces. These are driven by air supplied by three Roots blowers, and the cupolas are capable of melting 150 tons of iron in 12 hours. There is the usual Basic Shop for preparing the basic bricks made of burnt dolomite and tar for the linings and perforated bottoms of the Bessemer converters. The bulk of the iron from the furnaces is taken to the Bessemer department in the molten state in ladles supported on four-wheeled carriages, the ladle being capable of carrying 12 tons of molten iron.

The molten iron, as it arrives in the ladle from the furnaces, is weighed, the ladle is tipped, and the liquid iron is poured into a huge receiver or mixer, which is large enough to hold 120 tons of molten iron. The operation of pouring the iron into the mixer is very efficiently and quickly done by hydraulic power, and it is possible, though not usual, to empty a ladle containing 12 tons of molten iron in half-a-minute. The iron can be poured and weighed with equal facility from the mixer for use in the Bessemer converter. There is always a ready supply of uniformly regular iron for conversion into steel. At the beginning of the week the mixer is filled with molten iron on Sunday night ready for commencing work at six o'clock on Monday morning. During the week it is being continually filled as the iron is taken from it. A bath of 100 tons of molten iron is always available until the Saturday when the mixer is emptied. The iron remains liquid from its own initial heat, and no fuel is required to be burnt to keep it liquid. The iron remelted in the cupolas is also run into a ladle and poured into the receiver to mix with the molten iron taken direct from the blast-furnaces.

The conversion of the iron into steel by the Bessemer process is a refining operation effected by means of a powerful blast of air blown through the liquid iron with the result that the metalloids, carbon, phosphorus, etc., which are constituents of the pig iron, are oxidized or burnt out, and separated from the iron.

The analysis of the iron used is as follows :—

Combined carbon	2·85 per cent.
Graphitic carbon	Traces.
Silicon	0·60 per cent.
Sulphur	0·05 ,,
Phosphorus	2·60 ,,
Manganese	1·75 ,,
Iron (by difference) . . .	92·15 ,,

The converter is a strongly built egg-shaped furnace made of thick steel plates and lined with dolomite bricks. The bottom portion of this lining is perforated with about 70 holes, of about $\frac{3}{4}$-inch diameter to allow of a powerful blast of air at a pressure of 20 to 25 lbs. per square inch being driven through the liquid iron resting on the bottom ; the pressure of the blast is more than sufficient to prevent the liquid iron running down the holes. The converter is supported on pedestals at about its centre by means of a strong trunnion belt. It is capable of being rotated on its centre by means of a powerful hydraulic cylinder with ram actuating a rack and pinion. The converter is placed in the horizontal position to receive the charge of molten iron from the ladle, and is made of such shape and capacity that when set in the horizontal position the charge of metal rests quietly in the hollow of the converter, and below the level of the perforated bottom. The charge of metal being already in the converter when set in the horizontal position, the burning or refining is commenced by putting on the blast, and then rotating the vessel to bring it to a vertical position, when the molten metal flows on to the perforated bottom. During the process a strong flame issues from the mouth of the converter, arising from the combustion of the carbon contained in the iron. Other impurities, notably the silicon and phosphorus, pass into the slag.

The usual weight of iron treated at one operation in the converters is about 12 to 13 tons for the larger vessels, and to this is added about 20 per cent. of burnt lime, which enters into chemical combination with the silica and phosphoric acid, and thus aids in the elimination of the silicon and phosphorus originally present in the iron. The metal thus rid of carbon, silicon, and phosphorus then receives the necessary hardening additions, and is poured into a ladle to be cast into steel ingots weighing from 1 to 2 tons each. The slag is poured off separately, and is afterwards ground to a very fine powder, and in this condition is sold as a manure under the name of basic slag, phosphate slag, and phosphate meal. This material is now considered to be one of the most valuable and economical manures in use.

The blast for the Bessemer process is supplied by a horizontal engine made by Messrs. John Fowler and Co., Leeds. The air-cylinders are 4 feet 6 inches in diameter. The converters and cranes in this department are actuated by hydraulic power, for which the Galloway pumps feed two accumulators that are loaded to a working pressure of 620 lbs. per square inch. After the ingots have been cast in the moulds, the latter are stripped off the ingots promptly, and the red-hot ingots are lifted by a long jib hydraulic crane, and placed upon a hydraulic charger, which delivers the ingots into four reheating furnaces; there they are allowed to be heated uniformly throughout to pass through the next operation of rolling into various sections of girders, tram-rails, etc.

The Rolling Mill is a 32-inch mill and consists of three sets of standards placed in line across the mill. The cogging rolls are driven by a pair of engines made by Messrs. Galloways, of Manchester. The steam cylinders are 39 inches diameter by 4 feet 6 inches stroke, and are geared up by two sets of helical tooth-pinions to the main shaft, which can thus be driven either at the same speed as the engine, or at half the speed. The roughing and finishing rolls are driven by a powerful pair of engines made by Messrs. Davy Brothers. The cylinders are 51 inches diameter by 5 feet stroke. There are 19 boilers of Lancashire type to provide steam for both the Bessemer and rolling-mill engines, and these are worked at a pressure of 100 lbs. per square inch.

Owing to the number of various sections rolled in the mill, rolls have to be changed frequently, and this is done by an overhead electrically-driven travelling crane. This crane extends from the mill into a conveniently situated Roll Lathe Shop, thus facilitating rolls being taken direct from the lathe into the rolling mill. The ingots when rolled down to blooms have the ends cropped off by a pair of shears before passing to the roughing rolls. The mill is provided with roller gear and transfer arrangement to facilitate the handling of the heavy bars in the course of the rolling operations. The bar on leaving the mill passes on live rollers to the circular saw, and after being cut is transferred by mechanical power to the cooling bank; from there it is taken by the jib cranes to the finishing department, where the cold straightening and other operations which may be required on the bar are carried out. The mill is principally engaged in the manufacture of steel girders or joists, channels, flat bars, girder tram-rails for street tramways, patent bars, etc. The joists rolled vary in size from 16 inches by 6 inches, weighing 62 lbs. per foot to 4 inches by $1\frac{3}{4}$ inches at $7\frac{1}{2}$ lbs. per foot, and the sections of joists and channels rolled number 63. The number of tram-rail sections rolled is 44.

The Leeds Steel Works claim to be the largest makers of girder tram-rails in Great Britain. Forty-four different sections of these rails have been rolled, varying in size and weight from 5 inches by 5 inches at 65 lbs. per yard to 7 inches by $7\frac{1}{2}$ inches at 108 lbs. per yard. Orders have been executed for most of the important corporations in this country, including the London County Council, the rails and fastenings for the Tooting line recently inaugurated by the Prince and Princess of Wales having been supplied by these Works. Of late years there has been a demand for these rails in longer lengths to reduce the number of joints, and this firm was the first to undertake the supply of tram-rails in 60-foot lengths. This was in 1899, in compliance with the requirements of the Glasgow Corporation. The work on tram-rails is of a more complicated character than in ordinary permanent-way rails, and the Company has provided a very complete plant to do the necessary finishing work upon the rails, including machines for squaring the ends up to exact lengths, and for putting in holes for fishplates, sole-plates, tie-bars, and electrical bonds, also for milling recesses in the ends of the rails for the reception of the "Dicker" fish-plate.

A considerable area of ground is occupied by the constructional department for the building up and riveting of steel joists and plates to produce very strong compounds for building purposes. The company has also undertaken important contracts for steel piers, light bridges, etc., both at home and abroad. The plant for this class of work is extensive, and includes the usual drilling machines, hydraulic and pneumatic riveters. The sheds are provided with three overhead travelling-cranes, both electrical and rope-driven, for lifting heavy girders. Considerable stocks of various sections of joists, channels, flat bars, etc., are kept, and for the purpose of dealing readily with these an overhead gantry has been erected on columns, and covering an area of 640 feet by 95 feet span. Two steam-travellers are working on this span, putting into stock, also taking out of stock, and supplying the material to the cold saws and other machines as may be desired. The total quantity of finished steel turned out averages 1,400 tons per week.

The arrangements for testing steel supplied to customers are complete, including chemical analysis, falling-weight tests, and tensile tests as made in Buckton's 100-ton testing machine. The company has recognized the value of electricity for distribution of power to the various machines of small power spread over the Works, and has built a generating station in which two dynamos of 100 H.P. each are driven by Robey engines, and quite recently a 300-H.P. Fowler generator direct-coupled to a vertical cross compound Fowler engine working at 100 lbs. pressure, and this is now being duplicated by the same firm. Upwards of 50 motors, ranging from 1·5 to 50 H.P. each, have been supplied by Messrs.

Greenwood and Batley, of Leeds. These are of enclosed ventilated direct current type. The whole of the electrical plant is giving satisfaction, and arrangements are being made for the fixing of 50 other motors in place of small engines, whereby a considerable economy in fuel, greater efficiency, and less liability to stoppage for repairs will be realized. New repair shops, 250 feet long by 38 feet broad, have been erected for greater convenience in carrying out the repairs incidental to the iron and steel works.

The number of men employed is about 1,350.

MESSRS. TANNETT WALKER AND CO., GOODMAN STREET WORKS, HUNSLET, LEEDS.

These Works, founded more than 40 years ago, stand on an area of about 8 acres of freehold ground. There is an ample water supply, and all the engines are highly expansive and condensing. The Works are laid out for the manufacture of very heavy machinery used in iron and steel making, also dock and railway and works equipment, such as tilting furnaces, blast-furnace and iron and steel works plant, hydraulic forging and bending presses, rail, bar, plate, cogging, corrugating, and tyre machinery and engines; shearing, punching, straightening, bending, and blocking machines, for rails, plates, tyres, etc.; steam and hydraulic hammers, as supplied to His Majesty's Government, etc.; armour plate rolling machinery; hydraulic machinery for docks, railways, and works generally; hydraulic overhead travelling-cranes (a speciality); flanging machinery, hydraulic riveting plant, etc.; patent 3-cylinder blowing engines, for cupolas, foundries, forges, converters, and smiths' fires; hydraulic coal tips and cranes for coaling; and capstans.

This Company has recently supplied a 160-ton tilting furnace for Messrs. Guest, Keen and Nettlefolds, at their Dowlais-Cardiff Works, Cardiff. This is the largest furnace ever made in this

country, and it is tilted by hydraulic power, all other movements
being hydraulic. The 100-ton tilting furnace Talbot continuous
process at Frodingham was supplied by this Company. A 3,000-ton
hydraulic press has just been handed over to the authorities at
Woolwich Arsenal. In process of manufacture can be seen all kinds
of hydraulic machinery, and in use in the foundries and other
departments are hydraulic travelling cranes, hydraulic hammers, and
other appliances.

MESSRS. JOSHUA TETLEY AND SON,
THE BREWERY, HUNSLET ROAD, LEEDS.

The original Brewery was rented by Mr. Joshua Tetley in 1823, and
the present one, the building of which was commenced in 1855, now
covers an area of over six acres. Two new cast-iron mash-tuns have
been erected by Messrs. Kitson and Co. (in addition to four previously
in use): one of these is 17 feet diameter in a single piece, and the
other is 18 feet diameter in sections. The total mashing power is
225 quarters per day. There are two large steam coppers of 14 and
15 feet diameter, in addition to four older ones boiled by fire. The
fermenting vessels, 114 in number, are all on the Yorkshire square
system, most of them of slate, erected by Messrs. Alfred Carter
and Co. There are two sets of refrigerators, each set consisting of
four vertical refrigerators (Lawrence) and one Morton's horizontal
refrigerator, each set capable of cooling 160 barrels of wort per
hour, and each barrel of wort cooled requires by this method 1·3
barrels of water at 54° F. contrasted with 3 barrels of water when
the verticals were used by themselves.

Hydraulic machinery supplied by Sir W. G. Armstrong, Whitworth
and Co. is used for working six piston-hoists and five chain-hoists,
for passengers and goods, and five sack hoists. In the new maltings
the grain is distributed by elevators and by an india-rubber carrying
band, with discharging appliances, which were also supplied by the
same firm, and driven by a two-phase motor.

There are seven large boilers supplying steam to eleven engines of various sizes and of about 200 aggregate nominal horse-power. Steam is also largely used for heating, boiling, and washing purposes. The supply of water for brewing is obtained from a borehole 155 feet deep, and the water for cooling and washing purposes is obtained from a well 90 feet deep and 25 feet diameter at the bottom, and also two boreholes about 200 feet deep. The firm do the whole of their own malting, and manufacture and repair on the premises their own casks, drays, &c.; and employ altogether over 500 hands. The casks are all washed by machinery, which does away with the necessity of unheading and heading up again.

GREAT NORTHERN RAILWAY CO.'S LOCOMOTIVE, CARRIAGE, AND WAGON WORKS, DONCASTER.

The Locomotive and Carriage Works are situated between the south bank of the River Don and the Railway Station, Fig. 3 (page 618). The work of building and repairing wagons, originally carried

Fig. 2.

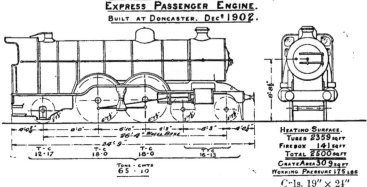

GREAT NORTHERN RAILWAY.
EXPRESS PASSENGER ENGINE.
BUILT AT DONCASTER. DEC.ᵗ 1902.

HEATING SURFACE.
TUBES 2359 sq ft
FIREBOX 141 sq ft
TOTAL 2500 sq ft
GRATE AREA 30·9 sq ft
WORKING PRESSURE 175 lbs
Cyls. 19″ × 24″

Fig. 3.—*Plan of Works.*

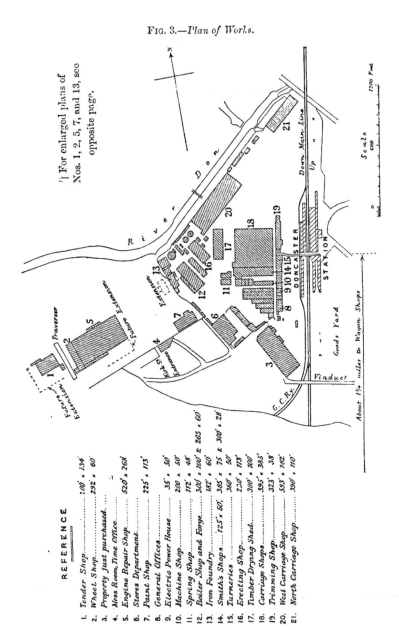

† For enlarged plans of Nos. 1, 2, 5, 7, and 13, see opposite page.

REFERENCE

1. Tender Shop................. 200' × 134'
2. Wheel Shop................. 292' × 60'
3. Property just purchased. ....
4. Mess Room, Time Office. ....
5. Engine Repair Shop........ 520' × 268'
6. Stores Department. ....
7. Paint Shop................. 225' × 113'
8. General Offices. ....
9. Electric Power House....... 35' × 50'
10. Machine Shop............. 200' × 50'
11. Spring Shop.............. 112' × 48'
12. Boiler Shop and Forge..... 300' × 100' & 265 × 60'
13. Iron Foundry............. 182' × 60'
14. Smith's Shops...... 125' × 60', 365' × 75'
15. Turneries .... 360' × 50'
16. Erecting Shop............ 220' × 113'
17. Timber Drying Shed....... 300' × 100'
18. Carriage Shops.......... 395' × 385'
19. Trimming Shop........... 323' × 38'
20. West Carriage Shop....... 593' × 142'
21. North Carriage Shop...... 390' × 110'

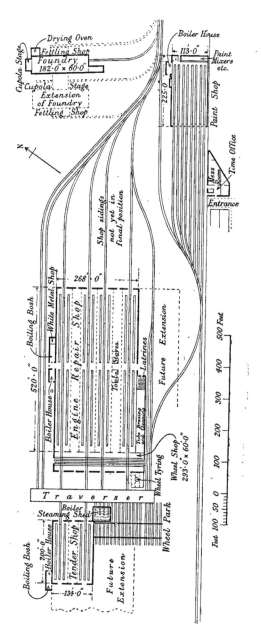

FIG. 4.

G. N. Ry. Doncaster Works. Plan of Engine Repair Shop, etc.

On upon the same site, was removed about 1890 to land alongside the main line, 1¾ miles south of Doncaster. About 100 acres of land along the bank of the River Don and alongside the old works has recently been acquired, and new engine, repair, and paint shops have been built on this site. The whole area occupied for Locomotive, Carriage, and Wagon Works (including sidings for the use of the shops) is at present about 200 acres. The total number of men employed in the department at Doncaster is 4,800. The rolling stock of the Company, in December 1902, numbered 1,279 engines, 2,175 carriages, and 39,000 wagons.

The Locomotive Shops comprise the following Departments:—

Engine Repair Shops.—The general arrangement of this block of buildings is shown on Fig. 4 (page 619). Accommodation for the repair of 100 engines is provided in the four bays, each 52 feet span by 520 feet length. The kindred trades, that is, copper-smiths, light boiler-repairs, fitting and turning, are carried on in the two smaller bays, 30 feet span. There are three roads in each of the larger bays, and each bay is equipped with two 35-ton 4-motor travelling cranes. The smaller bays have 8-ton 3-motor cranes, and are furnished with the necessary tools and machines to effect all the repairs required without the work leaving the shop. A complete system of 18-inch gauge roads run through the shops everywhere, both inside and out, for conveying the engine parts where required. Compressed air and electric current is conveyed round the shops for use where necessary. The shops are very lofty, spacious, and well lighted, half of each roof—that facing the north—is glazed. Electric light is installed, both arc and glow lamps being used. Hot-water pipes are used for heating the sheps. All the new shops are driven electrically; compressed air is largely used, and the electric mains and air pipes can be tapped for small portable motors anywhere along the pits and walls. Gas as well as electricity is provided for heating purposes or lighting where necessary. The offices, stores, boiling bosh, etc., are arranged near the central part of the shop.

The Wheel Shop (No. 2 on Fig. 3, page 618), 60 feet by 292 feet, is situated at the west end of the engine repair shops. The lathes, except two which have their own motors, are driven in groups by motors, and are fixed along the sides of the shop, and fed by electrically-driven walking-cranes each capable of lifting 6 tons. Two roads run through the shop for the manipulation, etc., of wheels. All wheels when completed in this shop are immediately returned to the engines or placed in a wheel park to the west of the Shop.

The Tender Shop at present forms the extreme end of the Works. The height and width of the bays and general equipment is the same as the Engine Repair Shops. Arrangements have been made in the building of this and the Repair Shop to admit of future extension where necessary. Between the Wheel Shop and the Tender Shop a 40-foot electric traverser is provided, and is capable of hauling on and traversing the heaviest locomotive.

Iron Foundry.—The weekly output is about 100 tons. A new foundry is shortly to be built adjacent to the existing one; alterations are to be made in the arrangement of cupolas, fettling shops, and roads leading to them. The enlarged plan, Fig. 4 (page 619), shows what the general arrangement will be when finished. The Pattern Shop is conveniently placed next to the Foundry.

Boiler Shop.—The output of this shop ranges between 100 and 150 boilers per annum. It is equipped with 14-ton overhead travelling-cranes. Extensive use is made of compressed air in this shop for pneumatic chipping, drilling, and caulking. Electric power is also laid on for the supply of current to drills, etc. Lubricant from an overhead tank is piped to all drilling machines, etc., in order to obtain the maximum output; a Tool Room also forms part of the equipment.

Forge.—This building is next to the Boiler Shop, and is provided with four steam-hammers. The average output is 610 tons per annum.

Engine-Erecting Shop, No. 16, Fig. 3.—This building consists of two bays 48 feet wide by 230 feet long. It is equipped with four 30-ton rope-driven cranes, shortly to be converted to electric motor cranes. One bay is used entirely for the building of new engines, the other for the manufacture of tanks, splashers, cabs, etc.

Smiths' Shops, No. 14, Fig. 3.—These shops contain 115 smiths' hearths, and are fitted with steam hammers, etc.

Turneries, No. 15, Fig. 3.—These comprise two storeys; the lighter parts of engines (motion, brake work, etc.), are manufactured in the upper one, and the heavier parts (such as cylinders, etc.) in the lower one. The shop marked No. 10 on Fig. 3 is used for similar class of work, and for repairs to cranes and out-station machinery.

Power House, No. 9, Fig. 3.—Power is provided for driving the Turneries from a battery of boilers which supply steam to a Robey 200-H.P. engine, also for the engines driving the electric generators. These at present consist of one 220 and two 88-kilowatt machines. Provision is made for two further sets of 220 kilowatts each, one of which is about to be fixed. When the installation is completed the smaller isolated stationary engines now working will be superseded by electric motors.

Carriage Works.—All new Great Northern Railway stock and a large proportion of East Coast Joint stock is built in these shops; there is accommodation for some 250 vehicles under repair and varnishing, etc. The Lifting Shop, No. 20, Fig. 3, is provided with two electric travelling-cranes, each capable of lifting 20 tons, for dealing with the larger coaches. Compressed air is largely used for lifting, etc., and the vacuum process for cleaning cushions and the inside lining of carriages is also employed.

Wagon and Wheelwrights' Shops.—These comprise two Wagon Shops, one capable of holding 210 wagons and the other 230 wagons, and two shops for dealing with the building and repair of floats, drays, omnibuses, etc. (technically known as "Road Goods"), and of which vehicles there are some 3,000 to be maintained. The shops are entirely lighted by electricity, and electric driving is gradually being introduced. There is little or no house accommodation near the works—a workman's train conveys the men to and from Doncaster, and a mess room, accommodating about 500 men, is provided.

The Running Shed is situated to the east of the main line. It has 12 roads and holds 96 engines. The general arrangement is completely "double ended," that is, sidings, coal stage, turntable, etc., are provided at either end. A small repair shop adjoins the shed, with overhead traveller, etc.

The Oil-Gas Works is just south of the shed. The capacity is about 12,000,000 cubic feet per annum.

FRODINGHAM IRON AND STEEL WORKS, FRODINGHAM, NEAR DONCASTER.

Ores.—The Lincolnshire ores are geologically secondary jurassie (lias) and oolitic. At present the surface ores only are worked, and are quarried in open workings, with a covering varying from nil to 30 feet. The ore seam has a depth varying from 10 to 20 feet. The ore field extends in a straight line from north to south, from the Humber to Kirton Lindsey, and with a width varying from about 150 yards to about a mile and a half. The ore contains iron from 18 to 30 per cent., the average being about 26 per cent., and it is therefore a very poor ore, very low in iron.

The reason why such low grade ore can be economically worked is due :—(1st) To its containing its own admixture of limestone ; (2nd) To its being quarried, therefore cheaply got ; (3rd) On account of the close proximity to coal and coke, viz., about 28 miles.

The pig-iron which is manufactured from the ore contains about 1·50 per cent. of phosphorus, and about 2·00 per cent. of manganese, and is therefore a pig-iron eminently suitable for the basic open-hearth process.

Blast-Furnaces.—There are at present four blast-furnaces, each 70 feet high, with 9 feet 6 inches diameter of hearth, height of boshes 18 feet, diameter of bosh 18 feet 6 inches, diameter of throat 15 feet. There are eight tuyeres to each furnace. Two of the furnaces are fed with material by means of bell and hopper, with standard beam and winch. The third furnace has an open top, but will be altered into a close top immediately. The ore and other material goes on a high level, emptying itself into bunkers, from

2 x

which it is loaded by hand into larries, which are wheeled to the hoist, and lifted up to the furnace gantry.

A new furnace is in course of erection, the foundations of which are just being put down. This furnace will be built entirely on the newest principles, with complete mechanical arrangements to do away with all labour in connection with filling of ores and feeding the furnace. All the present furnaces are served by eight stoves, six of which are Whitwell's, and two Massicks and Crooke's. They vary in height from 50 to 70 feet, and in diameter from 18 to 23 feet. The pressure of blast now maintained is 5 lbs. per square inch, and its temperature on entering the furnace is about 1,100° F. At the present time three furnaces are in blast; the fourth one is about to be rebuilt on modern principles, as mentioned previously. The furnaces make basic pig-iron for the open-hearth steel works.

The engines are five in number, of the vertical type, with steam cylinders on top, and blowing cylinders below, non-condensing, the sizes being as follows:—

No.	Blast.	Steam.	Stroke.	Max. Speed.	Max. Pressure.	Initial Steam.	Cut-off.
	Ins.	Ins.	Ins.		Lbs.		
2	72	24	48	36	6	90	½
2	78	26	48	34	6	90	½
1	100	38	54	28	6	90	⅓

For the new furnace new blowing-engines have been ordered of the Southgate type, compound, condensing, worked at a steam pressure of 150 lbs. per square inch, and built for a blast pressure of up to 20 lbs. per square inch. These engines are being built by Messrs. Richardsons, Westgarth and Co., of Middlesbrough. The steam required for the present blast-engines is supplied by twelve Lancashire boilers, 8 feet 6 inches diameter by 30 feet long, having 1,100 feet of heating surface. They evaporate 500 gallons per boiler per hour, and work at a steam pressure of 100 lbs. per square inch. For the new plant seven Babcock and Wilcox boilers are about to be put down, working at a pressure of 160 lbs. per square inch.

The waste gases from the blast-furnaces, after heating the air, are utilized for raising the steam, of which there is sufficient to drive 480 H.P. of electric power, which is utilized at the Steel Works. This power plant of 480 H.P. consists of one set of 250 H.P. driven by a Robey compound engine, and one direct-driven set of 230 H.P. The generators are of Westinghouse make, 220 volts pressure.

Steel Works.—The Melting Shop consists of eight open-hearth furnaces, as follows:—

  2 of 20 tons capacity each.
  2 of 25 tons capacity each.
  1 of 35 tons capacity.
  2 of 40 tons capacity each, and
  1 of 100 tons capacity.

The last-named furnace is worked on the continuous principle, Talbot patent. The seven ordinary furnaces work a mixture of 70 per cent. pig-iron, and 30 per cent. scrap, and make 1,600 tons of steel ingots per week. The 100-ton furnace, on the continuous principle, works liquid pig-iron only, without scrap, and makes 650 tons of ingots per week, so that the Melting Shop makes 2,250 tons of ingots per week.

The slag from the furnaces, containing from 13 to 14 per cent. of phosphoric acid is carried to the Manure Works, where it is ground down in three Krupp mills, built on the Ball principle, into a fine powder, and is sold as phosphate manure for agricultural purposes. The production of this manure is 25,000 tons per annum. The ingots are cast partly in straight and partly in circular pits, and are served direct to four coal-fired soakers, where they are reheated, and then supplied to the Cogging Mill.

The Cogging Mill consists of one pair of housings, 36-inch centres, steel rolls 7 feet 6 inches long, driven by a pair of two-cylinder reversing engines, geared 1 to 2. Diameter of cylinder is 42 inches, and stroke is 60 inches. These engines were built by

Messrs. Kitson and Co., of Airedale Foundry, Leeds. The capacity of this Mill is more than equal to the output of the Melting Shop, and is today dealing with the whole output, say 2,250 tons of ingots per week. The blooms, after leaving the cogging engines, are cropped by steam shears, built by Messrs. Joshua Buckton and Co., Leeds, and are capable of dealing with slabs up to 30 inches wide, and 10 inches thick. These shears are driven by a reversing steam-engine, and are supplied with live-roller tables.

There is a large re-heating furnace with producers attached, for reheating blooms for sections of larger sizes which are not capable of being rolled off in one heat, namely, all sections 12 inches high and over. This furnace is served by an electric charging machine, made at the Frodingham Works, which charges the blooms from the shears, and supplies them to the finishing mill. The ingots, after being cogged and cropped, are transferred, at the same heat, to the finishing mill, consisting of three pairs of housings, 30-inch centres, and rolled into all kinds of bridge and ship material, and also into material for wagon building, and structural work of all kinds.

The Finishing Mill is driven by a three-cylinder reversing engine, direct coupled, all high pressure; each cylinder is 45 inches diameter with 52-inch stroke, built by Messrs. Davy Brothers, Sheffield. At the end of this Mill are two hot saws, pendulum type, to crop the bars, which are pushed on to the hot bank, and lifted by 5-ton steam travelling-cranes either direct on to the trucks or into stock, or supplied to the straightening presses, of which there are two, built by Messrs. Craig and Donald, of Johnstone, worked by electric motors. There are also four cold saws, driven by electric motors. The Cogging and Finishing Mills are served by two electric overhead cranes, each of 15 tons capacity, one made by Messrs. Joseph Booth and Brothers, of Rodley, and the other by Messrs. Joseph Adamson and Co., Hyde. The overhead-crane track is lengthened, and connects with the repair shop and roll lathe shop, where there are different tools for the necessary repairs to the Works, and two roll lathes for dressing the rolls.

The 14-inch bar mill is driven by a high-pressure non-condensing flywheel engine, with cylinder 30 inches diameter, and stroke 36 inches, and consists of two pairs of housings for sectional work, and two stands of guide rolls for guide rounds and squares. This mill is served by a Siemens gas furnace, with producers attached, and the output of this mill is 350 tons per week. There is a hydraulic plant, with two high-pressure Worthington pumps, one accumulator and tank with automatic governor gear attached, working at a pressure of 600 lbs. per square inch.

There is a battery of 16 Lancashire boilers, fired by automatic stoking gear, Proctor's system. All the boilers are designed to carry 100 lbs. per square inch pressure, and are usually worked at 80 lbs. In connection with the Works there are the usual engineering shops, namely, fitting, blacksmiths, boiler smiths, pattern-making, joiner, and locomotive repairs, where all renewals and repairs to machinery and other plant are executed.

The Works employ about 1,500 men.

MESSRS. POPE AND PEARSON'S
WEST RIDING AND SILKSTONE COLLIERIES,
ALTOFTS, NEAR NORMANTON.

These Collieries have been working for the past 52 years, and at the present time the output of coal is about 3,000 tons per day. Last year 553,000 tons of coal were raised, of which 421,000 tons were got by coal-cutting machines. There are three shafts at present winding coal, with other shafts for ventilation and pumping.

The Silkstone coal (4 feet thick) is raised from a depth of 336 yards by means of a pair of horizontal winding-engines 32 inches diameter by 6 inches stroke, made by Messrs. R. Daglish and Co., St. Helens, with automatic cut-off valves; the drum is 16 feet diameter, and the rope-plough steel $4\frac{3}{4}$ inches in circumference. The wrought-iron head gear is 100 feet high, and was made by Messrs. Goddard and Massey, of Nottingham. Eight tubs of coal are raised at each lift, the weight of coal being about 3 tons 10 cwts., and the dead weight of loaded cage and rope is 10½ tons. About 1,300 tons of coal are raised per single shift.

Power is supplied by four Lancashire boilers 9 feet diameter and 30 feet long, working at a pressure of 120 lbs. per square inch, made by Messrs. Spurr and Inman, of Wakefield. The heapstead is built of steel girders 12 inches by 6 inches. The screening plant produces eleven qualities of coal, and the tipplers were made by Messrs. M. Coulson and Co., of Spennymoor, Durham.

The Silkstone plant also contains the following engines. A screen engine, 16 inches diameter by 3 feet stroke, made by Messrs. Marshall, Sons and Co., of Gainsborough. A pair of compound condensing air-compressing engines on the two-stage principle, by Messrs. Walker Brothers of Wigan. The high-pressure steam-cylinder is 20 inches

FIG. 5.—*Evolution of the Miner's Pick.*

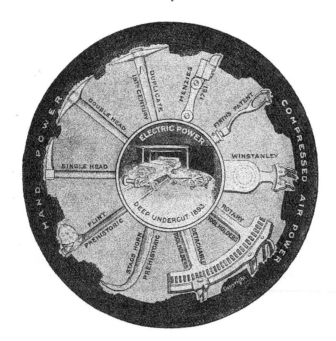

diameter, and the low-pressure cylinder 36 inches; the high-pressure air-cylinder is 22 inches diameter, and the low-pressure 34 inches by 4 feet stroke. A pair of 16-inch self-contained engines, by Messrs. Wood and Gee, of Wigan, were used for sinking the 16-foot pit 336 yards deep, and were subsequently converted to drive a strap rope which works an endless chain to convey the full and empty tubs to the pit bottom. The working places in-by are about 3 miles from the shaft, and the haulage is on the endless-chain system driven by compressed air.

The Diamond coal, $3\frac{1}{4}$ feet thick, is raised from a depth of 500 yards by a pair of vertical high-pressure engines, 40 inches diameter by 6 feet stroke, made by Messrs. Bradley and Craven, of Wakefield, fitted with Daglish's automatic cut-off. The weight of coal lifted each wind is about 3 tons 5 cwts., the dead weight of loaded cage and rope being 10 tons 15 cwts., and the rope is $4\frac{3}{4}$ inches circumference made of ploughed steel. The output per single shift is over 1,000 tons. The screening and banking plants are undergoing alteration, and when finished will be similar to those described at the Silkstone shaft. The Diamond plant also contains 9 Lancashire boilers working at 70 lbs. pressure, made by Messrs. Spurr and Inman, of Wakefield, and Messrs. Davy Brothers, of Sheffield. The ventilating Schiele fan is 15 feet diameter, and is capable of producing about 200,000 cubic feet of air per minute with a 6-inch water gauge; it is driven by a high-pressure engine 32 inches by 3 feet duplicated.

The West Riding pit is 80 yards deep, from which the Haigh Moor seam, 135 yards deep and 4 feet thick, is drawn, the coal face being distant about 3 miles underground. There are 8 Lancashire boilers, made by Messrs. Daniel Adamson, Thomas Beeley, and Spurr and Inman. Besides the above, the following plant may be seen: A washer capable of washing and separating 600 to 700 tons of coal per day into four different sizes, made by Herr F. Baum, Herne, Westphalia; bee-hive coke ovens; two pairs of air-compressors by Messrs. Walker Brothers, of Wigan, for supplying Diamond coal-cutting machines; and an air-compressor by Messrs. Ingersoll and Sergeant, London.

The electric lighting plant was made by Messrs. Greenwood and Batley, Leeds, supplying eleven hundred lamps of sixteen-candle power to underground and surface works, and consists of 3 dynamos driven by two Marshall engines. A Ledward condenser receives the exhaust steam from 10 cylinders. There are also smiths' and fitting shops, wagon shop and saw shed. Four locomotives, made by Hunslet Engine Co., Leeds, and a steam navvy made by Messrs. Whittaker Brothers, Horsforth, is used for filling up stacked coal.

At the Fox Lane pit (one mile distant), a Capell fan, made by Messrs. Thwaites Brothers of Bradford, driven by a 26-inch by 4-foot high-pressure engine by Messrs. Bradley and Craven, Wakefield, is capable of producing 250,000 cubic feet of air per minute with a 6-inch water-gauge. There are also a pair of winding engines, 6 Lancashire boilers made by Messrs. Spurr and Inman, of Wakefield, a polyphase electric power plant of 550 volts, made by the General Electric Co., Manchester, to supply coal-cutting machines, and two Diamond electric coal-cutting machines under-cutting 7 feet.

Two Diamond coal-cutting machines, one driven by compressed-air and one by electric continuous current, with a $5\frac{1}{2}$ feet undercut, were on view at the surface. (*See* Fig. 5, page 628, and Proceedings 1902, page 545.)

MESSRS. GEORGE CRADOCK AND CO., STEEL-ROPE WORKS, WAKEFIELD.

The founder of this firm had works at Darlington and Stockton for the manufacture of hemp rope, which were carried on for a considerable time. In 1853 he removed his business to Wakefield, and in 1854 commenced the manufacture of wire ropes, but the principal business then was hemp rope making. Until about 1881 both hemp and wire ropes were manufactured; the firm however gave up the manufacture of hemp ropes, excepting those for use in the centre of its own wire ropes. The wire-drawing mills were started in June 1885, and in January 1900 the firm commenced the manufacture of its own steel.

The Steel Works are in a building 270 feet long by 120 feet wide. At one end are two Siemens' furnaces, each capable of taking a charge of eight tons, and worked with producer gas. At the rear of the furnaces is the casting-pit, which is served by a 3-ton travelling jib-crane. The steel, as it comes from the furnace, is run into ingot moulds ready for use. Sometimes the rods are made direct from the ingots, at other times—depending on the nature of the ultimate product required—the ingots are first of all hammered under a steam-hammer. Generally, the weight of the ingots produced is $9\frac{1}{2}$ cwts., and the weight of the billets passed through the mill is from 100 lbs. to 110 lbs. when making rods of No. 5 gauge.

The rolling plant consists of a large cross-coupled horizontal engine, made by Messrs. Stevenson and Co., of Preston, having two cylinders of 36 inches diameter, and with 36 inches stroke. It runs at 90 revolutions per minute, and is supplied with steam at 70 lbs. per square inch. The crank-shaft carries a large fly-wheel grooved for twenty-one ropes, which drive by means of gearing a set of three cogging-rolls and two sets of ordinary rolls. Steam is supplied by two Babcock and Wilcox and two Lancashire boilers. Of the two Babcock boilers one receives the waste heat of a second re-heating furnace, and the other is fed by hand. Arranged at the rear of the boilers is a Green's economiser. All the engines are condensing.

The whole process of rolling the billets into finished rods is performed in less than two minutes, which entails three passages through the cogging mill, and no less than sixteen passes through rolls gradually decreasing in aperture, and then coiling up. The rods are allowed to cool sufficiently, and are then taken into the hemp rope walk where the various bundles are weighed and marked, and irregular ends cut off. The annealing process follows, which occupies about twenty-four hours. The bundles of rod are first dipped in a stone bath containing a diluted solution of sulphuric or hydrochloric acid. There they remain about ten minutes, when they are drained, and then washed for a time with water until a

" coat " is formed on the metal.   They are then dipped again in a
solution of lime, so that the acid used in cleaning may be neutralised.
After they have been well dried, the bundles of rods are ready for
drawing.

There are two drawing mills each arranged in two long cast-
iron tables placed end to end.   The driving mechanism consists of a
horizontal shaft running underneath the table and carrying on it
a bevel wheel for each block.   Each block consists of a die and a
revolving drum or block placed on a vertical shaft, the lower end of
which carries a bevel wheel which gears with one of the bevel wheels on
the horizontal shaft already mentioned.   There are two methods of
pulling sufficient wire through the die for it to be connected to the
revolving drum.   The number of times the wire is drawn through
the dies, and the number of times it is annealed during the process·
of drawing, depend upon the ultimate size of wire required, and the
nature of the material employed.   After the wire has been brought
down to the size required, it has several processes to go through
before being made into rope.   First it is tempered, and afterwards
finally cleaned and drawn, with very slight reduction in diameter,
several times.   Then it may require to be galvanised.   In that
case the wires are placed on vertical bobbins, and are then drawn
through a bath containing a cleansing composition and other materials,
whose object is to make the zinc take to the wire.   They are next
drawn through a bath of molten zinc, which is contained in an iron
pot over a coke furnace with a fire-brick setting.   As they emerge
from the bath the wires are drawn through a material, which
removes the surplus zinc, and are wound on reels.   Twelve wires
can be galvanised at once.

The next process is that of testing.   Of every coil of wire drawn
two pieces are tested.   As an example of the results obtained, a wire
for tramway rope should test as an average about 85 tons per square
inch tensile strain.   The wire is then ready for manufacturing into
rope.   There are twenty-six stranding machines of all sizes for
making rope from one-sixteenth inch diameter to any size required.
This rarely runs above one inch in a single strand, and hempen
ropes form the core.   The hemp ropes used for the cores

of the wire ropes are all made on the premises, and the firm also makes the machinery for the wire-rope department.

For the material used in wire-roping making, only two classes of pig-iron are used, namely, Swedish and Cumberland. For plough steel wire Swedish pig is alone used ; for all other purposes a mixture of Swedish and Cumberland is 'employed. Every delivery of iron is analysed before it is used, as is also each ingot as it comes from the furnace.

The company, in addition to making its own machines, carries out all its own structural work, and also does its own repairs. The lighting of the whole premises is by electricity produced on the site. The number of men employed is about 280.

MESSRS. E. GREEN AND SON, ECONOMISER WORKS, WAKEFIELD.

These Works were founded in 1821, rebuilt in 1864, and extended 1897 to 1900. They are occupied solely in the manufacture and building of Messrs. Green's specialities ; namely economisers, air heaters, boiler feed-pumps, and small economiser engines, and are the largest of their kind in the world, covering nearly 20 acres and employing about 1,000 men.

The demand for their economisers from every quarter of the globe has now assumed such proportions that an entire reorganization of the business has been necessary ; the plant has been re-constructed throughout, new special tools and labour-saving devices introduced, and additional facilities for transit arranged.

SEAMLESS STEEL BOAT CO., WAKEFIELD.

The Works of this company are situated in Calder Vale Road, Wakefield, being originally known as the old forge, when in the occupation of the late Mr. Samuel Whitham. The present company took them over in 1890, and laid down a plant for pressing steel boats in two halves.

The press shop in the centre of the yard, measuring 120 feet by 80 feet, contains a hydraulic press, 34 feet long by 12 feet wide, with six 16-inch rams, furnace, 32 feet by 10 feet, and electric welding plant. The boat fitting shop adjoining measures 120 feet by 80 feet, and, beyond a stand on which the two sides of the boat are punched and riveted by hydraulic pressure to a T-bulb bar, is used exclusively for fitting the woodwork on to the boats. Passing through this the galvanising department is seen, which contains an acid bath 32 feet by 6 feet 6 inches for pickling the plates, and a galvanising bath 44 feet by 7 feet 6 inches for electric deposition of zinc. The company in addition to galvanising all the boats, also undertakes the galvanising of water-tubes and such other work.

The engine-house contains a double-acting twin-cylinder hydraulic pump, having steam cylinders 21 inches diameter, rams 6 inches and 4 inches diameter with a stroke of 2 feet, feeding an accumulator having a ram $24\frac{1}{2}$ inches diameter, with a lift of 19 feet. The working pressure is 850 lbs. per square inch.

Other departments, such as the saw mill, blacksmiths' shop, tinsmiths' shop, occupy the west side of the yard.

Electrical driving has recently been installed, taking its power from the corporation mains, and one 50-H.P. and two 15-H.P. motors are running. Pumps for hydraulic press are, however, worked by steam. Since the company was started, nearly 3,000 boats have been sent out, varying in length from 15 feet to 36 feet 6 inches, the boats being chiefly used as ships' lifeboats and cutters, but many boats have also been fitted to meet special needs for Africa and other parts. The manufacture of launches has lately been added, and a 30-foot launch was recently supplied to H.H. The Khedive, for use of the war department of the Egyptian Government on the Nile; and among the other orders at present on hand, is one for a 34-foot launch for the Crown Agents for use on the Niger.

FOUNTAINS ABBEY.

The Abbey of Fountains originated in a small body of Benedictine Monks from St. Mary's Abbey, York, who desired to obey the rule of St. Benedict more strictly than it was there practised. They were compelled to leave their own abbey, and in 1132 received from Archbishop Thurstan the site of the present Abbey of Fountains. The buildings were laid out on the normal Cistercian plan, perhaps under the direction of a monk of Clairvaux, who came over to teach the brethren the Cistercian rule. Fountains Abbey therefore belongs to the great group of Cistercian abbeys founded in rapid succession in England in the twelfth century. The reason for the great popularity and success of this order is due to the fact that each daughter house, while submitting to the rule and the general directions of the mother house at Citeaux, was allowed to increase its position and work for its own benefit, and it is interesting that the organization of this system was due to an Englishman, by name Stephen Harding, who was the real founder of the order. It is known from the Customs that ornament was not allowed, that expensive materials were avoided, and that the services, dress and living were kept to a severe simplicity.

There were two classes of inmates in the Cistercian Abbeys, one consisted of the monks and the other of the conversi or lay brethren, and the common property of the house was managed by an abbot with a small number of officers. The duty of the monk was to devote himself to religious exercises without the interruption of worldly affairs. The conversi lived partly in the abbey and partly in outlying farms ; they attended certain services and did the work necessary to carry on the business of the abbey. The monk's days and nights were divided into twelve hours, which varied in length according to the time of the year. They began their services at about two o'clock in the morning, and they completed their offices at various hours during the remainder of the day. Each monk had a daily allowance of one pound of bread and a measure of drink, and at dinner in early times there were two cooked dishes, but no flesh meat or fish was to be eaten. While they were at their meals a chapter was read to them from a pulpit.

The buildings of a Cistercian Abbey can fortunately be identified, and the uses proper to each be confidently assigned, owing to the order in which they were traversed by the Sunday procession. The church of course is the important building to which all the others are subordinated, and on the south side of this is a great open square, in which the monks would take their exercises or sit and read, and this square is called the cloister. Around the cloister are arranged various buildings.

The dates at which different portions were built or altered are as follows: In 1146 or 1147 the church and buildings were burnt by the friends of William, Archbishop of York, out of revenge for the part taken by Abbot Murdac in the Archbishop's deposition. The repairs that followed were spread over many years, and it was not until about 1180 that the buildings round the cloister were again completed. The Chapter House and the monks' dormitory over the northern half of the western range were first rebuilt, probably in. place of wooden structures. The two guest houses in the outer court next followed, and finally the frater, warming house and kitchen, forming the southern range, and the remainder of the cellarium.

During the opening years of the thirteenth century, Abbot John of York (1203–1211) began to enlarge the church eastwards, but, at his death, only the foundations had been laid and some of the pillars built, and the work was carried on by his successor, another Abbot John (1211–1219). The next Abbot, John of Kent (1220–1247), completed the work, including the eastern transept, called the " Nine Altars." He also reconstructed the cloister, built the monk's infirmary, and the poor folk's guest house in the outer court. During the fourteenth century the infirmary kitchen and chapel were rebuilt, and the aisles of the great infirmary hall were cut up into chambers. Quite at the end of the fifteenth century large traceried windows were inserted in the east and west fronts of the church, and in the end gables of the " Nine Altars " by Abbot John Darnton (1479–1494). Abbot Marmaduke Huby (1494–1526) built the great tower at the end of the north transept in place of the original one over the crossing that had to be taken down on account of its instability. He also rebuilt and enlarged the abbot's lodging, which stood midway between the infirmary and the dormitory.

The Abbey was suppressed in 1539, but the buildings, after being unroofed and dismantled, were otherwise left intact. Some parts, such as the infirmary, the abbot's lodging, the gatehouse, and other buildings in the outer court, have since been more or less destroyed. Fountains Abbey, nevertheless, remains the most imposing and instructive ruin of its kind in this or any other country, and between 1848 and 1856 the entire Abbey was systematically explored and brought to its present orderly condition by the late Earl de Grey and Ripon.

A little to the west of the Abbey gatehouse stands the picturesque house known as Fountains Hall, with a quaint garden in front. The house was built, at the expense of some of the Abbey buildings, by Sir Stephen Proctor, who bought the site of the Abbey in 1597 from the representatives of Sir Richard Gresham, the grantee.

MESSRS. ROWNTREE AND CO.,
COCOA AND CHOCOLATE WORKS, YORK.

One of the chief developments of the modern and commercial aspect of the ancient City of York is the large Cocoa and Chocolate Works of Messrs. Rowntree and Co.—the home of Rowntree's " Elect Cocoa."

The building of this Firm's large modern Haxby Road Factory was begun in 1890, and every year since has seen a steady demand for increased accommodation. The business was founded in 1838— Queen Victoria's Coronation Year—and the development which it has experienced is evidenced in the extensive Haxby Road building which covers about seven acres. This factory is served by the firm's private railway, nearly a mile and a quarter in length, which is connected with the North Eastern Railway Co.'s York and Scarborough line. In 1900 the storeroom was doubled in size, the new box mills were built and new dining rooms to seat 1,500 of the employees have recently been opened. This centre of industry provides employment for about 2,500 workpeople, for whose use extensive gardens, allotments, and recreation grounds are provided, the total extent of the estate being about 56 acres. In the design and construction of the work rooms great care has been exercised in providing adequate ventilation, and in considering the comfort of the employees generally. The largest installation is for ventilating the offices and store-rooms. Three fans, two of 8 feet diameter and one of 6 feet, running about 120 revolutions a minute, draw the air through water screens to clean and humidify it. They then force it along 425 feet of excavated corridors and over nearly 4 miles of steam-pipe batteries, which can be turned on in sections according to the temperature required. The aggregate amount of fresh air supplied hourly by means of these three fans is over three million cubic feet.

First amongst the many of Messrs. Rowntree's manufactures is their " Elect Cocoa " which has in recent years taken such a high place in the opinion of the public. The popularity of the chocolates manufactured by this firm is everywhere recognised, and the constantly increasing demand which is being made for them has necessitated a recent addition to the chocolate manufacturing department, which will enable employment to be given to about 850 girls, making and packing chocolate creams. Interesting work rooms are where the pastilles and clear gums are made and packed. Messrs. Rowntree are the largest manufacturers in the kingdom of these gum sweetmeats. A huge cellar, containing over 20,000 giant pots of preserved fruit used for flavouring these sweetmeats, is evidence of the enormous extent of this branch of their business.

No doubt the most interesting feature to the Members will be that relative to the engineering department. In the centre of the

Factory is the Engine and Boiler House, &c. With the exception of a small amount of shafting which is driven by belts off a horizontal compound condensing engine, practically the whole of the machinery is driven electrically, and owing to the large ground space covered by the buildings, this method of driving is particularly suitable. The current is continuous and the voltage is 220.

The plant consists of two 90-kilowatt dynamos driven off the above horizontal compound engine and two 150 kilowatt direct-driven sets, one a Willans-Parker, the other an Allen-Parker ; the engines in each case are compound and condensing, and owing to the rapid extension of the factory all these machines are more than fully loaded at times. Some of the current is used for lighting, about 400–500 ampères being required for this purpose out of a current of rather more than 2,000 ampères. To avoid excessive fluctuations in the voltage, due to the varying motor load, all the machines are slightly over-compounded, the inner ends of the compound windings being connected in parallel when the machines are running. The cables and wire throughout are concentric ; the outer being uninsulated, but all the mains from the dynamo to the switchboard are insulated so as to enable the current of each dynamo to be measured when running in parallel.

At the back of the dynamo panels on the switchboard there are two alternative sets of bus-bars, one bar in each set being connected to the outer of the compound winding, the other bar in each set being connected to the inner of the compound winding; double change-over switches are provided on the dynamo panels so that the dynamos can be grouped in any two sets. At the back of the circuit panels there are two bus-bars for the insulated conductors of the feeders, so that the circuits can also be grouped in any two sets, change-over switches being provided on the circuit panels for that purpose. Any dynamo or dynamos can therefore supply any circuit or group of circuits, and by grouping together the steady motor load with the lighting circuits the fluctuation of the voltage is greatly diminished.

The switchboard is arranged for ease of extension ; each dynamo has a panel of its own, and two circuits go to one circuit panel.

The earth bus-bar is fixed on the wall at the back of the switchboard, all the earth conductors being permanently connected to it. The uninsulated outer conductor of all the cables consists of galvanised steel wire, and to avoid excessive size there is a further layer of stranded copper wire underneath; the inner conductor consists in the case of the cables, of lead-covered paper insulated cable. All wire of $7/21\frac{1}{2}$ size and below consists of an inner copper conductor insulated with vulcanised rubber and taped; the outer conductor consists of a layer of copper wire with a lead sheathing overall. The wire for the electric light throughout the buildings is on Messrs. Mavor and Coulsons C. C. system, and the various fittings are of their make. The whole of the distributing switch and fuseboards throughout the Factory (with the exception of one in the office block) whether for power or lighting, have cast-iron cases and covers so as to avoid risk of fire; and it has been found that even if a short circuit takes place inside these boxes, the damage done is entirely confined to the interior of the box.

The motors are of all sizes from 35 H.P. down to 1 H.P. and are mostly of Messrs. Thomas Parker's make. The number of units generated in the course of the year amounts to about 1,100,000, and the cost per unit, including all interest, depreciation and maintenance, is at the present time about $0 \cdot 66d.$, the lowness of the figure being largely due to the fact that there is a considerable night load.

In addition to the electrical machinery, the engine house also contains the pumps for supplying water to the Factory for fire and other purposes, and the refrigerators. The latter consist of a small ammonia absorption machine, which is now out of use, a CO_2 machine of 12 tons ice-making capacity by Messrs. Hall of Dartford, and an ammonia machine by Linde of 24 tons capacity. These machines are used entirely for cooling brine, which is circulated through different rooms for chilling and ventilating purposes. There is also a central condensing plant, capable of dealing with 14,000 pounds of steam per hour, with two cooling towers for the condensing water, made by the Klein Engineering Co., Manchester.

The Boiler House contains a range of five Lancashire boilers, 30 feet by 8 feet, working at 100 lbs. pressure, and fitted with Green's economisers, and mechanical stokers of the " compressed air" type, made by Messrs. Edward Bennis and Co., of Bolton. These stokers are given an efficiency in continuous work of 8·1 lbs. of water per lb. of coal from and at 212° F. with the cheapest form of coal to be obtained in the locality, and will easily evaporate 1,000 gallons of water per boiler per hour. Automatic coal-elevators to serve the stokers are now being fitted. The Boiler House at the time of the Meeting was undergoing alterations, with a view to increasing the number of boilers, two boilers, 8 feet 6 inches diameter for 150 lbs. pressure being on order. About three-fourths of the steam generated by the boilers is distributed over the Factory for boiling, drying and similar requirements, and nearly the whole of the condensed water is returned to the Boiler House.

Adjoining the Factory are the Mechanics' and Joiners' Shops, &c. In the former a part of the machinery used in the Factory is made, and also repairs are effected.

Not the least interesting feature of the firm's work is the provision that is made for the physical and social betterment of their employees. Several experienced persons are employed to give their special attention to this, both amongst the men and girls, and the opportunity of joining the many recreative and instructive clubs in connection with the works is largely taken advantage of by the employees.

In addition to their factory at York, Messrs. Rowntree have estates in the West Indies, giving employment to about 300 hands, where much of the cocoa used in their several manufactures is cultivated.

MEMOIRS.

CHARLES WILLIAM ADAMS, second son of the late Robert Adams, Superintendent of the Tilbury Docks, and great grandson of John Samuel Adams, Engineer of the East and West India Docks, London, was born on the 29th August 1875 at Bow, London. He was educated at Grays College, Essex, and left school at the age of fifteen when he was apprenticed for three years to Messrs. Henry Fletcher Son and Fearnall, millwrights, of Tilbury Docks. He carried out his indentures and remained in their employ until May 1894. In the following September he entered the service of Messrs. Harland and Wolff, of Belfast, as an improver, and was bound to them for two years in order to complete the five years required by the Board of Trade for marine engineers. On the completion of his time with this firm, he was employed for a short period by Messrs. Workman, Clark and Co., of Belfast, and then returned to London and entered the works and afterwards the drawing office of Messrs. Yarrow and Co., of Poplar. He left this firm at the end of 1902 to establish a motor engineering works at Brixton, but he was not destined to carry this on for long owing to failing health, which developed into disease of the kidneys and acute dysentery. His death ensued on 5th August 1903, in his twenty-eighth year. He was a proficient swimmer and had saved life from drowning. He became an Associate Member of this Institution in 1901.

WILLIAM FOULIS was born on 29th March 1838, at St. Andrews, where his father was manager of the gas works. In 1850 he went to Paisley along with his father, who had been appointed to take charge of the gas works in that town. At the age of fourteen the lad was apprenticed in the engineering works of Messrs. Craig, Fullerton and Co., and about this time he also attended Glasgow University. Shortly after the completion of his apprenticeship, he

2 Y 2

entered the service of the late Mr. W. M. Neilson, of the Hyde Park
Locomotive Works, Glasgow.   Then he proceeded to London, and
in the interest of an English firm he went to the Mediterranean,
superintending the erection of gas works at Malta and elsewhere.
Seven years were thus spent in different towns and cities in Italy
and Greece.   On returning to Glasgow, he entered into partnership
with Mr. W. R. Copland, and carried on with him the business of
civil and gas engineer.   When preparations were on foot to
municipalize the gas supply of Glasgow, which had previously been
in the hands of two private companies, he was employed to advise
the Corporation; and when the transfer of the works was
effected in 1869, he was appointed Gas Engineer.   His first work
was to erect new gas works at Dawsholm, and then to dismantle the
old works at Townhead and Partick.

Many improvements were introduced by him into the manufacture
of gas.   In conjunction with Sir William Arrol, he constructed a
machine, worked by hydraulic power, for charging and drawing
retorts, the water being also used to cool the shovel of the charger
and the rake of the drawer.   Another improvement was the
introduction of the gas-producer furnace invented by Sir William
Siemens; a double saving of fuel is effected by the device,
and a continuous high temperature is easily maintained.   In
addition to invention, his genius found scope in the adaptation of
plant and processes to perfect the production of gas and to cheapen
its supply to the consumer.   An apparatus for producing gas from
oil alone was designed by him, and put up in 1896 at the Temple
Farm Works, near Dawsholm, the resulting gas being used to
enrich the coal gas.   The question of the conveyance of material
early had his attention, and he installed at Dawsholm and
at Tradeston a system of conveying coke by means of tiny
locomotive engines, running upon narrow-gauge railways, for the
hauling of trucks from the basement floor of the works out to the
yard.   Another time- and labour-saving device of his was an
adaptation of wheels and rails for the removal of purifier covers,
taking the place of the overhead tackle almost universally in use.
He was the first gas-engineer in the kingdom to instal plant for

the recovery of cyanogen.   In 1887 he was elected President of the British Association of Gas Managers, afterwards the Gas Institute, and he was one of the founders of the Institution of Gas Engineers, of which he became a Member of Council, a Trustee, and President in 1896.   He was a Member of the Institution of Civil Engineers; and was President of the Institution of Engineers and Shipbuilders in Scotland at the time of his death.   He had been in failing health for some months from an internal complaint, and his death took place suddenly at his residence in Glasgow, on 29th June 1903, at the age of sixty-five.   He became a Member of this Institution in 1877.

HENRY PERCY HOLT was born in Wakefield on 14th August 1848. He began his engineering career in 1865 as a pupil of his father, Mr. Henry Holt, civil engineer, of Leeds and Wakefield, under whom he obtained experience in colliery branch railways, drainage and waterworks.   During 1868 he was an improver at the Yorkshire Engine Co.'s works, Sheffield, under Mr. A. L. Sacré, in charge of the erection and repairs of locomotive engines; and between 1869 and 1871 he was employed by that Company to superintend the running of new engines on the Great Northern Railway at Doncaster, and on the Midland Railway at Derby; and subsequently on the installation of engines of the Tanstoff-Kosloff Railway, and the Lemberg-Czernowitz Railway, and on the shipment of engines for the Poti-Tiflio Railway.

In April 1871 he took his father's place and share in the firm of Holt and Childe, civil and mining engineers, of Wakefield, and was engaged on various works until the following July, when the partnership was dissolved and he commenced business as a civil and consulting engineer in Leeds, his practice consisting in the design and superintendence of the construction of fireproof mills, warehouses, hydraulic hoists, and packing machinery; works for the drainage and purification of waste waters; well-sinking and shaft-driving reservoirs; mechanical cooling, heating and ventilating of large buildings; and the erection of large concrete buildings, iron roofs, girders, &c.   Since 1872 he also practised in London, and carried

out the following, amongst other works:—Queen's Parade Tramway,
Scarborough; Drill Shed Roof, Finsbury, for the Honourable Artillery
Company; combined iron and timber roof, truss construction, made
by Messrs. A. Handyside and Co. from his design for Messrs. J.
Fowler and Co., of Leeds; the Electric Construction Corporation,
Wolverhampton; and the Royal Engineers' and Home Office
Departments. In 1881 he joined the firm of Messrs. Crossley
Brothers and Co., and designed and superintended the erection of the
Otto Gas-Engine Works at Openshaw, becoming subsequently a
Director and Consulting Engineer of Messrs. Crossley Brothers,
Manchester. His death took place at his residence in Kensington
Court, London, on 15th July 1903, in his fifty-fifth year. He
became a Member of this Institution in 1873.

JAMES MACTEAR was born in Glasgow on 3rd April 1845, where
he was educated. From an early age he displayed a strong tendency
towards chemical science, and studied chemistry under Dr. M.
Wallace, of Glasgow. After a short time spent in acquiring a
knowledge of manufacturing chemistry, he was appointed assistant
manager to Mr. Edmund Stanford, whose processes for the preparation
of iodine from sea-weed were then being introduced into Scotland.
During this period he devised a method for obtaining the bromine
which had hitherto been lost. In 1864 he became assistant manager
in the chemical works of Messrs. C. A. Allhusen and Sons at
Newcastle-on-Tyne, and in 1867 he was appointed manager of the
works of Messrs. Charles Tennant and Co., of St. Rollox, Glasgow.
There he improved most of the old processes, and invented new
methods and appliances, which were so successful that within two
years he became managing partner, having the chief control of the
technical business of the Company. During the fifteen years of his
management he invented, designed, and erected a large amount of
chemical plant and mechanical furnaces. In 1884 he retired and
removed to London, to take up the profession of consulting chemical
expert and engineer.

   After settling in London, he took up mining and metallurgical
work as a speciality, and became an authority on the mining and

metallurgy of quicksilver, in connection with which subject he visited nearly all the countries in which that metal has been discovered, and invented appliances for the distillation and condensation of the metal from its ores. He also devoted much attention to the processes of gold extraction. He contributed many Papers on chemical subjects to the scientific literature of the times, and was awarded the silver medal of the Society of Arts in 1876 for his contribution on the new methods introduced by him into the alkali manufacture. He was appointed a Juror at the Paris Exhibition of 1878, and received the gold medal instituted in memory of Professor Graham for the best chemical research of the preceding three years. He was widely known as the discoverer of the method of producing artificial diamonds. For some years he commanded the 5th Lanark Rifle Volunteers. His death took place at his residence in London on 3rd June 1903, at the age of fifty-eight. He became a Member of this Institution in 1892; and was also a Member of the Institution of Civil Engineers, a Fellow of the Royal Society of Edinburgh, and a Fellow of the Institute of Chemistry and of the Chemical Society.

CHARLES GRAHAM PRICE was born at Hampton Wick on 27th February 1864. He was educated at the Institution, Mauchline, Ayrshire, and at Daniel Stewart's College, Edinburgh. In 1880 he commenced an apprenticeship with Messrs Mather, of Edinburgh, and completed it with Messrs. Umpherston, of Leith. From 1885 to 1889 he was engaged as draughtsman in the latter works, and also in the works of Messrs. Cran and Co., marine engineers, of Leith. In 1889 he was employed by Messrs James Currie and Co., Leith, as second and afterwards as chief Engineer, proceeding principally to Baltic Ports. In 1892 he obtained chief engineer's certificate from the Board of Trade, and in the same year became assistant to the chief engineer of the National Boiler and General Insurance Co., of Manchester, with whom he remained until 1896. He was then appointed by Messrs. Samuel, Samuel and Co., of London, to proceed to Japan as superintending engineer of their works and to extend the business. During this period he visited Corea for a similar purpose. On his return to England he was engaged by the same firm to proceed to Baleck Pappan on the east coast of Borneo, to take charge of their oil works and to visit their other works on the coast. While he was thus engaged, paralysis commenced, the result of a serious accident twelve years before when in the Baltic, and he went to the hospital at Singapore; there he was advised to return immediately to England, where his death took place at Wimbledon on 21st August 1903, in his fortieth year. He became a Member of this Institution in 1897.

LIONEL LINCOLN SMITH was born at Peterborough on 21st July 1872, and received his education at a school at Enfield. On its termination he studied at the City and Guilds of London Institute, where he obtained first-class honours in mechanical engineering. He served an apprenticeship from 1888 to 1892 at the Royal Small Arms Factory, Enfield, as a fitter, and for three months afterwards was employed as engineer fitter at the London Small Arms Co. In 1893 he went to Messrs. Thornycroft's works at Chiswick, where he remained nearly two years, being subsequently employed for a short time at the Birmingham Small Arms and Metal Co., and at the Royal Dockyard at Sheerness. In 1895 he was engaged at the Royal Laboratory, Woolwich, and was promoted in 1898 to be foreman of mechanics, a position in which he was immediately associated with the production of drawn-metal cartridge cases and the machinery in connection therewith. Subsequently he was appointed to the position of mechanical adviser to the manager of the Royal Laboratory Department; and just prior to his death he was recommended for the post of assistant-manager of the composition establishment. On three occasions he had saved life from drowning, and had received the medal of the Royal Humane Society. His death took place at Plumstead on 23rd July 1903, at the age of thirty-one. He became an Associate Member of this Institution in 1901.

WILLIAM WRIGHT was born at Dudley on 9th April 1836. He received his education at a private school in Tipton, and commenced an apprenticeship in 1849 at his father's engineering works at

Dudley. On its completion in 1856 he went into the drawing-office for two years, and then became a partner in the works with his brothers, superintending the manufacture and erection of steam-engines, boilers, and general machinery. In 1867 he was appointed manager of the engineering works of Messrs. Joseph Wright and Co., Tipton, where he remained until 1888, when he became manager to Wright's Heater Condenser Co., Westminster. In this position he constructed some of the largest feed-water heaters and water softeners in England. His death took place at Lambeth, London, on 26th July 1903, at the age of sixty-seven. He became a Member of this Institution in 1895.

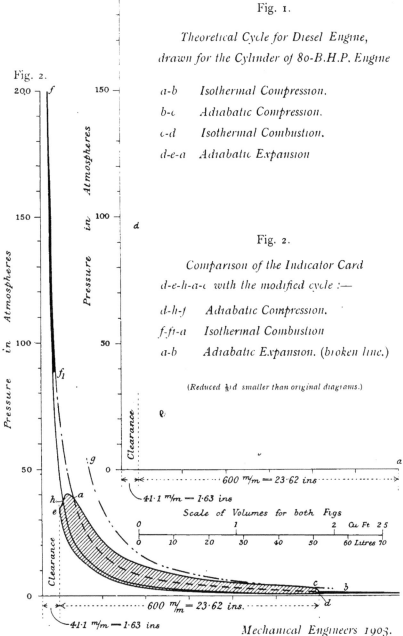

Fig. 1.

Theoretical Cycle for Diesel Engine,
drawn for the Cylinder of 80-B.H.P. Engine

a-b   Isothermal Compression.
b-c   Adiabatic Compression.
c-d   Isothermal Combustion.
d-e-a   Adiabatic Expansion

Fig. 2.

Comparison of the Indicator Card
d-e-h-a-c with the modified cycle :—

d-h-f   Adiabatic Compression.
f-f₁-a   Isothermal Combustion
a-b   Adiabatic Expansion. (broken line.)

(Reduced ⅓rd smaller than original diagrams.)

600 ᵐ/m = 23·62 ins
41·1 ᵐ/m = 1·63 ins
Scale of Volumes for both Figs

0   1   2   Cu Ft 2·5
0   10   20   30   40   50   60 Litres 70

600 ᵐ/m = 23·62 ins.
41·1 ᵐ/m = 1·63 ins

Mechanical Engineers 1903.

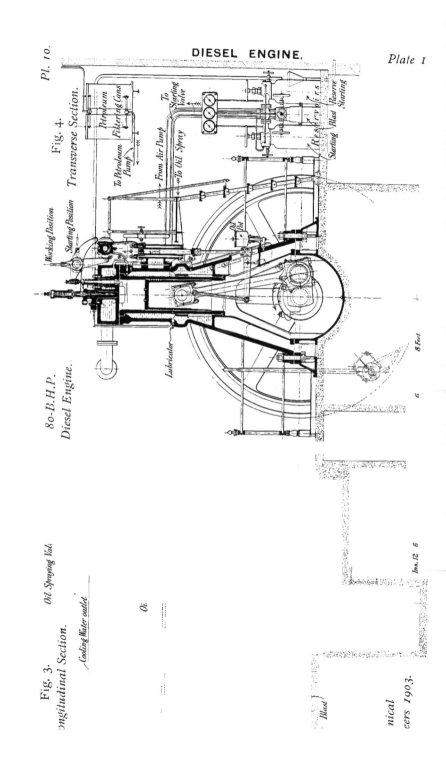

DIESEL ENGINE.

Pl. 10.

Plate I

Fig. 4.

Transverse Section.

80-B.H.P.
Diesel Engine.

Working Position

Starting Position

Petroleum
Filtering Cans

To Petroleum
Pump

To Starting
Valve

To Oil Spray

From Air Pump

Reservoirs

Starting Blast Reserve
Starting

Oil Pot

Lubricator

8 Feet

6

Fig. 3.
ongitudinal Section.

Oil Spraying Valve

Cooling Water outlet

Oi:

Blast

nical
eers 1903.

Ins. 12 6

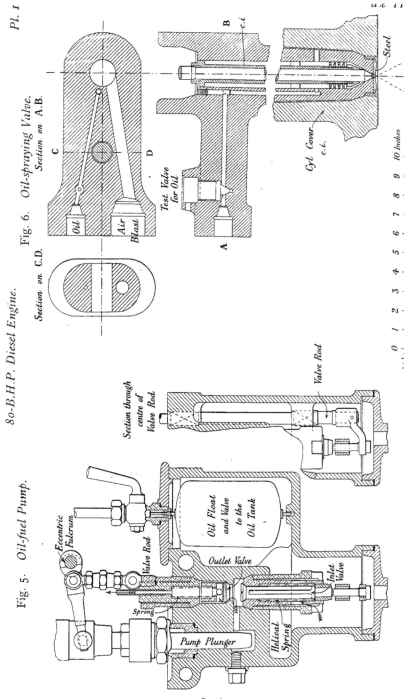

Pl. I

80-B.H.P. Diesel Engine.

Fig. 6. Oil-spraying Valve.

Section on A.B.

Section on C.D.

Fig. 5. Oil-fuel Pump.

Oil

Air Blast

C

D

B

c.i.

A

Test Valve for Oil

Cyl. Cover c.i.

Steel

0 1 2 3 4 5 6 7 8 9 10 Inches

Section through centre of Valve Rod

Valve Rod

Eccentric Fulcrum

Valve Rod

Oil Float and Valve to the Oil Tank

Outlet Valve

Inlet Valve

Spring

Helical Spring

Pump Plunger

(Author's remarks)

Fig. 23. *The upper line is for the 80-B.H.P. Engine (Fig 7)*

The 3 square dots are Oil Consumptions given in Table 7

The lower line is for the 35-B.H.P. Engine.

The 2 round dots show Oil Consumption after twelve months work.

$$0 \quad 10 \quad 20 \quad 30 \quad 40 \quad 50 \quad 60 \quad 70 \quad 80$$

B.H.P. *at which Engine is working.*

Fig. 24

(Prof John Goodman's remarks)

Diagram showing the effect of Compression on the Efficiency of an Oil-Engine

in weight of oil.

— Ordinary Oil Engine

$N = 1 \cdot 3$

Diesel Engine — approximately

Fig. 1. *Ordinary Tool Steel, turning Car-wheel Tyres.*

Speed 8 ft per min   Metal removed 8 lbs per hr.

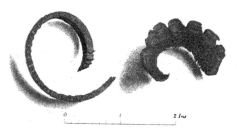

Midvale Tool Steel, turning Locomotive Tyres.

Fig. 2.
'eed 24 ft. per min.
Metal removed
120 lbs. per hr.

Fig. 3.
Speed 18 ft. per min.
Metal removed
450 lbs. per hr.

Fig. 4. *Car-wheel Lathe using High-speed Tool Steel on Steel Tyres.*

Fig. 5. *Planer working with High-speed Tool Steels.*

Mechanical Engineers 1903.

(*Mr Druitt Halpin's remarks*)

Fig. 1. *Thermal Storage of Feed Water for Varying Loads.*

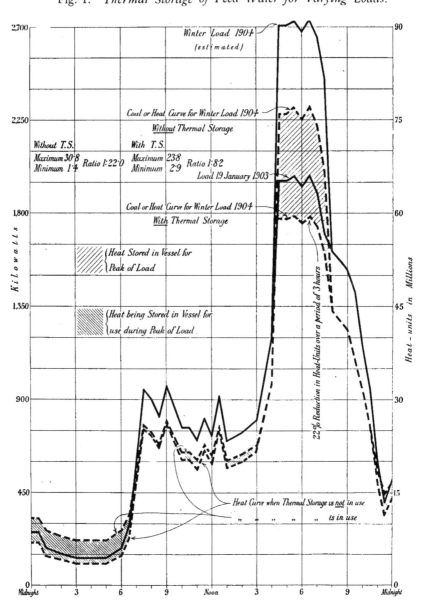

Mechanical Engineers 1903.

(*Mr Druitt Halpin's remarks*)

Thermal Storage.

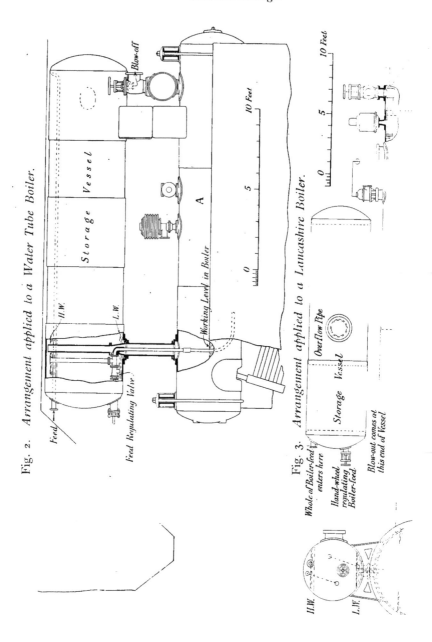

Fig. 2. *Arrangement applied to a Water Tube Boiler.*

Fig. 3. *Arrangement applied to a Lancashire Boiler.*

𝕿𝖍𝖊 𝕴𝖓𝖘𝖙𝖎𝖙𝖚𝖙𝖎𝖔𝖓 𝖔𝖋 𝕸𝖊𝖈𝖍𝖆𝖓𝖎𝖈𝖆𝖑 𝕰𝖓𝖌𝖎𝖓𝖊𝖊𝖗𝖘.

PROCEEDINGS.

OCTOBER 1903.

The first ORDINARY GENERAL MEETING of the Session was held at the Institution on Friday, 16th October 1903, at Eight o'clock p.m. ; J. HARTLEY WICKSTEED, Esq., President, in the chair.

The Minutes of the previous Meeting were read and confirmed.

The PRESIDENT announced that the Ballot Lists for the election of New Members had been opened by a Committee of the Council, and that the following fifty-three candidates were found to be duly elected :—

MEMBERS.

AULD, JAMES PATON, ...	Southern Nigeria.
BAYLISS, WILLIAM, ...	Manchester.
DEWAR, JAMES McKIE, . . .	London.
DOVE, JOHN CHARLES, . . .	Carlisle.
INGRAM, SAMUEL, ...	Exeter.
KETT, GEORGE, ...	London.
LIVOCK, HENRY ARTHUR, . . .	London.
MASSEY, HAROLD FLETCHER, . .	Manchester.
PRICE, ALAN, ...	Sydney.
ROBERTS, DAVID EVAN, . . .	Dowlais.
SHORT, ALFRED, ...	Durban.
SINCLAIR, HERBERT, ...	Leeds.

TULLY, WILLIAM, ...	London.
WICKS, GERALD HAMILTON, . . .	Puebla, Mexico.
YOUNG, FREDERICK WILLIAM, . .	Liverpool.

ASSOCIATE MEMBERS.

BEARE, HAROLD HARGREAVES, . .	Burton-on-Trent.
BENNETT, FREDERICK WILLIAM, . .	London.
BENTLEY, PERCY, ...	Leeds.
BOWDEN, JOHN, ...	London.
BROWN, CHARLES ANDERTON, . .	Belfast.
CLAUGHTON, GEORGE, ...	Newcastle-on-Tyne.
CROCKER, CLAUDE EDWARD, . .	Kalgoorlie.
FERNANDEZ, RAFAEL, ...	Tucuman.
FRYER, FREDERICK GEORGE, . .	York.
GEEN, HARRY, ...	Newton Abbot.
GROOM, ISAAC STEPHEN WEST, . .	Grantham.
HILL, FREDERICK JOHN, . . .	Leicester.
HUNTER, JOHN CHALMERS, . . .	Glasgow.
JOHNSON, HENRY HOWARD, . . .	Johannesburg.
LAMBERT, JOHN HENRY, . . .	Nottingham.
MCLAREN, WILLIAM DAVID, . . .	Roorkee.
MONTAGUE, GRAHAM, ...	Gloucester.
POWELL, CHARLES SKRYNE, . . .	London.
PROCTER, CHARLES GILBERT, . .	London.
RAINER, EDWARD ALLAN, . . .	Ipswich.
SEMARK, CHARLES HENRY, . . .	Faversham.
SHARR, FRANCIS JOSEPH, . . .	Rugby.
STEEL, WILLIAM BERTRAM, . . .	Cardiff.
WALKER, ARTHUR ERNEST, . . .	London.
WILSON, CHARLES AUGUSTUS, . .	Brighton.
WINDSOR, ARTHUR WHALESBY, . .	Birmingham.

GRADUATES.

BEALE, SAMUEL RICHARD, . . .	London.
BURRELL, CHARLES WILLIAM WILBERFORCE, ...	Thetford.

COBBALD, THOMAS ERNEST, .   .   . Thetford.
COOK, JOHN, ... Northwich.
CROSS, WILLIAM MARK,   .   .   . Bristol.
DAVIS, FRANK, ... Doncaster.
FRY, WILLIAM ELLERTON RYAN,   .   . Bloemfontein.
HOLMES, GEORGE NELSON, .   .   . Liverpool.
MACGREGOR, ALEXANDER JOHN,   .   . London.
POIGNAND, FRANCIS NESBITT,   .   . Leeds.
SHELMERDINE, ERIC DEAN SAATWEBER, . Liverpool.
WILSON, EWAIN MURRAY, .   .   . Thetford.

The PRESIDENT announced that the following three Transferences had been made by the Council since the last Meeting :—

Associate Members to Members.

MORTON, DUNCAN ANDERSON, .   .   . Errol.
PAYNE, HENRY, ... Cape Town.
WINSTON, HAROLD HOLMES,   .   .   . London.

The following Paper was given in the form of a Lecture, illustrated by lantern slides, and was followed by a brief discussion :—

" The Newcomen Engine "; by Mr. HENRY DAVEY, *Member of Council.*

The Meeting terminated shortly before Ten o'clock. The attendance was 187 Members and 76 Visitors.

THE NEWCOMEN ENGINE.

By MR. HENRY DAVEY, Member of Council.

A great deal has been written on the Steam-Engine generally, but the author has not met with any connected record of the invention and construction of the first steam-engine—the Atmospheric Engine of Newcomen. Unfortunately it does not appear that much detailed information is available; but the author has been able to bring together some facts, which, with the aid of appendices contributed by others, and with some illustrations of the engine itself, may be found to form a useful contribution to place on record in the Proceedings of the Institution. There are not many examples of the engine now in existence; and when they are consigned to the scrap heap, that receptacle of great efforts of the past, all will perhaps be forgotten.

Towards the end of the seventeenth century, philosophers and mathematicians searched for a new method of obtaining motive power. Mining was an important industry, requiring in most cases a new power, in order that the mines might be worked to greater depths. Water-power where available was often insufficient; and manual and animal power were altogether too small and too expensive for working any but shallow mines. Deep mining was and is possible only with pumping machinery. Water-wheels were used for working pumps. The construction of the common pump was known. Papin had proposed to transmit power by means of pistons moving in cylinders and acted on by the atmosphere, a vacuum having been formed under the pistons by the explosion of gunpowder; and he even hinted that it might be done by steam. It was claimed for Papin that he invented the steam-engine, because in 1685, in one of his letters, he illustrated what was known of the properties of steam, by saying that if water were put in the bottom of a cylinder under a piston, and the cylinder were put on a fire, the water would evaporate and raise the piston; and that if, after the piston had been raised, the cylinder were removed from the fire and cooled, the steam would condense and the piston would descend; but this was only an illustration of common knowledge. Sir Samuel Morland had, in 1683, stated * that steam occupied about two thousand times the space of the water from which it was produced; and had made some calculations as to the powers to be obtained from different sized cylinders, but had suggested no practical mode of operation. Experiments to determine the density of steam were made by John Payne in 1741, published in the Philosophical Transactions, vol. xli, page 821. As the result of these he concluded that one cubic inch of water formed 4,000 cubic inches of steam. Beighton calculated, from an experiment with the Griff engine, the second Newcomen engine erected, that the specific volume of steam was 2,893. The properties of steam were probably no better known to philosophers than to the ordinary observer who had seen the lid of a kettle dance under pressure, or steam issue from the spout. The only practical application of steam was made by Savery, who, in 1696, described his invention in a pamphlet entitled "The Miner's Friend."† Savery's engine was a pistonless steam-pump—in fact, the pulsometer of to-day without its automatic action. It remained for Newcomen to combine the bits of common knowledge in his mind for inventing the steam-engine. He was a blacksmith, probably accustomed to invent methods of construction in the prosecution of his art. At that time mechanics were more self-reliant than they are now. He knew from experience what a lever was, a pump, a piston, a cylinder, a boiler; and he knew that the atmosphere had pressure, and that steam possessed a far greater volume

* *See* Tredgold's "Steam Engine," 1853.

† "The Miner's Friend" was published in 1696; and a dialogue in answer to objections in 1699, and both together in 1702. Savery's Patent is dated 1698, or two years after the publication of "The Miner's-Friend." *See* R. Stuart's "Descriptive History of the Steam-Engine."

than the water which produced it. It did not require much more than common knowledge and observation to realise that. To produce the steam-engine from such known facts required invention. Philosophers probably conceived what might be done; but Newcomen had the advantage of seeing what could be done, and he did it. The engine when produced was imperfect; but defects became obvious to the designers and constructors of steam-engines, and the want of perfection at the present day is not from want of theory, but because of practical limitations and the want of practical invention. The theory of the steam-engine has followed its application, which is the natural order of things. Ascertained facts form the best foundation of theory. Watt's invention of the separate condenser was founded comparatively on no greater theoretical knowledge than Newcomen's invention of the engine itself. Watt was probably a better educated man than Newcomen; but no knowledge superior to that possessed by many others was necessary for the invention of the separate condenser. Watt roughly ascertained the specific volume of steam, and so did Morland, and the makers of Newcomen's engines, incorrectly. It was also given with even greater inaccuracy in a text-book edited by Dr. Brewster; * but quantitative results did not determine either invention. There are so-called inventions which are nothing more than obvious combinations of mechanism, but inventors such as Newcomen and Watt are pioneers who discover and occupy the country, whilst the theorists are surveyors who follow to measure it up.

At this distance of time it is difficult to appreciate the inventive power required to produce the atmospheric engine from the crude ideas of Papin and others. It appears, from papers in possession of the Royal Society, that Dr. Hooke had demonstrated the impracticability of Papin's scheme, and, in a letter addressed to Newcomen, had advised him not to attempt to make a machine on that principle : adding however, " Could Papin make a speedy vacuum under your second

---

* " Ferguson's Lectures on Mechanics, &c.," by Dr. Brewster, 1806. In this book the specific volume of steam is given as 14,000. Emerson, in " The Principles of Mechanics and the General Laws of Motion," 1758, gives it as 13,340.

piston, your work is done." *   A great deal of controversy hangs
about this, as about many things historical; and little is to be
gained by minute research into disputed claims.   What we do with
certainty know is that, with the common knowledge existing, and
with the mechanical contrivances available, Newcomen alone
succeeded in making a workable engine.

In 1698, Thomas Savery, of London, obtained a patent for
raising water by the elasticity of steam. †   It is stated in many
popular histories that in 1705 Thomas Newcomen and John Cawley,
of Dartmouth, in Devonshire, aud Thomas Savery, of London, secured
a patent for " Condensing the steam introduced under a piston, and
producing a reciprocating motion by attaching it to a lever," but no
record of such a patent exists in the Patent Office.   Stuart gives a
list of patents commencing with 1698, and in that list is one said to
have been granted in 1705.   Dr. Pole, author of "The Cornish
Engine," had a search made at the Patent Office, and no such
record could be found.   It is possible that Savery's patent was
thought to cover Newcomen's invention, as Savery was associated
with Newcomen.‡   This was sixty-four years before Watt invented
his separate condenser.   Very little is known of Newcomen.   It is
recorded that he was a blacksmith or ironmonger residing at
Dartmouth, and that he was employed by Savery to do some work
in connection with his water-raising engines.   In this way he had
some experience in the condensation of steam. §

---

* *See* R. Stuart's "Descriptive History of the Steam Engine," 1824.

† Savery was born at Shilstone, near Modbury, in Devonshire, in 1650; died
in London in May 1715.

‡ *See* Appendices I and II, from which it appears that there is every reason
to believe Newcomen had no patent, and that his invention was supposed to
be covered by Savery's patent of 1698, and that the latter was kept in force for
thirty-five years, the original patent having been extended for twenty-one years.

§ Newcomen was born at Dartmouth about the middle of the 17th century,
and died in London in 1729.   It is stated in Haydn's "Dictionary of Dates"
that at the time of his death he was in London, trying to secure a patent.
A sketch of the house in Dartmouth, occupied by Newcomen when he invented
the steam engine, is shown in a pamphlet published in 1869 for Mr. Thomas
Lidstone of Dartmouth.

Newcomen appears to have conceived the idea of using a piston for giving motion to pumps. He became associated with John Calley or Cawley, a glazier of Dartmouth, probably for business reasons. His connection with Savery was doubtless because of Savery's patent for condensing steam for raising water. He must however have been a good mechanic, because the construction of such an engine at a time when there was no previous experience or data to guide him was a task of no ordinary magnitude. He could not get workmen skilful enough to do his work, till, erecting an engine near Dudley in 1712, he secured the assistance of mechanics from Birmingham.*

In this age of mechanical developments we wonder at the slow progress made in early times, till we realise the condition of society, and the apathy with which all things technical and scientific were regarded by the public generally. That ignorance and prejudice are well illustrated by the following extract from the preface to a book published in 1758, entitled " The Principles of Mechanics," by W. Emerson, a copy of which was kindly lent to the author by Mr. H. W. Pearson, of Bristol :—

" I might have given the cuts of many more machines, but perhaps what I have already done may be thought too much in such a nation as this, where natural knowledge wants due encouragement, and where no Mecaenas appears to patronize and protect it, and where arts and sciences hang, as it were, in suspense whether they shall stand or fall, and where public spirit and English generosity are just expiring. This decline of arts and sciences is wholly or in a great measure owing to the ambition and most extreme avarice of the present age, where men, not being able to lift their eyes above this earth, think nothing worth their care but raking together the dross it affords, striving, like the toad, who shall die with the most earth in his paws. The duller part of mankind are entirely engaged in the pursuit of filthy lucre, and the brighter sort are wholly devoted to low, trifling, and often barbarous diversions. In such momentous concerns as these, it is no wonder

* *See* Appendix X. (page 695).

if arts and sciences flag, and natural knowledge meets with nothing
but contempt, and Minerva gives place to Pluto. And indeed, if
the general temper and disposition of men had been the same in all
ages, as it is in this, I am in doubt whether we had ever had any
such things as a mill to grind us corn for bread, or a pump to draw
us water. It is a trifling excuse for men of exalted station to urge
that they are unacquainted with such arts or sciences. For learning
has always been esteemed to be under the peculiar care and
superintendency of the great, who ought to protect and encourage
both that, and the professors of it, or else arts and sciences can
never flourish. And as the encouragement of these evidently tends
to the benefit of mankind and the promoting of the public good,
nothing can excuse so gross a neglect or such a manifest disregard
as they show for the happiness of their fellow-creatures. The
industrious students have only the fatigue, whilst all the world reaps
the advantage of their labours."

The Newcomen engine was soon brought into use ; for in 1712
Newcomen, through the acquaintance of Mr. Potter, of Bromsgrove,
erected an engine near Dudley Castle for Mr. Back, of
Wolverhampton. The cylinder of this engine was surrounded
with water. The piston was packed and had a water seal.
It is reported that by accident a hole in the piston admitted
water into the cylinder, and the condensation thereby became so
rapid, compared with that produced by cooling the cylinder from
the outside, that the engine worked much quicker. This may or
may not be correct ; but it is certain that, by accident or design,
the first improvement in the engine was condensation by injection in
the cylinder. It appears that the second engine was erected at the
Griff Colliery, near Nuneaton, in 1715. It had a 22-inch
cylinder. At this time the cocks and valves were all worked
by hand ; but automatic devices were soon introduced. The first
appears to be that for actuating the injection-cock by means of a
buoy in a pipe connected to the cylinder. Desaguliers thus describes
the apparatus : "They used to work with a buoy in the cylinder
enclosed in a pipe, which buoy rose when the steam was strong, and
opened the injection and made the stroke." It is said that a boy,

Humphrey Potter,* added a catch or "scoggan"† which was caught by a pin upon the plug-rod from the beam ; and by this means the speed of the engine was increased from 8 or 10 to 15 strokes per minute.

Thomas Barney's engraving of the engine ‡ erected in 1712 near Dudley Castle, was made in 1719, and contains the plug-frame and tumbling-weight device attributed to the invention of Beighton in 1718. It is possible that the tumbling-weight had just been added for actuating the steam-valve. The injection-valve is released by the buoy, said by Desaguliers to have been enclosed in a pipe attached to the cylinder, but here shown in a pipe attached to the boiler. The scoggan is also shown, and it is clear that the only thing Humphrey Potter added, if he added anything, was a cord for causing the plug-frame to actuate the scoggan, instead of the float doing it. The engraving shows such an addition, but the story of Humphrey Potter is discredited.

The scoggan appears in Belidor's illustration, § also in Stuart's illustration (Fig. 19, page 73, second edition, 1824), which is said by Stuart to be Beighton's gear with improved details. It also appears in the York engine, Plate 21, and the engine, Plate 23. Emerson says, " In some engines there is a catch held by a chain fixed to the great beam, and this catch holds the injection-cock lever or eff ‖ from falling back and opening the cold-water cock, till the rising of the beam pulls the catch up by the chain, and then the eff falls."

---

* It is curious to observe that the first engine was erected for Mr. Back through the influence of Mr. Potter of Bromsgrove. Mr. W. G. Norris writes that John and Abraham Potter were engineers in Durham, and erected an engine for Mr. Andrew Wauchope in Midlothian in or about 1725. *See also* Bald's "View of the Coal Trade of Scotland," pp. 18, &c, for a full account of this engine. He prints the contract in full, giving many interesting details.

† To scog is said to be a Yorkshire term meaning to skulk—"Mechanic's Magazine," p. 154; Stuart, p. 66. In Celtic Scotch "sgog" or " sgogan" is a fool or idler. In modern Scotch "scogie" is a kitchen drudge. In Barney's engraving the word is spelt "scoggen."

‡ Proceedings 1883, Plate 87, and page 623.

§ " Architecture Hydraulique," Paris, 1739.

‖ So called from its resemblance to the letter F.

In all the earliest engines the "eduction pipe" or drain pipe
from the cylinder was arranged with sufficient fall to prevent the
hot-well water from returning to the cylinder when the vacuum was
formed. It terminated below the water level of the hot-well. In
later engines it was shortened, and a non-return valve was placed
on the end, which was turned up in the hot-well, and in this way
had a water seal.

The next improvement consisted of the plug frame, invented by
Henry Beighton, who erected an engine at Newcastle-on-Tyne in
1718, in which a rod suspended from the beam actuated what was
called the hand-gear. This contrivance, with some slight modification,
remained till Watt's time.

In Plate 16 is reproduced from Price's "Mineralogia Cornubiensis,"
by William Pryce of Redruth, published in 1778, an illustration of
the Newcomen engine of that day. Also a section of the Bullen
Garden Mine, near Camborne, in Cornwall, Fig. 12, Plate 24, showing
the application of two atmospheric engines for enabling the mine to
be worked to a greater depth than was possible with only the water
wheels in use.

In 1758 Emerson described in detail the Newcomen engine of
that day. The description is so excellent that it is here reproduced,
omitting the letters of reference to an accompanying engraving;
the description is clear enough without the engraving. It will be
observed that the dimensions are given of an engine and pumps, and
the reported consumption of coal; neither the water pumped nor
the consumption of coal can be correct.

Description of the Engine.

" The fire-engine to raise water has a great beam or lever, about
24 feet long, 2 feet deep at least, and near 2 feet broad. It lies
through the end wall of the engine house, and moves round a centre
upon an iron axis. The steam cylinder is of iron, 40 inches diameter
or more, and 8 or 9 feet long, the piston sustained by the chain.
The fire-place is underground. The boiler is 12-feet diameter, which
communicates with the cylinder by a hole and throat-pipe 6 or

8 inches diameter. The boiler is of iron, and covered over close with lead; in this the water is boiled to raise steam. The regulator being a plate within the boiler which opens and shuts the hole of communication, this is fixed on an axis coming through the boiler, on which axis is fixed a horizontal piece called the spanner, so that moving the spanner forward and back moves the plate over the hole and back again. There is a horizontal rod of iron movable about a joint, and a piece of iron with several claws called the wye, moving about an axis in a fixed frame. The claw is cloven, and between the two parts passes the end of the spanner with two knobs to keep it in its place. There is the working-beam or plug-frame, in which is a slit, through which the claws pass and are kept there by a pin going between them. There is a leaden pipe called the injection-pipe, carrying cold water from the cistern into the cylinder, and turned up at the end within the cylinder. To the injection-cock is fixed an iron rod lying horizontal. The end goes through a slit in the end of another piece, and on the end is a knob screwed on to keep it in place. There is a piece of iron with several claws called the 'eff,' movable about an axis. The claw goes through the slit in the plug frame, and is kept there by the two pins; the claw goes over the piece. As the piece is moved back and forward, the injection cock opens and shuts. There are several holes in the plug frame that by shifting the pins serve to set the pieces higher or lower as occasion requires. There is the snifting clack balanced by a weight and opening outwards, to let out the air in the cylinder at the descent of the piston. In some engines a pipe goes from it to convey the steam out of the house. There is a leaden pipe called the sinking pipe or *eduction pipe* going from the cylinder to the hot-well; it is turned up at the end, and has a valve opening upwards; this carries away the water thrown in by the cold-water pipe or injection pipe. There is the feeding pipe, going from the hot-well to the boiler, to supply it with water by a cock opening at pleasure. There are two gauge pipes with cocks, one reaching a little under the surface of the water in the boiler, the other a little short of it. By opening these cocks, it is known when there is water enough in the boiler, for one cock will give steam and the other

water ; they stand in a plate which may be opened for a man to go
into the boiler to clean or mend it.   There is the puppet-clack or
safety-valve ; from this a wire comes through a small hole, to which is
fixed a thread going over a pulley, with a small weight at it; the
weight on the clack is about a pound for every square inch.   There
is a steam-pipe going from the clack out of the house.   When the
steam in the boiler is too strong, it lifts up the puppet-clack and
goes into the steam-pipe, by which it is conveyed away ; otherwise the
boiler would burst.   There is a pipe carrying water from the cistern
into the cylinder to cover the piston to a good depth ; also a cock
opening to any wideness that the water may run in a due quantity ;
and a hole to let it out through a pipe into the hot-well, when there
is too much.   There is a force pump, with a bucket and clack, and
two valves opening upwards.   This pump is closed at the top,
and being wrought by a lever it brings water out of a pit into a cistern
for supplying condensing water.   There are speers (spears), which
work in wooden pumps within the pit.   The cylinder is supported
by strong beams going through the engine-house.   There is the first
floor and the upper floor.   At the end of the beam there are two
pins, which strike against two springs of wood, fixed to two timbers
or spring beams lying on each side of the great lever ; these pins
serve to stop the beam, and hinder the piston from coming too low in
the cylinder.

   " When the engine is to be set to work, the water in the boiler
must be made to boil so long till the steam is strong enough, which
is known by opening the gauge-cocks.   Then the inlet hole is opened
by moving the spanner by hand; then the steam is let into the
cylinder which lets that end of the beam rise up; this raises the
working beam or plug-frame and moves the eff, which moves and
opens the cold-water cock ; at the same time is moved the wye, and
shuts the inlet steam-hole.   The injection-cock being opened, the
cold water rushing into the cylinder is thrown up against the piston,
and descending in small drops condenses the hot rarefied steam, and
makes a vacuum under the piston.   Consequently the weight of the
atmosphere pressing upon the piston brings down the inner end,
which raises the other end, which works the pumps.   As the inner
end descends, the working plug-frame descends, and moving the eff and

the wye shuts the cold-water cock and opens the hole inlet for steam, and the steam goes into the cylinder, which takes off the pressure of the atmosphere, and the outer end descends by the weight of the speers and the inner end ascends as before, which opens the injection cock and shuts the steam inlet. So by the motion of the beam up and down the cock and holes shut and open alternately; and by this means of condensing and rarefying the steam by turns within the cylinder, the lever or beam constantly moves up and down, by which motion the water is drawn up by the pumps, and delivered into troughs within the pit, and carried away by drifts or levels. At the same time the motion of the beam works the cold-water pumps, and raises water into the injection-water cistern.

" When the engine is to cease working, pins are taken out of the plug-frame, and the cold-water cock is kept close shut while the inner end is up.

" The diameter of the pumps within the pit is about 8 or 9 inches, and the bores of the pumps where the speers work should be made wide at the top, for if they be straight more time is required to make a stroke, and the barrels are in danger of bursting. Likewise, if water is to be raised from a greater depth at one lift, the pumps will be in danger of bursting; therefore it is better to make two or three lifts, placing cisterns to receive the water.

" The speers or rods that work in the pumps, consisting of several lengths, are joined thus:—each piece has a stud and hole, which are made to fit; and the studs of one being put close into the holes of the other, and an iron collar drove upon them to the middle, they are then firmly fixed together.

" There is never made a perfect vacuum in a cylinder; for as soon as the elastic force of the steam within is sufficiently diminished, the piston begins to descend. The vacuum is such that about 8 lbs. presses upon every square inch of the piston, or in some engines not above 6 lbs. This engine will make thirteen or fourteen strokes in a minute, and makes a 6-foot stroke; but the larger the boiler is, the faster she will work.

" A cubic inch of water in this engine will make 13,340 cubic inches of steam, which therefore is fifteen times rarer than common air. But its elastic force within the boiler is never stronger or weaker than common air; if stronger it would force the water out of the feeding pipe. This engine will deliver 300 hogsheads of water in an hour to the height of 60 fathoms. She consumes about thirty bushels of coal in twelve hours. In some engines there is a different contrivance to open and shut the regulator, which is performed thus :—As a beam ascends, it turns the wye about its axis, causing a weight to fall and strike a smart blow on a pin, and shuts the regulator. And when the beam descends, a similar action opens the regulator. There is a cord fixed at the top of the wye to hinder it from going too far on each side; likewise for opening and shutting the injection-cock, instead of the pieces described. Some engines have quadrant wheels with teeth, which, moving one, the other opens or shuts the cock of the injection pipe.*

"In some engines there is a catch held by a chain fixed to the great beam, and this catch holds the injection-cock lever or eff from falling back and opening the cold-water cock, till the rising of the beam pulls the catch up by the chain, and then the eff falls."

This last is the contrivance said to have been invented by Humphrey Potter, applied to the working of a buoy for opening and closing the cocks. The story probably arose from calling the cock-buoy " cock-boy." The contrivance is embodied in the Cornish engine gear. From the above description, written twenty-nine years after Newcomen's death, it is evident that the atmospheric engine was largely in use. The writer of it speaks of some engines differing from others in details. Engines were at work in most of the mining districts of England. In this year (1758) many engines were at work in Cornwall; and it appears that in that county Newcomen erected an engine in 1720, or two years after Beighton's invention of the hand-gear. It is said that his first engine in Cornwall was erected in 1814, but there are no definite records. In 1769 Smeaton computed the duty of fifteen engines in the Newcastle district, and

---

* This is the Beighton gear or tumbling-weight device, Plate 16. From the above description it is clear that the plug-frame was in use without Beighton's device, and probably before it.

one year later made note of eighteen large engines in Cornwall, eight of which had cylinders 60 to 70 inches diameter.

A calculation of the Cylinder and Pumps of the Fire-Engines.

" If it be required to make an engine to draw any given number of hogsheads of water in an hour from f fathoms deep, to make any number of strokes in a minute by a 6-foot stroke, find the ale gallons to be drawn at one stroke, which is easily found from the number of strokes being given.

" Let g = number of ale gallons to be drawn at one stroke.

„　p = pump's diameter (in inches).

„　c = cylinder's diameter (in inches).

Then $p = \sqrt{5g}$;

and supposing the pressure of the atmosphere on an inch of the piston to be 7 lbs.,

$$\text{then } c = p \sqrt{\frac{2 \cdot 614 f}{7}} = \sqrt{\frac{13 \cdot 07 fg}{7}}.$$

" Note, if instead of 7 lbs. you suppose the pressure of the atmosphere to be 1 lb., and instead of a 6-foot stroke to make an r feet stroke; then $p = \sqrt{\frac{6}{r} \times 5g}$, and $c = \sqrt{\frac{6}{r} \times \frac{13 \cdot 07 fg}{1}}$."

The author has here arranged the events in connection with the subject in their chronological order ; and it is interesting to observe that, though Watt had finished his labours in Cornwall in 1800, bringing the duty of his engines up to 20 millions, yet, between 1810 and 1821, three Newcomen engines were erected at the Farme Colliery in Scotland ; particulars of these engines are given in Appendix III (page 681). In those days the want of knowledge of what others had done formed a great bar to progress. By the publication of what has been accomplished, and by the knowledge thereby imparted, the progress of mechanical science and arts has been hastened during the last fifty years as much as by the theoretical teaching of schools.

The Newcomen Engine.

Events in chronological order :—

1696. Description of Savery's Fire Engine, published under the title of " The Miner's Friend."

1698. Thomas Savery, of London, obtained a patent for raising water by the elasticity of steam. His engine had no piston.

1702. Savery's "Miner's Friend" was republished. Savery's advertisement in " Post Man," 19–21 March, notified that his engine might be seen at work " at his workhouse in Salisbury Court, London." This advertisement was reproduced in " Notes and Queries," 27 January 1900 (9th series v. 64).*

1712. Newcomen erected an engine near Dudley Castle for Mr. Back of Wolverhampton. It had a water-jacket around the cylinder for condensing the steam; but afterwards injection in the cylinder was adopted. All valves were worked by hand.

1712 to 1718. A buoy was used to give automatic action to the injection-cock.

1714. A Newcomen engine is said to have been erected at Wheal Vor, near Breage, in Cornwall; and another at Ludgvan in 1720.

1715. Savery died in London.

1717. 29 December.—Calley (or Cawley) died whilst erecting an engine at or near Ansthorpe, Yorkshire. This is from the burial register of Whitkirk. *See* Farcy's " Steam Engine," p. 155.*

1718. Beighton invented the " hand-gear." The steel-yard safety-valve was introduced; also the snifting valve, and the shortened eduction-pipe with its non-return valve. All the essential features of the perfected engine were now present.

1720. Newcomen went into Cornwall and erected an engine at Wheal Fortune, St. Day. Another engine on the same model was erected at Pool Mine, Carn Brea, in 1746.

---

* These and some other notes have been contributed by Mr. Richard B. Prosser.

1721. An advertisement appeared in the " Daily Courant," 24 July
1721, beginning " Whereas an engine to raise water by Fire,
commonly called Savery's engine . . . .," and inviting
attention to a new form of engine. The above was printed
in " Notes and Queries," 27 January 1900 (9th series v. 64).
See also " Notes and Queries," 17 February 1900, for a
communication from J. E. Hodgkin.*

1725. Joseph Hornblower erected an engine at Wheal Rose, Truro; a
second engine was erected at Wheal Busy, Chasewater;
and a third at Polgooth, near St. Austell.

    8 April.—Steam engine at work at Tipton, Staffordshire.
[On this day the son of John Hilditch, " Manager of
ye Fire Engine at Tipton," was baptised in the Parish
Church of Bilston.—" The Engineer," 11 Nov., 1898.] *

1729. Newcomen died in London.

1733. 24 July.—Savery's patent expired, having been in existence
for 35 years.

1758. Many engines at work in Cornwall, one at Herland, near
Gwinear, having a 70-inch cylinder. Emerson describes in
detail in his " Principles of Mechanics " the Newcomen
engines as then used.

1767. Smeaton first turned his attention to the atmospheric engine.

1769. Smeaton computed the duty of fifteen engines in the Newcastle-
on-Tyne district, and found the average duty to be 5·59
millions of foot-lbs. per bushel or 84 lbs. of coal.†

    5 January.—Watt's first patent.

1770. Smeaton made note of eighteen large engines in Cornwall,
eight of which had cylinders from 60 to 70 inches
diameter.

---

   * These and some other notes have been contributed by Mr. Richard B.
Prosser.

   † The bushel was taken as 84 lbs. in Savery's time, and continued until
Watt adopted 94 lbs. The duty of steam-engines continued to be calculated on
94 lbs. until 1856, when the Cornish engineers adopted the cwt. (112 lbs.)
See Davey on " Pumping Machinery," page 20.

1772. Smeaton made improvements in details, not altering the general construction, and succeeded in obtaining a duty of 9·5 millions.

1775. Smeaton erected a Newcomen engine at Chasewater in Cornwall, the steam cylinder 72 inches diameter. Water load 7¾ lbs. per square inch. Lift of pumps 360 feet. This engine was altered by Watt to his system.

1776. Watt corresponded with Smeaton, and claimed 21·6 millions duty for his engines. Smeaton, after making experiments with Watt's engines, laid it down as a general rule that the Watt engines did double the duty of the Newcomen.

1777. Watt erected three more of his engines in Cornwall, his first having been erected the previous year. In these engines the load on the piston was increased from the 8 lbs. per square inch in the Newcomen to 11 or 12 lbs. in the Watt engines.

1778. Smeaton found that a Watt engine at the Birmingham Canal* did a duty of 18 millions, and one at the Hull Waterworks 18·5 millions. Two engines at Poldice, St. Day, were found to do a duty of 7 millions on one bushel of coal.

1781. 25 Oct.—Watt's second patent.

1800. Watt finished his labour in Cornwall, having raised the duty of his engines to 20 millions of foot-lbs. per bushel (94 lbs.) of coal. In 1798 the 70-inch engine at Herland, near Gwinear, was reported as giving a duty 27 millions. It was probably the best which at that time had ever been erected, and was under the care of William Murdock, who had the general charge for Boulton and Watt of their engines in Cornwall.

1810. A Newcomen engine was erected at the Farme Colliery, Rutherglen, near Glasgow, for winding and pumping; in 1820 another was added for winding; and in 1821 a third having a 60-inch cylinder for pumping.

---

* A drawing of almost the first Watt engine for the Birmingham Canal was illustrated in "The Engineer," 15th July 1898. This is now erected in the yard at Ocker Hill, near Wednesbury.

Newcomen had associated with him Cawley, a plumber and glazier ; and it will be observed that the pipes of the engines were at first made of lead with plumber's joints. In the early days the steam cylinders only were obtained from iron-founders, and the other parts of the engine were built by local blacksmiths, carpenters, and plumbers, under the direction of an engineer.

The engine cylinder was at first fixed on a boiler of haystack form ; but the vibration of the working so loosened the joints that it was found advisable to secure the cylinder to strong wooden beams above the boiler. At a later date the cylinder was fixed on a separate foundation by the side of the boiler, and as time went on iron pipes were substituted for lead, and the wagon-boiler was introduced to take the place of the haystack.

Among the first erectors of the Newcomen engine were the Hornblowers in Cornwall. Newcomen visited Mr. Potter, of Bromsgrove, and erected the engine near Dudley Castle in 1712. This is the historical engine in which injection in the cylinder was first used. In the vicinity lived Joseph Hornblower, an engineer who became acquainted with Newcomen's engine, and who was sent for into Cornwall about 1720 to 1725 to erect an atmospheric engine at Wheal Rose near Truro. It may be interesting here to observe on the authority of Cyrus Redding, a great grandson of Joseph Hornblower, and author of " Yesterday and To-day," &c., that the Newcomen engine was not such a simple machine as to require the attention only of boys, as stated in popular histories, but that it required the united exertion of three men to start.

A second engine it appears was erected by Hornblower at Wheal Busy or Chasewater Mine. A third at Polgooth near St. Austell. Joseph Hornblower then left the county, and his son Jonathan came down and erected his first engine at Wheal Virgin, St. Day, about 1743. The fourth son of Joseph was Jonathan Carter, the inventor of the compound engine * and the double-beat steam-valves, who died at Penryn in 1815.

---

* Proceedings 1862, page 243.

From 1720 to 1740 few engines were erected in Cornwall, because of the high duty on sea-borne coal. In 1741 an Act of Parliament was passed for the remission of the duty on coal used for fire-engines for draining tin and copper mines in the county of Cornwall.* The effect of the passing of this Act was that by the year 1758 many engines had been brought into use ; one engine at Herland had a 70-inch cylinder.

Rotative Atmospheric Engines.—It appears that attempts were made as early as 1768 to produce a rotative motion from a Newcomen engine ; but it was not till about 1780 that it was successfully accomplished by the use of the crank.

It does not appear that any attempt was made, before Watt's separate condenser was invented, to reduce the cooling effect of the injection-water in the cylinder by effecting the condensation in a small vessel attached to the cylinder. It is evident however from Plate 28 that, after Watt's patent, Newcomen engines were made with separate condensers without air-pumps, the air being discharged through a snifting-valve. Such condensers were known as "pickle-pots," and are shown clearly in both illustrations.

In Fig. 23, Plate 30, will be found a sketch of the "pickle-pot" condenser. Such condensers are indicated in both engines, Plate 28, and were operated without air-pumps, as already described. It is more than probable that such condensers were not known till after Watt's invention of the separate condenser, and that they were applied to improve the economy of the Newcomen engine and to evade Watt's patent.

In Fig. 27, Plate 30, will also be found a diagram constructed by the author to indicate the economy of fuel resulting from various improvements commencing with the earliest engines of Newcomen. A diagram above also indicates the increase in steam pressure corresponding to the increased economy.

---

* The Act referred to is the 14th Geo. II., Cap xli., and intituled :—An Act for granting to His Majesty the sum of one million out of the sinking fund, and for applying other sums therein mentioned for the service of the year 1741 ; *and for allowing a Draw-back of the Duties upon Coals used in Fire Engines for drawing Tin and Copper mines in the County of Cornwall, &c. . . .*

The steam-engine has held its own as a prime mover for two centuries. The gas-engine has now become a more efficient heat engine, and a powerful competitor, and electricity has become an economical transmitter of power.

Heat, electricity, and mechanical work are mutually convertible. The time may come when heat may be converted into electric current with as little loss as that involved in the conversion of electric current into mechanical work; when that time comes, the heat efficiency of the prime mover will exceed that of the gas-engine in a greater degree than the gas-engine has exceeded that of the steam-engine.

The Paper is illustrated by Plates 16–30, and is accompanied by the following ten Appendices:—

Appendix I.   Mr. W. G. NORRIS.
    II.   Sir FREDERICK BRAMWELL.
    III.   Mr. HENRY DAVEY.
    IV.   Mr. H. W. PEARSON.
    V.   Mr. JOHN H. CRABTREE.
    VI.   Mr. W. E. HIPKINS.
    VII.   Messrs. THORNEWILL and WARHAM.
    VIII.   Mr. W. B. COLLIS.
    IX.   Mr. HUGH S. DUNN.
    X.   INSTITUTION NOTES.

APPENDIX I.

BY MR. W. G. NORRIS,* Member, OF COALBROOKDALE.

Among the miscellaneous articles in the Works stores at Coalbrookdale at Midsummer, in 1718, was a brass fire-engine valued

---

* Mr. Norris has also contributed a list of large cylinders and other castings, &c., supplied to colliery owners and others between the years 1732 and 1792, which is invaluable as a means of tracing the dates of erection of many large engines during the eighteenth century.

at £7 3s. This is not likely to have been a utensil such as is now
known by the same designation; but probably was a model of the
newly introduced machine, which was then beginning to make itself
widely known under that name.

No record has been preserved of direct communication between
Thomas Newcomen with his partner John Cawley and Coalbrookdale,
if any took place; and the name which is first found associated with
"Fire-Engine" work in the account books at Coalbrookdale is
that of Stanier Parrot of Coventry, as though he were at the
time at Coalbrookdale.   In December and March 1718, some
castings and " pipes 6 inches bore " were charged to him for " yᵉ
fire-engine."

Thomas Newcomen however had in some previous year (about
1715), erected one of his engines for Sir Robert Newdigate at Griff
Colliery, near Nuneaton; and Stanier Parrot, resident at Coventry,
and evidently also interested in adjacent coalworks, may be assumed
to have made acquaintance with Newcomen and with his invention
and its results.   There are particulars in print * of the difficulties
connected with the erection of this first engine at Griff, and also of
the economic advantages of one or other of these early engines.

Evidently it was not difficult to obtain wrought-iron for the rods
and links connecting the wooden beam with the cast-iron articles;
and the work must have been done by local mechanics to suit each
particular engine at the time of its erection.   For years the necessary
cast-iron articles only were supplied from Coalbrookdale; and also
it was not until the middle of the century that boilers came
to be made entirely of wrought-iron, when " engine plates," or
small plates hammered at a "plating forge," feebly made their
appearance.

The earliest entry of a set of castings for an engine is in
October 1724, when there were sent to the order of Richard Beech, of

---

* Stuart's " Historical Account," etc.   Desaguliers' " Experimental
Philosophy," Vol. 2, 1734.

Walton near Stone, to "Harding beyond Chester," now known as Hawarden—

		£.	s.	d.
1 Cillinder and Bottom . . .	25 0 10 @ 32/6 =	40	15	5
1 Pistern	1 1 21 @ 32/6 =	1	17	7
2 Brass Pipes and Brass to s^d Pistern .	1 1 4 @ 14d. =	8	8	0
A parcel of Wrought Iron Screwpins.	0 0 40 @ 9d. =	1	10	0
Carriage of Do. to Harding . .		3	16	0
		£56	7	0

In 1725 there are entered :—

Aug. Stanier Parrot of Coventry.
  1 Cillinder, 1 Bottom, 1 Pistern and 4 Pipes weighing 74 1 11 @ 32/6 = £120 16s. 4d.
Oct. Sir Richard Newdigate of Griff.
  1 Cillinder, 1 Bottom, 1 Pistern with 3 Pieces to go round it, and 3 Pit Barrels (for pump) 51 3 25 @ 32/6 = £84 9s. 1d.

From this period there followed a continuous demand for sets of these castings, with working barrels and pipes for pump work ; the cylinders and castings increasing in dimensions and weight. Unhappily the volumes of "the Journal" account books for the years 1749–65 and 1771–78 are missing, so that a complete list of the cylinders made and sent away up to 1800 cannot be prepared.

The only reference that can be traced to a patent connected with this engine work occurs in a letter from Richard Ford to Thomas Goldney, of Bristol, both partners in the concern, under date 26 March 1733, * " and as ye patent for ye Fire Engine is about expiring, that business will consequently more increase."

There is no evidence in the books of difficulty in making the castings. With the increase in the size of the cylinders an important practical difficulty connected with the boring began to be experienced from the frequent breakage of the boring bars. It was desired in 1734 by the managing partner, Richard Ford, to have for this purpose "a wrought-iron spindle, 12 feet long and full 3 inches in

---

* Savery's patent dated July 1698 (extended for 21 years) expired 25th July 1733.

diameter; one end to be left square for 18 inches, and y^r other end to be left square for 6 inches, and the remaining 10 feet to be left round, but to be as true as may be, and to be made of right tuff iron and right sound." This was ordered from an anchor-smith in Bristol, and supplied at a cost of £26 10s. in September 1734. Another was obtained in 1745, ^{c}5·3·15 @ 8d. per lb. = £21 19s. 4d.

The motive power for the blowing apparatus for the blast furnaces, and for any machinery, in the works at Coalbrookdale was the water from the brook running through the valley in its course to the Severn, which necessarily was a variable supply. In order to utilise the limited quantity in the drier periods, it was decided in 1743 by the then acting partner, the second Abraham Darby, to erect an engine for pumping back the water after it had passed over two or more water-wheels, up into the reservoir from which it had first issued, and to make it proceed on its course of labour again. The difference in level through which the water was raised was about 120 feet. The dimensions of the cylinder and of the pumps have not been preserved. The total cost, as upheld in the year's accounts for 1745, appears to have been for—

	£.	s.	d.
Cylinder, Working Barrels and Cast Iron Articles . .	307	18	7
Brass Work and Lead	108	14	1
Plate Iron and Bricks	70	15	8
The Oak Regulating Beam	10	15	0
Various Labour in and with the Erection and the Pit .	356	0	3
	£854	3	7

Similar engines for the same object were subsequently erected at the other works belonging to the concern at Horsehay in 1754, with cylinders 10 feet by $47\frac{3}{4}$ inches, and in 1756 with cylinders 10 feet by 60 inches; and at Ketley about the same years with the same sizes of cylinders. With the Coalbrookdale engine the experiment was made of using spelter in place of brass for the working barrel of the pump, either entirely or as lining for an iron pipe. A second working barrel of the same size was sent with some engine work into Cornwall, but was quickly returned. The size of the working barrel was 9 feet long by $8\frac{1}{4}$ inches diameter, and the weight of spelter in each was 8 cwt.

The application of a pumping-engine to return water for recurring use by a water-wheel was subsequently adopted at several furnaces in England and Scotland, because of the advantage of a continuous blast upon the furnace.

From 1755 to 1765 James Brindley was frequently at Coalbrookdale. He superintended about 1756 the preparation of castings for an engine for Mr. Broade of Fenton near Stoke-on-Trent, but there is no record of experiments with wood for cylinders : * experiments which seem hardly probable, having regard to the experience already acquired with both steam and iron ; but probably experiments were largely made with easing the cylinder, condensation, and steam pressure. He obtained a patent for an arrangement of boiler to be made largely of cast-iron, upon which several experiments were made by the Coalbrookdale concern, but without satisfactory results. At last, in 1762, it was treated as old iron. He also superintended the castings for a larger engine sent to Newcastle-on-Tyne for the Walker Coal Co. in 1763. The cylinder was $10\frac{1}{2}$ feet by 74 inches, and was said to be the largest that had been sent into that district. It is stated also that the "bore was turned perfectly round and well polished," and "the whole a complete and noble piece of work."

The accurate boring of the large cylinders was still a source of anxiety. In 1751 a note is made against a "Journal" entry :— "The 54-inch cylinder for Lord Ward having been examined, was found to be true and exactly bored, viz., within $\frac{1}{16}$-inch in all the working part." Afterwards, in 1776, James Watt wrote † to James Smeaton : "Mr. Wilkinson has improved the art of boring cylinders, so that I promise upon a 72-inch cylinder being not

---

* *See* "Lives of Engineers—Brindley," by Dr. Smiles.

† *See* Farey's "History of the Steam Engine." Mr. John Wilkinson, of Bersham, near Wrexham, contrived about 1775 a new machine for more accurately boring the insides of cylinders. A straight bar was fixed in the axis of the cylinder, which was made to revolve slowly round it, while a sliding cutter was caused to travel along the central stationary bar.

further distant from absolute truth than the thickness of a thin
sixpence in the worst part."

After 1750, notices of " engine plates," or " boiler plates," are more
frequent. They were all hammered under a " Plating " hammer from
slabs or moulds. They seldom exceeded half a cwt. each. It was
not until after 1790 that boiler plates, about 4 feet by 8 inches by
$\frac{1}{2}$ inch were rolled at the Coalbrookdale Co.'s works at Horsehay :
and no other works in Shropshire then made them.

With the improvements in boilers consequent on the use of
wrought-iron boiler plates, the greater efficiency and economy of the
steam-engines conformable to the designs of James Watt were
becoming generally acknowledged. In 1780, Matthew Boulton visited
Coalbrookdale, in order to confer with the Coalbrookdale partners
upon the requirement for steam-engines for Cornwall, on which
Boulton and Watt had reports from their own agents. One of the
results was that the Coalbrookdale partners decided to supersede the
Newcomen engines in use at the works at Coalbrookdale and at
Ketley, by engines on the plans of James Watt. It was found that
the Newcomen engine in Coalbrookdale consumed 12 tons of small
coal, valued at 2s. 6d. per ton, beyond what would be required for
every 10,000 strokes by a Watt engine having a cylinder 66 inches
in diameter and 11 feet long, making 9 strokes a minute. It was
agreed, therefore, in 1781 to pay to Boulton and Watt one-third of
the estimated saving in fuel upon this basis, or 10s. for every 10,000
strokes, as recorded by a mechanical counter, during the unexpired
period of the term of Watt's renewed patent.

The difficulty and extreme delay involved in sending such large
castings into the North of England, all being forwarded to Bristol
and thence shipped to London and often transhipped, had led to the
cessation of business with the North.

At that time there were sixteen large engines at work in
connection with the coal works and furnaces of the Coalbrookdale
concern at Ketley, Dawley, Madeley, and Coalbrookdale, chiefly for
pumping.

The enlargement of the cylinder bottom, and the addition of a
" hot well," appear to have been improvements which were made

upon the earlier castings for the engine; but experiments were also made upon the expansion or extended use of the steam by means of two cylinders. Arising partly from these experiments Jonathan Hornblower, of Penryn, obtained a patent for an engine somewhat of this arrangement in 1781.* Similar experiments led to the patent in 1790 of Adam Heslop, who was then engaged at the Ketley Branch of the Coalbrookdale concern, for the application of two cylinders, one hot and the other cold. A pumping-engine to Heslop's designs † was made in 1791–2 and sent to Workington, where it worked for years at one of Lord Lonsdale's collieries. In 1878 an engine of this kind but arranged chiefly for winding, supposed to be the last, was presented to the Commissioners of Patents by Lord Lonsdale, and was erected at the Patent Museum. The notice of this donation in the "Times" of 24th May 1879 led to a letter from W. R. Anstice of the Madeley Wood Co., saying "that they still have three of the same description at work; they had five, and had had eight." One continued to work at a coal pit until 1890. The last of the single-cylinder Newcomen engines continued to work a helve in one of the Coalbrookdale Co.'s forges until 1879, when it gave place to a steam-hammer. A photograph of a winding engine built at Coalbrookdale about 1790 is shown in Plate 17.

The engines completed at Coalbrookdale must have been adapted in some manner to rotative action at least by 1780; but no detail drawings or descriptions are preserved there. Small engines working with cranks for coal winding were in frequent sale by 1790. Several office drawings of various designs made previous to 1800, which had been the property of William Reynolds, of the Ketley Branch of the Coalbrookdale concern, were sent to London in 1879; but recent enquiries at the Patent Office Library and at the Board of Education at South Kensington have failed to discover their present owners.

In Farcy's "History of the Steam Engine," 1827, and in Stuart's "Descriptive Anecdotes of Steam Engines," 1829, are engravings

---

* Proceedings 1862, page 243.

† See also H. A. Fletcher on the Heslop Engine, Proceedings, 1879, page 85.

representing Newcomen and other engines; also in the Transactions
of the Chesterfield and Derbyshire Institution of Engineers, Vol. IX,
1881-2, p. 37, is given an illustrated description of the Atmospheric
Engine at Handley Wood, built by the Coalbrookdale Iron Co. in
1776.

APPENDIX II.

By SIR FREDERICK BRAMWELL, BART., *Past-President.*

The writer wishes to draw attention to the details of Newcomen
engines given in Belidor's " Architecture Hydraulique," published in
Paris in 1739, in which a description of Newcomen's engine occurs
on page 309 of the first part, illustrated by three extremely good
engravings. These have not been referred to by the author, although
he has quoted so fully from Emerson's book.

In his own investigations * as to the inventions of Newcomen,
the writer found that Newcomen never took out a patent,† probably
because he joined forces with Savery, and the terms of Savery's
patent of 1698 were wide enough to cover Newcomen's invention.‡
A recent examination of the official lists of patents from the beginning
of 1617 to 1852 shows neither the name of Newcomen nor that of
Cawley. He has however found three patents in the name of
Thomas Savery, one of which is of 1698, above referred to. The year
1705, quoted by Mr. Davey (page 658) from Stuart, was the year
wherein only a single patent was taken out, and that was by a
gentleman of the name of Lydall for separating gold and silver
from tin.

---

* Life of Watt in " The Dictionary of National Biography."

† Mr. Norris writes (page 675) that, as contemporary evidence in the belief of
the grant of a patent, may be quoted an extract from a letter from Richard
Ford of Coalbrookdale to Thomas Goldney of Bristol, both partners in the
Coalbrookdale concern, under date 26 March 1733, "And as yᵉ patent for yᵉ
Fire Engine is about expiring, that business will consequently more increase."

‡ *See also* Transactions, British Association, 1877, Mechanical Science
Section, page 217.

APPENDIX III.

By MR. HENRY DAVEY, *Member of Council.*

Newcomen Engines at the Farme Colliery, Rutherglen, near Glasgow.

Plate 18.

The Newcomen engines at the Farme Colliery were three in number. The first was erected in 1810. In its present state it is illustrated in Plate 18 by photographs and drawing, together with indicator diagrams taken from it. It has a 42-inch cylinder with a stroke of 5 feet 8 inches, and has worked almost constantly since it was put up, being employed in winding coal. One rope draws from a depth of 30 fathoms, the other from a depth of 44 fathoms. Until recently, it also pumped water from a depth of 23 fathoms with an 8-inch bucket pump. It drew from 150 to 200 tons per day, and pumped for four or five hours in the twenty-four, on a consumption of 34 cwts. of "dross" coal per twenty-four hours.

In 1820 was erected another winding engine; and in 1821 a pumping engine, having a 60-inch cylinder, with a stroke of 7 feet. It worked three set of pumps. In the top set the pump was 20 fathoms deep, with $15\frac{1}{2}$-inch bucket; the second set was 30 fathoms below, with a 12-inch bucket; and the bottom set was 20 fathoms lower, with 10-inch bucket: making 70 fathoms total lift. Steam was supplied by two haystack boilers, 30 feet and 25 feet in diameter; the pressure of steam was $2\frac{1}{2}$ to 3 lbs. on the square inch. The piston was made as an ordinary air-pump bucket, and was packed with the old hemp ropes that had been used in the pit. A head of water 12 inches deep was always kept above it, so as to prevent air leaks. The engine worked night and day, driving two pumps, and sometimes three when they were hard pressed with water in the pits. Wood blocks were bolted to the beam to regulate the stroke. They were covered with old ropes to deaden the shock of the beam coming down, should the engine-man happen to give her a little extra steam.

The winding engine erected in 1810 is still working; it was constructed by John McIntyre, who made his own patterns, and obtained the castings from the Camlachie Foundry, three or four miles away. The two other engines were taken down in 1888.

APPENDIX IV.

BY MR. H. W. PEARSON, *Member*, OF BRISTOL.

The writer having offered to the Institution in 1901 a description of the Ashton Vale Newcomen Engine, the Council considered it would form an interesting contribution to the history of the subject; and Mr. Davey having offered to compile a general history of the Newcomen engine, the writer trusts the following short account of some early illustrations of pumping and winding engines, which in most cases have come directly under his notice, will be of some service to the Members of the Institution, and of sufficient interest to form a pleasing link between the past and the present. The existing developments, and the perfection to which the steam-engine and pump have been brought, will perhaps make the retrospect a still more interesting and instructive comparison in the study of their history, more especially in connection with mechanical works. The earlier and more ancient examples of steam as a power for working pumps and driving machinery were referred to by the writer in a pamphlet, published in 1896, entitled " The Creation and Development of the Steam-Engine," which is in the Institution Library.

*Newcomen Engine at Ashton Vale Iron Works, Bristol.**

Plates 19 and 20.

The earliest instance of a pumping engine for colliery purposes, which has come under the writer's notice, is the Newcomen engine at the South Liberty Colliery of the Ashton Vale Iron Company,

---

* Compiled by the writer from particulars taken by him from the engine itself at the South Liberty Colliery in 1895, at which date the engine was in regular work. It has since been dismantled in 1900. There is at South Kensington a photograph of this engine, taken many years ago, when it was in a more perfect condition than it was in 1895.

near Bristol. According to common statements Thomas Newcomen, of Devonshire, invented in 1705 the first successful cylinder and piston engines; and this example is considered one of the early large engines built by him, possibly about 1746–60. * The drawing in Plate 19 shows the general arrangement of the engine, and has been prepared from a sketch the writer had made in September 1895, when he had occasion to take some particulars on the spot for illustrating this old relic. The drawing, together with the three photographs, Plate 20,† conveys a good idea of the rude though successful working structure of Newcomen's age, with its open-topped cylinder and trussed oak beam having arch-heads at the ends. The cylinder was 5 feet 6 inches diameter, and being 10 feet long had clearance enough to allow of working 8-foot stroke if required. The piston was packed with rope ends, and loaded with pig-iron to help to balance the indoor stroke; and was worked with a covering of water on the top, so that, if the packing was not quite tight, only water leaked through and helped to maintain the vacuum. The valves fitted to the cylinders consisted of a circular lift steam-valve for admitting steam during the outdoor stroke; a slide-valve B for admitting injection water to condense the steam for forming the vacuum; and a flap valve C, kept closed by a weighted lever, for the escape of the injection and condensed water at the end of the indoor stroke. The Newcomen engine was utterly dependent on manual working of the valves, until that egregious youth Humphrey Potter devised self-acting catches whereby the working beam was made to perform the opening and closing of the valve.‡ This led to the introduction of the plug frame, fitted with catches to act on the valve

---

* Mr. W. G. Norris writes that he has little doubt the castings for this engine were made at Coalbrookdale about the year 1760, or perhaps a few years earlier; no castings so large could have been obtained much earlier than that date.

† Mr. W. G. Norris writes, in reference to these illustrations, that there must have been renovation in the early years of the last century, as the large adjusting screws on the arch-head at the pump end of the beam in Fig. 7, Plate 20, cannot have been of the same date as the cylinder.

‡ This story is quite discredited.—H. D.

levers—first put on a Newcastle engine by Henry Beighton in 1718.*
The engine was at first supplied with steam from a haystack boiler, but
latterly from the boilers working the winding engine; the steam
was passed into a receiver through a reducing valve by which the
pressure was brought down to 2½ lbs. to the square inch before
entering the cylinder. The beam was built of oak timber trussed
with rods, 24 feet long and 4 feet deep, and had V-shaped gudgeons
of cast-iron bolted on it in the form of a collar, which rocked in
cast-iron plummer blocks. Its weight with the gudgeons, trusses,
&c., was about 5 tons. The engine made from ten to eleven double
strokes per minute.

The pumps, which were three in number, were bucket lifts, and
9 inches in diameter, and were worked from the end of the beam
by chains secured to the beam, and passing over the wooden
arch-head at outer end of beam and thence down to the pump-rods,
which were of wood, bolted together with side straps. The pumps
were in three lifts, delivering one into another from the bottom of
the shaft, which was 700 feet deep. There was a jack-pump J
attached to the inner end of beam, which lifted water into a cistern
on the beam floor for supplying the injection water to the cylinder.
A vacuum was formed on the under side of the piston by the
admission of steam during the outdoor stroke, which was then
condensed by the admission of injection water; and the top of
cylinder being open to the atmosphere, the atmospheric pressure
on top of piston caused it to descend with an energy proportional
to the perfection of the vacuum formed, thereby producing the
indoor stroke, which lifted the pump rods and buckets at the outer
end of the beam, and so raised the water from the bottom of the
shaft. The outdoor or descending stroke of the pumps was made
chiefly by the weight of the rods, &c.; and thus the double stroke
was completed.

No indicator diagrams had ever been taken from the cylinder
until the writer took some in 1895, for which purpose the cylinder
had to be drilled and indicator connections and gear fitted. Fig. 5,

---

* "Imperial Cyclopædia of Machinery," by W. Johnson, 1852.

Plate 19, represents the indicator diagram, showing the amount of vacuum to be between 9 and 10 lbs. At a speed of 10 strokes per minute this diagram gives about 52 horse-power. The old driver seen in Fig. 6, Plate 20, near the gudgeon blocks, worked the engine up to the time of its being dismounted in 1900, and had driven it since he was a boy. His father and grandfather had worked it before him. No records could be obtained of the cost of the engine, neither was the writer able to gather any information as to who made the castings or erected the engine. He searched for a date on the cylinder and other parts of the engine, but could find no evidence of it. The gear for actuating the steam-valve and injection-valve was almost identical with that shown in the illustrations of the earliest engines: in confirmation of which the writer consulted an old book called " The Principles of Mechanics and the General Laws of Motion," by Emerson, published in 1758, kindly lent to him by Mr. T. Bush, of Bath, formerly of the firm Messrs. Bush and De Soyres of Bristol ; this he has shown to Mr. Davey. Mr. W. G. Norris believes that a second edition of this book was published some years later.

Engine at Buffery Old Colliery, near Stourbridge.

Fig. 15, Plate 26.

Mr. E. B. Marten, of Stourbridge, has lent the writer some sketches of early examples of open-topped cylinder engines of the Newcomen kind, which were worked in the neighbourhood of Stourbridge. These sketches * are from a collection of Mr. W. B. Collis, of Stourbridge, and are from the pen of the engineman who drove the engines, named William Patrick, whose father and grandfather had worked them before him.

"A Scribble of the Buffery old Pumping Engine," Fig. 15, Plate 26, drawn from memory in 1878, illustrates the Newcomen atmospheric engine, and shows the old haystack boiler with its

---

* See also Appendix VIII (page 691).

steam connections, and the admission of steam controlled by the levers worked from the tappet on the plug-rod from the beam. The engine as shown had a Watt condenser. There are also shown the jack-pump and hot-well tank, dealing with the injection water after it had done its work in the cylinder in forming the vacuum to give the indoor stroke. The outdoor stroke was made by the weight of the pump-rods, as in other instances of this class of engine. The oak trussed and tied beam, with quadrant ends and chain connections to pump and piston-rods, was similar to those previously described in connection with the engines of the Newcomen period.

Atmospheric Winding Engine,* near Stourbridge.

Fig. 16, Plate 26.

The other illustration is of a curious old atmospheric winding engine of the same kind as the one just described, with the exception that the quadrant ends to the beam were dispensed with; the couplings for the piston-rod and for the connecting-rod to the flywheel and crank-shaft were formed by a gudgeon or pin working in a sort of gudgeon block attached to the ends of the wooden beam, which was doubtless in two slabs, with a space between them to allow of the oscillation of the piston-rod and connecting-rod at the beam ends. An interesting note was made by the draughtsman, Mr. Patrick, stating that this is a representation of the atmospheric winding engine worked by him when he was ten years old in the year 1814; it was drawn by him in 1878 from memory. Several engines of somewhat similar description were to be seen working in Shropshire and South Staffordshire up to 1850.

York Water Works Pumping Engine.

Plate 21.

On occasion of being consulted by the York Water Company, in a visit to their works the writer came across an early illustration of

---

* *See* Indicator Diagrams, Plate 27.

their "Fire Engine," which is here reproduced by the kindness of Sir Jos. Sykes Rymer, the Chairman, and Mr. Humphreys, the engineer; Plate 21 is a copy of the original working drawing, reduced to one quarter the size of the original.

The action of this engine, it will be seen, is almost identical with that of the Ashton Vale Newcomen Engine previously described. The engine was designed by Smeaton, and erected in 1784. It was 18 horse-power, and raised 16 gallons of water per stroke, and could work up to 18 strokes per minute, and appeared to be used from 7 to 8 hours each day. The cylinder was placed immediately over the centre of the boiler, and the steam admitted at the bottom of the cylinder, the admission being regulated by the cataract; diameter of the steam-supply pipe 5 inches. The diameter of the cylinder was 27 inches, and its total length was 8 feet 3 inches. The length of the stroke was 6 feet; length of beam 23½ feet, with quadrant ends. Two pumps, one 9 inches and the other 7½ inches diameter. The rising main from the larger pump to the reservoir on the top of the tower was of flanged iron, and that from the smaller pump delivered into the reservoir through wooden pipes. The lift was about 70 feet. The supply pipe from the reservoir to the city was 7 inches diameter. A wooden overflow pipe was fixed for returning the water into the well below, in which the pumps were situated. A tell-tale worked by a float was placed in the engine-room, for showing the depth of water in the reservoir. The boiler was haystack; the body was made of copper, 9 feet 3 inches diameter at its broadest part, and 7 feet at the bottom, the crown was of iron-plate; total depth 8 feet 6 inches. For the most part the cataract appeared to have been adjusted to eleven strokes per minute.

An interesting trial of the engine was made by Smeaton in August 1785, and is recorded in Table 1 (page 688).

TABLE 1.

Date.	Number of Trial.	Strokes counted.	Length of Strokes.	Total Time of counted Strokes.	Weight of Coals.	Time of Consumption.	Reduced Strokes per minute.	Metts used per hour.†	Hogsheads raised per hour.	Hogsheads raised per mett.	
			ft.	min. sec.	stone. lbs.	hour. min.					
1785. 24 Aug.	1	10	6	1 3	11 4½	0 44	$9\frac{5}{10}$	$1\frac{28}{100}$	$328\frac{3}{10}$	$256\frac{5}{10}$	{ Slack.* Cataract.
	2	10	6	1 3	12 5¾	0 56½	$9\frac{5}{10}$	$1\frac{10}{100}$	$328\frac{3}{10}$	$298\frac{5}{10}$	{ Coals. Cataract.
	3	13	6	1 2	11 6	0 46¼	$12\frac{6}{10}$	$1\frac{24}{100}$	$412\frac{5}{10}$	$356\frac{9}{10}$	{ Coals. Self-worked.

* Same as Slack.

† Mett, or measure, is believed by Mr Humphreys, the present engineer of the York Water Works to mean a bushel, one bushel of coals weighing 84 lbs.

APPENDIX V.

By Mr. JOHN H. CRABTREE, of Oldham.

Newcomen Pumping Engine at Bardsley, near Ashton-under-Lyne.

Plate 22.

This engine, in a ruined condition, is still to be seen at Bardsley, near Ashton-under-Lyne, and was used for pumping water out of a coal mine there. It had a wooden beam with metal bracings, and is said to be of very early date. It had a cylinder, open at the top, about 28 inches in diameter, with a stroke of about 6 feet. The steam entered the cylinder at the bottom, and the condensation was effected by the injection of cold water into the cylinder. There was no separate condenser. The engine passed out of use about 1830. It pumped water from the "Camel" mine, which was 70 yards deep. The boiler was of the "wagon" pattern, and was placed in front of the beam-bed of the engine.*

* A Photograph of this engine is in South Kensington Museum.

APPENDIX VI.

By Mr. W. E. HIPKINS, *Member*, of BIRMINGHAM.

The drawing reproduced in Plate 23 represents a Newcomen mine-engine, but the writer has no knowledge at what mine it was worked. James Watt visited various mines for the purpose of making hand sketches and taking particulars of the engines and coal consumption, for enabling him to send in a report pointing out the saving to be effected by adopting his own plan of engine in lieu of the atmospheric. When the law cases with Hornblower, Bull, and other later infringers of Watt's patents, were proceeding, most of the sketches were properly drawn out with a view to their production in evidence ; and the drawing in Plate 23, dated 1826, was taken from a portfolio containing all these drawings in question, and is doubtless made from one of James Watt's sketches of a much earlier date, probably about 1796. This drawing represents a modern Newcomen engine arranged to work on the injection principle, the external cylinder or water-jacket having been discarded. In the position shown the engine is at rest, the weight of pump-rod U having drawn the outer end of the working beam R down, thereby bringing the piston up to the top of the cylinder. The manipulation necessary in working the engine was the alternate opening and closing of two valves ; the regulating or steam valve M and the condensing or injection valves H. When the piston reached the top of the cylinder, M had to be closed and H opened ; and on reaching the bottom M had to be opened, and H closed.

In starting * the engine from its position of rest shown in the drawing, the regulating valve M was opened, and the steam admitted was at first condensed by contact with the cold cylinder. After a short time however, the cylinder acquired the temperature of the steam, which then ceased to be condensed, and became mixed with the air that previously filled the cylinder. The steam and heated air, having a greater force than the atmospheric pressure, opened the

---

* With trifling verbal modifications this paragraph is reproduced from " The Steam Engine," by Dr. Dionysius Lardner, seventh edition, 1840, pages 69-70.

blowing or snifting valve V, which was placed at the end of a small pipe at the bottom of the cylinder, and opened outwards. From the snift the steam and air rushed out in a continuous stream until all the air had been expelled, and the cylinder was filled with pure steam. This process was called blowing the engine, preparatory to starting it. When it was about to be started, the engine-man closed the regulating valve M, and thereby stopped the supply of steam from the boiler. At the same time he opened the injection valve H, whereby a jet of cold water was thrown up into the cylinder. This immediately condensed the steam contained in the cylinder, and produced a vacuum. The atmosphere could not enter the snift valve V, because it opened outwards, so that no air could enter to destroy the vacuum. The atmospheric pressure above the piston now took effect, and forced it down in the cylinder. The descent being completed, the engine-man closed the condensing valve H, and opened again the regulating M. By this means he stopped the play of the jet within the cylinder, and admitted fresh steam from the boiler. The first effect of the steam was to expel the condensing water and condensed steam, collected in the bottom of the cylinder, through the eduction pipe N, containing a valve (called the eduction valve), which opened outwards ; the pipe led down to the hot cistern, into which the water was therefore discharged. When the steam admitted through M ceased to be condensed, it balanced the atmospheric pressure above the piston, and thus permitted the latter to be drawn up to the top of the cylinder by the weight of the pump-rod U. The ascent of the piston was also assisted by the circumstance of the steam being somewhat stronger than the atmosphere.

APPENDIX VII.

By Messrs. THORNEWILL AND WARHAM, OF BURTON-ON-TRENT.

Basset Atmospheric Engine, Denby Colliery, near Derby.

The atmospheric engine photographed in Fig. 13, Plate 24,* was erected in the position shown in 1817, but had been working

---

* *See also* " Engineering," 1887, page 285.

previously at another pit. It ran until 25th April 1886, working 24 hours a day in winter and 9 or 10 hours a day in summer, and in the opinion of the present colliery manager at Denby it was at that time working remarkably well and economically.

The stroke of the piston was 3 feet 3 inches, diameter of the cylinder 2 feet 2 inches, and stroke of the pump 3 feet. There were two lifts of buckets, each 7 inches diameter, with a lift of 42 yards, the total depth of the pit being 84 yards. The engine worked at 22 strokes a minute, and the safety-valve was loaded to a pressure of 2 lbs. per square inch.

Plate 25 shows a photograph and drawing of an Atmospheric Rotary Pumping Engine, made at the "Moira Furnaces" in 1821.

APPENDIX VIII.

BY MR. W. B. COLLIS, OF STOURBRIDGE.

Communicated through Mr. E. B. Marten, Member.

The writer sent to the Institution a number of sketches of old engines, made from memory by Mr. William Patrick (pages 685–6). From these have been selected for illustration Plates 26 and 28, on account of their being Mr. Patrick's personal recollections of definite engines of the atmospheric class.

Mr. Patrick was foreman over the colliery engine-men some twelve in number, at Messrs. Cochrane and Co.'s Woodside Collieries between the years 1854–60, and came at that time under the writer's direction. When he left Woodside he went to Middlesbrough, and subsequently returned to Dudley. The writer believes the whole of his sketches are from memory. None of them relates either to the present Buffery pumping engine or to the atmospheric winding engine, now being worked by the Mines Drainage Commissioners in the Old Hill district. The writer was informed that the former of these was erected in 1801, and the latter some thirty years later. He re-started them both, and put them to work about 1874. The pumping-engine works with about 12 lbs. steam pressure, and has an air-pump and condenser. The air-pump is on the outer end of the beam. The engine has no valve between the boiler and the closed top of the steam cylinder. The steam-pressure on the top of the piston with the vacuum on the underside is sufficient to raise the pump-rods and water. When the vacuum is destroyed by the admission of a small whiff of steam, the weight of the rods overpowers the steam, and drives it back into the boiler. The atmospheric (open-topped) engine receives the injection (condensing) water in the bottom of the cylinder, whence it escapes through a "pickle-pot" vessel beneath, similar to that shown on Fig. 20, Plate 28, provided with a "sinker," or self-acting flap-valve opening outwards. The writer has had to alter both engines by introducing equilibrium valves in place of the original mushroom valves, which had become too heavy for handling, on account of the use of a higher steam-pressure than that for which they were designed; but the principles of the engines remain the same.

APPENDIX IX.

By Mr. HUGH S. DUNN, *Member*,

OF THE CAPRINGTON COLLIERY, NEAR KILMARNOCK.

The writer draws attention to the description and illustration * of a pumping-engine set up at Caprington in 1806 (Plate 29), which, according to the late Mr. Bryan Donkin, was a Newcomen engine.† A short description of it has also been published in a recent number of the Transactions of the Institution of Mining Engineers.

---

* *See also* South Kensington Museum.

† *See* "The Engineer," 15 July, 1898, p C5, illustrated p. 57

Notes on a Newcomen Engine at Caprington Colliery, Ayrshire.

Plate 29.

This engine, erected at Caprington Colliery in the year 1806, is understood to have been built at Carron Iron Works in Stirlingshire, sometime between 1770 and 1780 ; it continued at work until July 1901, pumping from a pit, the exact distance from the surface of the water in the sump to the lip of the discharge being 165 feet 8 inches. As originally erected at Caprington Colliery the engine had the usual wooden beam with cast-iron gudgeons, and a rocker or cradle on each end. The pumps were worked at one end by a chain fastened to the upper end of the cradle or rocker, and the piston-rod was connected to the other end of the beam in the same manner. The cradles served the purpose of a parallel motion. About the year 1837 the wooden beam, having shown signs of distress, was removed, and a cast-iron beam with cast-iron gudgeons was substituted, and was fitted with a Watt parallel motion at each end. At the same time the original boiler, which was spherical, was removed, and an egg-ended boiler was substituted. These were the only alterations made on the engine during the ninety-five years it worked at Caprington Colliery. The engine is of the usual kind, made on the Newcomen principle. The cylinder is $30\frac{1}{2}$ inches internal diameter, and about 8 feet long. The working stroke however is only 4 feet 6 inches to 5 feet. The excess length of cylinder is necessary, owing to the lower part of it being used as a condenser, and also because there is no crank to regulate the stroke, which constantly varies with the least change of either steam or atmospheric pressure, or with the condition of the pump bucket. The steam is admitted to the cylinder at just over atmospheric pressure by a plain plate-and-pin valve, worked by a rod passing through an ordinary stuffing-box. The valve-rod is actuated by the tappet gear through a series of rods and levers, worked direct from the beam. The water for condensation is admitted at the bottom of the cylinder, also by a plate-and-pin valve similarly worked ; but here the valve-rod passes through

the side of the water valve-chest, and is kept tight by a cramp
arrangement, which presses a collar on the valve-rod against a small
brass collar screwed into the side of the water valve-chest. The
condensed steam and the condensing water are discharged through an
eduction port in the bottom of the cylinder, which is practically
the sole-plate of the engine. The discharge or eduction pipe, fitted
with a hinged non-return valve, is immersed in a cast-iron tank or
hot-well filled with water, which effectually secures the engine
against taking in air at the discharge valve. From this hot-well
the feed-water flows into the boiler by gravity, being regulated
by a float which opens the feed-valve according as water is
required in the boiler. The cycle of the engine is as follows:—
The pump-rods being heavier than the piston and other attachments
at the cylinder end of the beam, the piston is always kept at
the top end of the cylinder when the engine is stopped. To
start the engine, the steam-valve is opened, and the steam allowed
to blow through for a few seconds; it is then closed, and the
injection-valve opened; this produces a partial vacuum, and the
piston will move down perhaps a few inches. The water is then
at once shut off, and the same process is repeated until the air is
entirely expelled from the cylinder, and the engine continues to
work full stroke. Every time the steam-valve is opened, the water
which was admitted during the previous stroke is expelled by the
steam; and as soon as the cold water is admitted, the vacuum is
formed, and the non-return eduction-valve closes. The piston
is well worth study, as showing how the difficulties of construction
were got over in those early engineering times. The body is of
cast iron, and is fastened to the piston-rod by an old form of splice-
joint secured by a collar-ring and cottar-pin, Fig. 22, Plate 29. To
prevent leakage of air into the cylinder, the piston was packed
round the flange with old rope or junk, over which were placed
quadrant-shaped weights to hold the junk down. A jet of water
was always kept running upon the piston, to prevent air from
passing. The pump-rods were made of red pine, 4 inches square,
jointed with malleable iron plates, 7 feet long by 3 inches by $\frac{3}{8}$ inch;
and were carried down the pump pipes to the working barrel, which

was 9 inches in diameter. The stroke of the pump was equal to the stroke of the piston. The piston usually made from 11 to 13 strokes per minute. The water for condensation, and for keeping the piston covered was supplied from the jack-head tank, which was about 5 feet by 3 feet by 3 feet, and was kept full of water by the jack-head pump, which was an ordinary bucket-pump 4 inches diameter, making half the stroke of the engine piston. The spherical boiler was 8 feet 4 inches in diameter. The under half was made of malleable iron plates $\frac{3}{8}$ inch thick; the top half was cast iron. Each half had a flange about 4 inches broad for fastening them together, which was done with $\frac{3}{4}$-inch bolts spaced about 9 inches apart. The joint was made with gasket and white lead, and the outside stemmed with cast-iron turnings and sal-ammoniac. The whole of the engine castings are cold-blast iron; each casting has a number and its weight cut on it with a diamond-pointed chisel. The cylinder has cut on it

No. 31935

1—6—3—0

CARRON

The engine is to be erected by the Corporation of Kilmarnock in the Dick Museum.

APPENDIX X.

Additional Notes by the Secretary.

In connection with a Paper read before the Institution in November 1883, by Mr. E. A. Cowper, on " The Inventions of James Watt," an illustration (Plate 87, described on page 623) shows " The Steam Engine near Dudley Castle, Invented by Capt. Savery and Mr. Newcomen. Erected by ye later, 1712." See ante, page 661.

A reproduction of the original engraving of this engine may also be seen in " The Engineer," 28 November 1879.

A short Paper on " Thos. Newcomen's Steam-Engine," by Thomas Lidstone, was read at the British Association in 1877 (see page 680); and a pamphlet " Some account of the Residence of the Inventor of the Steam-Engine " was published (2nd edition) in 1869 by the same author.

In a series of articles published in " The Engineer " during 1880, on " Pumping Engines at Staveley, Old and New," will be found, on page 84, a full-page drawing of an Atmospheric Engine with separate condenser, built at Coalbrookdale in 1776.

An interesting model of the Newcomen engine, as known in 1720, may be seen at work in the South Kensington Museum. This was exhibited at the Meeting.

Discussion.

The PRESIDENT said he was sure that the Meeting might congratulate Mr. Henry Davey upon the great success of his labours in bringing the history of the Newcomen engine so vividly before them ; and they might also congratulate themselves upon having listened to so interesting a lecture. He thought that at this stage of the proceedings formal thanks might be given to the Lecturer and others, so that the whole of the evening might not be taken up by expressions of praise, but that the remainder might be spent in dealing with critical remarks or questions for obtaining information. He would therefore ask them to indicate their agreement with him when he proposed that their best thanks should be given to Mr. Davey for his Lecture and for his Paper, and also to those gentlemen who had made contributions to that Paper, of either drawings or supplemental Papers, namely :—Sir Frederick Bramwell; Mr. W. B. Collis ; Mr. John H. Crabtree ; Mr. Hugh S. Dunn ; Mr. W. E. Hipkins ; Mr. E. B. Marten ; Mr. W. G. Norris ; Mr. H. W. Pearson ; Messrs. Thornewill and Warham ; and the Board of Education, whose Senior Keeper of the Science Museum, Mr. W. I. Last, had so kindly arranged for the exhibition of the working model of the Newcomen Engine shown at the Meeting.

The vote of thanks was given by the Meeting with much applause.

Mr. VAUGHAN PENDRED said that the admirable Lecture which they had just heard appeared to him to present a suitable occasion for calling attention to the work that had been done by their predecessors. He was afraid that in the present day there was some tendency among younger members of the profession to underrate what was done by the pioneers of the profession. On coming into the hall he mentioned this matter to Mr. Davey, who said that in olden times a man had had not only to invent, but he had to invent something which could be made. The speaker did not think that engineers of the present day thoroughly realised the difficulties with which men like Newcomen and others had to contend. A keen controversy raged between Watt and Trevithick in their work. Trevithick advocated high-pressure, and Watt low-pressure, and he thought that it was quite possible that either of them would have killed the other if he could have got the chance, and it was perhaps as well that James Watt, who was rather a delicate man, never came in contact with Trevithick. Mr. Pendred then read an extract from the "Life of Trevithick," containing the Recollections of the late Captain Charles Thomas,* Manager of Dolcoath Mine, in which a brief description was given of Captain Trevithick's large high-pressure steam-puffer pumping engine, erected about 1814 at the Herland Mine, and worked at about 150 lbs. pressure (Vol. 2, pages 74–75). He, Mr. Pendred, thought they would agree with him that there was a large amount of courage in a man who could do this with the boilers of that day, which were made with rope yarn between the lap of the boiler plates to make the seams tight (page 78). Mr. Pendred quoted also from pages 90, 91 and 395,† which he thought illustrated very clearly the principles on which the men worked who made England what she was to-day. He also thought that, if a selection were made amongst the best of those present, having technical education and skill of all kinds, they would not be able to deal as successfully as their predecessors with the

---

* "Life of Trevithick," by Francis Trevithick. Published by E. and F. N. Spon, London, 1872.
† *Ibid.*

(Mr. Vaughan Pendred.)

materials and men available 100 years ago. There was an old
saying which possibly a great many of them had heard, that the
difference between an amateur and a regular workman was that the
amateur would attempt to do anything with hardly any tools at all,
whereas the workman was entirely helpless unless he had the precise
tools he thought necessary. All of the early engineers must have
been, in the first sense, amateurs. He thought that Newcomen,
producing an engine such as they had heard described, was a man
whose reputation should live through all time.

Mr. HENRY LEA, Member of Council, said that as a Birmingham
man he had been considerably cheered up by the reference to
Newcomen's obtaining the assistance of mechanics from Birmingham.
In 1712 Birmingham was the only place apparently in which
Newcomen could get two or three good mechanics to help him to build
his engines. He might mention that in Birmingham, up to within quite
recently, there was an old beam-engine at Cliffords' Rolling Mill, in
Fazeley Street, with a cylinder of about 4 feet in diameter, and with
a 9 or 10-foot stroke. That engine had a wooden beam built up of
baulks of timber, and a wooden connecting-rod. It drove that rolling
mill for about ten hours a day for ninety-five years, and the cost of
repairs was almost nil. He had taken some diagrams from it and worked
out the coal consumption per day, and, as far as he could recollect—
it was some eight or nine years ago—it worked out at about 7 lbs.
of engine slack per indicated horse-power. The users of that engine
wanted to know whether it would pay them to take it out and put
a more modern one in, and he reckoned that, with the cost of taking
it out and putting in a new engine and new high-pressure boilers,
it would be simply change for sixpence. As they would not get
more than 10 per cent. reward for their labour, and as the saving in
coal would not pay this 10 per cent., there was no financial reason
for making the change; and the engine remained, and he believed
was there to this day, but he understood the old wooden beam had
been taken out and an iron one put in. It was most curious to hear
this engine at work; it creaked like an old cane-bottomed chair.
One would think it would all come to grief, but it had stood for close

upon 100 years and did its duty well. He did not think that they sufficiently realised the difficulties that Newcomen and Boulton and Watt experienced in the old days in making engines of the size that they did. When one considered these people making cylinders 66 inches in diameter and with a 10-foot stroke, in the days when there were no lathes to bore them with, it made one appreciate very much the perseverance with which they went to work.

He had once looked through the old correspondence between Boulton and Watt, and through Watt's old drawings, and had come across a letter from Boulton to Watt, when Boulton was making the Watt engines at Soho, and Watt was still working in Edinburgh. " My dear friend, congratulate me," he said ; " I have just succeeded in making a cylinder " (for some engine) " nowhere more than a quarter of an inch out of truth." Of course Boulton had no lathe to bore out such a cylinder. It was 4 feet in diameter, and had a 10-foot stroke. The cylinder was laid on its side in the yard, and a block of lead, about 2 or 3 cwt., was cast inside it ; to this was added about a bucketful of emery and oil. Then two gangs of men pulled this block backwards and forwards, by means of chains, until they had scoured the cylinder smooth. The cylinder was then turned round a little on the ground, and another portion was scoured smooth. When one considered the means which they had for this purpose, he thought it was really a matter of much congratulation that they had produced a cylinder 4 feet in diameter and 10-foot stroke " nowhere more than a quarter of an inch out of truth."

Mr. MARK ROBINSON, Member of Council, drew this conclusion from the Paper: that it was a great pity that ancient and historical engineering records were allowed to fall into ruin and dilapidation. In any other country in the world he felt sure the Government would think it its duty to secure such relics. He went down with Mr. Davey a few months ago to see the old engine at Bardsley, illustrated on Plate 22. It was lamentable to see its neglected state. When they came back, he wrote to the owner of the property on which it was standing and begged to be allowed to

3 c

(Mr. Mark Robinson.)

remove it, adding that his Company would be delighted to remove it, together with the stone column which carried the beam, and to re-erect it without any cost to the owners. He proposed to give an undertaking that the engine should be properly taken care of, and that, if at any future time his Company desired to give it up, it should be available for the purposes of a museum. It was probably the oldest steam-engine in the world, and enough of it was still left to form a most interesting relic, yet he could not even get an answer, so little interest was taken in preserving that old relic. If any of the Members who were present, who had it in their power to influence people who had the control of such relics, would endeavour to get more attention paid to them by museum authorities, and would persuade them to put in claims for such relics, he felt sure that a very useful result would have followed from the reading of Mr. Davey's delightful Paper.

While upon ancient curiosities, he wished to point out that the open-top cylinder had come much nearer to their own day than he fancied many people were aware, for about thirty-five years ago he travelled on a steamer, which used to run between Jersey and St. Malo, with oscillating cylinders open at the top; they had no cover of any kind, and one saw the pistons moving up and down—pressed down, of course, by atmospheric pressure alone.

The PRESIDENT said he would now call upon Mr. Davey to conclude the Meeting with any remarks he might wish to make. He had shown how slowly improvements travelled a century ago from one end of the country to the other. In those days there were no such societies as this Institution; and he did not think it would take very long now to find out such a difference in economy between the Boulton and Watt engine and that of Newcomen, and to disseminate the knowledge from one end of the country to the other.

Mr. DAVEY said that he did not know that he could say very much in reply to the questions which had been raised. He was sure they could not realise the difficulties of the situation at the time when Newcomen set to work to make a steam-engine. He might mention that the first cylinders were made of brass. At that time cylinders of iron could not be cast sufficiently good and true. He had seen a copy of the invoice of the engine erected for Mr. Wauchope in Midlothian only a few years after the first Newcomen engine. The cylinder was made of brass somewhere near London. It was then put on board ship and sent to Scotland. Engines were, at that time, designed by Newcomen, and by engineers such as Hornblower and others, who went into the neighbourhood where the engines were required to find the materials for building them. The pipes at first were made of lead, and the wrought-iron work was made at local blacksmiths' shops.

He had listened with very much interest to the account Mr. Pendred had read of the starting of Trevithick's "pole" engine.* His (Mr. Davey's) father was present on the occasion of the first starting of that engine and had given him very graphic descriptions of what took place. Captain Dick (Richard Trevithick) was there with a crowbar which he inserted under a prop beneath the crosshead to make the plunger start. Then he called out to a man to come and hit the end of the crowbar with a sledge-hammer.

Referring to the lip round the outlet in the bottom of cylinders, Figs. 23 and 24, Plate 30, it was interesting to notice that in some later engines the bottoms of the cylinders were dished and brought down to a cone for the water to tumble into. The steam-valve was moved from underneath the cylinder to the side so that the discharge of the injection water did not at all interfere with the steam inlet. It was quite clear that the Newcomen engine was not ignorantly made. All those little points received consideration, and it was not the want of knowing what should be done as much as knowing how to do it. Therein lay the great difficulty. There was very often as much invention in being able to utilise the materials for the purpose for which they were required, as in the design of the mechanism itself.

* "Life of Trevithick," Vol. 2, page 78.

Communications.

Mr. ARTHUR TITLEY wrote that he thought all members interested in early mechanical matters would be grateful to Mr. Davey for collecting under one head so many facts and illustrations relating to steam-engine history. Francis Trevithick, in the life of his celebrated father, had recorded, among many other such facts, the existence in 1830, near Camborne, of the remains of what was known as the old moor-stone boiler. This was a floor about 12 feet square of granite blocks which, in the recollection of people then living, had walls several feet in height; and members of a previous generation were known to have taken a contract to cut out the copper pipes, &c., from it. He also recorded that cotter bolts and screw bolts were used to hold such blocks together. The only available iron in Cornwall in the middle of the 18th century was bar, up to 2 inches square, and all weights were carried between the mines and sea-ports on the pack-saddle. This seemed to point to an intimate connection between granite and boiler-making, but in precisely what manner it was hard now to say. The engine illustrated by Pryce, Plate 16, was described as the one purchased and re-erected by Richard Trevithick, Senior, at Bullen Garden (Dolcoath) in 1775 at a cost of £2,040. This engine was of great interest, as it remained for years at Dolcoath in competition with a Boulton and Watt engine, and also with one built by Richard Trevithick, the younger, who ultimately added to it a cylinder cover and plunger air and feed-pumps, retaining the timber beam and "horses' heads." It then (in 1812) took steam from his earliest cylindrical (Cornish) boilers, so that this engine, probably one of the Newcomen's own construction, survived to take part in the first practical experiments with high pressures.

Hornblower's valve, as illustrated by Tredgold, although double seated, gave only a single passage for steam, Fig. 28 (page 703), the present forms being probably much later. Tredgold also illustrated a modification of Hornblower's valve which had a cylindrical packing instead of the upper seat, Fig. 29 (page 703).

Mr. John Greene, Mining Manager to the Lilleshall Company, Shropshire, informed the writer that up to ten years ago they had had an atmospheric engine at work, and it was only broken up for scrap about three years since. It was used at their "Lawn" Pumping Station as a "Man Engine." It had to be "handled" and the levers set in motion at each stroke. It had, of course, an open-top cylinder, and was used about once a week to raise and lower the men employed in repairs and examination of the pumping shaft, the rope used being of thick round hemp 3 inches in diameter. The pit was 400 yards deep.

FIG. 28.—*Hornblower's Valve.*   FIG. 29.—*Modification.*

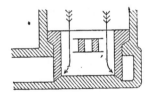
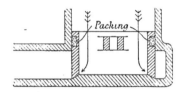

Professor W. CAWTHORNE UNWIN wrote that he thought Mr. Davey was a little hasty in calling the "pickle-pot" a "separate" condenser. It was not separate, being always open to the cylinder; and, as it increased the surface cooled during exhaust and reheated during admission, he would think it decreased rather than increased the efficiency. It looked merely like the arrangement of a draughtsman. There was a working drawing of a Newcomen engine in Smeaton's Reports.

Mr. DAVEY wrote in reply to the communications that he thought Professor Unwin had failed to realize the difference in the mode and conditions of working of the engine with the "pickle-pot" and without it. There was no doubt whatever that the "pickle-pot" had been added to reduce the fuel consumption of the engine. The engine without the "pickle-pot" might have water covering the bottom of the cylinder at the time of opening the steam-valve, and the incoming steam would come into contact with water having an exposed surface equal to the area of the

(Mr. Davey.)

cylinder. In the "pickle-pot" the water surface would be reduced to about one-fourth, and the total surface of metal exposed alternately to water and steam would also be reduced.

The great economy of the Watt condenser arose more *from the mode of removing the water and air from the condenser* than from the use of a separate vessel. The cooling effect of the condenser on the cylinder was nearly as much with the Watt as with the "pickle-pot." In the "pickle-pot," as in the Newcomen cylinder, the air and water were removed by direct contact of steam with the water, involving great condensation in the operation; but, in the Watt condenser, the work was done by an air-pump. The early Watt condensers were constantly open to one end of the cylinder, and in all double-acting engines the condenser was practically constantly open to one end or the other. No doubt the "pickle-pot" reduced the condensation due to the removal of the injection water. There would also be more cooling of the cylinder from direct contact of injection water than from contact of a mixture of steam and air at the same temperature. The economy might not have been very great.

Many letters had been received with reference to the existence of old engines, showing how great an interest was taken in the subject; but it had not been possible to make use of the information offered without unduly extending the limits, and going beyond the title, of the Paper.

The Institution of Mechanical Engineers.

PROCEEDINGS.

NOVEMBER 1903.

An ORDINARY GENERAL MEETING was held at the Institution [on Friday, 20th November 1903, at Eight o'clock p.m.; J. HARTLEY WICKSTEED, Esq., President, in the chair.

The Minutes of the previous Meeting were read and confirmed.

The PRESIDENT announced that the following three Transferences had been made by the Council since the last Meeting :—

Associate Members to Members.

BRADLEY, GODFREY THOMAS, . . . Colombo.
BROWNE, BENJAMIN CHAPMAN, JUN., . . Newcastle-on-Tyne.
HUGHES, GEORGE HENRY, ... London.

The PRESIDENT said that the Council had recently received from the American Society of Mechanical Engineers a most cordial invitation to hold a Joint Meeting of the two Societies in Chicago in the last week of May, 1904, the reason for selecting Chicago being that it was within convenient reach of the St. Louis Exhibition. At this Meeting the American Society desired to invite the members of the Institution " to be its guests during the days of such Joint Meeting, so far as participating in the business meetings, and in the social opportunities, and in the local excursions which will be planned as features of their Meeting." This invitation the Council had accepted; and he hoped that the response to the circular about to be issued to the Members would be of such a nature as to ensure the success of the Meeting, which was intended to fill the place of the usual annual Summer Meeting.

———————

The following Papers were read and discussed:—

"Roofing existing Shops while Work is proceeding"; by Mr. R. H. Fowler, *Member*, of Leeds.

"Experiments on the Efficiency of Centrifugal Pumps"; by Mr. Thomas E. Stanton, D.Sc., of Teddington.

———————

The Meeting terminated at Half-past Nine o'clock. The attendance was 150 Members and 80 Visitors.

ROOFING EXISTING SHOPS
WHILE WORK IS PROCEEDING.

By Mr. R. H. FOWLER, *Member*, of Leeds.

To many members of the Institution who are connected with old established works, a description of a method of re-roofing their shops without interfering with the work, may be of interest.

At the Steam Plough Works, Leeds, it became necessary to put roofs on the Boiler Shop and Black Store and Fettling Shop. Fig. 1, Plate 31, is a ground plan of the two shops, from which it will be seen that the Boiler Shop was divided into two bays, each 50 feet wide by about 270 feet long. The Fettling Shop was also 50 feet wide and about 280 feet long. Fig. 2 is a section through the shops showing the old roofs, and Fig. 3 is also a section through the shops, but showing the new roof. Fig. 4 is a combined plan showing the two roofs. In bay A of the boiler shop the usual bending rolls, punching and shearing machines, drills, etc., were fixed, driven from shafting attached to the beams of the roof. In bay B of the boiler shop a steam-crane travelled from end to end, and in the fettling shop an electric crane also travelled from end to end of the shop.

As practically the whole output of the Steam Plough Works depended on the supply of boilers, it was an absolute necessity that work in the boiler shop should be carried on without any stoppage.

It was therefore impossible to adopt any plan of raising the parts of the new roof from the floor inside the shop. Another important consideration was that the men working in the shops should be in absolute security, and feel that they were so, during the operations above. It was therefore decided to deal with one bay at a time and to construct a platform over the old roof for the full length of the shop, and to make this platform of sufficient strength to carry the tackle required in fixing the roof, and also the girders for carrying the roof, and to transfer this platform to the other two bays in succession.

The two shops are practically closed in on three sides, leaving only one end from which the work could be conducted. This available end is shown on Fig. 1, where is also shown the railway which brings material into the boiler shop, and is also connected with Messrs. Kitson's Works, and consequently had to be kept free while the operations connected with the roofs were carried on ; along this railway also the whole of the material for the roofs had to be brought. At the point C on Fig. 1, Plate 31, a steam derrick to lift 4 tons was fixed, this derrick being of sufficient height to place all the material for the roofs on to the platform, and with a jib of sufficient length to place the first girder D for carrying the platform in position. At the opposite end to this girder the first piece of platform was carried on an old brick wall of the shop marked E. When this girder was fixed, the first piece of the platform was erected, on which the cranes (to be described farther on) could be worked. Bay A of the boiler shop was the first one to be dealt with, and this necessitated a row of stanchions on each side, and as these stanchions had to be fixed in such positions as would not interfere with the existing walls and columns, the width of the shops had to be slightly altered. All the stanchions were made in two pieces for convenience of erection. To erect the stanchions, holes were made through the old roofs, and each stanchion was erected by means of a single pole, the power for lifting being obtained from a winding forward drum on a traction engine, and by this means all the stanchions were put up during meal times. When all the stanchions for one bay were in position, the next operation was to fix the crane and roof girders. To accomplish this a travelling crane, shown on Fig. 5, Plate 32, was constructed with an overhanging jib of sufficient length to carry the girders, 25 feet long, when suspended from the centre, from stanchion to stanchion. This travelling crane was mounted on a railway 8 feet 6 inches gauge, and of sufficient height above the platform to allow a narrow-gauge railway to run under it, on which the roof girders and crane girders were moved along on specially constructed lorries, and as the platform could not be constructed of sufficient width to allow this crane to be over the centres of the girders, it had to be made with a movable jib, as shown on the plan. As soon as a pair of girders, one on each side of the shop, were fixed, the trussed beams for carrying the platform were placed in position by means of a somewhat rough but very useful movable jib-crane, consisting simply of two baulks of timber mounted on some old traction wheels, with a small crab at the rear end and a fixed block at the end of the jib. This crane is illustrated by Fig. 6, Plate 33. The trussed beams were placed at a distance apart of 8 feet 4 inches centres, but this crane was made with such an overhang that the weight of the crane and the load it was lifting should be distributed over not less than two trussed beams. This crane ran on a wooden track in the centre of the platform. As soon as a trussed beam was fixed in position the space between it and the one behind it was covered with 3-inch planks all cut to dead lengths; the crane and track was then advanced and another trussed beam put in position and covered over as before. Three trussed beams were fixed on each girder, consequently when the platform from end to end was completed it was composed of about 30 trussed beams 8 feet 4 inches centres, carrying a platform or deck 3 inches thick. By this method two roof girders and two crane girders and two trussed beams and the platform were fixed in a working day.

In order to facilitate the making of the platform two 20-inch gauge railways were laid, one on each side of the centre crane. On these railways were carried all the planks and other material required. The roof principals were lifted on to the platform by the steam derrick, in two halves, each half was put on a trolley on the

20-inch gauge railway and conveyed to the far end where the halves were riveted together. When the platform was completed, the wooden crane was taken away and the crane used for lifting the girders was brought into the centre of the platform, and employed for lifting the roof principals and putting them in their proper positions. This crane was also used for lifting out the trussed beams as the work proceeded. The roof principals are placed at a distance of 12 feet 6 inches apart, consequently either one or two trussed beams had to be removed before each pair of principals could be fixed. The whole of the platform and all the principals for the roof of bay B were fixed in three weeks ordinary working time. As the principals of the new roof were erected, the planks that had formed the platform on the top of the trussed beams were then used for making a scaffold or platform inside each of the four divisions of each roof. On this platform was then deposited, as required, all the glass, slates, and timber, necessary for the new roof. The glazing, timbering and slating were then done with great rapidity, and by this means the whole of the last roof was glazed and slated in a fortnight.

Fig. 7, Plate 33, shows the travelling or swivelling crane with a girder suspended ready for dropping on to its stanchions.

Fig. 8 shows the same crane lifting a roof principal into its position.

Figs. 9 and 10 also show the same crane with a trussed beam that has just been lifted off the girders.

The stanchions for bay B of the boiler shop had to be erected between the existing longitudinal girders of the travelling crane and the wall of the shop, so as not to interfere with the old travelling-crane. It became necessary, therefore, to make these stanchions in such a way that the back leg K, Fig. 3, Plate 31, could be fixed after the old roof and wall were taken down. Details of the roofs are given on Plate 34. They are of the ridge and valley type, each roof divided into four ridges. They are glazed on each side with Heywood's patent glazing, with a slate ridge and ventilation for the whole length of each ridge. This type of roof is well adapted for shops where only top light can be obtained.

At the commencement of the Paper two points were mentioned that had to be considered, namely, the security of the workpeople and no interference with the output of the shop, and it is pleasing to be able to state that no accident occurred during the whole time the work was being carried out, and that the work of the shops went on without the slightest interruption.

After the erection of the swivelling crane the work on bay A was completed between the end of August and the end of October, and the two other bays were commenced the following spring and finished about the end of October.

The work was designed and carried out, and all the appliances designed and made by the staff at the Steam Plough Works, as it was quite impossible to employ any contractor on work of this nature.

The Paper is illustrated by Plates 31 to 34.

Discussion.

The PRESIDENT, in conveying the thanks of the Institution to the author, said that he thought it was a Paper which would be referred to many times by those who had to do similar work. The preparation that had been made for carrying out the work by a series of cranes had proved itself to be the right thing, and the author had given an excellent example of what a ridge-and-valley roof ought to be to cover travelling cranes. He noticed that it was not like a saw-tooth roof, which was generally made standing upon deep girders to support it across the shop, but that it was self-supporting, and that the crab of the travelling-cranes could go close up underneath the ridge and valley roof, or what was called in the Paper the roof principals, thus saving 4 or 5 feet of height to the building which was generally required to give head-room under the lattice girders supporting the roof. The Institution had the advantage of the presence of one of

(The President.)

its oldest members, Mr. Daniel, who perhaps might wish to add something to what had been said in the Paper, as he was there to represent the author.

Mr. WILLIAM DANIEL was afraid that the Paper would not be of much interest to a meeting composed chiefly of London members. It was prepared, as no doubt the members were aware, for the Summer Meeting held in Leeds, where it was thought many members would have the opportunity of seeing the work itself, and it might be of interest to them to know how the work had been carried out. It was a Paper that did not lend itself to much discussion; but if there was any point in connection with it that any member required an explanation, he would be very pleased to give it.

Communications.

Mr. W. H. MASSEY wrote that he was pleased to learn from the Paper that the work, which he had seen in hand some months previously, had been carried out successfully and without accident. Nothing was, however, said about the cost of the undertaking, but this was, he supposed, nothing to the convenience of being able to carry on the ordinary work of the shop without any interruption. He noted that all the "appliances" were made by the staff at the Steam Plough Works; but he presumed that the roof itself was made elsewhere?

Mr. LEWIS A. SMART wrote that he thought it might be of interest to the members to know that somewhat similar work had been carried out at St. Louis. The writer had been there last Spring, and at that time cement, though costing only about 70 cents per barrel, was selling at from 2 to $2\frac{1}{4}$ dollars delivered at the works carriage free. A new works called the St. Louis Portland Cement Works was just starting at the time of his visit. The plant and machinery were all ready, but through some hitch the buildings were not even started when the machinery was delivered.

With typical American energy they hastily constructed the finished cement store, and as the process, unlike the one in common use in this country, was a perfectly continuous one, the company erected the plant quite out in the open, started it, and were selling cement at the price already mentioned, while the actual buildings were being constructed over the machinery in motion. He thought he was correct in saying the works had an output of 5,000 barrels per day, or about 800 tons, which he believed compared favourably with some of the largest works in this country.

EXPERIMENTS ON THE EFFICIENCY OF CENTRIFUGAL PUMPS.

By Mr. THOMAS E. STANTON, D.Sc., of Teddington.

The determination of the actual Efficiency of Centrifugal Pumps has been the object of several careful investigations, the results of which have been published during the last thirty years, and it was to obtain if possible further information on the losses of efficiency in these pumps and the means of reducing them, that the following experiments were made.

The question possesses considerable interest for hydraulic engineers, because at the outset it may be somewhat difficult to account for the well-known fact, that although a centrifugal pump is merely a turbine reversed, and therefore, theoretically, should have the same efficiency as the turbine, yet in practice the efficiency of the pump is considerably below that of the turbine, a good turbine converting about 80 per cent. of the potential energy of the water into useful work, whereas the effective work in lifting water by centrifugal pumps rarely exceeds 55 per cent. of the work put into the shaft.

The explanation of this difference seems to be, that whereas the potential energy of the water in the turbine is converted into kinetic energy with very little loss, in the pump there is considerable waste of energy in converting the kinetic energy of the water leaving the

3 D

wheel into the potential or pressure form, even when the greatest care is taken to avoid shocks and sudden enlargements. The prejudicial effect of this loss on the efficiency is clear from the consideration that in a radial-vaned pump approximately half the work put into the shaft is in the form of the kinetic energy of the water leaving the wheel.

The methods adopted in pump design to make this loss as small as possible are :—(1.) To allow the water leaving the wheel to form a free spiral vortex in the pump casing, the pressure increasing radially outwards according to the known law, until the velocity has fallen to the value obtaining in the discharge pipe, as first suggested by Professor James Thomson. Although this method has been much used, there do not seem to be any published data of experiments on the efficiency of such a free vortex. From a study of the loss of energy in streams flowing in diverging channels the author was led to the conclusion that the efficiency of the vortex could not be a high one, and this was found to be the case.

(2.) To discharge the water leaving the wheel into guide passages of gradually increasing area until the velocity is sufficiently reduced in value. This is the method used by Professor Reynolds in the Mather-Reynolds pump. It is clearly not suitable for variable discharges unless the area of the guide passages can be regulated. No results of experiments on the efficiency of guide passages have been published as far as the author is aware.

(3) To recurve the vanes of the wheel at the outlet in a direction opposite to that of rotation, so that the velocity of the water leaving the wheel, which is the resultant of the velocity relative to the wheel and the velocity in common with the wheel, shall be comparatively small, and hence that any dissipation of its kinetic energy in shock will not effect the efficiency of the pump to an appreciable extent. This was the arrangement adopted by Mr. Appold in 1851 in his pump, the efficiency of which far exceeded that of the radial-vaned type, and the theory of which has been fully investigated by Professor W. Cawthorne Unwin.*

---

* Encyclopedia Britannica, Hydraulics.

It may be remarked that this reduction of resultant velocity can only be realised in moderately slow-speed pumps. In the case of high-speed pumps, it will be generally found that the velocity which the water has in common with the wheel at the outlet is so much greater than the velocity of the water relative to the wheel, that the recurving has very little effect in reducing the velocity of discharge from the wheel.

An objection to this method is that, by recurving the vanes of the wheel, the speed is increased at which the wheel must be run to give the same lift; and since the loss of work due to friction varies approximately as the cube of the speed, this will tend to diminish the gain due to diminished velocity of discharge. The superiority, however, of curved vanes over radial vanes, which is undoubted for the case of moderately slow-speed wheels, is not so clearly established for the case of high-speed wheels, the best practice on the Continent apparently being to revert to the radial-vaned type.

Previous Experimental Work.—Apart from Mr. Appold's experiments already referred to, the chief work in this country has been done by the Hon. R. C. Parsons, who has published the results of his experiments in the Proceedings of the Institution of Civil Engineers,* forming a most valuable addition to the literature of the subject.

In these experiments, which were made on wheels of 14 inches diameter with a vortex chamber, lifting water at heads varying from 6 feet to 18 feet at speeds varying from 165 to 430 revolutions per minute, it was found that the efficiency of the wheels with curved vanes was always higher than that of the wheels with radial vanes, and further that the efficiency depended on the form of the vortex chamber. To quote an example, with a discharge of 3,000 gallons per minute the efficiency of the wheel with curved vanes was 44·3 per cent., that of the wheel with radial vanes when discharging the same quantity being 37·9 per cent., and this although the speed in the latter case was 163 revolutions per minute as against 206 revolutions per minute for the curved wheel. With a specially

---

* Institution of Civil Engineers, Proceedings, 1876-77, vol. xlvii, page 267.

designed casing Mr. Parsons obtained efficiencies as high as 62·4 per cent. when lifting 1,753 gallons per minute to a height of 17·6 feet.

Mr. Parsons' experiments must be taken as proving conclusively that, for wheels of moderate speed, the curved vanes give better results than radial vanes when the wheels discharge into a vortex chamber. Our knowledge of the efficiency of high-speed pumps is chiefly derived from a Paper by M. Gerard Lavergne,* in which the results of a very complete series of trials on centrifugal pumps made from M. Schabaver's design are stated. M. Schabaver's object has been to construct a single-wheel pump which should be capable of raising water to heights which have only been previously attained by connecting several pumps in series (each pump discharging into the suction chamber of the one ahead of it), and which should also be capable of giving a high efficiency. These pumps have radial-vaned wheels discharging through a narrow, sharp-lipped slit at the centre of the wheel into a collecting chamber, and are fully described in the Paper referred to. According to M. Lavergne's curves of efficiency, the performance of these pumps is remarkably good, as, to take an example, he was able to pump 265 gallons a minute against a head of 164 feet with an efficiency of 50 per cent.

M. Lavergne's experiments show more clearly than has hitherto been done the fact that the efficiency of a centrifugal pump increases rapidly with the quantity of the discharge. This of course might have been predicted with a tolerable amount of certainty, since, for a given head and speed of wheel, the slip and friction will be nearly constant, so that an increase in the discharge will only affect the work supplied to the pump by an amount practically equal to the net work done, and so the efficiency will be improved. Another result of great interest brought out in M. Lavergne's experiments was that although the efficiency of the pump diminished with the lift as this became considerable, yet the maximum efficiency was not attained at the lowest lift. In the particular pump used the

---

* " Le Génie Civil," 21st April 1894.

maximum efficiency was reached when the lift was about 60 feet, a result which should modify the generally received opinion that centrifugal pumps are only efficient when working with lifts of under 20 feet.

Necessity of Further Investigation.—It will be seen from consideration of the previous experimental work that the questions which require further study are :—

1. The relative efficiency of curved vanes and radial vanes at high speeds.

2. The efficiency of the vortex chamber.

3. The efficiency of guide passages.

4. The possibility of high lifts by centrifugal pumps with a single wheel.

The Experimental Apparatus.—The experiments here described were carried out in the Hydraulic Laboratory of University College, Bristol, in August and September 1901. For the purpose of centrifugal pump and turbine trials this is equipped with two tanks each of 800 gallons capacity, the level of the upper tank being 48 feet above the laboratory floor and the lower tank 17 feet below it. There is a 2½-inch rising main connected to the pump for the purpose of filling the upper tank, and a 2½-inch falling main for working the turbine and the other hydraulic appliances. A branch pipe is taken from the rising main near the delivery side of the pump to the experimental tank in the laboratory, and by regulating the valve in this branch pipe the head in the rising main when pumping can be varied from 2 feet to 45 feet, the water passing over a 4-inch notch in the experimental tank, so that its amount can be calculated from the constants which were carefully determined by passing known quantities of water over it. In the high-lift experiments when the water was delivered into the upper tank, the discharge was estimated from the rise of water in the tank.

The pump casing, which is shown in section in Fig. 1 (page 720), consists of two parallel plates 1 inch apart, with a circular collecting chamber; the sides of the casing were 18 inches diameter, so that wheels from 4 inches to 18 inches diameter could be used in it. For the 7-inch wheels used in the experiments, the free vortex was formed in the space between the tips of the vanes and the collecting chamber. This was found quite sufficient for the purpose, as the pressure increased very little in the external part of this space. In cases where guide

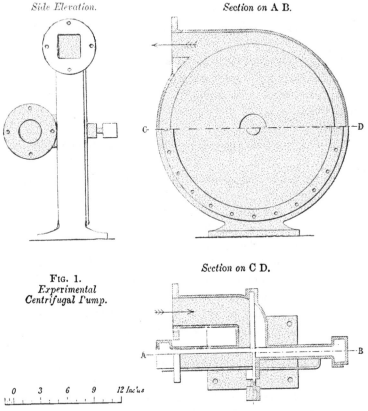

Side Elevation.

Section on A B.

C—   —D

Fᴵɢ. 1.
*Experimental
Centrifugal Pump.*

Section on C D.

A—   -B

0   3   6   9   *12 Inc'es*

passages were used, the guides were built up on a disc of thin sheet brass and bolted to the cover as shown in Fig. 2, Plate 35. It would, of course, have materially improved the performance of the pump if the inflow had been on both sides of the wheel, but this advantage was sacrificed to obtain greater simplicity in the arrangements for the trials, the object of the experiments being not so much to obtain a high efficiency as to make a comparison between various types of wheel and means of utilizing the kinetic energy of the discharge.

For driving the pump the wheel spindle was connected directly through a dynamometer to the shaft of a 4-H.P. motor, which was provided with a speed-regulating switch, the general arrangement being shown in Fig. 3, Plate 35. As the twisting moment on the shaft was in many cases quite small, *i.e.*, between 3 and 5 foot-pounds, some difficulty was experienced in finding a dynamometer sufficiently sensitive and accurate, and which would be preferable to any determination made by calibration of the motor.

The form of dynamometer finally adopted was of simple construction, but, in the author's opinion, susceptible of very considerable accuracy in cases when the twisting moment on a shaft

FIG. 4.—*Dynamometer.*

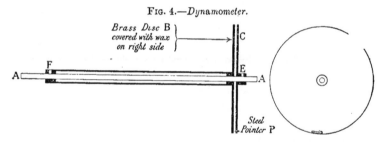

is practically constant, and therefore suitable for the experiments to be carried out. As the author has not seen this device used before, it may be interesting to describe it. In Fig. 4, AA is a steel shaft through which the power is transmitted. B and C are two brass discs secured to the shaft, C by a set pin at E, and B, together with the sleeve to which it is attached, by a set pin at F. The angular displacement of the two discs therefore measures directly the torsion of the shaft between the points E and F. To obtain a record of this displacement when the shaft is running, the inner face of the disc B is covered with a thin layer of paraffin wax. A small hole is drilled through C near the edge, and a sharp steel pointer P, at the end of a flat spring riveted to C, projects through the hole to within a small distance of B. A very slight pressure on the spring when the shaft is rotating is sufficient to make a clearly defined indentation on the wax. On stopping the motor the torsion can be measured by applying a known moment, until the pointer is in the same position as in the actual trial. The method is somewhat tedious, but the author considers that the indications were correct to 1 per cent., a result which would be difficult to obtain by any other means.

The pressures were taken by mercury gauges, all the readings being reduced to the centre line of the pump as a datum. By means of these gauges the pressures in the suction and discharge pipes, and at various points on the wheel casing and vortex chamber, were measured so that curves showing the variation of pressure in the wheel and in the vortex chamber could be plotted for purposes of comparison.

Results of the Trials.

(1.) *The relative efficiency of curved vanes and radial vanes in high-speed wheels.*—For this purpose the two small wheels shown in Fig. 2, Plate 35, and Figs. 5 and 6 (page 723), were made, the diameter of each being 7 inches, the width 1 inch, and the number of vanes twenty-two. A set of trials was carried out on each wheel at varying discharges and approximately constant speed. Tables 1 and 2 (page 724) give the results obtained in each case. The comparison between the two wheels is perhaps best seen in Fig. 7 (page 723), in which the curves of efficiency are plotted.

In these trials the water discharged from the wheel formed a free spiral vortex between the parallel plates of the casing, and then passed into the collecting chamber. In the Tables the horse-power of the water raised is calculated from the discharge, and the difference between the pressure on the delivery side and suction side of the pump, these pressures being reduced to the centre line as datum. The last two trials in Table 2 refer to trials made on a cast-iron wheel $6\frac{7}{8}$ inches diameter, with radial vanes twelve in number. As the efficiencies do not materially differ from those obtained when using the wheel with twenty-two vanes, it would seem that there is no particular advantage in increasing the number of the vanes beyond twelve.

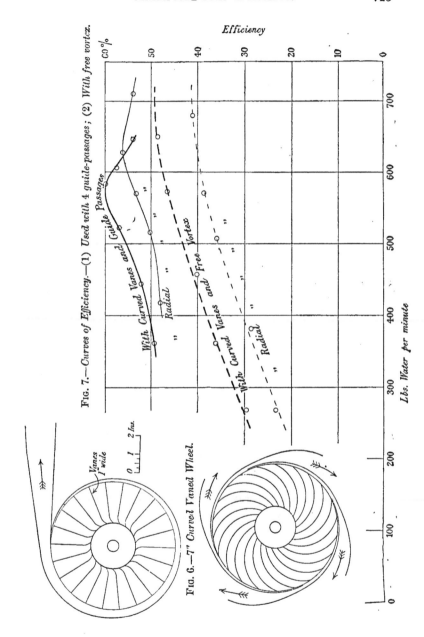

FIG. 7.—Curves of Efficiency.—(1) Used with 4 guide-passages; (2) With free vortex.

FIG. 6.—7" Curved Vaned Wheel.

TABLE 1.

7-inch Wheel with 22 curved vanes.

Revolutions per minute.	Discharge in lbs. per minute.	Moment on shaft in foot-lbs.	Pressure in discharge-pipes in feet of water.	Pressure in suction-pipes in feet of water.	Horse-power supplied to pump shaft.	Horse-power of the water raised.	Efficiency of the pump.
1358	656	4·64	+ 2·30	−27·10	1·20	0·579	48·3%
1341	574	4·19	+ 4·00	−24·50	1·07	0·496	46·4%
1343	457	3·96	+ 7·40	−21·75	1·01	0·404	40·0%
1343	362	3·48	+ 9·45	−19·80	0·89	0·321	36·1%
1343	268	3·17	+11·10	−18·45	0·81	0·240	29·6%
1345	0	2·49	+13·60	−15·90	—	—	—

TABLE 2.

7-inch Wheel with 22 radial vanes.

1295	680	6·27	+ 2·58	−27·90	1·54	0·628	40·8%
1300	572	5·52	+ 5·92	−24·50	1·36	0·527	38·7%
1311	508	5·44	+ 8·50	−22·95	1·35	0·485	35·9%
1330	382	4·86	+10·20	−20·00	1·23	0·350	28·5%
1336	268	4·31	+13·30	−18·30	1·09	0·256	23·5%
1340	0	3·46	+15·20	−15·84	—	—	—

6⅞-inch Wheel with 12 radial vanes.

1200	500	4·16	2·20	−22·65	0·95	0·377	39·7%
1175	451	3·96	2·10	−21·20	0·884	0·318	36·0%

From Tables 1 and 2 and the curves in Fig. 7 (page 723) it is seen that the efficiency of the curved vane is approximately 8 per cent. higher than that of the radial-vaned wheel at all discharges. It is difficult in this case to account for this increased efficiency by the reduction of the velocity of discharge of the water from the wheel, because this is practically the same in each wheel. Thus in Trial 1, Table 1, the velocity of discharge from the wheel is approximately 38·7 feet per second, the corresponding velocity in Trial 1, Table 2, being 39·6 feet per second.

The cause of the superiority of the curved vanes over the radial vanes would seem to be indicated in the last line in the two Tables, where are recorded the twisting moments on the shaft when the pump was full, but no discharge taking place. This is evidently the moment due to the slip and the friction, and is seen to be considerably less in the case of the curved vanes than in that of the radial vanes. Since the friction may naturally be supposed the same in both cases it follows that the slip or "internal circulation," as it may be termed, is much less in one case than in the other, and when the conditions of flow in the two wheels are carefully considered, this conclusion seems to be quite justifiable. For considering the path of the water in the curved vanes it is seen that the velocity is practically uniform, whereas in the radial vanes there is the case of a diverging channel in which the velocity continually diminishes. Now, if the phenomenon of the flow of water in a diverging channel be carefully studied by means of colour bands, it is observed that the streams have a strong tendency to flow back to the region of low pressure from which they have emerged, that is, there is an internal circulation or back flow set up which is greater the greater the divergence of the channel. A result of this is that there is a considerable loss of energy due to this circulation, and on the assumption that this takes place in the case of the wheel with radial vanes, it is evident that the work absorbed will be greater than when this internal circulation does not take place.

This, in the author's opinion, is the chief reason of the greater efficiency of the curved vanes, and leads at once to the conclusion that for maximum efficiency the vanes of the wheel should be so designed that the velocity of flow through the wheel is practically constant.

(2.) *The Efficiency of the Vortex Chamber.*

The suggestion of Professor James Thomson that the water leaving the wheel should be discharged into a chamber in which a free spiral vortex could be set up, thereby partially converting the kinetic energy of the water leaving the wheel into pressure head, has been generally adopted by designers of centrifugal pumps.

To test the efficiency of such an arrangement a wheel with radial vanes 11 inches diameter was fitted to the pump, and trials made in which the rise of pressure of the water after leaving the wheel was carefully observed, and the curve of actual pressure plotted. The values obtained are given in the following Table :—

TABLE 3.

Revolutions per minute.	Velocity of discharge from tips of vanes.	Pressures (in feet of water) at points distant from the axis.					Pressure in Collecting Chamber.	
		5·5″ (tips)	6 38″	7·33″	8·28″	9″		
1203	58·04	22·92	29·31	34·73	37·45	38·76	43·5	{ actual values
—	—	22·92	36·32	46·02	52·32	55·82	75·42	{ theoretical values

Below these for purposes of comparison are given the values on the assumption of no loss of any kind. The curves in Fig. 8 are plotted from these values, and show the loss of efficiency very clearly. In the above case the total rise of pressure from the tips of the vanes to the discharge pipe was 20·58 feet of water. If there had been no loss the rise of pressure would have been 52·5 feet, so that the efficiency of the vortex chamber is 39 per cent.

Experiments with a 7-inch wheel and a smaller velocity of discharge gave similar results, a set of four trials with this wheel giving values varying from 37 per cent. to 41·5 per cent. as the efficiency of the vortex chamber. The cause of this low efficiency is doubtless to be found in the instability of diverging streams of water, the energy being absorbed in the eddies set up in flow of this kind. These experiments refer of course to a vortex in which the flow takes place between parallel plates, which is not the most usual arrangement adopted in pumps, which are usually provided with a spiral casing of circular cross-section. It seems reasonable to suppose, however, that in this latter case the loss of energy is even greater than in the case of parallel plates, which by limiting the

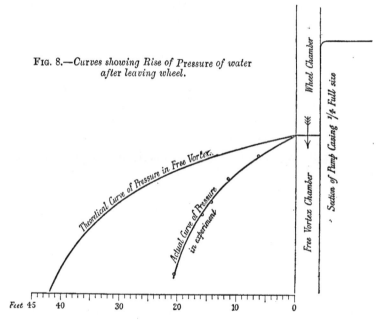

FIG. 8.—*Curves showing Rise of Pressure of water after leaving wheel.*

divergence of the stream to one plane, make for the greater stability of the motion.

(3.) *The Efficiency of Guide Passages.*

In pumps fitted with guide passages the water is discharged into a certain number of narrow channels distributed round the circumference of the wheel, these channels diverging outwards so that the velocity of the water is gradually reduced to such a value that its kinetic energy is small compared to the total lift.

A radial-vaned wheel fitted with guide passages may be worked as a pump or as an inward flow tui bine.

In order to obtain information as to the efficiency of such a channel in converting the kinetic energy of the water at the least section into the potential form, some preliminary experiments were made on a specially constructed rectangular channel, the diverging angle of which was about five degrees. By passing a known quantity of water through the channel and measuring the rise of pressure which took place above the value at the least section, it was possible, from a calculation of the velocities at entrance and exit, to compare this gain of pressure with the theoretical gain on the assumption of no loss in friction and eddying. This was done for various discharges with the result that the loss due to friction and eddying did not exceed 20 per cent. of the initial energy of the water, indicating that a pump fitted with such guide passages would probably be more efficient than one in which the water formed a free vortex, and on trial this was found to be the case. For the purpose of the experiments the same wheels were used as in the previous trials, the guide blades, four in number, being placed symmetrically round the wheel, the area of each channel being 0·18 square inch at its smallest section. Fig 2, Plate 35, shows the method of arrangement of the guide passages for a 12-inch wheel.

The following Tables give the results obtained, the curves of efficiency plotted from these values being shown in Fig. 7 (page 723). A somewhat remarkable feature of these trials is that the maximum efficiency is not obtained with the maximum discharge, the probable explanation of which is that the guide-blade angles, having been calculated for a given discharge, any discharge greater than this would lose part of its energy in shock at the entrance to the guide passages. Evidence of this was found in the fact that for the trials at maximum discharge the rise of pressure in the guide passages was considerably lower than that measured in the more efficient trials. It is clear from these results that, in designing pump wheels which discharge into guide passages, great care should be taken in the determination of the guide-blade angle and also of the area of the passages.

TABLE 4.

Wheel 7 inches diameter, radial vanes and guide passages.

Revolutions per minute.	Discharge in lbs. per minute.	Moment on shaft in foot-lbs.	Pressures in feet of water.		Horse Power.		Efficiency.
			Discharge.	Suction.	Supplied to shaft.	Of water raised.	
1312	710	4·83	+ 3·13	−26·80	1·20	0·645	53·8%
1336	629	4·69	+10·80	−24·20	1·19	0·667	56·0%
1342	564	4·58	+12·60	−23·50	1·17	0·618	53·0%
1347	517	4·44	+14·50	−21·90	1·16	0·570	50·2%
1358	418	3·94	+18·00	−19·70	1·02	0·476	46·7%
1362	0	2·10	+24·70	−15·30	−	−	−

TABLE 5.

Wheel 7 inches diameter, curved vanes and guide passages.

1313	647	4·07	+ 2 45	−25·20	1·01	0 542	53·7%
1343	607	3·94	+ 7·15	−23 95	1·00	0·572	57·2%
1370	577	3·86	+10·80	−23·40	1·00	0·598	59·8%
1363	523	3·71	+12·45	−21·90	0·962	0·545	56 7%
1380	447	3 62	+16·12	−20·40	0·950	0·495	52·1%
1383	362	3·16	+18·30	−19·20	0·832	0 414	49·5%
1385	0	2·05	+22 40	−16·50	−	−	−

The advantage of the use of guide passages over that of the free vortex chamber is thus seen to be considerable, amounting ˙to approximately an increase of 10 per cent. in the efficiency.

The‾reason for this increase in efficiency seems to be twofold, for from the value of the moment when the discharge was nil the slip and friction are shown to be less than in the previous trials, and further it was observed that the gain of pressure in the guide passages was greater than that obtained when discharging into a free vortex.

To: estimate the effect of the guide passages, a pressure-gauge hole was made at the narrowest part of the channel, and the pressure taken at each trial. The quantity of the discharge, the area of the channels and the initial and final pressures being known, it was possible to estimate the efficiency of the guide passages. These calculations are given in Table 6, in which the first column gives the velocity of discharge from the wheel calculated from the speed and the quantity of water pumped. The second column gives the velocity head $\frac{V^2}{2g}$ due to this, and the third column the measured pressure on entering the guide passage, this being the same as the pressure at the tips of the vanes.

Wheel 7 inches diameter discharging into guide passages.

TABLE 6.

RADIAL VANES.

Velocity of Discharge.	Velocity Head of Discharge.	Pressure of Water leaving Wheel.	Pressure of Water leaving Guides.	Gain of Pressure in Guides.	Efficiency of Guides.
41·1	26·3	− 1·70	+14·50	16·2	61·6%
40·8	26·0	+ 2·70	+18·00	15·3	58·8%
39·8	24·7	−13·46	+ 3·75	17·2	69·7%

TABLE 7.

CURVED VANES.

39·9	24·8	+ 3·27	+16·12	12·85	51·8%
39·4	24·3	+ 0·90	+12·45	11·55	47·4%

In these trials it will be noticed that the efficiency of the guide passages is much lower for the curved vanes than for the straight vanes, and this was found to be the case after repeated adjustment of the guide-blade angles. A possible explanation is that the velocity of the water relative to the vanes was greater than that calculated from the discharge, which would make the calculated velocity of discharge less than that given in the Tables, and consequently raise the calculated value of the efficiency. This assumption may be justified by the fact that in spite of the apparent inefficiency of the guide passages in the case of the curved vanes, these latter give a better result for the pump trials than the radial vanes.

Experiments with Wheel discharging into a single Guide Passage.— As the friction in the guide passages is considerable, it was deemed advisable to discover if a reduction in the number of guide passages would improve the efficiency. The arrangement adopted is shown in Fig. 5 (page 723), the area of the single passage being equal to the sum of the areas of the four passages in the previous trials. The results obtained are given in Table 8 (page 732).

In this case it was found that the single guide passage did not give such good results as the four passages, the difference being most marked in the case of the wheel with curved vanes.

It was also noticed that the single guide passage did not convert the kinetic energy of the discharged water into pressure head very efficiently, which doubtless accounts for the results not being so good. In fact there can be no doubt that, neglecting the losses due to friction, maximum efficiency would be obtained by using the largest possible number of guide passages.

(4.) *The Possibility of the Efficient Working of a Single-Wheel Pump for High Lifts.*—As previously mentioned, single-wheel pumps for high lifts have been made by M. Schabaver, which have an efficiency of over 50 per cent. when raising water 160 feet, and it is likely that this performance has been exceeded in the best Continental practice.

For the purpose of the experiments here described, and for lifts of over 50 feet, the wheel and guide blades shown in Fig. 2, Plate 35,

3 E

TABLE 8.

7-inch Wheel, curved vanes, discharging into single guide passage.

Revolutions per minute.	Discharge in lbs. per minute.	Moment on shaft in foot-lbs.	Pressure in discharge pipe.	Pressure in suction pipe.	Horse-power supplied to pump shaft.	Horse-power of the water raised.	Efficiency of the pump.
1331	583	3·63	+ 2·25	−23·30	0·92	0·451	49·0
1370	445	2·81	+ 7·80	−19·40	0·73	0·364	50 0

7-inch Wheel, radial vanes, discharging into a single guide passage.

1324	671	4·86	+ 3 47	−25 60	1·22	0·591	48·4%
1316	650	4·24	+ 2·30	−25·00	1·06	0·538	50·7%
1347	561	4·21	+11·60	−23·00	1·08	0·588	54·5%
1318	526	4·00	+11·00	−22·40	1·00	0·532	53·2%

TABLE 9.

12-inch Wheel, curved vanes, discharging into four guide passages.

1190	626	15·20	33·3	−27·6	3·44	1·15	33·5%
1226	501	14·40	39·9	−24·1	3 36	0·972	29·0%
1219	401	12 40	46·5	−20·2	2·84	0·811	28·6%
1230	411	11·83	46·3	−21·5	2·77	0·845	30·5%

12-inch Wheel, curved vanes, discharging into a free vortex.

1180	553	21·95	45·0	−24·8	4·92	1·17	23·8
1192	453	19·30	45·7	−22·1	4·36	0·932	21·4

were designed and made. The results obtained from them were quite satisfactory, but owing to the power of the motor not being great enough to lift larger quantities of water, the pump efficiencies actually obtained were small, the values being given in Table 9. From comparison with the previous trials when using the 7-inch wheels, it may fairly be assumed that the efficiency of this wheel when discharging 120 gallons a minute would be over 50 per cent., although for the above reasons this was unable to be verified.

The four guide passages used with this wheel were found to be very efficient, converting from 70 to 75 per cent. of the kinetic energy of the water leaving the wheel into pressure head at discharge.

Summary.—The general conclusions arrived at from the results of the experiments described in this Paper may now be stated.

1. In high-speed wheels, that is, wheels in which the velocity of the tips of the vanes exceeds 40 feet per second, the effect of moderately recurving the vanes at the outlet is beneficial (especially when the water is discharged into a free vortex), the curvature being such that the velocity of flow of water through the wheel is uniform.

2. Wheels which discharge the water into guide passages give a higher efficiency than those which discharge the water into a free vortex, this advantage being more marked in the case of wheels with radial vanes than in the other type.

3. The number of guide passages should be not less than four, and the areas at inlet should be such that the velocity of flow into the passages should be equal to the velocity of discharge from the wheel, to avoid losses from sudden changes of velocity.

As regards any modification of the existing design of centrifugal pumps, the author ventures to suggest that increased efficiency and considerable economy in material and space occupied would result in the adoption of a high-speed pump driven direct from the motor and designed on the principles laid down above. For dealing with very large quantities of water at a low lift, no doubt the present slow-speed pump driven direct from a steam-engine has great advantages, but for supplying moderate quantities of water at high—or moderately

3 E 2

high—lifts the high-speed motor-driven pump is very suitable. When the simplicity of construction and small cost of centrifugal pumps are considered, it seems remarkable that the use of them should be so limited. This is perhaps due to the impression which has till recent years prevailed, that these pumps were useless at other than quite low lifts. Now that their adaptability and satisfactory working at high lifts is being recognised, there seems every reason to suppose that they will come into use in many cases where reciprocating pumps have hitherto been solely employed.

The Paper is illustrated by Plate 35 and six Figs. in the letterpress.

Discussion.

The PRESIDENT moved that a hearty vote of thanks be given to Dr. Stanton for his interesting and suggestive Paper, and this was passed with applause.

Dr. STANTON said, that if he might venture to anticipate a little criticism, he would point out that the Paper was written two years ago, and since that time the practice in centrifugal pumps had gone ahead very much. He expected that his statement that the maximum efficiency of a centrifugal pump did not exceed 55 per cent. would be severely criticised, and he might explain that at the time of writing he was thinking of single-wheel pumps lifting 100 feet, and not, of course, of low-lift pumps.

With regard to the curves of efficiency, it was obvious that the values given were very low, but that was on account of the small quantity of water which was put through the pumps. No doubt, with an increase in the quantity of water, the efficiency would be considerably

higher.   As to the wheel and casing, every one would see they were
not designed for high efficiency in any way; they were made simply
for analysing the various losses of efficiency that took place in the
pump.

Mr. F. HOWARD LIVENS said that, in referring to these interesting
experiments, which illustrated a subject on which there was not too
much accurate information, it had struck him that it would be of
interest to the members if he gave the results of very similar
experiments upon pumps made for practical work on lifts up to
40 feet.   The pump made by Dr. Stanton was evidently designed in
the form shown for mechanical reasons, and in such a manner that
the guide blades could be easily added; but in itself it was an
exceedingly bad form, and the efficiencies obtained were low in
consequence.   Whilst therefore Dr. Stanton had been able to prove
by its aid that curved blades were better than straight blades, and
also that guide blades might considerably increase the efficiency of a
pump, it was quite questionable whether the latter results were of
real value.   That guide blades had improved Dr. Stanton's pump
was evident, but with a pump having the efficiency shown on some
of the accompanying diagrams, it was doubtful whether guide blades
could add an efficiency commensurate to their cost; and, for reasons
which would be given later on, it might happen that such blades
would be rather a disadvantage than otherwise.

Referring to the form of fan blades, the company, of which
the speaker was engineer, obtained orders some time ago for
large irrigation pumps, and as the average lift was known, in order
to settle the best form of pump, a 3-inch model (that is, with a
3-inch diameter delivery pipe) was constructed.   This had a fan
$7\frac{1}{2}$ inches diameter, and the width of the opening round the
periphery of the fan was $\frac{1}{2}$ inch.

Fig. 9 (page 736) showed these vanes, and Fig. 10 (page 737)
showed the efficiency curves obtained from them, the lift in all cases
being 29 feet; the pump was driven by a motor, and the current
used taken by a voltmeter and ammeter.   These curves were correct
from a comparative point of view, although incomplete, and the

(Mr. F. Howard Livens.)

actual percentages of efficiency might be slightly out. The quantities of water delivered, however, were absolutely correct, as in each experiment it was diverted by a swinging shoot into a tank fixed on scales, and the weight delivered therein was timed by a stop-watch. It would be of interest to note that this pump, the fan of which could be carried in the coat pocket, delivered up to about 240 gallons per minute.

Turning now to the fans themselves, No. 1, Fig. 9 (page 736) showed a fan with curved vanes well inclined backwards; No. 2 had vanes which finished radially, and No. 3 vanes only slightly curved backwards. For all general purposes No. 1 was undoubtedly the best, the efficiency curve of this and No. 2 crossing each other at

FIG. 9.—*Flyers for Model Centrifugal Pump.*

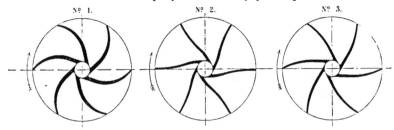

an output of about 140 gallons per minute. No. 2 fell much below in efficiency when the output was raised to 200 gallons per minute. No. 1 permitted of forcing, while No. 3, although giving the highest efficiency at 190 gallons per minute, when forced to 240 gallons showed a loss in efficiency, and therefore this type must be very carefully speeded so that forcing was prevented.

One of the pumps, actually constructed after the experiments, was shown in Fig. 11 (page 738), the fan being 6 feet diameter; at a lift of 27 feet 4 inches it was computed that the pump delivered about 135 tons per minute. The indicated horse-power of the Corliss compound engine, driving each pump by belt, was about 435 I.H.P., and the percentage of water horse-power to indicated horse-power 66 per cent. The pump was made with a cut-water, as shown on Fig. 11, brought close to the path of

FIG 10.

Tests of Model Pump.   Lift 29 feet constant.

Comparison between Flyers Nos. 1, 2 and 3.   Fig. 9.

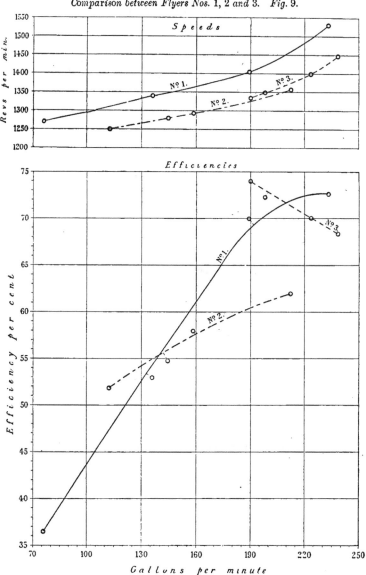

(Mr. F. Howard Livens.)

the fan, with as little clearance as possible ; in fact, a guide blade exactly similar to one of those shown on Dr. Stanton's illustration. When, however, the fan ran at full speed, a violent vibration was set up in the case, and this was attributed at first to air accumulating in the upper part ; an outlet was made for this air without improvement, and after careful consideration the speaker then came to the conclusion that the vibration was caused by the momentary breaking

FIG. 11.—*Diagram of 36-inch Irrigation Pump.*

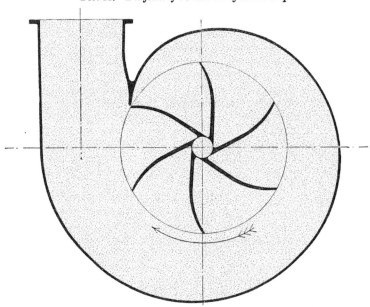

of the current as each vane passed the face of the guide. The cut-water therefore in the first place was drilled and a small notch made, with some slight improvement, and, finally, a large notch was made and the vibration permanently ceased. It might be fairly questioned therefore whether the addition of guide blades might not carry a practical disadvantage. It might also be noted that, in the experiments with the model, no appreciable loss in efficiency was noted when the cut-water was taken out altogether.

A curious point that resulted from alterations made in the attempt to discover the cause of the vibration was that the advancing edges of the fan blades in one pump were slightly rounded off, probably to about a $\frac{5}{8}$-inch radius; the second pump in the installation had the cut-water notched straight away without the

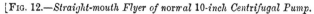

[FIG. 12.—*Straight-mouth Flyer of normal 10-inch Centrifugal Pump.*

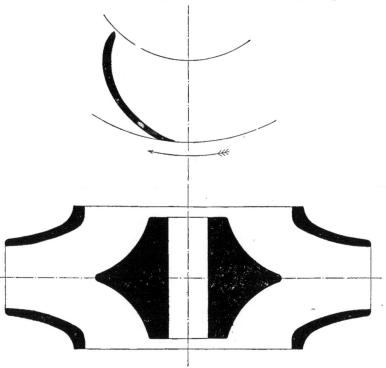

fan being touched, and consequent upon this rounding the normal speed of the first pump was increased by about two revolutions per minute over the pump which was not touched.

Fig. 12 showed a fan of the normal type with which experiments had been made during the last few months, and Fig. 13 (page 740) showed the results obtained from it. The pump was driven by a motor, of which the efficiency curves were first of all carefully determined,

(Mr. F. Howard Livens)

FIG 13.

Tests of 10-inch Centrifugal Pump, with Flyer, Fig 12

Forced — · — · — · — · -   Full Bore — — — — — —   Just Pumps ————

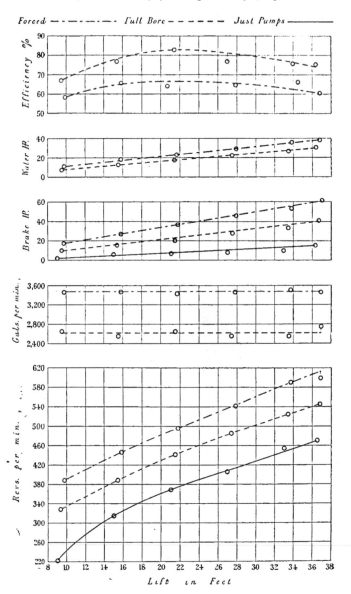

Lift in Feet

as shown on diagram Fig. 14 by the full line. A large tank 16·636 feet by 22·626 feet in size, holding about 11,640 gallons at a depth of 4·5 feet, was very carefully measured, and its capacity calculated, and the same checked over at each 3 inches of

FIG. 14.—*Efficiency Curve of 60 H.P. Motor. Used in Tests, Fig. 13.*

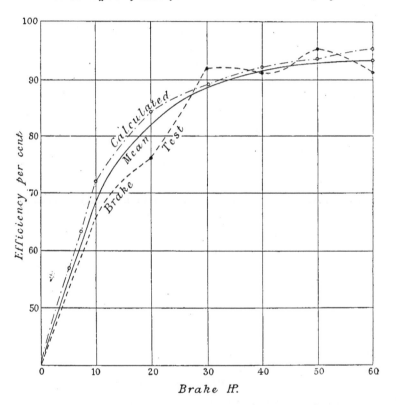

depth by a Kennedy positive-piston water-meter. The electrical instruments were tested as follows:—The voltmeters were checked by placing three different instruments in parallel and comparing with a recording Richard wattmeter, and three ammeters each of different make were placed in series and so checked, so that the readings might be considered as correct within 1 per cent.

(Mr. F. Howard Livens)

For each experiment a depth of from 12 to 15 inches was pumped out of the tank, the water-level being carefully determined before and after each experiment. The mean depth was considered as halfway between; the lift was calculated as the height from tho mean depth to the centre of delivery bend. Each experiment was of from one to two minutes' duration, and was repeated three times. As it was found that the time required to reach full delivery after closing the switch was as nearly as possible the same as the time the pump continued pumping after breaking the circuit, the actual time the current was on was considered as the pumping time.

The results on Fig. 13 (page 740) therefore could be looked upon as reliable. It will be seen that at a lift of about 22 feet an efficiency of 83 per cent. was obtained, and at 36-feet lift an efficiency ot 74 per cent. with an output of water of 2,600 gallons per minute; when the output was forced to 3,400 gallons per minute the efficiencies at these two lifts were about 67 and 64 per cent. respectively.

As regards the speed curves, these were as nearly as possible straight lines, showing that the generally accepted rule for the revolutions of a centrifugal pump—based on the square root of the height pumped—was departed from, the reason apparently being that the line showing the quantity of water delivered was almost horizontal; and consequently the speed of the water in the delivery pipe was more uniform than a formula based simply on the velocity due to a falling body could permit. The speed for the full bore or proper output at any lift from 9 to 37 feet, in this particular pump was :—

Revolutions per minute $= 260 + 7 \cdot 8$ H in feet, which for the above reason of course was different to the text-book formula; the output was 50 per cent. in excess (with a possible of 110 per cent.) of the usual normal output for this size of pump.

Referring to Fig. 15 (page 743), this was an interesting diagram of another pump of 8 inches diameter of delivery which emphasised this point; in this case the speed of the fan was kept as nearly as possible constant, whilst the height the water was lifted varied from 10 feet to 40 feet. It was instructive to notice that the greatest

FIG. 15.—*Test of 8-inch Centrifugal Pump. Constant Speed.*

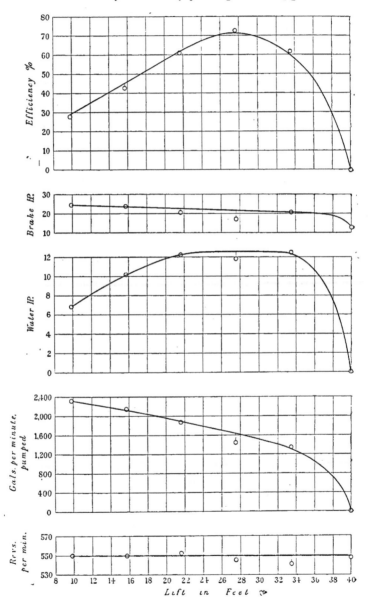

(Mr. F. Howard Livens)

efficiency was about 72 per cent. at a lift of $27\frac{1}{2}$ feet, and an output of 1,420 gallons, showing this to be the best speed for this pump at that height. The efficiency dropped to 28 per cent. at 10 feet with an output of 2,320 gallons per minute, whilst at 34 feet the output was only 1,350 gallons and the efficiency 61 per cent. At 40 feet the speed was only sufficient to keep the pipe full without delivering any water, and the efficiency therefore dropped to zero.

It was very commonly thought among users that a centrifugal pump delivering well at a given lift could accommodate itself to any reduced lift without reduction in speed, but the diagram showed clearly that there was of course a limit, because of the excessive forcing involved ; for a similar reason fans with the vanes inclined backwards were preferable for the pumps of commerce, as, on account of their greater " slip," they naturally allowed the wider range in speed without so serious effect on their efficiency.

In the ideal pump the fan should be designed in the most suitable form for a given output and lift, and the speed fixed accordingly, a beautiful mathematical formula being constructed afterwards to suit; but in practice with the majority of pumps the height at which a given pump had to work might continually vary with the tide or river-level, and was often an unknown quantity until after the pump had left the maker's hands. At times also the user would set it to run at any speed that happened to be convenient. The designer therefore must produce a pump which should be very accommodating, and yet keep up its efficiency whilst delivering the most advantageous quantity of water—a compromise difficult to attain.

Professor W. CAWTHORNE UNWIN had read the Paper with a great deal of interest, and had come to the conclusion—perhaps it was unnecessary to say it—that the experiments had been made with extreme care, that the results were extremely consistent, and that they did, if taken as results applicable to the particular pump experimented on, give a good deal of guidance. But he thought the author had not confined himself to that. He had gone on to state some conclusions in a very general form, and had reached conclusions which might be misleading.

As he was going to touch on two or three theoretical points, he would prepare the ground by saying, that in the "sixties" of the last century, he had built a very large number of centrifugal pumps, one of which was probably one of the largest centrifugal pumps ever built, partly to his own design and partly to the design of Professor James Thomson, a pump with a pump disc 16 feet in diameter and a whirlpool chamber 32 feet in diameter. He very much doubted whether, if the pump had to be designed at this moment, a better pump could be designed than that was. The author said that the whirlpool arrangement had been very generally adopted by centrifugal pump makers, but at that time they were the only people who used a whirlpool chamber, and they did so at the instance of Professor James Thomson ; and his (Professor Unwin's) experience was that it had been very seldom adopted since.

He desired to call attention to one or two points on which he thought there could be no sort of doubt, and first of all to draw attention to the result at the top of page 717, where the author said : " It may be remarked that this reduction of resultant velocity can only be realised in moderately slow-speed pumps." He had thought over that, and he had had some experience in designing pumps, and he could not see why the author had made that statement. He saw absolutely no difficulty in getting the same reduction of resultant velocity in pumps at any speed, unless perhaps in the case of pumps lifting hundreds of feet, which was not the case in the author's tests. That point was not an unimportant one, because a little later on in the Paper when the author was discussing the cause of the advantage due to recurving the vanes, he came to the conclusion that the only advantage of recurving the vanes of the pump was the reduction of slip and friction. Of course, if that statement was right, he could understand the author's conclusion ; but if the statement on page 717 had to go, he was afraid the other conclusion had to go also. The designs of centrifugal pumps made by the author were very largely based on experience with the inward-flow turbine, which was the exact converse of the centrifugal pump. In designing the inward-flow turbine, very careful estimates had to be made of the various losses which occurred in the turbine. There was the loss at the eye

(Professor W. Cawthorne Unwin.)

of the turbine, in change of direction, and certain losses due to shock, and there was the friction of the discs, residual energy, and so on. He did not suppose quantitatively those results could be got very right, but he thought relatively they did get them right; and one conclusion that came out of the investigation was that the designer was very definitely limited indeed as to the size of the pump which should be used for a given quantity of water at a given lift. It was that which he thought had been generally neglected by ordinary centrifugal pump makers, and was the reason why a very large number of centrifugal pumps were giving very poor efficiencies.

Reading the Paper, he thought one could say that the author began his investigation with a prejudice. It was quite true that it was more difficult to prevent any losses in diverging streams than in converging streams, and it surprised him the more that the author seemed to have designed his pump so as to have the maximum possible of divergence in the streams, and then he got a low efficiency, and no doubt properly explained the low efficiency by saying that it was due in great measure to the large divergence of the streams passing through the pump and the whirlpool chamber. The pump the author used he no doubt had reasons of convenience for adopting, and he, Professor Unwin, was only urging his statement to show where he thought the author's results must be used with caution. But he thought the pump erred as a centrifugal pump in being of the wrong type, of the wrong proportions, and of the wrong size. It was of the wrong type, because it received water only on the one side, and because the author had adopted what was a very bad form of centrifugal pump, the uncased centrifugal pump in which there was a great deal of slip. It was of the wrong proportions because it was too narrow for the size of its eye, and the diameter was too large. In the first place it should be noticed that the pump disc was of uniform width. A centrifugal pump should be coned, the reason being that by coning the disc the divergence of the streams passing through the pump was considerably decreased and the residual energy was reduced. The non-coning of the pump discs was one reason why the recurving of the vanes was not of much service in high-speed pumps. If it was desired to increase the backward relative velocity and reduce the absolute velocity of discharge, it was necessary to cone the pump. In the next place, the whirlpool chamber itself was of uniform width, which again made a very unnecessary amount of divergence of the streams flowing through the whirlpool chamber. But there was a worse fault about the arrangement of the whirlpool chamber in the author's pump. Round the whirlpool chamber, as he understood the drawings in the Paper, there was a discharge chamber of uniform section, and the result was that if the whirlpool was discharging uniformly over its circumference, there must have been a varying velocity all round that chamber.

FIG. 16.

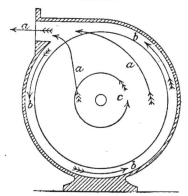

Therefore, there would be a difference of head between the whirlpool chamber and the supply chamber at different points which would tend to prevent the whirlpool chamber discharging uniformly, and would tend to prevent any proper rotation of the water in the whirlpool chamber. There was a still worse fault, that on one side there was that large discharge pipe.

It was obvious that a great deal of the water would find it very much easier to go off from the whirpool chamber straight to the discharge pipe, where the pressure was lower in the direction of the arrows a a, Fig. 16, instead of going round the supply chamber in the direction of the b b, and that would prevent the proper rotation of the water in the whirlpool chamber in the direction of the arrow c on which its efficiency depended. It was because of that very bad arrangement

3 F

(Professor W. Cawthorne Unwin.)

that he thought the author got a very much lower increase of pressure in the whirlpool chamber than he ought to have done. If he (Professor Unwin) was right at all in that, he believed some light was thrown on another point. In the model pump the guide blades very greatly improved the action of the pump; they improved it so much he believed, mainly because they prevented the cross flow of the water and forced the water in the whirlpool chamber to act a little more as it ought to have done. If he might make a suggestion, he would say he believed that the efficiency of that pump instead of being diminished would have been improved, if the aperture from the whirlpool chamber to the discharge chamber had been narrowed, forcing the water to go out through a comparatively narrow aperture, so as to force the whirlpool chamber to discharge pretty nearly uniformly over its circumference. Of course that involved a loss, but he thought it would have been more than replaced by the better action of the whirlpool chamber. In all the whirlpool chambers constructed in the " sixties," he first coned the wheel, and in the next place coned the whirlpool chamber as well, and in the third place, either had an expanding chamber round the whirlpool chamber, so that there was a uniform velocity in the discharge chamber, or a good deal contracted the discharge from the whirlpool chamber so as to force it to discharge uniformly all round its circumference. Those were two or three points of criticism of the Paper, and he hoped he was not doing more than use scientific criticism, because he was quite sure the author's experiments were made very carefully, and he thought they ought to be thankful to him for them, even if they could not apply them very widely.

With regard to the whirlpool chamber, it was quite true it was difficult to make the whirlpool chamber act quite as it ought to do. It was possible to try and get too much out of the whirlpool chamber, and therefore he preferred first to use recurving vanes so as to diminish the energy of the discharge from the wheel, and secondly use the whirlpool chamber, and allow it to take up some of the energy which was left after using the recurving vanes. He thought something might be got out of the whirlpool chamber even with recurved vanes as good as they could be made, especially when the volume of discharge varied. With an uncased pump, no doubt a very large loss would be due to slip past the blades of the wheel in revolving. Of course, if a cased pump was used there was no slip, but in place of it there was friction on the casing, but that was a calculable quantity, whereas the slip was an incalculable quantity. He preferred in designing machines to deal with quantities that could be calculated instead of quantities that could not, and he very much preferred the cased pump, and he thought experience showed that it was better. The author spoke of the efficiency increasing very fast with the quantity discharged. That was quite true up to a limit: but it should be quite clearly recognised that it was up to a limit; beyond a certain limit the efficiency would decrease again. It did so in some of the author's curves, and it would also do so in all if the experiments were pushed far enough. The author attributed that to the fact that the friction and slip were constant, while the useful work done as the pump increased was an increasing quantity. The losses due to the slip might be constant, but he was surprised it did not occur to the author that in a properly designed centrifugal pump, at any rate one considerable part of the loss was fluid friction, and the work expended in fluid friction increased as the cube of the quantity discharged.

Mr. MICHAEL LONGRIDGE, Member of Council, said he wished to add a few words to the explanations given by Professor Unwin, but before doing so he would ask Dr. Stanton to be good enough to add to the figures given in the Paper the angles at which the vanes cut the inside and outside circumferences of the pump wheel, the thickness of the vanes, the angles at which the guides in the discharge chamber cut its inner and outer circumferences, and the dimensions of the discharge chamber, if the wider ring round the outside of the pump might be so called. These dimensions would add much to the value of the Paper because, having them, it would be possible by using the formula given in the excellent Paper * by Professor Unwin, to calculate what the pump

---

* Proceedings, The Institution of Civil Engineers, vol. liii, 1877-78, page 249.

(Mr. Michael Longridge.)

ought to do under the various conditions imposed upon it by the author of the Paper, and to compare the calculated results with those actually obtained. In this way it would be possible to form a much sounder opinion upon the merits of curved vanes, guide blades, and whirlpool chambers, than by simply looking at the figures now given, without knowing how the conditions of working affected them.

To attempt to estimate the hydrodynamic efficiency of a centrifugal pump, without reference to such standards of comparison as Professor Unwin had laid down, was like trying to estimate the thermodynamic efficiency of a steam-engine, without reference to the possible thermodynamic efficiency of a theoretically perfect engine

FIG. 17.   FIG. 18.

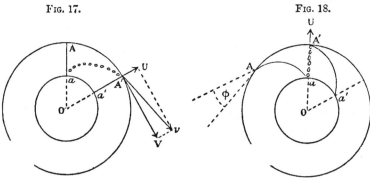

working under like conditions. The comparison in the case of the pump would enable them to ascertain how nearly the hydraulic efficiency approached the maximum possible under the imposed conditions of working, and to evaluate the amounts of the losses from eddies and skin friction, just as in the case of the steam-engine it enabled them to ascertain how nearly the heat expenditure approached the minimum possible under the conditions of temperature imposed, and the amounts of the losses from cylinder condensation and wire-drawing in ports and passages. Such a comparison of experimental results with theoretical standards would enormously increase the value of every Paper of this kind, and he thought that if it had been made in this case, a much clearer idea would have been gained as to why curved vanes had given higher efficiencies than the radial vanes in these experiments.

Let them consider for a moment the action of radial and curved vanes upon the water ; first of radial vanes. Let Fig. 17 be the centre of the pump-wheel, a A one of the vanes, and suppose that while any globule of water was passing through the wheel from a to A the vane moved from a A to a' A'. Then the globule would traverse the wheel by the path a A', and would issue from it at A' with a velocity v, which was the resultant of the radial velocity U ft. per second (equal to the discharge of the pump in cubic feet per second divided by the area of the opening round the circumference of the pump-wheel in square feet) and the linear velocity of the tips of the vanes V.

In the high-lift pump, in which V must necessarily be great compared to U, while the velocity v would approach V in magnitude and direction, and the kinetic energy in each pound of water leaving the pump-wheel would be $\frac{v^2}{2g}$ approximating in a high-lift pump to $\frac{V^2}{2g}$. Unless this energy could be utilised after the water had left the pump-wheel, it was shown in Professor Unwin's Paper, already referred to, that the hydrodynamic efficiency of the pump, which was $= \frac{V^2 - U^2}{2V^2}$, could not exceed 50 per cent.

Taking next the case of curved vanes, Fig. 18 (page 750), and again supposing the vane a A to have moved to a' A', while the globule of water moved through the wheel, the water would pass through the wheel by the path a A', and issue at A in a radial direction without any tangential velocity. Such a case of course could not occur in practice, but he had put it to them merely as an illustration to point the distinction between the action of the curved and radial vanes. By reducing the angle ϕ between the tangent to the tip of the vanes, and the tangent to the circumference of the wheel as much as structural conditions permitted, the tangential velocity of the water leaving the wheel could, however, be reduced to a comparatively small value.

In this case the hydrodynamic efficiency of the pump given by the expression $\frac{V + U \cosec \phi}{2V}$ might be made to approach unity, but the two facts,

(Mr. Michael Longridge.)

(1) that as ϕ diminished and U cosec ϕ approached V, the speed of the wheel had to be increased to obtain a given lift; and

(2) that the skin friction increased rapidly with the speed, militated against any great recurvature of vanes for high-lift pumps.

Now the low hydrodynamic efficiency of the wheel with radial vanes was due to the high tangential velocity of the water leaving the wheel. They did not require this velocity. What they wanted at the bottom of this discharge pipe was a velocity V_1, equal to the discharge in cubic feet per second divided by the area of the discharge pipe in square feet and a pressure P_1 lbs. per square foot, equal to the pressure of the head in the discharge pipe. The vortex chamber supplied the means of reducing the kinetic energy of the water, leaving the wheel from $\dfrac{v^2}{2g}$ to $\dfrac{V_1^2}{2g}$ and at the same time increasing the pressure from P, whatever that might be at the circumference of the wheel, to P_1, equal to the pressure at the bottom of the discharge pipe.

But to accomplish the transformation efficiently, the vortex chamber must be of proper size, and the motion of the water in it must be regular. The relations between PP_1, the pressures, and $v\ V_1$, the velocities, at the circumference of the wheel, distant R ft. from centre of the pump, and at the outer circumference of the vortex, distant R_1 ft. from the same point, were,

$$\frac{P}{62 \cdot 2} + \frac{v^2}{2g} = \frac{P_1}{62 \cdot 2} + \frac{V_1^2}{2g}, \text{and } v\,R_1 = V_1\,R.$$

By making provision for a regular vortex of radius R_1 and conducting the water from it through a properly designed spiral collecting chamber to the discharge pipe, the energy $\dfrac{v^2}{2g}$ could be utilized which otherwise would be largely wasted by shock and eddies in the pump casing. With such a vortex, the hydrodynamic efficiency of a pump having a wheel with the outer ends of the vanes radial might theoretically be increased from 50 per cent. almost to unity. When a vortex was provided for, it was probably best to place the spindle of the pump vertical, as the motion of the water in the vortex would be symmetrically affected by gravity.

Dr. STANTON, in reply, said Mr. Livens made a criticism which he had anticipated at the beginning, namely, that the pump was fitted with a bad form of wheel and was inefficient. He quite agreed with that, but would point out that it was necessary for the purpose with which he had set out to have a wheel of a simple kind, because the main thing was to find the relative efficiency of the various means of using the kinetic energy of the water leaving the wheel. The type was also necessary because he hoped at the time to use different-sized wheels, which he had done, and it would have been very difficult to do that in any other way. Mr. Livens had referred to the very high efficiencies obtained in his own pump trials, results of which were given in his remarks (page 735). Unfortunately these efficiencies seemed to be calculated from a combined efficiency, as there was no transmission dynamometer attached to the shaft ; that was, the work put into the pump was calculated from the known or supposed efficiency of the motor. He contended that in all serious work on the subject there ought to be a transmission dynamometer used. There were such things, and it was possible to get the exact value of the work put into the shaft.

With regard to Professor Unwin's criticisms, he was very glad to have them, and he agreed to a great extent with them. Professor Unwin was the greatest authority on centrifugal pumps, and he was afraid that he himself knew very little about them compared with the Professor. There were one or two points in which he ventured to disagree with him. One that Professor Unwin had referred to was the possibility of getting a low value in the velocity of discharge for high-speed pumps. He was still of opinion that it was very difficult to get this low value, not only in pumps lifting hundreds of feet, which Professor Unwin admitted, but in pumps raising small quantities of water 40 or 50 feet, which was the case in his experiments. In practice there was a limiting value of the discharge, consistent with moderate efficiency, which in the case of high-speed wheels prevented this low value being obtained. He referred to a very modern case ; the pumps which had been put in by Messrs. Sulzer Brothers at the Horcajo mines in Spain, which he thought were examples of the most modern kind of centrifugal

(Dr. Stanton.)

pumps, and in these the velocity of the tips of the vanes was 78 feet a second; and he calculated that the relative velocity of the water and vane at outlet was something like 11 feet a second, even with a very high discharge. The consequent reduction in the velocity of discharge due to the re-curving would not be anything considerable.

With reference to what Professor Unwin had termed the worst fault of the pump—that, owing to the circular form of the discharge chamber, and the discharge pipe being placed at one side of it, the water would go straight from the outside of the whirlpool chamber to this pipe, instead of circulating in the discharge chamber—he failed to see that this could possibly have affected the efficiency of the vortex chamber. The extreme difference of head at the various points of the discharge chamber owing to its circular form would not amount to more than 9 inches of water, a quantity considerably less than the difference of head at the top and bottom due to gravity. He did not believe that this would affect the proper rotation of the water in the vortex chamber, there being no indications of this in the curves of Fig. 8 (page 727), which showed a nearly constant loss of 50 per cent. of the kinetic head throughout the vortex chamber.

In reply to the questions asked by Mr. Longridge (page 749), the angles were as follows :—

  Inclination of vanes to tangent at outlet = $30°$
     „   „   „   inlet = $15°$
   „ guide blades to tangent at circumference of wheel = $8°$
  Thickness of vanes = $0·05$ inch.

In his remarks, Mr. Longridge had made a very complete statement of the theoretical conditions necessary to secure a high efficiency, finally arriving at the conclusion that a vortex chamber was necessary, and stating that by its means the efficiency could be made nearly equal to unity. It would be seen by referring to the Paper that it was mainly for the purpose of testing this assumption that the experiments had been undertaken, and he claimed that his experiments showed unmistakably that the vortex chamber was very inefficient, and that to discharge the water leaving the wheel into suitably diverging passages was a much better plan.

Communications.

Mr. W. H. ATHERTON wrote that the Paper was interesting, inasmuch as it clearly pointed out the main features which influenced the efficiency of centrifugal pumps, and gave some idea of their relative importance in the case of the small experimental pump employed. The author's main conclusions, however, though no doubt arrived at independently, were not new, but had been known for a number of years; and the suggestions put forward had been for some time embodied in the everyday practice of the more advanced makers of centrifugal pumps. Nevertheless one was glad to have a clear statement and independent experimental confirmation of the **principles** underlying modern practice in high-speed centrifugal **pump** design.

Mention had been made of the Mather-Reynolds pump, the pioneer of high-lift centrifugal pumps, which was brought out so long ago as 1875. Large numbers of these pumps had been made, many of them coupled direct to electric motors running at high speeds, and delivering against heads up to 150 feet, in the multiple **chamber** or series type.

A further development took place about five years ago, when Messrs. Sulzer Brothers on the continent, and Messrs. Mather and Platt in England, brought out their new high-lift centrifugal pump, the success of which was immediate and pronounced. Two pumps of this type were shown at the Glasgow Exhibition of 1911, where they attracted much attention. This design embodied all the features that made for high efficiency, as pointed out by Dr. Stanton, namely, diverging guide passages, curved vanes, and a large vortex or **whirlpool** chamber. The actual efficiency of such pumps ranged from 65 to 73 per cent. at the normal speed and delivery, according to the size of pump and the conditions of working. An interesting installation of two such pumps was that at the Newcastle-on-Tyne Corporation electric traction station, described by Mr. Charles

(Dr. Stanton)

pumps, and in t̶ ...locity of the tips of the vanes was 78 feet
a second; and ...d that the relative velocity of the water
and vane at ou ...ething like 11 feet a second, even with a
very high dis ...e consequent reduction in the velocity of
discharge due ...rving would not be anything considerable.

With ref... ...at Professor Unwin had termed the worst
fault of the ... owing to the circular form of the discharge
chamber, a... ...rge pipe being placed at one side of it, the
water wou... ...from the outside of the whirlpool chamber
to this p... ...circulating in the discharge chamber—he
failed to ...ould possibly have affected the efficiency of
the vorte ...he extreme difference of head at the various
points o ...chamber owing to its circular form would not
amount ... inches of water, a quantity considerably less
than th ...ead at the top and bottom due to gravity. He
did no... ...is would affect the proper rotation of the water
in th ...r, there bein... ...ns of this in the
curv... ...727), w... ...constant loss of
50 ... ...x chamber.

...ge 749),

th

Co

Mr. W. H. ATHERTON
inasmuch as it clearly pointe
the efficiency of centrifugal
relative importance in the
employed. The author's ma
arrived at independently, wei
number of years ; and the su
time embodied in the everyda
of centrifugal pumps. Ne
clear statement and independ
principles underlying modern
pump design.

Mention had been made
pioneer of high-lift centrifug
long ago as 1875. Large n
many of them coupled di
speeds, and delivering ag
chamber or series type.

A further de

the Paper was interesting,
in features which influenced
gave some idea of their
small experimental pump
however, though no doubt
t had been known for a
orward had been for some
e more advanced makers
was glad to have a
tal confirmation of the
high-speed centrifugal

olds pump, the
ught out so
made,
gh

(Dr Stanton.)

pumps, and in these the velocity of the tips of the vanes was 78 feet a second; and he calculated that the relative velocity of the water and vane at outlet was something like 11 feet a second, even with a very high discharge. The consequent reduction in the velocity of discharge due to the re-curving would not be anything considerable.

With reference to what Professor Unwin had termed the worst fault of the pump—that, owing to the circular form of the discharge chamber, and the discharge pipe being placed at one side of it, the water would go straight from the outside of the whirlpool chamber to this pipe, instead of circulating in the discharge chamber—he failed to see that this could possibly have affected the efficiency of the vortex chamber. The extreme difference of head at the various points of the discharge chamber owing to its circular form would not amount to more than 9 inches of water, a quantity considerably less than the difference of head at the top and bottom due to gravity. He did not believe that this would affect the proper rotation of the water in the vortex chamber, there being no indications of this in the curves of Fig. 8 (page 727), which showed a nearly constant loss of 50 per cent. of the kinetic head throughout the vortex chamber.

In reply to the questions asked by Mr. Longridge (page 749), the angles were as follows :—

Inclination of vanes to tangent at outlet = 30°
   „   „   „ inlet = 15°
   „ guide blades to tangent at circumference of wheel = 3°
Thickness of vanes = 0·05 inch.

In his remarks, Mr. Longridge had made a very complete statement of the theoretical conditions necessary to secure a high efficiency, finally arriving at the conclusion that a vortex chamber was necessary, and stating that by its means the efficiency could be made nearly equal to unity. It would be seen by referring to the Paper that it was mainly for the purpose of testing this assumption that the experiments had been undertaken, and he claimed that his experiments showed unmistakably that the vortex chamber was very inefficient, and that to discharge the water leaving the wheel into suitably diverging passages was a much better plan.

Communications.

Mr. W. H. ATHERTON wrote that the Paper was interesting, inasmuch as it clearly pointed out the main features which influenced the efficiency of centrifugal pumps, and gave some idea of their relative importance in the case of the small experimental pump employed. The author's main conclusions, however, though no doubt arrived at independently, were not new, but had been known for a number of years ; and the suggestions put forward had been for some time embodied in the everyday practice of the more advanced makers of ` centrifugal pumps. Nevertheless one was glad to have a clear statement and independent experimental confirmation of the principles underlying modern practice in high-speed centrifugal pump design.

Mention had been made of the Mather-Reynolds pump, the pioneer of high-lift centrifugal pumps, which was brought out so long ago as 1875. Large numbers of these pumps had been made, many of them coupled direct to electric motors running at high speeds, and delivering against heads up to 150 feet, in the multiple chamber or series type.

A further development took place about five years ago, when Messrs. Sulzer Brothers on the continent, and Messrs. Mather and Platt in England, brought out their new high-lift centrifugal pump, the success of which was immediate and pronounced. Two pumps of this type were shown at the Glasgow Exhibition of 1901, where they attracted much attention. This design embodied all the features that made for high efficiency, as pointed out by Dr. Stanton, namely, diverging guide passages, curved vanes, and a large vortex or whirlpool chamber. The actual efficiency of such pumps ranged from 65 to 73 per cent. at the normal speed and delivery, according to the size of pump and the conditions of working. An interesting installation of two such pumps was that at the Newcastle-on-Tyne Corporation electric traction station, described by Mr. Charles

(Mr. W. H Atherton.)

Hopkinson in a Paper * read before the Institution. A larger pump of the same type had since been installed. '

Another noteworthy application of a large high-lift centrifugal pump was that at the Sydney Water Works, New South Wales, described in " The Engineer " of 12th December, 1902. This pump was designed to deliver 2,800 gallons of water per minute against a head of 265 feet. It was not a series pump, but had a single revolving wheel or impeller, which was driven direct by a 4-pole continuous current motor at the high speed of 720 revolutions per minute, the power being about 340 B.H.P. During the last twelve months the multiple-chamber design of pump had been modified by Messrs. Mather and Platt, by casting the body in interchangeable sections instead of in one piece, and by omitting the vortex chambers, while still retaining the diverging guide passages and curved vanes. The good points of both the former types of high-speed centrifugal pumps had been combined, and an improved form of pump thereby produced. This simplification had not resulted in any loss of efficiency, a quadruple or four-wheel centrifugal pump of this type having recently shown under test a maximum efficiency of 75 per cent. when delivering 1000 gallons per minute against a head of 320 feet. In testing these high-speed pumps, it had been found that radial vanes were less efficient than curved vanes.

Thanks were certainly due to the author for a careful series of experiments to determine the conditions of maximum efficiency of high-speed centrifugal pumps; though at the same time one could not but regret that it was not found possible to employ much larger apparatus, and also to make comparative tests of a number of pumps of a less experimental type than that described.

Professor ALFRED CHATTERTON, of Madras, wrote that for some time past he had been employing centrifugal pumps, driven by small oil-engines for raising water from wells for irrigation, and his attention had been very forcibly drawn to the low efficiency of small centrifugal pumps, and the desirability of an improved design with

* Proceedings 1902, Part 3, page 437.

a higher efficiency. Dr. Stanton's Paper very opportunely brought forward the question, and clearly indicated the direction in which improvements could be expected. There was, he believed, a very large field of employment for small centrifugal pumps in India; and the prospects of successful water-lifting by mechanical means would be greatly enhanced if it was possible to obtain a 3-inch or 4-inch pump, with an efficiency of 70 per cent., instead of from 35 to 45 per cent. which was about the range of efficiency at the present time.

In the Madras Presidency water for irrigation was drawn from more than 600,000 wells, and probably in the whole of India the number of wells sunk for this purpose was about three million. The native appliances for lifting water were very effective, and had hitherto been sufficient for the needs of the country, but the time had now arrived when a much greater draught on the stores of underground water was desirable, and cheaper and more effective methods of lifting water were necessary. Where the continuous inflow to a well was anything like 100 gallons per minute, it would pay to use a small oil-engine and a centrifugal pump, working the pump about 16 hours a day in two shifts of 8 hours each, with an interval of 4 hours between them, during which time the inflow might accumulate in the well. If the well was 20 feet in diameter and 40 feet deep, the surface level of the water would fluctuate about 12 feet, and the lift at the commencement of pumping would be 27 feet, and at the end nearly 40 feet. An oil-engine developing from 3 to $3\frac{1}{2}$ B.H.P. should be sufficient; and what was wanted was a pump which would lift the quantity of water mentioned with this amount of power. Dr. Stanton obtained (Table 5, page 729) an efficiency of 59·8 per cent. with a small wheel delivering only 57·7 gallons per minute on a lift 34·2 feet. It should therefore be possible to obtain centrifugal pumps of three times this capacity with an efficiency of possibly as much as 70 per cent.

A scheme was under consideration in Southern India for irrigating a considerable tract of land by water drawn from a number of wells by centrifugal pumps by alternate-current motors. The electric

(Prof. Alfred Chatterton.)

current would be supplied from a central station to 40 wells scattered over a tract of country 10 miles long, and from 1 to 2 miles wide. The engines and alternators would be kept running continuously when water was wanted for irrigation, and the wells would be practically underground reservoirs into which the water would percolate and which would be emptied at intervals. It was calculated that the losses in distributing power were more than compensated by the economy of working in a central station, but it was felt that the success of the scheme mainly depended upon obtaining efficient alternators and pumps. At least 50 per cent. efficiency in the motor and pumps was necessary, and the prospects would be more satisfactory if a combined efficiency of something like 60 per cent. could be obtained, divided as follows :—alternator, 85 per cent. ; pump, 70 per cent. The pumps would have to deliver from 200 to 250 gallons per minute on a lift from 25 to 35 feet, and would require about 4-inch suction pipes. The centrifugal pump was at present a comparatively simple, cheap, and inefficient machine, and in the total outlay on a pumping project the cost of the pumps was a small item, yet the efficiency of the whole system depended upon them ; and it would be advantageous to pay more for centrifugals, if, by making them more complex, they could have their efficiency materially increased. For irrigation in India large centrifugal pumps were not likely to be much used, but for a small and efficient centrifugal pump, ranging from 3 inches to 6 inches diameter of suction pipe, there were prospects of a good demand, and it was to be hoped that the experiments of Dr. Stanton would serve as a starting point for fresh advance.

Mr. ISAAC SMITH wrote that, from numerous experiments he had made in turbines, he ventured to state that practical test confirmed what the author stated with regard to the difference in efficiency between radial and curved turbines. The difference, in the writer's opinion, would have been still wider if, instead of using the type of turbine shown on the left-hand side of Plate 35, he had employed one incorporating the principle of curved channels as illustrated and described by the writer in the discussion on Mr. Schönheyder's

Paper on " Water Meters." *   The Smith turbine therein described possessed great power, and this might be attributed mainly to the way in which the curves were described and mathematically calculated. It was now used as a turbine for measuring water, but it would make a capital pump by reversing the passages in the drum. Both curved and radial turbines would continue to be used for the special work for which they were best adapted, but data appeared to be wanted for calculating the friction of water on the walls of the passages. Dr. Stanton had given results of certain experiments which would be of service to those who were interested in this department of Mechanics, and he desired to thank the author for the Paper.

The radial vane turbine was now used for the purpose of producing sound, and its principle applied in the construction of the Smith syren. He was not in favour of applying the principle in the construction of pumps, as the slip was too great; the principle of curves was much better, owing to the great energy derived. He himself had now adopted it in the turbine of the Smith water meter. He tried a 6-inch diameter drum on this curved principle under a steam pressure of 60 lbs., and it was surprising the amount of power that was given out. A circular saw was first connected to it, and it quickly cut its way through a $1\frac{1}{2}$-inch gate ; and in various ways he had tested it in dressing brass castings, etc.

Mr. JAMES A. SMITH, of Melbourne, wrote that he could endorse the view that transmission dynamometer based upon the tension of a relatively slender shaft—a principle first brought into prominence by Hirn in 1867—had many points of excellence. It would be found that readings might be made during the course of an experiment, if a graduated arc and pointer were substituted for the two discs illustrated in Fig. 4 (page 721), and their relative positions noted by the use of a system of interrupted illumination. This was effected by the rotation, with the shaft, of a radially slotted disc, running in proximity to a correspondingly slotted fixed plate screening a source of light ; the slots should not exceed $\frac{1}{30}$ inch in width.

---

* Proceedings 1900, Part I., pages 65-68.

(Mr James A. Smith)

The general effect was that the arc and pointer appeared constantly illuminated, and at rest completely so far as axial rotation was concerned ; but use in a darkened room was necessary.

Relation of Turbine and Pump.—The view that " a centrifugal pump is merely a turbine reversed " (page 715), although usually accepted, was deserving of close scrutiny, since it involved fundamental assumptions which, if incorrect, must lead to false inferences when applied to actual design. There might be differences of kind as well as of degree. In making the comparison it was implied that an ideal fluid medium was used, possessing, amongst other properties, an infinite resistance to tendency to rupture of continuity under tension. It might be shown that in the pump there were forces in action tending to produce the tensional state, therefore the analogy should be applied with caution pending the acquirement of further experimental data.

Method of Visual Examination.—To clear away some of the complexity arising from the consideration of the problem on a basis of partial data, and to arrive at an approach to basic causes from a consideration of new facts, or old facts presented graphically and in new lights, an apparatus was designed by the writer * to enable the interior of a pump to be observed as though rotary actions were absent. He now proposed to give some of the results of the application of the method to deductions in Dr. Stanton's Paper. Figs. 19 to 21 were diagrams showing graphically some of the derived results. Fig. 20 was produced thus :—A card templet $\frac{1}{10}$ inch thick was set out (to Fig. 5, page 723, on a six-sevenths scale, a reduction required by the size of the apparatus), enclosed between glass discs rotating at a predetermined speed, and supplied centrally with a coloured fluid ; the whole constituted an analogy to a 7-inch " Radial-Vaned

---

* "Notes on Some Experimental Researches upon the Internal Flow in Centrifugal Pumps and Allied Machines," a Paper read before the Victorian Institute of Engineers, 4th June 1902, reprinted in "Engineering," 5th December 1902.

Wheel." By suitable means observations were made as though motion of rotation had been non-existent, and the effects measured and plotted as shown. Fig. 21 similarly represented the 7-inch " Curved-Vane Wheel." Fig. 19 will be dealt with hereafter.

Velocity of Flow through Runner.—" The explanation of this difference [efficiency in pump *v.* turbine] seems to be . . . in the pump there is considerable waste of energy in converting the kinetic

FIG 19.

The original photo was obtained by passing the streams under pressure through a stationary " runner " *a.*

Velocity is inversely as widths of dark streams

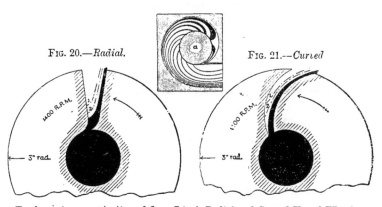

FIG. 20.—*Radial.*   FIG. 21.--*Curved*

Tendency to concentration of flow, 7-inch Radial and Curved Vaned Wheels

energy of the water leaving the wheel into potential pressure form " (page 715). Although this was supported by the curves in Fig. 8 (page 727), showing, in that particular experiment, heavy loss in the casing through disturbed vortex action, yet it was an explanation in part only, since those effects were subsequent to, and partially dependent upon, pre-existing action in the runners. These latter might, in pumps of other proportions, either reduce or accentuate the losses referred to.

This led to the consideration of:—" . . . the conclusion that for maximum efficiency the vanes of the wheel should be so designed

(Mr. James A. Smith.)

that velocity of flow through the wheel is practically constant" (page 725).

That a parallel passage might be superior to, for instance, the radial or curved channels described, was a fair inference from the experimental results, but it did not follow that it was the perfection of type.

Premises were involved :—

(a) That the full velocity of progression through the runner—not velocity in space—must either exist prior to entering the runner, or must be instantaneously imparted.

(b) That, given a parallel passage, of whatever contour, and whatever the speed of rotation, it could be assumed that the ·fluid would at all times fill the section as a sufficiently effective flow.

Unless these premises were proved, the condition specified (page 725) was arbitrary, and might be incorrect.

The experimental results, Fig. 7 and Tables, demonstrated an increase of efficiency following a decrease of divergence in the channels, but did not indicate necessarily that parallelism was the limit of desirable convergence.

Figs. 20 and 21 (page 761) related to this phase, each represented a single passage through the runner. At relatively low speeds the passages ran full, but as the velocity increased the edge of the flow, not constrained by the wall of the "driving" vane, overcame viscosity and gradually leasing the "leading" non-controlling vane, passed through the intermediate states, 1, 2, etc., and when the speed reached and became stationary at 1,400 revolutions per minute, the flow assumed the sharply defined stable curves denoted by the full black shading.

The deductions were :—

(a) That the velocity of flow in relation to the channel was an increasing quantity, therefore that the channel should be, not parallel, but should, to conform to the flow, correspondingly diminish in section to that point (which might be at the periphery) where it was desired that velocity-pressure conversion should begin.

(b) That as a consequence, the form of the second, or "leading" vane must, to avoid inert or detrimental spaces, be complementary to the driving vane suited to the working conditions.

As a corollary, it followed, that under varying conditions of lift or discharge, with non-adjustable vanes, even if otherwise theoretically perfect conditions existed, the highest efficiency could not be constantly attained.

Internal Circulation.—The tendency to slip and eddy in the vanes (page 725) could not be truly inferred from observations of the flow in stationary channels. Rotation of the passage introduced effects of quite another order, but these were by suitable apparatus rendered visually evident, if one or both—as in the preceding instances, Figs. 5 and 6 (page 723), 20 and 21 (page 761)—of the walls of a passage were out of accord with the working conditions.

Quantitative analysis of component parts of an integral stream was carried out by injecting thin filaments of coloured fluid into a general clear flow, moving in a given passage rotating as in a pump, and observed as before.

Cavitation was pronounced when the wall curves were unsuitable. As the speed rose a state of instability—confined to that part of the passage ultimately found to be in excess—arose ; small, irregular, transient cavities formed, and eventually, as already described, the stream concentrated, leaving, when the discharge was into air,* a space void of fluid, but potentially an eddy-forming, energy-consuming area when the runner was in actual use in a pump.

Number of Vanes.—The preceding section appeared to point to the advisability of securing more direct transfer of energy by the impulse of solid to fluid, and simultaneously of decreasing disturbances due to considerable paths of transmission transversely across the fluid, by augmenting the number of the vanes and thus reducing the widths of the spaces. As before, skin-frictional effects

---

* The condition throughout these tests.

3 g

(Mr. James A. Smith.)

(mentioned in one aspect, page 717) would ultimately impose a limit to number. This would appear to negative the application, as a general principle, of the deduction (page 722) that there was " no particular advantage in increasing the number of the vanes beyond twelve."

As a corollary, it was apparent that in experimental work relating to comparison of form, the number of vanes should be—as in the Paper under discussion—equal in all cases, or due weight should be attached to the differences or relation of frictional surface and section of flow.

Guide Blades.—It is usual to consider the problem from the point of view of associated velocities and pressures in a frictionless fluid medium. This section was limited to an endeavour to show that some of the results obtained by the methods described (page 731) discounted the value of that treatment, and were in consonance with those which must ensue when streams of water, originally moving as diverging spirals in space, were collected into common channels.

Those streams nearest the discharge were least constrained; those furthest therefrom suffered the most deflection and resistance, therefore retardation. Hence an unequal velocity of the flow would arise in passages in the runners, differently situated as regards the discharge opening; and such differences would constantly vary, as the vanes changed their angular position. This necessarily implied departure from the condition of greatest efficiency—a secondary result which the writer believed had hitherto escaped notice.

Fig. 19 (page 761) * illustrated analogous action, but in that case the engraving was from a photograph taken with the "runner" (a) at rest, and the clear and coloured streams passing, under pressure, between glass plates as in stream-line methods. Velocities were inversely as the widths of the dark streams. The great retardation of the flow in the streams from the channels at the greatest distances from the outlet was very marked.

---

* Reproduced from the " Notes..." previously referred to.

It would be noted that the position of the test-gauges might have a considerable effect upon the curve of apparent vortex efficiency in Fig. 8 (page 727).

By the application of guide blades, eddy was localised; solid non-eddy-producing skin-friction was substituted for the unstable action of fluid-stream upon fluid-stream, and relieved of these disturbing factors, the channels could be better designed to satisfy the purely velocity-pressure requirements ; further, the conditions of flow in the runner could be better equilibrated.

The general gain, under the experimental conditions, was very clearly brought out in Fig. 7 (page 723) and the Tables. Although four blades were mentioned (page 728 and Fig. 6), the results did not disclose that the efficient limit of number had been reached, a matter fully conceded (page 731).

In concluding, the writer would express his appreciation of the numerical and graphic data made available for comparison, data of a type sadly lacking hitherto, and when—as in the Paper under discussion—accompanied by a clear explanation enabling the methods of derivation to be considered, they should prove of very considerable value.

Dr. STANTON wrote, in reply to the communications, that he was extremely glad to have Mr. Atherton's confirmation of the conditions of maximum efficiency laid down in the Paper. The conditions themselves, as Mr. Atherton had pointed out, were not new, but that they were the conditions of maximum efficiency was not well known, as had been apparent in the discussion on the Paper. It was gratifying to learn that in pumps of recent design the vortex chamber had been done away with altogether, and the efficiency improved, as this was in exact accordance with the results of his experiments.

With reference to Mr. James A. Smith's communication, while expressing his admiration of the beauty of Mr. Smith's experimental results and the ingenious method of obtaining them, he ventured to doubt if they threw any considerable light on the actual nature of the flow on a centrifugal pump wheel. It must be remembered that

3 G 2

(Dr. Stanton.)

the curves in the diagrams represented the flow of water between plates $\frac{1}{70}$ of an inch apart—that is, the flow was dominated by the viscosity of the water. It was indeed probable that the radial-vaned wheel did not run full, as Mr. Smith's experiment indicated, but experiments on diverging channels showed that this depended on the quantity of the flow.

Institution of Mechanical Engineers.

PROCEEDINGS.

December 1903.

An Ordinary General Meeting was held at the Institution on Friday, 18th December 1903, at Eight o'clock p.m.; J. Hartley Wicksteed, Esq., President, in the chair.

Before proceeding to business, the President said that, since they had last met, their oldest and most staunch Member, Sir Frederick Bramwell, had been gathered to his fathers. He had been a Member of this Institution close upon fifty years, and he contributed a Paper to the Institution three years before he joined it as a Member. Nearly thirty years ago he was their President, but, since occupying the Chair, he had never abated in the work that he had done on Committees and at the Council Meetings, and in the interest he had taken in the Institution. Although his form had been familiar to most of them, he had not been able, of late years, to attend as many of the General Meetings as he would have liked to do, owing to his connection, as Honorary Secretary, with the Royal Institution, which held its meetings on the same evenings as their own. This Institution was, however, his first love and, although since he joined it he had become the valued Member of all the principal Engineering and Scientific Societies in London, including the Fellowship of the Royal Society, yet, as the Council knew, even better than the Members, he had never slackened in interest for the welfare of this Institution, and had attended, within the last few months, both General Purposes Committees, and prolonged Council Meetings.

(The President.)

He contributed to this Institution a Paper in 1851 on a Vacuum Gauge; in 1867 on Floating Docks; in 1872 on Marine Engines and Fuel Economy. In 1874 he delivered a Presidential Address, and in that he said that " the talent of the Mechanical Engineer may be likened to the trunk of the elephant, which, as we all know, has been said to be competent to root up a tree, or to pick up a pin." In 1892 he gave a Paper on the Sewage Outfall Works at Portsmouth, of which he was the Engineer; and, so lately as 1899, he read a Paper on Atmospheric Railways.

His description of the talent of the Mechanical Engineer well described his own. He had a great memory and a powerful mind to grapple with large problems, and his brain worked extremely quickly. His wit was delightful and perennial. He was an extremely good speaker ; he was a good listener, and, after hearing the description of complicated machinery, he could gather together all the points, arrange them in their due proportion, and epitomise the whole description in a clear-cut memorable form, made picturesque, in many cases, by some unexpected analogy. In 1895 he gave evidence before the Select Committee on Weights and Measures, and exhibited one of the finest displays of mental arithmetic ever recorded, while he was upholding our existing system of weights and measures and numeration against a severe cross-examination on behalf of the advocates of the metric system.

During the thirty years that have elapsed since his Presidency, the Presidents who succeeded him, one and all, have felt that, while they commanded the ship, they had a pilot by their side in Sir Frederick Bramwell. To his widow and daughters a letter had been addressed on behalf of the Institution, expressing condolence with them upon his death.

The Minutes of the previous Meeting were read and confirmed.

The PRESIDENT announced that over two hundred Members had signified their intention of attending the Joint Meeting of this Institution and the American Society of Mechanical Engineers in Chicago next May, and that some of the Members would be accompanied by ladies.

The PRESIDENT announced that the Ballot Lists for the election of New Members had been opened by a committee of the Council, and that the following one hundred candidates were found to be duly elected:—

MEMBERS.

ADIE, ALFRED EDMOND, ...	Calcutta.
AMOR, WALTER, ...	Calcutta.
BELLAMY, CHARLES VINCENT, . . .	Nicosia, Cyprus.
CHARLTON, FRANCIS JAMES, Engineer Lieut. R.N.	London.
CHORLTON, ALAN ERNEST LEOFRIC, . .	Manchester.
DEAN, JOSEPH, ...	Bradford.
DONALDSON, PETER, ...	London.
EDMONDS, RICHARD JAMES, Capt. A.O.D., .	Portsmouth.
HILDRED, CHARLES WILLIAM, . . .	London.
HUMPAGE, THOMAS, ...	Bristol.
LILLY, WALTER ELSWORTHY, . . .	Dublin.
MCREDDIE, CHARLES EDWARD BRYSON, . .	Buenos Aires.
MORRISON, RALPH PERCY, ...	Leeds.

ASSOCIATE MEMBERS.

AUSTIN, SAMUEL, JUN., ...	Manchester.
BARKER, LAURENCE CHECKLEY, . . .	Taunton.
BITHEL-JONES, HARRY, ...	South Shields.
BOX, WILFRID LEONARD, ...	London.
BROWNJOHN, WILLIAM HENRY, . . .	Gravesend.
CHILD, JOHN, ...	Liverpool.
CHRISTIANSEN, ADOLPH GOTHARD, JUN., . .	Bombay.
CHRISTIANSEN, ALBERT GEORGE, . . .	Bombay.
CLARK, HUBERT CHARLES, ...	Birmingham.
DARBY, ARTHUR ERNEST, ...	Wolverhampton.
EADES, JOHN, JUN., ...	Manchester.
EDGE, THOMAS, ...	London.
ETLINGER, GEORGE ERNEST, . . .	Bombay.
FORSTER, RICHARD, ...	Horsehay, Salop.

GASKELL, HOLBROOK, JUN., ...	Widnes.
GEE, THOMAS JOHN, ...	Buenos Aires.
GULLIVER, GILBERT HENRY, . . .	Edinburgh.
HANDY, CHARLES EDWARD, ...	Preston.
HARRIS, FRANCIS GRAHAM REYNOLDS, . .	Cardiff.
HAYWARD, GEORGE, ...	London.
HEATH, JAMES FOX, ...	Leeds.
HOULT, WILFRED, ...	Dublin.
JOHNSON, JOHN HENRY, ...	Chelmsford.
JOLLIFFE, CHRISTOPHER HUBERT, . . .	Colombo.
KNOCKER, GEORGE STODART, . . .	Lowestoft.
LAKIN-SMITH, CLIFFORD, ...	Dorking.
LE CLAIR, LOUIS JEAN, . . , .	Dublin.
LIMOZIN, FERDINAND LOUIS JOSEPH, . .	Braintree.
LISTER, ARTHUR JAMES, ...	Manchester.
LONDON, WILLIAM JAMES ALBERT, . .	Manchester.
LOVERIDGE, ARTHUR WALTER, . . .	Swindon.
MACLENNAN, HOPE VERE, ...	Kimberley.
MARGETSON, ALFRED JAMES, . . .	Bristol.
MARPLES, STEPHEN ARNOLD, . . .	London.
McGREGOR, JAMES, ...	Durban.
MEDLEY, CHARLES POWIS, ...	Surbiton.
MILLER, JOHN, ...	Dublin.
MILLS, JAMES JESSE, Lieut. A.O.D., . .	Karachi.
MOON, JAMES GEORGE, ...	London.
MORGAN, DOUGLAS HOWARD, . . .	Cheltenham.
MORGAN, FREDERICK JOHN, ...	Landore, Glam.
NICHOLAS, DAVID CURTIS, ...	Obuassi, W. Africa.
PATERSON, SYDNEY, ...	Old Charlton.
PEARSON, HARRY, ...	Stockport.
PLAISTER, WILLIAM EDWARD, . . .	London.
PORTER, EUSTACE WILLIAM, . . .	Devonport.
POVER, GEORGE ALFRED FRANKLIN, . .	Slough.
RIX, GAYFORD, ...	Woolwich.
ROWCLIFFE, JOHN ARTHUR, ...	Manchester.
SANDERSON, HERBERT WILLIAM, . . .	Mansfield.

SMITH, FREDERICK GEORGE,	. . .	Chelmsford.
SPURR, GILBERT RICHARD, .	. . .	Ilford.
WATSON, GEORGE WILLIAM,	. . .	London.
WELLS, JOSEPH MALCOLM, .	. . .	London.
WIDDOWSON, ERNEST LEEDHAM,	. . .	Halifax.
WILDING, JAMES ARMSTRONG,	. . .	London.
WOLSTENHOLME, WALTER, .	. . .	Bolton.
WOOD, SAMUEL,	. . . .	Sheffield.
WORNUM, JOHN RUSKIN,	. . .	Hartlepool.
YATES, DONALD RUSSELL MARTIN,	. .	Newark.

GRADUATES.

ADAM, BENJAMIN, JUN.,	. . . .	Hull.
ALDRIDGE, LIONEL SHUTTLEWOOD,	. .	Birmingham.
BATES, HERBERT HENRY, .	. . .	Kidderminster.
BLACKMORE, WILLIAM ROBERT,	. . .	Accrington.
FARLOW, CHARLES FITZROY,	. . .	London.
FERRIER, SAMUEL KEY,	. . .	London.
GIBBONS, NORMAN BARRINGTON,	. . .	London.
GRIGGS, JOHN WILLIAM,	. . .	London.
HARLISS, WILLIAM LEWIS, .	. . .	London.
HAWKINS, EUSTACE FELLOWES,	. . .	Doncaster.
HODGSON, HERBERT EDWARD,	. . .	London.
HOLTHOUSE, CHARLES SCRAFTON, .	. .	London.
HOWARD, PERCY,	. . . .	Erith.
KNOWLES, GUY JOHN FENTON,	. . .	London.
LI FOKI,	. . . . .	London.
LLOYD-PARTON, FRED,	. . .	Wolverhampton.
OGG, EASTON LEWIS, .	. . .	London.
PELLOW, JAMES RICHARDS, .	. . .	London.
PORTER, GEOFFREY, .	. . .	London.
RICHARDS, SAMUEL HOPE, .	. . .	London.
SAYE, KENNETH NOEL,	. . .	Sutton, Surrey.
SMITH, LOUIS WILLIAM,	. . .	Birmingham.
THORNE, DOUGLAS STUART, .	- . .	Crewe.

TURNER, JOE, ... Preston.
TURNER, WILLIAM LAWRENCE, . . . London.
WARD, MONTAGUE WESNEY, . . . London.
WRIGHT, FRANK THOMAS, ... London.

———

TRANSFERENCE.

The PRESIDENT announced that the following Transference had been made by the Council since the last Meeting:—

Associate Member to Member.

WILLIAMS, ARTHUR, ... Mexico.

———

The following Paper was read and discussed:—

"An Inquiry into the working of various Water-Softeners"; by Mr. C. E. STROMEYER, *Member,* and Mr. W. B. BARON, of Manchester.

———

The Meeting terminated at Ten o'clock. The attendance was 138 Members and 134 Visitors.

AN INQUIRY INTO THE WORKING OF VARIOUS WATER-SOFTENERS.

By Mr. C. E. STROMEYER, *Member*, of MANCHESTER, and
Mr. W. B. BARON, of MANCHESTER.

It has frequently been stated that scale seriously reduces the heat efficiency of boilers, and experiments have been made which seem to prove this assertion, but it will be found that they have been carried out on wrong lines, and they only prove that scale very seriously interferes with the transmission of heat if the heat source, usually a flame, is of equal temperature over the whole surface. In a boiler the temperatures vary from $3,000°$ to $4,000°$ F. at the furnace, down to $500°$ to $1,000°$ F., where the gases leave the boiler. Let us take a simple case, assuming for convenience of calculation that the heat transmission from flame to boiler-plate is proportional to the difference of temperatures. Let the ratio of air to fuel be as 20 to 1; let the air temperature be $80°$ F., then the flame-temperature will be $3,000°$ F. If the steam-temperature is $380°$ F., the maximum temperature of the furnace-plates will, in the above example, be only $20°$ F. higher than that of the steam, viz., $400°$ F. Now let it be assumed that the heating surface is covered with scale $\frac{1}{8}$ inch thick; then, if the same quantity of heat were transmitted through the coated furnace-plates as through the clean ones, the temperature difference between one side of the scale and the fire side of the furnace would be 350° F., and the temperature of the plate would be 730° F. It is, however, clear that as the boiler-plate is hotter than in the first example, less heat will be transmitted to it and the temperature gradient in the scale will be less steep. Naturally also the flame will not get cooled so rapidly, and its temperature, as it reaches the next portion of heating surface, will be higher than before. The temperature distributions will therefore be roughly as follows :—

TABLE 1.
Temperature Distribution in a Boiler.

Square feet of heating surface per pound of fuel per hour . . .	0	$\frac{1}{4}$	$\frac{1}{2}$	1	2	4	8
	Boiler with plates free from scale.						
Flame and flue temperature . °F.	3000	2421	1961	1335	728	426	381
Maximum plate temperature . °F.	400	396	392	387	383	381	380
Total heat transmitted . per cent.	0	19·8	35·6	57·0	77·8	89 2	89·7
	Boiler with scale $\frac{1}{8}$ inch thick.						
Flame and flue temperature . °F.	3000	2484	2070	1471	835	459	384
Maximum plate temperature . °F.	691	630	581	510	434	389	382
Total heat transmitted . per cent.	0	17·5	31·8	52·4	74·2	87·0	89·5

Boilers are generally designed to have $1\frac{1}{2}$ to 2 square feet of heating surface per pound of fuel burnt per hour under ordinary working conditions. Four square feet in Table 1 represents a lightly worked boiler, and $\frac{1}{4}$ square foot represents the heating surface usually swept by the flame.

It will be seen from the Table that, even for this small value of heating surface to fuel, the addition of scale $\frac{1}{8}$ inch in thickness only reduces the transmitted heat by 11·6 per cent., whereas when the gases reach the end of a lightly worked boiler, where this value is 4, it is found that the total reduction has fallen to only 2·5 per cent. It may therefore be safely said that even a thick coating of scale does not materially reduce the efficiency of a boiler. On the other hand, Table 1 shows very clearly that even $\frac{1}{8}$ inch of wet scale raises the temperature of the furnace-plate in this case by nearly 300° F. In this particular instance there is therefore danger that the furnace has been sufficiently weakened by heat to be nearly collapsing. Scale is thus a serious danger, and, as is well known, has frequently caused accidents. It will be noticed that in clean boilers the temperature of the furnace-plate is nearly the same as that of the water, whereas in the scaly boiler the excess temperature is about one-tenth of that of the flame. If, therefore, in the first case the furnace door is opened and cold air is admitted, the excess temperature of the furnace-plate can at most be reduced 20° F., causing a contraction of only $\frac{1}{90}$ inch in 8 feet, whereas cold air admitted to the furnace of the scaly boiler will effect a rapid reduction of about 311° F. accompanied by a contraction of $\frac{1}{6}$ inch in 8 feet, which is a very serious matter. In fact, in a rigid structure these two strains would be accompanied by stresses of $1\cdot5$ and 22 tons respectively. Boilers are elastic, but a large fraction of these stresses certainly make their appearance every time that a furnace door is opened, say once every half-hour, or 6,000 times a year. No wonder, therefore, that in high-pressure boilers, which are necessarily more rigid than low-pressure ones, this constant straining leads to grooving at the furnace flanges. With scale $\frac{1}{4}$ inch thick the stresses would be nearly doubled.

It is thus seen that scale does not materially reduce the efficiency of a boiler, but it seriously increases its wear and tear, whereby its life is considerably reduced. It also endangers the safety of boilers. The same remarks apply to coatings of grease due to feeding the boiler with water containing condensed steam from the engine. It is, therefore, desirable to remove all scale-forming impurities from the feed-water. These impurities are suspended matter, carbonate of lime, sulphate of lime, magnesium salts and grease. One has also to be on one's guard against introducing large quantities of soluble salts, as these concentrate when the water evaporates, until thick scales of crystals are formed.

Let us examine the behaviour of these various impurities.

Suspended Matter is often organic, and appears to have a beneficial effect on such mineral precipitates as may be formed by combining with them and forming loose sediment, which is easily dealt with in water softeners, or it settles down as mud in boilers. Some suspended matter, such as fine sand and more particularly paper pulp, settles on the furnaces and leads to collapses.

Carbonate of Lime is the chief cause of Temporary Hardness.—Its chemical formula is $CaCO_3$. It is practically insoluble in water, but it is easily converted into bicarbonate of lime, having the formula $CaH_2 (CO_3)_2$, which is fairly soluble in cold water, and is a constituent of most natural waters. Its second equivalent of carbonic acid is easily removed on boiling, when of course the carbonate of lime is precipitated, forming a scale. The slower this reaction is carried out, the slower the heating, the more chance is there for this carbonate of lime to form crystals—called calcite in mineralogy—which constitute a fairly hard scale, as found in economisers, where the conditions are very favourable to its production. This explains why, in the south of England, where the waters contain much temporary hardness, economisers are very little used, for they would get choked unless the waters be first softened. Waters poor in carbonate of lime but rich in sulphates do not choke the economiser pipes. If the heating is effected rapidly, as for instance when natural waters are pumped direct into the steam space of a boiler, the precipitation of carbonate of lime is so rapid, that only a mud is formed. Another means of converting bicarbonate of lime into carbonate of lime is to add solutions of caustic lime, burnt lime, or caustic soda, in water. For solubilities see Appendix I (page 789). The chemical reaction is as follows:—

$$\left. \begin{array}{c} \text{Bicarbonate} \\ \text{of Calcium} \end{array} \right\} + \left\{ \begin{array}{c} \text{Caustic} \\ \text{Lime} \end{array} \right\} = \left\{ \begin{array}{c} \text{Carbonate} \\ \text{of Lime} \end{array} \right\} + \text{Water}$$

$$\underset{\text{Soluble}}{Ca\,H_2(CO_3)_2} + \underset{\text{Soluble}}{Ca\,O} = \underset{\text{Insoluble}}{2\,Ca\,CO_3} + \underset{\text{Water}}{H_2O}$$

The second equivalent of the carbonic acid in the bicarbonates of lime and of magnesia is generally called half-bound. Any excess is called free carbonic acid. Of course sufficient lime is required to neutralise both.

In the Table of Chemical Analysis the bicarbonate of lime is split up into carbonate of lime and carbonic acid, because the latter disappears on boiling. One grain of carbonic acid combines with 2·27 grains of carbonate of lime, forming 3·27 grains of bicarbonate of lime. Temporary hardness cannot be entirely removed by lime treatment, because carbonate of lime is not absolutely insoluble.

Sulphate of Lime.—The chemical formula is $CaSO_4$. It is fairly soluble in waters up to the boiling point at atmospheric pressure, but it is less soluble at temperatures corresponding to high pressures of steam, Appendix I (page 789). Because this sulphate of lime cannot be removed by ordinary boiling, it is called permanent hardness. One grain of sulphate of lime equals 0·7350 permanent hardness. The result is that if waters containing sulphate of lime are pumped into a boiler, the constant evaporation effects a slow concentration until saturation point is reached, when the sulphate of lime is precipitated so slowly that it crystallises and adheres to all parts of the boiler, but more particularly to the hottest parts; then, whenever the pressure drops—and with it the temperature—part of the precipitate is redissolved, the spaces between the crystals are filled with concentrated solutions of sulphate of lime which crystallises out again on heating, whereby the originally small and loose crystals are enlarged and are firmly cemented together, forming a very hard scale, which can only be removed by chipping or by heating; it is practically gypsum or selenite. If any carbonate of lime is mixed up with the scale, this too gets thoroughly cemented with it.

At the temperatures corresponding to high pressures, water can dissolve only about 20 grains of sulphate of lime per gallon. On cooling down a boiler and letting it stand for some time, the water will dissolve parts of the crystals until there are about 170 grains to the gallon. This dissolving action loosens the scale, which can be easily removed as long as it is wet. If the scale be allowed to dry, all the 170 grains of sulphate crystallise and thereby cement together the loose parts of the scale, making it hard and difficult to remove.

The principle of pumping feed-water into a trough in the steam space of a boiler has several times been patented, but as boiler water can hold up to 20 grains of sulphate of lime in solution, this method is of no benefit with waters having less permanent hardness than 15°.

The sulphate of lime can always be entirely removed by conversion into carbonate of lime. The reaction is as follows :—

$$\left. \begin{array}{c} \text{Sulphate} \\ \text{of Lime} \end{array} \right\} + \left\{ \begin{array}{c} \text{Carbonate} \\ \text{of Soda} \end{array} \right\} = \left\{ \begin{array}{c} \text{Carbonate} \\ \text{of Lime} \end{array} \right\} + \left\{ \begin{array}{c} \text{Sulphate} \\ \text{of Soda} \end{array} \right.$$

$$\begin{array}{cccc} Ca\,SO_4 & + & Na_2\,CO_3 & = & Ca\,CO_3 & + & Na_2\,SO_4 \\ \text{Soluble} & & \text{Soluble} & & \text{Insoluble} & & \text{Soluble} \end{array}$$

If there is any free or half-bound carbonic acid in the water, and this is generally the case, caustic soda, if available as a waste product, may be used. The caustic soda combines with the carbonic acid to form carbonate of soda, and the reaction is then as sketched above; and the lime which has lost its one equivalent of half-bound carbonic acid is also precipitated. Caustic soda without carbonic acid will not precipitate the sulphate of lime. It is, therefore, wrong to introduce caustic soda into boilers which are fed with water having only permanent hardness. In fact caustic soda should not be pumped into a boiler except together with the cold feed, for as long as the water is cold, the carbonic acid can be fixed, and additions of caustic soda would come too late if made in the boiler.

The above sketched chemical reaction can easily be carried out in water softeners, but the separation of the precipitate is as slow as in the last case, and unless heat is applied, the newly-formed carbonate of lime has to settle for a long time.

Magnesium Salts.—These are found in natural waters as nitrates, chlorides, sulphates, and cause permanent hardness, or as bicarbonate which causes temporary hardness; they are very soluble except the last-named. They all react on soap, and heat will not precipitate them, except the bicarbonate by driving off its carbonic acid. At the high temperatures to be found in boilers a reaction resulting in precipitation of magnesia takes place between the carbonate of lime and the soluble magnesium salts. The bicarbonate of magnesia can be partly removed by converting it into carbonate of magnesia by the addition of caustic lime.

The reaction in water softeners is as follows —:

$$\left\{\begin{array}{c}\text{Bicarbonate of}\\ \text{Magnesia}\end{array}\right\} + \left\{\begin{array}{c}\text{Caustic}\\ \text{Lime}\end{array}\right\} = \left\{\begin{array}{c}\text{Carbonate of}\\ \text{Magnesia}\end{array}\right\} + \left\{\begin{array}{c}\text{Carbonate}\\ \text{of Lime}\end{array}\right\} + \text{Water}$$

Mg H$_2$ (CO$_3$)$_2$	+	Ca O	=	Mg CO$_3$	+	Ca CO$_3$	+ H$_2$ O
Soluble		Soluble		Slightly Soluble		Insoluble	Water

The last traces of magnesium carbonate are removed by adding an excess of lime, the reaction being as follows:—

$$\left\{\begin{array}{c}\text{Carbonate of}\\ \text{Magnesia}\end{array}\right\} + \left\{\begin{array}{c}\text{Caustic}\\ \text{Lime}\end{array}\right\} + \text{Water} = \left\{\begin{array}{c}\text{Hydrate of}\\ \text{Magnesia}\end{array}\right\} + \left\{\begin{array}{c}\text{Carbonate}\\ \text{of Lime}\end{array}\right\}$$

Mg CO$_3$	+	Ca O	+ H$_2$ O	=	Mg (OH)$_2$	+	Ca CO$_3$
Slightly Soluble		Soluble	Water		Insoluble		Insoluble

The other magnesium salts react in a similar way, being easily converted into carbonate and hydrate by the addition of burnt lime and soda ash.

The carbonate of magnesia, $MgCO_3$, is very slightly soluble in water, whereas the hydrate of magnesia $Mg(HO)_2$, is practically insoluble, but being a gelatinous mass it ·causes serious difficulties in water softeners, more especially by clogging the filters.

Grease.—This is, of course, only found in the waters discharged from jet or surface condensers or from feed-heaters in which steam is condensed. Vegetable and animal fats cause corrosion in boilers, but as they are unsuitable for cylinder lubrication, they are now rarely used and heavy mineral oils have been found to be more suitable. Modern marine practice tends towards the running of engines without any cylinder oil, but factory engines still consume large quantities, and thereby contaminate their feed-water if drawn from the condenser. Part of this grease floats on the water and could be removed by filtration; probably it does little harm if it gets into the boiler, but a small trace is emulsified in the water and is very difficult to deal with. It cannot be separated from water by boiling at atmospheric pressure, but it is completely removed by the conditions which exist in a boiler, and, as it adheres to the heating

3 H

surfaces, it causes overheating which may result in a collapse and certainly increases the wear and tear.

The peculiarity of grease deposits in boilers is that their effect is out of all proportion to their thicknesses. It has been seen that scale of $\frac{1}{8}$ inch thickness will raise the temperature of furnace plates about 300° F. As grease offers ten times more resistance to heat, one would expect that $\frac{1}{80}$ inch would have the same effect as this thickness of scale, but experience shows that the merest trace of grease, certainly less than 1–1000th inch, or one-tenth of the above, can cause far more serious injury than scale. Various explanations have been attempted. According to one of these, thin films of grease form tough bubbles on the heating surface and prevent the water from keeping it cool. Another view is that grease, either alone or joined to mineral matter, forms an impalpable powder like oxalate of lime and other precipitates, and, like these, retards ebulition. In support of these views we find a fairly well-founded belief that grease in boilers is more injurious if these boilers are clean than if they are coated with mineral scale, and against this view we have the undoubted experience that land boilers with scale at once give trouble if condensed water is used instead of natural water. Increase of pressure above 110 lbs. seems to accentuate this evil; perhaps this may be due to decomposition of magnesium carbonate when this temperature is reached. In any case it is highly desirable to remove every trace of grease from the feed-water. As already stated, this cannot be done by filters; and grease separators, which appear to be rather more efficient, do not remove the last trace of grease.

As yet the only effective method for doing this is to add mineral matter in solution to the condensed water, and then to cause precipitation by chemical means. The grease then adheres to the precipitate, and can easily be removed by settlement or filtration. Water from jet condensers contains the necessary mineral matter, but this has to be added, if the waters are drawn from surface condensers.

The preceding remarks may be here briefly summarised.

Carbonate of lime forms hard scale in economisers and a soft mud in boilers, unless sulphate of lime is present when it also is cemented into a scale. Carbonate of lime can be removed by boiling, or by adding enough caustic lime or caustic soda to combine with the free and half-bound carbonic acid which holds it in solution.

Sulphate of lime forms no scale in economiser pipes, but it forms a very hard scale in boilers and also cements the carbonate of lime deposits. Its deposition in boilers is due to slow concentration of the water, and it is, therefore, desirable to remove the salt entirely. This can always be done by adding carbonate of soda, which converts it into carbonate of lime.

Carbonate of magnesia generally behaves like carbonate of lime, except that it is slightly more soluble.

The other salts of magnesia are very soluble. They seem to cause corrosion and should be removed. This can be done by treating them like the sulphate of lime by adding carbonate of soda.

Grease should, if possible, be kept out of a boiler. Neither separators nor filters will remove it entirely from feed-waters, but this can be done by mixing them with impure waters and treating them in water softeners.

The cost of raising the temperature of 1,000 gallons of water from 60° to 212° F. is about one shilling when the price of coal is 12s. per ton, no matter what the hardness of the water may be.

With waters of 10° temporary hardness 1·14 lbs. of caustic soda of 77 per cent. strength, or 0·8 lb. of burnt lime, will suffice for 1,000 gallons of water. The same quantities will also neutralise all the free carbonic acid represented by 6 grains per gallon. As the prices of caustic soda and lime stand in the ratio of 12s. 6d. to 1s. per cwt., the relative costs of the two treatments would be 1·53 and 0·09 penny per 1,000 gallons, and naturally one would use caustic soda only if it is a waste product and if there is permanent hardness also present. Large proportions of soda, either carbonate or caustic, affect brass fittings.

It requires 1·5 lbs. soda ash of 58 per cent. strength to remove 10° permanent hardness out of 1,000 gallons; as the price of this soda is about 5s. 6d. per cwt. the cost would be about 0·88 penny per 1,000 gallons. The strength of washing soda is about 20 to 22 per cent.

3 H 2

In some water softeners, such as the Archbutt-Deeley, Lassen and Hjort, caustic lime and carbonate of soda are mixed, producing a milky fluid, which consists of caustic soda, free lime or free soda ash, and insoluble carbonate of lime, the latter substance being of course a useless constituent. The cost of the caustic soda so produced would be about 6s. per cwt.

The interest and depreciation of the softening plant, the attendant's time, generally only half-an-hour a day, have to be added to the above expenses. If the waters are treated in the boilers— when burnt lime may, of course, not be used—the cost of removal of the scale and the wear and tear of the boiler, due to overheating, have to be added. The following Table 2 shows the relative cost of working installations of from one to seven boilers. Each boiler is supposed to be 8 feet diameter, and to have cost £800, including setting, etc. The interest on the first outlay and on the sums annually set aside for depreciation and renewal is taken at 3 per cent. The best worked boiler using pure or softened water is supposed to last fifty years, and the last boilers on the Table are supposed to last only fifteen years, and have to be opened out and scaled every three months. The other boilers, having each a spare one in the set, are supposed to last the number of years shown in the Table. In each case, except the first, the water is supposed to be very sedimentary.

It will be seen that according to this estimate, boilers using pure water should not cost more than about £35 per annum, including chemicals in a water softener, whereas the other boilers using sedimentary water and boiler compositions may cost twice and three times as much. The minimum saving, say £30, capitalised at 5 per cent. represents £600, whereas the cost of a water softener per boiler would amount to about £100 to £200.

Turning one's attention to some of the practical difficulties which water softeners have to overcome, the most important of these are the proper adjustment of the supply of chemicals and the removal of the precipitates from the treated waters. If, by suitable chemicals, carbonate of lime is precipitated out of water it is at first in a colloidal condition, its first appearance

TABLE 2.

Hypothetical.

Nature of Feed.	Pure.	Very Sedimentary Water.						
Boilers at work . . .	1	1	2	3	4	5	6	6
Spare boilers . . .	0	1	1	1	1	1	1	0
Assumed life of boilers . .	50	40	40	40	40	30	20	15
Interest on first cost . £	24	48	72	96	120	144	168	144
Depreciation . . £	7	21	32	42	52	100	209	258
Scaling and cleaning @ 30s. £	2	20	39	59	78	78	78	36
Chemicals . . . £	2	10	20	30	40	50	60	60
Total . . . £	35	99	153	227	290	372	515	498
Totals per working boiler £	35	99	71·5	75·7	72·5	74·4	85·8	83*

* The actual cost would be much greater, as the works would be closed down for 4 weeks per annum.

being that of a bluish-white thin starch, which can freely pass through the best chemical filters, and could never be arrested by wood wool, cloth or sponges. If this fluid is allowed to stand for a considerable time, or if it is heated a little, the precipitate settles down, changing to a yellowish colour, and no amount of shaking will again convert it back to its original condition; but even now this precipitate is so fine that it settles down very slowly. According to Professor Wanklyn's experiments, this precipitate will settle down in twenty-five minutes through $\frac{3}{4}$ inch of water, which will then be quite clear, whereas it takes eight hours to clear 20 inches of water charged with this precipitate. The rate of settlement is about 1·8 to 2·5 inches per hour. No experiments seem to have been made on hot water.

It is thus understood why very large tanks are necessary when working with cold water, and why even filters will not remove all the precipitated carbonate of lime.

When the softening operation is carried out hot, these settling tanks, as can be seen by comparing various softeners, may be much reduced in size, and filters now become far more effective than before.

In most of the cold-water softeners the treated water moves upwards against the descending sediment, which, being coarse, assists the newly-formed precipitate in settling down. In other softeners, notably in that of Desrumaux, the treated water is made to travel in comparatively thin sheets. Thus by dividing the tower into narrow layers, the precipitate has not to fall very far. This principle may be carried too far, for the thinner and longer the sheet of water, the greater the velocity and the stronger the eddies.

In spite of the use of very large settling tanks, supplemented by filters, there are very few cold-water softeners which can be relied upon to remove all the solid matter which the added chemicals have precipitated, and serious complaints are not infrequently heard that after the installation of a water softener, the injectors or the feed-pipes get choked with scale. Evidently all the chemical precipitates do not settle down thoroughly until the comparatively hot pipes are reached.

To overcome this difficulty the Archbutt-Deeley process is arranged so that the carbonic-acid gas from a stove comes in contact with the treated water and dissolves the trace of precipitate which would have appeared in the pipes. Steam users, however, have a prejudice against deliberately introducing carbonic acid into their boilers, and some at least add very large settling tanks, and do not use the stove. Unquestionably the most effective way of removing the residual precipitate would be to heat the treated water, but as already shown, that is a very expensive remedy.

In practice, magnesia precipitates are still more difficult to deal with than the lime precipitates, because of their gelatinous nature, which although it assists in separating out the carbonate of lime, also retards the settling process, and seriously interferes with the working of filters. Here again heat is a remedy, but only a partial one, and is very costly. Nevertheless, several of the softeners dealt with in this Paper seem to have removed large quantities of magnesia without choking their filters.

In the Archbutt-Deeley softener, which, due to a sufficiency of lime, has effectively removed nearly all magnesia salts, each new tankful of treated water is mixed with the sediment of the previously treated waters, the settlement being expedited by the adhesion of the old and coarse particles to the new flocculent and gelatinous precipitates. The same principle is adopted in those softeners in which the treated water is led to the bottoms of the settling tanks, and has to pass through previously precipitated mud. Use is also made of the precipitated mud for the removal of grease in the apparatus of Babcock and Wilcox, Boby, Maxim, and Wollaston, in which exhaust steam is introduced. The explanation of the cause of this removal is doubtless to be found in the coagulation of the first-formed colloidal precipitate, which on separating effectively removes both the oil in suspension and in the emulsified state. Wollaston finds that the greasy scum rises upwards, and at the top of his reaction tower a scum tap is fitted, and the exit from this tower to the settling tank is placed some distance below, thereby preventing this scum from contaminating the softened water.

Out of seventeen continuous-water softeners dealt with in this Paper fourteen are fitted with filters. Two of these, Bell's and Reisert's, are sand filters. Porter-Clark and the Atkins Co. have cloth filters, and the others have wood wool or sponge filters. Wood wool is cheap and can be renewed, say, twice a year, while sponges have to be cleaned. These removal or cleaning operations are rather tedious, but cannot be entirely obviated. Those filters through which the water passes downward, and on to which the sediment falls, have, of course, to be cleaned much more frequently than those through which the water passes upward.

Cloth filters have to be cleaned by hand. They generally consist of a number of wooden or iron frames, each frame being separated from its neighbour by a piece of cloth. The frames are pressed together by two long bolts. The water enters every alternate frame through a hole in the top, side, or middle, and is discharged out of holes in the bottom of the adjoining frames. The cloths are easily removed by slacking back the nuts on the long bolts. The Atkins Co.'s filters are circular discs which are brushed clean without removing them.

The second difficulty encountered in practice is that of properly apportioning the chemicals.

It appears that in all water softeners the chemicals are dissolved in water or at least mixed with it, and the introduction of small measured quantities of powdered slaked lime and of powdered soda does not appear to have been tried, and numerous complicated and ingenious contrivances have been devised for overcoming the difficulties which are encountered when using fluids.

Thus, caustic lime, as will be seen by Appendix I (page 789), dissolves only very sparingly in water, the quantity decreasing with rising temperature from 91 grains at 59° F. down to 40 grains at 212° F. At about 70° F. the solubility decreases at the rate of about $\frac{1}{2}$ per cent. for every 1° F. rise of temperature. It requires $1\frac{1}{4}$ grain of caustic lime, or about $\frac{1}{70}$ gallon of lime water, to precipitate the carbonate of lime, which is dissolved in water containing 1 grain of a free carbonic acid; and, as this free carbonic acid is rarely less than 7 grains per gallon, it is generally necessary to add quite 10 per cent. of lime water. This lime water is prepared by throwing slaked lime into a deep tank and letting unsoftened water flow slowly through the milk of lime in the bottom. If the flow is slow enough, and the tower high enough, the overflow is clear saturated lime water. The relative quantities of lime water to untreated water is regulated by letting both streams emerge from a single tank having adjustable nozzles or weirs. Except that the lime tank must be high, and that it ought to be provided with stirrers near the bottom, this arrangement is a fairly convenient one. It is used by the Atkins Co., Desrumaux, Reisert, and Stanhope.

The apparatus can, of course, be made much smaller if milk of lime, containing say 10 per cent. caustic lime, is used instead of clear lime water, but then it is imperative that this milk of lime should be perpetually stirred, as in the Bell, Doulton, Harris-Anderson and Porter-Clark softeners, nor is it permissible to add water during the working day, as appears to be the case with the Harris-Anderson softener. The mixture is thereby made weaker and weaker.

In the Lassen and Hjort softener the quantity of milk of lime and soda solution, injected with each tilt of the water trough, is regulated by the duration of the opening of a small valve, but as the head of the re-agent diminishes during the working day, so does the quantity of injected fluid, and it has been found necessary to keep the chemical tank half full. These two examples will suffice to show that, in spite of some advantages, the use of milk of lime is perhaps attended with more difficulties than is the use of lime water.

Most water softeners are provided with small lime-slaking tanks, but this is not an invariable practice, nor does any continuous softener seem to be fitted with a tank in which either soda or caustic soda is dissolved before use ; yet it takes some time to dissolve these salts and thoroughly mix the solutions. The addition of the soda to the untreated water is perhaps most easily effected by a small pump, which must be of iron, working in unison with the feed-pump. Scoops and bucket-chains are also used. In all these arrangements the ratio of soda to water is best regulated by varying the strength of the soda solution. The Harris-Anderson softener has a very ingenious arrangement for delivering soda of the right density. It is, however, questionable whether all these extremely ingenious contrivances are at all necessary. They are intended to replace the pump, but have grown so complicated that they are unquestionably now more costly. Pumps for supplying the chemicals are used by Bell, Boby, Maxim, Porter-Clark, and Wollaston. Weirs, whose relative widths can be regulated, are used by Harris-Anderson and Wright. Carefully-gauged holes or taps, whose openings can be regulated, are used by Babcock and Wilcox, Carrod, Desrumaux, Doulton, and ball taps by the Atkins Co. and Tyacke.

Assisted by the above general remarks the sketches of the water softeners, together with their descriptions, Appendix II (page 790), and the analyses of the unsoftened and softened waters (Appendix III, page 826) permit of a fair comparison being made as to the suitability of the various types for special purposes.

Glancing over the analyses, it will be noticed that, with two exceptions, the various users have added too little lime. This cannot be on the ground of expense as lime is very cheap, nor on account of increasing the amount of dissolved solids, for, unlike soda ash or caustic soda, it both decreases the temporary hardness and also the amount of dissolved solids. The cause of the insufficiency of lime is most probably due to an incorrect idea of the functions of the added lime. They are three in number :—

(i.) To combine with the free carbonic acid.

(ii.) To combine with the half-bound carbonic acid.

(iii.) To reduce all the magnesia to hydrate.

In most of the analyses only that represented by (ii.) has been approached. A simple differential soap and colour test, devised by Mr. Baron (Appendix IV, page 834), will permit of better softening-effects being easily attained.

In nine cases out of nineteen an excess of soda has been added. This excess should be avoided, as it tends to corrosion of the brass fittings, especially when in the form of caustic soda or even sodium carbonate.

With regard to the apparatus in which the temporary hardness is supposed to precipitate by the use of steam, it would seem that only prolonged boiling completely evolves the free and half-bound carbonic acid. The result is that caustic soda is generally required to precipitate the permanent hardness and neutralise the still unexpelled carbonic acid. The presence of any magnesia salts also calls for the use of caustic soda or lime as explained above.

The Paper is accompanied by 4 Appendices, which are illustrated by 17 Figs. in the letterpress.

APPENDIX I.

Solubilities.

Temperatures.	Corresponding Steam Pressure.	Chemicals and Experimenters.			
		Calcium Oxide. (Caustic lime.)	Calcium Sulphate.		
		Lanny 1878	Marignac 1874	Poggiale 1879	Tilden and Shenstone —
°Fahr.		Grains per Gallon.			
32	—	96·7	133·0	143·5	—
40	—	94·0	—	—	—
59	—	91·0	—	—	—
64·4	—	—	143·0	—	—
68	—	—	—	168·7	—
75·2	—	—	146·5	—	—
86	—	81·4	—	—	—
95	—	—	—	177·8	—
101·5	—	—	150·3	—	—
111	—	70·1	—	—	—
127·4	—	—	147·0	—	—
140	—	70·8	—	—	—
158	—	—	—	170·8	—
161·6	—	—	140·8	—	—
210·2	—	—	122·6	—	—
212	0	40·2	—	151·9	—
284	37	—	—	—	54·6
323·6	79	—	—	—	39·2
356	131	—	—	—	18·9
464	484	—	—	—	12·6
482	575	—	—	—	12·6

APPENDIX II.

ARCHBUTT-DEELEY WATER SOFTENER.*

The apparatus, Fig. 1 (page 791), consists of a small chemical tank, a re-action and settling tank combined, a storage tank and a coke stove. The inlet for the untreated water is at the side of the tank about two or three feet above the bottom. The treated water is drawn off the surface by means of a floating discharge. This pipe is connected to the chimney of a coke stove. Steam pipes are led into the chemical tank and an air blast actuated by a steam jet is led into the re-action tank.

Capacity.—2,500 gallons at a time or 1,000 gallons per hour.

Dimensions.—Re-action and settling tank, 8 feet cube; the storage tank is 22 feet long, 17 feet wide, 3 feet 9 inches high, and holds 5,500 gallons; floor space of softening tank 64 square feet. Total floor space, 438 square feet. About 180 of these softeners of various sizes are in use.

Working.—The re-action tank is filled with the untreated water, a definite quantity of burnt lime is put into the chemical tank, slaked, mixed with water and boiled by admitting live steam. The necessary quantity of soda is then added and again boiled. The untreated water is introduced, by means of a steam injector, through the upper one of two sets of pipes near the bottom of the tank. On its passage through the pipes the re-agent is slowly added. By turning a cock, the steam injector is made to drive air through the perforations in the lower pipe in the bottom of the re-action tank. This air effects a thorough mixing of the water and re-agent, and also stirs up the sediment remaining from previous operations, the newly formed sediment readily adheres to it and is quickly settled. This precipitate is occasionally removed by suitable means. The softened water is drawn off through the swivelled floating discharge pipe. While passing down this pipe the water should come in contact with the waste gases from the coke stove and should absorb some carbonic acid. The sample analysed was not carbonated.

---

* Proceedings 1898, Part 3, page 404.

Supply.—Partly town's water, partly brook.

Chemicals used per 1,000 gallons.—2·5 lbs. lime, 0·6 lb. soda, 0·15 lb. alumino ferric.

The user reports that this apparatus is easily worked and gives satisfactory results. The old boiler scale came off after six weeks' use. Now there is only a slight mud in the boilers. The carbonating is done for every alternate tank.

FIG. 1.—*Archbutt-Deeley.*

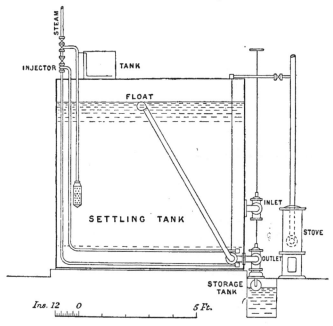

Result of Analysis.—The supply is the same as that for the Desrumaux apparatus, Fig. 7 (page 803), but was not collected at the same time. A slight excess of lime as well as of soda was added, which had the effect of almost entirely removing the magnesia salts. This sample of water was not carbonated, but the treated water had absorbed 0·63 grain of carbonic acid, probably by contact with the atmosphere.

The Atkins Co. Water Softener.

The apparatus, Fig. 2 (page 793), consists of a lime-slaking tank, a lime-water tank, a settling tank and a tank with revolving filter discs. The settling tank has a central division plate, and over it are placed two troughs, one for mixing the supply with the lime water, the other for collecting the treated water. The supply pipes to the mixing trough and to the bottom of the lime-water tank are controlled by ball-taps.

The filters consist of hollow cast-iron discs covered with perforated sheet zinc and filtering cloths. They are bolted together by means of a hollow shaft which acts as a discharge pipe. Brushes are placed between the discs.

Capacity.—2,500 gallons per hour.

Dimensions.—Settling tank, 7 feet wide, 15 feet long, 5 feet deep. Lime-water tank, 3 feet 6 inches diameter, 8 feet deep. Filter tank, 5 feet 6 inches wide, 6 feet long, 6 feet deep. Floor space, exclusive of lime-slaking tank, 138 square feet.

Working.—Slaked lime and water are poured into the lime-water tank, and a small stream of water, regulated by a ball-tap in this tank, enters the bottom of the tank, and clear lime water flows into a mixing trough, on to which the remainder of the water to be treated is poured. The supply and the outflow of the lime water are regulated by ball-taps in the settling tank. The mixed water passes along the settling tank and overflows to the filters. These are cleansed occasionally by giving them a turn, when the brushes remove the sediment. No information was obtained as regards supply, chemicals used or effects on boiler.

Results of Chemical Analysis.—In this case water which had only 3·165° of permanent hardness was only treated for temporary hardness, which was successfully reduced to 2·32°.

Babcock and Wilcox Water Softener (Guttman).

The apparatus, Fig. 3 (page 795), consists of a chemical tank, a re-action tank, filters and soft-water tank. The supply pipe passes down

Fig. 2.—*The Atkins Co.*

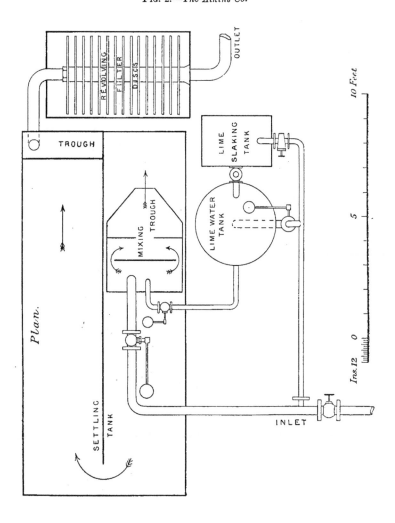

the side of the chemical tank and ends in the bottom of the re-action tank. It has a branch pipe and cock over the chemical tank, and is fitted with valve V, shown on larger scale, which is regulated by the float in the soft-water tank. Live steam enters the re-action tank as shown, through a silent injector; a branch from the steam pipe for occasional use is led into the bottom of the chemical tank. Where exhaust steam is available an additional injector is fitted as shown. Attached to the bottom of the chemical tank is a pipe with two cocks, C_1 for regulating the ratio of soda solution to the feed, and C_2 for regulating the quantity in accordance with the feed drawn off. At the top of the re-action tank is a weir over which the treated water flows into the filter, which consists of a series of boxes filled with wood-wool. The bottoms of these boxes are provided with cocks through which the sludge is drawn off occasionally. The filtered water passes into the soft-water tank, whence it is pumped into the boiler.

Capacity.—5,000 gallons per hour.

Dimensions.—Chemical and re-action tanks, 5 feet 5 inches square, 13 feet high. Filters and soft-water tank, 5 feet 5 inches wide, 17 feet long and 8 feet high, measured over flanges. Total floor space, 122 square feet.

About thirty-five of these softeners of various sizes are in use.

Working.—Enough soda to last for one day is placed in the chemical tank, water is added and steam turned on till all the soda is dissolved. The cock C_1 is opened to suit the density of the soda solution and the impurity of the feed water, and has to be adjusted during working, according to whether the softened water is alkaline or not. The cock C_2 and the feed valve V are actuated by levers from one float in the settling tank, so that the ratio of the amount of chemical and of the water is maintained. The mixture flows into the re-action chamber, where it is mixed with waste steam if available. It is here brought to the boiling point by live steam, which should remove the temporary hardness; the soda removes the permanent hardness. The hot water overflows to the filters, which it passes in half-an-hour, and here the sediment is removed. Some

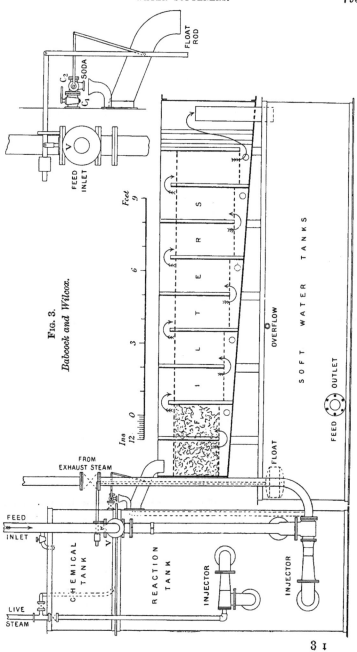

FIG. 3.

Babcock and Wilcox.

sediment collects in the lower tank. The grease from the exhaust steam adheres to the sediment and does not pass the filters.

Supply.—From a reservoir.

Chemicals used.—Soda ash, quantity not stated.

The users report that this softener has worked well for three years and has given every satisfaction, although the water is hard and sometimes acid. No trouble has been experienced with their three water-tube boilers with $1\frac{1}{2}$ in. diameter tubes. All old scale is now removed, and only a little soft mud is formed. On one occasion a little grease was carried into the boilers, but it was found that the filter had not been properly packed.

Result of Chemical Analysis.—The analysis shows that about 60 per cent. condensed steam have been added to the supply. Neither the high temperature nor the chemicals have produced a beneficial effect, for on subtracting the above-mentioned condensed water (see third column of Table of Analyses, page 827), the various constituents and the hardness are practically the same as the supply, except that some added soda has converted all the permanent hardness into temporary hardness.

BELL BROTHERS' WATER SOFTENER.

The apparatus, Fig. 4 (page 797), consists of a mixing tank, a re-action tank, a settling tank and two sand filters. The mixing tank and chemical tanks have mechanically-driven paddles. A pipe leads from the mixing tank through a small pump to the inlet pipe of the untreated water. Another pipe leads from the top of the settling tank to the two filters. The sand filters are provided internally with revolving hollow arms, which on occasion break up and wash the filtering sand.

Capacity.—2,500 gallons per hour.

Dimensions.—Two chemical tanks, 4 feet wide, 8 feet long, 4 feet high. Settling tank, 5 feet diameter, 9 feet above ground to top of handwheel. Each sand filter is 3 feet diameter and 8 feet 6 inches high. The floor space is 172 square feet.

About twelve of these softeners of various sizes are in use.

Working.—The chemicals, lime and soda, are placed in their chemical tanks and well stirred.   They are then passed in correct proportions into the mixing tank with its stirrer, and pumped into the feed inlet pipe, thence into the settling tank and then into the sand filters.   When the pressure gauge indicates that the filters are

FIG. 4.—*Bell Brothers'.*

Plan

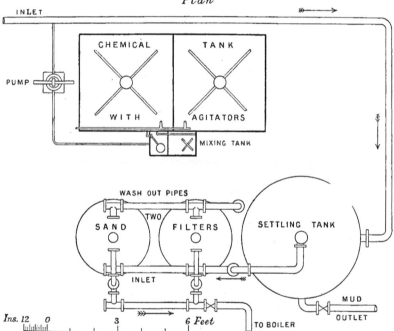

nearly choked, the sand is loosened by water pressure from below, and then more water is injected through revolving arms into the body of the sand bed, which is thereby thoroughly washed out.   The filters are under a pressure of 160 lbs. per square inch.

Supply.—Well water.

Chemicals used per 1,000 gallons.—2 lbs. alkali and 1·5 lbs. lime. The users report that the softener works fairly well, but the

3 I 2

pipes get choked too often.   The makers affirm this to be due to
faulty manipulation.

Result of Chemical Analysis.—By adding an excess of soda the
permanent hardness was reduced from 30·9° to nothing.   The 2 lbs.
of lime, said to have been added, do not appear to have been properly
mixed with the supply, and the temporary hardness has for this
reason been increased from 5·2° to 6·8°, which may account for the
trouble.

BOBY WATER SOFTENER : CHEVALET DETARTARISER.

The apparatus, Fig. 5 (page 799), consists of several perforated trays
placed one above another and surrounded by a cylindrical shell, the
lower part of which is the softened-water tank.   At the side of this
tank is placed a closed vessel with a float.   The latter is connected
to the inlet valve and regulates the supply.   A pipe for injecting
soda solution enters the untreated water-supply pipe.   The exhaust
pipe from the engines ends in a grease separator attached to the
softened-water tank.   The outlet for the uncondensed exhaust
steam into the atmosphere is at the top of the apparatus, and can
be connected to a main condenser which would deliver water free
from grease.

Capacity.—3,400 gallons per hour.

Dimensions.—Diameter of softened-water tank, 6 feet 2 inches.
Height to top flange, 15 feet.   Total floor space, exclusive of float
and soda tank, 37 square feet.

About five hundred of these softeners of various sizes are in use.

Working.—Soda is dissolved in a tank, not shown on the drawing,
and the solution is pumped into the water supply which falls on the
top trays of the softener.   There it comes in contact with the
escaping steam and gradually reaches the soft-water tank at the
bottom.   In its downward passage the water is in contact with the
exhaust steam, acquires its temperature, and partly condenses it.   The
heat effects a precipitation of the carbonate of lime, and the soda
precipitates the sulphate of lime.   These precipitates, mixed with
grease, adhere to the trays.   Occasionally the apparatus is taken to
pieces and the deposit removed.   The joints are easily remade with

FIG. 5.

Boby, Chevalet Detartariser.

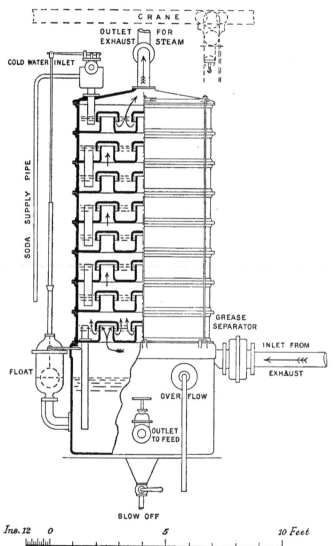

putty. The feed-pump, drawing from the softened-water tank, actuates the soda pump, and thus the ratio of the chemical to the feed-water is kept constant.

The exhaust steam, after passing the grease extractor, comes first in contact with the softened water, and must inevitably impart to it traces of grease.

Supply.—Well water.

Chemicals used.—Soda, quantity not stated.

The users report that after three months' use it takes three days to clean the apparatus and to remove 2 tons of scale. No grease has been noticed in the feed, and no scale is formed in the boiler.

Result of Chemical Analysis.—The analysis shows that 27 per cent. condensed steam was added to the supply. Subtracting this addition, the effect of the heat and the carbonate of soda is shown in the third column of the Table of Analyses (page 827), where it will be seen that the temporary and permanent hardnesses have been reduced from 22·7° and 12·6° to 4·6° and 4·22°. Only a small part of the magnesia has been removed. Better results would have been obtained by the use of caustic soda instead of carbonate of soda. Heat alone does not appear to be able to reduce the temporary hardness to a minimum.

CARROD WATER SOFTENER.

The apparatus, Fig. 6 (page 801), consists of two chemical tanks, one mixing tank, and a settling tank and filter combined, and a distributor. The water-supply pipe has a hand-cock over each re-agent tank, and a ball-tap over the distributor, which discharges to the bottom of the settling tank through a valve regulated by a float in the settling tank. A trunk conveys the mixed water to the bottom of the settling tank where several baffles are fitted. The upper part of this tank is a filter. The settling tank has a drain cock for the sediment, and the chemical tanks have also drain cocks to remove the impurities of the lime.

Capacity.—600 gallons per hour.

Dimensions.—8 feet long, 4 feet wide, 12 feet 9 inches high. Floor space, 32 square feet.

About sixty of these softeners of various sizes are in use in England and about 2,000 on the Continent.

Working.—The re-agent is alternately prepared in the left or right-hand tank by mixing together the correct amounts of lime, soda and water and stirring these by hand. The re-agent is drawn

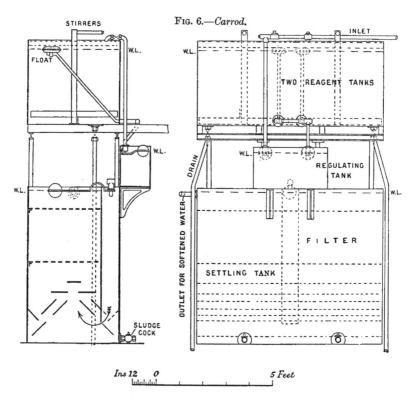

FIG. 6.—*Carrod.*

off near the surface through an india-rubber tube, supported by a float, and passes through a ball-tap into the mixing tank. The untreated water also enters this tank through a ball-tap, and the two quantities should, therefore, always be in a definite ratio. The water is let out of the mixing tank through another ball-tap and descends to the bottom of the settling tank, where baffles arranged in a peculiar manner are intended to assist precipitation. The precipitate is occasionally drawn off through a cock. The treated water now passes through the filter and thence to the feed-pump. The filtering material, wood-wool, which costs 4s. 6d. per cwt., is said to last two years.

Supply.—Well water.

Chemicals used.—Caustic soda, quantity not stated.

The users report that the softening (*sic*) is not carried quite so far for the quarry locomotives as for the fixed boiler, because of priming. The effect on the boilers is good, and the brass work or copper fire-boxes are not injured, but the gauge-glass, taps, &c., give more trouble than when ordinary water is used. The sediment from the apparatus is easily removed.

Result of the Chemical Analysis.—This apparatus is designed to use both lime and soda, but waste caustic soda and no lime was used, and the result is that the temporary hardness has been increased 10° and the permanent hardness decreased. This bad result is not due to the apparatus.

DESRUMAUX WATER SOFTENER.

The apparatus, Fig. 7 (page 803), consists of a cylindrical settling tank and a cylindrical lime tank. Over these is placed a distributor and a waterwheel which turns a hollow shaft reaching to the bottom of the lime tank, where several stirrers are attached to it. The same waterwheel also actuates a crank-shaft with a small scoop, which ejects the necessary quantity of soda solution out of the soda tank, which is placed over the settling tank. The settling tank has a central mixing trunk. Attached to the circumference of the settling tank are several screw-shaped inclined plates, the inner edges of which touch the mixing trunk. Higher up is a filter. At the discharge-water level there is a float which controls the water supply. There is also a lime slaking tank over the lime tank.

Capacity.—8,000 gallons per hour, worked at a rate of 5,000 gallons.

Dimensions.—Settling tank, 12 feet 6 inches diameter over angles; lime tank, 4 feet 3 inches over angles. Height of tanks, 34 feet.

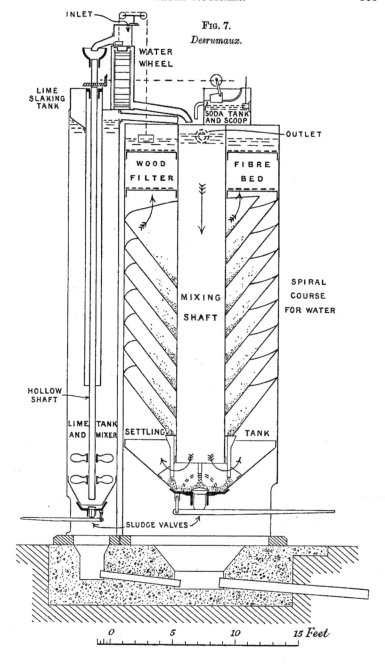

FIG. 7.
Desrumaux.

Total height, 40 feet. Floor space, 16 feet 9 inches by 12 feet 6 inches = 210 square feet.

These softeners are said to be working in all parts of the world.

Working.—The water supply, which is controlled by a float at the delivery, enters the distributor and through two carefully adjusted openings. One stream of water passes through the hollow shaft to the bottom of the lime tank, and gets converted into lime water which overflows into the mixing shaft. The other water stream descends over a water-wheel to the mixing shaft. The water-wheel rotates the stirrers in the lime tower, and also causes the small scoop to throw definite quantities of soda solution into the mixing shaft. The chemical re-action takes place here and the precipitate is formed. The partially cleared water then ascends the settling tank, being carried round and round the helical plates. The sediment which is here formed slips towards the centre and down the steep helical junction line. Before being discharged the water is filtered.

Supply.—Partly town's water, partly brook.

Chemicals used per 1,000 gallons.—2·56 lbs. lime, 0·59 lb. soda.

The analysed water was taken from a reservoir holding 730,000 gallons which was filled in one week, viz. 144 hours, equal to 5,000 gallons an hour.

The users report that the apparatus has been giving satisfaction since 1901. The boilers are now opened out only after steaming 1,000 hours, equal to about 4 months at 60 hours per week. Growths of soda appear on the fittings, but the brass is not attacked.

Result of Chemical Analysis.—The supply is the same as that for the Archbutt-Deeley apparatus, but being collected at another time its composition is slightly different. The hardness of the delivery is 4·3° as against 2·3° in the other case. This is chiefly due to a deficiency of the chemicals added.

DOULTON WATER SOFTENER.

The apparatus, Fig. 8 (page 805), consists of two cylindrical settling tanks with a filter in the upper part of one tank and a mixing funnel and tube in the other. The two tanks are connected near the bottom. Resting on these two tanks are three smaller ones, one for

FIG. 8.—*Doulton.*

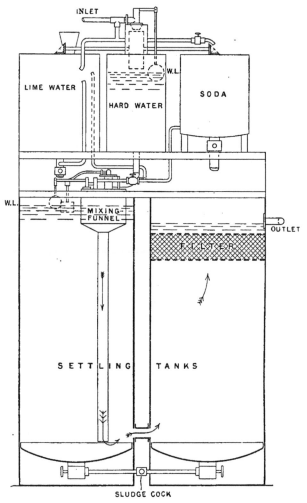

Scale, 5 feet = 1 inch.

the milk of lime, one for the untreated water, and one for the soda
solution. The lime and the soda tanks have stirrers driven by a
water-wheel. Each tank has a pipe leading to the mixing funnel.
Above these three tanks is a shaft with a water-wheel and bevel-
wheels which actuate the stirrer in the chemical tanks.

Capacity.—4,000 gallons per hour.

Dimensions.—Tank diameters, 6 feet; height of tanks, 14 feet
6 inches. Total height, 24 feet 3 inches. Floor space of tanks,
75 square feet.

About forty-five of these softeners of various sizes are in use.

Working.—Correctly determined quantities of soda and of burnt
lime are mixed with hot and cold water respectively, and poured into
their tanks. The water supply is then turned on, whereby the
water-wheel is rotated and the fluids mixed. The connection from
the centre tank to the mixing funnel is now opened, and as the water
rises in the mixing tank, the ball-regulator valves of the chemical
tanks open, and their contents also pass slowly into the mixing
funnel. Should the demand for purified water stop, the water rises
in the funnel and causes the three ball-taps to close, and after a time
the supply is also cut off automatically over the hard-water tank.
Should the water supply be cut off, then, by the sinking of the balls
of the taps, the chemical tanks will close automatically. A supply
of hard water, regulated by a ball-tap, flows into the lime tank.
The sediment which collects in the two settling tanks is run off through
two taps. The filtering material is occasionally renewed or washed.

Supply.—Well water.

Chemicals now used are caustic soda and carbonate of soda.

The users report that the water has not yet been used for feeding
the boilers. The water does not injure brass fittings. The sponge
filters give some trouble, and it is intended to put down two or three
settling tanks.

Result of Chemical Analysis.—The water treated by this apparatus
was heavily charged with mineral matter, and although the chemical
treatment was not correct the temporary hardness has been reduced
very low, namely to 3·1°. The water is, however, very alkaline, due to
the excess of soda, but there is no free carbonic acid.

HARRIS-ANDERSON WATER SOFTENER.*

The apparatus, Fig. 9 (page 809), consists of a distributor, a lime tank, soda tank, air vessel and re-action tank, and settling tank with a filter. The distributor has a circular weir, the water flowing from the circumference to three central compartments, and thence to the air vessel, to the lime tank and to the soda tank respectively. The air vessel, by means of its syphon, acts as an air pump and agitates the milk of lime in the lime tank, in which pipes and cones are arranged for this purpose. The soda tank consists of four cylinders, the solid soda being poured into a cage, the centre one, while the water flows into the second inner cylinder, and, according to the strength of the low-lying soda solution, either passes straight down or overflows into the second cylinder.

Capacity.—1,000 gallons per hour.

Dimensions.—Diameters of the re-action chamber and settling tank are 5 feet 6 inches. The diameter over platform round re-action chamber is 11¾ feet. Total height from bottom of settling tank to top of distributor, 17 feet 3 inches. Floor space of platform and settling tank, 158 square feet.

Working.—The water supply passes through a vertical turbine to the circumference of the distributor, thence over the circular weir, whose circumference is properly subdivided, to the air vessel, to the lime tank and to the soda tank. Alternately the water rises and sinks in the air vessel and drives the air into the lime tank, agitating the milk of lime, and then the water in the air tank is syphoned out and the action is repeated. The amount of water added to the lime regulates its supply, and can easily be altered by shifting the subdivision on the circular weir. The central cylinder of the soda tank is filled with soda crystals which partly dissolve in the water, the heavy soda solution settling at the bottom as shown. A small but constant supply of water is run from the distributing weir into the second cylinder, passing through the dense fluid at the bottom and up outside the third cylinder. Should, however, this rising soda solution be very dense, then the added water will flow over the top

* *Engineering*, 10 July 1903, page 52.

of the second cylinder and under the third cylinder. Thus, by adjusting the height of the second inner cylinder, a soda solution of a definite strength is obtained. The sediment formed in the re-action chamber and in the settling tank is run out through a sludge cock.

Supply.—Pit water.

Chemicals used per 1,000 gallons.—2·25 lbs. lime, 2·0 lbs. soda crystals.

The users report that the apparatus works satisfactorily, and that no scale, only mud, is formed in the boiler. The soda solution maintains its strength, but the milk of lime loses about 4 per cent. of its strength per hour.

Result of Chemical Analysis.—The water treated by this apparatus was also heavily charged with mineral matter, and, although an excess of soda was added, the deficiency of added lime resulted in a fairly high temporary hardness, viz. 6·20.

LASSEN AND HJORT WATER SOFTENER (BRUUN LOWENER).

The apparatus, Fig. 10 (page 811), consists of one large tank subdivided into three compartments; a re-action tank, settling tank, and filter. Immediately over the re-action chamber is a pivoted double-V-shaped oscillating trough, and above this is a semi-cylindrical chemical tank with stirrer and a valve, both actuated by the oscillating trough. Steam is employed in the re-action chamber for heating and mixing. The untreated water is discharged immediately over the double-V trough, and when one side is full it tips over and empties itself. In doing so a fin, projecting down into the re-action chamber, agitates the mixed water, and a small wheel on the top of the trough knocks against the bottom of a valve spindle and raises it for an instant. The valve is situated in the bottom of the semi-cylindrical chemical tank.

Capacity.—1,000 gallons per hour.

Dimensions.—Length over angle flange, 8 feet 8 inches. Width over angles, 4 feet. Height of tank, 4 feet. Total floor space, 48 square feet.

Working.—The chemicals, about 10 per cent. lime and the necessary soda, are mixed with water, and are poured into the semi-

Detail of Distributor.

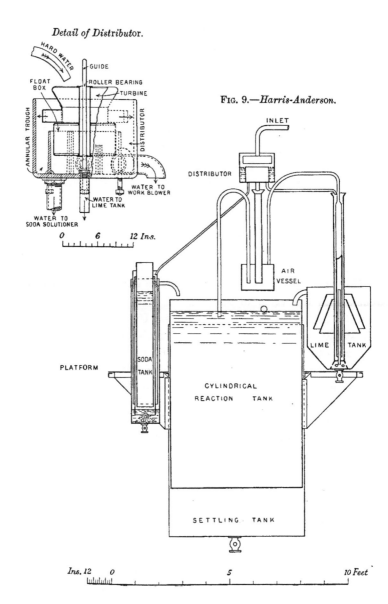

FIG. 9.—*Harris-Anderson.*

cylindrical tank, and the supply, which is regulated by a ball-tap at the discharge, is turned on, and runs into one of the pivoted V chambers of the oscillating trough. When one side is full, the trough tips over, discharges its water, opens and closes the valve in the bottom of the chemical tank, and by means of blades agitates the chemicals and the mixed water. If no steam is used, the settling tanks have to be larger than shown. The quantity of chemicals discharged through the valve with each tilt of the trough is regulated by the length of the valve rod. The chemicals, assisted by heat from the steam, cause the mineral matter to precipitate in the re-action and settling chamber. The water then passes through the filter to the discharge.

Exhaust steam can be used.

Supply.—No. 1, town's water; No. 2, well water.

Chemicals used per 1,000 gallons.—No. 1, 2 lbs. lime, 1 lb. soda; No. 2, 3·6 lbs. burnt lime, 2·8 lbs. calcined soda.

The users report that this softener works well. There is a little trouble with the valve in the chemical tank, which sticks if the apparatus has been resting for some time. Unless this tank is kept at least half full, the flow of chemicals diminishes.

Result of Chemical Analysis.—Two softeners were dealt with, but only one, referred to as No. 1 in the Table of Analyses (page 829), is illustrated. In this case the proper chemicals were used, and a considerable total hardness has been reduced to a very low limit. No. 2 apparatus dealt with a still harder water, in which the temporary and permanent hardness were about equal, and both have been reduced to nearly the lowest limit, but, a large excess of soda having been added, the alkalinity is high.

MAXIM'S WATER SOFTENER, "WARWICK" DETARTARISER.

The apparatus, Fig. 11 (page 813), consists of two cylindrical vessels placed one above the other. The untreated water-supply pipe enters the top of the upper vessel and ends in a spraying arrangement, while an exhaust steam-pipe, into which a grease separator is introduced, enters the bottom of the upper vessel. The lower vessel acts partly as a settling tank and partly as a filter.

Capacity.—3,000 gallons per hour.

Dimensions.—Diameter of lower cylinder over angles, 7 feet 6 inches. Total height, exclusive of sludge cocks and exhaust pipe, 24 feet. Floor space, exclusive of grease separator, is 56 square feet.

Working.—The hard water enters the upper chamber through a valve which is regulated by a float on the discharge. The necessary amount of soda solution is injected into this water by means of a pump connected to the feed. This water comes in contact with exhaust steam from the engine, from which most of the grease has

FIG. 10.—*Lassen and Hjort.* (No. 1.)

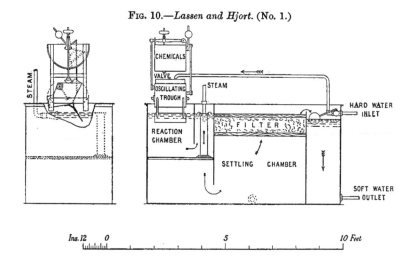

already been removed by a grease separator. The water thereby acquires a temperature of about 212° Fahr. The condensed steam is added to it, and it then falls into the lower tank. The heat and the soda cause the mineral matter to be precipitated, and the grease adheres to the sediment. As a safeguard a filter is fitted. The sludge is occasionally run out of the cocks in the lower tank.

Supply.—Pit water.

Chemicals used.—Soda, quantity not stated.

The users report that they are well pleased with the softener, and now have two at work. Partly because the boilers are now free from

3 к

scale, partly because their feed is now heated to 200° F., the softeners have paid for themselves in 12 months' time.

Result of Chemical Analysis.—The analysis shows that only 6 to 11 per cent. condensed steam were added to the supply. By subtracting 6 per cent. of pure water, the values in the third column of the Table of Analyses (page 830) are obtained. It will be seen that the slightly deficient carbonate of soda has converted nearly all the permanent hardness into temporary hardness, and that the small quantity of steam combined with a probably low temperature had no power to drive off the carbonic acid, which now held nearly 26 grains of carbonate of lime and magnesia in solution. Caustic lime or caustic soda ought to have been added to this water.

PORTER-CLARK WATER SOFTENER.

The apparatus, Fig. 12 (page 815), consists of several tanks placed side by side. The surface heater fitted at these works is not an essential feature of the softener. At the right-hand corner is the lime tank, next to it is the soda tank. The large tank is subdivided into a mixing tank, a settling tank, and next to it are the cloth filters. An engine-driven shaft passes over the lime, soda and mixing tank, and by means of bevel wheels actuates the stirrers. It also drives two small iron pumps which draw from the chemical tanks. The lime solution discharge is near the inlet of the untreated water. The soda discharge is at the outlet of the mixing tank. There is a collecting trough at one end of the settling tank, and adjoining it are eighteen cloth filters.

Capacity.—3,000 gallons per hour.

Dimensions.—Lime mixer, 4 feet 9 inches square, 5 feet high. Soda tank, 3 feet by 2 feet by 3 feet high. Mixing tank and settling tank, 10 feet by 6 feet by 5 feet high. Filter, 10 feet by 5 feet by 5 feet high. Floor space, 152 square feet. Heater, 3 feet diameter, 21 feet long.

Working.—The water, after passing the surface heater, enters the mixing tank, where it gets mixed with milk of lime which is pumped

Fig. 11.—*Maxim's, "Warwick."*   *Detartariser.*

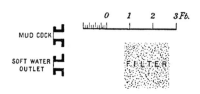

into this tank in the correct proportion. The partly treated water while flowing out of this tank is mixed with a regulated quantity of soda solution and slowly passes along the settling tank where most of the precipitate is deposited. It then overflows into a distributing tank, and thence on to the cloth filters and to the feed pump.

Supply.—Not stated.

Chemicals used per 1,000 gallons, 7 lbs. lime putty and 1·5 lbs. soda crystals.

The users report that the water was intended more for washing purposes than for boilers; in these some pitting was noticed. Subsequently the softener was replaced by another patent.

Result of Chemical Analysis.—In this case the supply had much temporary and permanent hardness, but as only half the necessary quantity of soda was introduced the result was naturally very unsatisfactory, and it is impossible to judge how the apparatus would have acted if properly worked.

PULSOMETER WATER SOFTENER. CRITON.

Working.—Details from which to make a sketch of this softener were not obtainable, for this company have redesigned their softener since the samples were analysed. The flow of the water and chemicals was regulated by syphon and ball-taps.

Supply.—Well water.

Chemicals used.—Lime, but the quantity is not stated.

The users were apparently not aware that they had no permanent hardness in their supply, and they report that they formerly added soda; this naturally caused their boiler fittings to give trouble, which has now ceased. No scale, only mud, is formed in the boiler.

Result of Chemical Analysis.—The water treated was alkaline and had no permanent hardness. No soda was, therefore, added. Unfortunately an excess of lime was used, which produced 2 grains of caustic soda, and which accounts for the resultant high temporary hardness and high alkalinity. An addition of water having permanent hardness would have improved the product.

FIG. 12.—*Porter-Clark.*
Plan.

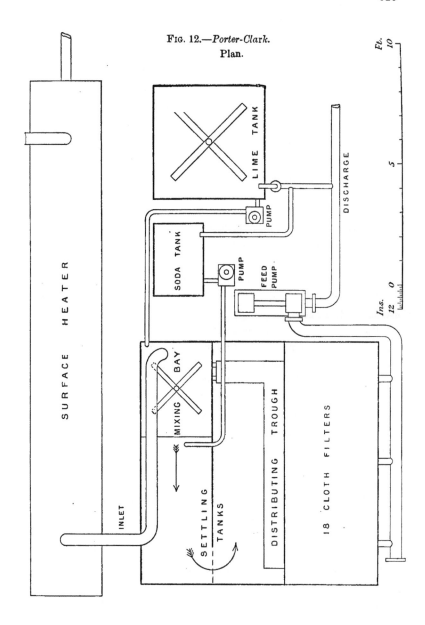

REISERT WATER SOFTENER. TYPE E.

The apparatus, Fig. 13 (page 817), consists of a large settling tank and filter combined, over which is placed a distributor. In addition there is a conical lime tank and a cylindrical soda tank. Pipes with carefully gauged apertures lead the water from the distributor to the top of the soda tank and to the bottom of the lime tank and to the bottom of the settling tank. A syphon from the soda tank and an overflow from the lime tank are also led into the bottom of the re-action tank. A pipe passes from the supply under the filter for automatically washing away the precipitate on the filter when the resistance is increased so much as to raise the water level. The mud in the settling tank is drawn off through a sludge cock.

Capacity.—1,300 gallons per hour.

Dimensions.—Settling tank, 5 feet 3 inches diameter. Lime tank, 3 feet 9 inches diameter at top to 6 inches at bottom. Total height, 22 feet 6 inches. Floor space, 41 square feet.

About 2,000 of these softeners of various sizes are said to be in use on the Continent.

Working.—The correct quantity of lime is placed in a small tank, slaked and converted into milk of lime, which is then run into the bottom of the conical lime tank. The right quantity of soda and water is placed in the cylindrical soda tank, and steam is turned on until the soda is dissolved and the contents of the tank are well mixed and of uniform density. The water supply, which has been temporarily stopped, is now re-started, and passes from the distributor through three carefully gauged holes. One portion goes direct to the bottom of the settling tank into a little mixing chamber. Another portion flows to the bottom of the lime tank, the same quantity of concentrated lime water flowing off the top of this tank into the mixing chamber. The third portion of water is led to the top of the soda tank, and being lighter than the soda solution it floats on its surface, and by depressing it, causes it to rise up the syphon pipe to the mixing chamber. The three fluids being well mixed precipitation takes place, the sludge settling in the bottom of the settling tank and being drawn off occasionally. The partly clarified water passes

FIG. 13.—*Reisert.*

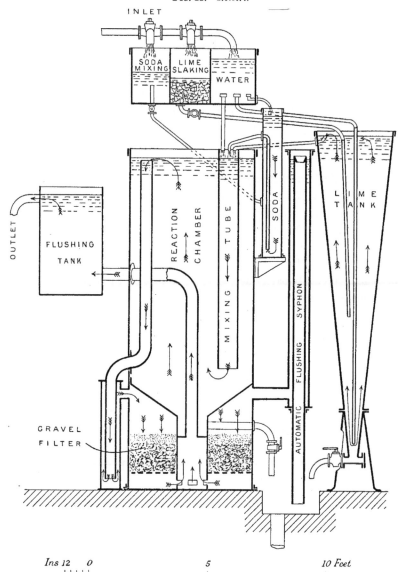

Ins 12   0   5   10 Feet

through a pipe to the bottom of the gravel filter where the remaining
precipitate is removed, and the purified water then passes to the feed-
pump. The filter is occasionally cleaned by an automatic reversal
of the flow, whenever the filter is fouled. The mud passes down the
waste pipe.

Supply.—Well water.

Chemicals used per 1,000 gallons.—2·7 lbs. lime and 1·4 lbs. soda.

The users report that the apparatus has worked satisfactorily for
two and a half years, and the boilers have given no trouble. It
takes one man half-an-hour every twelve hours to charge the
chemicals.

Result of Chemical Analysis.—This is one of the few softeners in
which an excess of lime was added, with the result that nearly all
the magnesia was removed, but the excess of lime has also caused
the temporary hardness to be fairly high, namely 4·4°. The treated
water also shows 3·5° of permanent hardness, which could have been
entirely removed if enough soda had been added. This insufficiency
of soda is probably partly due to the presence of sulphate of lime
as an impurity in the burnt lime.

STANHOPE CONTINUOUS WATER SOFTENER.

The apparatus, Fig. 14 (page 819), consists of a lime-mixing tank,
with a small compartment for the mixing chain, a re-action tank and a
settling tank with inverted perforated funnels and a filter. Above
these cylindrical tanks is a distributor, a soda tank and a lime-
slaking tank. The distributor has three carefully adjusted outlets,
one discharging into the re-action tank, one into the lime tank,
and another into the soda tank. The chain in the lime mixing tank
is driven by a water wheel.

Capacity.—2,500 gallons per hour, but worked at 1,000 gallons
per hour.

Dimensions.—Settling tank, 7 feet 6 inches diameter over angles.
Re-action and settling tower, 5 feet 6 inches by 2 feet 6 inches over
angles. Height, 26 feet. Floor space, 70 square feet.

About ninety of these softeners of various sizes are in use.

FIG. 14.—*Stanhope Continuous.*

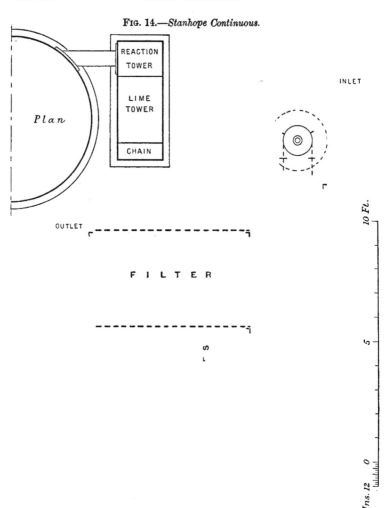

Working.—The water enters the distributor, the supply being regulated by a ball-tap at the discharge. Three carefully gauged outlets lead this water respectively into the mixing tank, into the lime-mixing tank, and into the soda tank. The slaking of the lime is done in a separate tank. The re-action between the lime, soda, and untreated water takes place at the same time. The treated water passes upwards through perforations in the cones and then through the filter. The sludge falls down through the openings in the centre of the cones and is run out of the sludge cock.

Supply.—Town's water.

Chemicals used per 1,000 gallons.—0·5 lb. lime, 0·5 lb. soda crystals.

The users report that the sediment in the softener is easily removed. The filters are cleared twice a year, which takes about two days. Hardness of treated water varies from 2° to 4·5°.

Result of Chemical Analysis.—Insufficient lime and soda were added, but as the water contained little magnesia salts, the result was a good one, the combined hardness being reduced to 2·6°.

TYACKE WATER SOFTENER.

Working.—The apparatus, Fig. 15 (page 821), consists of a cylindrical settling tank with a central funnel and mixing tank. It has a sludge cock at the bottom. Above the funnel are two regulating tanks with ball-taps; above them are two chemical tanks with floating discharge pipes. Only one tank is used at a time. Sufficient lime and water to produce clear lime water is put alternately into one of the two chemical tanks; the fluid is drawn off through a floating discharge, the quantity being regulated by a ball-tap in a lower tank. The water to be treated is discharged into a similar tank with a ball-tap, and the mixture passes to the bottom of a settling tank and is drawn off at the surface. The apparatus has been in use since 1894.

Result of Chemical Analysis.—This apparatus was supplied with the hardest of the waters submitted, there being 55° permanent hardness and 21° temporary hardness. Unfortunately far too little lime was added, so that the 18 grains of magnesium salts were hardly reduced at all. Only half the necessary quantity of soda was added, and the general result was most unsatisfactory. It is

FIG. 15.—*Tyacke.*

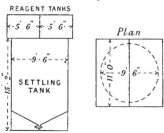

impossible to judge from this case how the apparatus would have behaved with less difficult water, and if the chemicals had been properly porportioned.

WOLLASTON WATER SOFTENER.

The apparatus, Fig. 16 (page 823), consists of a settling tank and a re-action tank, surmounted by a feed-heater with numerous trays. The inlet for the chemicals is at the bottom of this heater, from which a pipe (shown in dotted lines) leads to the bottom of the first re-action tank. An external overflow pipe leads from this tank to the bottom of the settling tank, which is provided with settling planes. The pump for the chemicals is actuated by the feed-pump.

Capacity.—4,000 gallons per hour.

Dimensions.—Re-action tank, 3 feet 4 inches diameter over angles. Settling tank, 5 feet 5 inches by 10 feet 5 inches; height, 13 feet. Feed-heater, 2 feet 7 inches diameter, 8 feet high. Floor space, 66 square feet.

About fifteen of these softeners of various sizes are in use.

Working.—The water is admitted at the top of the feed-heater, exhaust steam which has passed through a grease separator being blown into the bottom. The descending water passes over several trays and then meets with a stream of soda solution which completes the re-action initiated by the heat of the steam. The grease and precipitated lime adhere to each other and mutually increase their tendency to separate from the water. Most of the precipitate remains in the re-action tank, and only a small quantity is carried into the settling tank. No filter was used.

Supply.—Well water.

Chemicals used.—Caustic soda, quantity not stated.

The users report that the effect on the boiler is satisfactory except that old scale is not loosened. With recent installations, in which there is no excess of alkalinity, the old scale is rapidly coming away, even from within the Makin's boiler cones.

Result of Chemical Analysis.—The analysis shows that about 33 per cent. of condensed steam was added to the supply which has to be subtracted from the delivery. This has been done in the third column of Table of Analyses (page 833), where it will be seen that a total hardness of $20 \cdot 5°$ had been reduced to $0 \cdot 86°$, but the alkalinity of the treated water is so high, due to 18 grains of carbonate of soda, that one might expect that the concentrated boiler water would attack the brass fittings. The users make no complaint on this subject. At these works waste caustic soda from the mercerisers was available, of which an excess of 2 lbs. per 1,000 gallons was used. It would have been better to use only $0 \cdot 809$ lb. of caustic soda and $1 \cdot 076$ lbs. of lime. The lime would have dealt with 13° of temporary hardness, and the caustic soda, after combining with the carbonic acids, would have dealt with the permanent hardness, and practically no carbonate of soda would have been found in the delivery.

WRIGHT WATER SOFTENER.

The apparatus, Fig. 17 (page 824), consists of a rectangular settling tank, filter, mixing tank, and lime tank combined, surmounted by a distributor. At the side of the lime tank is a lime-slaking tank and a soda tank, not shown. The settling tank has a large number of 6-inch baffle plates, inclined at an angle of 45°, on which the mud settles and slips down. Above these baffles is a filter, and at the discharge level is a float attached to a leather hose in the distributor, which causes the two water levels to rise and fall together. Another float in the distributor regulates the supply of unsoftened water.

This arrangement has been superseded. The distributor has a weir with two adjustable overflows to the mixing tank and lime tank. A mixing chain reaches to the bottom of the lime tank, and a small bucket chain draws from the soda tank. These chains are actuated by a water-wheel.

Fig. 16.—*Wollaston.*

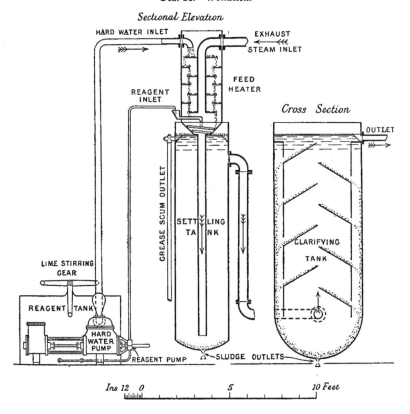

Capacity.—10,000 gallons per hour, but only worked up to 5,000 gallons.

Dimensions.—Combined tank, not including lime-slaking tank and platform, 6 feet 5 inches wide, 13 feet 6 inches long, 23 feet high.

In addition there are two cylindrical settling tanks, not shown, 7 feet 6 inches diameter, 27 feet high.   Floor space of apparatus, 86 square feet ; ditto, including two tanks, 112 square feet.

FIG. 17.—*Wright.*

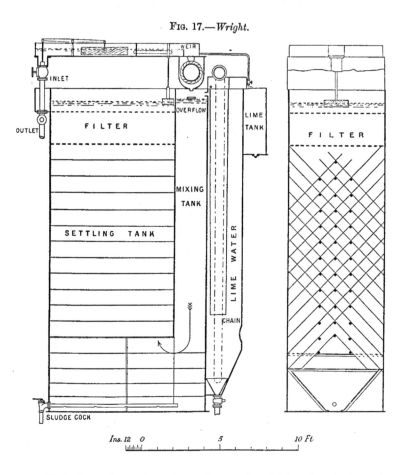

This firm also make water softeners in which exhaust steam is utilised.

Working.—The untreated water flows over a weir, one part descending over a water-wheel direct into the mixing tank, the other part passes to the bottom of the lime tower, forcing the same quantity of clear lime water into the mixing tank. A small, but definite, quantity of soda solution is also tipped into the mixing tank, the bucket chain being actuated by the water-wheel. From the bottom of the mixing tank the mixed water passes up through the settling tank, most of the mud settling on the large number of baffles and slipping down. It is now and then removed through the sludge cock. The mud which gets arrested at the bottom side of the filter is easily washed into the settling tank by opening the sludge cock.

The users report that the boilers are fed partly by softened, partly by unsoftened, water, and the effect on the boilers has not been noted. The baffle plates were cleaned twice in twelve years. Half the filter stockings are changed every day. The mud is easily run off. There used to be a deposit in the pipes; now two boiler shells, 27 feet high by 7 feet 6 inches diameter, are used as extra settling tanks.

Result of Chemical Analysis.—The water treated by this apparatus presented no serious difficulties, but here again the mistake was made of adding too little lime and too much soda, and although the delivery is fairly soft its alkalinity is rather high.

(*Continued to page 833.*)   APPENDIX I

Chemical Analyses.

Name of Water Softener.			ARCHBUTT-DEELEY.		THE ATKINS C	
Chemicals, &c.		Chemical Symbols.	Supply.	Delivery.	Supply.	Delive
			Grains per Gallon.			
Carbonic Acid Gas.		CO_2	7·161	0·63	6·795	1·2
Scale-forming Constituents.	Calcium	Sulphate $Ca\ SO_4$	2·639	0·000	0·715	0·
		Carbonate $Ca\ CO_3$	6·659	1·704	13·909	1·
		Silicate $Ca\ Si\ O_3$	1·733	1·175	1·110	1·
	Magnesium.	Hydrate $Mg\ (OH)_2$	—	—	—	
		Carbonate $Mg\ CO_3$	5·418	0·073	—	
		Silicate $Mg\ Si\ O_3$	—	—	—	
	Ferric Oxide		0·330	0·148	0·024	
	Total		16·779	3·100	15·758	
Non-Scale-forming Constituents.	Sodium.	Nitrate $Na\ NO_3$	—	—	0·716	0·7
		Chloride $Na\ Cl$	2·675	3·211	2·881	2·9
		Sulphate $Na_2\ SO_4$	0·000	5·402	—	—
		Carbonate $Na_2\ CO_3$	—	—	—	—
		Bi-Carbonate $Na\ H\ CO_3$	0·000	1·401	—	—
		Silicate $Na_2\ Si\ O_3$	—	—	—	—
		Hydrate $Na\ HO$	—	—	—	—
	Calcium.	Nitrate $(Ca\ NO_3)_2$	—	—	2·903	2
		Chloride $Ca\ Cl_2$	—	—	—	
	Magnesium.	Nitrate $Mg\ (NO_3)_2$	—	—	0·904	0
		Chloride $Mg\ Cl_2$	0·251	0·000	—	
		Sulphate $Mg\ SO_4$	—	—	—	
	Total		2·926	9·814	7·404	
Total Mineral Matter . . .			19·705	12·914	23·162	
Alkalinity			14·582	3·632	14·680	
Temporary Hardness			14·582	2·799	14·680	
Permanent Hardness			2·203	0·000	3·165	
Total Hardness			16·785	2·799	17·845	

Chemicals required in pounds per 1,000 gallons *

* The minus sign (-) stands for excess.

ANALYSES.   (Continued on next page.)

Chemical Analyses.

BABCOCK AND WILCOX.			BELL.		BOBY.		
Supply.	Delivery.	Delivery less Steam.	Supply.	Delivery.	Supply.	Delivery.	Delivery less Steam
Grains per Gallon.							
3·940	4·070	—	2·39	3·22	10·36	1·40	—
8·001	0·000	0·000	29·918	0·000	17·126	4·532	5·734
2·549	6·319	9·972	0·000	1·706	17·271	0·205	0·259
2·615	1·184	1·868	—	—	2·176	1·560	1·974
3·339	1·365	2·154	3·619	3·598	2·966	1·768	2·237
—	—	—	0·913	0·800	--	—	—
0·611	0·008	0·013	0·008	0·005	0·044	0·011	0·014
17·115	8·876	14·007	34·458	6·109	39·583	8·076	10·218
16·676	10·612	16·745	8·690	8·669	5·423	4 739	5·995
32·736	26·045	41·100	12·491	56·700	1·157	10·322	13·059
0·000	11·923	18·815	0·000	9·844	—	—	—
—	—	—	—	—	—	—	—
—	—	—	—	—	—	—	—
—	—	—	10·662	0·000	—	—	—
49·412	48·580	76·660	31·843	75·213	6·580	15·061	19·054
66·527	57·456	90·667	66·301	81·322	46·163	23·137	29·272
8·757	16·050	28·960	5·199	12·628	22·661	3·642	4·606
8·757	.8·954	14·131	5·199	6·770	22·661	3·642	4·606
5·883	0·000	0·000	30·858	0·000	12·590	3·332	4·216
14·640	8·954	14·131	36·057	6·770	35·251	6·974	8·822
Chemicals required in pounds per 1,000 gallons.*							
1·034	0·000	—	1·599	1·459	2·166	0·000	—
0·892	- 1·075	—	4·676	- 0·888	1·908	0·500	—
0·635	—	—	2·922	—	1·354	—	—

* The minus sign (-) stands for excess.

3 L

(Continued from previous page.)   APPENDIX I

Chemical Analyses.

Name of Water Softener.				CARROD.		DESRUMAUX.	
Chemicals, &c.			Chemical Symbols.	Supply.	Delivery.	Supply.	Delive
				Grains per Gallon.			
Carbonic Acid Gas.			CO_2	6·21	1·14	5·83	1·25
Scale-forming Constituents.	Calcium.	Sulphate	$Ca\,SO_4$	33·105	3·430	2·963	0·6
		Carbonate	$Ca\,CO_3$	0·000	4·780	5·990	1·4
		Silicate	$Ca\,Si\,O_3$	2·893	3·902	1·909	0·1
	Magnesium.	Hydrate	$Mg\,(OH)_2$	—	—	—	—
		Carbonate	$Mg\,CO_3$	10·999	14·562	5·115	1·8⁻
		Silicate	$Mg\,Si\,O_3$	—	—	—	—
	Ferric Oxide			0·113	0·155	0·063	0·05
	Total . . .			47·110	26·829	16·040	4·2
Non-Scale-forming Constituents.	Sodium.	Nitrate	$Na\,NO_3$	—	—	—	—
		Chloride	$Na\,Cl$	3·380	3·671	2·818	3·60
		Sulphate	$Na_2\,SO_4$	5·449	38·528	0·000	3·15
		Carbonate	$Na_2\,CO_3$	—	—	—	—
		Bi-Carbonate	$Na\,H\,CO_3$	—	—	—	—
		Silicate	$Na_2\,Si\,O_3$	—	—	—	—
		Hydrate	$Na\,HO$	—	—	—	—
	Calcium.	Nitrate	$Ca\,(NO_3)_2$	—	—	—	—
		Chloride	$Ca\,Cl_2$	—	—	—	—
	Magnesium.	Nitrate	$Mg\,(NO_3)_2$	—	—	—	—
		Chloride	$Mg\,Cl_2$	—	—	0·228	0·0
		Sulphate	$Mg\,SO_4$	1·204	0·000	—	—
	Total			10·033	42·199	3·046	6·
Total Mineral Matter . . .				57·143	69·028	19·086	10·
Alkalinity				15·54	25·410	13·701	3·
Temporary Hardness				15·54	25·410	13·701	
Permanent Hardness				25·34	2·522	2·419	
Total Hardness				40·88	27·932	16·120	
Chemicals required							

* The minus sign (-) stands for excess.

ANALYSES.   (Continued on next page.)

Chemical Analyses.

DOULTON.		HARRIS-ANDERSON.		LASSEN AND HJORT. (1)		LASSEN AND HJORT. (2)	
Supply.	Delivery.	Supply.	Delivery.	Supply.	Delivery.	Supply.	Delivery.
Grains per Gallon.							
4·78	0·00	8·856	0·000	6·84	1·39	10·000	1·87
22·772	0·00	18·680	0·000	1·99	—	20·472	0·000
12·305	0·814	} 6·53	1·938 {	14·701	1·009	5·588	0·311
—	—			2·308	1·951	1·338	0·000
—	—	0·291	0·000	—	—	—	—
0·000	1·896	9·177	3·235	0·000	0·365	13·110	2·841
—	—	—	—	—	—	—	0·398
0·147	0·112	0·244	0·111	0·017	0·091	0·067	0·047
35·224	2·822	34·571	5·575	18·116	3·416	40·575	3·597
0·000	25·932	—	—	—	—	—	—
8·238	10·755	23·971	23·971	1·145	3·161	4·586	8·887
0·000	26·462	30·717	49·462	0·000	1·881	0·000	19·980
0·000	13·921	0·000	0·671	—	—	—	—
0·000	30·502	—	—	0·000	1·783	0·000	19·128
—	—	—	—	—	—	—	—
14·626	0·000	—	—	—	—	—	—
—	—	—	—	0·707	0·000	—	—
9·380	0·000	—	—	—	—	—	—
1·177	0·000	—	—	0·803	0·000	2·961	0·000
33·421	107·572	54·688	74·104	2·655	6·825	7·547	47·995
68·645	110·394	89·259	79·679	21·671	10·241	48·122	51·592
12·303	34·33	17·350	6·907	16·685	4·179	22·297	15·459
12·303	3·064	17·350	6·275	16·685	3·118	22·297	4·077
33·215	0·000	13·731	0·000	2·945	0·000	18·160	0·000
45·518	3·064	31·081	6·275	19·630	3·118	40·457	4·077
Chemicals required in pounds per thousand gallons.*							
1·5	1·5	2·476	0·312	1·31	0·000	3·31	1·56
5·0	-4·274	2·081	-0·095	0·44	-0·16	2·75	-1·69
3·11	—	1·475	—	0·4	—	1·965	—

* The minus sign (-) stands for excess.

3 L 2

(*Continued from previous page.*)   APPENDIX I

Chemical Analyses.

Name of Water Softener.				MAXIM.		
Chemicals, &c.			Chemical Symbols.	Grains per Gallon		
Carbonic Acid Gas			CO_2	6·33	5·67	
Scale-forming Constituents.	Calcium	Sulphate	$Ca\ SO_4$	38·600	4·746	5·0
		Carbonate	$Ca\ CO_3$	0·000	10·149	12·3
		Silicate	$Ca\ Si\ O_3$	0·984	1·170	1·2
	Magnesium	Hydrate	$Mg\ (OH)_2$	—	—	—
		Carbonate	$Mg\ CO_3$	12·032	15·333	16·3
		Silicate	$Mg\ Si\ O_3$	—	—	—
	Ferric Oxide			0·367	0·112	
	Total			51·983	31·510	
Non-Scale-forming Constituents.	Sodium	Nitrate	$Na\ NO_3$	—	—	—
		Chloride	$Na\ Cl$	130·790	117·05	124·740
		Sulphate	$Na_2\ SO_4$	14·743	55·772	59·382
		Carbonate	$Na_2\ CO_3$	—	—	—
		Bi-Carbonate	$NaH\ CO_3$	—	—	—
		Silicate	$Na_2\ Si\ O_3$	—	—	—
		Hydrate	$Na\ HO$	—	—	—
	Calcium	Nitrate	$Ca\ (NO_3)_2$	—	—	—
		Chloride	$Ca\ Cl_2$	—	—	—
	Magnesium	Nitrate	$Mg\ (NO_3)_2$	—	—	—
		Chloride	$Mg\ Cl_2$	—	—	—
		Sulphate	$Mg\ SO_4$	8·122	0·000	0·0
	Total			153·655	172·772	184·1
Total Mineral Matter . . .				205·635	204·282	219·2
Alkalinity				15·128	29·346	31·2
Temporary Hardness				15·128	29·346	31·2
Permanent Hardness				35·128	3·490	3·7
Total Hardness				50·256	32·836	34·9
Lime			$Ca\ O$	2·834		
Soda-Ash			$Na_2\ CO_3$	5·323		
Cost per 1,000 gallons . . .			Pence	3·44		

* The minus sign (-) stands for excess.

ANALYSES.

(*Continued on next page.*)

Chemical Analyses.

PORTER-CLARK.		PULSOMETER.		REISERT.		STANHOPE.	
Supply.	Delivery.	Supply.	Delivery.	Supply	Delivery.	Supply.	Delivery.
Grains per Gallon.							
5·03	0·43	5·54	0·00	6·68	0·000	5·70	0·32
12·178	10·074	—	—	4·696	4·762	4·084	1·192
8·971	0·000	5·193	1·551	10·423	1·456	1·259	0·516
1·694	0·493	3·671	2·541	2·050	1·355	1·371	1·184
—	—	0·000	0·376	0·000	0·284	—	—
0·109	0·918	4·270	0·000	1·453	0·000	0·049	0·938
—	—	—	—	—	0·737†	—	—
0·246	0·057	0·047	0·034	0·251	0·054	0·082	0·036
23·198	11·542	13·181	4·502	18·873	8·647	6·845	3·866
0·000	12·261	—	—	0·106	6·247	—	—
5·463	5·576	10·556	11·037	3·750	3·773	1·443	2·801
0·000	1·933	8·729	7·186	0·000	0·943	0·00	2·743
—	—	0·000	3·484	—	—	—	—
—	—	7·336	0·000	—	—	—	—
—	—	—	—	—	—	—	—
—	—	0·000	2·158	—	—	—	—
—	—	—	—	5·003	0·000	—	—
—	—	—	—	—	—	—	—
12·309	1·621	—	—	—	—	—	—
—	—	—	—	—	—	1·010	0·00
—	—	—	—	—	—	—	—
17·772	21·391	26·621	23·865	8·859	10·963	2·453	5·544
40·970	32·933	39·802	28·367	27·732	19·610	9·298	9·410
11·432	1·512	17·780	10·356	13·910	3 501	13·830	2·656
11·432	1·512	13·416	4·379	13·910	3·501	13·830	2·656
17·248	8·498	0·000	0·000	6·502	4·424	4·063	0·867
28·68	10·01	13·416	4·379	20·412	7·925	17·893	3·523
Chemicals required in pounds per 1,000 gallons.*							
1·658	0·253	1·412	− 0·218	1·353	− 0·105	1·127	0·146
2·614	1·288	—	—	0·985	0·530	0·616	0·132
1·71	—	0·15	—	—	—	0·49	—

* The minus sign (-) stands for excess.   † (Reisert delivery) calcium oxide, CaO.

(Continued from previous page.) APPENDI

Chemical Analyses.

Name of Water Softener.				TYACKE.	
Chemicals, &c.			Chemical Symbols.	Supply.	Deli\
				Grains per Gallo	
Carbonic Acid Gas.			CO_2	9·29	4·'
Scale-forming Constituents.	Calcium	Sulphate	$Ca\,SO_4$	66·576	32·
		Carbonate	$Ca\,CO_3$	7 649	0·
		Silicate	$Ca\,SiO_3$	2·458	1·
	Magnesium	Hydrate	$Mg\,(OH)_2$	—	
		Carbonate	$Mg\,CO_3$	9·586	8
		Silicate	$Mg\,SiO_3$	—	
	Ferric Oxide			0·314	0·
	Total			86·583	42·
Non-Scale-forming Constituents.	Sodium	Nitrate	$Na\,NO_3$	—	
		Chloride	$Na\,Cl$	19·415	20
		Sulphate	$Na_2\,SO_4$	0·00	35
		Carbonate	$Na_2\,CO_3$	—	
		Bi-Carbonate	$Na\,HCO_3$	—	
		Silicate	$Na_2\,SiO_3$	—	
		Hydrate	$Na\,HO$	—	
	Calcium	Nitrate	$Ca\,(NO_2)_2$	—	
		Chloride	$Ca\,Cl_2$	—	
	Magnesium	Nitrate	$Mg\,(NO_3)_2$	—	
		Chloride	$Mg\,Cl_2$	0·978	0·
		Sulphate	$Mg\,SO_4$	6·061	6·
	Total			26·454	62·
Total Mineral Matter				113·037	105·
Alkalinity				21·139	1:
Temporary Hardness				21·139	
Permanent Hardness				55·008	
Total Hardness				76·147	

Chemicals required in pounds

Lime   .   .   .
Soda-Ash .   .   .
Cost per 1000 gallons .

* The minus signs (-) stands for excess.

ANALYSES.　　　　　　　　　　　　　　　(*Concluded from page 82*

Chemical Analyses.

	WOLLASTON.		WRIGHT.	
Supply.	Delivery.	Delivery less Steam.	Supply.	Delivery.
Grains per Gallon.				
6·21	0·00	—	3·406	0·000
9·610	0·000	0·000	14·174	—
5·592	0·355	0·471	2·033	0·162
1·737	0·000	0·000	1·574	1·778
0·000	0·171	0·227	—	—
5·400	0·000	0·000	3·201	1·635
—	—	—	—	—
0·371	0·024	0·032	—	0·002
22·710	0·550	0·730	20·982	3·577
—	—	—	—	—
7·503	8·013	10·635	1·351	3·897
0·551	7·982	10·594	0·000	12·703
0·000	13·702	18·183	0·000	6·142
—	—	—	0·000	4·630
0·000	1·187	1·575	—	—
0·000	0·502	0·666	—	—
—	—	—	—	—
—	—	—	—	—
—	—	—	1·737	0·000
—	—	—	—	—
8·054	31 386	41·653	3·088	27·372
30·764	31·936	42·383	24·070	30·949
13·493	15·161	20·121	7·184	12·173
13·493	0·650	0·862	7·184	3·628
7·065	0·000	0·000	12·246	0·000
20·558	0·650	0·862	19·430	3·628
Chemicals required in pounds per 1,000 gallons.*				
1·641	⎰ excess of	—	1·069	0·625
1·071	⎱ caustic soda	—	1·856	- 1·295
0·806	⎰ - 1·661	—	1·205	—

* The minus signs (-) stands for excess.

APPENDIX IV.

INSTRUCTIONS FOR CONTROLLING CHEMICAL TREATMENT.

Chemicals Required. — Standard Soap solution, Wanklyn's strength (1 cm^3 * = 0·001 $gr.$ $CaCO_3$), Standard $\frac{N}{50}$ acid solution (1 cm^3 = 0·001 $gr.$ $CaCO_3$), Standard $\frac{N}{50}$ sodium carbonate solution, must be free from bicarbonate. Powdered ammonium oxalate crystals, methyl orange and phenol-phthalein indicators.

Apparatus. — Three 50 cm^3 burettes graduated to $\frac{1}{10}$ cm^3. Three 8-oz. glass-stoppered bottles, funnel and filter papers, two white porcelain basins, 4½ inches diameter, and one 70 cm^3 pipette.

I.—70 cm^3 of the water are placed in a clean bottle, and the soap solution run in in small quantities at a time, and the bottle vigorously shaken after each addition. As soon as lather appears, lay the bottle on its side, and observe if the lather persists for five minutes. When this is the case, read off the quantity of soap solution used and, because of some uncombined soap in the lather, subtract 1 cm^3, the remainder is the total hardness, A. When the water requires more than 16 cm^3 of soap solution, it must be diluted with 70 cm^3 of distilled water, and 2 cm^3 must be subtracted from the reading.

II.—Take about 120 cm^3 of the water, add about 4 grains of ammonium oxalate, shake and let it stand for ¼ hour ; filter, testing the filtrate with ammonium oxalate to ensure complete precipitation, and take 70 cm^3 of this filtrated water and conduct the soap test as before. The reading less 1 cm^3 is the reduced hardness, B.

III.—70 cm^3 of the water are placed in a porcelain basin, one drop of the methyl orange indicator added, and standard acid dropped in from the burette until the yellow tint just changes to rose pink. Read off the alkalinity (sometimes called temporary hardness), C.

---

* cm^3 is used instead of *c.c.*, in conformity with the arrangement recommended (Proceedings 1900, June, page 394).

IV.—70 cm^3 of the water are placed in a porcelain basin and 3 or 4 drops of phenol phthalein added. If the water is acid it remains milky (generally so with natural waters). In this case (1) add the standard sodium carbonate until a pale persistent pink colour is obtained. Read off the phenol phthalein acidity, D.

The water—especially softened water—may turn the phenol phthalein pink. .

In this case (2) the 70 cm^3 are titrated with the standard acid until the pink colour is almost destroyed. Read off the phenol phthalein alkalinity, E.

Tests on Hard Water.—Re-agents required in lbs. for softening 1,000 gallons:—

Soda ash.—$0 \cdot 15 \ (A - \frac{1}{3} B - C)$ lb.

Lime.—Case IV (1). (Water is acid to phenol phthalein) $0 \cdot 085 \ (\frac{2}{3} B + C + D)$ lb.

Lime.—Case IV (2). (Water is alkaline to phenol phthalein) $0 \cdot 085 \ (\frac{2}{3} B + C - 2 E)$ lb.

Using these values the ratio of lime water to total supply is $\dfrac{\text{Lime}}{13 + \text{Lime}}$.

Tests on the Softened Water.—If an excess of soda ash has been added, the amount in excess, $0 \cdot 15 \ (A - \frac{1}{3} B - C)$ lb., is negative.

When too little soda ash has been added, the insufficiency is $0 \cdot 15 \ (A - \frac{1}{3} B - C)$ lb., as in the untreated water.

If too much lime has been added, the amount in excess is $0 \cdot 085 \ (2 C - 2 E - \frac{2}{3} B)$ lb.

If too little lime has been added, the insufficiency is $0 \cdot 085 \ (\frac{2}{3} B + C - D)$ lb., as in the untreated water.

Discussion.

Mr. C. E. STROMEYER drew the attention of the members to the samples which had been placed upon the table, containing the scale or mud which had come out of boilers using respectively unsoftened and softened water. He also exhibited various crystals and deposits, some of which are illustrated in Fig. 18, Plate 36, and a piece of an economiser pipe, Fig. 19, fed with unsoftened water which was so encrusted, that there was not even a small hole through the centre for the water to pass through. It would be found mentioned in Appendix II that the scale in economisers was practically nothing else but carbonate of lime. Just before coming to the Meeting, a case had been mentioned to him of two boilers, which had been fed with the same water; the low-pressure boiler (30 lbs.) had no scale and the high-pressure boiler (60 lbs.) had a hard sulphate of lime scale. The change of solubility with a rise of temperature, as given in Appendix I (page 789) might readily explain this difference, the saturation point in the low-pressure boiler not being low enough to cause the precipitation of the sulphate of lime.

The PRESIDENT, in presenting the thanks of the Institution to the authors, said he thought Mr. Stromeyer would like him to acknowledge the thanks that were due to the Manchester Steam Users' Association for having given him the power to use a great deal of time, a large amount of money, and a great many opportunities that were at his disposal for the purpose of preparing the Paper for the Institution. As an indication of the direction in which he wished to see some little scientific light brought into the discussion, he would be glad if any gentleman who took part in the discussion could speak as to whether pressure purely took the place of heat in facilitating precipitation. If anyone had had experience on that point it would be valuable, because it cost a considerable amount to heat a thousand gallons of water from 40° to 200°. According to the Paper, Bell's softener described in Appendix II (page 797) was practically under a pressure of 160 lbs. per square inch.

Mr. W. G. ATKINS asked whether he understood the question to mean that pressure of the boiler affected incrustation.

The PRESIDENT said the question had been handed to him by Mr. Longridge, whom he would ask to explain the point further.

Mr. MICHAEL LONGRIDGE, Member of Council, said the question he had suggested was whether purification under pressure—pumping the water through the purifying plant and filter under pressure— affected the speed of precipitation. He had heard from one or two sources that the precipitation was facilitated and hastened by pumping the water through the apparatus under pressure, that was to say, connecting the apparatus to the boiler and pumping through it to the boiler, and that the pressure had the same effect as heat in facilitating and hastening the precipitation of the impurities.

Mr. W. G. ATKINS said he had misunderstood the question, which he thought was whether the pressure of the boiler affected incrustation.

Mr. W. T. HATCH said, that with regard to the query as to filtering under pressure, he had not had any such experience with feed-water for boilers, but he had had considerable experience with filtration under pressure of softened water for domestic purposes, where the filtration was carried out under a pressure of 40 lbs. per square inch, and also with filters which were under practically only atmospheric pressure. His experience was that filtration under atmospheric pressure was more thoroughly carried out than filtration under pressure. One reason was that pressure filters were liable to get choked up, the water apparently always attempting to make a short cut through them, and when the flow was reversed for cleansing purposes, the water returned by the same short cut; instead of completely cleansing the filters throughout as one would wish, the were only cleansed along the so-called "short-cut." In addition, there was a great deal of labour entailed in clearing out filters worked under pressure; the lids (which required to be moved away)

(Mr. W. T. Hatch.)

were very heavy, and often it was very difficult to get at the filters, whilst those working under atmospheric pressure needed very little time to be spent upon them and were always open to view. Personally he favoured those that were worked under atmospheric pressure.

With regard to the results given in the Paper, he wished to know whether the tests had been carried out by pre-arrangement, or whether surprise visits had been paid to the works. He had several softening plants under his care, and found at times, when he did not send word that he was coming down, and took samples of the softened water, that the results were not always equal to those taken when he had sent word. If the tests that were given in the Paper corresponded to the best that the machines could do, the results were rather disappointing. Out of the eighteen types given there were six results fairly satisfactory, but the other twelve were unsatisfactory. Four he noticed increased the amount of carbonate of lime above what was present originally, and five increased the amount of magnesium carbonates. Some simple method was desirable of dealing with the "softened water" question in connection with steam boilers, because the apparatus was generally attended by stokers who were not often of a very intelligent class, whilst the resident engineer usually looked upon it as something which might look after itself.

With regard to the types illustrated in the Paper, some were admittedly of a very simple nature whilst others were indeed "fearfully and wonderfully made." All who had worked with water-softening plants knew that the simpler they were the better they worked, and the more easily they were cleaned. With regard to one type where the inclined plates were supposed to allow the deposit to slide down them into the "blow-off," they worked all right when new, but he knew two cases of similar plants working at the rate of 9,000 gallons per hour, where, after a couple of years' time, numerous hand holes had to be made in the sides to provide facilities for clearing the sediment from the inclined plates, or otherwise it did not get away.

With regard to "filters," he noticed that the author stated that "even filters would not remove all the precipitate of carbonate of lime," and he thought that statement would be generally agreed with. He had a case in mind where a softening plant was working for some years and doing something between 9,000 or 10,000 gallons an hour, with a 6-inch main taking the supply some distance. That 6-inch main, although on a cold water service, became encrusted with a deposit of carbonate of lime to such an extent, that it only had a 3-inch hole through it, due to the filters not acting as they should have acted theoretically, but which they were inherently not able to do, as stated in the Paper. If some method could be found of carrying out that work without using filters, he thought many engineers would be much more satisfied than they were at present.

With regard to boiling water, he was laying down a plant for dealing with 9,000 gallons per hour. That plant was not originally intended for boiler use, but was required to sterilise water for drinking purposes. The supply was from a well which generally gave good water, but at times showed symptoms of containing something it ought not to. The plant was designed for 9,000 gallons and when completed he made a test of it, but it would only deliver 6,000 gallons, and hence some little alterations were being made. Although, however, it was not working up to its full capacity, he used a large independent boiler for the sake of testing the quantity of fuel required, and found that the cost, with coal at 15s. 6d. a ton, worked out at $2\frac{1}{4}d.$ per thousand gallons. In that case the water was not allowed to escape hot. It entered at, say, 60° F., and on its way it was heated (after the manner of a surface condenser, the outgoing water not coming into actual contact with the incoming water) by the water which was coming away from the boiling vessel, so that it absorbed a large amount of heat, which was given up by the outgoing water. The effluent was 15° F. higher temperature than the intake, and in that case softened and sterilised water was obtained which had the temporary hardness reduced from 19° down to 5° at a cost of $2\frac{1}{4}d.$ per 1,000 gallons; although the cost in this case exceeded that of the lime and soda process, yet good potable water free from contamination was obtained, whilst the removal of

(Mr. W. T. Hatch.)

such a large proportion of the temporary hardness, although of secondary importance, was welcomed. The precipitation of deposit was so rapid however, that simple flat plates only were provided to collect the carbonates, no filters being required; these plates were easily removable and needed only tapping with a hammer, when they became freed from deposit and were ready to be used again. When the apparatus was working at its full capacity of 9,000 gallons, he hoped the results would be even more favourable than that given.

He desired to add that he would like the author, if possible, to add a few words to the title of the Paper, because in referring to the matter, those who were interested in softening plants, used other than for boiler-feed purposes, might possibly be misled. The title was " An Inquiry into the working of various Water-Softeners," whilst all the types mentioned in the Paper referred to softening plants using soda for boiler-feed purposes, and were of course unfit for dealing with water-softening for drinking purposes. If there was added to the title " for Boiler-Feed purposes," it would help engineers when making references in the future.

Mr. WILLIAM BROWN, of Woolwich, referred to a thermal storage vessel which Messrs. Siemens Brothers and Co. had found to be a very efficient water softener, being of great value in eliminating the carbonate of lime from the feed-water. It was attached to a battery of seven large boilers, and the amount of water treated in a run of seven weeks was about 2,600,000 gallons, the vessel being then laid off one week for cleaning. The boilers were on the Kent water service which averaged 18 to 19 degrees of hardness. Before the storage vessel was erected, each boiler had to be laid off every seven weeks and thoroughly chipped and scraped. Now after seven weeks the water was simply followed down by the cleaners and brushed out, again at fourteen weeks the same thing was done, and at twenty-one weeks was done what before had taken place every seven weeks, namely chipping and scraping. The vessel, Fig. 20, Plate 37, was a plain cylinder 9 feet in diameter and 35 feet in length, designed with dished ends to do away with all internal stays. The inside of the vessel was kept as free from obstructions as possible, the only fittings consisting of a diaphragm across at one end and a series of sloping trays.

The method of working was as follows : steam was turned into the vessel, and at the same time water was pumped from a feed-heater heated with exhaust steam (not an economiser) at about 130° F. into it through a valve close to the steam one, the water falling on an inclined tray, which passed backward and forward in the steam space, finally falling into the small space cut off by the diaphragm. It was in the small space and on the trays where the majority of the precipitate was found.

In view of the Paper now being discussed, the last time the vessel was cleaned they had the whole of the precipitate collected and carefully dried and weighed, which amounted to about 2,200 lbs. for a run of seven weeks. A certain quantity was precipitated in the feed-heater before it reached the vessel. A photograph of the interior was taken looking along a tray toward the end where the water entered the vessel, this photograph showed the precipitate undisturbed, it having heaped up as shown, Fig. 21. The interior of the vessel was kept at the same pressure and temperature as the battery of boilers, the feed being pumped into the vessel and from there it was distributed to the boilers by gravity.

Mr. EDWARD STEPHENSON said the Paper was certainly a very practical one, practical from two points of view. First of all, the subject was of very great importance. There was undoubted economy to be derived from turning one's attention to the water employed in the boiler. A great deal of care was given to the design and working of engines in the matter of their economy for the production of power, but he considered that, as only a comparatively small proportion of the power in the coal was delivered to the engine, amounting in many cases to not more than one half, there was a great scope for economy on the boiler side as well, and the Paper dealt with one of the points which would produce a marked improvement in that direction.

Further, the subject had been very practically treated. The deductions appeared all to have been derived from the actual working results of water softeners in use, without very much theory ; and

(Mr. Edward Stephenson.)

nearly all the water softeners mentioned in the Paper had hard knocks given to them all round, when they did not appear to have answered the purposes thoroughly for which they were intended. In a great number of cases he noticed the inefficiency was to some extent due to misproportion in the chemicals that were used, a matter that required continual and constant attention. Questions had been asked as to whether a water softener could not be made as simple as possible. Well, the simpler it was the more inefficient it might be, and that was generally because a water softener was really a chemical apparatus which performed laboratory experiments, so to speak, on a large scale, and the adjustment had to be absolutely exact; the machine which could adjust the quantity of chemicals exactly, and keep them acting day after day in a more or less automatic manner, was the softener which was likely to produce the best results in water passed on to the boiler. It must be borne in mind that they frequently dealt with very large quantities of water in one apparatus, as much as 40,000 and 50,000 gallons per hour, and such could not be treated in a mere tank into which the attendant had to measure off the water and chemicals, and mingle them from time to time, as someone had suggested. A plant of this size made on the " Desrumaux" system was working at Valenciennes, on the Chemin de Fer du Nord; the North Eastern Railway Co. also had one at Hessle.

With regard to economy, he was inclined to think that the authors of the Paper had hardly given as much credit to the effect of pure soft water, free from scale, on the efficiency of the boiler as might be given. The loss of efficiency due to scale had been calculated from certain temperatures, and the possible economy had been deducted therefrom; but he thought the most practical way of looking at that point was the actual saving which occurred in the case of a softener, as against the same boilers under precisely identical circumstances not using the softener. He had a case before him in which a softener of the same system, producing about 3,000 gallons of water an hour, had been in use for a year on a train of locomotive type multitubular boilers working at a high pressure, and the saving in fuel, from the books, had actually amounted to 5 tons of coal a week. That was a very considerable amount. It was in the London district, and coal would be worth say at least £1 a ton, so that the saving of coal in the course of a year might amount to £300. There had been a further saving in the case of that water softener, actually proved by facts, in six months, of £150 under the head of repairs, due to the constant tinkering and patching up that used to go on with the boilers when in a foul condition. This was a second cause of efficiency, which also amounted to £300 a year. Therefore, from those two sources the amount of saving had really reached about the same, and in twelve months had more than paid for the cost of the whole installation of the water softener. That was a very important and remarkable result, and one worthy the attention of all users of steam power.

With regard to the treatment of oily water, this was even a more important subject than that of hard water. He thought the question of oily water should not be mixed up too much with that of the treatment of hard water. His system was to treat the make-up water chemically for its hardness, and treat the oily water for the oil that was in it, and mix them together subsequently in the apparatus for the sake of filtration and deposition. Two bottles were shown on the table, one containing thick oily water and the other the clear translucid product of treatment by the system he advised; the finished sample was quite soft, and absolutely free from oil.

Mr. J. A. SMEETON thanked the authors particularly for bringing forward the theory that a scaly boiler was not inefficient from the coal consumption point of view. It had been stated in the discussion that, by using pure water from a softener, £5 per week was saved in coal consumption; and it would be rather interesting if these two divergent views were explained. He had never understood the question thoroughly from a cost point of view, because it had been so thoroughly impressed upon his mind by various coal users and boiler owners, mostly in collieries and big steel works, and in connection with condensing plant particularly, that if a plant of any kind could be obtained which would give perfectly pure boiler feed-water, an enormous amount of coal would

3 M

(Mr. J. A. Smeeton.)
bo saved, by the mere fact that scale would not be deposited in the
boilers. It seemed to him that the mistake, if it were a mistake,
might be brought about in the following way. For every $\frac{1}{4}$ inch of
scale formed on the boiler plates, furnaces, tubes, etc., the water
capacity and steam-raising efficiency of every boiler would be
decreased in ratio, and more coal would have to be used per hour to
raise the necessary amount of steam required during that period to
drive the engines; and probably that was giving the engineer the
idea that he was using more coal per lb. of steam with scaly boilers.
It appeared to him that too much had been made in the Paper of the
question of the cost of raising the temperature of the water. He
held that for every 1 lb. of steam or for every unit of heat put into
the water, that heat was utilised in raising the temperature of the
water in the boilers, less the amount lost in transit, so that the
actual additional cost of heating the water in a softener was not so
great as it really appeared to be on paper. For instance, taking the
Boby machine, which was principally and essentially a hot-water
softener, a considerable amount of heat or steam had to be used to
raise the temperature of the water to a maximum temperature of say
200° F. That heat, if the water was put straight into the boiler and
not into a settling tank, would reduce the amount of coal required in
heating the water in the boiler.

After looking through the various points of the Paper, it seemed
to him that sufficient care was not taken in any one of the softeners
described, to get the actual ratio of chemicals into the water or into
the softener; and he thought that if some suitable arrangement could
be made which would automatically control all those arrangements,
without making the apparatus particularly complicated, so that the
amount of water going into the apparatus to be softened would
exactly regulate the inflow of the soda and lime solutions, an
apparatus would be obtained which would be mechanically perfect.
With a mechanically perfect softener the mere question of treating
the water was a subject only for the chemist. Boiler-fluid makers
came to boiler users and told them that they had a fluid which would
prevent scale in their boilers, and that there would never be any
more scale if that particular fluid was used; but each one of the

circumstances must be treated individually; it was not possible to take one boiler fluid or one ratio of re-agent in a softener to treat all troubles. Each man must deal with his own trouble in his own particular way, according to the necessities of the case. The opinion of the lime saturators mentioned in the Paper seemed to him open to some considerable doubt. If a certain quantity of water were put into an apparatus, and if a sufficient quantity of lime for a certain number of hours' work were put into the saturator, and the lime water were taken out in equal quantities, and the make-up water let in at another point in equal quantities, the water coming in would act as a stirring arrangement; and the water going out would be consistent lime water, because the amount of lime put into the softener would be sufficient to saturate, not only the water put in the first instance to fill the saturator, but the make-up water as well.

There was another point he particularly noticed in the Paper, namely, that very few softeners seemed to use soda in solution for injecting into their main chambers. He thought it was an extremely simple matter to put soda solution into the softener in specific quantities. It was much easier to regulate a liquid than to regulate a powder or ash. If the re-agents were put into the main tank satisfactorily and stirred together with the water, either by a steam coil, or a steam and air injector, a very feasible and successful apparatus for softening would be obtained.

The weak point he had noticed particularly, when inspecting the various softeners of different makers in Great Britain, had been the filters. Influenced by the flow of the water, a portion of the precipitated carbonate of lime would rise to the top of the apparatus if the outlet was at the top, and would naturally sink if the outlet was at the bottom, and it was apt to clog the filters. It seemed to him to be distinctly necessary that there should be suction on the discharge side of the softener, so that in case the carbonate of lime got into the filters and clogged them, there was something sucking a passage through the filters. He found that in quite a number of cases the inlet side of the perforated filter plates was completely coated with a sticky yellow substance, which in many cases completely clogged them up. Thus eventually a point was reached when it was

3 M 2

(Mr. J. A. Smeeton.)

absolutely impossible for the water to flow through the filters. A perfectly satisfactory filter, and one which could be easily regulated, was certainly not one which required the shutting down of the apparatus three or four days at a time at periods of three or even a less number of months. Filters should be so arranged that they could be cleaned out while the apparatus was working, so that it would be in reality a continuous softener, and not a softener which could only be worked intermittently.

Mr. LEONARD ARCHBUTT said that having studied the chemistry of water softening for many years, and having had the honour of reading a Paper * on the subject before the Institution in 1898, he had been much interested in the Paper that was before the meeting that evening. He thought, however, that it might have been more valuable if the different makers of apparatus had been asked to select apparatus or plants working in certain districts, and satisfying certain specified conditions, if they had also been given the opportunity of seeing that those plants were doing the best they were capable of, and if samples had then been taken from each apparatus at random intervals during a certain period of time. That would have given a better idea of what could be done in the way of softening water with an efficient apparatus than by taking single samples, as he imagined had been done in many of the cases, without preliminary arrangements being made. He was quite aware, however, that such an investigation, although more thorough, would have been much more expensive, and would have taken much more time, and therefore he did not wish to blame the authors for what they had done. In fact, he was extremly pleased to find a gentleman in Mr. Stromeyer's position taking such a great interest in the subject of water-softening, because it was a subject which had been too much neglected in the past by engineers and manufacturers, who had not really known what advantages could be gained *by it, and that neglect had been partly due to the fact that those who had undertaken to soften water in the past had not always understood

* Proceedings, 1898, Part 3, page 404.

the difficulties. It had not been sufficiently realized that some waters were much more difficult to treat than others: that whilst certain waters could be efficiently treated by an intelligent labourer, and used afterwards for almost any purpose, other waters required more skilled superintendence, and possessed properties which must be taken into account when the softened water was required for certain purposes, and that unless the necessary skilled superintendence were given, and the special properties were allowed for, the result would be a failure. It should be clearly understood that the chemical principles involved in softening water were well understood by competent chemists, and that there was no water which could not be efficiently softened in a suitable water-softener.

In the first part of the Paper the authors went into the question of the transmission of heat, and came to the conclusion that even a thick coating of scale did not materially reduce the efficiency of the boiler. He would be greatly obliged to Mr. Stromeyer, if, in his reply, he would give a little more information as to how the figures were arrived at. They were perhaps well understood by some present who were trained engineers, but he could not quite follow them himself. Very varied statements had been made. Some had estimated that $\frac{1}{2}$ inch of scale caused 150 per cent. more fuel to be consumed, whilst another stated that it was possible to have the water spaces nearly choked with scale without greatly increasing the fuel consumption. It was very important that the question should be cleared up, and if Mr. Stromeyer could make the figures so plain that everybody could understand them and agree with him, very useful service would have been done. Apart from that, however, the reduction of wear and tear, the reduced cost of cleaning, and the increased safety of working boilers kept free from scale, were most important advantages effected by water-softeners.

He was pleased to see that the authors had again directed attention to the fact that sulphate of lime was not insoluble in water at any ordinary working boiler temperature (page 777). He had pointed out that in his own Paper in 1898, and the statement which used to be made and copied from book to book that sulphate of lime became insoluble at about 45 lbs. per square inch steam

(Mr. Leonard Archbutt.)

pressure was quite erroneous. The authors referred (page 779) to the last traces of magnesium carbonate being removed by adding "an excess of lime." He did not agree with the use of the word "excess" in that place, because no more lime was needed than was required to turn all the magnesium carbonate into magnesium hydroxide, and unless that amount were added it was impossible to remove the magnesia from the water. Further on the authors stated that "the hydrate of magnesia is practically insoluble, but being a gelatinous mass it causes serious difficulties in water softeners, more especially by clogging the filters." The fact was that magnesium hydroxide, if properly precipitated, was not gelatinous, but crystalline; it settled very rapidly indeed, and needed no filtration.

The authors referred (page 784) to the process of passing carbonic acid into the water in order to prevent a deposit taking place in hot pipes leading to the boiler, and they stated that "steam users, however, have a prejudice against deliberately introducing carbonic acid into their boilers, and some at least add very large settling tanks, and do not use the stove." He was not acquainted with any single instance where that was the case. He presumed Mr. Stromeyer had found it to be so; but whoever did that had not done it on his advice, and were making a great mistake. Water softened in the cold might be kept for 24 hours or even longer than that settling; but, even when perfectly bright, it would still deposit a small quantity of precipitate when heated, unless it had absorbed sufficient carbonic acid from the air. Carbonic acid must be added; but an excess was not necessary. The process of carbonating was very simple; and if the users would only take the trouble to understand it and look after it, no harm could be done to the boilers.

Mr. SIDNEY TEBBUTT said he had one of Archbutt and Deeley's softening apparatuses at work, and he had been fairly successful with it. He was treating a water of about 25° of hardness and using that water, amongst other purposes, in the boilers for some years with practically no hard deposit on the boilers at all. He did not suggest it was because the apparatus itself would soften water

better than any other apparatus, but the whole point was, as had
been said many times, that something was required extremely simple
in its action ; and it was because of the simplicity of the apparatus,
in which it was possible actually to measure out the amount of
chemicals required each time, put them into a tank and boil them, and
then run the solution into a fixed quantity of water, that made it so
valuable.　　Before he adopted that apparatus he saw others,
involving, for instance, mixing saturated solutions by means of a
running pipe, with the water to be softened (also running from a
pipe), and he found that whenever he inspected them casually
the results were very much inferior to those which were said to
have been obtained originally, chiefly because of the difficulty of
mixing in correct proportion. He believed that many of the
apparatus could be made to work successfully, if sufficient time and
proper care were given to them.

　　With apparatus for feeding boilers, which was of course a very
small one and not to be compared with those used for water
for public purposes, unless the service of chemists were available
one could not afford the labour and knowledge required to work
successfully, and therefore it was necessary to have something of
the simplicity of Archbutt and Deeley's system. He did not
consider it was a chemist's question entirely. The information
Mr. Archbutt gave him at the time of starting his plant to enable
him to soften his water was hardly sufficient, and he experimented
a great deal with the chemicals before he was able to get them right.
The difficulty was with regard to the question of settling, and
he found out he could not settle his water sufficiently quickly
until such time as he used a " commercially supplied " caustic soda
instead of a cheaper alkali ; and when he used caustic soda in
a small proportion, he found the lime all settled down very much
quicker, perhaps in half the time, and made a practical working.
He also found with caustic soda, if much were needed, that using it
alone it worked through the joints of the fittings on the boiler, though
it did not apparently do any harm to them. When the quantity of
caustic soda was reduced, the joints kept up without leaking at all
and without needing remaking. The chemicals recommended

(Mr. Sidney Tebbutt.)

produced a caustic solution, but he was convinced it acted differently to the use of a 98 per cent. caustic soda as supplied ready manufactured.

As regards softened water for use in laundries, it was not a matter so much of how it was softened as that it was soft, except where softened water was required for washing flannels. The caustic soda softening was very injurious to these goods as well as to coloured goods. But an excess of alkali was strongly to be avoided, and it was not to be recommended to soften lower than 5 per cent. of hardness. An automatic continual measuring system for water and solution separately appeared to him the only satisfactory method as against measuring in bulk, but there still remained the question as to the difficulty of keeping the right strength of the solution.

Mr. G. H. READ said it had become necessary for him to ascertain what was being done in dealing with the problem of using waters of both temporary and permanent hardness, in steam boilers. He had to thank the authors of the Paper for many instructive points. He had been in correspondence with friends whose experience in different parts of the world, and particularly in the United States of America, peculiarly fitted them to give him advice upon this difficult subject. The great railroad companies of America were entering deeply into the subject, as indeed they were compelled to do, through the enormous wear and tear upon boiler shells, tubes, and fire-boxes, owing to the very bad water they were often compelled to use over great stretches of country. In one case that had come to his notice, a very hard water was being used, containing 89 grains of incrusting solids. It became possible however, to mix this water with that from other wells of greater purity, so as to bring the final mixture down to 21 grains to the gallon. He was sorry that the authors did not seem to deal, to any great extent, with what was being done in America, but in the case just referred to, he would like to say that this extreme hardness was reduced to $3\frac{1}{2}$ grains per gallon. This was done under a system extensively used in the United States, and known as the Kennicott system. If any member present had had an extended experience of this system, he would be glad indeed to hear what the result had been. He had been advised that the advantage of this system was the fact that, after an analysis of the water to be used had been made, the machine was so set that the proportion of re-agents employed, whether lime or soda or both, was exactly in accordance with the volume of water delivered to the plant. This seemed to him to be a very great feature, as he had come across a great many instances where the quantity of re-agents used had been largely in excess, principally owing to carelessness, and additional troubles had been introduced. It would seem to be a very great advantage to have the exact proportion of re-agents automatically controlled, and where this was done it seemed to be easily possible to bring the worst water down to a hardness of only 5 grains per gallon.

He did not know whether the members would be particularly interested in waters for locomotive boilers, but he might mention that on the Atchison, Topeka and Santa Fe Railroad, in Kansas, it was almost impossible, with the water originally used, to run an engine many hundred miles without having trouble of some sort. Tubes, sheets, and staybolts were incrusted with scale $\frac{3}{8}$ inch thick, which had to be removed every three weeks. Great trouble was experienced in handling heavy traffic on this division, owing to leaky tubes through over-heating. Since the plant, already referred to, had been taken into use it was claimed that a reduction of 75 per cent. under this head had been affected. It was superfluous to point out the loss of fuel resulting from scale even $\frac{1}{4}$ inch in thickness. On a question of cost it appeared to the speaker that, given the same location and the same water, any system that did its work perfectly must cost practically the same in working, as in any case, with a given water, only a certain proportion of re-agents was required, the cost of which would be the same for any apparatus. The whole thing depended upon the exact distribution of re-agents to water supply.

Some of the members would doubtless have an opportunity, when in the United States, of investigating the system so largely used there. He understood that so large a quantity as 60,000 gallons per hour was continuously treated for twelve hours at a

(Mr. G. H. Read.)

station on the Pittsburg and Lake Erie Railroad. The water there contained, he was advised, 21 grains of incrusting solids per gallon, and this was brought down to 5 grains without variation from day to day. It was generally found that locomotives having boilers not just as they should be, as regards scale, had the shells and tubes wonderfully cleaned after using water that had been treated in in this way.

The subject of suitable water was still more vital in the case of water-tube boilers, and was being exhaustively studied in the United States. He had made great efforts to obtain accurate particulars upon the subject of the treatment of water for water-tube boilers. The most satisfactory reply he had received referred also to the system already named, in which the cost was $1\frac{1}{2}d.$ per 1,000 gallons, but this was for the utmost purification of the water. With a purification satisfactory for all ordinary purposes, and leaving only a slight deposit of mud, a cost of only $\frac{9}{16}$ths penny per thousand gallons had been arrived at.

He had also been obliged to give some study to the question of waters impregnated with acids. In the course of conversation with a professional acquaintance recently returned from the United States, and who was also interested in the water-softening problem, he had brought to his notice two if not three cases in Pennsylvania, where water had to be drawn from ponds, and water which varied with almost every rain, and frequently contained a very considerable proportion of acid. The variations were taken note of, and soda and lime introduced in the right proportions to ensure the required reduction in hardness, and to neutralize the acid. Before the introduction of this plant there had been a great deal of corrosion and pitting. Leakage was the inevitable result, and a very hard scale was also formed. Since softening the water, these had been entirely done away with. Where the water to be used in locomotives was muddy and thick, in addition to being hard, there was great advantage in using a softener, provided there was adequate settling surface, as the particles of mud seemed to be mechanically enclosed by the precipitate of carbonate of lime.

He regretted that he could not give particulars of more systems than the one referred to, but he could vouch with certainty for the accuracy of the statements which he had made, as they were given to him by engineers whose business it had been to discover the best systems for reducing the cost of fuel, and wear and tear upon boilers.

Mr. VILHELM HJORT, referring to the question of a secondary tank for mixing chemicals before they were placed in the softener, stated that he was interested in one softener, largely in use in this country, using such a tank, and by means of that tank the difficulty, to which the authors referred, namely, the varying height of chemical re-agents in the reservoir of the softener, was entirely overcome. That tank was provided with a small pump, and the trough vessel was always kept full. It was only right he should mention this matter, as one of the plants referred to in Appendix III (page 829) and also on page 810, was provided with the secondary tank, and had been working for three years.

He also wished, if possible, to have some information as to why the scale in a boiler increased the amount of fuel used.

Mr. STEPHEN H. TERRY said that he had had, in his early days and since, a considerable experience of steam boilers. As a young man he had helped to scale them, and he therefore did not agree with the first part of the Paper, in which the author seemed to attach less importance to the reduction of steam production which took place by scale than he had been always taught to do. From his early experience he had always supposed that the less a boiler was invited to be a chemical retort the better for those who had to feed it with fuel, repair it from time to time, renew it when worn out, and use the steam it produced. One could not help thinking that the nearer the contents of a boiler consisted of H_2O the better for the users of the steam, with the presence possibly of just sufficient lime to form that scale which might and did do a great deal to prevent the combination of the oxygen with the iron and so rust it away. Certainly the presence of any other

(Mr. Stephen H. Terry.)

matter which lime combined with or associated with, namely, oil,
whether in minute drops or in an emulsion, was very fatal to the
life of the boiler. It was known very well that all those things
had caused very serious damage not only at sea but on shore.

His late friend, Mr. Samson Fox, did a great deal for the
advancement of steam pressures at sea, whereby speed was also
increased, and it was he who found out the necessity for the
corrugated furnaces, largely through furnaces collapsing which
were not made corrugated. Why did they collapse? Not merely
from pressure but from undue heating of the surfaces; that undue
heating was caused by the solid salts, not necessarily lime, but by
solid matter mingled with oil which adhered to the crowns of the
furnaces. These together formed a sort of kid glove which scorched
like a burnt pancake in a badly managed frying-pan, and produced
something which would not let the heat through, but which did as
much damage as the similar condition did in the frying-pan if the
product had to be eaten. He thought electrical engineers had for
many years devoted great and proper attention to the subject of
economy, and had compelled mechanical engineers to improve their
high-speed engines, both in durability and in power of continuous
running. They had given immense attention to steam consumption,
but much less attention to steam production. It was about time to
turn more attention to the steam economy of the boilers, not only of
the boiler itself, but to other collateral economies which followed
very closely the production of pure steam unmixed with a lot of
chemicals which came over in priming in a more or less
crystallised form. These chemicals cut the valve faces and cylinder
faces to bits, with a subsequent loss from steam leakage, the high
efficiency of trial trips sometimes ceasing after a few months'
running. He thought those who had practical charge of engines
would agree with him, that it was very important that scale should
not be formed in the boilers.

The various nostrums in use for boilers had for their object the
prevention of the formation of hard scale; this they to some extent
succeeded in doing. Some of them seriously injured the plates and
stays, and assisted in causing pitting, others were directly

responsible for priming with all its wasteful dangers. To admit
that a "boiler-fluid" was necessary was to admit that a water
softener was necessary. Yet taking locomotives only, there were in
this country thousands of locomotives using the hardest water and
incurring enormous expense in scaling burning tubes and fire-boxes,
and (as an indirect result) smoke-boxes also; using contracted
nozzles to exhaust pipes and steam-jet blowers for intensifying the
draught; burning out firebars, and producing pyrotechnic displays
at night; causing rail-side fires and great waste of fuel and
destruction of rolling stock, paint and varnish and discomfort to
passengers. (The extra fuel involved was probably at least a
shovelful of coal per mile.) There was, in addition to all this, the
great loss of tractive power due to the unnecessarily high back-
pressure in the cylinders, whereby 10 to 15 per cent. of the
tractive force was not forthcoming. In other words, out of the total
of 30 lbs. to 35 lbs. coal per mile, burned by a passenger locomotive,
the same work with a clean boiler, clean fire-box, and tubes and hot-
feed would be obtained with 25 lbs. to 30 lbs. of coal, instead of
30 lbs. to 35 lbs. and hot feed with all its attendant advantages of
uniformity of boiler temperature and increased dryness of steam.
This would be easily achieved without the continuous trouble of
choked feed-pipes due to heating hard feed-water, for it was an
undoubted fact that many feed-heaters had been condemned through
the fault of the hard water. Much of the so-called economy of
"economisers" was due to the fact that many boilers had just
sufficient heating surface in proportion to grate area and fuel burnt
to make the required amount of steam, when the surfaces were clean,
but had a wholly inadequate amount of surface to entrap and pass
the heat to the water when these surfaces were covered with scale.
High flue and chimney temperature resulted, and the remarkable
economy said to result from fitting economisers, the said economisers
being really an apology for a foul boiler. Much smaller economisers
would be sufficient to heat the feed, if the economisers did not have
to do part duty as water-tube boilers, and in this capacity to provide
sufficient surface to permit the passage of heat from the hot waste
gases to the water.

(Mr. Stephen H. Terry.)

In the early days of portable engines,'it was found that heating the feed with exhaust steam produced 15 per cent. economy. It also frequently furred up the feed-heater, by precipitating such of the lime as constituted temporary hardness, namely, that which could be thrown down by the application of heat. There were numerous installations of elaborate " economisers " and " feed-heaters," working to-day with very poor results, because they were choked with lime. He knew of a case in which the feed-heater was disconnected because it got so constantly choked with lime, and the boilers of a large electric-light station were fed with cold water, the ratepayers making up the deficiency of " thermal-units " in the feed by some fraction (nearly a whole number) of a penny in the pound. All this was an object lesson to show that, if economy was aimed at, the boiler and what went into it demanded closer attention than they had generally received in England.

He had been engaged at one time as managing partner in works in the Midlands, where numbers of water softeners up to 4,000 gallons per hour had been constructed on a patent system now lapsed. These softeners had been freely made by the works before he joined, and he well remembered the continual complaints which arose of leakage and bulging due to bad design. Immense flat unstayed and insufficiently stiffened surfaces were exposed to some 15 feet head of water with the natural result of bulging and leakage at riveted seams. It fell to the speaker's lot to rectify some of these installations, but in some cases no sooner were they rectified than further orders followed, showing the economy experienced by steam users in softening their water.

His experience with softeners of this description with flat sides showed that such softeners required (when of large dimensions) efficient cross-staying by rods or bars, and stiffening with angle-iron, all such staying interfering with the cleansing of the apparatus. A far better design was the cylindrical type with conical bottoms, such the " Desrumaux." This type required no staying, and was easily got at both inside and out for cleaning, repairs, and painting. In France and in Germany a water softener, generally of the " Desrumaux " pattern, was a necessary adjunct to all works, factories and shipyards, where steam power was required if the water was hard.

When visiting Messrs. Bloom and Voss's Shipyard and Marine Engine Works in Hamburg, he was specially struck with the architectural proportions of the cylindrical water softeners, which in some cases were built round a chimney for warmth and prevention of trouble from frost in severe weather. The large list of firms, using this particular type of water softener supplying water for steam raising of an aggregate of nearly one million horse-power in France alone, showed the importance which that thrifty country attached to having pure water for boiler feed. It was only in wasteful England that they were quite happy in continuing the softening of water in boilers constructed for the purpose of raising steam, and not intended as chemical retorts. He was convinced that a far greater field for economy in steam production was available for engineers in seeing that their boilers were fed with pure water, than in any other portion of the heat sieve, between the combustion of fuel, at the beginning, and the rotations of the crank-shaft at the other end.

Having learned quite recently that a Desrumaux Water Softener was installed at Deptford at the London Brighton and South Coast Railway Company's Wharf, and being anxious to see its working with the view of adopting the same in two cases where he was acting as consulting engineer, he paid a surprise visit to these works and the following conditions were found:—A water softener capable of dealing with 4,000 gallons per hour was in operation, having been there installed by Messrs. Bowes, Scott and Western, the manufacturers of the Desrumaux system. The water dealt with was very hard, being supplied by the West Kent Co.; it was chiefly derived from deep wells, and producing (when not treated) a calcareous deposit in pipes and boilers of a hard limestone character. The water as delivered had the following hardness:—$20 \cdot 75°$ total hardness, and $16 \cdot 65°$ alkalinity, in terms of grains of $CaCo_3$ per gallon.

The quantities of reagents used per 1,000 gallons of water were:—160 gallons saturated lime water containing 2 lbs. $2\frac{1}{4}$ ozs. soluble lime, and $13\frac{3}{4}$ ozs. of refined soda ash of 58 per cent. The proportion of the plant was such that each particle of water took

(Mr. Stephen H. Terry.)

2½ hours to travel through, and at the end of that time the maximum hardness was 4°, with an alkalinity of about 6°, although the water drawn from the storage tanks went down as low as 3° and even 2½° in hardness.

Many engineers began a conversation on the merits of some type of engine under discussion with " What is the water consumption per B.H.P. ? " " What is your cut-off ? " " What compression have you ? "—while they left the boiler as before, or attempted at economy by substituting water-tube boilers for cylindrical, and were still filling their boilers with water highly charged with lime (and other even more injurious mineral matter in solution and in suspension). It would be well if locomotive superintendents, factory managers, and all employers of steam, would spend no more time in designing fancy valves and fancy valve-gears to economise steam, until they had examined the boiler feed-water to see if the true place for economy (which would show in whole numbers, not decimals) lay in the efficiency of the boilers.

Mr. WILLIAM H. MAW, Past-President, wished to refer to one great advantage of water-softeners not mentioned in the Paper, but which was of very great importance in many instances, namely, the way in which the use of efficient softeners enlarged the choice of the type of boiler available for a given service. In the case of works in the country where there was ample room for putting down Lancashire boilers, or boilers occupying considerable space, that was not a matter of very great importance; but in cities like London where boilers of the Lancashire and kindred types were often very difficult and expensive to get into the boiler-room available, and where in consequence boilers which were too small for their work were very often put in, it was a very great thing to have the command of pure water, and to be able to use boilers of a tubular type, either locomotive or Cornish multitubular or water-tube boilers, which could be readily got in, and which would have ample heating surface for their work. In many cases the fact of being able to use boilers of that kind was an advantage quite out-weighing any question of the cost of softening.

With regard to the first part of the Paper referring to the effect of scale on heating surfaces, he thought some of the gentlemen who had spoken had somewhat misunderstood Mr. Stromeyer's position. He did not think Mr. Stromeyer meant that a very dirty boiler was as efficient as a clean boiler, he understood him to mean—and, perhaps, Mr. Stromeyer would correct him in his reply if it were not so—that a boiler which was well looked after and cleaned at suitable intervals would not vary much in economy between the times of being cleaned and being just ready for cleaning. When Mr. Archbutt read his Paper at the Derby Meeting in 1898, he himself had occasion to speak on this point,* and he then mentioned that in the case of the London water, which was a very fair sample of town water, he had found that if the boilers were not worked longer than about 1,000 hours between cleanings— a period which corresponded to an evaporation of some 250 to 350 gallons per square foot of surface—the efficiency did not fall off appreciably owing to the formation of scale. In Mr. Stromeyer's case he mentioned certain percentages of loss of efficiency due to the formation of $\frac{1}{8}$ inch of scale. In estimating the effect of that loss on fuel economy, it was necessary to halve the percentages. He inferred that that diminution of efficiency of 2·5 per cent., which was referred to at the bottom of page 774, would be the loss of efficiency when the boiler was just ready to be cleaned again, and the loss of efficiency would vary from that to nothing when one started with a clean boiler, so that the average loss of efficiency would be half the loss mentioned. He thought that many boiler owners were deterred from using softeners by a want of knowledge of what it cost them to clean their boilers. In the discussion on Mr. Archbutt's Paper, he mentioned that in London he found the cost of cleaning boilers varied from about 1d. to 2d. for every 1,000 gallons evaporated. That was in the case of boilers which were thoroughly accessible. In the case of boilers of complicated type the cost would very often run to twice and three times that sum, and he knew of cases in which it cost as much as 8d. per 1,000 gallons

---

* Proceedings 1898, Part 3, page 445.

3 N

(Mr. William H. Maw.)

evaporated to clean the boilers. In Table 2 (page 783) the author had given the cost in the last column as £6 per year per boiler for cleaning, namely, four cleanings at 30s. each. It was, perhaps, convenient to remember that a boiler working ordinary factory hours, and evaporating from $3\frac{1}{2}$ to $3\frac{3}{4}$ lbs. per square foot of surface per hour, would evaporate almost exactly 1,000 gallons per square foot per annum. In the case of the boilers Mr. Stromeyer referred to as costing about £800, including setting, he thought it might be assumed that the heating surface would not be very far off 1,000 square feet. If that was so, there would be 1,000 times 1,000, or a million gallons of water evaporated per annum per boiler, and the £6 per annum which Mr. Stromeyer put down amounted to 1·44d. per 1,000 gallons for cleaning, which was a mean between the figures he gave at Derby of the London experience.

There was a point in Table 2 (page 783) that he should like to ask a question about. In the last column, where a group of six boilers (without any spare one) was referred to, the total cost of cleaning was put down at £36, or £6 per boiler per annum, but in the column preceding that, where there were seven boilers of which six were at work, the cost was put down at £78 for the group. He did not know why the fact of having a spare boiler should more than double the cost of cleaning. The same enquiry referred to some of the other columns in that line, the cost of cleaning a single boiler, for instance, being stated at £20 per annum.

He noticed that in the assumed life of the boilers Mr. Stromeyer put down the life of a boiler using pure water at fifty years, and in the next column the pair of boilers, of which one was a spare, were allotted a life of forty years. He did not know whether that was the joint life of the two boilers, or the life of each boiler. If it was the life of each boiler, of course the joint life would be eighty years, and the depreciation allowance, which was really a sinking fund, would be decreased accordingly. The same remark applied to all the other groups in which there was a spare boiler provided. He thought the amount Mr. Stromeyer took for interest was too low for any investment on plant; it should be quite one-and-a-half times that assumed. He also noticed that the authors put down as depreciation on the amount which would have to be invested annually at 3 per cent. to repay the cost of the boiler in the number of years set down for its life; but in addition to that sinking fund there were other depreciation expenses which would have to be met, and which would affect the total charges given in the Table.

Mr. WILLIAM BOBY was glad of the opportunity of saying a few words, particularly on the subject of purification of water by the application of heat. He presumed that the question of the value of any purifier hinged, first of all, on its convenience and efficiency, and secondly on what it would cost to carry out the process. There seemed to be an idea that there might be a good deal of cost involved in purifying water by heat, but the members had heard the testimony of Mr. Hatch (page 839) that he did not find it expensive. In most cases the effect of heating the water was found to conduce to economy, and not to be expensive. For instance, there was a large number of electric stations now in which there were not many auxiliary engines, but in most cases there was enough exhaust-steam to purify the feed-water, even though the installations might be condensing; in that case the heating did not cost anything, and heat was utilised which would otherwise go to waste. After all, the main point was what could be done by heating the water. He found on page 776, in speaking of bicarbonate of lime, the authors said: "Its second equivalent of carbonic acid is easily removed on boiling, when of course the carbonate of lime is precipitated." It was further stated (page 800) that heat alone did not appear to be able to reduce the temporary hardness to a minimum. He thought there was a discrepancy in those statements, and he was disposed to think that the last statement was based on a single sample of water which had been drawn from a machine; and with reference to that sample as shown in the chemical analyses, he would like to point out that the carbonate of lime, which was the largest constituent of temporary hardness, was reduced almost to nothing. He found also that the silica was reduced as much as it was with any system, and there was simply a question of a small amount of magnesia. There was nothing shown to indicate

3 N 2

(Mr. William H. Maw

evaporated to cle

had given the cc

cleaning, namel

convenient to

hours, and eva

per hour, wor

foot per annv

to as costing

assumed th

square feel

million gal

£6 per ann

per 1,000 ç

he gave at

There

a questic

(without

was put

precedi

work,

know

the c

other

insta

F

put

the

alle

the

th

a

ers. In Table 2 (page 783) the author

st column as £6 per year per boiler for

anings at 30s. each. It was, perhaps,

that a boiler working ordinary factory

m 3½ to 3¾ lbs. per square foot of surface

almost exactly 1,000 gallons per square

ase of the boilers Mr. Stromeyer referred

including setting, he thought it might be

surface would not be very far off 1,000;

so, there would be 1,000 times 1,000, or a

evaporated per annum per boiler, and the

Stromeyer put down amounted to 1·44d.

ning, which was a mean between the figures

ondon experience.

able 2 (page 783) that he should like to ask

last column, wher   of six boilers

s refer   of cleaning

he

were at

ot

would have to be invested
st of the boiler in the number
addition to that sinking fund
ɜ which would have to be met,
ɟes given in the Table.

e opportunity of saying a few
purification of water by the
t the question of the value of
ɔnvenience and efficiency, and
ry out the process. There
ι good deal of cost involved in
·s had heard the testimony of
find it expensive. In most
was found to conduce to
instance, there was a large
here were not many
xhaust-steam
might be
ing, and

(Mr. William H. Maw.)

evaporated to clean the boilers. In Table 2 (page 783) the author had given the cost in the last column as £6 per year per boiler for cleaning, namely, four cleanings at 30s. each. It was, perhaps, convenient to remember that a boiler working ordinary factory hours, and evaporating from $3\frac{1}{2}$ to $3\frac{3}{4}$ lbs. per square foot of surface per hour, would evaporate almost exactly 1,000 gallons per square foot per annum. In the case of the boilers Mr. Stromeyer referred to as costing about £800, including setting, he thought it might be assumed that the heating surface would not be very far off 1,000 square feet. If that was so, there would be 1,000 times 1,000, or a million gallons of water evaporated per annum per boiler, and the £6 per annum which Mr. Stromeyer put down amounted to 1·44d. per 1,000 gallons for cleaning, which was a mean between the figures he gave at Derby of the London experience.

There was a point in Table 2 (page 783) that he should like to ask a question about. In the last column, where a group of six boilers (without any spare one) was referred to, the total cost of cleaning was put down at £36, or £6 per boiler per annum, but in the column preceding that, where there were seven boilers of which six were at work, the cost was put down at £78 for the group. He did not know why the fact of having a spare boiler should more than double the cost of cleaning. The same enquiry referred to some of the other columns in that line, the cost of cleaning a single boiler, for instance, being stated at £20 per annum.

He noticed that in the assumed life of the boilers Mr. Stromeyer put down the life of a boiler using pure water at fifty years, and in the next column the pair of boilers, of which one was a spare, were allotted a life of forty years. He did not know whether that was the joint life of the two boilers, or the life of each boiler. If it was the life of each boiler, of course the joint life would be eighty years, and the depreciation allowance, which was really a sinking fund, would be decreased accordingly. The same remark applied to all the other groups in which there was a spare boiler provided. He thought the amount Mr. Stromeyer took for interest was too low for any investment on plant; it should be quite one-and-a-half times that assumed. He also noticed that the authors put down as depreciation on the amount which would have to be invested annually at 3 per cent. to repay the cost of the boiler in the number of years set down for its life ; but in addition to that sinking fund there were other depreciation expenses which would have to be met, and which would affect the total charges given in the Table.

Mr. WILLIAM BOBY was glad of the opportunity of saying a few words, particularly on the subject of purification of water by the application of heat. He presumed that the question of the value of any purifier hinged, first of all, on its convenience and efficiency, and secondly on what it would cost to carry out the process. There seemed to be an idea that there might be a good deal of cost involved in purifying water by heat, but the members had heard the testimony of Mr. Hatch (page 839) that he did not find it expensive. In most cases the effect of heating the water was found to conduce to economy, and not to be expensive. For instance, there was a large number of electric stations now in which there were not many auxiliary engines, but in most cases there was enough exhaust-steam to purify the feed-water, even though the installations might be condensing ; in that case the heating did not cost anything, and heat was utilised which would otherwise go to waste. After all, the main point was what could be done by heating the water. He found on page 776, in speaking of bicarbonate of lime, the authors said : " Its second equivalent of carbonic acid is easily removed on boiling, when of course the carbonate of lime is precipitated." It was further stated (page 800) that heat alone did not appear to be able to reduce the temporary hardness to a minimum. He thought there was a discrepancy in those statements, and he was disposed to think that the last statement was based on a single sample of water which had been drawn from a machine ; and with reference to that sample as shown in the chemical analyses, he would like to point out that the carbonate of lime, which was the largest constituent of temporary hardness, was reduced almost to nothing. He found also that the silica was reduced as much as it was with any system, and there was simply a question of a small amount of magnesia. There was nothing shown to indicate

3 N 2

(Mr. William Boby.)

that the carbonates could not be reduced to a minimum by
the system of applying heat. The sample in reference to which
the remark was made showed 4·6 of temporary hardness, owing
to the presence of a certain amount of magnesia, and it should
be borne in mind that the assistance of soda was required to
precipitate this impurity entirely, whether in cold or hot water. As
mentioned by the authors, the dose of soda in this case had been
insufficient; had it been rather larger, the magnesia would have
gone down. He did not however agree with the authors that caustic
soda was required, but simply more carbonate of soda ; caustic soda
would not in this case have had the desired effect.

Assuming then that the results showed that heat would reduce
the carbonates to a minimum, it followed that sulphate hardness
could also be reduced to a minimum, since it was in all cases reduced
by first inverting it so as to form carbonate. He had recently put
down a large installation where the hardness of the water
consisted of :—

Temporary hardness	.	.	.	.	.	3·5
Sulphate	,,	.	.	.	.	18·5
Total	.	.	.	.	.	22·0

and tests of the purified water showed a total hardness of 2·2 grains
per gallon.

Purification by the application of heat gave an almost entirely
automatic operation, dispensing with the employment of lime and
giving as complete purification as was possible. It should however
be borne in mind that the heat must be applied effectually, and that
it was not sufficient to drop water through steam at atmospheric
pressure, which would not cause it to give up all its lime, and no
amount of filtering would assist matters. Heat, if properly applied,
caused precipitation of almost every particle of lime.

Mr. W. B. BARON said that a comment had been made as to
collection of samples, and he would say a few words on that subject.
The makers of the various water-softeners were asked to submit the
names of people who used their apparatus, and who were using
waters which might be considered representative of the various forms

of hardness, and by preference those which were modern types. From those were taken the names of people from whom it was thought samples of water might be obtained, both hard and soft, and those samples were analysed, and the results given in Appendix III (pages 826–833). It was desired to get at what the apparatus was practically doing; and if it was working satisfactorily, so much the better. Mr. Stromeyer and himself were quite unbiassed: they simply wanted to see exactly for their own information what purification was being effected, and to embody in the Paper a statement of what they found. The results of examination were before the members, and a summary of their views was on page 788. [See also page 879.]

Mr. STROMEYER, in reply to Mr. Maw and other members who enquired about the data on which Tables 1 and 2 (pages 774 and 783) were based, remarked that these Tables did not represent actual conditions; they were calculated merely for the purpose of enabling comparisons to be made. Thus anybody who desired to go into the question, as in Table 1, of the efficiences of the different parts of the boilers, or into the question, as in Table 2, of the relative cost of working a boiler with unpurified and with purified water, could do this if they followed out the system as laid down in the Tables, but using their own data.

With regard to Table 2 (page 783), he had been asked why it was cheaper to clean the six boilers having no spare boiler than to clean the six boilers having one spare boiler. This was due to the assumption underlying the Tables that the six boilers without a spare boiler could only be cleaned four times a year, during the holiday times, and that meant there was an interval of three months between each cleaning period. These boilers would therefore get very dirty and have a short life, whereas the assumed condition for the other boilers was that they should be cleaned every five weeks, or such longer time as was possible with the number of spare boilers available. Naturally, a boiler which was cleaned more often than another was likely to last longer than the other one. As regards the assumed life of boilers noted in Table 2, no reliable estimate

(Mr. Stromeyer.)

could yet be formed, some of the first boilers built of steel being still in use and as good as new. He had some iron boilers which were certainly fifty years old, and he had heard of others sixty years old stated to be doing good service. He hoped that the Tables would be simply looked upon as models on which to base comparisons; they did not pretend to be scientifically accurate.

With regard to Table 1, he had used the best determined co-efficients of transmission of heat through plates, and through scale, and of heat from the flame to the plates. To facilitate calculations he had taken those co-efficients which were based on the assumption that the transmission of heat from the flame to plate was proportional to the difference of temperature. Some people believed that it was proportional to the square of this difference. Calculations carried out on this assumption would lead to smaller differences between a scaled and a clean boiler than those given in Table 1. The Table showed that a hard-worked boiler did suffer considerably if it had a thick scale, whereas a very lightly worked boiler would be practically as economical with scale as without scale. The difference would be so slight that any change of condition in the firing would quite mask any effect of scale. He knew that a bad stoker could bring down the efficiency of a boiler to 50 per cent., and a good stoker could bring it up to 65 or 70 per cent.

With regard to the chemical analyses, Appendix III (pages 826–833), he wished to say that the amount of labour expended on them by Mr. Baron had been very great. He had heard it said by his Professor that these analyses were practically as accurate as atomic weight determinations. Each column represented a full week's work, which would be about 38 weeks altogether.

With regard to the question of mixing tanks mentioned by Mr. Hjort (page 853), he was in this Paper of course only dealing with the actual softeners which were submitted to him. Some were comparatively old and not fitted with the inventor's newest appliances. The Lassen and Hjort softeners on which they experimented had not the mixing tank, but, if added, it would remove the only feature of this softener to which he had taken any exception.'

With regard to Mr. Halpin's Thermal Storer (page 840), of course it could only be used as a water softener in districts where the water contained chiefly carbonate of lime, and practically no sulphate of lime.

He hoped the discussion would not be looked upon as telling the members which was the best and which was the worst of the softeners, but he hoped it would lead makers of water softeners to reconsider their designs in the light of the information which had now been laid before them, and that they would introduce such improvements as were possible to meet the requests which had been made by speakers for simplification and reliability. In connection with that point, he believed they would be doing themselves and everybody else a great service if they would give very strict instructions with regard to the chemical testing. None of the softeners could work at all unless the water was analysed, for it was perpetually changing its composition. Some users were not aware of these changes, and would have to be convinced. Even the chemicals which were used varied. Thus it often happened that burnt lime contained sulphate of lime, which required an additional amount of soda, and some sodas were not pure. It was therefore quite necessary every now and then to check both the hard and the softened water. [*See also page* 883.]

Communications.

Mr. Frederic A. Anderson wrote that, with regard to the reaction in water softeners (page 779), carbonate of magnesia was very much more soluble in pure water than carbonate of lime. The most recent figures he had found were given in the "Chemische Zeitschrift" (15th October 1903, page 52), in a communication by Herr Mach-Marburg "Über die Löslichkeit der Boden-konstituenten." The solubility of carbonate of lime in pure water was given as 2·08 grains per gallon, and that of carbonate of magnesia as 11·64 grains per gallon.

(Mr. Frederic A. Anderson.)

With regard to the grease in water (page 779), he differed from the statement that a small trace only was present in the emulsified form. In many modern land installations nearly all the grease occurred emulsified, and only a small trace was free. The absolute amount of such oil was often considerable. His firm was constantly engaged in examination of such samples, and from 4 to 7 grains per gallon was about the average amount found emulsified. In connection with this subject, and referring to page 780, he would like to add that his firm were the first to put on the market a successful apparatus for the purification of emulsified condenser-water by precipitation, and that without any addition of impure water. This was about five years ago, and the apparatus, known as the " Harris-Anderson Purifier," was rapidly coming into general use. Within the last few days they had received an order for plant to deal with over 30,000 gallons of condensed water per hour, for the new London County Council Tramways Station at Greenwich.

He would like to correct one or two slight inaccuracies which occurred in the description of the " Harris-Anderson Softener " (page 807). This apparatus did not supply the lime in the form of " milk of lime," but as saturated lime water. It was therefore to be classed with the Atkins Co., Desrumaux, Reisert, and Stanhope machines, and not with the Bell, Doulton, and Porter-Clark. The continual addition of water, to which the authors would very properly take exception in the class in which they had placed the machine, was therefore a fundamental necessity.

The progressive weakening of the lime water (page 808) was quite exceptional. He might remark that the first intimation of it in this case was given by reading the statement in the Paper. The users of the plant had never complained of this to him, and the analytical results and the effect on the boilers were sufficient evidence of the efficiency of the plant. The cause in this instance was probably the employment of an unsuitable lime.

With regard to the automatic soda apparatus, there was an important advantage in the use of this which seemed to have escaped the authors' notice. In the first place, any weighing out of soda was unnecessary ; the apparatus might be charged with an indefinite quantity of soda, and mere inspection showed when more was required. As a corollary from this, it was unnecessary to know how much actual soda there was in the product in use. The apparatus worked indifferently with the ordinary soda crystals of 36 per cent. strength, or with alkali of 98 per cent. strength, or any other grade. The advantage of this apparatus was still more marked when using *two* very soluble reagents for water treatment, as was the case with the " Harris-Anderson Purifier " already referred to.

The distributing device (page 809) called for a word of explanation. The word " weir " was used more than once in referring to it. There were, however, no weirs at all, properly speaking. The principle was briefly—the uniform distribution of the whole water by means of a revolving turbine over the surface of a ring-shaped trough, and this ring-shaped trough being divided by radial partitions into compartments in proper proportions, each compartment received a fraction of the whole water in proportion to its size He would not go at length into the question of what constituted a satisfactory softening, but would only say that the authors throughout referred to *complete* softening as the ideal. Practically this was by no means necessary in every case, and he did not think it always advisable to reduce the soap-destroying salts to the lowest limit.

He would add a word in conclusion upon the very full and valuable Tables of analytical results, Appendix III (pages 826–833). He regretted that, in addition to the constituents of the waters, the authors had not given separately the amounts of the various basic and acid radicals found. This would of course have greatly increased the mass of tabular matter, but would have been a great convenience to those who might not regard the arrangement of salts given as really the most probable.

Mr. HENRY ATKINSON wrote that he would be glad if the authors would give some explanation of the following:—In the first place, Mr. Baron made a statement at the close of the Paper that the waters dealt with in the various machines were *representative* waters ; but if this were the case, who chose the waters? If the authors chose them, why did they select the most difficult waters for the

(Mr. Henry Atkinson.)

simplest machines? If the manufacturers of the machines selected the waters, it seemed strange that in the latter cases they should have chosen very difficult waters as representative ones. The cases he referred to were the Tyacke (hardness 76·147), Doulton (hardness 45·518), and Carrod (hardness 40·88). There was only one other water of a higher degree of hardness dealt with, namely, the one for the Maxim machine (50·256), and this was not as hard as the one dealt with by the Tyacke softener. The Maxim machine was a somewhat complicated one, and his remarks therefore did not refer to it. The Archbutt-Deeley (16·785), Atkins Co. (17·845), Babcock and Wilcox (14·64), Desrumaux (16·12), Lassen and Hjort [1] (19·63), Reisert (20·412), Stanhope (17·893), Wollaston (20·558), and Wright (19·43), all dealt with waters under 21° of hardness, and all these softeners were more or less complicated compared with the first three mentioned. He could not reconcile Mr. Baron's statement with these figures, which seemed to the writer rather unfair, unless, of course, the makers chose the most difficult waters, for reasons not stated, in the first three cases.

There was, however, a still more important matter. On the occasion of the meeting, there were exhibited the results of softening in a few cases only. One was scale from the boilers which used water after treatment by the Archbutt-Deeley process, and another from boilers which used water after treatment by the Carrod process. Now, the Archbutt-Deeley softener was a rather complicated machine (he only used the word "complicated" comparatively), in which it was necessary to use live steam and a carbonating apparatus; the Carrod was, it seemed to him, the simplest of all the machines mentioned. The former reduced a simple water from 16·785° to 2·799° of hardness, which the authors said was a satisfactory result; the latter reduced a very difficult water from 40·88° to 27·932°, which was said to be an unsatisfactory result. This did not strike him particularly, except that the simplest softener was treating the most difficult water, and *vice versâ*; but to his great astonishment, the samples of scale showed that the water in each case deposited almost precisely the *same amount of scale*, the difference, if any, being on the side of the Carrod machine. How was this accounted for? And why, if the result was the same, was the one satisfactory and the other not? He thought the extreme importance of this case would be manifest to every one who was interested in the subject, though it was overlooked by those who took part in the discussion.

It was a pity that more samples of scale were not obtainable, as practical results were more useful than anything else in such cases, and doubtless the explanation would have been forthcoming if other results had been shown. A water-softener was required in connection with a large mill which he was erecting, but he really did not know what to do about it. The undoubtedly good result which was obtained by the Archbutt-Deeley machine favourably inclined him to it; but the water he had to deal with was a very difficult one, and contained a large amount of salt. Judging from actual results, the Carrod would suit his purpose better, and it was very much simpler to work, and the cost would be much less.

He would be glad if the authors of the Paper would give some explanation of the above-mentioned matters, and if they would also say how, and for what reason, the chemicals required were obtained and given, as these in almost every case differed considerably from those in actual use, and especially in the amount.

Mr. MICHAEL M. BROPHY wrote that his firm (Messrs. James Slater and Co.) had used a Chevalet-Boby Detartariser for about twelve months, and found that it heated the water to boiling point by means of the exhaust steam, and, in the process of heating, softened it. They also could not find the slightest trace of oil in the boiler; they had been very careful to look for this, as had also the officers of the Manchester Steam Users' Association, in whose care the boiler was placed. Their works' manager frequently tested the water both in the detartariser and in the boiler for softness and for soda. Some of these tests for the last few weeks were given in Table 3 (page 870). The makers advised the addition of a small amount of soda solution for the New River Company's water, which advice had been followed.

TABLE 3.

Date 1903.	Points of Hardness in Detartariser.	Points of Hardness in Boiler.	Soda Colour.
27 November	3½	2	Faint
4 December	3½	1	Strong
11	3	0·5	
18	„	1	
24	„	1·2	„

A little water was blown out from the boiler four times daily, lowering the water in the gauge-glass about half-an-inch each time. The boiler was cleaned out three times a year, and found all but free from deposit; the very little that was found might be due to the fact that they laid off the detartariser for cleaning for a couple of days in each four months. This was done because they then could have the use of their electric crane, which they could not have if the engine and dynamo were not running. The boiler was practically as clean as new.

Mr. OSCAR GUTTMANN concurred in praising the general excellence of the Paper and its intentions, but feared that a wrong impression would be obtained from its results. The Paper meant to deal with the working of various water softeners, but it drew its conclusions from the chemical analysis of the water before and after treatment. Now the chemical reactions taking place in a water softener were well known and unalterable, and they could only be influenced by an apparatus, if it were badly designed or out of order. The former was certainly not the case with most softeners, and the latter could not be taken as a measure for the working. Had the inquiry been divided into two parts, the one dealing with the construction of softeners and their work under certain definite conditions, and the other dealing with the efficiency of the chemical treatment for different waters, then very useful comparisons could have been made.

Another drawback of the Paper was that its conclusions were drawn from single specimens of apparatus and single observations with them, which was bound to give rise to misunderstandings. Thus in the case of his own softener an apparatus was inspected, which had to work under peculiar conditions. The water came from a pond, which collected all the rain water from the roofs and other contaminated sources, and thus varied considerably. This water contained a large quantity of sodium salts, and it was a well-known property of sodium-salt solutions that they were capable of dissolving and keeping in solution other salts. The result in this case was, that the calcium carbonate formed by the reaction of the sodium carbonate on the calcium sulphate could not be precipitated, but was held in solution. No other treatment could get a different result, and therefore the working of the apparatus could not be judged from the chemical analysis. The essential feature in this case was, that the apparatus had to serve as a condenser and oil-catcher for the large quantities of exhaust-steam available in these works, which it performed very well, and for the prevention of scale in the boilers, which it did quite satisfactorily, according to the testimonial of the owners given by Mr. Stromeyer. The fact was that this factory spent formerly £21 per week on labour for cleaning their many boilers, and had bulged tubes regularly every fortnight in a boiler working at 300 lbs. pressure, which was now no more the case. There were many other softeners at work on normal waters, which gave excellent results both chemically and mechanically.

Mr. WILLIAM PATERSON wrote in reference to the President's suggestion as a subject for discussion, "Does pressure facilitate precipitation?" In the writer's opinion this depended upon the nature of the suspended impurities. Should these be free from all air or other compressible gas-bubbles, then precipitation was not facilitated by increase of pressure, but where the precipitate had entangled with it minute bubbles of air or other gas, then precipitation was undoubtedly facilitated by increase of pressure. The decreased misplacement, due to the decreased volume of the gas bubbles under pressure destroyed the buoyancy of the attached

(Mr. William Paterson.)

suspended matter. The Cartesian Diver was a scientific toy which illustrated this phenomenon ; by depressing the stretched membrane across the mouth of the vessel, one increased the pressure throughout the various liquid layers, compressing the air in the bell of the diver and causing entrance of a portion of the liquid into the bell. The floating body thus became heavier, and the increased weight caused the diver to sink to the bottom. On releasing the pressure the air in the bell assumed its former volume, decreasing the amount of water in the bell, and so increasing its displacement and buoyancy,

FIG. 22.—*Paterson Softener.*

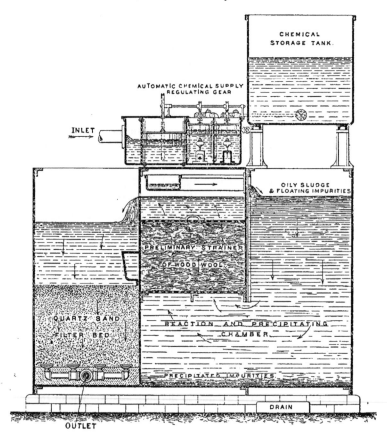

and causing it to rise to the surface of the liquid. In short, the increased pressure had facilitated the precipitation of the diver by decreasing its buoyancy, while the decrease in pressure had caused its refloatation by restoring its buoyancy. In the same manner a precipitate which contained entangled air or other compressible gas might be precipitated by increase of pressure.

In practice it would not be satisfactory to rely upon such means for clarification of water, as a reduction in the pressure might cause the refloatation of a large accumulation of the precipitate. In his opinion satisfactory purification of softened water from its fine precipitate could only be satisfactorily obtained by filtration through a quartz sand filter bed, which could be cleansed periodically by the reversal of the current of water through it.

In the Paterson softener, Fig. 22 (page 872), the hard water was measured continuously in its passage through a narrow vertical discharge slot or tumbling bay. A float in this chamber rose and fell upon variation of the amount of supply, and so controlled the quantity of chemical re-agents discharged through the long taper valves in the chemical re-agents storage-tanks alongside. These were kept furnished at a constant level by ball cocks, piped to the storage tank.

After mixture with the re-agents, the water passed through a re-action and precipitating chamber, and wood-wool strainer, and overflowed into the quartz sand-filter bed for the final purification. The quartz sand was cleansed by the reversal of the current through the bed, the separated impurities overflowing through the waste gutter to the drain.

Mr. J. H. PAUL, from a considerable experience of water-softeners, and the analysis of softened waters, came to the conclusion that water-softeners, like other machines, required intelligent attention, and they must not be expected to work automatically when once the valves were set and the supply turned on. There were many details to be attended to, and the feed-water was so constantly changing that the amount of chemicals to be used required frequent adjustment. This was specially the case with softeners using lime, either as milk of

(Mr. J. H. Paul.)

lime or lime water, for, as had been pointed out by Mr. Stromeyer, the quantity required was dependent upon the gaseous carbonic acid in the water; this was very variable and changed from day to day, so that attention had to be paid to this. As an excess of lime in the water if allowed to continue would form a very hard impervious scale, this was probably the reason why the users of softeners preferred to err on the safe side and rather use too little than too much lime. With soda, on the other hand, a slight excess did no

FIG. 23 —*Continuous Working of a Softener, observed 3 times daily.*

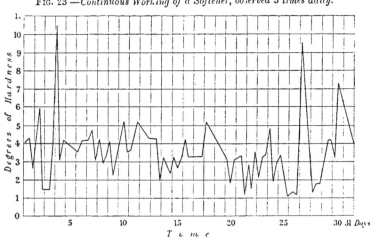

harm, and would effectually prevent any sulphate of lime scale should any improperly softened water get into the boiler. Of course, any continued excess was bound to become concentrated in the boiler water, and if allowed to get too concentrated, would undoubtedly do harm by attacking the fittings, or even cause leaky seams and joints, and at high pressures would develop caustic soda, which would aggravate the evil.

With regard to the authors' results, the conclusions they drew about the particular softeners were taken from one, or at most two, particular instances, and were drawn on particular samples. Fig. 23 showed a diagram of the working of a softener for a month. The samples were taken three times every day, tested, and results marked on the diagram. It would be seen that at certain times the hardness was very high, while in the majority of instances it was quite low. Suppose the authors had had a sample when the hardness was high, the result of their test would be to condemn the softener, while the average of the month would show it to be a very efficient machine. Of course it would have been almost impossible for the authors to have made so many tests, and therefore the conclusions they had drawn from the tests must be greatly modified by the remarks they quoted as to the expeiience of the users. In one case which they gave, namely, the Babcock and Wilcox softener, the results of the test showed that the water had not been softened at all, while the users stated that they were able to use the softened water in their small-tubed water-tube boilers without any formation of scale. This could hardly be the case, if the sample the authors had analysed was a fair average sample of the water discharged by this particular softener.

Further, as to the analysis of the delivery after it had been adjusted for the excess of condensed steam, it was hard to understand how it was possible for the water to contain the amount of bicarbonate returned in the analysis. This bicarbonate could only have been formed by the combination of the gaseous carbonic acid in the feed-water with the added carbonate of soda. Now, if the whole of this 3·94 grains of carbonic-acid gas was converted into sodium bicarbonate, it could only yield 15·04 grains, so that according to the analysis there were neaily 4 grains of sodium bicarbonate more than could pcssibly have been formed; and this was the more remarkable, when it was found that the sulphate of soda returned in the delivery corresponded to the second place of decimals with what one would expect to find if the whole of sulphate of lime had been converted into sulphate of soda, and this had been added to that already existing in the water. As the writer happened to have made two analyses of waters treated by the Babcock and Wilcox softeners, he gave the results below.

In the first case the total hardness had been reduced from 18·38° to 5·1°, with an excess of 4·4 grains of carbonate of soda; and in the second case a total hardness of 19·15° had been reduced to 3·78°,

3 o

(Mr. J. II. Paul.)

with an excess of carbonate of soda of 13·49 grains, and when it was considered that the solubility of carbonate of lime in water was about 3 grains per gallon, it must be conceded that the softener had done remarkably well.

Water No. 1.

	Untreated.	Treated.
Silica	0·48	0·53
Oxide of Iron	0·19	0·26
Sulphate of Lime	6·49	—
Carbonate of Lime	13·06	3·25
Carbonate of Magnesia . . .	1·11	0·45
Sulphate of Sodium	—	6·85
Nitrate of Sodium	2·97	2·96
Chloride of Sodium	2·45	3·26
Carbonate of Sodium	—	13·49
	26·75	31·05

Water No. 2.

Silica	0·42	0·37
Oxide of Iron	0·65	0·43
Sulphate of Lime	6·64	—
Carbonate of Lime	12·91	4·83
Carbonate of Magnesia . . .	0·50	0·23
Sulphate of Sodium	—	6·49
Nitrate of Sodium	2·98	2·67
Chloride of Sodium	3·72	2·39
Carbonate of Sodium	—	4·41
	27·82	21·82

Mr. T. ROLAND WOLLASTON wrote that he thought the Paper would prove of great value to steam users, and that it had been written opportunely, as the necessity for water-softening in connection with perhaps 95 per cent. of the boiler installations in the country was becoming recognised, while the majority of engineers and others who had to select plant had the crudest notions regarding the nature of the process, and required such assistance as the Paper provided. He welcomed the figures relating to heat losses due to boiler scale. He had direct evidence that the conclusions were reasonably in accord with practice, and the figures should at once put an end to the many ridiculous estimates in circulation. At the same time he thought that the figures given were only intended to apply so long as it was possible to turn the flue gases away to the chimney at a low temperature. If with scaled heating surface the gases were turned away at say 200° F. higher temperature than when the surface was clean, the rate of combustions being normal, there was obviously a greater loss. Such cases were frequent in practice.

There were two matters in connection with practical water-softening concerning which he thought further remarks from the authors would be interesting :—

(1) He had invariably found that scale, resulting from the use of hard and *organically foul* water, was more readily removed when that water was softened than was the scale formed from hard water in which the dissolved impurities were *only mineral*. He would like to know whether this experience corresponded with general experience, and to have the scientific explanation.

(2) It was common knowledge that with some feed-waters the softening need be only partial to prevent scale in boilers. Some of the results named in the Paper illustrated this. He was anxious to know if there were any laboratory test, chemical or physical, which would indicate how far it was necessary to soften any water to prevent scale. This matter was one of vital importance to makers of water-softening plant. Guarantees that scale should be prevented were generally required by purchasers, and the cost of full softening was often prohibitive, whereas it was probable that in many cases a very mild and cheap partial treatment would be quite satisfactory. He had heard it stated that the sodium-chloride of sea water would not form a hard scale, if the small amount of calcium and magnesium salts were eliminated.

The authors' remarks in regard to the cost of heating to assist water-softening might, he suggested, be misleading to many. The advantages of using exhaust-steam, when available, were too obvious to need discussion. When live steam was used, as a rule nearly all its heat could be returned to the boiler. The following was, in his experience, the most common case in practice : Hard water was taken from the hot-well of a jet-condenser at say 100° F. It was heated

(Mr. T. Roland Wollaston.)

to about 150° F. by live steam in the softener, at which temperature a nice flocculent precipitate would form, then there was a loss from 5° to 10° of temperature before the softened water reached the boiler. In many cases this procedure had led to an apparent saving in fuel. His experience therefore was similar to Mr. Boby's (page 861), but he did not pretend to account for it.

While in many softeners, his own among them, the control of the chemical treatment was by geared pumps for hard water and re-agent respectively, there were, as was noted by the authors, several which depended upon bored nozzles, adjustable weirs, etc., for regulation. These latter were, it was claimed, entirely automatic, and this claim was perhaps fair enough when the hard water supply reached the machine under sufficient head. But he thought in the great majority of cases the water had to be first raised to the machines, for which purpose pumps must be used. Thus these so-called automatic machines were on the same footing as the "pump controlled" machines in this respect. The question of the relative accuracy of the two systems was perhaps too delicate to be raised, although an expression of the authors' opinion would be very valuable.

The authors had given Tables of the results of working with several water-softeners of different types. He questioned whether these figures conveyed information in any way commensurate with the labour evidently involved in obtaining them, for probably any one of the machines described was capable of producing the best result named, if under skilled control. He thought that the best water-softener was the one which, in unskilled hands, would, with the minimum amount of attention, give the best and most consistent results. It was unfortunate that the authors had not felt themselves justified in expressing their views on this important question.

The authors, in exhibiting a sample of scale taken from a boiler where the writer's water-softener was in use, had inadvertently done him serious injustice. The sample was labelled in such a manner as to lead one to suppose that the scale had formed while the softened water was being used, whereas it was really old scale, formed previously to the installation of the water-softener, and removed thereby. That the wrong impression was formed by visitors was proved by their questions to him after the meeting.

Mr. BARON, in replying to the discussion and communications, wrote that perhaps the chief reason why the general result of the inquiry was not better was that the makers of water softeners had confined themselves too much to the mechanical side of the question, nor did he think that the subject had received at the hands of chemists the care and consideration it deserved. The facility with which magnesia formed basic salts, necessitating the employment of a proportionate amount of caustic soda or lime, if the purification from lime and magnesia was to be the maximum, seemed especially to call for the attention of the chemists.

This amount of caustic lime was the quantity referred to (page 779) as "excess," to which Mr. Archbutt took exception; the equation which followed the statement showed that its amount was the chemical equivalent of the magnesium carbonate.

The fact that magnesium hydrate was a gelatinous precipitate, and caused trouble by clogging the filters, was confirmed by the following experiment. A very dilute solution of magnesium chloride was precipitated with caustic potash and filtered through shredded asbestos on a Gooch crucible, the air being extracted underneath; observations were made on the quantities filtered and vacuum throughout the day. The readings were as follows:—

TABLE 4.

Time.	Volume filtered.	Vacuum Manometer.	Time.	Volume filtered.	Vacuum Manometer.
A.M.	Cm³.	Cm.	P.M.	Cm³.	Cm.
10.15½	—	—	12.0	95	24·0
10.16	15	—	12.15	99	28·8
10.16½	20	37	1.0	118	23·4
10.22½	30	38	2.0	146	18·5
10.30	45	23·2	2.30	157	19·0
10.45	58	24·9	3.0	168	19·0
11.0	65	27·0	3.30	176	20·0
11.15	80	24·8	4.0	185	21·0
11.30	88	18·8	4.30	193	21·8
11.45	92	20·8	5.0	200	23·0

(Mr. Baron)

205 cm³ only were filtered in the 7 hours, the mean rate being
29·3 cm³ per hour.   For the first minute the mean rate was 600 cm³
per hour; it dropped gradually to 25 cm³ per hour between 12 and 2
o'clock, and at 5 o'clock was only filtering at the rate of 14 cm³ per
hour.   1,000 cm³ of Manchester town water was filtered in 20 minutes
by the same filter, the mean rate being 3,000 cm³ per hour.   The
diameter of the perforated bottom of the crucible was 2·5 cm, and
the weight of magnesia collected on it in one day 0·0591 grm.
MgO from 212 cm³ of the above solution.

The gelatinous nature of magnesium hydrate was also shown by
the reproduction from the microscopic slides which he had prepared,
Fig. 24, Plate 38.

It was possible to obtain it in a more compact form by
precipitation along with calcium carbonate, the gradual transition
of the latter from the amorphous condition to the crystalline
"weighting" and shrinking the latter, and making it readily
separable.

There was no discrepancy in the two statements pointed out by
Mr. William Boby (page 861).   The first statement referred to
carbonate of lime, and the latter one was a general inference from
the results of the investigation, and referred to carbonate of
magnesia as well, which was soluble to some considerable extent.
This magnesium carbonate could not be removed completely by
boiling or by any reasonable increase in the carbonate of soda
added.   Caustic soda or lime was necessary for this, and must be
added to the boiled water.   The carbonic acid in the steam of
some boiler waters must not be forgotten in this connection.   Of
course in the chalky districts of the South of England the magnesia
present in water was small, and the need for its removal was
scarcely felt.

He could not agree with the explanation Mr. Oscar Guttmann
gave (page 871) of the retention of lime salts in solution, in the case
he mentioned, by soda salts.   It would be seen by referring to the
analyses that the softened water of Lassen and Hjort (2) contained
nearly the same amount of soluble salts, and the scale-forming
matter had been reduced to 3·597 grains per gallon.   The reason it

had not been precipitated was probably that the temperature had not been high enough.

He quite agreed with Mr. J. H. Paul (page 874) that variations in the chemical constituents of the water could only be met by regular testing. When once the adjustment of re-agents to feed-water was made, it should remain the same throughout the day, otherwise the apparatus was mechanically imperfect. The more simple and direct such adjustments were, the more certainly could the water be treated by the class of workmen to whose lot the softening of the water generally fell. He could not agree with Mr. Paul that there were great variations in the quantities of lime required through the day. His own experience was that very little variation in the quantity of lime occurred. Considerable latitude could also be made in the quantity of lime between the higher limit represented by that required for the total carbonic acid, and for the precipitation of magnesia as hydrate, and the lower limit of that required for the free carbonic acid alone. Further, an excess of lime in the water was very easily detected, and if due to variations in the water must almost invariably be accompanied by soda, the result being not a scale but an excess of caustic soda.

There were several reasons for the increase in the amount of sodium bicarbonate, in the case mentioned by Mr. Paul, beyond the amount equivalent to the free carbonic acid existing in the hard water. The steam with which it was heated contained free carbonic acid. Magnesium salts also, when precipitated with sodium carbonate, formed basic salts containing relatively more hydroxide the higher the temperature, a more acid salt (with carbonic acid) being left in solution.

The following reaction occurred when mixed cold:— $5MgCl_2 + 6Na_2CO_3 + 2H_2O = \{MgCO_3\}_4 Mg(OH)_2 + 10NaCl + 2NaHCO_3$ When the solutions were boiling, the following occurred:—$4 MgCl_2 + 5Na_2CO_3 + 2H_2O = \{MgCO_3\}_3 Mg(OH)_2 + 8NaCl + 2NaHCO_3$. The normal carbonate re-acted in the same way, thus:—In the cold, $5MgCO_3 + Na_2CO_3 + 2H_2O = \{MgCO_3\}_4 Mg(OH)_2 + 2NaHCO_3$. It would thus be seen that even with a considerable excess of soda, the magnesia and lime could not be as efficiently precipitated as by the

(Mr. Baron)

use of caustic soda or lime, and at the same time the excessive alkalinity was avoided.

With regard to Mr. T. Roland Wollaston's first question (page 877), the presence of organic matter in water was generally in the direction of increasing the rate of settlement by aggregating the small particles of precipitate, which would otherwise remain in suspension or settle very slowly. This was especially the case when heated, as it generally caused the coagulation of organic matter.

In reply to his second question and the point raised by Mr. Anderson, he himself did not know of any simple laboratory test which would give the amount of softening necessary to prevent scale. At low pressures, say up to 80 lbs., it was not very necessary to add soda equivalent to the soluble salts of magnesia present. Above this pressure the case was different, as magnesium hydrate was produced by the interaction of soluble magnesia salts and any calcium carbonate scale which was almost invariably present in the boiler :—$CaCO_3 + MgSO_4 + H_2O = CaSO_4 + Mg(OH)_2 + CO_2$, and hence all the permanent hardness must be converted into temporary for high pressures.

When the whole of the permanent hardness was converted into temporary in this way, there was not much fear of a hard scale being produced in the boiler. It became the reaction chamber of the "detartariser" class of softeners, the CO_2 being driven off, and the carbonates precipitated, and by an intelligent use of the blow-off cock these might be to a large extent removed. This method was the one attempted by most boiler compositions, and was open to the same objection as scale, unless the precipitate was got rid of in the way indicated.

The gelatinous nature of precipitated magnesium hydrate was shown on Fig. 24, Plate 18. The cracks in the semi-transparent mass were due to shrinkage, and the square holes were due to crystals of potassium chloride which had been removed subsequent to drying by washing in distilled water.

The crystalline nature of carbonate of lime precipitate was shown Fig. 25. It would be readily understood that these compact grains could settle fairly quickly. When magnesium hydrate was present in the water the crystals of carbonate of lime formed on it, and, by increasing the density of the flakes, increased the rate of settlement.

Fig. 26 showed crystals of selenite (sulphate of lime). Their lengths readily suggested that when close together they must bind each other, and thus form a compact scale.

Mr. STROMEYER wrote, in further reply to the discussion and communications, that Mr. Hatch had suggested the title of the Paper should indicate that only feed-water softening was being dealt with, but as all the inventions dealt with were capable of purifying waters for manufacturing or domestic purposes equally, serious complaints would have been made if the title had indicated a restricted inquiry. The above-mentioned subjects might have been included in this Paper, but only at the expense of very much space, and they would also have confused the issue; although the effect of scale in boilers was discussed, the wider question as to the permissible amount of impurities was not touched upon. It must therefore be left to specialists who knew the influence of soda, lime, and magnesia carbonates and sulphates on silks, flannels, linens, cottons, and on health and taste, to find out which system of softening was likely to give the desired qualities to the waters with which they had to deal.

Another reason for the somewhat restricted nature of the inquiry was the present impossibility of doing what so many speakers had suggested, namely, making continuous comparisons of various softeners fed from the same source. This could only be done on the occasion of an exhibition, or when, as he hoped at no distant future, water-softeners of every type would be found in every district and counted by thousands instead of as now by tens or by hundreds. It was the scarcity of water-softeners throughout the country which made it impossible to compare the various softeners in the manner suggested. The several inventors were asked for lists of installations from which one might hope to obtain samples of water, and of which one could get detailed drawings. Very few of these lists named a dozen works, and

(Mr. Stromeyer.)

generally these were all in one district, and the authors' choice was consequently very limited. Wherever possible, they selected localities where they suspected sulphate of lime and magnesia salts, and it was a mere coincidence that, as Mr. Atkinson had pointed out, only the simpler and older apparatus were met with in these districts. Under these conditions of limited selection, it would clearly have been unfair to make the rigorous comparisons asked for, and although the analyses could not be used for this purpose they threw considerable light on the possibilities of all water-softeners, if supplied with the proper quantities of chemicals. The analyses also indicated how easily one could go wrong, and how desirable it was to test the supply and delivery frequently. The fairly general deficiency of lime (*see* last 2 lines of Appendix III, pages 826–833) was probably due to the imperfections of the ordinary tests.

It had been suggested that certain bad results were due to the improved dissolving powers of water containing salt, but according to experiments made by M. Étard, who did not however deal with calcium salts, the reverse was the case. Experiments had also been made on the rapidity of settlement of fine powders, and it had been found that it was increased by acids added to the water, and retarded by alkalies. Waters containing lime could not of course be acidified, but their alkalinity could be kept at a minimum. This experience suggested that waters treated with caustic lime would clear more rapidly than those treated with caustic soda, and also that waters containing much carbonate of soda or requiring much of this salt would settle more slowly than others, and should be treated in larger softeners.

An apparent discrepancy had been pointed out between their remarks (on page 781) that boiling removed carbonic acid and their comments; the latter were made on several softeners using steam for this purpose but without complete success, and Mr. Brown (page 841) supported the latter view when he showed that the thermal storage vessel removed only about one third of the mineral matter in the water. This apparent discrepancy was simply a case of so-called theory (of test-books) not agreeing with practical experiences on a large scale; possibly improvements might be introduced, more especially in such a case as that mentioned by Mr. Paul (page 875), where, in spite of a large addition of condensed steam, the carbonic acid had actually increased. Here the most reasonable explanation seemed to be that the condensed steam contained carbonic acid, and that under the condition of the treatment heat alone was unable to remove it. In view of this difficulty it seemed desirable to add limewater to the supply of softeners using steam, the additional cost, less than 1 per cent., being almost insignificant.

FIG. 27.—*Temperature Distribution in a Boiler.*

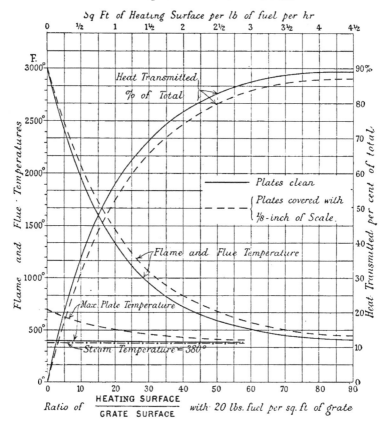

Sq Ft of Heating Surface per lb of fuel per hr

Ratio of $\dfrac{\text{HEATING SURFACE}}{\text{GRATE SURFACE}}$ with 20 lbs. fuel per sq. ft of grate

(Mr. Stromeyer.)

Mr. Maw mentioned (page 858) that water-softeners enlarged one's opportunities as regards choice of boiler type, but there was another advantage that boilers of any type using softened water could be worked harder, with higher flame temperatures, and therefore more economically than other boilers using sedimentary waters. At the request of Mr. Maw some values from Table 1 (page 774) had been plotted in Fig. 27 (page 885). As already pointed out, the decreased efficiency due to scale was but slight, and would appear even less than shown if, as was probably more correct, the heat transmission had been assumed to be proportional to a higher power than one of temperature difference. The diagram showed very clearly how impossible it was to attain the same efficiency in a scaly boiler as in a clean one, if the condition should be insisted upon of keeping the plate temperature in both cases equally low.

GRADUATES' ASSOCIATION.

DEEP WELL PUMPING MACHINERY.*

By Mr. W. PERCY GAUVAIN, *Graduate*, of London.

Wells are, and have been for many centuries, an important means of water supply for large districts, and it may be interesting to go briefly through the methods which have been adopted for raising water from them. All the first methods employed were by winding, for although pumps were in use to a small extent, they were made of wood, and unable to hold water with a greater head than a few feet. Wooden winches were employed, the water being drawn up in small buckets. At present there are numbers of such wells scattered about the country, supplying hamlets or groups of houses situated at so great a distance from the nearest town that it would be impossible to connect them with the town supply. Windmills were largely employed in driving two and three-throw pumps, and are still used in certain districts and in country houses; but, as their action is uncertain, it is usual to adopt a small oil-engine, or some such motor, as a standby.

The earliest type of steam-engine that was successfully applied to raising water (other than the Newcomen) was the Cornish engine, which was first used to drain the tin mines in Cornwall. Engines used for deep well pumping may be divided into two classes :—

    (1) Rotative.

    (2) Non-Rotative.

the non-rotative class being subdivided into :—

    (2 a) Cornish.

    (2 b) Those other than Cornish.

---

* This Paper, read on 9th March 1903, has been selected by the Council for publication.

Rotative.—A large variety of different types of rotative engines have been used. A very common type is the overhead beam-engine. It is cumbersome, the first cost is heavy, and it is subject to the disadvantage open to all classes of rotative engines employed in deep well pumping, namely, that there is no distinct pause at the end of the stroke, which is so essential to the efficient and economical working of the pump-valves. Even though a large fly-wheel be employed, it is impossible to run such engines very slowly, such as can be done with the Cornish engine and the Davey engine described below. Gas- and oil-engines are usually employed to drive three-throw pumps. The speed has to be reduced, and the gearing involved not only reduces the mechanical efficiency, but makes the upkeep very heavy. Where small quantities of water have to be dealt with however, the upkeep is probably less than that of a steam plant, the reduction being in labour cost, rather than in fuel. Electric motors are also used to drive well pumps, and in this case also a certain amount of power is lost in gearing.

Cornish.—For those who are not conversant with this type of engine, it will be advisable to explain upon what principle it works. The engine has to lift the pump-rods (which are of timber, usually pitch-pine) and then let them fall by their own weight and force up the water. In the case of an engine forcing into the mains and having no pump-rods, a balance weight or weighted plunger is used. The mode of working is as follows :—Suppose the steam-piston to be at the top of its stroke. Steam is admitted at the top of the cylinder and the piston moves downwards, raising the pump-rods and drawing into the pump a charge of water. Steam is cut off at about one-quarter to one-third stroke, the piston continuing its stroke, partly by the expansive force of the steam and partly by the momentum of the pump-rods. At the end of the stroke, the equilibrium valve opens, allowing the steam, which was at the top of the piston, to come round underneath it, putting the piston in equilibrium. The pump-rods then fall by their own weight, forcing water up by means of the plunger. The pumps-rods should be at least 20 per cent. heavier than the water load, but in many cases they far exceed this, the extra weight having to be counterbalanced by balance bobs. Before the end of the stroke is reached the equilibrium valve is closed, thereby cushioning steam in the top of the cylinder and bringing the rods gradually to rest. The exhaust- and steam-valves now open, the exhaust-valve remaining open nearly the whole of the stroke. The Cornish engine takes advantage of the momentum of the pump-rods to enable it to use steam expansively; in fact, the pump-rods are to it what a fly-wheel is to a rotative engine.

The energy stored up in the pump-rods is expressed by the formula $\dfrac{W\,V^2}{2\,g}$, where W is the weight of the moving parts, V the velocity in feet per second, and $g = 32 \cdot 2 \cdot$ It will be seen, therefore, that the degree of expansion depends on (1) the weight of the moving parts, and (2) the velocity which they attain; and as the weight is constant for a given engine, any increase in the degree of expansion employed increases the value of V. For the purpose of expansive working the speed of a Cornish engine during the steam stroke (that is, when it is lifting the pump-rods) is about twice the speed of the water stroke; and for this reason the suction-pipes and valves should be large, generally two suction-valves and one delivery-valve being employed. The Cornish engine employed as a mining engine is usually more economical than when it is used for waterworks purposes, for the reason that as a mining engine the weight of the pump-rods usually far exceeds the water load, whilst the balance weight on a waterworks engine is very little in excess of the load on the plunger. There is a limit however to the number of expansions, for increasing the weight and the velocity of the moving parts puts greater stresses on the rods and connections, so that steam of a higher pressure than about 40 lbs. cannot be used effectively. The consumption of steam by a high-class Cornish engine is about 26 to 30 lbs. per I.H.P. hour, and the duty about 60 million foot-lbs.

Non-Rotative Engines other than Cornish.—Of non-rotative engines for deep-well pumping, probably the best known type is the Davey high-duty engine. The following is a description of such an engine which was erected at the Hooton pumping station of the West Cheshire Waterworks in 1901, the author being present during the whole time of erection. Two boreholes were sunk 22 feet apart and to a depth of 450 and 500 feet respectively. For the first 200 feet they were 30 inches diameter, and from there onward 18 inches diameter, it being decided to place the pumps at a depth of 200 feet from the surface. The engine had to lift water from the boreholes into a sump in the bottom of the engine-house, from which sump the force-pump drew and forced the water into the mains against a head of about 400 feet. Running at 15 double-strokes per minute, equivalent to a piston speed of 150 feet, 2,160,000 gallons had to be pumped per 24 hours.

The engine, Fig. 1, Plate 39, is of the horizontal compound tandem type, having cylinders 30 inches and 56 inches diameter by 5 feet stroke, actuating a pair of single-acting borehole pumps and a double-acting force-pump. From the engine crosshead two timber rods go to the first pump quadrant, where they couple on at the bottom and from thence a single timber rod goes to the second pump quadrant. These rods are made of pitch-pine, and are strapped the whole length with mild steel plates. Timber is much better than steel for this purpose, there being no vibration such as would occur in steel rods of a great length. The piston-rod is continued through the low-pressure back cover, and is cottered on to the pump-rod of the force-pump. Both high- and low-pressure cylinders are steam-jacketed on the bodies and on the covers, the jacket space on the bodies being about 1 inch. The steam for supplying the steam-jackets is taken from the boiler side of the engine stop-valve, so that the engine can be thoroughly warmed up before starting. Jacket steam passes first round the high-pressure cylinder body, and from thence to the two high-pressure covers and to the receiver. The receiver is an enlarged steam-pipe in which is placed a copper coil, the effect of the jacket steam passing through this coil being to superheat slightly the steam on its way to the low-pressure cylinder. The jacket steam then passes round the low-pressure cylinder and the two covers. All cylinder jackets and covers are connected to steam traps.

In an engine of this type some special governing arrangement has to be adopted, otherwise it would be impossible to obtain a constant stroke; sometimes the pistons might be striking the cylinder covers and at other times the stroke might be several inches short, owing to a variety of circumstances, such as rise and fall of steam-pressure, pump-valves sticking, etc. It is of great importance that a full stroke should be made every time, for if the stroke is two or three inches short at each end, the clearance volume is thereby greatly increased. In small duplex pumps such as are used for boiler feeding, the bringing to rest of the piston at the end of the stroke is performed by means of a double steam-port which enables the piston to cushion steam. These pumps, however, have no pretence to economy, and, being of small size, would do little or no damage if the covers were struck occasionally; but in the case of a large engine a heavy bump might be a very serious matter. To overcome these difficulties the differential valve-gear was invented by Mr. Henry Davey.

At first sight the action of the gear may perhaps be a little difficult to follow. The following explanation should, however, make it clear. The gear is a small engine which is run at a definite speed, and which is used to open the valves of the main engine; the main engine itself closes its own valves. The motion given to the valves is thus a differential one, made up of a definite motion from the gear and a varying motion from the main engine, so that the engine automatically adjusts its own cut-off. For example, if the engine is very heavily loaded the gear will open the valves, and the engine will move off slowly and cut off late in the stroke; if, on the other hand, it is lightly loaded as soon as the gear starts opening the valves, the engine starts off and commences closing them overtakes the gear and cuts off early. In the case of the engine racing, for example, if a pipe bursts and the load is suddenly thrown off, the engine starts off at a high velocity and overtakes the gear, so that the valves are closed soon after they commence to open.

The usual form of differential gear is that in which a steam cylinder, controlled by a cataract, works the engine-valves. In the case of the Hooton engine, however, the gear was worked by water-

3 P

pressure derived from the force-pump of the engine. This also has a great advantage, for in the case of the pressure falling by reason of a burst pipe, the gear ceases working. Both high and low-pressure cylinders have slide-valves, on the backs of which work Meyer expansion-plates. The cut-off can be adjusted whilst the engine is in motion, in the case of the high-pressure cylinder from 0·3 to 0·7, and in the low-pressure from 0·3 to 0·9. The main slide-valves receive their motion from the gear through a rocking-shaft, and the expansion-valves through a similar rocking-shaft worked from one of the pump quadrants. Both valves move in the same direction, and the faster the engine goes the earlier the expansion-valve cuts off. The gear consists of a subsidiary double-acting cylinder ; the valve controlling admission of water to this cylinder is of special type, and is itself worked by a second and smaller cylinder known as the pausing cylinder. This cylinder, as its name implies, regulates the amount of pause the engine makes at the end of its stroke. The valve controlling admission of water to this cylinder is worked from the engine. By throttling the supply of water to this cylinder it can be arranged that its piston takes several seconds to complete its stroke, so that the valve which admits water to the subsidiary cylinder is not opened till two or three seconds after the engine has reached the end of its stroke. Consequently the engine pauses, as the main valves are not open. The amount of pause given to the engine can be regulated to a nicety by means of this cylinder. Allowing a pumping-engine to pause at the end of its stroke is very beneficial ; it enables the valves to fall to their seats by their own weight instead of being forced down, the wear and tear on the beats being much less. This, of course, is impossible in a rotative engine, and the valves of pumps driven by that type of engine require renewing more frequently than in the non-rotative type.

In the high-pressure steam ports are placed two Corliss valves receiving their motion from the engine, the object of these valves being to close the port before the engine gets to the end of its stroke and thereby cushion steam. These are capable of adjustment whilst the engine is running, from three-quarters to nineteen-twentieths of the stroke. The exhaust steam, after leaving the low-pressure cylinder, passes through a feed-water heater, through which water, on its way to the boiler, passes. The condenser is of an open type standing in the sump into which the borehole pumps deliver, and contains 530 square feet of cooling surface. There are two single-acting air-pumps driven from one of the pump quadrants.

Pumps.—The borehole pumps receive the engine motion by means of quadrants or compensating levers. The pump-rods not being of sufficient weight to allow of an early cut-off (as in the case of the Cornish engine), a compensating method was devised. The pumps are connected to the engine in such a way that, as the engine advances in its stroke, it gains a mechanical advantage over the pumps, or, in other words, the pump resistance gets less as it proceeds in its stroke ; Fig. 2 illustrates this. Let A B C D be the pump

FIG. 2.

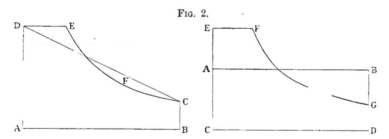

resistance and C E F G D the resultant engine diagram, then A D E C B is the pump resistance when the compensating levers are used, the momentum of the moving parts of the engine and pumps being sufficient to equate the two diagrams. By this means it is possible to cut off early in the stroke, and a great economy is thereby effected. In Fig. 3 (page 894) the author shows energy-diagrams of various classes of engines. The rectangle represents the pump resistance, whilst the curve is that of engine pressures. The percentage of excess work at the beginning of the stroke is given in each case.

The pumps are of the single-acting bucket type, the pump-rods working inside the raising main. The action is as follows :—On the

upstroke the bucket draws in a charge of water through the suction
valve, at the same time delivering water over the top of the well; on
the down, which is the idle stroke, the suction valve is closed, the
bucket coming back to its position for the working stroke, water
passing up through its valve. During the idle stroke it is possible
to ascertain if the suction valve is holding water, and as the bucket
has to be drawn before the suction valve can be got at, it (the

FIG. 3.   *Energy-Diagrams.*

Non-Rotative Pumping Engines.

Variation per cent. from Actual Engines.

Compound Forcing Engine.   *Cornish Mining Engine.*   *Cornish Waterworks Engine.*
Steam Pressure 70 lbs.   *Steam Pressure 45 lbs.*   *Steam Pressure 32 lbs.*
15·4 per cent.   *26 per cent.*   *21·1 per cent.*

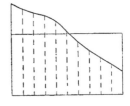

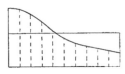

Cylinder Ratios 1 : 3·3.   *No. of Expansions* 5.   *No. of Expansions* 3·5.
Cut-off in H.P. Cylinder ¼.
Total number of Expansions 6·6.

bucket) can be examined at the same time. The rising main consists
of a number of mild steel lap-welded pipes, 18 inches diameter and
$\frac{3}{8}$-inch thick, in 15 feet lengths, having wrought-iron flanges screwed
on and riveted through. Fig. 4 shows a section of one of the pipes,
and illustrates the methods of attaching the flanges and making the
joints. The working barrel and clack piece, which are of cast-iron,
are attached to the pipes by a screwed joint. A steam-winch is
supplied, which makes the operation of lowering in the pumps quite
a simple matter. The suction pipe is first put into the borehole, and
clamps somewhat similar to those used by smiths are placed round
it, supporting it at the borehole mouth. The working barrel and
clack piece are next lifted up by the winch and screwed on to the

suction pipe. The winch then lifts the whole lot slightly, enabling the men to free the clamps. The working barrel clack piece and suction pipe are next lowered into the borehole and the clamps fixed near the top of the working barrel. Then another pipe is added, and so on. To the top pipe is screwed an iron casting which rests upon, and is bolted down to, a second casting firmly bedded in concrete. In this position the rising main and working barrel hang. The suction valve is then lowered down to its seat. The valve, Fig. 5, Plate 40, is of the double-beat type, having gutta-percha beats.

It will be noticed that a long weight is cottered on to the valve-spindle. This is necessary in order to keep the valve from lifting

FIG. 4.

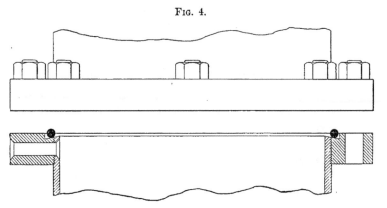

on the suction stroke. In some cases taper seatings have been used, but these are troublesome, as the valve seating gets firmly bedded in, and difficulty is experienced in drawing it. It will be observed that the ribs on the valve are cast at an angle. This is done so that the water impinges on them, and causes the valve to revolve and continually bed itself on to its beats.

The fishing tackle used for lowering in the suction valve is also shown in Fig. 5, Plate 40. It is of a very simple construction and most effective in " fishing " for the valve when it is required to put in new beats or for other causes. The valve spindle is made with a mushroom head under which the wings of the tackle grip. Fig. 6,

Plate 40 is a drawing of the bucket, which has a double-beat valve similar to the suction valve. The bucket packing consists of gutta-percha rings of stout section. Three or four small holes about ¼ inch diameter are drilled on the top of the bucket, so that the pressure of water gets to the back of the rings and makes a tight joint. The first rod is cottered on to the bucket spindle and is then lowered, the method adopted being the same as with the pipes, clamps being used. The rods are of steel 4 inches diameter and in 20-feet lengths. At the end of each rod a guide is cottered, which consists of a cast-steel wheel, Fig. 6, and is made about 1 inch less in diameter than the pipes. There is no thrust on the rods, all the work being done on the upstroke; on the downstroke the rods simply fall by their own weight, but in the case of the bucket sticking and the weight of the rods being insufficient to overcome the friction, the engine then having to help them, the guides would stay the rods very materially. The pump quadrants are connected to the pump crossheads by two steel connecting-rods. The pins on the quadrants to which these connecting-rods are attached are prolonged about 10 inches each side, and are made to act as bumpers. In the event of the engine overstepping its stroke, these pins come on to buffers and prevent the pistons striking the covers. The buffers consist of blocks of hard wood about 7 inches deep, lying on top of about 3 inches of indiarubber. There are similar buffers in the engine slide-bars for the slippers to bump against.

In starting this engine great difficulty was experienced, owing to the large amount of sand which found its way into the boreholes. Sand was pumped in such large quantities that it had to be ladled away as soon as it rose to the surface, and finally it was decided to stop, as the sump in which the borehole pumps delivered became so full of sand that nearly all the rows of condenser tubes were covered and the vacuum fell away. The rods and bucket of one borehole were drawn, and after some difficulty the suction-valve also, but when it came to drawing the suction-valve of the second borehole, it was found that the mushroom head was completely covered with sand by about 12 inches, the sand in the 200 feet of water having settled. The fishing tackle was tried several times but without avail, and it looked very much as though the tubes would have to be drawn, when it was decided to try the effect of putting a small pipe down the tubes with water under pressure. A 1-inch wrought-iron pipe was connected to the mains under a pressure of about 150 feet, and then led down the tubes. It had the effect of stirring up the sand which gradually rose and flowed over the top with the water. After running several hours enough sand was cleared away to enable the fishing tackle to grip the mushroom head, and the suction-valve was drawn.

The valves of the force pump are of the same type as those described, the piston being 17 inches diameter and double acting.

By-passes are provided, making connections from above to below the delivery valves, and above to below the suction valves. By means of these by-passes the engine can be moved over, provided there is pressure on the main, the force-pump acting as a power cylinder. The engine can also be run with only one borehole pump in action, or the force pump single-acting, or any such combination, by the adjustment of the expansion plates, which is a very simple matter. This is extremely important, for in the case of a breakdown, one set of pumps can be run whilst the other is being repaired. This type of engine can be run dead slow, and the pause may be as long as ten to fifteen seconds; running at a slow speed the steam consumption is not much more than at the normal, for steam is not throttled, the piston speed during each stroke being the same.

Unfortunately the author is unable to give any results as to the steam consumption of this engine, as at the time of writing this Paper no trial had been carried out. A similar type of engine, but working with a steam pressure of only 70 lbs. per square inch, gave a duty of 124 millions on a ten to one evaporation, the steam consumption per I.H.P. being 15·6 lbs. and the mechanical efficiency 87 per cent.

Taking into consideration the fact that the steam pressure used was only 70 lbs. per square inch, the above result is extremely good. It will be noticed also that the mechanical efficiency is high. The Hooton engine may reasonably be expected to give an even better result, as the steam pressure in this case is 100 lbs. per square inch, and various improvements, all tending to reduce steam consumption, have been added.

It would take too long to go into the various classes of pumps and pump valves used for deep wells, but the engine and pumps already described may be taken as an example of the best practice of the day.

The Paper is illustrated by Plates 39 and 4C, and by 3 Figs. in the letterpress.

GRADUATES' ASSOCIATION.

AUTOMATIC COUPLERS.*

By Mr. FREDERICK C. HIBBERD, *Graduate,* OF MANCHESTER.

The subject of Automatic Couplers is important for all connected with the working of railways, the question of their adoption having been of late years the cause of a considerable amount of agitation in this and other countries. It is, however, an extremely wide one, and it will be impossible in this short Paper to do more than touch upon the most important points, namely :—

(1) The history of the automatic coupler question.

(2) The various systems of automatic couplers in use, with brief descriptions of typical examples of each system.

(3) The requirements of an ideal automatic coupler.

(4) The question of the adoption of the automatic coupler in Great Britain.

History.—In 1874 the Master Car-Builders' Association, a subsidiary association of technical men representing the various railways in the United States, turned their attention to the subject of an efficient automatic coupler, that is, one which would couple by impact. Various Committees were appointed, but it was not until 1884 that the Massachusetts Legislature passed an Act requiring that "as freight cars were constructed or purchased, or when cars were repaired, they should be fitted with such form or forms of automatic or other safety couplers as the Board of Railroad Commissioners may

---

* This Paper, read on 12th January 1903, has been selected by the Council for publication.

prescribe after examination and test of the same, to take full effect after 1st March 1885." In 1884 the Commissioners prescribed certain forms of couplers, of which 5,000 had been applied in 1888. Meanwhile the Master Car-Builders' Association in 1886 and 1887 undertook a series of competitive tests, and as the result of these tests they recommended a standard type known as the "Janney" coupler, a description of which will be found on page 904. It is important to notice that it was not the Janney coupler that was adopted, but one of the Janney type. The Master Car-Builders' Association, in fact, fixed what was known as the "Contour Line," Fig. 1, Plate 41, and determined the principal dimensions, Fig..2, which had to be followed by all automatic couplers. They left the railway companies and inventors to fix the form and method of hinging the knuckle and the arrangement of the locking device. The outcome of this was that in 1891 a convention of the State Railroad Commissioners resolved that the States Legislature should require freight-cars to be equipped with automatic couplers of the Master Car-Builders' type.

Later, in 1893, the United States Congress passed an Act dealing, amongst other things, with automatic couplers, which says, " That on and after the 1st day of January 1898, it shall be unlawful for any such common carrier to haul or permit to be hauled or used on its line any car used in the inter-State traffic not equipped with couplers coupling automatically by impact, and which can be uncoupled without the necessity of men going between the ends of the cars. At the end of the period allowed, however, a large number of companies applied for an extension of time and were allowed a further two years, so that from January, 1900, all cars in the United States were to have been fitted with automatic couplers.

The progress of the automatic coupler in the United States of America (up to July 1898), as a result of this legislation, can be seen from the Table on page 901.

It will be noted that this Act of 1893 did not, like the Massachusetts Legislature, specify the particular form of coupler, but left it to the companies to decide upon one which would be suitable to the requirements of their traffic.

Date.	Total No. of Cars reported.	Fitted with Automatic Couplers.	Percentage.
1st January 1897 . .	1,054,815	517,617	49·1
1st July 1897 . . .	1,054,022	573,296	54·4
1st January 1898 . .	1,110,045	674,675	60·8
1st July 1898 . . .	1,113,745	784,596	70·4

Description of Couplers in use.—At present there are about eighty different variations of the M.C.B.* coupler in use in the United States; but owing to the fact that they are all on the lines of the Master Car-Builders' Standard, conforming with the leading dimensions shown in Fig. 2, and having the contour line shown in Fig. 1, any one coupler will engage with any other, although the locking mechanism may vary; moreover, of the total number of couplers in use in the States about 70 per cent. are of the Janney type.

It may be asked how it is that the rolling stock of the United States is almost exclusively fitted with automatic couplers, whilst in this country they are comparatively rare. To understand this fully, it must be remembered that the old form of American coupler was the "link-and-pin" coupler, Fig. 3, which did not require side buffers. When the Master Car-Builders' coupler was introduced, it had the merit of being readily made to couple with the old attachment. This link-and-pin arrangement, moreover, was the cause of such an enormous number of accidents that the railway companies in America decided to look out for an improved coupler. The number of accidents in this country never reached that in the United States of America; in fact, in 1898, when the majority of vehicles in the States were fitted with automatic couplers, 1 in 486 servants employed on the railways was killed, whilst in this country

* The Master Car-Builders' Coupler is generally known as the "M.C.B." coupler.

the proportion was only 1 in 1060, and in the United States 1 in 30 was injured as against 1 in 128 in this country. It is partly due to these facts that railway experts in this country have not given so much consideration to the question of automatic couplers as the American engineers.

In 1882, however, an automatic coupling was exhibited at Darlington, and in 1886 some trials took place at Nine Elms and Derby. At the former place the "Brockelbank" coupler was exhibited, Figs. 4 and 5, Plate 41, which made a favourable impression on the minds of some of those present. A full account of the performances of this coupler up to this time may be seen in the Journal of the Society of Arts, in a lecture * delivered by Mr. Brockelbank in 1876.

Since then some of the railway companies have expressed themselves as being on the look-out for a suitable coupler. On some of the principal railways automatic couplers have been and are now employed on vehicles used in passenger trains. One of these is the Gould coupler, which is of the M.O.B. type, somewhat similar to the Janney. In 1899, legislation on the subject was attempted in this country. A Bill was introduced by Mr. Ritchie, President of the Board of Trade, giving this body power, after five years had expired, to make an order upon the railway companies to apply a suitable form of automatic coupler after another term of years— provided such suitable coupler could be found during the first five years. This Bill was subsequently rejected.

In Australia, France, and other countries, experiments have been and are being made with automatic couplers; the question may be said to be in an experimental stage on the Continent. In India automatic couplers are used to a considerable extent, the form adopted being the "Jones" type; this coupler is of an entirely different form from that adopted by the Master Car-Builders' Association, being of the "Norwegian hook" type, and a description is given on the next page. It was first introduced on the Indian State Railways in 1883. The couplers used on many of the

* Journal, Society of Arts, 1876, vol. 24, page 415.

Government Railways in Australia, South Africa , and Natal, are semi-automatic, and are known as the " Johnston " couplers ; they are of the " link-and-pin " type.

Description of various types of Couplers.—An automatic coupler is one which is capable of self-coupling by impact. It is usually connected to a central buffing and draw gear, although coupling devices have been applied to side buffers. Cheesewright's coupler is an example of this, consisting of the Norwegian hook adapted to side buffers. There is also a later design, which consists of two Master Car-Builders' couplers in the position of side buffers—each acting as a buffing and draw gear. The automatic couplers in general use are of the following kinds :—

(1) Norwegian Hook — the Standard of the Indian States Railways, also used in Australia.

(2) The Master Car-Builders' or American type.

The best example of the Norwegian hook type of automatic coupler is the "Jones" arrangement Fig. 9, Plate 42, which was selected out of many, and adopted by the Indian Government for application to the States Railways. In this arrangement, when the vehicles are buffed together, the hook A, which is pivoted at B on one head C, strikes against a bar D on the other head E, and, due to the angle of inclination of the face F of the hook A, it rises, and passing forwards it engages with the bar D. In order to effect tight coupling, the bar D is drawn backwards towards the headstock H by means of the right- and left-handed screw K and the link-work L L_1 L_2. In this manner the two buffer faces, M and N, are fixed so tightly together that the two draw-bars become practically one solid rod. Referring to Fig. 6, Plate 41, it will be seen that when the vehicles are on a curve, the centre line of the coupling a–b will stand at an angle with the centre lines of the vehicles c–d, e–f ; consequently allowance must be made for the necessary radial motion of the draw-bar. It was found that, when the vehicles stood on a curve, the friction against the knee P, Fig. 9, due to the draw-bar being at an inclination to the axis of the vehicle, was sufficient to wear the bar down, so that its strength was materially reduced. To obviate this the rod R was made perfectly free of the knee P as shown.

When the vehicles are of different length, and are put on a sharp curve, the coupler of the longer vehicle will stand farther out from the rail centre than that of the shorter, so much so that it was found impossible to couple on sharp curves. To overcome this difficulty the guide vanes V_1 V_2 V_3 V_4 were introduced. They are really extensions of the buffer faces in an outward direction, the top half to the right and the lower to the left (looking from the vehicle), and are inclined forward at an angle of 45°. The result is that when the couplers meet on a curve at different inclinations to and different distances from the rail centre, they mutually engage and guide each other into the central position as shown on Fig. 9. But in order to effect this desired result, it is necessary to give the couplers a greater degree of flexibility than before; for this purpose the india-rubber pads, X and Y, are introduced, securely bolted to the knee P, the casing Z, and the guide casting S. The buffing and draw springs are the india-rubber springs G, which fit fairly tightly in the casing Z, and are clear of the draw-bar. In this manner the whole arrangement is flexible, and is capable of being turned through a considerable angle relatively to the headstock. The original "Jones" flexible buffer and coupling was designed—(a) to remove all wear by abrasion—abrasion being removed by tight coupling which necessitates flexibility ; (b) to enable the Indian State Railway to run double trains. When loose coupled it was found that the snatch at the rear end of the train broke the couplings. Hence tight coupling was necessary, and tight coupling is not possible in a centre buffer which is not flexible, owing to the curves upon which the vehicles have to run.

The great advantage this coupler possesses over others is that it will couple automatically under almost any circumstances, a 43-foot vehicle with 25-foot wheel-base coupling easily with a 12-foot wagon, 5-foot wheel-base on a 5-chain curve.

The "Janney" coupler may be taken as the representative of the Master Car-Builders' type, for, as previously mentioned, it is that found upon the majority of railways in the States. Fig. 7, Plate 41,

shows a sectional elevation and inverted plan of this coupler. A is
the main casting of the coupler, having a "guard arm" B, C is the
knuckle pivoted by means of the pin D, and capable of turning through
an angle of 90°. When uncoupled, the knuckle is in the position
indicated by the chain lines, and upon two of the couplers coming
together the tail of the knuckle is turned through an angle of 90°
and locked in position, in the case of the Passenger Car coupler by
the "catch lever" H, which in the later forms is operated from the
end of the coach by means of a connecting lever. In the case of the
freight-car coupler, this catch lever is replaced by a "locking
spring," which is also operated from the end of the vehicle. The
buffing and draw spring is shown at E. It should be noticed that the
buffers F 1 and F 2 do not touch when the cars are coupled, but only
come into play after the main buffing spring E is partly compressed.
One great difficulty met with in the early days of these couplers in
the United States of America was that wagons were sometimes
buffed together when the knuckles were in the "coupled" position.
This of course necessitated men going between the wagons to open
the knuckles, and consequently improved forms were devised in
which the knuckle could be opened from the side of the vehicle by
means of a lever. The "Gould" coupler has this attachment.

The "Perry-Brown" Tandem Spring Coupler is another example
of the M.C.B. type. In this case, the lever on being pulled over
raises the locking bolt and the knuckle flies open. This coupler is
composed of three main parts : — (1) The head and shank ;
(2) Knuckle or Jaw ; (3) Locking bolt. The shank is of cast steel,
cast either in two halves or one piece, and is fitted with two springs
in the inside to take up the thrust and pull motion. These springs
fit between three plates fitted in the shank through slots, and the
plates are removable, allowing the springs to be taken out to be
replaced. The knuckle is of cast steel, fitted to the shank-head with
a steel pin. The nose of the knuckle is fitted with a hardened steel
wearing plate which can be replaced when the old one is worn.
The knuckle is so formed that when the coupler is closed the strain
is not on the pin, and in the chance of the knuckle pin breaking or
falling out the knuckle is still held fast by a "locking bolt." The

latter is an arrangement for locking and unlocking the coupler, and is worked from the side of the vehicle by means of a lever. The locking bolt is provided with a small trigger which, when the bolt falls, locks the mechanism so that the bolt cannot jump out. To uncouple the vehicle the shunter pulls over the lever, thus lifting the locking bolt and causing the knuckle to fly open.

Another difficulty that has been experienced with couplers of the M.C.B. type was that it was not easy to uncouple vehicles standing on a curve. During the last visit of Messrs. Barnum and Bailey to this country, they brought special trains over with them, which were, in accordance with American practice, fitted with Master Car Builders' couplers. It was found impossible to uncouple these trains whilst standing at Willesden Junction Station without putting an engine on to the end of the train and backing slightly to relieve the tension on the draw springs. In the early days of these couplers it was found that trains parted, due to the couplers becoming disengaged in a mysterious manner without opening the knuckles. After considerable investigation it was found that the pin was disposed to creep up, owing to the friction of one knuckle upon the other; this was remedied by putting a little tooth into the locking spring.

In a report of the Master Car-Builders' Association dated June 1897, it is shown that of 5,755 cases of trains parting, 2,155 were due to failures in the automatic coupler. Many of these were due to defective material, some to the creeping of the locks, and some to other points of weakness. The chief points of wear on these couplers are the heel of the knuckle which comes in contact with the guard arm of the opposing coupler, the inside of the guard arm, the inside face of the knuckle where the pull is most severe, the pivot pin-hole in the knuckle and the lug of draw-bar, the pivot pin, and the locking surface of the tail of the knuckle, any of which might cause a break away.

Another coupler, used to some extent in the United States of America, is the "Miller" coupler, Fig. 8, Plate 41. In this case the buffer B and draw-hook A are entirely separate, and are controlled by separate springs. The great objection to the "Miller" form is the trouble experienced in uncoupling, which is difficult at all times and may become impossible on curves. This coupler dates from before the time of the "Janney" coupler, and is generally acknowledged to be inferior to it. The "Miller" arrangement is really the origin of the American automatic coupler, a development of it having a knuckle release, which was further developed by the "Janney" coupler.

Ideal Automatic Coupler.—Some of the requirements of an ideal automatic coupler are:—

1. It must be capable of coupling automatically by impact, and should uncouple without the necessity of men going between the vehicles. This, as already pointed out, has been accomplished by many couplers.

2. It should be reliable and inexpensive. In the early days of the "Jannoy" coupler, cases of trains parting, due to failure of the couplers, were very frequent. The question of cost is of course very important. It is difficult to conceive a form of coupling which would be less expensive than our three-link coupling. The introduction of the Master Car-Builders' coupler in the United States of America has caused a considerable increase in expenditure.

3. It should be capable of coupling not only the draw-bars of the vehicles, but also where necessary the brake-pipes, and in some cases the heating pipes, and passenger communication. Experiments have been made in this direction, but, so far as the author is aware, without any satisfactory result. A coupling recently invented by Mr. George Westinghouse for use on electrical coaches is said to couple not only the draw-bars but also the air-brake pipes and electrical connections.

4. It should be capable of coupling the longest vehicle with the shortest whilst standing on the sharpest curve, and of uncoupling easily under the same conditions.

5. A coupler which is to be adopted in Great Britain must be capable of coupling during the transition stage with all the forms of couplings now in use in this country. Figs. 10 and 11, Plate 42, are photographs of an arrangement of the "Buckeye" coupler, which can be used as a Master Car-Builders' coupler, Fig. 11, or as an ordinary

3 Q

draw-hook, Fig. 10. It is worked in conjunction with side buffers, which are of the "Spencer" form, Fig. 12, being so arranged that the distance between the face of the buffer A and the headstock H can be varied, by means of the insertion or omission of a "slipper" S. When the automatic coupler is in use, the slipper S is removed, and the length between the buffer face A and headstock H in each vehicle is reduced by sliding the buffer rod C back and fixing it in this position by the set screw B, so that the buffer faces of the opposing vehicles do not touch. When the draw-hook is used, the slippers S are inserted, and the buffers thus packed out so that they act as side buffers in the ordinary way.

Adoption of the Automatic Coupler in Great Britain.—It is by no means easy to predict the future of the automatic coupler in this country. Although the railway companies have been trying to discover an automatic coupler suitable for use on the railways of this country, there appears to be in the minds of some of the leading mechanical engineers in the railway world considerable doubt as to the advantages to be obtained from its use. Mr. S. W. Johnson, of the Midland Railway, in stating his opinion "as an engineer" before the Royal Commission of Accidents to Railway Servants in 1899, said that he did not approve of the automatic coupler, nor did he see what advantage there was to be derived from it. He thought that the "Gedge" coupling and the buffers were the best, and would prefer to keep to them. He also remarked it was possible that some arrangement would be made whereby the "Gedge" coupling could be worked by a lever at the side, and that there was not sufficient warranty for the adoption of the automatic coupler when they had a coupling like that particular one.

The railways of the United States can scarcely be said to have made a mistake in introducing their automatic coupler, as their previous system—the "link-and-pin" coupling, Fig. 3, Plate 41—was so bad. They could not adopt the British system as they had no side buffers, but the opinion of many American railway officials agrees with that of Mr. Johnson in asserting that in this country there is no case for an automatic coupling. On the other hand, the Hon. E. A. Moseley,

Secretary to the Inter-State Commission at Washington, says :—
"There can be no objection to the adoption of the coupler
now in use here (United States, America), namely, the Master
Car-Builders' type, by reason of the difference either in length
or weight of the freight-cars now used in Great Britain and
those in the United States." But Sir Francis Hopwood, in
his report to the Board of Trade on Automatic Couplers, says :—
" Mr. Moseley . . . naturally takes a hopeful view of the American
coupling." At any rate, there can be no doubt that the ordinary
Master Car-Builders' coupler is not suitable for application to
the existing British stock, of which vehicles other than bogie
stock form a large proportion. There is not sufficient play in the
ordinary Master Car-Builders' coupler to enable such vehicles to go
round curves, although in some forms the coupler may be made
flexible. It is a fact worthy of notice that the Act of Congress,
which in 1893 made the adoption of the automatic coupler in the
United States of America compulsory, made an exception in the case
of four-wheel vehicles. It should be remembered that the successful
employment of automatic couplers in the United States does not
necessarily establish that they can be advantageously adapted to
wagons in use in this country. The curves, in the United States,
are neither so numerous nor so sharp as those in the shunting yards
of the United Kingdom. The wagons of the United States vary in
capacity from 30 to 50 tons and are bogie vehicles, whilst our
wagons have until recently rarely exceeded 10 tons, having four
wheels. This being so, an entirely flexible coupler, such as the
" Jones, " Fig. 9, Plate 42, is far more suitable than the Master
Car-Builders' coupler. But the great objection to the adoption of
automatic couplers in this country is that, although they can be worked
in conjunction with the side buffers now in use, they are in themselves
centre buffers ; and it is only necessary to compare the cost of an
automatic coupler (of either the Norwegian Hook or Master Car-
Builders' type) with that of the ordinary Gedge hook and three-link
coupling, to see that until the railway men in this country become
firmly cónvinced that an automatic coupler is a necessity, the
progress of these couplers will not be rapid.

3 Q 2

The question of expense is not the only consideration which hinders the adoption of the automatic coupler in this country, although the cost of such would in many cases be a large percentage of the cost of the vehicle. But there are also the apparently insurmountable difficulties of the transition period to be overcome. Many devices have been tried, but all have their faults. Amongst these are the arrangement of the " Buckeye " coupler as illustrated in Fig. 10 and 11, Plate 42, and the Westinghouse couplers. In the case of the latter the link on the coupler, which engages with the ordinary draw-hook which opposes it, is so placed that the pull is not directly through the draw-bar, but in an upwardly oblique direction. The conditions which an automatic coupler in this country should fulfil are :—

1. It should be applicable to spring buffer wagons.

2. It should be applicable to wagons on a curve.

3. It should be capable of coupling when one wagon is light and the other loaded.

4. There should be no tendency on the part of the combination to jump off the drawbar-hook if the vehicles are shunted together.

5. There should be no chance of the combination becoming jammed, if the vehicles are buffed sharply.

6. There should be no chance of the uncoupling gear of one wagon becoming entangled with the other.

7. The apparatus should be out of the way of any capstan ropes and brake gear.

8. The apparatus should have as few joints as possible.

9. The apparatus should provide means of coupling with the Gedge-hook.

10. It should not be necessary for a man to go between the vehicles to couple or uncouple.

11. It should be easy of access and renewal at any part of the line.

It is undoubtedly a fact that the leading men of the railway world of Great Britain are very anxious to do all they can to protect their servants from injury, and if it were an individual matter they would spare no effort in finding a coupler suitable to their requirements. Having done so, the automatic couplers would be given a fair trial, with a view to finding out whether the number of accidents was reduced. It must be remembered however that the coupler must be suitable to universal application, and consequently the matter does not rest with individual effort.

It may be mentioned that, since the adoption of the automatic coupler in the United States, another source of danger has arisen. Knowing that there will not be a man between the cars, the shunters " bash " the cars together to ensure their coupling, and sometimes the end timbers are thus started ; the fault is not discovered until an accident occurs. Then again there is also a tendency in uncoupling to throw over the levers at the end of the cars and pull, without first uncoupling the train-pipe connection. The result is that the latter are uncoupled forcibly and strained, causing leaking train-pipes and insufficient brake pressure.

The author wishes it to be distinctly understood that he does not profess to have treated this subject in any other than a most superficial manner, for the obvious reason that in 1900 about 550 patents had been filed at the British Patent Office for automatic couplers. By this time the number has probably reached 800, whilst there are several thousands in the United States; consequently a full treatment of the subject would fill several volumes.

The author wishes to express his best thanks to Messrs. George Spencer, Moulton and Co., Mr. O. Winder, Assistant Carriage and Wagon Superintendent, Lancashire and Yorkshire Railway, Mr. W. R. S. Jones, formerly Carriage and Wagon Superintendent, Rajputana Malwa Railway, and other gentlemen who have favoured him with information and advice.

The Paper is illustrated by Plates 41 and 42.

MEMOIRS.

Sir FREDERICK BRAMWELL, Bart, was born on the 7th March 1818.
He was the third son of Mr. George Bramwell, of the Firm of
Dorrien and Co., Bankers, of Finch Lane, London, (who eventually
became amalgamated with the present Firm of Glyn, Mills, Currie
and Co.), and was the younger brother of Baron Bramwell, the Judge,
who died in 1892.

In 1834 he was apprenticed to John Hague, a mechanical
engineer, of Cable Street, Wellclose Square, London, with whom, on the
expiration of his apprenticeship, he remained for some years as chief
draughtsman and manager. Amongst the matters on which he was
engaged was that of the vacuum system for distributing power, with
the prospects of which he was so impressed that, between the years
1846 and 1850, he, in conjunction with his fellow apprentice,
Mr. Samuel Collett Homersham, worked out a proposal for a
Subterranean Atmospheric Railway, between Hyde Park Corner and
the Bank. In connection with this subject, he read an interesting
Paper at the Plymouth Meeting of this Institution, in 1899, on
" The South Devon Atmospheric Railway."*

In 1843 he became associated with the Fairfield Railway Carriage
Works at Bow, at that time under the management of the late
Mr. W. Bridges Adams. While there he devised a means for making
tyres without welding, but found that, as regards " subject matter"
for a patent, his invention had been forestalled. A similar fate
attended another of his projects—that of constructing an endless
band-saw. He, however, persevered with the idea, and designed a
multiple arrangement of four band-saws for breaking down timber.
His early connection with the motor-car movement has been

---

* Proceedings 1899, Part 3, page 299.

described by him in a Paper read before Section G of the British Association, at the Oxford Meeting in 1894.

To the present generation, Sir Frederick was known principally as a scientific witness, and it was in this particular branch of the profession that he made his reputation. As a witness, either in matters connected with engineering or with patent litigation, in the Law Courts, or in connection with Bills before Parliamentary Committees, his services were much sought after, and, as an Arbitrator, his judgments were sound, equitable, and marked with rare legal acumen.

He was elected a Member of this Institution in 1854, entering the Council in 1864, and being elected a Vice-President in 1868, and President in 1874. In the latter year he delivered his Presidential Address * at the Cardiff Meeting—an interesting and instructive Address, wherein, from the storehouse of his marvellous memory, he set forth, with the skill of one who appreciated their significance, a statement of engineering facts dealing with the progress of civilisation. It was at the York Meeting of the British Association, in 1881, that, during the discussion of Mr. Emerson Dowson's Paper on the subject of using cheap gas for gas motors, he prophesied that in 1931 the steam-engine would only be seen in museums as an interesting relic of a past age, having been superseded by the internal-combustion engine. As long ago as 1851 he contributed a Paper to this Institution on an Improved Vacuum Gauge for Condensing Engines (Proceedings 1851, page 37); and, in 1867, he gave one on Floating Docks (Proceedings, page 80). At the Liverpool Meeting, in 1872, he contributed a Paper on the Progress effected in the Economy of Fuel in Steam Navigation, being the first of a series of Papers † which have appeared decennially by various authors. On the occasion of the Summer Meeting, at Portsmouth, in 1892, Sir Frederick contributed a Paper on the Sewage Outfall Works of that town (Proceedings, page 319).

---

* Proceedings 1874, page 103.

† Proceedings 1872, page 125; 1881, page 449; 1891, page 306; and 1901, page 607.

He joined the Institution of Civil Engineers in 1856 as an Associate, being transferred to full membership in 1862. From the year 1867 he served continuously on the Council, becoming a Vice-President in 1880, and President in 1884. In 1873 he was elected a Fellow of the Royal Society, on the Council of which he served in the years 1877–78. He was one of the most prominent of the Members of the Royal Institution, being elected one of its Managers in 1879, and acting as Honorary Secretary from 1885 to 1900. He was also a constant attendant at the Annual Meetings of the British Association, being selected as President of the Mechanical Science Section in 1872 and in 1884, and President of the Association itself at the Bath Meeting, in 1888. He joined the Society of Arts in 1874, and was for many years a Member of Council and a Vice-President, and at intervals filled the posts of Treasurer and of Chairman of the Council, and for a short period was President.

Always lamenting the fact that in his younger days there were practically no facilities for obtaining Technical Education, he was a warm supporter of the movement in the earlier stages of its development in this country. On the establishment of the City and Guilds of London Institute, he was appointed by the Goldsmiths' Company (of which he was Past-Prime Warden) one of its representatives on the Governing Body.

In 1881 he received the honour of Knighthood. On the formation of the present Ordnance Committee in 1881, he was appointed one of its two Civilian Members, and served on this Committee continuously until his death. In 1889, he was, in recognition of special services in connection with war material, created a Baronet. The success of the Inventions Exhibition, at South Kensington in 1885, was due, in a great measure, to the indefatigable energy and tact which he displayed. He was a D.C.L., of Oxford and of Durham, and an LL.D., of Cambridge and of McGill University, Montreal. His death occurred at his residence in Kensington, after a few days' illness, from cerebral hemorrhage, on the 30th November 1903, and in the eighty-sixth year of his age.

HENRY MARC BRUNEL was born in Westminster on 27th June
1842, being the second son of Isambard Kingdom Brunel, the
celebrated Engineer of the Great Western Railway and of the
steamship " Great Eastern," and consequently grandson of Sir Marc
Isambard Brunel, the Engineer of the Thames Tunnel. He was
educated at Harrow and at King's College, London. Although only
seventeen years old when his father died, he had taken a personal
part in some of his later engineering works, notably in the launch
of the " Great Eastern," and his contributions to the life of
" I. K. Brunel " (published in 1870) which his brother (the late
Dr. Isambard Brunel) and he had been engaged for some years in
compiling, are in some measure descriptions at first hand. He
served his time as a premium apprentice at Messrs. Armstrong's
Works at Elswick, and subsequently became a pupil and assistant of
Mr. (afterwards Sir John) Hawkshaw, with whom he remained until
about 1870. During his service with Sir John Hawkshaw he was
engaged, amongst other things, in an elaborate series of soundings in
the English Channel, which were undertaken for the purpose of
selecting the best route for the early scheme of a channel tunnel
proposed by Sir John Hawkshaw. He was subsequently engaged
with other members of Sir John's staff in a very elaborate
examination into the condition of the Caledonian Railway with the
lines affiliated therewith. He also acted as assistant to the resident
engineers in the construction of Penarth Dock near Cardiff and of
the Albert Dock at Hull. His strictly professional work during the
period between 1870 and 1878 included the construction of a large
reservoir for the water-supply of Torquay and a careful and
complete investigation of methods for the prevention of waste ; also
a visit to Brazil for the purpose of examining and reporting on an
important installation of hydraulic hoists at Bahia, and engagements
in connection with a variety of parliamentary proposals. Apart
from these subjects, being a friend and devoted disciple of the late
Mr. William Froude, F.R.S., formerly a member of his father's
engineering staff, he took a lively interest in the scientific
researches bearing on Naval Architecture. In the experiments on
ships of H.M. Navy which were made during this period, in

connection with these researches of Mr. Froude, he placed his time and talents for several years at Mr. Froude's disposal. In 1878 he entered into partnership with Mr. (now Sir John) Wolfe Barry, and his subsequent professional career is bound up with that of Sir John and other partners. He was intimately concerned with all that they undertook, but in particular may be mentioned the important Barry Dock in South Wales, the St. Paul's Station and railway bridge over the Thames at Blackfriars, The Tower Bridge, and the bridge recently completed at Connel Ferry, near Oban. In the autumn of 1901 he had a slight apoplectic stroke, followed a few months later by the bursting of a blood vessel in the brain, from the effect of which he never fully recovered. His death took place in Westminster, where he had resided all his life, on 7th October 1903, at the age of sixty-one. He became a Member of this Institution in 1873; and was also a Member of the Institution of Civil Engineers, and of the Institution of Naval Architects.

JOHN BULMER was born at Newburn, a small village near Newcastle-on-Tyne, on 7th April 1834. At eight years of age he was put to work at one of the local collieries and from there to various engineering works, until in 1873 he commenced business with the late Mr. Wardhaugh, the firm being known as Wardhaugh and Bulmer. In 1880 he entered upon the Spring Garden Engineering Works, Newcastle-on-Tyne, on his own behalf, where he made a speciality of wire and hemp rope-making machines, also machinery for making electrical wires and cables. He gained by his inventive genius and excellent workmanship a world-wide reputation for his machines, which are now working not only in nearly all the rope-works in this country, but on the Continent and in America, Japan, and other countries. His death took place at his residence in Newcastle-on-Tyne on 31st October 1903, after a day's illness, in his seventieth year. He became a Member of this Institution in 1882.

WILLIAM COCHRANE was born at Blackbrook, near Dudley, on 23rd January 1837, being the second son of the late Mr. A. Brodie

Cochrane* of Stourbridge. He was educated at Hawthorn Hall, Wilmslow, and at King's College, London, with a view to proceeding to Cambridge and taking up the Bar as a profession, but, owing to his father's failing health, he was obliged to relinquish this intention, and entered business in his father's iron works and collieries in Staffordshire. In 1857 he went north to superintend the development of Elswick, Tursdale, and New Brancepeth Collieries. His abilities as a mining engineer were soon recognized, and he was frequently engaged as a professional witness in important mining cases and arbitrations. He took a leading part in the foundation of the Durham College of Science at Newcastle-on-Tyne, and was a member of its Council from the commencement. In recognition of his work on behalf of the College, the University of Durham in 1901 conferred upon him the honorary degree of Master of Science. He was instrumental in introducing the Guibal ventilating fan into this country; and, in conjunction with the late Professor Marreco, conducted a series of important experiments on the explosive nature of coal dust in mines. He took a prominent part in the organization and management of the Exhibition held at Newcastle-on-Tyne in 1887. In 1898 he succeeded his brother, Mr. Charles Cochrane,† as chairman of Messrs. Cochrane and Co. His death took place, after a long illness, at his residence in Newcastle-on-Tyne, on 25th November 1903, in his sixty-seventh year. He became a Member of this Institution in 1868; he was also a Vice-President of the Institution of Mining Engineers, and a Past-President of the North of England Institute of Mining Engineers.

Engineer-Lieut. RAYNER DAVIS was born at Wallingford, Berks, on 20th February 1871. He was educated at the County School, Bedford, and at the Crypt Grammar School, Gloucester. In 1887 he commenced a four years' apprenticeship with Messrs. Strachan and Henshaw, of Bristol, during which time he worked at the evening classes of the Merchant Venturers' Schools, and obtained

---

* Proceedings 1864, page 13.

† Proceedings 1898, Part 2, page 309.

many certificates in the various branches of engineering. He obtained a free studentship at University College, Bristol, given by the Bristol County Council; and while studying at the College he gained a Whitworth Exhibition, for which Messrs. Strachan and Henshaw kindly released him from the last year of his apprenticeship. In 1894 he joined the Royal Navy, serving in H.M.SS. " Benbow," " Revenge," " Diana," " Cressy," and in December 1902 he was appointed to H.M.S. " Coquette," one of the destroyer squadron at Malta. In August 1903 he was seized with Mediterranean fever, to which he succumbed at the Royal Naval Hospital at Bighi, on 3rd September 1903, in his thirty-third year. He became a Member of this Institution in 1900.

CHARLES RALPH DÜBS was the second son of the late Mr. Henry Dübs, founder of the well-known firm of locomotive builders in Glasgow. He served his time in his father's works, and on its completion in 1869 he assisted in the management of the works, succeeding to full management upon the death of his brother. He became a partner in 1875. Under his direction, and that of Mr. William Lorimer, who subsequently joined him in partnership, the business of the firm gradually extended to all. parts of the world. In 1902 the firm amalgamated with Messrs. Neilson, Reid and Co., and Messrs. Sharp, Stewart and Co., under the title of the North British Locomotive Co. His death took place at his country residence at Moniaive, Dumfriesshire, on 19th July 1903. He became a Member of this Institution in 1877.

SAMSON FOX was born on 11th July 1838, in Bradford, Yorkshire. At about ten years of age he went to work at a cloth mill in Leeds, where his father was also employed. Showing an early aptitude for mechanics, he was afterwards apprenticed to the firm of Messrs. Smith, Beacock and Tannett, Machine Tool Makers, Round Foundry, Leeds, where he became foreman and ultimately traveller. Whilst in the employ of this firm he designed several special tools for the machine-cutting of bevelled gear, for the manufacture of trenails,

920   MEMOIRS.   DEC. 1903.

and for several of these machines he took out patents. In 1862 he
was in charge of the machine tool exhibit of Messrs. Smith, Beacock
and Tannett at the London Exhibition. Later, he started a small
engineering works in Leeds for the manufacture of special machine
tools at the Silver Cross Works, Leeds. In 1874 he founded the
Leeds Forge Co. for the manufacture of "Best Yorkshire" Iron and
boiler plates and general forging work, and in 1877 he brought out
his first corrugated boiler furnaces, the material for which originally
was "Best Yorkshire" Iron and the corrugations of the furnaces were
hammered by means of swage blocks under a steam-hammer. The
advantages of these corrugated furnaces led to the practical
application of triple-expansion engines, and machinery for rolling
the furnaces in place of hammering them was undertaken in 1882 ;
and a Siemens steel plant was laid down for manufacturing the material
for the manufacture of plates for the production of the furnaces.
In 1877 he took out a large number of patents for various details
connected with the manufacture of corrugated furnaces, and in 1887
and 1888 he took out patents for the manufacture of pressed-steel
underframes for railway wagons, &c. The works of the Leeds
Forge Co. in 1889 were again further extended for the manufacture
of this new form of railway rolling stock. In 1888 he started works
for the manufacture of steel frame rolling stock at Joliet, near
Chicago, and made there the first pressed-steel cars which had been
used in America, and large numbers of the Fox pressed-steel bogie
truck which was principally used for freight cars ; both of these
inventions met with great success. The extension of the business
in America led to other works being built at Pittsburg, and
these were in 1889 merged into the Pressed Steel Car Co.
In 1894 he produced at Leeds, in works specially erected,
the first carbide of calcium for the production of acetylene
gas which was made in England, and founded the industry
which is now manufacturing its supplies of carbide of calcium
at Foyers, N.B. For the greater part of his business life at
Leeds he was the managing director of the Leeds Forge Co.,
and had succeeded to the chairmanship of the company just before
his death, which took place at Walsall on 24th October 1903,

at the age of sixty-five. He resided at Grove House, Harrogate. He became a Member of this Institution in 1875; and was also a Member of the Legion of Honour of France.

ALBERT FRY was born in Bristol on 27th July 1830. After being educated at a private school in Clifton, he entered the engineering works of Messrs. Gilkes, Wilson and Co., Middlesbrough. Subsequently he became proprietor of the Bristol Wagon Works, and on its conversion into a company he became managing director. He ·was greatly interested in educational work, especially in the University College, Bristol, which largely owed its inception and successful career to his ability as Chairman of the Council. His death took place at his residence in Bristol on 22nd April 1903, in his seventy-third year. He became a Member of this Institution in 1866.

WILLIAM ARTHUR LEA was born at Templemore, Ireland, on 22nd August 1864, being the eldest son of the late Major W. R. Welch Lea, of the Bedfordshire Regiment. He was educated at Blundell's School, Tiverton, which he left in 1882 to serve a pupilage of three years under Mr. William Dean at the Great Western Railway Works at Swindon. In 1886 he was employed for three years in the locomotive department of the Canadian Pacific Railway at Winnipeg, and in 1890 he underwent a course of training in the University of Toronto, where he took his degree in 1893. Having been employed for nearly two years as an electrician on the City and Suburban Electric Railway of Toronto, he was appointed in 1894 superintendent of the Galt and Preston Street Railway, a small line connecting with the Canadian Pacific Railway. In 1896 he accepted the position of civil engineer in the construction department of the Mexican District Railway Co., in the City of Mexico, and on the construction company being dissolved he was transferred to the operating department. His death took place from typhoid fever in Mexico City on 3rd June 1903, in his thirty-ninth year. He became a Member of this Institution in 1895.

DENIS RALPH LOVELL was born in Glastonbury on 25th August 1861, and was educated at a private school in Weston-super-Mare. In 1879 he was articled to Earle's Engineering and Shipbuilding Co., Hull, where he went through the various shops and the drawing office. On the expiration of his term in 1884, he went to sea for two years as marine engineer on the Wilson Line of steamers, and then returned to Earle's Works for one year as draughtsman. In 1887 he again went to sea for four years, and obtained a first-class Board of Trade Engineer's certificate. On the recommendation of Mr. A. E. Seaton, general manager of Earle's Works, he went out to Brazil in 1891 to build a steam-wheel paddle-steamer on the River São Francisco. On its completion, he was engaged as general shop foreman on the Alagoas Railway, having full charge of all the rolling stock ; and in 1896 he was promoted to be locomotive superintendent of the same railway. On the sale of the line to the Government in 1902, he accepted a more lucrative position on the Leopoldina Railway, but after a few months' time he was compelled to return to England on account of ill-health, the result of an accident. His death took place at Weston-super-Mare on 3rd April 1903, in his forty-second year. He became a Member of this Institution in 1900.

JOSEPH MOORE was born at Tranent, Haddingtonshire, on 29th April 1829, and received such education as that neighbourhood afforded. At the age of fifteen he moved to Glasgow to serve an apprenticeship as an engineer at the Hill Street Foundry of Messrs. Murdoch and Aitken, Glasgow. In 1849 he went to San Francisco, where he followed the engineering profession, and was instrumental in assisting in the development of the resources of a new country. His position as one of the foremost engineers in California for a lengthy period caused him to apply himself to a great variety of engineering work. The first cable railway in California was developed by him, as was also the use of sheet-iron pipe, and machinery for making the same, for water works and mining purposes. He devised many new appliances for utilizing the power of water in connection with mines, and introduced improvements in the machinery for extracting the juice from cane. In 1883 he was obliged to retire from business through illness caused by overwork, and since then he lived in England. His death took place in London on 31st March 1901, in his seventy-second year. He became a Member of this Institution in 1876.

CHARLES O'CONNOR was born in Manchester on 23rd November 1823, and was educated in the same city. After leaving school he was apprenticed to Messrs. Sharp, Roberts and Co., Falkner Street, Manchester, machine manufacturers, but owing to the dissolution of partnership about 1840, he went to Messrs. Sharp Brothers, Oxford Road, Manchester, locomotive builders, where he finished his apprenticeship in 1844. In 1846 he was sent to Paris by the same firm to establish a locomotive works there for Messrs. Ernest Gouin, where he remained for about two and a half years, until after the Revolution broke out in 1848. He, then returned to Manchester, and remained with Messrs. Sharp Brothers until 1852, when he was offered the position of crank turner, and after two years was made foreman over the machinery department of Messrs. Beyer, Peacock and Co., Gorton, Manchester, and remained in this position for seven years. Shortly after this he was appointed manager at the works of Messrs. Slaughter and Grunning, Avonside Engine Works, Bristol, which position he occupied for seven years, when he became works manager for Messrs. Stothert and Pitt, Bath. Upon relinquishing this position in 1870, he was appointed Superintendent of the Machinery and Fitters Department, at Messrs. John Elder and Co., Fairfield Engine Works, Govan, Glasgow, where he remained for nine years. He next became works manager to Messrs. Maudslay, Sons and Field, London, where he remained for a considerable time. He was next offered the position of manager of the Mersey Forge, Liverpool, where he was actively occupied in the production of several important works, amongst which may be mentioned the manufacture of the largest solid crank (in those days), for the S.S. "Servia," and also the largest iron stern-frame, weighing over 40 tons, for the S.S. "City of Rome." He continued in this position until the firm dissolved. Upon the closing of these works, he commenced business in Liverpool on his own behalf, as consulting

3 R

engineer, which occupation he continued to follow. His death took place at his residence in Liverpool on 18th October 1903, in his eightieth year. He became a Member of this Institution in 1868.

THOMAS PARKER was born in Ayrshire on 11th July 1829, and commenced his apprenticeship in 1847 at the Greenock Works of the Caledonian Railway under the late Mr. Robert Sinclair. In 1851 he served for one year with the London and South Western Railway under Mr. Beattie, after which he returned to the Caledonian Railway, and was engaged for some considerable time on the inspection of material and rolling stock supplied by contract to that company. In 1858 he was appointed carriage and wagon superintendent on the Manchester, Sheffield, and Lincolnshire Railway, and during his term of office he constructed the whole of the now existing carriage and wagon shops. He introduced on the same railway a six-wheel bogie coach, and the dining-car built by him in 1885 was one of the first constructed in this country. In 1886, on the resignation of Mr. Sacré he was appointed locomotive, carriage, and wagon superintendent. At the Manchester Jubilee Exhibition he exhibited engine No. 561, which was an entirely new type of locomotive on the line, and subsequently used Belpaire fire-boxes extensively. In 1892 he remodelled the locomotive shops, and built a large new erecting shop. He retired in 1893 after a railway career of nearly fifty years ; and during the time he was with the Manchester, Sheffield, and Lincolnshire Co., the wagon stock was increased from 4,182 to 17,933 vehicles, and the carriage stock from 338 to 1,106. His death took place at his residence at Gorton, near Manchester, on 25th November 1903, at the age of seventy-four. He became a Member of this Institution in 1872; and was also a Member of the Institution of Civil Engineers.

HENRY WILLIAM PEARSON was born at Forest Gate, Essex, on 6th January 1846, and was educated first at a private school and later at Bow Grammar School. In 1862 he was apprenticed to the Eastern Counties Railway Co. (afterwards the Great Eastern

Railway Co.) under Mr. Robert Sinclair, then locomotive engineer. On the completion of his term in 1867, he was engaged in the drawing office under the then locomotive engineer, Mr. Samuel W. Johnson,* and, on leaving this department, he was appointed in 1870 assistant to Mr. H. W. Davis, with whom he was engaged in civil engineering and surveying.   In 1872 he became assistant engineer to the Bristol Water Works Co., under Mr. T. Bell, whom, on his death two years later, he succeeded as resident engineer, which position he held until his death.   During the period he was at Bristol the supply was more than doubled, the daily supply in 1903 being eight million gallons, with a population of 350,000. Two large wells have also been sunk in the Nailsea Valley, and large pumps erected, the engine-power during his tenure of office having been trebled.   The storage has also been largely increased, for which two reservoirs were constructed, in addition to great extensions in the filter area.   Besides being a Member of this Institution, he was also a Member of the Institution of Civil Engineers, the Iron and Steel Institute, a Fellow of the Royal Geographical Society, and a Member of the British Association of Waterworks Engineers; and was also Past-President of the Bristol Engineers Association.   In conjunction with Mr. Henry Davey he aided in the preparation of his Paper on the Newcomen Engine.†   He was the inventor of several appliances mainly for use on waterworks, amongst which was an apparatus for making branch connections to mains under pressure; his waste-preventing devices were also well known.   His death took place at his residence in Bristol on 20th October 1903, in his fifty-eighth year.   He became a Member of this Institution in 1885.

JOHN PENN, M.P., was born at Lewisham, Kent, on 30th March 1848, being the son of the late Mr. John Penn and grandson of the founder of the firm of John Penn and Sons, marine engineers, of Greenwich and Deptford.   He was educated at Harrow

---

* Locomotive Engineer of the Midland Railway from 1873 to 1903; and President of this Institution in 1898.

† Proceedings 1903, Part 4, page 682.

3 R 2

School, subsequently proceeding to the University of Cambridge.
On leaving the University about 1867 he proceeded to take charge
of his father's works, and directed his attention principally to the
design and manufacture of warship machinery which should
embody the latest improvements, as instanced in H.M.SS.
" Crescent," " Magnificent," and " Goliath." In 1888 the firm
was converted into a private company, but has since been absorbed
in the Thames Ironworks Shipbuilding and Engineering Co. He
was a director of the Great Eastern Railway Co., and also of the
Kent Waterworks Co. After an illness lasting for more than a
year, his death took place on 21st November 1903, in his fifty-sixth
year. He became a Member of this Institution in 1873; and was
also a Member of the Institution of Civil Engineers.

FREDERICK WILLIAM RAFAREL was born at Barnstaple on 26th
February 1837, and was educated at the Grammar School at
Colyton. About the age of fifteen, he entered the employment of
the Great Western Railway Co. at Paddington under Mr. Gibson,
the superintendent of the carriage department. In the spring of
1863, he obtained the appointment of paymaster of a fleet under
command of Captain Sherard Osborn, which the British Government
sent out to assist the Chinese Government in suppressing piracy on
the coast of China. On his return to England about two years later,
he joined the Patent Nut and Bolt Co. at their London Works,
Smethwick, near Birmingham, and was placed in charge of the
prime-cost department. He was promoted to the position of
assistant to the managing director, Mr. F. G. Grice, at the
Cwmbran Iron Works and Collieries of the Company in 1867,
and continued in the position under Mr. E. J. Grice, who succeeded
his brother as managing director. On the death of Mr. E. J.
Grice in March 1899, he was appointed general manager of the
Cwmbran Iron Works and Collieries, a position which he held up
to the time of his death, retaining the appointment on the works
being taken over by Messrs. Guest, Keen and Co. in July 1900, and
later by Messrs. Guest, Keen and Nettlefolds in February 1902.
During his thirty-six years' service with the company at Cwmbran

Works the works were very largely extended and developed. In 1885 he invented a metallic railway sleeper and method of fixing rails. He was a member of the Monmouthshire and South Wales Coal Owners' Association and of the Newport District Board of the same Association, and a member of the Hartley Colliery Trust. He was a magistrate for the County of Monmouth, and he took a large interest in local matters, being a member of the Pontypool Board of Guardians, and Chairman of numerous local public bodies for a number of years. His death took place at his residence at Cwmbran, Monmouthshire, on 2nd June 1903, at the age of sixty-six, after a long illness, from bronchial asthma and an affection of the heart. He became a Member of this Institution in 1868.

JOHN WARDEN WATSON was born in Glasgow on 8th July 1867. After an elementary education, he attended evening classes at the Glasgow and West of Scotland Technical College from 1882 to 1890, and was successful in obtaining first class Science and Art Certificates in all his subjects. In the former year he commenced an apprenticeship with Messrs. Duncan Stewart and Co., at their London Road Iron Works, Glasgow, and served the usual time in the shops and drawing office. On its completion in 1887 he was engaged as draughtsman by the same firm, being promoted in 1891 to the post of chief draughtsman. During that period he also superintended the erection of large steam-engines, steel works' plant, and sugar machinery, &c. In 1897 he was appointed engineering manager of the London office of Messrs. Duncan Stewart and Co., and remained in this position until 1900, when he entered into partnership with Mr. F. W. Scott as engineers and contractors. His death took place in London from consumption, after several months' illness, on 4th December 1903, at the age of thirty-six. He became a Member of this Institution in 1898.

JOHN CHARLES GRANT WILSON was born at West Park, Leslie, Fifeshire, on 7th September 1857. He received his education at Kennaway Public School; Madras College, St. Andrews; and at the University College 'at St. Andrews. He served an apprenticeship

from 1874 to 1879 under his father at the works of the Avonside
Engine Co., Bristol. In 1880 he went to Manila as assistant
engineer to Messrs. Barlow and Wilson; and in 1885 was
appointed superintendent of the Manila Tramway. In 1888 he
was employed as sectional engineer on the Manila and Dagupan
Railway, having one of the most difficult sections of their line in
his charge; and on the completion of the line was appointed
locomotive superintendent, retaining his post until the outbreak of
the Spanish-American War, when he came home on account of ill-
health. He returned to Manila in July 1903, as estate manager to
Messrs. Ayala and Co.'s provincial distilleries, and died suddenly of
cholera on 14th August 1903, in his forty-sixth year. He became a
Member of this Institution in 1892.

WILLIAM HENRY WILSON was born in Manchester on 27th April
1850, and was a son of the late Mr. George Wilson, chairman of the
Anti-Corn Law League. He was educated at a private school in
Manchester, and afterwards attended the engineering classes at
Owens College. In 1867 he was apprenticed to the firm of Messrs.
Wren and Hopkinson, engineers, of Manchester, and on its completion
in 1874 he became assistant manager at the works of Messrs. James
Dodge and Co., of Newton Heath. In 1876 he was appointed
assistant manager to the Phoenix Engineering Co., of Ghent,
Belgium, remaining in that position until 1881, during which time
he designed several important improvements in flax-spinning
machinery. Owing to ill-health, he was compelled to return to
England; and then became manager to Messrs. W. F. Mason,
engineers, of Manchester. On the business being converted into a
company he was elected a director. He was closely connected with
the large improvements made by the firm in gas producers,
destructors, bakers' machinery, &c. In 1899 he started in business
for himself as engineer and ironfounder, but was unable to devote
his time to it on account of ill-health. His death took place on
26th December 1902, in his fifty-third year. He became a Member
of this Institution in 1897.

INDEX.

1903.

Parts 3-4.

930   INDEX.   Dec. 1903.

934 INDEX. DEC. 1903.

3 s

Fig. 1. *Newcomen Steam Fire Engine.*

From " A Treatise on Minerals, Mines, and Mining," by Wm. Pryce, folio, 1778, p. 160.

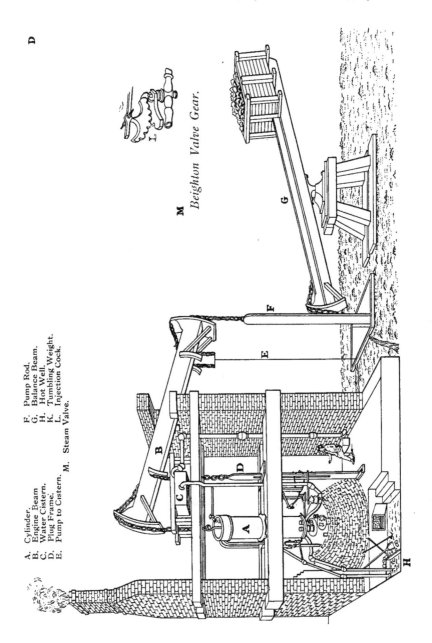

D

Beighton Valve Gear.

M

G

F

E

B

D

C

A

H

A. Cylinder.
B. Engine Beam
C. Water Cistern.
D. Plug Frame.
E. Pump to Cistern.

F. Pump Rod.
G. Balance Beam.
H. Hot Well.
K. Tumbling Weight.
L. Injection Cock.

M. Steam Valve.

NEWCOMEN ENGINE.

Plate 17.

Appendix I.

Fig. 2.   *Colliery Winding Engine, Coalbrookdale.*
Made about 1790.   Boilers much later.

Appendix III.

Fig. 3　*Newcomen Winding Engine at the Farme Colliery, Rutherglen.*
Erected 1810.   Still working.

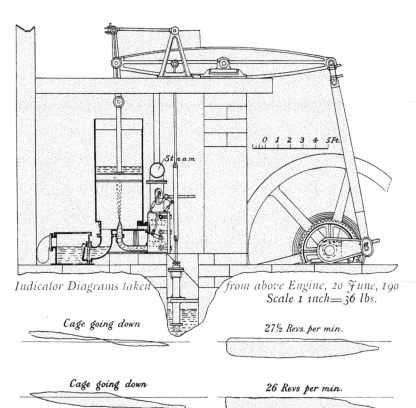

Indicator Diagrams taken . . . . from above Engine, 20 June, 190
Scale 1 inch = 36 lbs.

Cage going down　　　　　　　*27½ Revs. per min.*

Cage going down　　　　　　　*26 Revs per min.*

NEWCOMEN ENGINE.   *Plate 19.*

Appendix IV.

Fig 4.   *Newcomen Engine at Ashton Vale Iron Works, Bristol.*

From a Sketch made in September, 1895.   Erected about 1746-60   Dismounted 1900.

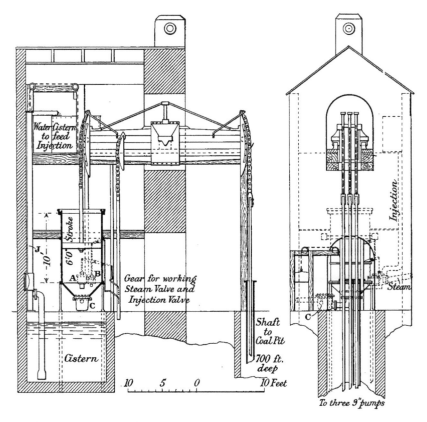

Fig. 5.   *Indicator Diagram taken from above Engine, 27th May, 1895.*

Dia of Cyl 5' 6"   Boiler Pressure 2 3 lbs.
Stroke   6' 0" about   Vacuum Gauge, none fixed.
No. of Strokes per min 10.   Time 3 p m. .

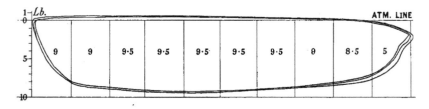

Mechanical Engineers 1903.

NEWCOMEN ENGINE. *Plate 20.*

Appendix IV.

Newcomen Engine at Ashton Vale Iron Works, Bristol.

Fig. 7. Side View of Pump End of Beam.

Fig. 8. Top of Cylinder.

Fig. 6. Beam and Driver.

Plate 21

Appendix IV.

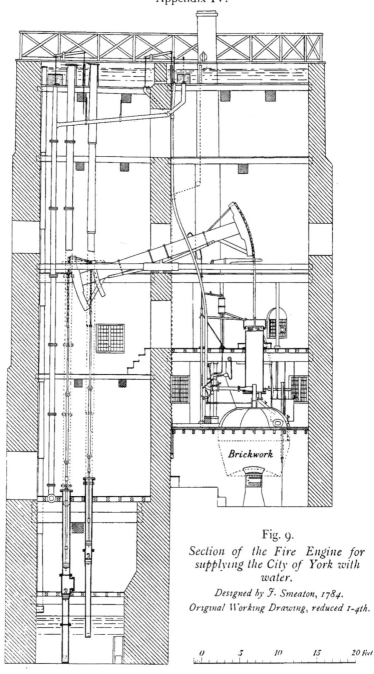

Fig. 9.

Section of the Fire Engine for supplying the City of York with water.

Designed by J. Smeaton, 1784.

Original Working Drawing, reduced 1-4th.

0   5   10   15   20 Feet

Out of use 1830.

The following three Views are from recent Photographs.

Mechanical Engineers 1903.

ᴘ ρρᴄ ᴜ

Fig. 11. *Newcomen Mine Engine.*

From a Drawing dated 12th January, 1826

Q

K

A	Cylinder.	H	Injection Pipe and Valve	P.	Injection Pump.	
B.	Piston	I	Injection	Q	Piston Chain.	
C	Piston Stems.	K.	Boiler.	R	Working Beam.	
D	The Cup.	L.	Steam Pipe	S	Plugtree which opens and	
E.	To pour water on piston	M.	Regulator Valve		Regulator and Injection	
F.	Overflow pipe from ditto	N.	Eduction Pipe	T	Gudgeon of Beam	
G.	Injection Cistern	O.	Its Valve.	U.	Pump Rod	
				V.	Snift	

Mechanical Engineers 1903.

From " A Treatise on Minerals, Mines, and Mining," by Wm. Pryce, folio, 1778, p. 172.

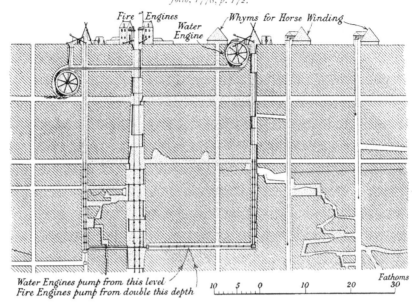

Fire Engines
Water Engine
Whyms for Horse Winding

Water Engines pump from this level
Fire Engines pump from double this depth

10 5 0   10   20   Fathoms 30

Appendix VII.
Fig. 13. Basset Atmospheric Engine, Denby Colliery, near Derby.
Erected 1817 in position shown, but had been working previously at another pit.
Dismounted 25th April, 1886.

Mechanical Engineers 1903.

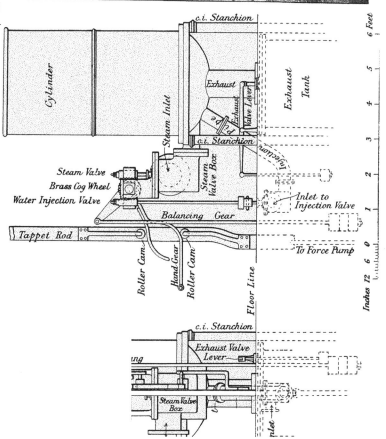

nospheric Pumping Engine, made at the "Moira Furnaces," 1821. Max. boiler press. 4 lbs.

Appendix VIII. (Described in Appendix IV.)

Figs. 15, 16, 19 & 20 are from memory drawings by Mr Wm. Patrick.

Fig 15.   *The Buffery Old Pumping Engine, near Stourbridge.*

(Being a Newcomen Engine with Watt's separate condenser added)

*Reproduced
⅔ths. original.*

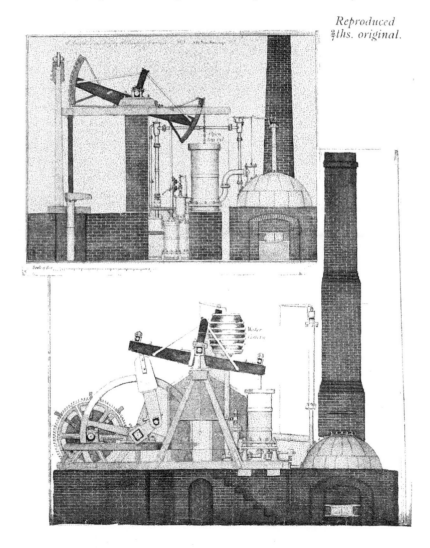

Fig. 16.   *Atmospheric Winding Engine, near Stourbridge*

NEWCOMEN ENGINE. Plate 27.

Indicator Diagrams, taken 8th Dec., 1903,

at Windmill End, near Netherton, Dudley, from Newcomen Winding Engine
with "Pickle-pot" Condenser.

(Similar to Engine, Fig 16, Plate 26)

Fig. 17   *Valves worked by hand.*

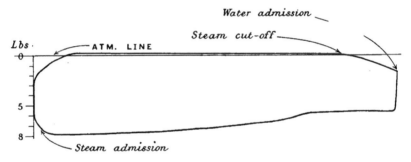

Fig. 18   *Valves worked by self-acting tappet.*

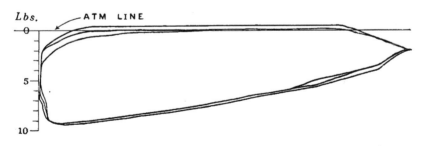

Dia of Cyl , 28¾ ins.   Length of Stroke, 5 feet.   Mean Pressure, 6 8 lbs. per sq. in.

Revs per min , 30   I H P , 19 7

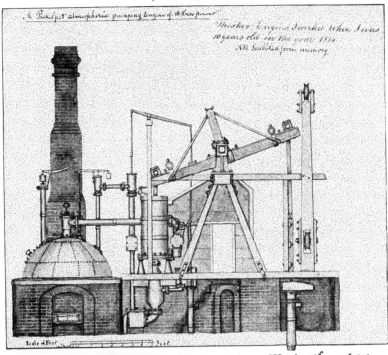

Fig. 20. *" Pickle-pot " Atmospheric Engine. Winding from 6 pits.*
Bore of Cyl. 28". Stroke 4 feet. ⅓rd. original.

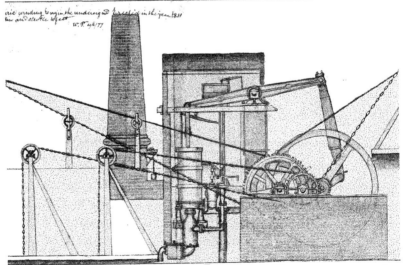

Mechanical Engineers 1903.

Fig. 21. *Newcomen Engine (King o' dery).*

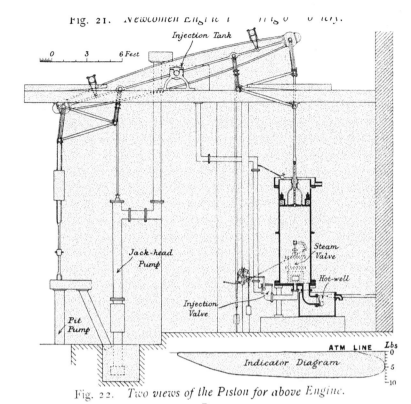

Injection Tank

0   3   6 Feet

Jack-head
Pump

Steam
Valve

Hot-well

Injection
Valve

Pit
Pump

ATM LINE   Lbs
           ⌐0
Indicator Diagram
           ⌐5
           ⌐10

Fig. 22.   *Two views of the Piston for above Engine.*

Feet.

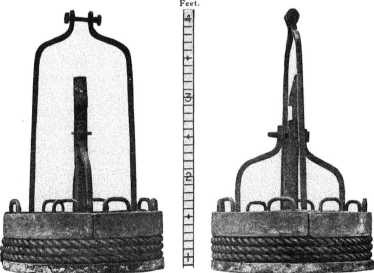

Mechanical Engineers 1903.

Fig. 23.
" Pickle-pot."

Fig 24. Sketch illustrating
the Newcomen Engine before
it was made automatic,
1712-1718.

Fig 25
Illustration of the
" Buoy " and the
"Scroggan" in the
Newcomen Engine.

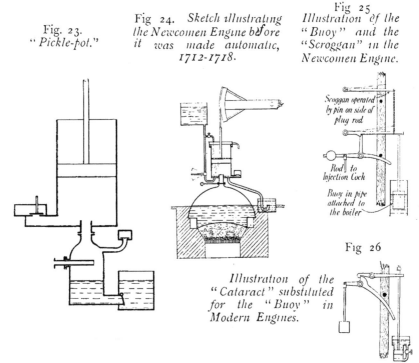

Scroggan operated
by pin on side of
plug rod

Rod to
Injection Cock

Buoy in pipe
attached to
the boiler

Fig 26

Illustration of the
" Cataract " substituted
for the " Buoy " in
Modern Engines.

Fig. 27.   Pumping Engines.

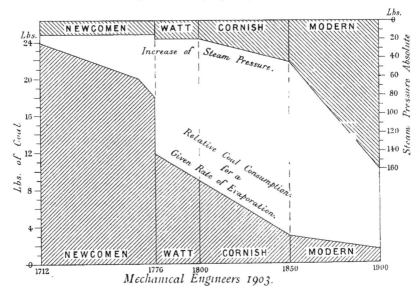

Mechanical Engineers 1903.

Fig 1.   *Boiler and Fettling Shops.*

F. *Swing Jib*   H. *Hydraulic Crane*
G. *Hand Crane*   I. *Steam Crane*

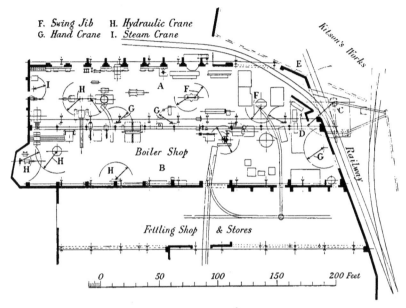

Fig. 2.   *Section of Old Roofs of Boiler Shop.*

Fig 3.   *Section of New Roofs of Boiler Shop.*

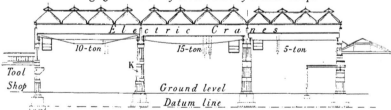

Fig. 4.   *Section of Old and New Roofs of Boiler Shop.*

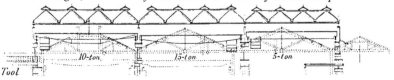

Plate 32.

Plate 32

RE-ROOFING EXISTING SHOPS.

Fig. 5. *Travelling Crane for lifting roof girders.*

Balance Weight

Centre line of Crane

0 5 10 15 20 Feet

Mechanical Engineers 1903.

Figs. 9 and 10. *Crane lifting trussed beam.*

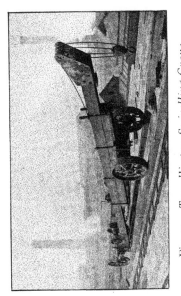

Fig. 7. *Travelling or Swivelling Crane.*

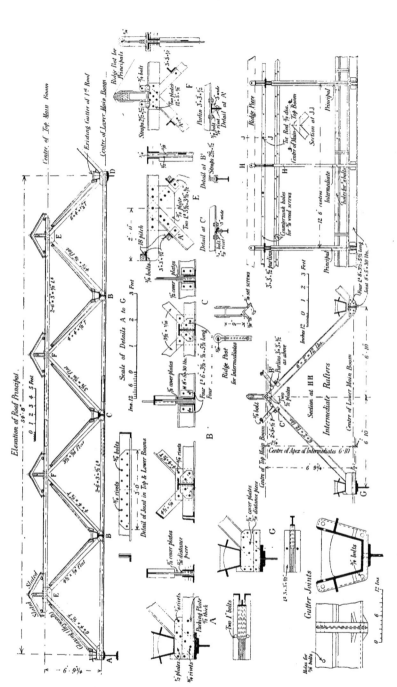

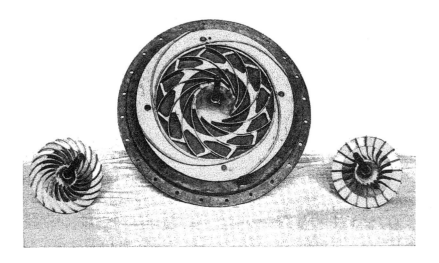

Fig. 3. *4-H.P. Motor driving Pump.*

Mechanical Engineers 1903.

Specimens of Crystals, etc., exhibited at Meeting.

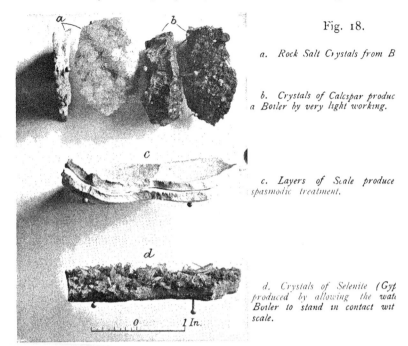

Fig. 18.

a. Rock Salt Crystals from B

b. Crystals of Calcspar produc
a Boiler by very light working.

c. Layers of Scale produce
spasmodic treatment.

d. Crystals of Selenite (Gyp
produced by allowing the wate
Boiler to stand in contact wit
scale.

Fig. 19. Chocked Economiser Pipe.

Mechanical Engineers 1903.

Fig. 20. Thermal Storage Vessel.

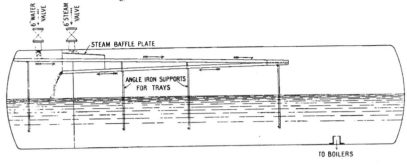

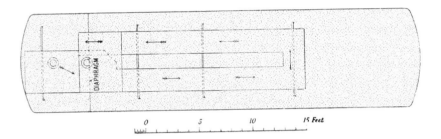

Fig. 21. Precipitate in Interior of Vessel,
looking along a tray toward the end where water enters.

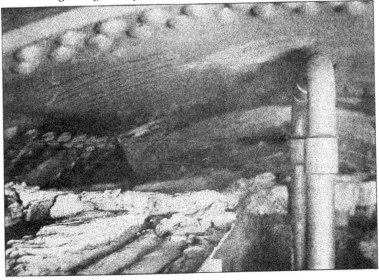

Mechanical Engineers 1903.

SOME CONSTITUENTS OF BOILER SCALE.

Fig 24 $Mg\,(OH)_2$ $K\,Cl$ crystals washed away.   × 200 diams

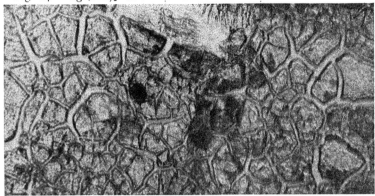

Fig 25. $Ca\,Co_3$.   Washed with distilled water.   × 200 diams

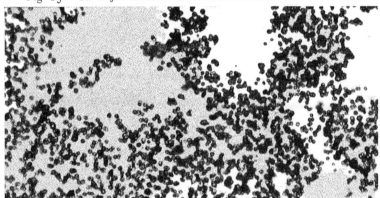

Fig. 26.   Crystals of Selenite.   × 200 diams.

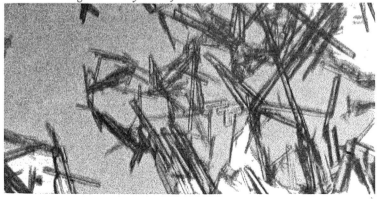

Mechanical Engineers 190 ;.

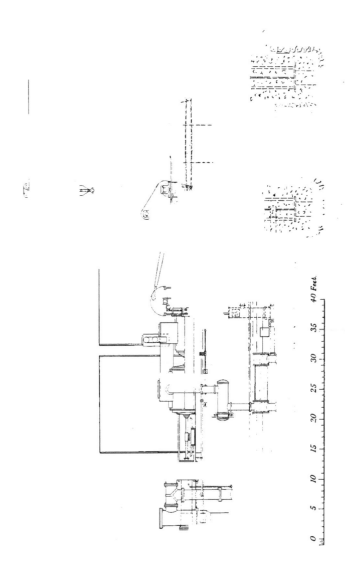

Mechanical Engineers 1903.

ig. 5. *Suction Valve and*
Fishing Tackle.

Fig. 6. *Bucket and*
Guide.

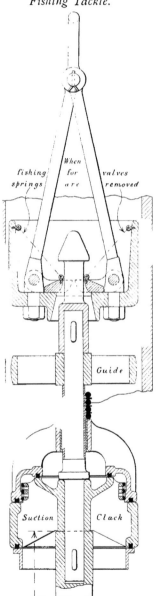

fishing
springs

When
for
are

valves
removed

Guide

Suction

Clack

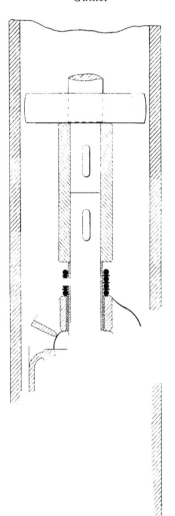

Fig. 7. Coupler (Janney).

Sectional Elevation

Inverted Plan

Fig. 8. Coupler (Miller).

Coupler (Brockelbank).

Fig. 4. Applied to 8 and 10-ton Wagons.

Fig. 5. The same, viewed from above.

Fig. 6 Diagram showing the Inclination of the centre line of the Coupler to the centre lines of two 50-feet Coaches on a 5-chain Curve.

Fig. 1. Form of "Contour Lines" for M.C.B. Standard Coupler.

Fig. 2. General View of the M.C.B. Standard Coupler.

Fig. 3. "Link and Pin" Coupler.

Fig 9 Flexible Centre Buffing and Draw Gear and Self-centreing
Coupler (Jones')

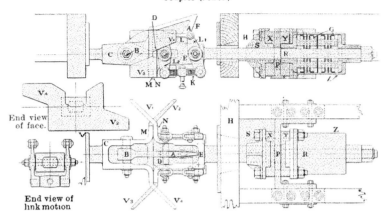

End view
of face.

End view of
link motion

Coupler (Buckeye)

Fig 10. Used with a 3-link
or screw coupling.

Fig 11.
Used as a M C. B Coupler

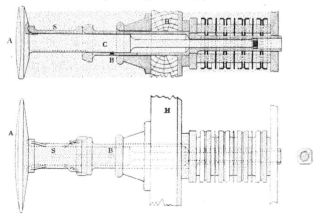

Fig 12 Slipper Buffer (Spencer)

Lightning Source UK Ltd.
Milton Keynes UK
UKHW02f1926060418

320655UK00007B/360/P